ENCYCLOPEDIA OF EVOLUTION

ENCYCLOPEDIA OF
EVOLUTION

STANLEY A. RICE, PH.D.
Associate Professor, Department of Biological Sciences,
Southeastern Oklahoma State University, Durant

Foreword by
MASSIMO PIGLIUCCI, PH.D.
Professor, Department of Ecology and Evolution,
State University of New York at Stony Brook

Checkmark Books
An imprint of Infobase Publishing

Encyclopedia of Evolution

For information contact:

Checkmark Books
An imprint of Infobase Publishing
132 West 31st Street
New York NY 10001

ISBN-10: 0-8160-7121-7
ISBN-13: 978-0-8160-7121-0

Library of Congress Cataloging-in-Publication Data

Rice, Stanley A.
Encyclopedia of evolution / author, Stanley A. Rice.
p. cm.
Includes index.
ISBN 0-8160-5515-7 (hc)—ISBN 0-8160-7121-7 (pbk)
1. Evolution (Biology)—Encyclopedia. I. Title.
QH360.2.R53 2006
576.803—dc22 2005031646

Text design by Joan M. Toro
Cover design by Salvatore Luongo
Photo research by Tobi Zausner
Illustrations by Dale Williams

Printed in the United States of America

VB Hermitage 10 9 8 7 6 5 4 3 2 1

This book is printed on acid-free paper.

This encyclopedia, not surprisingly, is dedicated to Darwin, but not to the Darwin you might expect. The author would like to dedicate this volume to Emma Wedgwood Darwin, wife of Charles Darwin, who remained Charles Darwin's supportive and loving companion throughout his life of illness and hard work, even though she believed that he was wrong about evolution and about religion. Without Emma Darwin's help, Charles Darwin would almost certainly have failed to complete his most famous works. Of course, some other scientist would eventually have discovered the evidence and mechanism of evolution. Indeed, Alfred Russel Wallace figured it out independently of Charles Darwin. Some other scientists who championed evolution, such as Ernst Haeckel and Herbert Spencer, proclaimed a violent and oppressive version of evolution, in contrast to the Darwinian version that allows for the evolution of altruism. It is possible that evolution would have been an ugly and oppressive theory if it had been presented to the world as Spencerism or Haeckelism. Instead, evolution is Darwinism, and we have Emma Darwin to thank for it.

CONTENTS

FOREWORD

The theory and facts of evolution have been part of modern science for almost 150 years, since the publication of Charles Darwin's *On the Origin of Species,* and yet, the public at large—especially, but not only, in the United States—still has little understanding of evolution, with almost 50 percent of Americans rejecting the Darwinian view of the history of life outright.

It is for this reason that Stanley Rice's *Encyclopedia of Evolution* is a particularly welcome addition to the popular literature on evolution. Until recently, most professional scientists have simply not deemed it worth their time and effort to talk to the public about the nature and importance of what they do. Even some of the notable exceptions have been somewhat mixed blessings, from Stephen Jay Gould's tiresome politically motivated crusades against biological "interpretations" of the human condition to Richard Dawkins's equally misconceived all-out attacks on religion.

It is not that science has no political, religious, or philosophical implications. On the contrary, it is precisely because of such implications that it is important for people to understand both the nature of scientific claims and the specific content of such claims. It is for a similar reason that while we do not want a nation of, say, economists or lawyers, we do want people to be able to understand enough about the economy and the law to make informed decisions in their own lives.

Accordingly, one of the encyclopedia's interesting features is a set of essays interspersed among the entries. Rice writes in a balanced and informative way about several "hot" issues where an understanding of evolutionary biology is relevant, although it cannot by itself be sufficient. In particular, the essays on genetic determinism, on the (biological) reasons for death, and whether an evolutionary scientist can also be a religious person, ought to stimulate some healthy level of thinking in the interested reader. Of course, within the scope of this reference book, such essays can only whet the appetite and provide a window on the often huge and intimidating literature concerned with the issues, but that is exactly what a good book is—a bait to the reader to delve more deeply, to begin a journey that will hopefully continue with twists and turns for her whole life.

Socrates claimed that the root of all human evil is simply ignorance: If only people knew better, if they spent a bit more time inquiring and examining their lives, all would be good. I am not quite as optimistic (or naive) as Socrates, but I do think that part of the reason we see a surge of fundamentalism around the world, with science being attacked by people ranging from the local preacher to the president of the United States, is (partly) because there are so few books like Rice's *Encyclopedia of Evolution.*

Massimo Pigliucci, Ph.D.
Professor of Ecology and Evolution
SUNY–Stony Brook
www.genotypebyenvironment.org

ACKNOWLEDGMENTS

The author gratefully acknowledges reviews of some of the entries by the following scientists:

Simon Conway Morris, Cambridge University, Cambridge, England
Richard Dawkins, Oxford University, Oxford, England
Niles Eldredge, American Museum of Natural History, New York
Victor Hutchison, University of Oklahoma, Norman
Lynn Margulis, University of Massachusetts, Amherst
Mark McMenamin, Mount Holyoke College, South Hadley, Massachusetts
Karl Niklas, Cornell University, Ithaca, New York
Ian Tattersall, American Museum of Natural History, New York
Michael Turnage, Southeastern Oklahoma State University, Durant
Douglas Wood, Southeastern Oklahoma State University, Durant

The author is also grateful to Edward O. Wilson of Harvard University, Lynn Margulis of the University of Massachusetts at Amherst, and Ian Tattersall of the American Museum of Natural History in New York for meeting with the author to discuss some of the material that eventually went into this encyclopedia. The author is also indebted to A. Lisette Rice for historical information, particularly about Thomas Jefferson, and also gratefully acknowledges the time, dedicated efforts, and skill of editor Frank K. Darmstadt, photo researcher Tobi Zausner, and literary agent Jodie Rhodes. The contributions of Mr. Darmstadt and Ms. Zausner to this volume exceeded those of editor and photo researcher and included alerting the author to valuable new information sources.

INTRODUCTION

What you do not know about evolution can kill you. Here are just two examples of how human survival depends on an understanding of evolution (see RESISTANCE, EVOLUTION OF):

- Evolution happens in hospitals, as populations of bacteria evolve resistance to antibiotics. Evolutionary science will continue to help medical professionals minimize the risk that super-germs will evolve.
- Evolution also happens down on the farm. Populations of insect pests and weeds evolve resistance to pesticides and herbicides through the same evolutionary process by which bacteria evolve resistance to drugs. Agricultural scientists now need to take evolutionary principles into consideration when using pesticides, or when introducing genetically modified insect-resistant crops, to minimize the evolution of super-pests. Insecticide-resistant pests, which were once controlled by spraying, can also spread diseases.

There are many other ways in which evolutionary science has transformed medical science (see EVOLUTIONARY MEDICINE). Evolutionary analysis has revealed the multiple origin of HIV, the virus that causes AIDS (see AIDS, EVOLUTION OF), and has also revealed the origins of different strains of influenza. This knowledge will help prevent the transmission of viruses from nonhuman animals to humans. Evolutionary science, as applied to human beings, helps us to understand our own behavior. The result (we hope) will be that we can better control the behavior patterns, such as aggression, that evolution favored in our prehistoric days but which evolution has now given us the flexibility to avoid (see EVOLUTIONARY ETHICS; SOCIOBIOLOGY). Evolutionary theory has even helped computer-aided design (see EVOLUTIONARY ALGORITHMS).

Encyclopedia of Evolution contains 215 entries, which span modern evolutionary science and the history of its development. Even though the entries use some technical terminology, I intend them to be concise and useful to nonspecialist readers. By reading these entries, you may discover that many of the things that you have heard all of your life about evolution are wrong. For example, many people have grown up being told that the fossil record does not demonstrate an evolutionary pattern, and that there are many "missing links." In fact, most of these "missing links" have been found, and their modern representatives are often still alive today (see MISSING LINKS). The biographical entries represent evolutionary scientists who have had and continue to have a major impact on the broad outline of evolutionary science. My choice of which biographies to include is partly subjective, and it reflects the viewpoint of scientists working in the United States. I have also included five essays that explore interesting questions that result from evolutionary science:

- How much do genes control human behavior?
- What are the "ghosts of evolution"?
- Can an evolutionary scientist be religious?
- Why do humans die?
- Are humans alone in the universe?

The appendix of *Encyclopedia of Evolution* consists of my summary of *On the Origin of Species,* by Charles Darwin, which is widely considered to be the foundational work of evolutionary science, and one of the most important books in human history, but which is seldom read.

Encyclopedia of Evolution adopts a stance, nearly universal among scientists, that is critical of creationism and its variations such as intelligent design theory. Criticism of creationism is not by any means the major purpose of this encyclopedia. Many scientists believe in God, as indicated in the essay "Can an Evolutionary Scientist Be Religious?" Conversely, many religious people (including many Christians) fully accept evolutionary science. Therefore the alarm that scientists feel at the continued and growing popularity of creationism is not based in a fear of or antipathy toward belief in God. Rather, the opposition that scientists have to creationism is largely based on two issues:

- Creationism is an indicator of the low value that many people place on scientific evidence, and of the partial failure of science education. Scientists and educators are alarmed when much of the public is more willing to listen to the assertions of preachers or politicians than to evidence from scientific research. The widespread public disregard for science evidences itself in many ways besides the popularity of creationism. People who smoke disregard the facts of human physiology; people who are skeptical of global warming disregard the facts of chemistry and geology; people who drive recklessly disregard the laws of physics. It is not just that many people do not believe in Darwin; they also do not act as if they believe in Newton.
- Creationism is usually part of a larger package of causes that serve political purposes. Creationism is a tool sometimes used by politicians to gain support for their platforms that also include opposition to environmental policies. In my home state, Oklahoma, the viewpoint of many people can be summarized as follows: As regards the natural world of forests and prairies, "It's okay to pour oil on it, it's okay to drive your truck over it, it's okay to bulldoze it, it's okay to chop it down and let it erode away, it's okay to spit on it, so long as you don't believe that it evolved." To believe in a Creator but to trash the creation is a fatally inconsistent position. The seriousness that I would accord a creationist viewpoint depends largely on whether that creationist is willing to declare himself or herself an environmentalist, something most creationists appear unwilling to do.

Among the many examples of the association of antievolutionism with antienvironmentalism are these:

- The conservative religious group Focus on the Family promotes an increase in the exploitation of natural resources, although they do not advocate the destruction of natural resources or habitats; this group is also one of the main proponents of intelligent design theory.
- A high-ranking official in the Interior Department proposed, in the same August 2005 document, that the National Park Service remove nearly all restrictions on off-road vehicle use in national parks and remove nearly every reference to evolution in Park Service educational signage and publications. The National Park Service rejected these proposals, but the official who wrote them ranks even higher than the director of the National Park Service.

It appears that, for many of its proponents, creationism is more a political tool than a sincere pursuit of scientific truth. Some creationists make outlandish claims, such as when U.S. Representative and former House Majority Leader Tom DeLay (R-TX) claimed that the Columbine High School shootings resulted in part from the teaching of evolution in public schools. Can he really believe that nobody killed anybody else before Darwin published his book? (See DARWIN, CHARLES; *ORIGIN OF SPECIES* [book]). What else is one to believe but that in such cases creationism is a political tool? In August 2005 President Bush began using intelligent design theory as a way of gaining support for his political platform.

When a scientist encounters challenges from creationists, the scientist sees not just an attack on evolution but on science in general and also sees a truckload of other issues coming along with it. I hope that the information in this encyclopedia will help readers to recognize the incorrect assertions upon which such politically motivated attacks are made.

In this encyclopedia, I have tried to avoid taking sides on most controversial issues within evolutionary science, of which there are many, including selfish genes, punctuated equilibria, symbiogenesis, and sociobiology. I have also limited myself to evolutionary science. For example, the discussion of the evolution of religion (see RELIGION, EVOLUTION OF) deals only with the evolution of the mental capacity for religion; it does not address whether particular religious beliefs, such as my own Christian beliefs, are or are not true.

My entry into the world of science might have been in 1976. A sophomore biology major at the University of California at Santa Barbara, I was a creationist. To me, there were only two ways of looking at the world. There was only one kind of creationism, and one kind of evolution, and I had chosen the former. The first scientific seminar I attended was a presentation by Lynn Margulis, who had only recently convinced most scientists that the mitochondria and chloroplasts in cells were the evolutionary descendants of symbiotic bacteria (see MARGULIS, LYNN; SYMBIOGENESIS). As I watched Margulis's film of spirochete bacteria embedded in the membrane of a protist, waving exactly like cilia, a light went on in my head. I realized that there were more than two ways of looking at the world, and more than one way of looking at evolution. I realized that evolution was not a doctrine chiseled in stone to which all scholars had to give religious assent. I saw that evolution was an exciting field of exploration in which not only many discoveries of fact but perhaps whole new concepts awaited scientific researchers.

The evolutionary view of the world helps us to appreciate our place in it and also to be grateful for the beauty of the world. When we see ourselves as existing, as highly intelligent creatures, for only a tiny fraction of the history of the universe, and as part of an ecological and evolutionary web of the life of this planet, rather than as recently created masters of a dead world, we should, I hope, be inspired to take better care of our planet and ourselves.

Entries A–Z

adaptation Adaptation is the fit of an organism to its environment, which allows successful survival and reproduction. An adaptation is often described as having the appearance of design, as if the characteristics had the purpose of allowing the organism to more successfully survive and reproduce. The eye, for example, appears to have been designed for vision. However, scientists consider adaptations to be the product of NATURAL SELECTION, rather than of a higher intelligence (see INTELLIGENT DESIGN). Adaptation can refer either to the characteristic itself or to the process that produced it.

The word *adaptation* is used in so many different ways that the meaning is often obscure. Physiologists, as explained below, often use the word to describe processes that occur within an individual organism. Evolutionary scientists, in contrast, restrict the use of the word to evolutionary processes and products. To say that an organism is adapted to its environment is a truism, since if the organism were not it would be dead. Evolutionary scientists are interested in going beyond this truism, to study how and why successful adaptations have evolved.

The concept of adaptation grew from the concept of causation. The ancient Greek philosopher Aristotle made reference to four different kinds of causation, in which he used the metaphor of building a house:

- The material cause is the materials from which the house is built.
- The efficient cause is the set of processes by which it is built.
- The formal cause is the abstract design of the house.
- The final cause is the purpose for which the house was built.

Modern science uses causation, therefore adaptation, only in the sense of Aristotle's efficient cause (see SCIENTIFIC METHOD). Although the success with which an organism is adapted to its environment may result from how closely its structure fits that of an abstract design for the perfect organism under those circumstances, scientists do not focus on the category of formal causes separate from the efficient causes.

The idea that organisms are perfectly adapted to their environments predates evolutionary theory. This idea was central to NATURAL THEOLOGY in which the perfect fit of organism and environment was considered evidence of divine creation. As British chemist Robert Boyle wrote in the 17th century, "There is incomparably more art expressed in the structure of a dog's foot than in that of the famous clock at Strasbourg." English theologian William Paley, in his famous 1802 book *Natural Theology,* used perfect adaptation of organisms as evidence that a Supreme Being had created them. At the time Paley wrote his book there was no credible evolutionary theory that could challenge this view. Charles Darwin's theory of evolution, published in 1859 (see *ORIGIN OF SPECIES* [book]), provided what most scientists consider the first credible natural explanation of adaptation.

Natural theology is wrong in one particularly important way. Scientific investigation has found numerous examples of adaptation that are far from being a perfect fit between organism and environment. One example is the digestive system of the panda. The immediate ancestors of pandas were carnivores, but pandas are herbivores, living exclusively on leaves. Pandas have intestines that are better suited to a carnivore. In well-adapted herbivores, such as sheep, the intestines are up to 35 times as long as the body, while in well-adapted carnivores, the intestines are much shorter, only four to seven times as long as the body. With the help of bacteria, longer intestines allow herbivores more time to digest coarse plant materials, such as cellulose. Meat, in contrast, requires less digestive breakdown. Pandas have intestines that are in the carnivore, not the herbivore, range. The panda's digestive system is, therefore, not well adapted to its function. Modern evolutionary scientists would attribute this to the recent evolutionary shift from meat to leaves in the diet of the panda's ancestors: There has not been time for better adaptation in this case. A natural theologian, in contrast, would have a difficult time explaining this example of imperfection. Evolutionary biologist STEPHEN JAY GOULD used the sixth digit, or thumb, of the panda as another example of the imperfection of adaptation produced by the ongoing process of evolution rather than by the Supreme Being invoked by natural theology.

Another example of an imperfect adaptation is pain. The function of pain is to alert an animal to danger or possible damage. Victims of some kinds of strokes, and of leprosy, lose much of their ability to feel pain and cannot feel the damage that they may do to their extremities. Healthy individuals feel pain and avoid movements that would damage their extremities. Excessive and prolonged pain, however, serves no useful purpose, and much of modern medicine is devoted to the control of excessive pain. Pain is therefore an adaptation, but an imperfect one.

There are many different ways in which an organism can vary with or be adapted to its environment. Below are presented two categories of causes for which the term *adaptation* has been used, with four examples within each (see table). Proximate causes are the immediate events that adjust an organism to its environment, while ultimate causes are the long-term evolutionary events that adjust organisms to their environments. Evolutionary scientists restrict the use of the word *adaptation* to ultimate causation.

Proximate Causes

This category refers to the immediate physical and chemical events that induce an organism's characteristics. Although physiologists often call these events adaptive, evolutionary scientists would not consider them to be true examples of adaptation, because they do not involve evolutionary changes.

1. *Immediate effects of environmental conditions.* On a hot day, a leaf becomes hot, and more water evaporates from the leaf. Evolutionary scientists do not consider this to be an adaptation, because the evaporation is the direct effect of the environment.

2. *Transient responses of the organism to changes in environmental conditions.* On a hot day, the stomata (pores) of a leaf open, allowing more water to evaporate from the leaf, making it cooler. A little mammal, on a cold night, curls into a ball, reducing its heat loss. These responses typically do not involve changes in gene expression (see DNA [RAW MATERIAL OF EVOLUTION]); instead, the molecules and structures that are already present behave differently under the changed conditions. Evolutionary scientists do not consider these events to be adaptation, because they are the immediate response of an organism to changes in its environment.

3. *Changes in gene expression that occur when the environment changes.* Consider again the example of a leaf on a hot day. If the leaf cells accumulate or manufacture more dissolved molecules, the leaf becomes more resistant to wilting. Or consider humans at high elevations. When humans live at high elevations for a week or more, their bone marrow begins to manufacture more red blood cells, which compensates for the lower availability of oxygen at high elevations. These are not immediate, transient responses; they require the genetically controlled manufacture of new materials. This is often called *acclimation* for a single environmental factor or *acclimatization* for an entire set of factors. Many of the features of an organism cannot be changed, after they have initially developed, and cannot acclimatize. In animals, acclimatory changes can also occur as a result of behavioral changes (e.g., muscle building). Evolutionary scientists do not consider acclimation to be adaptation, because acclimation involves changes in gene expression but not changes in the genes themselves.

4. *Differences in gene expression that occur when individuals develop under different environmental conditions.* The leaves of plants that grow in cool, shady, moist conditions are often larger than those of plants that grow in warm, bright, dry conditions, even if all the plants are genetically identical. The characteristics of both kinds of leaf result from the same genes, but the genes are expressed differently in each group of plants. Or consider humans that grow up from childhood at high elevations. These people may develop larger lungs than they would have had if they had lived at sea level; this compensates for the lower oxygen availability at high elevations. These developmental changes are often called *phenotypic plasticity.* Plasticity is rarely reversible and rarely changes after development of the organ or organism is complete. Although leaves can change the amount of dissolved substances in their cells, they cannot change their size once their development is complete; that is, acclimatization is reversible and plasticity is not, in this case. Red blood cell count can change, but lung size cannot, after childhood; in this case also, acclimatization is reversible and plasticity is not. Evolutionary scientists do not consider plasticity to be adaptation, because plasticity involves developmental differences in gene expression but not changes in the genes themselves.

Not all differences between individuals in different conditions are plasticity; in some cases, they may result simply

Eight Processes Often Called Adaptation

Category	Example
Proximate	Direct and transient effect of environment on physiology
	Physiological response of organism to environment
	Acclimation: changes in gene expression induced by environment
	Plasticity: developmental differences induced by environment
Ultimate	Genetic drift: genetic changes in populations without selection
	Contingency: historical accidents
	Constraints and exaptations
	Natural selection acting directly upon a characteristic*

*Evolutionary biologists restrict "adaptation" to this meaning.

from the allometric patterns of growth (see ALLOMETRY). For example, as herbaceous plants grow, they accumulate stems and roots but shed their old leaves. A bigger plant will therefore have a relatively lesser amount of leaf material. Plants that grow in the shade are smaller and have relatively more leaf material than plants that grow in bright sunlight. The greater amount of leaf material is not necessarily plasticity; it may be due solely to the smaller size of the shade plants. Scientists consider such allometric differences to be neither plasticity nor adaptation.

In practice, most plastic responses are observed within populations of organisms, and these individuals also differ in genetic makeup. When quantifying the plasticity of such individuals, the scientist must distinguish between the variability due to genetic differences, the variability due to the response to environment, and the interaction of the two (see POPULATION GENETICS).

Ultimate Causes

These are the causes that have affected the ultimate source of variation, the genes themselves. But even the processes or products of genetic change are not always adaptation.

1. *Genetic drift.* Some of the genetic differences among individual organisms may have resulted from genetic drift (see FOUNDER EFFECT). These genetic changes are random with respect to the environment in which the organisms live. Therefore evolutionary scientists do not consider these differences, although genetic, to be adaptations.

2. *Historical accidents.* Some of the genetic differences among individuals may have resulted from historical contingencies that affected the course of evolution. The most common genes are the ones that got lucky, rather than being selected. One example might be the five-digit pattern in the hands and feet of all living vertebrates. Why five? Why not four or six? There are clear reasons why vertebrates do not have just one digit (they could not grasp anything) or 20 (the digits would be too small and would get broken), but the number five may be arbitrary. Although several primitive terrestrial vertebrates had more than five digits (see AMPHIBIANS, EVOLUTION OF), the ancestor of all modern terrestrial vertebrates had five digits on each foot, and this ancestor is the one that happened to survive and flourish. Its success may have been accidental at first, and its success may have had nothing to do with having five, rather than four or six, digits.

Evolutionary biologist Stephen Jay Gould frequently criticized what he called the *Panglossian paradigm,* named after Dr. Pangloss in French author Voltaire's novel *Candide.* This paradigm claims that all biological characteristics are as good and perfect as they could possibly be, just as Dr. Pangloss claimed that everything on Earth is just as it should be in "the best of all possible worlds." In contrast, Gould said, many so-called adaptations were simply the best that could be achieved with the material that was provided by historical accident.

3. *Exaptations.* Some of the genetic differences among individuals may have resulted because of natural selection on traits other than the ones being considered. That is, natural selection may not have acted directly on the trait being considered; the trait may have been a side effect of the real story of natural selection. Gould and geneticist RICHARD LEWONTIN realized this as they looked at the spaces between the arches of a cathedral, and the artwork contained therein. The artwork was constrained by the spaces. The design of the cathedral was focused on the arches; the spaces between them, or spandrels, were a side effect. Later, Gould and paleontologist Elisabeth Vrba expanded this concept. Some evolutionary changes occurred, like artwork in spandrels, within necessary constraints. Other evolutionary changes in traits resulted from natural selection acting on other traits; they called these changes exaptations rather than adaptations.

Some characteristics of organisms are, like spandrels, structurally inevitable. Many allometric patterns fall into this category. Large plants must have relatively thick trunks, and large animals must have relatively thick legs.

Some exaptations are the side effects of natural selection acting upon developmental patterns (see DEVELOPMENTAL EVOLUTION). One possible example of such an exaptation is the human chin. Modern humans have chins, but earlier hominids did not, and other ape species did and do not. Some scientists tried to imagine how the chin might have aided the survival and reproduction of the human species. One scientist claimed the chin, when stuck out, constitutes a threat gesture. One of the major characteristics of human evolution has been NEOTENY, the retention of juvenile characteristics into adulthood. Neoteny affected the growth patterns of the bones of the skull, and natural selection favored jaws and teeth suitable for an omnivorous diet in human ancestors. The chin was a side effect of different facial bone and jaw growth patterns. The chin was probably not, itself, selected for anything; it could be considered an exaptation rather than an adaptation, because natural selection acted upon neoteny, not directly upon the chin.

Exaptations can also result when a characteristic evolves for one reason and then turns out to be useful for a quite different reason. One possible example of such an exaptation is the feathers of birds (see BIRDS, EVOLUTION OF). The major use of feathers in modern birds is for flight, but the earliest feathers may have functioned primarily to allow the bird to retain body heat. Feathers, which were adaptations for conserving body heat, turned out to be useful exaptations for flight. Such exaptations have been called *preadaptations,* but this term is now seldom used, since it implies that the evolutionary process can adapt organisms to future events, which is impossible. Gould considered many human characteristics, including major aspects of human intelligence, to be exaptations.

It can be very difficult to distinguish adaptations from exaptations. The classic example of an adaptation that has appeared in numerous popular science books and textbooks for over a century is the neck of the giraffe. To most observers, it would appear obvious that long necks are an adaptation that allows giraffes to feed at the tops of trees. If this is so, why do female giraffes have shorter necks than male giraffes, and why are giraffes so frequently observed actually bending down to eat leaves? Careful observations have shown that males with longer necks prevail in male–male competition, and females chose the males with longer necks (see SEXUAL SELECTION). Now that giraffes have long necks, they can use them for eating leaves from treetops, but that is not the original reason that long necks evolved in these animals. The long neck of the giraffe, therefore, is an adaptation to combat, and an exaptation for reaching leaves at the tops of trees.

4. *True adaptations.* The only true adaptations, according to many evolutionary scientists, are the genetic changes that result from natural selection favoring those actual traits.

For each of the preceding eight definitions, the term *adaptation* can refer either to the evolutionary process (natural selection) or to the product. This leads to 16 meanings of adaptation. As explained above, however, much clarity is achieved by restricting the use of the term to just two: the process and product of evolution acting directly on a characteristic.

In most cases, a characteristic is not an adaptation with just a single function. Most characteristics are adaptations with numerous evolutionary causes. For example, the glands within an animal's epidermis are an adaptation with several functions, each of which may have provided a separate evolutionary advantage. Some of the glands produce sweat, which allows the animal to become cooler as the sweat evaporates. Sweat also contains dissolved molecules. The body of the animal uses sweat as one of its methods of disposing of excess or toxic materials. Molecules in sweat can also serve as chemical communication between animals. Mammary glands are modified sweat glands. Sweat glands, therefore, are an adaptation to at least four different functions.

Once a true adaptation has been identified, an evolutionary scientist may attempt to test a hypothesis to explain how and why it evolved. To test a hypothesis, scientists need to obtain as large a sample of data as possible (see SCIENTIFIC METHOD). These data must be independent of one another, rather than repeated measures on the same phenomenon. Recently evolutionary scientists noticed that testing hypotheses about adaptation requires observations in which the adaptation has evolved independently. They refer to these observations as *phylogenetically independent.* As an example, consider small leaves as an adaptation for bushes that live in dry conditions, and large leaves as an adaptation for bushes that live in moist conditions. The scientist determines the average leaf area for each of 12 species that live in different moisture conditions and finds that there is a positive correlation between leaf size and moisture. If these 12 species represent six species from a small-leaved genus that lives in dry climates, and six species from a large-leaved genus that lives in moist climates, the number of independent observations is two, not 12. These 12 species evolved from just two ancestral species, one with small leaves, one with large leaves. If the investigator claims that bushes that live in dry climates have smaller leaves than those in moist climates, the investigator has only two, not 12, data to test the claim. The 12 species are not phylogenetically independent of one another. This phenomenon is called the *phylogenetic effect.*

Evolutionary biologist Joseph Felsenstein determined a way around the problem of the phylogenetic effect. The investigator first identifies pairs of closely related species and then compares the two members of each pair with one another. If, within most of the pairs, the species from the drier climate has smaller leaves, then the investigator has six observations that test the claim.

An excellent example of testing a hypothesis about adaptation, incorporating the phylogenetic effect, is the study by ecologists Angela Moles, David Ackerly, and colleagues. The hypothesis is that large seed size in plants is an adaptation to enhance the survival of seedlings in plant species that live a long time and grow to a large size (see LIFE HISTORY, EVOLUTION OF). Rather than calculating a simple correlation between seed size and body size in plants, the investigators produced a phylogenetic tree (see CLADISTICS) of 12,987 plant species and determined the evolutionary events in which significant changes in seed size occurred. They found that increases in seed size occurred along with evolutionary shifts toward large plant size. These investigators could conclude, with a great degree of confidence, that large seed size is an adaptation to large body size in plants.

In order for adaptation to be a useful concept in evolutionary science, investigators restrict the use of the word to characteristics that result from the direct effects of natural selection, and they investigate adaptation using phylogenetically independent comparisons.

Further Reading

Dawkins, Richard. *The Blind Watchmaker: Why the Evidence of Evolution Reveals a Universe Without Design.* New York: Norton, 1996.

Felsenstein, Joseph. "Phylogenies and the comparative method." *The American Naturalist* 125 (1985): 1–15.

Gould, Stephen Jay. *The Panda's Thumb: More Reflections in Natural History.* New York: Norton, 1980.

———, and Richard Lewontin. "The spandrels of San Marco and the Panglossian paradigm: A critique of the adaptationist programme." *Proceedings of the Royal Society of London B* 205 (1979): 581–598.

———, and Elisabeth Vrba. "Exaptation—a missing term in the science of form. *Paleobiology* 8 (1982): 4–15.

Moles, Angela T., et al. "A brief history of seed size." *Science* 307 (2005): 576–580.

Sultan, Sonia E. "Phenotypic plasticity for plant development, function, and life history." *Trends in Plant Science* 5 (2000): 363–383.

Van Kleunen, M. and M. Fischer. "Constraints on the evolution of adaptive phenotypic plasticity in plants." *New Phytologist* 166 (2005): 49–60.

adaptive radiation Adaptive radiation is the evolution of many species from a single, ancestral population. Throughout the history of life, many species have become extinct, either by natural selection or simply bad luck (see EXTINCTION; MASS EXTINCTIONS), while others have produced many new species by adaptive radiation. The net result has been a steady increase in BIODIVERSITY through evolutionary time. Adaptive radiation occurs because the single ancestral population separates into distinct populations that do not interbreed, thus allowing separate directions of evolution to occur in each (see SPECIATION).

The world is full of numerous examples of adaptive radiation. Nearly every genus that contains more than one species, or any family that contains more than one genus, can be

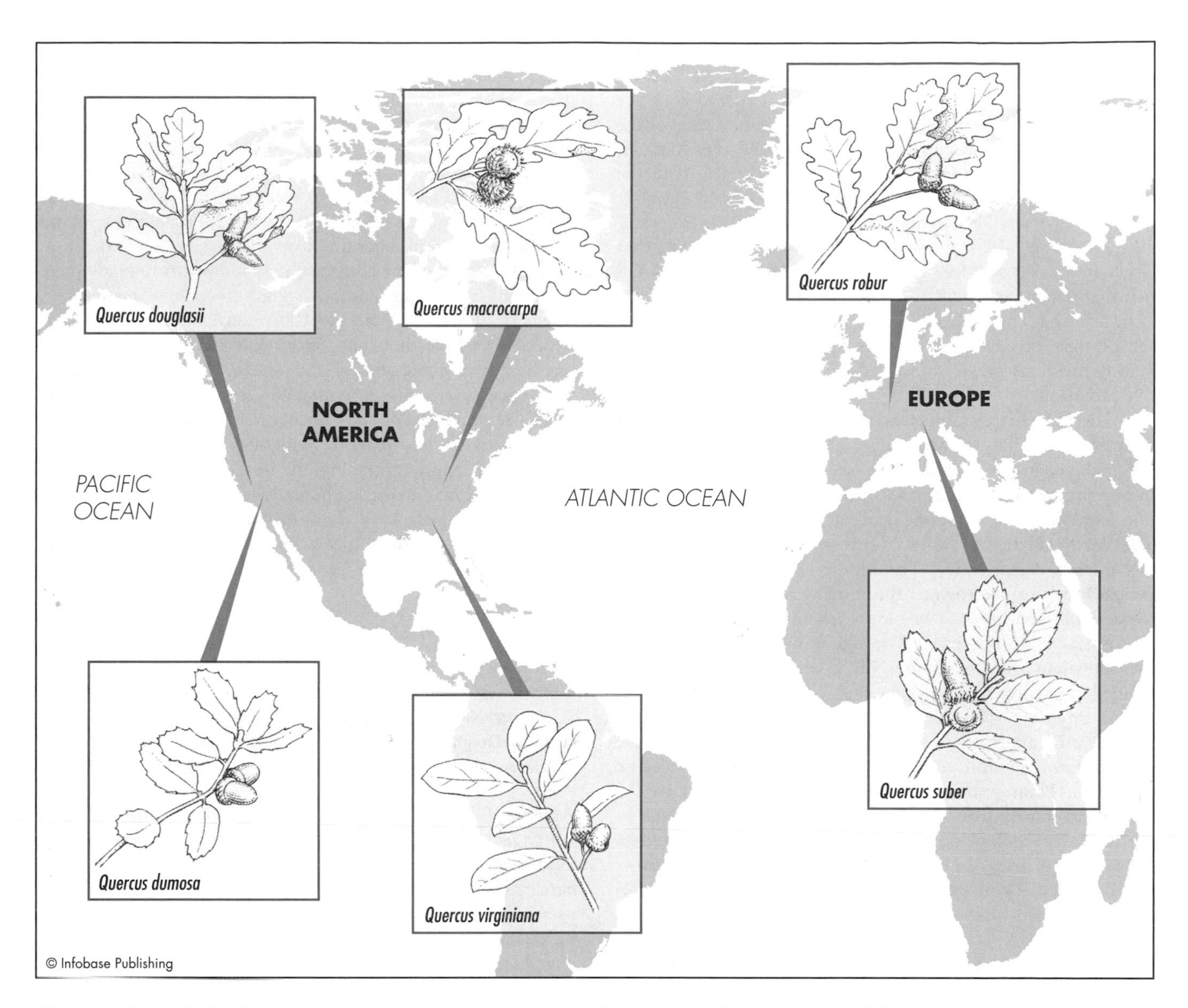

Adaptive radiation of oaks. Species of white oaks have evolved separately in Europe, eastern North America, and California. Also, evergreen and deciduous species of white oaks have evolved separately in these three locations. All of these oaks evolved (radiated) from a single ancestral species.

considered as an example. (The taxonomic hierarchy of classification used by scientists—species, genera, families, etc.—reflects the evolutionary history of species; see LINNAEAN SYSTEM). The following are two examples:

Consider the white oaks, part of the genus *Quercus*. At one time, probably before 80 million years ago, there was a single species of white oaks. In the late MESOZOIC ERA and early CENOZOIC ERA, warm conditions were widespread on the Earth. Oaks spread throughout the Northern Hemisphere. Three of the locations in which they are now found are Europe, eastern North America, and California.

North America then separated from Eurasia (see CONTINENTAL DRIFT), new mountain ranges such as the Rocky Mountains arose, and conditions became cooler and drier during the Cenozoic era. These events separated the white oaks of Europe, eastern North America, and California. Separate species evolved in each of these locations, including *Quercus robur* (English oak) in Europe, *Quercus macrocarpa* (bur oak) in eastern North America, and *Quercus douglasii* (blue oak) in California (see figure above). Furthermore, in each of these locations, cooler conditions allowed the evolution of the deciduous pattern of leaf production in some areas, while warmer conditions encouraged the evolution of the evergreen pattern in other areas. The previously mentioned species are all deciduous. *Quercus suber* (cork oak) is an evergreen white oak of Europe, *Quercus virginiana* (live oak) is the evergreen white oak of eastern North America, and *Quercus dumosa* (scrub oak) is an evergreen white oak of California. In all three areas, both deciduous and evergreen (live) oaks evolved (with some exceptions). These six oak species, and many others, constitute an example of adaptive radiation within the white oak subgenus.

While the adaptive radiation of oaks occurred over the course of about 80 million years, the adaptive radiation of the approximately 500 species of cichlid fishes of Lakes Malawi, Tanganyika, and Victoria in Africa has occurred just in the last few thousand years. From what was probably a single ancestral species, the cichlids in Lake Victoria have radiated into such different species as: fish with heavy jaws that crush mollusks; slender, swift fish that eat plankton; small fish that eat parasites off of the skins of larger cichlid fishes; large fish that eat other fishes; and fish with sharp teeth that scrape algae off of rocks. There is very little genetic variation among these species, so rapid and recent has been their adaptive radiation.

The Hawaiian archipelago also provides an example of adaptive radiation, in both time and space. The Hawaiian archipelago, in the Pacific Plate (see PLATE TECTONICS), sits over a plume of lava that rises from the mantle, known as a hot spot. Volcanic eruptions occur at the site of the hot spot. The Pacific Plate has moved northwest, while the hot spot has remained stationary. The oldest islands in the Hawaiian archipelago are toward the northwest and the youngest toward the southeast. The oldest islands have now eroded and are beneath sea level. The oldest major islands still in the Hawaiian archipelago are Niihau and Kauai, which were formed about five million years ago. They have no active volcanoes anymore. The islands of Oahu, Molokai, Lanai, and Maui are of intermediate age. The youngest island, Hawaii, is less than a million years old and still has active volcanic eruptions. Plant and animal species have dispersed to the islands, primarily from the southwest, and species unique to the Hawaiian archipelago have evolved (see BIOGEOGRAPHY). Species have also dispersed from older islands to younger islands once the younger islands became suitable for life, radiating into new species on each island. Adaptive radiation has occurred in space (unique species on each island) and in time (from older to younger islands) in the Hawaiian archipelago.

Perhaps the most famous adaptive radiation has been the evolution of DARWIN'S FINCHES in the GALÁPAGOS ISLANDS. A single ancestral species, which may have come from the South American mainland, has radiated out into 13 species, in four genera. Finches in three of the genera live in trees. Some of them have some amazing adaptations, such as the species that uses sticks to pry insects out of holes in branches. The largest genus, with six species, consists of finches that hop on the ground, eating mostly seeds.

Sometimes speciation occurs as one species responds to the presence of another (see COEVOLUTION). However, the presence of other species can preclude some directions of speciation. On the Galápagos Islands, predators are largely absent, and some of the finches have figured a way to drink blood from the legs of boobies (large marine birds) by pecking them. Also, woodpeckers are absent, and some of the finches, as mentioned above, use sticks to get insects out of holes. In a habitat that was already filled with efficient predators and woodpeckers, the amateurish adaptations displayed by the blood-drinking and insect-prying finches would probably not have succeeded. Therefore, adaptive radiation can occur rapidly when a population invades a new habitat—if, that is, the population is adapted to the conditions of the new habitat, and if the new habitat does not have potential competitors.

Release from competition is the reason that adaptive radiation has occurred most spectacularly in the history of the Earth after large disasters. During the Mesozoic era, mammals were few in number and low in diversity (see MAMMALS, EVOLUTION OF). Their constant body temperature made them successful nocturnal animals. The dinosaurs, meanwhile, ruled the daytime. After the CRETACEOUS EXTINCTION that cleared away the dinosaurs, mammals underwent an explosive adaptive radiation. Large mammals did not exist before the Cenozoic era; any mammal species that began to evolve large size would have been unsuccessful in competition with large dinosaurs, but many huge mammals evolved in the early Cenozoic. Along with the big mammals, many of the modern categories of mammals, including whales, primates, and bats, evolved. Mammals have continued to have adaptive radiations, but none quite as spectacular as the adaptive radiation that occurred early in the Cenozoic era.

Further Reading

Gillespie, Rosemary G. "The ecology and evolution of Hawaiian spider communities." *American Scientist* 93 (2005): 122–131.

Schilthuizen, Menno. *Frogs, Flies, and Dandelions: The Making of Species.* New York: Oxford University Press, 2001.

Schluter, Dolph. *The Ecology of Adaptive Radiation.* New York: Oxford University Press, 2000.

Agassiz, Louis (1807–1873) Swiss-American *Paleontologist, Geologist* Louis Agassiz, born May 28, 1807, in Switzerland, relocated to the United States to complete his scientific career. He is most famous as the discoverer of the ICE AGES, as a proponent of catastrophism, and as an opponent of evolutionary theory.

As a student, Agassiz left Switzerland to study in Heidelberg, Germany. As he studied the anatomy of animals, learning from the French scientist Cuvier (see CUVIER, GEORGES), and as he read what philosophers had written about the natural world, he became convinced that each species of animal was a reflection of an ideal type and connected with all other species structurally but not by origin. He was convinced that species had been created separately and never changed.

While a professor of geology at Neuchâtel Academy in Switzerland, he began a study of the geological formations that catastrophists believed were the product of the most recent worldwide catastrophe, the Flood of Noah (see CATASTROPHISM). He concluded that the formations were the product not of liquid water but of ice: He discovered the Ice Ages. Not only did his theory of glaciations explain much of geology, but it also made sense of the distribution of many plant and animal species.

Agassiz came to the United States for a speaking tour in 1846. Besides his attraction to the New World, Agassiz needed the money and got it: $6,000 in a half year on the speaking circuit. His lectures (even with his hastily acquired

English) were a resounding success, and Harvard offered him a professorship in 1848. Agassiz saw in the United States a land of opportunity for intellectual growth as well as for the financial possibilities that drew most immigrants, and he accepted the Harvard professorship. Agassiz played a major role in building Harvard's famous Museum of Comparative Zoology.

Agassiz's reaction to the thunderbolt of Charles Darwin's *Origin of Species* (see *ORIGIN OF SPECIES* [book]) was to defend the eternal separateness of species even more vigorously. He also, but less vigorously, opposed Lyellian UNIFORMITARIANISM. In 1860, just after the release of Darwin's *Origin of Species,* Agassiz's younger colleague at Harvard, the botanist Asa Gray, debated him publicly about evolution (see GRAY, ASA). Agassiz's defense of the fixity of species spilled over into his social convictions. He was offended by having to be close to African Americans. He considered non-European races to have been separately created by God. Agassiz defended slavery and opposed interracial marriage as being an unnatural mixing of what God had intended to keep separate.

Today, Agassiz might be classified as a creationist (see CREATIONISM), but he would not be accepted by the creationists as one of their associates, because (1) he was a Unitarian, rather than what would today be called a fundamentalist, and (2) he believed in a dozen separate creations of humans, rejecting the biblical story of Adam and Eve. Interestingly, it was his opponent, the evolutionist Asa Gray, who was the orthodox Christian and who accepted Africans as his fellow humans.

It is impossible to tell whether his continued opposition, in his later years, to evolution was due to stubbornness or sincerity. During a voyage he actually visited the Galápagos Islands and saw the same evidence that Charles Darwin had seen, but he refused to admit their evolutionary implications. Nevertheless his studies of zoology and of the Ice Ages have contributed substantially to the modern scientific understanding of the evolutionary history of the Earth. He died on December 12, 1873.

Further Reading

Lurie, Edward. *Louis Agassiz: A Life in Science.* Baltimore, Md.: Johns Hopkins University Press, 1988.

age of Earth Most mythologies contain a nearly instantaneous creation of the Earth by one or more gods, followed shortly thereafter by the creation of humankind. The modern evolutionary insight that humankind has existed only briefly in the long history of the Earth is a product of only the last two centuries and is an insight most people even today do not appreciate. The 5,000 years of human history is only one-twentieth of the approximately 100,000 years that *HOMO SAPIENS* has existed as a species in fully modern form, and only one-millionth of the 4.6 billion years that the Earth has existed. According to science writer John McPhee, if a person stretches out his or her arms to represent the age of the Earth, all of civilization corresponds to a day's growth of fingernail.

The predominant source of Earth history that was available to scholars in the Western world until just a few centuries ago was the Bible. By adding up the genealogies of the Old Testament, Bible scholars calculated that the Earth was about 6,000 years old. Some Bible scholars, early in the age of science, got caught up with a desire for precision, without having any more information from which to work. The most famous example of this is James Ussher, Archbishop of Armagh and Anglican Primate of All Ireland. He published *Annals of the Old Testament, Deduced from the First Origin of the World* in 1650, in which he identified the moment of creation as beginning at noon on October 23, 4004 B.C.E. Earlier scholars had concluded that the Supreme Being's plan for the history of the world fit into precise millennia. This is an assumption shared by many modern people, who expected Earth-shattering events to occur in the year 2000 (called "Y2K" in modern jargon). Ussher knew that King Herod, a Jewish king who ruled Israel under Roman occupation, died in 4 B.C.E. yet, according to the Bible, was alive when Jesus was born. Therefore, Ussher concluded that the Western calendar is off by four years, and he chose 4004 B.C.E. rather than 4000 B.C.E. as the year of creation. Biblical scholars disagreed as to whether there may have been "pre-Adamite" humans that existed, possibly for long periods of time, before the creation described in the first chapter of Genesis. Even from a study of the Bible, such precision as exhibited by Archbishop Ussher was not accepted by all scholars. Evolutionary scientist Stephen Jay Gould (see GOULD, STEPHEN JAY) has written more extensively on the topic of how Ussher and other biblical scholars calculated the age of the Earth and its major historical events.

If the Earth was only 6,000 years old, then it must be virtually unchanged since the moment of creation, with the exception of the Noachian Deluge (the Flood described in the Old Testament), according to these biblical scholars. As scientific investigation of the Earth began, many observations did not fit with the concept of a recently created Earth. Most famously, geologist James Hutton (see HUTTON, JAMES) observed the sedimentary layers of rock at Siccar Point in Scotland and realized that vast stretches of time were necessary for their formation (see UNCONFORMITY). Geologist Charles Lyell (see LYELL, CHARLES) expanded Hutton's view into a geological model of UNIFORMITARIANISM in which the Earth was, as far as scientific inquiry could determine, eternal. As Hutton wrote in a 1788 treatise, "The result, therefore, of our present enquiry is, that we find no vestige of a beginning,—no prospect of an end."

As the documentation of fossils continued (see FOSSILS AND FOSSILIZATION), it became apparent that there was a time early in the history of the Earth when complex life-forms did not yet exist (see PRECAMBRIAN TIME). Although this implied that there must have been a beginning, this beginning might have been many billions of years in the past. Geologists were able to determine the relative order in which the different assemblages of species had existed on the Earth (see SMITH, WILLIAM). The present sequence of GEOLOGICAL TIME SCALE was largely established by geologists during the 19th century, although they had no way to calculate the absolute periods of time in which these organisms had lived.

Physicist William Thomson (see KELVIN, LORD) calculated that it may have taken about 100 million years for the Earth to have cooled from a molten state to its present temperature. He also calculated a similarly brief period during which the Sun would have been able to burn without exhausting its supply of fuel. These calculations, which Kelvin published in 1866, seemed to place an upper limit of 100 million years for the age of the Earth, in only the last part of which the Earth was cool enough for life to survive and evolve upon it. To many scholars, this was insufficient time for evolution to occur by means of NATURAL SELECTION as proposed by Charles Darwin (see DARWIN, CHARLES). Some, like the zoologist Huxley (see HUXLEY, THOMAS HENRY) simply accepted the figure and proposed that evolution, occurring occasionally by leaps rather than by gradual change, had in fact occurred during just the final portion of that 100 million year time span. The problems with Kelvin's calculations became apparent early in the 20th century. Kelvin did not know, until possibly at the end of his life, about radioactivity (from the fission of elements such as uranium) as a source of heat energy in the Earth. Radioactivity could therefore have kept the Earth warm far longer than the 100 million years required for an Earth-sized ball of lava to cool. Further, he knew only of combustion as a possible source of energy for the Sun, not realizing that the fusion of hydrogen atoms would allow the Sun to have existed for several billion years.

Radioactivity provided a source of energy that had kept the inside of the Earth warm for billions of years. Radioactive elements degenerated on a precise and calculable schedule of half-lives. This fact allowed geologists to develop techniques of RADIOMETRIC DATING. Radiometric dating techniques contained some sources of error, but geologists continue to find ways to avoid these errors, and radiometric dating has become a very precise method of determining the periods of time in which different fossilized species lived, the times at which major Earth catastrophes occurred (see CRETACEOUS EXTINCTION; PERMIAN EXTINCTION), and of the age of the Earth.

The Earth formed about 4.6 billion years ago. Its initial heat, plus the impacts of extraterrestrial debris (see ASTEROIDS AND COMETS) prevented the formation of oceans until about 3.8 billion years ago. According to most evolutionary scientists, life began shortly after the oceans formed (see ORIGIN OF LIFE).

Further Reading

Dalrymple, G. Brent. *The Age of the Earth*. Stanford, Calif.: Stanford University Press, 1994.

Gould, Stephen Jay. "Hutton's purpose." Chap. 6 in *Hen's Teeth and Horse's Toes: Further Reflections in Natural History*. New York: Norton, 1983.

———. *Questioning the Millennium: A Rationalist's Guide to a Precisely Arbitrary Countdown*. New York: Harmony Books, 1997.

agriculture, evolution of Agriculture is the process by which animals cultivate plants (or occasionally fungi or protists) for food or other resources. More broadly defined, agriculture also includes the breeding and raising of livestock animals. In most cases of agriculture, the crop or livestock species and the animal species that raises it are mutually dependent upon one another for survival.

Several species of ants carry out activities that bear striking parallels to human agriculture. For example, leaf-cutter ants (genus *Atta*) cultivate gardens of fungus. Massive foraging parties of leaf-cutter ants gather pieces of leaf from many species of tropical plants and carry them back to their nests. They do not eat the leaves, which contain many toxins. Instead they chew them up into compost, on which a kind of fungus grows. The fungus grows nowhere else except in the mounds of leaf-cutter ants; when the ants disperse, they take fungus tissue with them. The ants eat fungus tissue and almost nothing else. Beneficial bacteria that grow on the bodies of the ants produce chemicals that inhibit the growth of other bacteria in the compost. For this reason, some biologists consider these and other ants to be a promising source of new antibiotics. Because the ants deliberately prepare compost for the fungus, and because of the mutual dependence of ants and fungus upon one another, the ant-fungus relationship can be considered an example of agriculture.

Several species of ants in the seasonal tropics of Central America live on and in trees of the genus *Acacia*. The ants do not eat the leaves of the acacia; instead they consume nectar that is produced by glands on the stems (not in the flowers) of the trees, and they eat globules of protein and fat, called Beltian bodies, that grow on the tips of immature leaves. The ants chew out the insides of the acacias' unusually large thorns, and live inside the thorns. In some cases, experimental manipulation has shown the ants to be dependent upon the acacias for survival. The ants attack and kill other insects and drive away larger animals that attempt to feed on the acacias. When vines or other plants begin to grow in the immediate vicinity of the acacias, the ants sting them or chew them down. In most cases, the acacias are completely dependent upon the ants; when the ants are experimentally removed, vines overgrow the acacias, and animals browse the leaves heavily. The acacias remain green during the dry season, when most of the other trees lose their leaves; but these green targets go undisturbed by herbivores, because of their protective ant army. Because the ants weed out other plants from the vicinity of the acacias, and defend their crops, and because of the mutual dependence of ant and acacia, the ant-acacia relationship can be considered an example of agriculture. When, about 10,000 years ago, agriculture evolved in the human species, it was not the first time that agriculture had evolved on this planet. Hereinafter, "agriculture" refers to human agriculture.

Some scholars used to believe that human agriculture was invented by a brilliant man in a tribal society of hunter-gatherers. Other scholars pointed out that, since women gathered most of the plant materials, agriculture was probably invented by a woman. Both the brilliant-man theory and the brilliant-woman theory are incorrect, however, because agriculture could not have been invented in a single step by anyone. It had to evolve. Agriculture had to evolve because unmodified wild plants are unsuitable for agriculture. There are four reasons:

Some Centers of the Evolution of Agriculture and Their Major Crops

Center of origin	Examples of major crops
Mesopotamia	Wheat, barley, lentils
China	Rice, soybeans
Central America	Maize, beans, tomatoes
South America	Potatoes, quinoa

- The seeds of most wild plant species have dormancy when they are mature. That is, when planted, the seeds will not germinate. Many require a period of exposure to cool, moist conditions before the inhibitors within the plant embryo break down and the seed germinates. If a brilliant man or woman planted the first agricultural seed from a wild plant, it would not have grown, and they would have rightly concluded that agriculture was not a good idea.
- The seeds of many wild grains shatter or fall off of the stem as soon as they are mature. Since the whole point of the seed is to grow in a new location, shattering is beneficial to the plant; but it is extremely inconvenient for a human harvester.
- The seeds of many wild plant species contain toxins.
- The seeds of many wild grains are small.

Furthermore, the advantages of primitive agriculture would not have been immediately apparent to intelligent hunter-gatherers. Agriculture requires intense labor. Modern hunter-gatherers often barely eke out an existence in marginal habitats such as the Kalahari Desert or Great Outback; but these are the habitats to which tribes and nations with more advanced tools have driven them. Before agriculture, many tribes lived in rich habitats in which hunting and gathering in many cases provided a comfortable level of existence. For these reasons also, agriculture had to evolve gradually.

Agriculture originated separately in several parts of the world. It originated at least twice in the New World (Central America and South America), in Mesopotamia, and in China. Agriculture may have had several separate origins in each of these areas, as well as in New Guinea. The only inhabited part of the world in which agriculture did not evolve was Europe; European agriculture was imported from the Middle East. Scientists and historians know that agriculture originated separately in these locations because the staple crops were different in each (see table). The domesticated sumpweeds and sunflowers grown by inhabitants of North America were replaced by Mexican maize and beans in pre-Columbian times.

Agriculture began earliest (about 10,000 years ago) in the Middle East, especially in the Tigris-Euphrates floodplain of Mesopotamia. Agriculture began there first, apparently because there were many wild species of plants that were almost suited for agriculture. Of the 56 species of wild grains that have large seeds, 32 grow in the Middle East. These wild grains needed little evolutionary transformation to become crops. The transformation from gathering to agriculture would have been a relatively quick and easy process in the Middle East.

At first, the transformation of wild into domesticated plants was effected unconsciously by the gatherers. Within each species of grain (for example, wild wheat or wild barley), the largest seeds, the seeds that shattered the least, and the ones that tasted the best were the ones that people preferentially gathered. The gatherers would also take grain seeds with them when they traveled and may also have cast them onto the ground in the new location. In this way, the seeds with the least dormancy were the ones most frequently chosen by the gatherers. Small, shattering, less palatable, dormant grains (the wild type) were thus gradually transformed, by unconscious NATURAL SELECTION and by deliberate ARTIFICIAL SELECTION, into large, non-shattering, palatable, non-dormant grains similar to today's crops. A similar process occurred in other wild food plants.

This same process occurred in all the other places where agriculture originated but took longer. There were fewer suitable wild food plants in Mexico and even fewer in the Andes. It took longer for wild teosinte to evolve into maize, as a greater evolutionary transformation was needed; it took even longer for poisonous wild potatoes to evolve into edible ones. As evolutionary biologist Jared Diamond points out, the delayed development of agriculture in other continents compared to the Middle East had nothing to do with any ethnic inferiority but was due to the lesser availability of suitable wild plants.

Meanwhile, the human population was growing. By 10,000 years ago, all available, favorable habitats were occupied by humans. Hunters and gatherers were much less free to move to a new location when resources became scarce. About that same time, the weather became cooler and drier in the Middle East. Hunting and gathering became a much less desirable way of life. When certain tribes then tried deliberate cultivation, it was worth the extra work, and the plants were suitable.

Once agriculture had evolved, societies that depended upon it could not easily revert to hunting and gathering, for several reasons:

- Agriculture allowed greater food production and greater population growth. A large population could not revert to hunting and gathering. This is obvious today, for the Earth cannot support six billion hunters and gatherers; but it was also true in all local regions in which agriculture evolved, thousands of years ago.
- Since agriculture allowed greater food production, it was no longer necessary for nearly all tribal members to participate in food procurement. Farmers raised enough food for everybody, which allowed other people to be soldiers and priests. A hunting, gathering tribe was ill-equipped to fight an agricultural tribe with a dedicated army. A world trapped in agriculture was now trapped into war. Agriculture allowed the rise of religious and governmental hierarchies as well as of armies.
- With the evolution of agriculture, productive farmland became valuable. People settled into cities, because they had

to stay in one place at least long enough for one harvest. Agriculture promoted the rise of civilization. Civilization rose earlier in Mesopotamia than in other places because agriculture began earlier there. With civilization came advanced technology. Because they needed to defend particular tracts of territory, the armies now had a lot more to fight about. The cultural groups that developed agriculture first were the first to be civilized and to have advanced technology, which allowed them to conquer the cultural groups in which this process had not progressed as far. Biologist Jared Diamond explains that this is why Europeans conquered America, driving natives into reservations, rather than Native Americans conquering Europe and driving Europeans into remote corners of the Alps and Pyrenees.

The evolution of agriculture is a perfect illustration that evolution does not operate for the good of the species (see GROUP SELECTION). Agriculture did not improve the average health of human beings. In fact, the average life span in early agricultural societies was less than that in contemporaneous hunter-gatherer societies. This occurred for two reasons:

- Diseases spread more rapidly in cities in which people were trapped with one another's wastes, garbage, and germs.
- Agriculture actually decreased the quality of human nutrition by making people dependent upon a few crop plants rather than a diversity of wild foods. In particular, the human body evolved under conditions in which ascorbic acid (vitamin C) was readily available from wild fruits. When entire populations became dependent upon crops with little vitamin C, scurvy became a way of life.

Along with agriculture, herding began, starting with the wild animal species most amenable to survival and breeding under captivity. Unconscious natural selection, then conscious artificial selection, resulted in the evolution of livestock species, such as the cow, which evolved in western Eurasia and Southeast Asia from two distinct wild species. Once again, the earlier development of herding in the Middle East than in other areas occurred because many wild animal species of the Middle East (such as goats and sheep) were amenable to herding, while wild animal species such as deer in North America were not. Livestock animals provided high quality food (meat and milk), often by consuming wild foods that humans could not eat. This was especially true of goats. Pigs, on the other hand, eat many of the same foods as humans. This competition between pigs and humans for food may be one reason that pigs were considered undesirable ("unclean") by some cultures in the arid regions of the Middle East. As with agriculture, herding caused a narrowing of the food diversity base from many wild animal species to a few livestock species.

During the migrations of people during the past few thousand years, new assortments of genes have occurred, producing racial diversification. The human species as a whole has undergone no significant evolutionary changes during the time since the origin of agriculture. Humans a hundred thousand years ago were physically indistinguishable (as far as is known from fossil remains) from modern humans. Nine-tenths of the history of the human species occurred under hunter-gatherer conditions, without significant genetic evolution. Therefore, while the evolution of agriculture is a prime example of the evolution of mutualism, it is not an example of COEVOLUTION, because the crop species genetically evolved, and the humans did not.

Further Reading

Diamond, Jared. *Guns, Germs, and Steel: The Fates of Human Societies.* New York: Norton, 1997.

Kislev, Mordechai E., Anat Hartmann, and Ofer Bar-Yosef. "Early domesticated fig in the Jordan Valley." *Science* 312 (2006): 1,372–1,374.

Konishi, Saeko, et al. "An SNP caused loss of seed shattering during rice domestication." *Science* 312 (2006): 1,392–1,396.

Rindos, David. *The Origins of Agriculture: An Evolutionary Perspective.* San Diego: Academic Press, 1984.

Tenno, Ken-ichi, and George Willcox. "How fast was wild wheat domesticated?" *Science* 311 (2006): 1,886.

AIDS, evolution of HIV, the human immunodeficiency virus, can ultimately result in Acquired Immune Deficiency Syndrome (AIDS) in humans. Many scientists and public health professionals consider AIDS to be the major epidemic of modern times. Evolutionary science contributes greatly to an understanding of the origin of and changes within this disease.

The human immunodeficiency virus (HIV) infects some human white blood cells and eventually causes them to die. This partial disabling of the human immune system allows opportunistic infections to occur as normally harmless microbes become parasitic (see COEVOLUTION). Because this immune deficiency is acquired through infection, and results eventually in a whole set of symptoms, it is called Acquired Immune Deficiency Syndrome (AIDS). Most people, when not infected with HIV, can easily resist microbes such as the cytomegalovirus, the bacterium *Mycobacterium avium,* or the protist *Pneumocystis carinii,* but these microbes can cause complications such as fatal pneumonia if HIV partially disables the immune system. The immune system apparently also protects the human body from viruses that can induce certain kinds of cancer, such as Kaposi's sarcoma, which also afflict people infected with HIV.

HIV infection spreads slowly within the body of the victim. After an initial acute phase of infection, during which the immune system launches a partly effective response, a chronic phase of invisible spread follows. It may be several years after HIV infection, during which individuals are HIV-positive, before those individuals exhibit symptoms resulting from the loss of immune competence, at which time they are said to have AIDS. There are different strains of HIV, of which HIV-1 is the most common worldwide.

HIV is a retrovirus, which means that it stores its genetic information in the form of RNA and is able to transcribe this information backward into DNA. In contrast, most viruses and all living cells store genetic information in DNA (see DNA [RAW MATERIAL OF EVOLUTION]). HIV carries, inside its protein coat, two molecules of an enzyme known as reverse transcriptase, which makes DNA from the genetic information in its RNA. HIV also carries an integrase enzyme, which helps the DNA insert into the host chromosomes, and a special protease enzyme that helps process the protein coat of a newly synthesized virus. Because of reverse transcriptase and integrase, HIV can insert DNA copies of itself into the chromosomes of victims. This is similar to the way some transposable genetic elements operate within the genomes of complex organisms (see SELFISH GENETIC ELEMENTS). Human chromosomes contain many copies of the reverse transcriptase gene, most of which are nonfunctional parts of what is sometimes called "junk DNA" (see NONCODING DNA) but some of which may function in the movement of transposons. Many biologists speculate that retroviruses were the evolutionary ancestors of transposons, and that retroviruses may be one of the ways in which segments of nucleic acids can move from one species to another unrelated species without sexual reproduction (see HORIZONTAL GENE TRANSFER). The practical consequence of this is that HIV can remain inside the chromosomes of some of the victim's cells for the rest of the victim's life. Even if viruses can no longer be detected in the blood, HIV can reemerge from its latent form inside the genome. For this reason, most researchers believe that no way will ever be found to eliminate HIV from a victim's body. Instead they focus on therapies to achieve a permanent state of remission.

About 40 million people worldwide are living with HIV infection. Five million new cases of HIV infection occurred in 2005. Over twenty million people have already died of AIDS, and most of the rest of those who are infected will develop AIDS, because 90 percent of the victims live in poor countries, two-thirds of these in sub-Saharan Africa. Medications that prevent HIV from spreading within the body are expensive and widely available only in the United States, Europe, and a few wealthier countries of eastern Asia. Sub-Saharan Africa had (as of 2005) more than 25 million people with HIV infection, compared to 950,000 in the United States. Since 1986, the prevalence of HIV-1 infection in South Africa and in Zimbabwe increased from less than one percent to its current rate of 22 to 25 percent. Almost 5 percent of human deaths are caused by AIDS. The AIDS death rate is below that of heart disease and cancer but, unlike those diseases, AIDS afflicts people in their young and middle-aged years. In countries with high AIDS prevalence, the AIDS epidemic is the single most significant factor in political and economic life. The massive amount of death and sickness has devastated productivity in these countries and produced a generation of "AIDS orphans," both of whose parents have died of AIDS. The latest information about AIDS incidence and its consequences can be obtained from the United Nations Web site.

HIV spreads primarily through sexual contact, and secondarily through contaminated needles. This is because the virus is inactivated by contact with dry conditions or atmospheric levels of oxygen. In Africa, HIV spreads primarily through sexual contact between men and women. In North America it originally spread through homosexual contact among males, primarily because of the way it was first introduced. Its spread is no longer limited to a single mode of sexual contact, and its prevalence among American women has increased. In Africa the contaminated needles that spread HIV are often used in inadequately equipped hospitals, while in North America this mode of spread is primarily through needles used by drug addicts. It can also, like some other viruses such as rubella, spread through the placenta. Therefore a pregnant HIV-positive woman can infect a gestating fetus. Thousands of children are born HIV-positive. Because sexual contact is the main form of HIV spread, condoms have proven effective at reducing the incidence of new HIV infections in populations that have adopted their use. The National Institutes of Health in the United States determined that condoms were more than 85 percent effective at preventing HIV infection during sex. Higher officials in the federal government during the presidency of George W. Bush prevented the publication of this information on government Web sites. These Web sites implied that abstinence from sexual contact was the only effective way to prevent HIV infection. Although abstinence, which is 100 percent effective at preventing sexual transmission of anything, is clearly more effective than condoms, the refusal of the federal government to report the high effectiveness of condoms aroused a storm of controversy among American medical professionals.

When HIV enters the bloodstream, it adheres to surface molecules on certain white blood cells and thereby enters the white blood cells. One approach to the treatment of AIDS focuses on blocking these surface proteins so that viruses cannot enter new white blood cells. This would allow the bone marrow to produce a new population of white blood cells. Another approach involves the use of alternative nucleotides. Reverse transcriptase cannot tell the difference between AZT (azidothymidine) and regular thymidine nucleotides and incorporates either one of them into the viral DNA that it produces. However, it cannot attach new nucleotides to the DNA strand with AZT. In this way AZT stops the reverse transcription of new viral DNA.

Many researchers consider HIV/AIDS to be the best example of a medical subject in which an understanding of evolution is essential (see EVOLUTIONARY MEDICINE). The following are three major reasons that evolutionary science is important in understanding AIDS: first, the evolutionary origin of HIV/AIDS; second, evolutionary differences among different strains of the virus and different genotypes of the human host; and third, evolutionary changes within populations of the virus after it infects an individual.

Evolutionary Origin of HIV

Where did HIV come from? HIV is very similar to SIV (simian immunodeficiency virus). SIV infects several different kinds of higher primates, for example mangabeys, mandrills, green monkeys, and chimps. SIV does not seem to make these primates ill. If HIV evolved from SIV, which SIV was the evolutionary ancestor? Nucleotide sequences of the different strains of HIV and SIV can now be compared by CLADISTICS, which clusters similar genotypes together. The various strains of HIV-1 cluster together with the SIV strains from chimpanzees. HIV-1 and chimp SIV are so mixed together in the analysis that it appears the virus jumped from chimps to humans on at least three occasions. Moreover, HIV-2 clusters together with the version of SIV found in sooty mangabeys. Therefore, HIV had multiple origins from SIV that came from chimps and mangabeys. The original suspects from the 1980s, African green monkeys, are hosts to SIV that are not closely related to any strains of HIV. The method by which the virus entered human populations has not been determined but did not necessarily involve sexual contact. SIV could have infected the first human host by blood contact by a hunter with an infected animal he had killed.

Evolutionary scientists have even been able to estimate a time of origin for HIV. Evolutionary biologist Bette Korber and colleagues limited their study to group M viruses. They estimated the degree of nucleotide difference between a strain of HIV and the common ancestor of all group M viruses. For each year between about 1983 and 1998, each of the strains became more and more different from the common ancestor. The researchers calculated a statistical line through these data, then extrapolated the line all the way back to a time when there would have been zero difference between the strains and their common ancestor. The line crossed zero for the year 1931. Especially with extrapolation, error ranges become very large, so the estimated age of group M viruses is between 1918 and 1941. It appears that HIV has been present in human populations for only about 70 years, while SIV has been in other primate populations for many millennia. Parasites often have their most severe effects upon first infecting a host species; thereafter, coevolution may result in less severe disease, both because of more resistant hosts and also because of milder parasites.

Evolutionary Diversity of HIV and of Human Hosts

There are several strains of HIV. One reason for this is, as noted above, HIV-1 and HIV-2, as well as different strains of HIV-1, had distinct evolutionary origins. A second reason is that natural selection among HIV variants occurs differently in each victim's body. Thus the genetic strain that a person passes on to the next host is not necessarily the same strain with which he or she was originally infected. Some strains of HIV reproduce more slowly than others. A strain of HIV from Australia, the Sydney Bloodbank Cohort, recognizes a slightly different class of white blood cells, which reproduce themselves more slowly, causing the virus to propagate more slowly. A slower virus would be at a disadvantage in the presence of viruses that spread more rapidly, but some individuals were infected only with the slow form of the virus. These individuals have few symptoms, even after two decades.

There are also differences among individual humans in their ability to resist HIV. Apparently some individuals, who have not developed AIDS even after exposure to HIV, have a slightly different set of surface proteins on their white blood cells. HIV cannot bind to these mutant proteins and therefore cannot get into the white blood cells. Interestingly, 9 percent of Europeans (more than 14 percent from Scandinavian areas) have the mutant protein form that conferred resistance to HIV infection, while less than 1 percent of Asians and Africans have this mutant protein. Nobody knows why this geographical pattern exists. Two explanations have been suggested. The first proposal is that the mutant protein was produced by NATURAL SELECTION, because the mutant proteins also conferred resistance to other kinds of infection that had struck the populations in earlier centuries. Resistance to bubonic plague has been suggested, since plague also spreads in conjunction with white blood cells, and because the Black Death struck especially hard in 1347–50 in the areas of Europe that today have the most people that resist HIV infection. The second proposal was that the mutant protein was produced by genetic drift (see FOUNDER EFFECT), because just by accident the Vikings had these mutant proteins, and they spread them whenever they went on raids. Genetic drift does not explain why the highest allele frequency for the mutant protein is found among Ashkenazi Jews. Estimates from POPULATION GENETICS equations suggest that the mutation apparently occurred about 700 years ago. This would be right at the time of the Black Death, but a little later than the heyday of Viking expansion.

Evolutionary Changes in HIV after It Infects an Individual

Evolutionary changes occur in populations of the viruses within an individual victim. In a typical victim, for example, a very small amount of AZT is all that is necessary to inactivate a large proportion of the viruses during early infection. By the second year of infection, much larger doses are needed to achieve the same effect. This occurs because the percentage of viruses that can resist AZT increase in the population of viruses. The figure on page 13 relates dosage of AZT to effectiveness; the horizontal axis is in powers of 10, which means that almost 10,000 times as much AZT was needed to kill about half the viruses in the second year of infection as in the second month in this particular person.

The most likely reason for the evolution of AZT-resistant viruses within an individual is that random mutations in the viral genes resulted in reverse transcriptase molecules that would not recognize AZT as a nucleotide. While a mutant reverse transcriptase molecule would normally be detrimental to a virus, in the presence of AZT this mutant enzyme, though somewhat defective, is at least able to operate. Therefore, mutant viruses thrive in the presence, but not in the absence, of AZT. This pattern, in which resistant organisms thrive in the presence of the chemical agent used against them but are otherwise inferior to the susceptible organisms, is general among the many cases of the evolution of resistance to antibiotics, pesticides, and herbicides (see RESISTANCE, EVOLUTION OF).

Resistance is less likely to evolve if several different chemical agents are used together. This is the reason that many different antibiotics, pesticides, and herbicides are in use and more are being developed. Populations of HIV can evolve resistance to any of the chemical treatments against it (reverse transcriptase inhibitors such as AZT; chemicals that inhibit proteases; chemicals that block the entry of HIV into white blood cells; chemicals that block the integration of viral DNA into host chromosomes) but a combination or "cocktail" of different chemicals has proven effective at stopping the spread of HIV within a victim. It is much less likely that any virus will happen to possess mutations that render it resistant to all four means of chemical control than that it will possess a mutation against any one of them. In fact, mathematical calculations show that a cocktail of three chemicals is much more than three times as effective as each chemical individually.

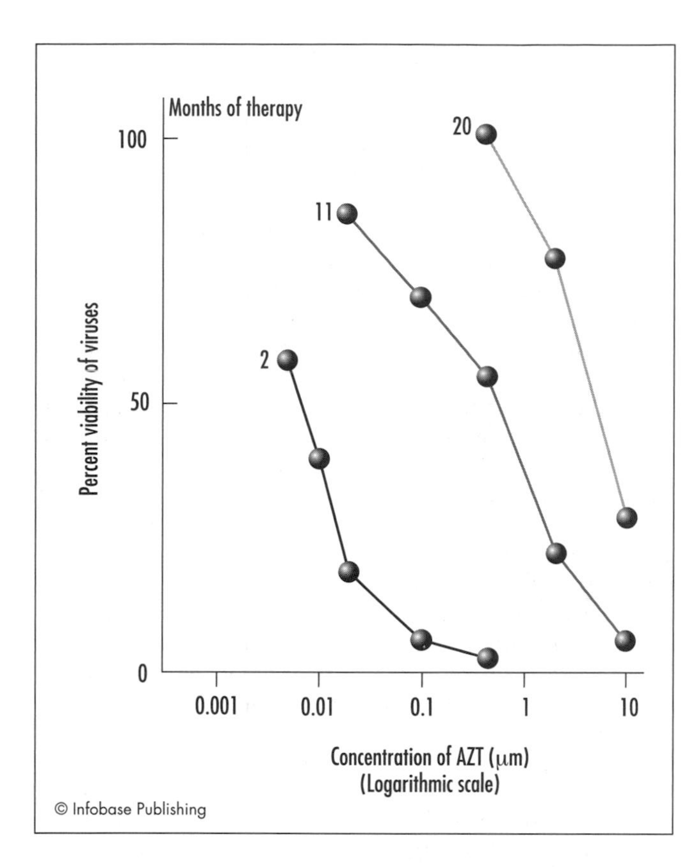

Even within a single patient, HIV evolves resistance to AZT, a drug used against it. Each line (connecting the data points) represents the death of viruses in response to increasing amounts of AZT. The three lines represent the death of viruses in the patient after two, 11, and 20 months of therapy. After two months of AZT therapy, small concentrations of AZT were able to kill most of the viruses. After 20 months of AZT therapy, all the viruses survived at an AZT concentration that would have killed nearly all of them earlier. *(Redrawn from Larder, et al.)*

RNA viruses, in general, have a high mutation rate. When a mutation occurs in double-stranded DNA, an enzyme can consult the intact strand in order to repair the mutated strand, but RNA is single-stranded. RNA viruses, such as influenza and cold viruses, mutate rapidly into many forms, which is why they can never be eradicated. DNA viruses evolve slowly enough that they can, in theory, be eradicated. Natural populations of the smallpox (variola) virus, a DNA virus, were eradicated by the World Health Organization of the United Nations by 1977; only laboratory samples of the virus remain. Until recently, international health organizations held out hope for the imminent eradication of another DNA virus, poliomyelitis. In contrast, HIV mutates rapidly. This rapid mutation rate has been measured. Over the course of seven years, populations of HIV from a patient evolved, becoming more and more different from the original population that had infected the patient. By the seventh year the average virus in the patient differed from the average original virus at 8 percent of its nucleotides for the gp120 gene. (This gene controls the protein that the virus uses for gaining access to white blood cells.) Eight percent nucleotide divergence in seven years is a phenomenal rate of evolution; in contrast, after six to seven million years of divergence, the DNA of humans and chimps differ by only 2 percent of the nucleotides in corresponding genes. After about the seventh year, however, the further evolution of the viruses in the patient virtually stopped. The reason appears to be that after the seventh year the patient had so few of that type of white blood cell remaining that there was very little opportunity for the virus to reproduce.

The tremendous variability of HIV is a major reason that medical researchers are skeptical that a vaccine will be developed. It would be difficult to develop a vaccine that would be effective in everybody, or even a single vaccine for HIV-1 and another for HIV-2. Even if such a vaccine could be developed, the rapid evolution of HIV would probably render it obsolete. Even in the world of vaccine development, which would seem to be as far away as one could get from Darwin and the struggle for existence in the natural world, the evolutionary process is at work.

Further Reading

Freeman, Scott, and Jon C. Herron. "A case for evolutionary thinking: Understanding HIV." Chap. 1 in *Evolutionary Analysis,* 3rd ed. Upper Saddle River, N.J.: Pearson Prentice Hall, 2004.

Heeney, Jonathan L., Angus G. Dalgleish, and Robin A. Weiss. "Origins of HIV and the evolution of resistance to AIDS." *Science* 313 (2006): 462–446.

Hillis, Daniel M. "AIDS: Origin of HIV." *Science* 288 (2000): 1,757–1,759.

Joint United Nations Programme on HIV/AIDS. "Uniting the world against AIDS." Available online. URL: http://www.unaids.org/en/. Accessed April 11, 2006.

Keele, Brandon F., et al. "Chimpanzee reservoirs of pandemic and nonpandemic HIV-1." *Science* 313 (2006): 523–526.

Korber, Bette, et al. "Timing the ancestor of the HIV-1 pandemic strains." *Science* 288 (2000): 1,789–1,796.

Larder, B.A., et al. HIV with reduced sensitivity to Zidovudine (AZT) isolated during prolonged therapy. *Science* 243 (1989): 1,731–1,734.

Quinn, Thomas C., and Julie Overbaugh. "HIV/AIDS in women: An expanding epidemic." *Science* 308 (2005): 1,582–1,583.

Rambaut, A., et al. "Human immunodeficiency virus: Phylogeny and the origin of HIV-1." *Nature* 410 (2001): 1,047–1,048.

Shliekelman, P., C. Garner, and Montgomery Slatkin. "Natural selection and resistance to HIV." *Nature* 411 (2001): 545–546.

allometry Allometry is the study of dimensional scaling as it applies to the comparison of big organisms to small organisms. Allometry is relevant to evolution because it allows the identification of which characteristics are the necessary results of body size, and which were free to evolve as evolutionary adaptations (see ADAPTATION).

Allometry has important biological consequences. If a large animal were structurally the same as a small one, the large animal would have less internal and external surface area relative to its volume. This would occur because surface areas increase as the square, and volume increases as the cube, of linear dimensions. That is, the small animal would have a higher surface-to-volume ratio.

Since animals lose body heat through surfaces, a higher surface-to-volume ratio would result in greater loss of body heat. This is why smaller animals lose more body heat. As a result, a mouse or a hummingbird must burn a lot of calories to maintain a constant body temperature. This is the main reason that, among warm-blooded animals, metabolic rate declines with increasing body size. Small mammals have evolved to produce relatively more body heat. Their faster metabolic rate means that they do not live as long. Small animals tend to have shorter lives than large animals, and warm-blooded animals shorter lives than cold-blooded animals. Conversely, large animals can have warm bodies just because they are large. Large dinosaurs, for example, were probably warm-blooded simply because of their low surface-to-volume ratio. Warm-bloodedness that is not controlled by the body but is simply due to large size is called *gigantothermy* and is different from the *homeothermy* of mammals and birds, which is the maintenance of a constant (warm) body temperature. Therefore homeothermy is an evolutionary adaptation; gigantothermy was not but resulted passively from the evolution of large body size. This also helps scientists understand why one of the evolutionary adaptations of mammals to cold temperatures was large body size. Many of the largest mammals are found not in the tropical rain forests but in colder regions of the Earth. Behavioral modifications allow animals to alter their surface-to-volume ratios. Mammals curl up into a little ball, thus reducing their surface-to-volume ratio, to keep warm.

Other processes that involve surfaces and volumes are also influenced by allometric relationships. One example is the exchange of molecules between a body and its environment. The rate at which molecules diffuse is related to the square of the distance the molecules move. A molecule takes 100 times as long to diffuse 10 times as far. Diffusion of molecules occurs very rapidly over small distances. Very small organisms, such as single-celled microbes, can rapidly absorb and excrete molecules through their surfaces. Larger organisms need to have internal, as well as external, surfaces. That is, in order to keep exchanging molecules with their environments (bringing in food and oxygen, getting rid of carbon dioxide and wastes, in the case of animals), a large animal simply cannot survive with just external surfaces. Large animals such as humans have a tremendous amount of surface area devoted to bringing in food and oxygen and getting rid of carbon dioxide and wastes. These surfaces are folded up inside of lungs, intestines, and kidneys. Human lungs have the equivalent of a tennis court of exchange surface hidden inside of them, in the form of air sacs called alveoli. Intestinal walls are fuzzy with villi (multicellular projections) and microvilli (projections from each cell). Aquatic animals have gills, which perform the same function as lungs but are on the outside of the body rather than the inside. Plants have a tremendous amount of external surface area (leaves and roots), which is why plants do not need lungs or intestines.

Some forms of surface area work better than others, depending on the scale at which they operate. Insects do not have lungs. They have little passageways called tracheae that penetrate throughout their bodies. These tracheae open to the outside world through little holes called spiracles. Oxygen can diffuse from the atmosphere into these tubes, and

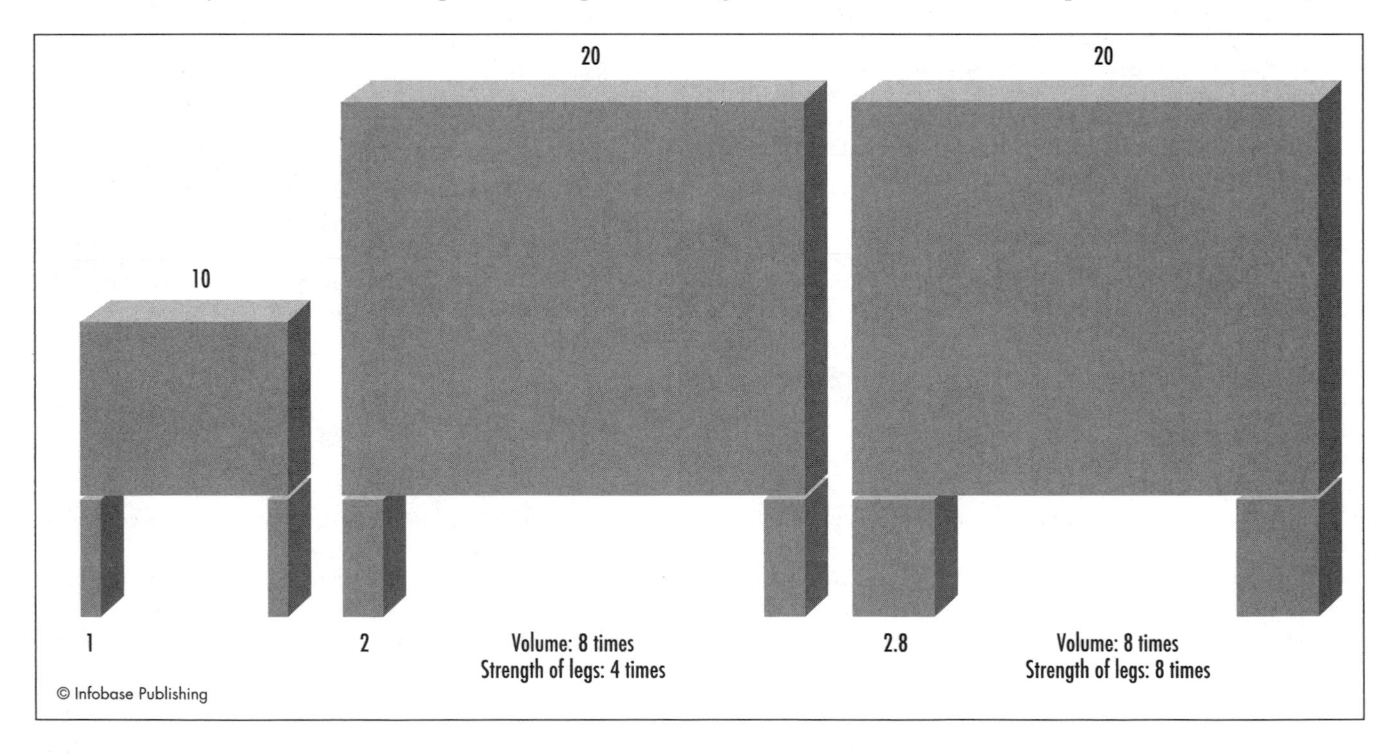

The boxes and columns represent animal bodies supported by animal legs. The middle animal has twice the linear dimensions of the small animal on the left; therefore it weighs eight times as much but its legs are only four times as strong. The legs of the middle animal may be unable to support the body weight. The animal on the right has a body twice as large in all three linear dimensions, but legs that have 2.8 times the diameter as the small animal. Both its weight and the strength of its legs are eight times as great.

then into the insect body; carbon dioxide can diffuse out. The bending of the insect's body as it moves, and the recently discovered muscular action of the spiracles, creates a limited amount of air movement. This arrangement works only for small organisms. A large animal could not obtain enough oxygen by means of tracheae. This is one reason why monster-sized insects could never evolve.

The allometric relationship of surface area and volume is only one of many allometric relationships in organisms. Another such relationship involves the strength of support structures. The strength of a cylinder (such as a tree trunk or an animal leg) is a function of its cross-sectional area. As indicated in the figure on page 14, an animal that is larger by a linear factor of two would weigh eight times as much but its legs would be only four times as strong. Larger animals, of course, need thicker legs, as everybody knows; what allometry indicates is that larger animals need relatively thicker legs. An animal twice as big, in linear terms, needs legs more than twice as thick. That is why it is possible to tell whether an animal is large or small by looking at its picture without being told the magnification of the picture. Something that looks like an elephant would be a large animal because its legs are thick relative to its other body dimensions; but something that looks like a spider, with spindly legs, could not possibly be a large animal. Galileo was one of the first to appreciate this mathematical principle in organisms. The same principle applies to tree trunks: Large trees will have thick-looking trunks, while small plants will appear delicate. Evolution cannot violate these constraints. These constraints put an upper limit on animal size. At about 140 tons, an animal would need legs so thick that they would touch one another, and the animal could not walk. The largest dinosaurs, at about 100 tons, came close to this theoretical limit.

Allometry also helps to explain animal function. Larger animals need larger brains, because they have more sensory information to process and more muscles to control. However, brain size in mammals does not increase to the same extent as body size. The allometric equation that relates brain size to body size in mammals is as follows:

$$\text{Brain weight} = 11.2\ (\text{body weight})^{0.74}$$

This formula allows the calculation of the *encephalization quotient* or EQ (see INTELLIGENCE, EVOLUTION OF). An EQ greater than 1 indicates a brain size greater than the typical mammal of that size. Primates have EQ greater than 1. The pre-Neandertal hominids of Atapuerca in Spain had EQ in the range of 3.1–3.8; the average Neandertal EQ was 4.8; the average modern human EQ is 5.3.

Organisms follow the same allometric principles as nonliving objects. Thus allometry is a mathematical bridge between biology and engineering. Very few biologists study engineering, and many allometric insights are lost in biological research. Allometric principles also control the adaptations that organisms have in terms of stress imposed by walking and running, or by the wind, or by horizontal support of a weight. In small buildings, windows provide enough ventilation, but large buildings need ventilation systems. Very large skyscrapers, if they had the same structure of elevator service as smaller buildings, would be 80 percent elevator; accordingly, these buildings have a streamlined system of express vs. local elevators. The genes have built buildings, called bodies, in which to live, and natural selection has been the engineer.

Evolutionary biologist W. Anthony Frankino and colleagues studied the sizes of wings in different butterflies. Different butterflies can have different sizes of wings for two reasons. First, larger butterflies need larger wings, for allometric reasons as described above. Second, natural selection or SEXUAL SELECTION can influence butterfly wing sizes within allometric constraints. Frankino and colleagues bred different butterfly strains that had different allometric relationships between wing size and body size and concluded that allometry did not greatly constrain wing size. Most of the differences in butterfly wing sizes observed in the natural world have resulted from natural and sexual selection, not from allometric constraints.

The classic work about allometry in organisms was the 1917 book *On Growth and Form* by Scottish biologist D'Arcy Wentworth Thompson, which scientists now mostly consult in its abridged format. Although Thompson's concept of the evolutionary process differed from that later adopted (see MODERN SYNTHESIS), the allometric insights remain important for modern evolutionary research.

Further Reading

Frankino, W. Anthony, et al. "Natural selection and developmental constraints in the evolution of allometries." *Science* 307 (2005): 718–720.

LaBarbera, M. "Analyzing body size as a factor in ecology and evolution." *Annual Review of Ecology and Systematics* 20 (1989): 97–117.

Martin, Robert D. "Brain size and basal metabolic rate in terrestrial vertebrates." *Nature* 293 (1981): 57–60.

McGowan, Chris. *Diatoms to Dinosaurs: The Size and Scale of Living Things.* Washington, D.C.: Island Press, 1994.

Niklas, Karl. *Plant Allometry: The Scaling of Form and Process.* Chicago: University of Chicago Press, 1994.

Rice, Stanley A. "Tree measurements: An outdoor activity to teach principles of scaling." *American Biology Teacher* 61 (1999): 677–679.

———, and F. A. Bazzaz. "Plasticity to light conditions in *Abutilon theophrasti*: Comparing phenotypes at a common weight." *Oecologia* 78 (1989): 502–507.

Thompson, D'Arcy. *On Growth and Form.* Abridged edition by John Tyler Bonner. Cambridge: Cambridge University Press, 1961.

altruism Altruism is an animal activity that benefits another animal, usually of the same species, but which incurs a cost to the altruistic individual. Since only genetically based characteristics can be the product of evolution, the individual acts of altruism are not the product of evolution; but most animal species appear to have a genetically based tendency to perform such acts. Scientists have spent much effort to find evolutionary explanations for altruism.

Altruism is a pervasive aspect of animal behavior. Perhaps the most visible example is when worker honeybees sacrifice their lives in the defense of the hive. In every animal species with social behavior, there are individuals that perform

Three Processes That Can Lead to the Evolution of Altruism

Term	Meaning
Kin selection	Individual sacrifices for close genetic relatives
Reciprocal altruism	Individual sacrifices for another individual that is likely to reciprocate in the future
Indirect reciprocity	Individual gains social status by being conspicuously altruistic

services and take risks for what appears to be the good of the population.

Like many rodents, Belding's ground squirrels of the Sierra Nevada Mountains of California have alarm calls that alert the entire group to the presence of predators. The rodent that sounds the alarm may or may not put itself in greater danger of predation by doing so. Animals that sound an alarm against predatory birds such as hawks actually reduce their own risk; alarm calls against hawks are therefore not altruism. An animal that sounds an alarm against predatory mammals such as cougars, however, puts itself at greater risk of being killed by that predator; alarm calls against cougars are therefore altruism.

The problem with explaining altruism relates to evolutionary fitness. NATURAL SELECTION will eliminate any tendency to perform altruistic acts that result in a net loss of fitness (successful transmission of genes into the next generation). This is because the costs of altruism, unless they are very minor, will deplete the resources that an individual would use to produce or provision its own offspring and may very well put the sacrificial altruist at risk of danger or death. Evolutionary scientists must find individual benefits that result from altruism; altruism cannot rely on a group benefit (see GROUP SELECTION). The benefits, furthermore, must outweigh the costs of altruism. At least three processes by which altruism can evolve have been suggested: kin selection, reciprocal altruism, and indirect reciprocity (see table).

Kin Selection

Evolutionary biologist W. D. Hamilton explained that the total fitness of an individual includes not only those genes passed on by the individual itself, through the successful production of offspring, but also by the offspring produced by its genetic relatives. He called this *inclusive fitness.*

Any two relatives share what Hamilton called a *coefficient of relatedness.* Consider two half-sibs, such as two children with the same mother but two different fathers. They share half their parents, and from the shared parent there is a 50 percent chance that they will inherit the same allele (see MENDELIAN GENETICS). Their coefficient of relatedness is $\frac{1}{2} \times \frac{1}{2} = \frac{1}{4}$. Since full sibs share both parents, they share $\frac{1}{4} + \frac{1}{4}$ of their alleles, resulting in a coefficient of ½. This is a matter of probability; in actuality, the two sibs may share more, or less, than half of their alleles. On the average, though, they will share one half of their alleles. Coefficients of relatedness are reciprocal, which means they are equal in both directions of comparison. An animal shares half of its genes with its offspring (coefficient of relatedness = ½), as well as with its parents. An animal shares one-fourth of its genes with its nephews and nieces (coefficient of relatedness = ¼), as well as with its aunts and uncles. An animal shares one-eighth of its genes with its cousins. (Such a measure of relatedness is not the same as a measure of DNA similarity; see DNA [EVIDENCE FOR EVOLUTION]). An animal can get its genes into the next generation half as efficiently by devoting itself to its siblings and enabling them to reproduce, as it would to produce its own offspring; devoting itself to its cousins would be one-eighth as efficient. As early 20th-century biologist J. B. S. Haldane said, "I would die for two brothers or ten cousins" (see HALDANE, J. B. S.). A good mathematician, Haldane said 10 rather than eight just to be on the safe side.

Self-sacrifice, even to the point of death, can be favored by natural selection, so long as it benefits the transmission of genes through one's relatives. Natural selection through inclusive fitness is appropriately called *kin selection.* Kin selection is also sometimes called *nepotism,* borrowing a term from human interactions. *Hamilton's rule* indicates that kin selection can favor altruism if $Br > C$, in which B is the benefit, C is the cost, and r is the coefficient of relatedness, all measured in terms of fitness, the number of offspring. Altruism may evolve if the benefit is large, the cost is low, and/or the altruists and their recipients are close relatives.

According to this reasoning, an individual animal should discriminate as to which other individual animals receive the benefits of its altruism. But can an animal distinguish between different degrees of relatedness? Species with intelligence and complex social behavior can learn the identities of different relatives. Mice can distinguish full from half-sibs on the basis of body chemistry, particularly the major histocompatibility complex proteins that function in the immune system. Research by zoologist Paul Sherman shows that the Belding's ground squirrels mentioned above issue alarm calls against cougars (true altruism) more often when close relatives are present than when more distant relatives are present.

In some cases, fledgling birds expend their efforts defending and providing food to the nests of other birds rather than starting their own. This altruism can be explained by the fact that the best territories have already been taken, and the fledgling would be unlikely to establish a nesting territory that would allow successful reproduction. The young bird does the next best thing: it assists other birds in their reproduction. In almost all cases, it is the close relatives that benefit from the altruism of these birds, as one would expect from kin selection.

Kin selection has been particularly successful at explaining the evolution of altruism in social insects, particularly ants, bees, and wasps (order Hymenoptera) and termites (order Isoptera). Social insects live in large colonies in which the reproduction is carried out by one or a few dominant individuals. Hymenopterans have a type of sexual reproduction known as *haplodiploidy,* in which females have pairs of chromosomes (they are diploid) while males have unpaired chromosomes (they are haploid). Females in these species produce eggs by MEIOSIS, which reduces the chromosome number by half, while males do not need to do this (see MENDELIAN GENETICS).

Social insects are famous for the tendency of worker females to sacrifice themselves to protect the hive. Consider a beehive in which all of the workers have the same mother (the queen) and the same father (a lucky drone). Since drones are haploid, the worker offspring receive not half but all of his alleles. Since the workers receive only half of the queen's alleles, they are more closely related to their fathers than to their mothers. Their coefficient of relatedness to one another is $(1 \times \frac{1}{2}) + (\frac{1}{2} \times \frac{1}{2}) = \frac{3}{4}$, in which the first term is their relatedness through their father and the second is their relatedness through the queen. Because siblings in most other species are related to one another only by $r = \frac{1}{2}$, a relatedness of $r = \frac{3}{4}$ indicates that sibling worker ants, bees, and wasps should be much more altruistic toward one another than siblings usually are. (A queen typically mates with several drones and stores their sperm. Many of the worker daughters have different fathers. Some workers have a coefficient of relatedness $r = \frac{1}{2}$ while others have $r = \frac{3}{4}$. The average relatedness among workers in most social insect colonies is therefore somewhere between one-half and three-fourths, which is still high enough to allow strong altruism to evolve.) Parents and offspring are usually related to one another by $r = \frac{1}{2}$. Worker insects are more altruistic toward one another even than parents and offspring. Anyone who has experienced an attack by a swarm of bees or wasps can attest to the way the worker sisters sacrifice themselves for their common welfare.

Since the worker insects are more closely related to one another than any of them are to their queen, the workers are in control of the nest. Even though the term *queen* implies rulership, queens in social insect colonies are mere egg-laying machines. It would be in the best interest of the queen to produce equal numbers of male and female offspring, but the workers will not allow this to happen: The workers kill most of the drone larvae. It is also the workers that decide which female larvae should receive the "royal jelly," which, unlike regular larval food, causes a female to develop into a queen. Workers may destroy some queens if there are too many.

Haplodiploidy is not the only evolutionary precondition for the altruism of social insects. All ants, bees, and wasps have haplodiploidy, yet the only ones that have evolved sociality are those that have also evolved a life cycle in which the larvae are helpless grubs (see LIFE HISTORY, EVOLUTION OF) and in which nesting behavior has evolved. Solitary bees have haplodiploid genetics but do not sacrifice themselves for one another.

Kin selection may also explain why animals, including humans, tend to behave more altruistically toward their true biological offspring than toward their stepchildren. In many mammal species, such as lions, a newly arrived dominant male will kill the offspring of the previous male, as shown in the photo above. These juveniles, while perfectly good for the prosperity of the species, have a zero percent genetic relatedness to the new dominant male. If these offspring represent any cost at all to these males, Hamilton's rule would predict that there would be no altruism at all. The unrelated juveniles do represent a cost, because while the females are feeding and protecting them they cannot produce offspring for the new dominant male.

This male lion has just killed a lion cub. When male lions take a new mate, they often kill the female's previous offspring, thus making the resources of the female lion available for raising the new male's offspring. ***(Courtesy of George Schaller)***

Kin selection helps evolutionary scientists to understand why humans are less solicitous toward stepchildren than toward biological offspring. This behavior pattern is a nearly universal feature among human societies: biologists Martin Daly and Margo Wilson call it "the truth about Cinderella." Crime data from Canada show that, while men very seldom kill children in their families, they are 70 times as likely to kill stepchildren as biological children. Stepchildren also have higher levels of blood cortisol (an indicator of stress) than do biological children. This indicates that the fathers and stepchildren both behave as though altruism is often missing from the father-stepchild relationship. In blended families with both biological and adopted offspring, fathers spend more time with their biological offspring than with their stepchildren. But is this due to kin selection, or simply due to the fact that stepchildren are older before their stepfather first becomes acquainted with them? Researchers have found that fathers were less solicitous of stepchildren than of biological children even if the stepchildren were born after the stepfather and the mother had begun living together.

Reciprocal Altruism

Evolutionary biologist Robert Trivers pointed out another way in which altruism could be favored by natural selection. An animal might perform some costly act of help to another animal if the recipient was likely to return the favor at some time in the future. Because the recipient may reciprocate in the future, this behavior is called *reciprocal altruism*. Reciprocal altruism helps to explain altruism toward individuals to which the animal is not closely related. Reciprocal altruism will not work unless there is a minimal level of intelligence. This is because reciprocal altruism is susceptible to cheaters. There must be some punishment for the individual that keeps all its resources while accepting the help of others. The other animals need to have enough intelligence to remember who the cheaters are. This could be one of the major contributing factors in the evolution of human language and human intelligence. Evolutionary biologist Robin Dunbar suggests that language evolved largely because it allowed humans to keep track of the intricacies of social structure, which would include the ostracism of cheaters.

In animal species with strong social hierarchies, the subordinate males receive no benefit for being altruistic toward the dominant males. Altruism between social classes conveys no benefit in those species. The males may, however, benefit greatly from carrying out acts of reciprocal altruism that win allies from their social peers. Because of the need both for paying back the altruism and for punishing cheaters, reciprocal altruism works best in animal species that are intelligent, social, and long-lived.

Indirect Reciprocity

Neither kin selection nor reciprocal altruism can explain altruism toward individuals who are unlikely or unable to reciprocate. While such altruism is rare in nonhuman species, it is very common among humans. Evolutionary biologist Geoffrey Miller points out that, in modern human charities, the recipient is often indigent and unable to reciprocate, and the recipient seldom knows who the donor is anyway. The donors often are not interested in the efficiency of resource transfer to the recipient. It would be much more efficient if a rich person continued to earn money, then donated that money, rather than working the equivalent number of hours in a soup kitchen. What, then, could be an evolutionary explanation for this kind of altruism?

The key to this kind of altruism may be whether or not another animal is observing it. The altruist can obtain greater social stature by being altruistic toward individuals that are unrelated or that cannot repay. Mathematicians Karl Sigmund and Martin Nowak have produced calculations that demonstrate this advantage. Human donations of time and/or money to charities, says Geoffrey Miller, more closely resemble a display of wealth than a calculated plot to get reciprocal benefits. Altruism, like conspicuous consumption, may constitute a message to the population at large. Conspicuous charity proclaims, "I am rich enough to give away some of my resources. This tells you that I am not only rich but also that I am a good person." Conspicuous consumption tells the observers only the first of those two things. The reputation of being a good person might yield enough social benefits to compensate for the cost of the altruism.

In particular, the altruist may gain advantages in mate choice. SEXUAL SELECTION could favor a conspicuous display of altruism, whether through charity or through a heroic deed to benefit the community. It is usually the males that display and the females that choose. Although among humans sexual selection has been more mutual, it is still the males who show off, and the females who choose, more often than the other way around. Conspicuous altruism is not merely showing off; it is actually useful information to the individual (usually the woman) making the choice of a mate: Such a man must have good resources and must be a good person who will be good to her. Displays of altruism need to be excessive, or prolonged, or both, in order for the woman to know the man is not faking it. Geoffrey Miller uses the example of Ebenezer Scrooge, the character in a novel by British writer Charles Dickens. Before his transformation, Scrooge not only did not participate in kin selection (he was not generous to his nephew) or in reciprocal altruism (he was not generous to Bob Cratchett), but also it is no surprise that he was single. Sexual selection can, and routinely does, produce adaptations that are costly to the individual, whether it is human altruism or the tail of a bird of paradise. Miller uses sexual selection as an explanation not only of the peculiarly human excesses of altruism but all aspects of human intelligence (see INTELLIGENCE, EVOLUTION OF).

Because altruism can provide fitness benefits, natural selection has also favored the evolution of emotions that reinforce altruism. Altruism feels good, in a number of ways, including feelings of satisfaction for being altruistic, gratitude toward donors, and rage toward cheaters. Neurobiologists have measured brain activity in human subjects involved in simulated situations of cooperation. They found that altruistic cooperation activated the same brain regions (such as the anteroventral striatum, also known as the pleasure center) as cocaine, beautiful faces, good food, and other pleasures. They also found this response when the subjects participated in sweet revenge against cheaters. The idea that the enjoyment of altruism has a natural basis is not new. American president Thomas Jefferson wrote in a letter to John Law in 1814, "These good acts give pleasure, but how it happens that they give us pleasure? Because nature hath implanted in our breasts a love of others, a sense of duty to them, a moral instinct, in short, which prompts us irresistibly to feel and to succor their distresses."

Evolution can therefore explain the tendency toward altruism in three ways: kin selection, reciprocal altruism, and indirect reciprocity. Since it is the proclivity, rather than the act itself, which evolution explains, humans can perform individual acts of self-sacrifice that yield no fitness benefit. But if such acts were common, the tendency to perform them would be selected against. A person can sacrifice herself or himself in a totally unselfish fashion—and there are numerous examples of such saints and heroes—because the behavior pattern evolved in the human species as a result of people sacrificing themselves in a selfish fashion.

Evolutionary altruism has also influenced the evolution of ethical systems (see EVOLUTIONARY ETHICS). Humans not only behave altruistically but believe that this is the right way to act. Evolutionary ethicist Michael Shermer indicates that, during the course of human evolution, feelings of affiliation with others and affection for others have evolved as reinforcements of altruism, first through kin selection within extended families and then through reciprocal altruism within communities. These feelings, being the product of natural selection, have a genetic basis. About 35,000 years ago, at a time Shermer calls the bio-cultural transition, cultural evolution became more important than biological evolution. The feelings of altruism that had already evolved were now applied beyond the community, to include society as a whole (altruism toward people who could not reciprocate), the entire human species, and even the entire biosphere of species. Today many humans choose to extend altruism to the whole world. The behavioral and emotional basis of this altruism evolved by means of kin selection, reciprocal altruism, and sexual selection.

Further Reading

Axelrod, Robert, and William D. Hamilton. "The evolution of cooperation." *Science* 211 (1981): 1,390–1,396.

Daly, Martin, and Margo Wilson. *The Truth about Cinderella: A Darwinian View of Parental Love.* New Haven, Conn.: Yale University Press, 1999.

DeWaal, Frans B. M. "How animals do business." *Scientific American,* April 2005, 72–79.

Dugatkin, Lee Alan. *The Altruism Equation: Seven Scientists Search for the Origins of Goodness.* Princeton, N.J.: Princeton University Press, 2006.

Hamilton, William D. "Altruism and related phenomena, mainly in the social insects." *Annual Review of Ecology and Systematics* 3 (1972): 193–232.

Miller, Geoffrey. *The Mating Mind: How Sexual Choice Shaped the Evolution of Human Nature.* New York: Doubleday, 2000.

Nowak, Martin A. "Five rules for the evolution of cooperation." *Science* 314 (2006): 1560–1563.

Sherman, P. W. "Nepotism and the evolution of alarm calls." *Science* 197 (1977): 1,246–1,253.

Shermer, Michael. "The soul of science." *American Scientist* 93 (2005): 101–103.

Singer, Peter. *A Darwinian Left: Politics, Evolution, and Cooperation.* New Haven, Conn.: Yale University Press, 1999.

Trivers, Robert L. "The evolution of reciprocal altruism." *Quarterly Review of Biology* 46 (1971): 35–37.

Warneken, Felix, and Michael Tomasello. "Altruistic helping in human infants and young chimpanzees." *Science* 311 (2006): 1,301–1,303.

amphibians, evolution of Amphibians are vertebrates that usually have an aquatic juvenile and terrestrial adult stage. Modern amphibians include animals such as frogs, salamanders, and caecilians. Frogs have an aquatic juvenile form (the *tadpole*) that swims with fins and a tail and breathes with gills, while the adult frog has no tail, has legs, and breathes with lungs. Most newts and salamanders, which are long and tailed, also have aquatic juvenile and terrestrial adult forms, although in some, such as the axolotl, aquatic juveniles become sexually mature (see NEOTENY). Caecilians, often mistaken for snakes, are legless and live in burrows. The term *amphibian* refers to the fact that most of them live (bio-) both (amphi-) on land and in water. Genetic analyses suggest that all amphibians share a common evolutionary ancestor that lived on the earth during the DEVONIAN PERIOD. Amphibians were the first *tetrapods,* or animals that walked upon four (tetra-) feet (-pod).

The common evolutionary ancestor of all modern amphibians has not been identified, and it certainly was not the only amphibian alive at the time. Nine genera of Devonian amphibians have been found, spanning a 20-million-year period. Many of these fish-amphibian animals lived at the same time, and scientists cannot determine with certainty which if any of them was the ancestor of all modern amphibians. There is no doubt of the evolutionary transition from fish (see FISHES, EVOLUTION OF) to amphibian, as it was occurring simultaneously in many different lineages. Animals intermediate between fishes and modern amphibians included:

- *Eusthenopteron foordi* lived in the late Devonian period. It had all the same fins that modern fishes still possess, rather than hands or feet. However, at the bases of the fins, it had bones analogous to the arm and leg bones of terrestrial vertebrates. *Eusthenopteron* had no bones that corresponded to the digits of modern tetrapods. This organism looked very much like an ordinary fish, and it probably spent nearly all of its time in shallow water.
- *Panderichthys rhombolepis* also lived during the late Devonian period. It lacked some of the fish fins and had thicker ribs than fishes possess. The thicker ribs were important in supporting the weight of the body when on land and away from the buoyancy of water. However, lacking legs, this species must have spent nearly all of its time in water.
- *Ichthyostega stensioei* and *Acanthostega gunnari* looked like fishes with legs. Their skulls and skeletons looked fishlike, but they had hands and feet and ribs even thicker than those of *Panderichthys. Ichthyostega* probably spent more time on land than *Acanthostega.* Ichthyostega moved like a seal, dragging itself by its forelimbs. It was the first vertebrate to have a non-swimming locomotion. In fishes, the hyomandibular bone helps to support the gills; this bone corresponds to the stapes, the ear bone of tetrapods (innermost ear bone of mammals). The stapes of *Acanthostega* has been found, and it resembles a fish hyomandibular bone. The stapes, however, was not free to vibrate and therefore could not have functioned in hearing on land. Later amphibians, in the Carboniferous period, had stapes that functioned in hearing.

In April 2006, paleontologist Neil Shubin announced the discovery of a new transitional form between fishes and amphibians. It was named *Tiktaalik roseae* from a word in the language of the Nunavat, the Canadian First Nations community that owns the fossil. This animal had a skull, neck, ribs, elbows, and wrists that resemble those of later amphibians, but had fishlike fins.

By the early Carboniferous period, there were many different amphibian lineages. All have become extinct except for two: the branch that led to modern amphibians, and the branch that led to reptiles (see REPTILES, EVOLUTION OF).

There has been debate regarding the reasons for the evolution of legs. Suggestions include:

- One early proposal was that fishes had to walk on land to get from one pond to another if their home pond began to dry up. There are fishes today that crawl on land and even climb trees, entirely without legs.

- Modern salamanders have legs yet many of them live underwater. They use their legs for walking underwater on rock surfaces against the current. It is possible that the first legged amphibians evolved in rushing water.
- Early amphibians may have used their legs to drag themselves around in shallow water, where they would be safe from deep water predators.
- The shallow water in which early amphibians lived may have been deficient in oxygen due to decomposition of leaf litter. If the amphibians lifted themselves up and breathed air, they could overcome this problem.
- Evolutionary biologist Robert A. Martin suggests that legs assisted in clasping during sexual reproduction, a function they still possess in many modern amphibians.

It is likely that legs proved useful for several different functions over a long period of time. Whatever combination of advantages may have selected for the evolution of legs, it had to be something that worked in a primitive condition. The earliest amphibians with legs could scarcely lift themselves at all.

The evolution of limbs would not require the acquisition of many new genes. Hox genes control the pattern of body part development in most animals (see DEVELOPMENTAL EVOLUTION). Some of these Hox genes (numbers 9–13) control limb development in mice, from shoulder to feet. Analogs of these genes are found in all vertebrates. Activation of specific Hox genes can produce limbs; evolution would then work out the structural details of these limbs, rather than produce them from scratch.

While the arm and leg bones have analogs in some fishes, the digits were an amphibian invention. The earliest tetrapods had more than five digits (*Ichthyostega* had seven, *Acanthostega* had eight). Though many tetrapods today have fewer than five digits (horses, for example, have just one; see HORSES, EVOLUTION OF), all surviving tetrapods have a five-digit fundamental pattern.

Many fossil species that are intermediate between fishes and modern amphibians have been discovered. In addition, amphibians (as their name implies) are themselves intermediate between fishes and fully terrestrial animals.

Further Reading

Benton, Michael. "Four feet on the ground." In *The Book of Life: An Illustrated History of the Evolution of Life on Earth,* edited by Stephen Jay Gould, 79–126. New York: Norton, 1993.

Clack, Jennifer. *Gaining Ground: The Origin and Evolution of Tetrapods.* Bloomington, Ind.: Indiana University Press, 2002.

———. "Getting a leg up on land." *Scientific American,* December 2005, 100–107.

Martin, Robert A. "Fishes with fingers?" Chap. 10 in *Missing Links: Evolutionary Concepts and Transitions through Time.* Sudbury, Mass.: Jones and Bartlett, 2004.

Shubin, Neil H., et al. "The pectoral fin of *Tiktaalik roseae* and the origin of the tetrapod limb." *Nature* 440 (2006): 764–771.

angiosperms, evolution of Angiosperms (the flowering plants) are one of the largest groups of plants, with at least 260,000 living species in 453 families. Angiosperms live in nearly every habitat except the deep oceans. They range in size from large trees to tiny floating duckweeds. Despite their tremendous diversity, angiosperms are a monophyletic group, which means that they all evolved from a common ancestral species (see CLADISTICS).

Most species of eukaryotes have life cycles in which MEIOSIS alternates with fertilization. Meiosis eventually produces haploid eggs and sperm that fuse together during fertilization (see MENDELIAN GENETICS). Plant life cycles differ from those of animals, fungi, and most protists (see EUKARYOTES, EVOLUTION OF) by having a multicellular haploid phase. The multicellular haploid structures produce eggs and/or sperm. The haploid structures of the simplest land plants, which evolved earliest, live in water or moist soil (see SEEDLESS PLANTS, EVOLUTION OF). In *seed plants,* however, the male haploid structures are pollen grains, which contain sperm and travel through the air from one plant to another; and the female haploid structures remain within the immature seed. Seeds contain, feed, and protect embryonic plants. Angiosperms and gymnosperms are the seed plants (see GYMNOSPERMS, EVOLUTION OF).

The shared derived features of angiosperms include flowers, fruits, and double fertilization, which all angiosperms (and no other plants) possess. Flowers produce fruits and consist of the following parts:

- *Sepals.* These are leaflike structures that protect unopened flowers
- *Petals.* These attract pollinators (see COEVOLUTION). In some cases, the petals cannot be distinguished from the sepals.
- *Stamens.* Pollen develops inside of anthers, each of which has two pairs of pollen sacs. Filaments hold the anthers up from the base of the flower. Stamens are the male component of a flower.
- *Carpels.* Tissue of the female parent completely surrounds the seeds during their development, forming a carpel. Pollen grains attach to the stigma, which is a surface at the top of the carpel; the immature seeds are protected within an ovary at the base of the carpel. Carpels may be fused together into pistils. The carpel and pistil tissue develops into a fruit, which assists in the dispersal of the seed to a new location. Carpels are the female component of a flower.

Not all flowers have all of these parts. A flower may lack sepals or petals or both. A flower may have only stamens or only carpels, rather than both.

The other unique feature shared by all angiosperms is *double fertilization,* in which each pollen grain contains two sperm nuclei. One of the sperm nuclei fertilizes the egg nucleus, and the other fertilizes female polar nuclei that develop into endosperm, a nutritive tissue inside the seed.

The earliest undisputed fossils of angiosperms date back about 130 million years (see CRETACEOUS PERIOD). Some researchers interpret a few earlier fossils to be those of angiosperms. Unlike the dinosaurs, the angiosperms as a group survived and recuperated from the asteroid impact of the CRETACEOUS EXTINCTION, perhaps because their seeds remained dormant and sprouted after the Earth recovered from this event.

Botanists understand that the angiosperms evolved from a gymnosperm ancestor. This had to occur, unless seeds evolved more than once, which is unlikely. One major difference between gymnosperms and angiosperms is that the wood of most angiosperms contains large water-conducting vessels, while most gymnosperms have only small conducting tubes. A few gymnosperms, including the now extinct Bennettitales and the modern group often classified as Gnetales, have wood with large vessels, while a few angiosperms have wood without large vessels. The Bennettitales and Gnetales had or have reproductive structures that are not flowers but resemble them in some ways. The angiosperms that have wood without vessels are those that have also been identified, for other reasons, as the most primitive angiosperms.

One major difference between angiosperms and gymnosperms is that most angiosperms have a much faster life cycle. In gymnosperms, a year may be required for the pollen nucleus to reach the egg and bring about fertilization. There are therefore no annual gymnosperms (i.e., that live only for a year, completing their reproduction at the end of the same year in which they germinated). In angiosperms, fertilization can happen in a few days. This observation has led many researchers to believe that the earliest angiosperms lived in disturbed environments. These disturbed areas may have included places that experienced a forest fire or landslide, seasonal drought, or even damage by dinosaurs. Rapid growth provides an evolutionary advantage in disturbed areas. If the earliest angiosperms lived in such environments, it may not be surprising that no fossils of them have been found, as such environments are very unsuitable for fossilization (see FOSSILS AND FOSSILIZATION).

There is also widespread recognition among botanists that angiosperms have proliferated largely because of COEVOLUTION with animals:

- Although many lineages of angiosperms have re-evolved wind pollination, most angiosperms have coevolved with the insects and other animals that pollinate their flowers. Insects as a whole did not experience an increased burst of diversity during the Cretaceous period. But insect lineages in which the adults pollinate flowers have proliferated, along with the angiosperms, since the Cretaceous period.
- Although many lineages of angiosperms have re-evolved wind dispersal of seeds, most angiosperms have coevolved with mammals and other animals that disperse their seeds. The fact that many mammals ate fruits, while dinosaurs appeared to rarely do so, paralleled the decline of dinosaurs during the Cretaceous and the proliferation of mammals since the Cretaceous.

The earliest angiosperm fossils do not provide a clear indication of what the earliest flowers were like, because by the time of these earliest fossils, angiosperms had already evolved a great diversity of flowers, from small, simple flowers to large magnolia-like flowers. Angiosperms radiated into many diverse forms very early in their evolutionary history. Since botanists have no fossils of the ancestral angiosperm, they have attempted to reconstruct the earliest flower by a phylogenetic analysis of existing groups of angiosperms. Most of these analyses have clearly identified a small woody plant that today exists only in the cloud forests of New Caledonia, *Amborella trichopoda,* as the only surviving member of the lineage that most closely resembles the ancestor of all flowering plants. Characteristics of *Amborella* include the following:

- Five to eight sepals and petals that are small, greenish-yellow, and largely separate from one another
- Numerous stamens, largely separate from one another
- Five to six separate carpels, each of which has an ovary with a single seed. The carpel does not completely close with tissue during growth; it is sealed by a secreted liquid.

If these characteristics represent what the first flowers were like, then the following major changes have occurred during the evolution of the angiosperms:

- In many but not all lineages, the floral parts have fused together. Two kinds of floral fusion are *connate fusion,* in which similar parts have fused, and *adnate fusion,* in which different parts have fused. In some lineages, petals have fused together to form a tube, or to form a landing platform that accommodates pollinators; in many lineages, carpels have fused into pistils; in a few, stamens have fused together. All these are examples of connate fusion. In some lineages, stamens have fused to petals, an example of adnate fusion.
- In many but not all lineages, the numbers of floral parts have changed. In many lineages, the numerous stamens have been replaced by a small number of stamens, usually the same as the number of petals. In some lineages, the number of parts has increased, as in the lineages in which the flowers have dozens or even hundreds of petals.

In many lineages of flowers, a mosaic of such changes has occurred. In roses, for example, all of the floral parts are still separate, the stamens are still numerous, but there are only five petals. In geraniums, all of the floral parts are still separate, but there are only five stamens and five petals. Mallow flowers have five separate petals but have numerous stamens that have fused together.

The traditional classification of angiosperms divides them into monocots and dicots. *Monocots* have characteristics such as these:

- The embryo has one leaf (mono- for one, -cot for cotyledon).
- The leaves usually have parallel veins.
- The stems usually have scattered bundles of conducting tubes.
- The flowers usually have parts in threes or multiples of three.

Dicots have characteristics such as these:

- The embryo has two leaves (di- for two, -cot for cotyledon).
- The leaves usually have netlike veins.

- The conducting tubes usually form a ring just underneath the stem surface.
- The flowers usually have parts that are numerous, or in fours, fives, or multiples thereof.

Phylogenetic analysis has allowed the classification of angiosperms into the following lineages:

- *Basal angiosperms,* which include *Amborella,* plants with flowers resembling those of water lilies, plants with flowers resembling those of magnolias, and the monocots
- *Eudicots,* or true dicots

Therefore botanists now generally recognize that the monocots are a monophyletic group, while the dicots are not.

The 260,000 species of angiosperms far surpasses the species diversity of any other plant group. Their importance in the evolutionary history and the current ecological function of the Earth cannot be overestimated (see BIODIVERSITY).

Further Reading

Ingrouille, Martin J., and William Eddie. *Plants: Evolution and Diversity.* New York: Cambridge University Press, 2006.

Klesius, Michael. "The big bloom: How flowering plants changed the world." *National Geographic,* July 2002, 102–121.

Soltis, Douglas E., et al. *Phylogeny and Evolution of Angiosperms.* Sunderland, Mass.: Sinauer Associates, 2005.

Soltis, Pamela S., and Douglas E. Soltis. "The origin and diversification of angiosperms." *American Journal of Botany* 91 (2004): 1,614–1,626.

animal rights Nonhuman animal species have some legal protections, and there is controversy about extending these protections even further. Modern democratic societies recognize the equal rights of all of their citizens. They also recognize that all human beings have equal rights to fairness and justice, even though noncitizens do not share in privileges such as the right to vote. Widespread belief in the equality of all people is quite recent in human history and, while widespread, is far from universal. In most societies, nonhuman animal species also have some rights, though far fewer than do humans. These rights are generally accorded on the basis of the mental capacity, and the capacity for suffering, that a species possesses. For example, intelligent animals, such as chimpanzees, have a greater capacity for suffering than do dogs and cats, which far surpass mice, which far surpass insects. Laws govern the use of chimpanzees, dogs, and cats for medical research; people get into legal trouble not only for abusing but even for neglecting confined dogs and cats; nobody gets in trouble for stomping on bugs. In contrast, the mental capacity of a human being is not used as a basis for rights and privileges.

Some religions recognize animal rights; for example, Jainism, a sect of Hinduism, reveres all animal life so much that people must sweep the ground before them to avoid stepping on insects. However, the major western religions still accept a binary definition of rights: All humans have infinite and eternal value, while animals do not and will simply die. Although many western people do not recognize the existence of animal rights, they do have feelings of empathy toward animals. Many religious people are appalled by cruelty to animals, but mainly because cruelty to animals reflects a disturbing attitude within the minds of the perpetrators, rather than because they have zeal for the rights of the animals themselves.

The more humans learn about other animal species, the more they recognize that evolutionary relatedness is not a sufficient basis for animal rights. Whales and dolphins, for example, show very high intelligence but evolved separately from the mammalian ancestors they share with humans. The controversy over whether it is ever ethical for humans to kill whales was one of the more strident ethical conflicts of the 20th century. For the last 20,000 years, *HOMO SAPIENS* has been the only human species in the world. However, when modern humans first evolved about 100,000 years ago and began to spread out of Africa, they were not alone. They met other species of humans in the Middle East and Europe (see NEANDERTALS) and perhaps even encountered other human species in southeast Asia (see FLORES ISLAND PEOPLE; *HOMO ERECTUS*). It is likely that the ancestors of modern humans showed these other human species no "humanity" but treated them as competitors who had no more rights than any other animal species. This is hardly surprising, for it is only very recently that the human species has shown kindness to members of a different race, religion, or culture. It is interesting to speculate what would happen in the modern age of ethical enlightenment if Neandertals or *H. erectus* still existed.

Even before the 20th century, scholars and statesmen thought about extending rights beyond humans to even include trees. Francis Hopkinson, a signer of the American Declaration of Independence, wrote, "Trees, as well as men, are capable of enjoying the rights of citizenship and therefore ought to be protected." The inability of the new American democracy to preserve the forests was one of the few things that made President Thomas Jefferson have second thoughts about it. He wrote, "I wish that I was a despot that I might save the noble, the beautiful trees, that are daily falling sacrifice to the cupidity of their owners, or the necessity of the poor ... the unnecessary felling of a tree, perhaps the growth of centuries, seems to me a crime little short of murder." Beginning in the 20th century, some scientists began to call for the extension of some rights to entire ecological communities of species. Ecologist Aldo Leopold, who once shot as many wolves as he could find, gradually came to realize that he had been not merely unwise but unethical in so doing. He was an early proponent of environmental or ecosystem ethics, which he called "thinking like a mountain." As environmental awareness spread in the middle of the 20th century, there were even some court decisions that recognized the rights of the natural world. One American Supreme Court decision was called "The Rights of Rocks." This represented a rare viewpoint in a society whose economic system defined the natural world as mere resources, trees as mere sources of wood, bears as mere sources of rugs, and evolution as irrelevant.

Further Reading

Wise, Steven M. *Rattling the Cage: Toward Legal Rights for Animals.* Boston: Perseus, 2000.

———. *Science and the Case for Animal Rights.* Boston: Perseus, 2002.

anthropic principle The anthropic principle is the concept that the universe has characteristics that, beyond coincidence, have allowed the evolution of intelligence. Because the proponents (mainly cosmologists) of this concept do not specify that a higher being is responsible for creating these characteristics, the anthropic principle differs from CREATIONISM and INTELLIGENT DESIGN, which attribute all complexity to a creative intelligence. First named in 1973 by cosmologist Brandon Carter, the anthropic principle maintains that human observers can deduce certain characteristics about the universe simply from the fact that they exist and are capable of studying it. In its strongest form, the anthropic principle insists that no universe could exist that did not have characteristics that would allow the evolution of intelligence at some point in its history. This principle relates to evolutionary science in that, if the principle is true, the evolution of intelligence could not have been a contingent (or chance) event.

Many cosmologists, such as the British Astronomer Royal Sir Martin Rees, consider the anthropic principle to be a tautology: Humans are here because humans are here. Others claim that it provides useful insights. From the evolutionary perspective, to say "the universe is just right for life" is backward; instead, life has evolved to fit the conditions of the universe. If the conditions had been different, life would have evolved differently. Anthropic theorists point out that physical constants of matter and energy took form during the first brief moments after the big bang (see UNIVERSE, ORIGIN OF) and that if these constants had been just a little bit different, nothing remotely resembling life could have evolved at all. They present examples such as these:

- During the expansion of the universe right after the big bang, small deviations from the uniform field of energy precipitated the formation of galaxies and stars. If those deviations had been slightly less, galaxies and stars would never have condensed; if those deviations had been slightly more, all the matter in the universe would have condensed into massive black holes.
- A force known as the strong nuclear force, one of the four fundamental forces, holds protons and neutrons together. If the strong nuclear force had been even 2 percent stronger than it is, protons would have bound into diproton pairs, which would have become helium atoms, and the universe would have had no hydrogen atoms. Helium atoms are very nonreactive, and nothing much would have happened in the universe, which might have been a helium cloud for all eternity. Hydrogen, on the other hand, undergoes thermonuclear reactions (producing helium) that allow stars to ignite.
- During atomic fusion, seven one-thousandths of the mass of the hydrogen becomes energy, while the rest of the mass becomes helium. If more energy were lost during fusion, all of the energy would have been used up by now. If just slightly less energy were lost, fusion would not have occurred at all and the universe would consist only of hydrogen.
- A slight difference in atomic forces might also have prevented the formation of carbon atoms in the interior of large stars. Without carbon, life may not have been possible (see ORIGIN OF LIFE).
- The force of gravity seems to be just right. If the force of gravity had been slightly more, stars such as the Sun would have burned out in less than a year; if the force had been slightly less, stars would never have condensed and ignited at all.

Because the physical constants are "just right" for the evolution of intelligence, the anthropic principle has been called the "Goldilocks principle," both by admirers and detractors. Anthropic theorists ask, could all this just be good luck?

Another possible example of a cosmic coincidence is that the cosmological constant is much smaller than it might have been. Physicist Albert Einstein posited a cosmological constant in his 1917 theory of the history of the universe. This constant is a number in the equation that accounts for the presence of dark energy, which is invisible but which causes the universe to expand faster as time goes on. Later, Einstein concluded that there was no empirical evidence that this was occurring, and he called the cosmological constant his "greatest blunder." Some evidence from the late 1990s suggests that the rate of expansion of the universe may itself be increasing, making it the first evidence to be consistent with a cosmological constant. Cosmologists remain unsure whether there even is a cosmological constant, but if there is, humans are pretty lucky that it is not any bigger than it is, or the universe would have expanded so fast that galaxies and stars would never have formed.

Life is possible on Earth because of water. Water is a molecule with highly unusual properties, mostly because the molecules stick together with hydrogen bonds. If the laws of chemistry had been slightly different, such that hydrogen bonds could not form, there would have been no life on Earth, or presumably anywhere else in the universe. This represents a non-cosmological example of the anthropic argument.

Rees and other cosmologists point out that if Earth had been a little different, life could not have evolved. If it had been a little larger, or a little smaller; a little closer to the Sun, or a little further away, no life of any kind could have evolved. So unusual are the conditions of the Earth that some scientists have suggested that, although bacteria-like lifeforms may be abundant on planets throughout the universe, complex life may be exceedingly rare and may not even exist anywhere else (see essay, "Are Humans Alone in the Universe?"). Given the possibility that there may be an uncountable number of Earth-like planets in the universe, it may not be unlikely (indeed, it may be inevitable) that intelligent life would have evolved.

The anthropic principle suggests that there is only one universe, and it is one that is strangely suitable for intelligent life. Some cosmologists, such as Rees, have suggested that there are many (perhaps an infinite number of) universes, each one with different cosmological properties. According to this *multiverse* model, most of these universes have characteristics in which the evolution of intelligence was impossible (they expand too fast, or they form into clouds of helium, or they collapse into black holes). But in one of them—the one humans know—the conditions were just right. In an infinite number of universes, Goldilocks has to exist somewhere. Since there can be no evidence of other universes, this possibility remains forever outside the realm of scientific investigation.

Further Reading

Goldsmith, Donald. "The best of all possible worlds." *Natural History,* July–August 2004, 44–49.

apes *See* PRIMATES.

archaebacteria Archaebacteria (now called *archaea* by scientists) are single-celled organisms that resemble bacteria. Most of them live in conditions of extreme acidity, temperature, salinity, or pressure that biologists at one time considered impossible for organisms to endure. *Pyrolobus fumarii,* for example, lives in deep ocean vents where the water reaches 235°F (113°C). Water boils at 212°F (100°C) at the air pressure of sea level, but the pressure at the bottom of the ocean allows superheated water to exist, in which this archaebacterium lives.

Like eubacteria (see BACTERIA, EVOLUTION OF), archaebacteria are *prokaryotic,* which means that their DNA consists of circular strands that float freely in the cell fluid rather than being enclosed in a nucleus (see DNA [RAW MATERIAL OF EVOLUTION]). In contrast, *eukaryotic* cells have DNA in the form of linear chromosomes, inside of a membrane-bound nucleus (see EUKARYOTES, EVOLUTION OF).

Because very small organisms generally come in just a few fundamental shapes, archaebacteria look like eubacteria and for decades were classified with them. Archaebacteria have some characteristics that distinguish them from eubacteria, including the following:

- All cells have membranes that are built of phospholipid-type molecules. In eubacteria and eukaryotic cells, the phospholipids are built from unbranched fatty acids linked to glycerol. In archaebacteria, the phospholipids are built from branched isoprenoids linked differently to a chemically backward form of glycerol. This is one of the most fundamental differences that are known to occur among cells.
- Archaebacteria have, in addition to cell membranes, cell walls that are chemically different from those of eubacteria.
- The transfer RNA and ribosomes of archaebacteria differ from those of eubacteria.
- The DNA of archaebacteria is associated with histone proteins, which is a characteristic that they share with eukaryotic cells, but not with eubacteria.

Archaebacteria resemble what most evolutionary scientists consider to be the most primitive life-forms on Earth. Not only are they structurally simple, but the extreme conditions in which they live resemble those of the earliest oceans. During the earliest PRECAMBRIAN TIME, the atmosphere and oceans contained no oxygen gas. All modern archaebacteria are obligate anaerobes, which means that they cannot live in the presence of oxygen gas. Many of the eubacteria that are identified by DNA analysis as most primitive also live in extreme conditions.

Carl Woese (see WOESE, CARL R.) began making nucleic acid comparisons among many different species in the 1970s (see DNA [EVIDENCE FOR EVOLUTION]). At that time, all bacteria were lumped together into one category. Woese found that certain bacteria that lived in extreme conditions had nucleic acid sequences as different from those of other bacteria as they were from those of eukaryotes. It was from this discovery that evolutionary scientists began distinguishing the archaebacteria as a distinct branch of life from the eukaryotes and the eubacteria (see TREE OF LIFE). Scientists prefer the term *Archaea,* because archaebacteria are as genetically different from the more familiar bacteria as they are from eukaryotes.

DNA studies since the time of Woese's original work have continued to confirm the uniqueness of archaebacteria. The complete genome of *Methanococcus janaschii* was published in 1996. Only 11–17 percent of its genome matched that of known eubacteria, and over half of its genes are unknown in either bacteria or eukaryotes. Moreover, some of the DNA similarity between archaebacteria and eubacteria may be due to HORIZONTAL GENE TRANSFER, in which bacteria exchange small segments of DNA. Although most horizontal gene transfer occurs between bacteria closely related to one another, exchange between archaebacteria and eubacteria has occurred frequently during the history of life.

Three major groups of archaebacteria are recognized by many evolutionary scientists on the basis of DNA sequences:

- *Euryarcheota* include the methanogens and the halophiles. Methanogens convert hydrogen gas (H_2) and carbon dioxide gas (CO_2) into methane gas (CH_4). They live in marshes and in the intestinal tracts of animals such as cows and humans. They are a principal source of methane in the atmosphere (see GAIA HYPOTHESIS). Halophiles live in extremely salty conditions such as the Dead Sea and the Great Salt Lake. Halophiles may have evolved more recently than other archaebacteria. First, they obtain their carbon from organic molecules produced by the decay of other organisms, which implies that they evolved after these other organisms were already in existence. Second, they use a molecule that resembles the rhodopsin visual pigment in the vertebrate eye in order to produce energy from sunlight. Third, DNA analyses place halophiles out toward the branches of the archaebacterial lineage, rather than near the primitive base of the lineage.
- *Crenarcheota* include thermophiles that live in very hot environments, but some crenarcheotes live in soil and water at moderate temperatures.
- *Korarcheota* have been identified only by their DNA sequences and little is yet known about them.

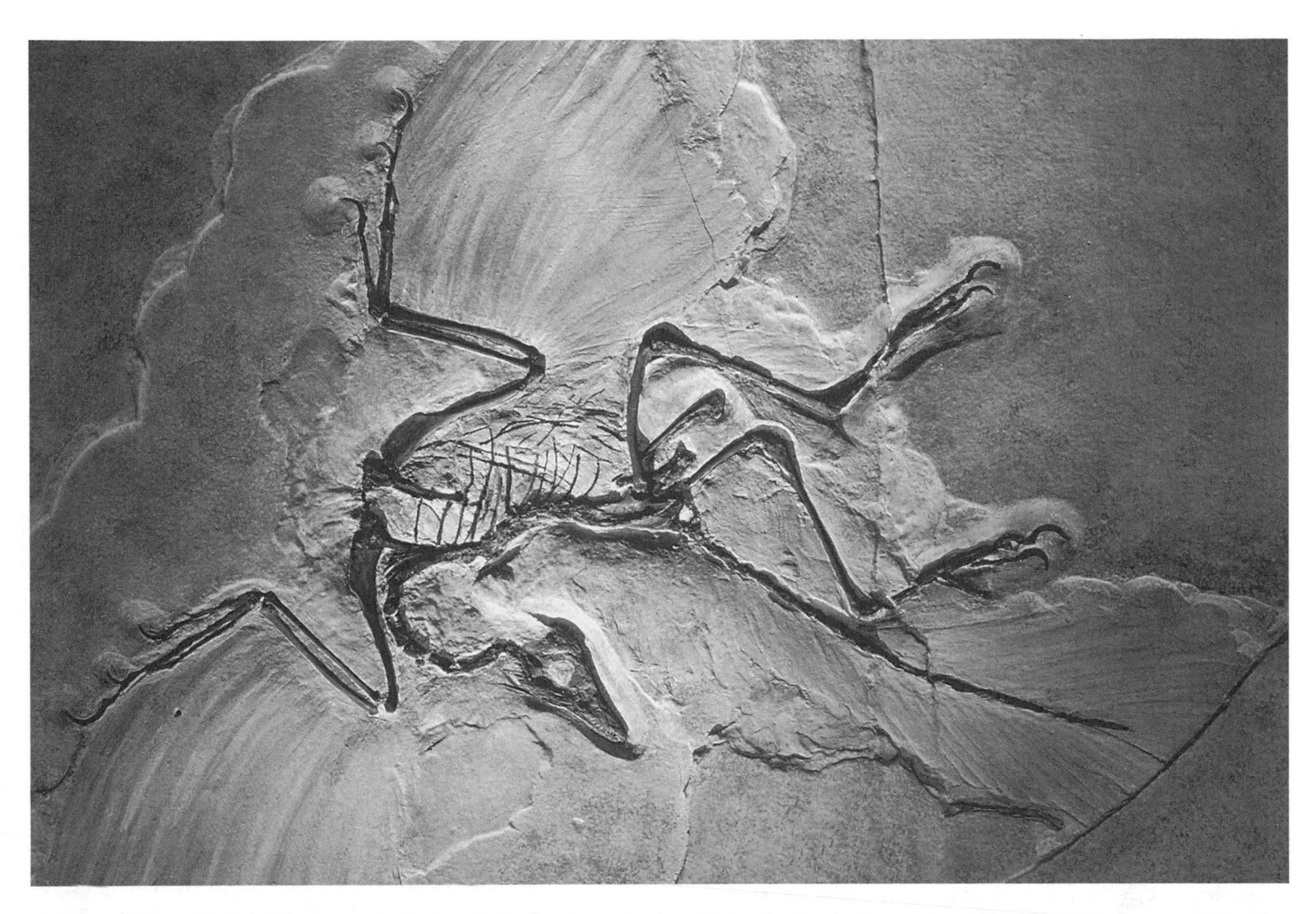

This is one of the original *Archaeopteryx* fossils found in the limestone deposits of Solnhofen, in what is now Germany. The skeleton is very reptilian; the impressions of feathers are clearly visible. *(Courtesy of James L. Amos/Photo Researchers, Inc.)*

Because the DNA of many archaebacteria is associated with histone proteins, just like the DNA of eukaryotes, some evolutionary scientists have suggested that the eukaryotic nucleus originated from an archaebacterium (see EUKARYOTES, EVOLUTION OF). In this scenario, an archaebacterium invaded a eubacterium and formed a mutually beneficial association with it. The archaebacterium acquired much of the DNA of the eubacterial host and became its nucleus. This would have been the first step in a series of mutually beneficial bacterial invasions that has produced the modern eukaryotic cells of plants and animals (see SYMBIOGENESIS). If this is true, then archaebacteria are not an evolutionary sideshow or mere leftover from the distant past but are the partial ancestors of all complex life on Earth, including humans.

Further Reading

Museum of Paleontology, University of California, Berkeley. "Introduction to the Archaea." Available online. URL: http://www.ucmp.berkeley.edu/archaea/archaea.html. Accessed March 22, 2005.

Archaeopteryx *Archaeopteryx* is the earliest undisputed bird fossil, intermediate between reptiles and modern birds.

When it was first discovered, *Archaeopteryx* looked like just another skeleton of a coelurosaurian dinosaur, until the discoverers saw the imprint of feathers in the rock. The discovery of the first fossil of *Archaeopteryx lithographica* by paleontologist Hermann von Mayer in a limestone quarry at Solnhofen, Bavaria, in 1861 could not have been more auspicious. Like the discovery of the first NEANDERTAL skull, it was just in time for Charles Darwin to use it in some editions of his famous book (see DARWIN, CHARLES; *ORIGIN OF SPECIES* [book]). Named after the Greek for "ancient wing," *Archaeopteryx* was just the kind of intermediate species that Darwin needed as evidence that entire modern groups of organisms (such as birds) had come into existence by evolution. One hundred and fifty million years ago, a bird fell into a shallow anaerobic pool and died. The anaerobic conditions inhibited decomposition, and microscopic marine organisms produced lime sediments, resulting in a vast number of fossils preserved in the limestone of what is today Solnhofen. The bones had transformed into stone; the feathers had decomposed but left their detailed imprint (see photo above) (see FOSSILS AND FOSSILIZATION). Six other fossilized *Archaeopteryx* individuals have been found in the same quarry. Some authorities recognize some of the specimens as a different species, *Archaeoptemx bavarica.*

Archaeopteryx is an almost perfect intermediate between modern birds and coelurosaurian dinosaurs, with a mixture of features characteristic of both. Its modern bird features include:

- Feathers
- Large brain
- Large eyes
- Clavicles fused into a furcula (wishbone)

Its reptilian features include:

- Long tail of vertebrae. Tails of modern birds are reduced to a small pygostyle from which tail feathers radiate.
- Digits in forearm. Wings of modern birds are not reinforced by long fingers.
- Conical reptilian teeth. Modern birds are toothless; the beaks of modern birds are hardened lips.
- Small sternum (breastbone). Modern birds have large keeled breastbones that allow attachment of flight muscles.

Archaeopteryx seems like something of a patchwork, since the features listed above are mostly either bird or reptilian, rather than intermediate. It is the organism as a whole that is intermediate between birds and reptiles. This is an example of *mosaic evolution,* in which different parts of an organism evolve at different rates. In particular, the feathers appear fully modern. A considerable amount of bird evolution must have preceded *Archaeopteryx.* Paleontologists have found other feathered dinosaurs, including a tyrannosaur discovered in 2004 that had been covered with feathers that appear to have a more primitive structure (see BIRDS, EVOLUTION OF). Because birds with much more modern features existed just a few million years later than *Archaeopteryx,* it is unlikely that *Archaeopteryx* was the actual ancestor of modern birds. It is more likely to have been a side branch of bird evolution. It clearly indicates the evolutionary transition that was occurring in several related branches of reptiles at that time. There were feathered dinosaurs running around all over the place during the JURASSIC PERIOD; *Archaeopteryx* is simply the first to be discovered, the most famous, and still the oldest.

Archaeopteryx had fully modern flight feathers but a small sternum. Scientists conclude that *Archaeopteryx* could fly but was not a strong flier. This was part of the reason that evolutionary scientists debated about whether bird flight had begun from the ground up (small motile dinosaurs running along the ground and eventually launching into the air) or from the treetops down (small dinosaurs gliding from treetops). The fact that the dinosaur group that contained feathered species was known for rapid pursuit of prey on the ground (as in *Velociraptor*) suggests the former possibility, while the weak flight of *Archaeopteryx* suggests the latter.

Further Reading

Mayr, Gerald, Burkhard Pohl, and D. Stefan Peters. "A well-preserved Archaeopteryx specimen with theropod features." *Science* 310 (2005): 1,483–1,486.

Nedin, Chris. "All about Archaeopteryx." Available online. URL: http://www.talkorigins.org/faqs/archaeopteryx/info.html. Accessed July 11, 2005.

Shipman, Pat. *Taking Wing: Archaeopteryx and the Evolution of Bird Flight.* New York: Simon and Schuster, 1998.

Ardipithecus *See* AUSTRALOPITHECINES.

arthropods *See* INVERTEBRATES, EVOLUTION OF.

artificial selection Artificial selection refers to genetic changes in populations that result from human choice. Artificial selection resembles NATURAL SELECTION in these ways:

- It is a process that occurs in populations.
- The population must have genetic variability (see POPULATION GENETICS) from which selection, natural or artificial, can choose.
- As a result of their genetic characteristics, some individuals in the population reproduce more than other individuals.
- As a result of differential reproduction of individuals, the characteristics of the population change.

In natural selection, the individuals that reproduce the best are those that are best suited to the natural conditions of their habitat, and the social conditions of their populations. In artificial selection, success depends entirely on human will.

Charles Darwin (see DARWIN, CHARLES) used artificial selection as a model for natural selection. Some writers have claimed that artificial selection was just a metaphor for natural selection, but it was more. Natural populations contain genetic variability, on which either natural or artificial selection can act. It was artificial selection that allowed Darwin to demonstrate that this variability did, in fact, exist. His most striking example was the domestic pigeon. Different breeds of domestic pigeon had different outlandish characteristics: Some had inflatable throats; some had ornate feathers; others could tumble in the air while flying. Darwin cited the historical fact that all of these pigeons had been bred, within recent centuries, from wild rock doves *(Columba livia).* Crossing any two of the domesticated breeds of pigeons produced offspring that resembled the wild rock dove. Artificial selection proved what extreme changes were possible, starting with just the genetic variability that exists in wild populations.

Artificial selection can have practical value or appear almost whimsical. Artificial selection was entirely responsible for producing the many species of crop plants and livestock animals from wild ancestors (see AGRICULTURE, EVOLUTION OF). Silviculturists have bred trees that rapidly produce strong wood, and horticulturists have bred garden flowers, bushes, and fruit and ornamental trees that are very useful in the human landscape. Some crop and garden plants have been so altered by artificial selection that they cannot survive in the wild. Garden tomatoes, for example, cannot even hold their heavy fruits off of the ground, where they rot if left unattended. Seedless fruits, from oranges to watermelons, can reproduce only through cuttings, which would seldom occur without human intervention. Some livestock animals, and many pets, have been bred into outlandish forms. Tiny hairless dogs and cats are an extreme example. A few centuries ago, breeders produced sheep whose tails were so fat that they had to drag a little cart behind them as they walked.

Koi and goldfish have been bred that are garishly colored and sometimes grossly misshapen. Few of these organisms would survive in the wild.

Perhaps the extreme form of artificial selection is genetic engineering. Biotechnologists can copy and move genes from one species into another and produce gene combinations that could never occur in the wild: corn with a bacterial gene, bacteria or pigs with a human gene, strawberries with a flounder gene. Although HORIZONTAL GENE TRANSFER occurs in the wild, it is much less common than biotechnology. Biotechnology is expensive, and all of these genetically modified organisms were believed by their creators to have some practical value (corn that makes its own pesticide, bacteria that synthesize human insulin, pigs that grow big but are not fat, and strawberries that resist frost damage). But there is still a selection process: the production of genetic variability, followed by reproductive success for some individuals when a human selects them.

Sometimes the changes that result from artificial selection are relatively minor. For example, most flowering plants have the ability to produce the red pigment anthocyanin. Wild plants usually do so only under certain conditions: for example, the red color in flowers or autumn leaves, or under conditions of deep shade or phosphorus deficiency. Most leaves produce the red color only after the green chlorophyll has been removed in the autumn. A simple genetic mutation causes some individuals to produce anthocyanin in their leaves all the time. The result is trees such as cherry, maple, and beech that produce green and red pigments at the same time; their leaves appear reddish black. Such trees survive quite well but would probably die out in a wild population in which they had to compete with other trees that did not waste their resources making anthocyanin at times or in tissues where it is not useful to their growth or reproduction.

Natural selection goes on in human-altered environments, right alongside artificial selection. Natural selection can occur as a result of human activity. The difference is that artificial selection is deliberate, while natural selection can occur as an unintended side effect. The most important examples are the evolution of weeds, of pesticide-resistant pests, and of antibiotic-resistant bacteria. Wild plant populations adapted to the cycle of plowing that farmers used for their crops; the wild plants evolved into weeds by natural selection, because the farmers did not intend it. It was certainly not the intention of farmers or doctors that the use of pesticides, herbicides, and antibiotics should result in the evolution of resistant populations of insects and weeds (see RESISTANCE, EVOLUTION OF). Many bird species have evolved, at least in urban populations, some characteristics that allow them to make use of civilized landscapes, such as mockingbirds that follow lawn mowers to catch insects that are scared by them, and blue tits that learned how to open milk bottles (see GENE-CULTURE COEVOLUTION).

Artificial selection of teosinte by Native Americans produced the first varieties of maize. Continued selection in modern agriculture has produced ears of maize that are longer and have more rows of larger seeds. A single grain of modern maize is shown to the right. *(Courtesy of Science VU/Visuals Unlimited)*

Humans have created artificial conditions that have stimulated the evolution of many other species. It has frequently been observed that two species will diverge from one another during conditions of scarcity (see CHARACTER DISPLACEMENT), but when life is easy the characteristics of the two species may overlap (as has been observed in DARWIN'S FINCHES). When hybridization occurs between species, the hybrid offspring are usually intermediate between the parents and may be inferior under conditions of scarcity, but they may survive and reproduce well under conditions of abundance. When humans cut down a forest or plow a grassland, they create conditions in which weeds can abundantly grow. Closely related species of weeds can also hybridize, and genes from one species can flow into another species as a result of introgression (see HYBRIDIZATION). Human activities not only encourage the growth of weeds but influence their further evolution. Humans inflict drastic and rapid changes in the environment, thereby speeding up the course of evolution in those species that can coexist with humans and driving to extinction those that cannot (see BIODIVERSITY).

Artificial selection might seem like the ultimate human conquest of wild species. From the viewpoint of the domesticated species, they may benefit at least as much as humans do. Science writer Michael Pollan points out that many kinds of plants (he uses apples, marijuana, tulips, and potatoes as examples) have greatly enhanced their reproductive success by taking advantage of the fact that humans like them. These plants have, in effect, used humans to spread around the world and build up huge populations.

Further Reading

Pollan, Michael. *The Botany of Desire: A Plant's-Eye View of the World.* New York: Random House, 2002.

asteroids and comets Asteroids and comets are small fragments of material that did not become major planets, left over from the formation of the solar system. When, rarely, comets or asteroids collide with the Earth, they can influence the course of evolution. Most of them, like the planets, revolve around the Sun.

Asteroids are bodies of metal and rock, without atmospheres. If all the asteroids in the solar system accreted into one planet, it would be less than half the size of the Moon. There was brief controversy over whether Pluto was a mere asteroid rather than a planet. Since Pluto is about as large as all the asteroids combined, most astronomers still consider it to be a planet. The largest known asteroid is Ceres, about 600 miles (1,000 km) in diameter, while the smallest asteroids are pebbles. Sixteen asteroids exceed 150 miles (240 km) in diameter. Most asteroids have a nearly circular orbit around the Sun. Asteroids are classified into three categories:

- More than three-quarters of the asteroids are *carbonaceous (C-type) asteroids.* They are found predominantly in the outer portion of the asteroid belt. Their composition is similar to that of the Sun, minus the volatile materials (such as hydrogen and helium) that have been lost. The *carbonaceous chondrite* meteors, such as the meteorite that fell near Murchison, Australia, in 1869, appear to be C-type asteroids. They contain organic molecules similar to those that may have formed on the primordial Earth and from which life is believed to have developed (see ORIGIN OF LIFE).
- A little less than one-fifth of the asteroids are *silicaceous (S-type) asteroids.* They are composed of metallic iron mixed with iron-magnesium silicates. They are found predominantly in the inner portion of the asteroid belt. Many astronomers believe that S-type asteroids are the source of most meteorites, but this issue remains unresolved.
- The other asteroids are *metallic (M-type) asteroids,* composed mostly of metallic iron and found in the middle region of the asteroid belt.

Comets consist mainly of wet and dry ice (H_2O and CO_2) but also contain many organic chemicals that are similar to those involved in the origin of life. Many comets revolve around the Sun in extremely slow and very elliptical circuits, in one of two main groups:

- The Kuiper Belt comets reside (along with some asteroids) in a band beyond the orbit of Pluto.
- The Oort cloud comets are beyond the Kuiper Belt, many of them nearly two light-years from the Sun. At this distance, they are nearly halfway to the nearest star, and the Sun itself would appear as merely a star.

Solar gravitation is weak, especially in the Oort cloud, but only the tiniest perturbation in gravity is enough to send the comets in a path toward the Sun. Comets can die (for example, when they collide with a planet), but there is a virtually unlimited supply of them in the Kuiper Belt and Oort cloud. The tail of the comet consists of water and CO_2 molecules that vaporize as the comet approaches the Sun; therefore, the tail of the comet points away from the Sun rather than trailing behind the comet.

A large number of comets and asteroids roamed in the paths of planets and moons early in the history of the solar system. Many of these objects collided with planets and moons until about four billion years ago. The craters from these impacts can still be seen on the Moon. The influence of the tremendous gravitational field of Jupiter has profoundly affected the asteroids of the solar system: Jupiter cleared many of them out of the plane of the ecliptic and stabilized a band of asteroids between its orbit and that of Mars that might otherwise have formed into a planet. Humans have Jupiter to thank for the fact that asteroid collisions with Earth are so rare.

Asteroids with orbits that bring them within about 125 million miles (200 million km) of the Sun are considered NEAs (near-Earth asteroids). Most of these asteroids were jarred from the asteroid belt by collisions and/or nudged by interactions of gravitation. The approximately 250 NEAs that have been found to date probably represent only a tiny fraction of the total. It is estimated that there may be a thousand NEAs that are large enough (1 km or more in diameter) to threaten mass extinction, as at the end of the Cretaceous period (see CRETACEOUS EXTINCTION) and at the end of the Permian period (see PERMIAN EXTINCTION).

On March 23, 1989, an asteroid one-quarter mile (0.4 km) in diameter came within 400,000 miles (640,000 km) of the Earth. Its existence had not been previously known. Scientists estimated that the asteroid and the Earth had passed the same point in space a mere six hours apart. A similar near-miss occurred on February 23, 2004, in which a previously unknown asteroid, large enough to destroy life on the planet, appeared near the Earth. An approximately 1,500-foot (500 m) asteroid passed within 250,000 miles (400,000 km) of the Earth on July 3, 2006. Near misses with smaller asteroids that came closer to Earth had occurred in 1991, 1993, and 1994. All three of these asteroids were about 26 feet (8 m) in diameter, enough to create local devastation but not a threat to life. If a large NEA appeared on a collision course with Earth, humans would not be able to send up space cowboys to save the Earth as in the *Armageddon* movie, because splitting the asteroid would only create smaller asteroids whose smaller impacts may have the same cumulative effect on the planet.

Asteroid and comet collisions have occurred numerous times during the history of the planet. The most famous impact occurred 65 million years ago at the end of the Cretaceous period (see table on page 30). Most of the craters listed in the table are found on dry land, in regions that are cold or dry enough that erosion has not erased evidence of them. The Barringer Crater in Arizona is still easily recognizable because the impact was recent (50,000 years ago) and the weather is dry (see figure on page 29). Chicxulub, the crater from the end of the Cretaceous period, has largely eroded and filled with sediment and would probably not have been discovered had scientists not been looking for it.

These impacts have had an important effect on the course of evolution. While NATURAL SELECTION explains most evolutionary patterns, asteroid impacts cause largely random extinctions. Paleontologist David Raup indicates that MASS EXTINCTIONS are due to bad luck rather than bad genes. In the period following an impact, a wide world of possibility is open to many new types of organisms, some of them with characteristics that would have been eliminated by competition in a more heavily populated world. In this way, asteroid impacts have created punctuations of evolutionary novelty.

Barringer Crater, Arizona, was formed about 50,000 years ago when a relatively small asteroid hit the Earth. ***(Courtesy of François Gohier/Photo Researchers, Inc.)***

The tremendous devastation that can result from an asteroid or comet impact was vividly illustrated in July 1994, when the comet Shoemaker-Levy 9 collided with Jupiter. Jupiter's gravity split the comet into 20 fragments, which produced prodigious plumes of hot gas when they collided with Jupiter's liquid and gas surface (see photo at right). Dark scar-like holes remained on the Jovian atmospheric surface for weeks afterward.

When an asteroid struck the Earth at what is now Manson, Iowa, 74 million years ago, it created a hole three miles (five km) deep and 20 miles (32 km) across, making it the biggest disaster ever to occur in what is now the United States. Glaciers during the Ice Ages filled it in, so no trace of it remains aboveground. One second after entering the atmosphere, an asteroid such as the one that made Manson Crater would hit, instantly vaporize, and throw out more than 200 cubic miles (a thousand cubic kilometers) of rock, causing devastation of life for 150 miles (240 km) around. A bright light (brighter than almost any other that humans have seen) would be followed

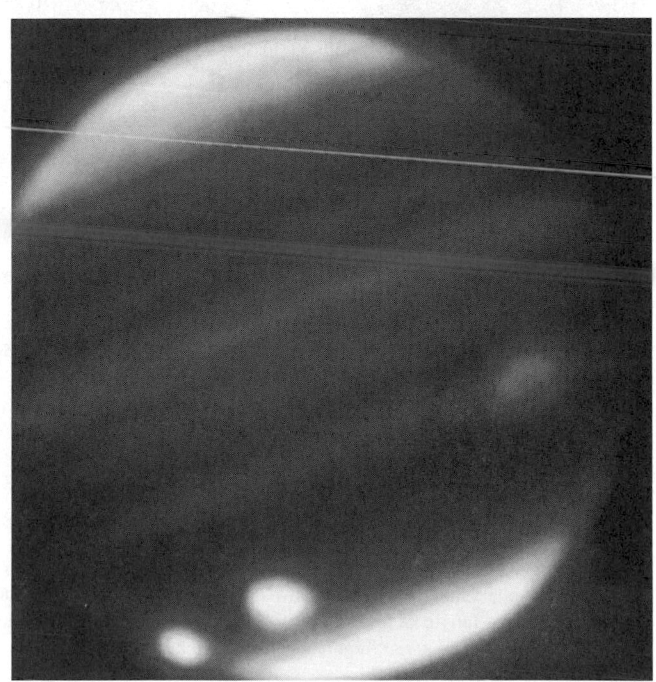

The comet Shoemaker-Levy 9 broke apart, and the fragments were drawn by Jupiter's gravity. They collided in 1994. ***(Courtesy of NASA)***

Large Impact Craters on Earth

Name	Location	Age (million years)	Period	Diameter (km)
Vredefort	South Africa	2023	Precambrian	300
Sudbury	Canada	1850	Precambrian	250
Bedout	Australia	250	Permian	200
Chicxulub	Mexico	65	Cretaceous	170
Manicougan	Canada	214	Triassic	100
Popigai	Russia	35	Tertiary	100
Chesapeake Bay	United States	36	Tertiary	90
Acraman	Australia	590	Precambrian	90
Puchezh-Katunki	Russia	175	Jurassic	80
Morokweng	South Africa	145	Jurassic	70
Kara	Russia	73	Cretaceous	65
Beaverhead	United States	600	Precambrian	60
Tookoonooka	Australia	128	Cretaceous	55
Charlevoix	Canada	357	Carboniferous	54
Kara-kul	Tajikistan	5	Tertiary	52
Siljan	Sweden	368	Devonian	52

Damage to trees caused by the Tunguska event. On June 30, 1908, a huge fireball exploded in the sky over western Siberia, followed by an enormous explosion. The explosion flattened almost 1,200 square miles (more than 3,000 square km) of forest. The explosion occurred near the town of Vanavara, in the Podkamennaya Tunguska River valley. The explosion was probably caused by a meteorite about 150 feet (about 50 m) in diameter that broke up in the atmosphere, although no major fragments or craters have been found. This photo was taken in 1910. ***(Courtesy of Novosti Press Agency/Science Photo Library/Photo Researchers, Inc.)***

by a silent wall of darkness advancing faster than sound. The impact would set off tsunamis, earthquakes, and volcanoes, and the hot ashes would ignite fires all around. If such an asteroid hit today in Manson, a billion and a half people would die in the first day. Despite the magnitude of this impact, there are no known extinctions associated with the Manson event.

Some extraterrestrial impacts have occurred during recent history. An impact crater in southeastern Germany is only 2,200 years old, a date that matches descriptions in Roman records of fire falling from the sky. The Kaali event, a natural disaster that occurred about 350 B.C.E. in what is now Estonia and was preserved in legend, may have been an asteroid. An asteroid or comet, about 300 feet (100 m) in diameter, exploded above a remote area of Siberia on June 30, 1908. The object did not hit the ground but devastated over a half million acres of forest in the Tunguska region (see photo on page 30). Scientists refer to this impact as the *Tunguska Event.* While it is not known whether humans died in the explosion, a man's clothing caught fire from it over 30 miles (50 km) away. Trees were flattened, radiating outward from the impact center. Debris from the explosion extended far into the upper atmosphere, reflecting sunlight for great distances. In Western Europe people were able to play tennis outside far into the night without artificial lighting because of it. If the impact had occurred over a densely populated region, the loss of life and property could scarcely be calculated.

Further Reading

Beatty, J. Kelly, ed. *The New Solar System.* Cambridge University Press, 1998.

Department of Geological Sciences, University of California, Santa Barbara. "Evidence of meteor impact found off Australian coast." Available online. URL: http://beckeraustralia.crustal.ucsb.edu/. Accessed March 22, 2005.

Stone, Richard. "The last great impact on Earth." *Discover,* September 1996, 60–61.

australopithecines Australopithecines were HOMININ apes that lived in Africa between about four million and one million years ago. Scientists universally recognize that some member of this group, perhaps yet to be discovered, was ancestral to modern humans.

Two features distinguished australopithecines from other apes. First, unlike all other apes except the genus *Homo,* they walked upright on two legs (see BIPEDALISM). Second, they differed from the genus *Homo* in that their brains were not consistently larger than those of modern apes such as chimpanzees and gorillas. Australopithecines may have used tools, such as sticks, or used rocks as tools; both of these are behaviors seen in modern chimps. There is no evidence that they deliberately altered stones into tools.

Different anthropologists classify extinct hominins into different numbers of species and genera. In this entry, the term *australopithecine* is used to indicate hominins intermediate between their pre-bipedal ancestors and the larger-brained genus *Homo.* As here used, the term *australopithecine* incorporates the genera *Ardipithecus, Australopithecus, Kenyanthropus, Orrorin, Paranthropus,* and *Sahelanthropus.* The paragraphs below review the genera of australopithecines and related hominins that have been thus far discovered, in the approximate chronological order of their evolutionary appearance.

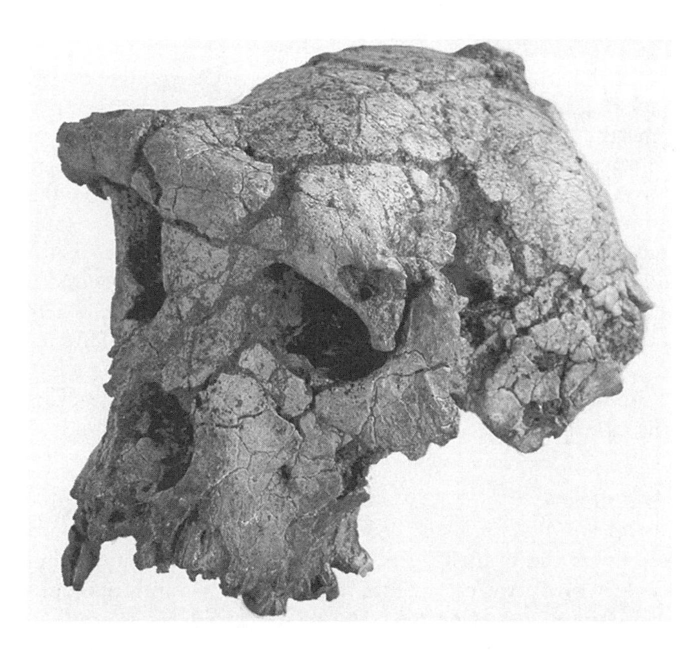

Skull of *Sahelanthropus*. Found in western Africa, this skull represents one of the earliest apes that was adapted to upright walking. *(Courtesy of Michel Brunet, Laboratoire de Géobiologie, Biochronologie et Paléontologie Humaine, Faculté des Sciences Fondamentales et Appliquées, Université de Poitiers, France)*

Sahelanthropus

The earliest bipedal ape currently known is *Sahelanthropus tchadensis* ("Sahel man that lived in Chad"), fossils of which were discovered in 2002 by a member of the research team of French anthropologist Michel Brunet in Chad, western Africa. *Sahelanthropus* lived six to seven million years ago (see photo above). DNA studies (see DNA [EVIDENCE FOR EVOLUTION]) suggest that chimpanzees and humans diverged from a common ancestor five million years ago. The structure of the fossil skull of *Sahelanthropus* suggests that this species walked upright. Evolutionary scientists have not yet determined whether *Sahelanthropus* was a direct human ancestor. If in fact *Sahelanthropus* represents the common ancestral condition of both humans and chimpanzees, it is difficult to understand how chimpanzees would have lost bipedalism. Alternatively, the hominin and chimpanzee lines may have diverged earlier than the five million years suggested by the DNA studies. The discoverers claim that *Sahelanthropus* was a true human ancestor, and other australopithecines were not.

Orrorin

Orrorin tugenensis ("ancestor that lived in the Tugen Hills") was discovered in Kenya, eastern Africa, in 2000 by paleontologists Martin Pickford and Brigitte Senut. The fossils are about six million years old. A leg bone, which includes the expanded end that connected with the knee joint, suggests that this hominin walked upright. Other limb bones, jaw fragments, and teeth were found as well. *Orrorin* presents the same challenge to the interpretation of human evolutionary history as does *Sahelanthropus:* Bipedalism may have been present in the common ancestor of humans and chimpanzees, and it calls into question whether the later australopithecines were human ancestors. The place of origin of the human lineage is likewise unclear: was it western Africa, where *Sahelanthropus* lived, or eastern Africa, where *Orrorin* lived? At the time, the extensive deserts of Africa had not yet formed, and the forests in which both *Sahelanthropus* and *Orrorin* lived may have been continuous across the continent.

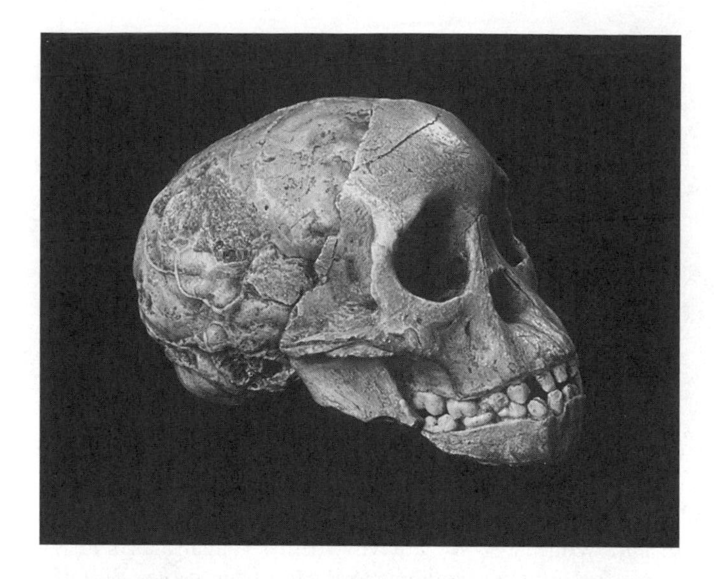

Skull of the "Taung child" *Australopithecus africanus.* Discovered by Raymond Dart in 1924, this eventually directed attention of researchers to Africa as the place where humans evolved. The specimen consists of part of a juvenile skull and mandible and an endocast of the right half of the brain (seen at center left). It is about 1.7 million years old. *(Courtesy of Pascal Goetgheluck/Photo Researchers, Inc.)*

Ardipithecus

Ardipithecus ramidus (called "basic root ape" because it may represent the branch point or beginning of the human lineage) was discovered in Ethiopia in 1994 by anthropologist Tim White and associates. The fossils are about 4.4 million years old. The bones suggest that *Ardipithecus* was bipedal. Anthropologists have not agreed whether *Ardipithecus* was the ancestor of later australopithecines.

Australopithecus

The genus *Australopithecus* ("southern ape") is represented by several species that lived between about four million and about two million years ago in various parts of Africa. The three earliest species were *A. anamensis* and *A. afarensis,* both named after the eastern African regions in which their fossils were found, and *A. bahrelghazali* ("Bahr el Ghazal," the western African location at which the fossils were found). These species lived between four and three million years ago. Two other species, *A. garhi* ("surprise") from eastern Africa and *A. africanus* ("African") from southern Africa, lived between three million and two million years ago. The evolutionary relationships among these species are unclear. Genus *Australopithecus* is called the "gracile australopithecines" because they were small (three to four feet [one m] in height), had small faces, and jaws adapted to an omnivorous diet.

Australopithecus africanus was the earliest prehuman hominin to be discovered. A South African anatomist (see DART, RAYMOND) discovered this species in the 1920s. The especially striking fossil of the Taung child (see photo at right) showed a mixture of human and apelike features. Anthropologist Robert Broom found many other *A. africanus* specimens in the field. At first, British anthropologists disregarded Dart's and Broom's discoveries, primarily because they relied on PILTDOWN MAN as a guide to understanding human evolution. The characteristics of Piltdown man suggested that human brain size had begun to increase very early in human evolution, and that this transition had occurred in Europe. The Taung specimen suggested that ape skulls had begun assuming modern characteristics before any increase in brain size, and that this transition had occurred in Africa. Dart lived to see Piltdown man revealed as a hoax, and the general acceptance of *Australopithecus* as a valid human ancestral genus, even if *A. afarensis* may not have been on the main line of human evolution.

Since the bones of *A. africanus* were found in association with those of many prey mammals whose bones had been crushed, it was at first thought that *A. africanus* was a hunter. This is the image presented by writer Robert Ardrey, upon which writer Arthur C. Clarke based the opening of the movie *2001: A Space Odyssey.* However, a skull of a child of this species was found that had holes that exactly matched the species of leopard that was present in southern Africa at that time. According to the research of South African paleontologist C. K. Brain, the piles of crushed and broken bones, found in limestone caves, were apparently leftovers from leopard meals, rather than australopithecine hunts. Modern leopards eat their kills in trees, which often grow out of limestone caves in the arid South African landscape. *A. africanus* was the prey, not the predator.

One of the most famous australopithecine fossils, "Lucy," was a nearly complete *A. afarensis* skeleton found in the early 1970s by Donald Johanson (see JOHANSON, DONALD) (see figure on page 33). Since the discovery of Lucy, both small and large individuals of this species have been found, which may represent females and males of this species or may indicate that there was more than one species of australopithecine at that place and time. Because it was so nearly complete, the Lucy skeleton revealed a great deal about the movements of this species. The major adaptations for bipedalism were present in *Australopithecus afarensis:* The opening for the spinal cord was underneath the skull,

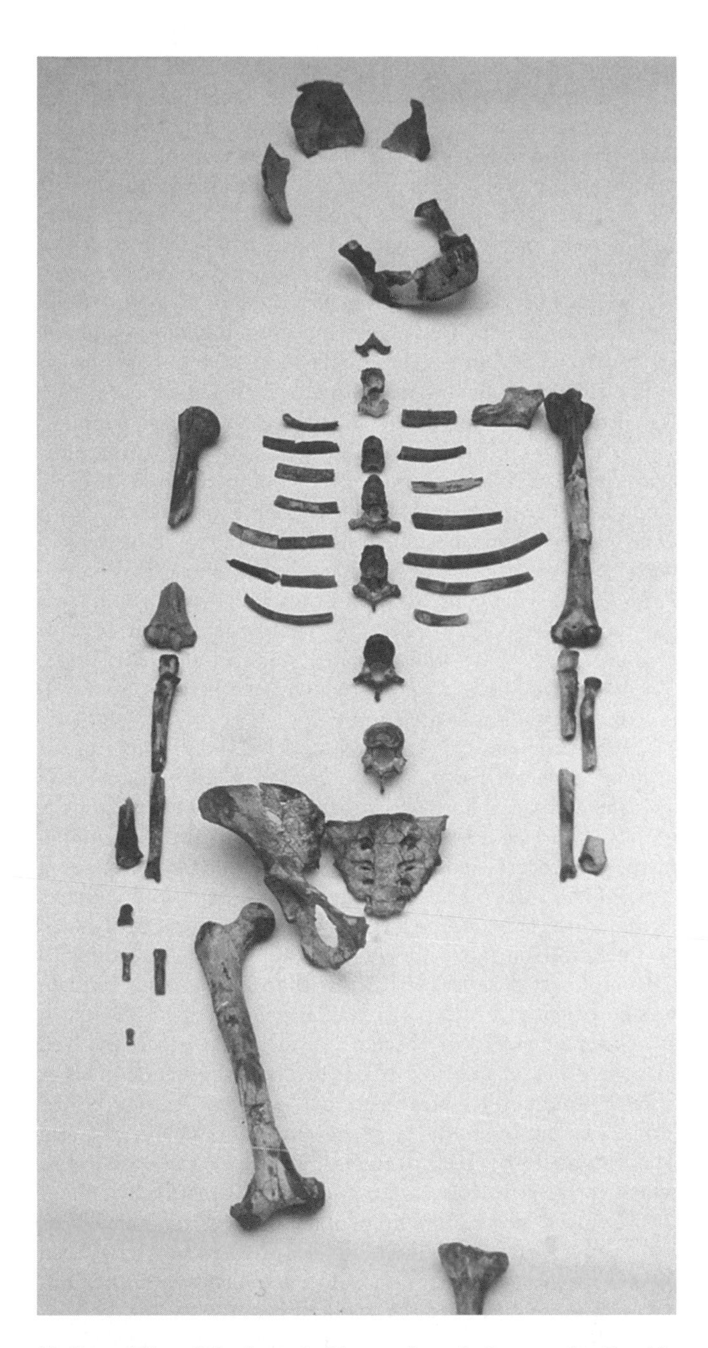

Skeleton of "Lucy," the *Australopithecus afarensis* discovered by Donald Johanson and associates. It is one of the most complete australopithecine skeletons that has been found, and it demonstrated that upright posture preceded increase in brain size in human evolution. *(Courtesy of Science VU/Visuals Unlimited)*

and the pelvis was suitable for walking. The feet, however, appear to have been only partly adapted to upright walking: the big toe was still at a noticeable angle. This suggests that *Australopithecus* lived mostly on the ground, walking upright, but frequently scrambled back into the trees when danger threatened. The relatively long arms and short legs of *A. afarensis* further suggest a mixture between arboreal and ground life. It even suggests the possibility that upright posture began as an adaptation for shinnying up trees from brief visits to the ground and later allowed full adaptation to ground life.

The primarily upright locomotion of *A. afarensis* and/or related australopithecines was confirmed by famous fossilized footprints found by anthropologist Mary Leakey (see LEAKEY, MARY) at Laetoli in Tanzania (see photo below). At about three million years of age, the footprints were undoubtedly produced by *A. afarensis* or a closely related species. Their date is fairly certain, as the footprints were imbedded in volcanic dust, on which RADIOMETRIC DATING can be used.

Trail of hominin footprints in volcanic ash. This 200-foot (70-m) trail was discovered by Mary Leakey's expedition at Laetoli, Tanzania, in 1978. The footprints were probably made by *Australopithecus afarensis* individuals 3.6 million years ago. They demonstrate that hominins had already acquired the upright, free-striding gait of modern humans. The footprints have well-developed arches and the big toe does not diverge noticeably. They are of two adults with possibly a third set belonging to a child who walked in the footsteps of one of the adults. *(Courtesy of John Reader/Photo Researchers, Inc.)*

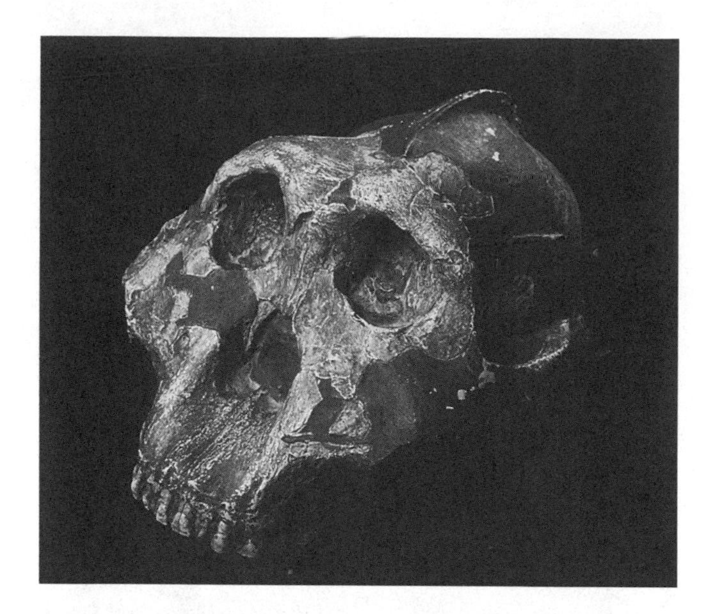

This skull of a robust australopithecine *(Paranthropus boisei)* that lived in Africa about 1.8 million years ago has a sagittal crest across the top, which served as an attachment surface for large chewing muscles. The sagittal crest and the large molars indicate that the robust australopithecines primarily ate coarse vegetation. *(Courtesy of Pascal Goetgheluck/Science Photo Library)*

In late 2006, two significant new *Australopithecus* discoveries were announced. A baby *A. afarensis* was found in Ethiopia by anthropologist Zeresenay Alemseged. The "Dikika Baby" fossil is even older than the Lucy fossil, and its shoulder blades indicated that this species retained some ability to swing in trees. The "Little Foot" australopithecine fossil, found in Sterkfontein, South Africa, had previously been represented only by a foot. In late 2006, the discovery of more of this same individual was announced.

Kenyanthropus

Kenyanthropus platyops ("flat-faced man of Kenya") was discovered by anthropologist Meave Leakey. This species lived about three and a half million years ago and resembled *Australopithecus* in many ways, including upright posture. As its name suggests, it had a much flatter face than any other australopithecine, a feature it shares with *Homo*. For this reason, some anthropologists suggest that *Kenyanthropus*, rather than any species of *Australopithecus*, was the ancestor of humans.

Paranthropus

Three species of the genus *Paranthropus* ("alongside human") were the "robust australopithecines." They all lived in eastern Africa and eventually became extinct without leaving evolutionary descendants. They contrasted with the gracile australopithecines in having huge jaws that allowed a diet of coarse vegetable materials. The first robust australopithecine to be discovered, by anthropologist Louis Leakey (see LEAKEY, LOUIS), was even called "Nutcracker man." The large teeth and prominent sagittal crest of robust australopithecine skulls indicated that they could chew powerfully. The sagittal crest, across the top of the skull, allowed attachment sites for large chewing muscles (see photo above). Chemical analysis of the tooth enamel of robust australopithecines suggests that they ate plants with the C_4 photosynthetic pathway, probably grasses, or termites that had eaten grass (see ISOTOPES). Detailed laser analysis of layers of tooth enamel in *P. robustus* specimens indicate that they were able to switch between food sources based on C_4 plants and those based on C_3 plants such as fruits and nuts. The conclusion is that robust australopithecines had dietary flexibility, and that their extinction was probably not due to the disappearance of food supplies. This conclusion raises the possibility that robust australopithecines were driven to extinction by competition with early modern humans, rather than by environmental changes. With their big jaws, the robust australopithecines may have looked fierce but were probably gentle, as are most herbivores. *Paranthropus aethiopicus* ("of Ethiopia") lived about two million years ago. This species may have been the ancestor of *P. robustus* ("robust") and *P. boisei* (named after Mr. Boise, a benefactor of Louis Leakey), which lived between two million and one million years ago.

When climates change, species evolve and proliferate, but in many different ways. About two and a half million years ago, the climate of east Africa became even drier, with more grassland and less forest cover. This occurred at the beginning of the Pleistocene epoch, when the Earth began its ongoing cycle of ICE AGES. While Ice Ages caused very cold climates in the north, they caused dry climates in tropical areas such as those inhabited by the African hominins. According to paleontologist Elisabeth Vrba, this change, while not sudden, was sufficiently severe that many species of mammals (such as species of antelopes) became extinct, and others evolved that were better able to survive in open grasslands. This is also approximately when baboons evolved from arboreal monkeys. The hominins were no exception to this pattern. This was about the time of the last gracile australopithecines, which apparently evolved in two different directions. Some of them evolved into a more humanlike form, with larger brains and use of tools. These early humans, which may have been several species, are often referred to as *HOMO HABILIS*. Others evolved into the robust australopithecines.

A half dozen or more hominin species may have lived in Africa all at the same time, about two million years ago. It is presently impossible to determine which of them may have been ancestral to modern humans, although some (such as *Paranthropus*) can be dismissed. The last of the australopithecines were the robust australopithecines, which did not become extinct until after the evolution of *Homo*. *Homo* and *Paranthropus* lived at the same time and general location. Evolutionary biologist Stephen Jay Gould suggested that, rather than a ladder of upward progress, the human evolutionary story more closely resembles a bush, with branches leading in different directions, only one of which became the modern genus *Homo* (see PROGRESS, CONCEPT OF).

Further Reading

Alemseged, Zerxenay, et al. "A juvenile early hominin skeleton from Dikaka, Ethiopia." *Nature* 443 (2006): 296–301.

Johanson, Donald, and Maitland Edey. *Lucy: The Beginnings of Humankind.* New York: Warner Books, 1982.

———, Lenora Johanson, and Blake Edgar. *Ancestors: In Search of Human Origins.* New York: Villard Books, 1994.

Johnson, Patrick, and Scott Bjelland. "Kenyanthropus." Available online. URL: http://www.kenyanthropus.com. Accessed March 22, 2005.

———. "Toumai Sahelanthropus tchadensis." Available online. URL: http://www.sahelanthropus.com. Accessed March 22, 2005.

Klein, Richard G., and Blake Edgar. *The Dawn of Human Culture: A Bold New Theory on What Sparked the "Big Bang" of Human Consciousness.* New York: John Wiley, 2002.

Kreger, C. David. "Australopithecus/Ardipithecus ramidus." Available online. URL: http://www.modernhumanorigins.com/ramidus.html. Accessed March 22, 2005.

———. "Orrorin tugenensis." Available online. URL: http://www.modernhumanorigins.com/lukeino.html. Accessed March 22, 2005.

Leakey, Richard, and Roger Lewin. *Origins Reconsidered: In Search of What Makes Us Human.* New York: Doubleday, 1992.

Sloan, Christopher P. "Found: Earliest Child—3.3 million year old bones discovered. *National Geographic,* November 2006, 148–159.

Smithsonian Institution, Museum of Natural History, Washington, D.C., USA. "Australopithecus afarensis: Composite reconstruction." Available online. URL: http://www.mnh.si.edu/anthro/humanorigins/ha/afarcomp.htm. Accessed March 22, 2005.

Sponheimer, Matt, et al. "Isotopic evidence for dietary variability in the early hominin Paranthropus robustus." *Science* 314 (2006): 980–982.

Tattersall, Ian, and Jeffrey H. Schwartz. *Extinct Humans.* New York: Westview Press, 2000.

Walker, Alan, and Pat Shipman. *The Wisdom of the Bones: In Search of Human Origins.* New York: Knopf, 1996.

Walker, Joanne et al. "U-Pb isotopic age of the StW 573 hominid from Sterkfontein, South Africa." *Science* 314 (2006): 1592–1594.

B

bacteria, evolution of Of all the cells and organisms in the world, bacteria and ARCHAEBACTERIA are the: (1) smallest, (2) simplest, (3) most numerous, (4) most metabolically diverse, and (5) oldest. Although for many decades scientists classified archaebacteria in the same category as bacteria, most now recognize archaebacteria to be an evolutionary lineage as distinct from the bacteria (also called *eubacteria,* or true bacteria) as they are from the complex organisms (see TREE OF LIFE).

Smallest. The typical eubacterial or archaebacterial cell is about 10 times smaller than the typical cell of complex organisms, in each dimension; therefore the typical bacterium has a volume a thousand times smaller than that of a typical cell of a complex organism. There are exceptions. The largest known bacterium is *Thiomargarita namibiensis,* which is one-twenty-fifth of an inch (1 mm) in diameter, which is larger than many complex cells.

Simplest. Eubacteria and archaebacteria are *prokaryotic,* which means they do not have a nucleus (Greek, *karyo*). Human, plant, fungus, and protist cells are *eukaryotic* (see EUKARYOTES, EVOLUTION OF) because they have a nucleus. Most eubacteria and archaebacteria have:

- a cell wall. In eubacteria, the wall consists mostly of a combination of protein and polysaccharide called peptidoglycan. Because the Gram stain indicates the presence of peptidoglycan, the *Gram positive bacteria* are those with a thick cell wall, and the *Gram negative bacteria* have thinner cell walls. Archaebacteria do not have peptidoglycan walls.
- a cell membrane, which regulates the molecules that can enter and leave the cell. The cell membranes of bacteria resemble those of eukaryotic cells, but the cell membranes of archaebacteria have a different chemical composition from those of all other cells.
- cytoplasm, the liquid portion in which the molecules are dissolved
- ribosomes, which make proteins. Bacteria and archaebacteria have smaller ribosomes than those of eukaryotic cells.

Although bacteria and archaebacteria do not have the complex structures found in the eukaryotic cell, they do carry out most of the same chemical processes. Prokaryotes do not have nuclei, but they do have DNA and use it in a manner similar to eukaryotic cells (see DNA [RAW MATERIAL OF EVOLUTION]). Prokaryotes do not have mitochondria, but many bacteria carry out the same reactions that mitochondria do. Some bacteria have numerous internal membranes and others have simple flagella that allow them to swim. Most bacteria function as single cells, although some can form multicellular aggregates. The shapes of bacteria are relatively simple: round ones are called *cocci,* rod-shaped ones are called *bacilli,* and spiral ones are called *spirilla* (if they have a central axis, they are *spirochetes*).

Most numerous. Bacteria and archaebacteria are the most numerous cells in the world. A typical spoonful of soil can contain billions of bacteria.

Most metabolically diverse. Bacteria and archaebacteria are the most diverse cells in the world, in terms of the metabolic reactions that they perform. Even bacteria that have similar appearances can have very different types of metabolism (see table on page 38):

- Many bacteria obtain energy the same way that animal cells do: by digesting organic molecules. Some of them live in the presence of oxygen and release energy through aerobic respiration, like most animal cells; some live in the absence of oxygen and rely on fermentation (see RESPIRATION, EVOLUTION OF).
- Some bacteria (in particular, the *cyanobacteria*) use sunlight energy to make food molecules in the same way that plants do. They obtain electrons by splitting water molecules (H_2O), and release oxygen gas into the air (see PHOTOSYNTHESIS, EVOLUTION OF).
- Some bacteria obtain energy using processes of which no eukaryotic cell is capable. Some photosynthetic bacteria use hydrogen sulfide (H_2S) rather than water (H_2O) as an

Examples of the Diversity of Eubacterial Metabolism

Metabolism	Category	Source of Energy or Electrons	Habitat
Heterotrophic	Aerobic	Obtain food from organic molecules	Live in presence of oxygen gas
	Anaerobic	Obtain food from organic molecules	Live in absence of oxygen gas
Autotrophic photosynthetic (energy from sunlight)	Cyanobacteria	Split H_2O molecules	Live in aerated water
	Sulfur bacteria	Split H_2S molecules	Live in anaerobic water, low light
Autotrophic chemosynthetic (energy from inorganic molecules)		Energy from H_2S	Deep ocean vents
		Energy from metals	Rocks
		Energy from NH_3	Anaerobic soil

electron source, releasing sulfur atoms instead of oxygen gas. *Chemosynthetic* bacteria obtain energy from inorganic chemical reactions. Some sulfur bacteria oxidize hydrogen sulfide. Others oxidize ions, such as ferrous (Fe^{++}) into ferric (Fe^{+++}), ammonia (NH_4^+) into nitrite (NO_2^-), or nitrite into nitrate (NO_3^-). The hydrogen sulfide bacteria are often found at the deep-sea vents that spew out water that contains this gas; in such places chemosynthetic bacteria are the basis of the entire food chain. Bacterial metabolism has been important in the production of important mineral deposits.

Oldest. Bacteria and archaebacteria were the first lifeforms on the Earth. They probably existed 3.5 billion years ago, and it was nearly two billion years later that the first eukaryotic cells evolved.

The total number of bacterial species is impossible to calculate. Some studies suggest that a spoonful of soil contains thousands of kinds of bacteria, while a different spoonful of soil from a different location contains thousands of other kinds of bacteria. Because they have simple structures, they can be distinguished mainly by their metabolism; but scientists have to know what that metabolism is in order to grow them and study them. There are many kinds of bacteria whose presence researchers overlook because they cannot grow them properly. Recently, biotechnology has provided new ways of identifying the bacteria on the basis of their DNA without having to grow them in a laboratory.

Many scientists speculate that there is more life under the Earth than on top of it, and this life is all bacterial and archaebacterial. These bacteria and archaebacteria obtain energy from minerals and live in the wet cracks in rocks up to two miles below the surface, metabolizing very slowly (dividing perhaps once a century). Some scientists estimate that the bacteria of these "subsurface lithoautotrophic microbial ecosystems" (abbreviated SliME) would cover the surface of the Earth to a depth of five feet (almost two m) if they were concentrated.

Many kinds of bacteria can grow under extreme conditions: high temperatures, strong acid, or concentrated salt, all conditions that would kill eukaryotic cells. For example, *Thiobacillus concretivorans* grows in sulfuric acid at concentrations that would dissolve metal. Other bacteria can tolerate extremely basic conditions. *Deinococcus radiodurans* can live in the water tanks of nuclear reactors. It resists nuclear radiation because it has a unique ability to reform its DNA after damage by radiation. *Streptococcus mitis* was found inside the lens of a camera that was left on the Moon for two years. Bacteria live even in deep ocean vents so hot that the temperature probes start to melt. *Thermus aquaticus* (shortened to Taq), which lives in hot springs, is the source of an enzyme that copies DNA at high temperatures and is therefore very useful in genetic engineering.

Many of the bacteria that live in extreme conditions are archaebacteria. However, the most primitive lineages of eubacteria, represented by *Aquifex* and *Thermotoga,* thrive under extreme conditions as well. The earliest eubacteria, like the earliest archaebacteria, must therefore have evolved under extreme conditions on the early Earth.

The earliest cell fossils, such as those found in the 3.5-billion-year-old Apex chert, were of types that, if they lived today, would be classified as bacteria (see ORIGIN OF LIFE). Throughout the Archaean era, the entire world was bacterial. Throughout the Proterozoic, the world was mostly microbial, shared between prokaryotes and eukaryotes (see PRECAMBRIAN TIME). Only in the last 540 million years (the last 14 percent of Earth history since the beginning of the Archaean) have multicellular organisms been numerous. Evolutionary biologist Stephen Jay Gould points out that in every age of Earth history, bacteria have been the modal type of organism (see PROGRESS, CONCEPT OF).

Bacteria freely exchange genetic material (see HORIZONTAL GENE TRANSFER). Throughout history, bacterial cells have accumulated genetic capabilities from one another, and from other types of cells. For example, the green sulfur bacteria (which use hydrogen sulfide as an electron source in photosynthesis) have only one set of reactions for absorbing and using light energy. The cyanobacteria, on the other hand, have two; one of them is very similar to that of the green sulfur bacteria, and that is where it may have come from (see PHOTOSYNTHESIS, EVOLUTION OF). Few bacteria produce cholesterol; the ones that do may have obtained it from eukaryotic cells.

Bacteria have also proven to be the champions of SYMBIOGENESIS by moving into, and remaining inside of, eukaryotic cells for billions of years. The mitochondria and chloroplasts found in eukaryotic cells are both the simplified evolutionary descendants of bacteria.

Bacteria can evolve rapidly. Their populations can evolve antibiotic resistance within a few years (see RESISTANCE, EVOLUTION OF). They can also remain the same for millions of years. Bacteria are very good at evolving, or staying the same; they are good at metabolizing rapidly, but also at doing nothing at all. Spores of *Bacillus permians* from a 250-million-year-old salt deposit near Carlsbad Caverns have been resuscitated, although scientists are uncertain whether the bacteria themselves are that old. A type of cyanobacteria began to grow after spending a century as a dried specimen in a museum.

Bacteria, while they are as simple as they were three billion years ago, have contributed enormously to the evolutionary history of organisms. Humans survive because their cells are full of the descendants of bacteria.

Further Reading

Dexter-Dyer, Betsey. *A Field Guide to Bacteria.* Ithaca, N.Y.: Cornell University Press, 2003.

Stevens T. O., and J. P. McKinley. "Lithoautotrophic microbial ecosystems in deep basalt aquifers." *Science* 270 (1995): 450–454.

Wassenaar, Trudy M. "Virtual Museum of Bacteria." Available online. URL: http://www.bacteriamuseum.org/main1.shtml. Accessed March 24, 2005.

balanced polymorphism Balanced polymorphism is the maintenance of genes in a population because of a balance between the damage they cause to individuals in homozygous form and the benefit they confer on carriers. Often, defective recessive alleles of a gene (see MENDELIAN GENETICS) can cause individuals that carry two copies of the allele (homozygous individuals) to die. Carriers (heterozygous individuals) are often unaffected by having one copy of the defective allele, because they have a dominant allele that functions normally; the defective allele, in that case, has no effect on the organism. Because the recessive allele can hide in the heterozygotes, NATURAL SELECTION may not be able to eliminate these alleles from populations (see POPULATION GENETICS). Balanced polymorphism occurs when the heterozygotes are not only unharmed but actually benefit from the defective allele.

The most famous example of balanced polymorphism involves sickle-cell anemia (SCA). The MUTATION that causes this genetic disease produces abnormal hemoglobin. Hemoglobin is the protein that carries oxygen in the red blood cells. The mutant hemoglobin differs from normal hemoglobin in only one amino acid, but this difference occurs near the crucial spot at which oxygen molecules bind. The mutant form of hemoglobin crystallizes under acidic conditions, such as during physical exertion, causing the red blood cell to collapse into a sickle shape. When this occurs, the white blood cells attack the sickled red cells, forming clumps that tend to clog the blood vessels. Therefore a person who is homozygous for the production of the mutant hemoglobin suffers anemia, first because of the collapse of the red blood cells into a form that cannot carry oxygen, and second because of impaired circulation. People with SCA frequently die young. In a person who is heterozygous for this mutation, half of the red blood cells have the mutant hemoglobin, therefore the heterozygous person is mildly anemic. SCA is very common in tropical regions of Africa and among black Americans descended from tribes that live in those areas.

People who are homozygous for the normal type of hemoglobin are susceptible to malaria, which is infection by a protist (see EUKARYOTES, EVOLUTION OF) that lives inside of red blood cells and is spread by mosquito bites. When, on a regular cycle, the protists emerge from red blood cells, the victim experiences fevers and chills. Millions of people die from malaria. The malaria parasite, however, cannot reproduce in a red blood cell that contains SCA mutant hemoglobin. The parasite's waste products cause the red blood cell to collapse, and the cell is destroyed, before the parasite has a chance to reproduce. Therefore, people with SCA cannot also have malaria. In people who are heterozygous for the mutant hemoglobin, the parasites can survive in only half of the red blood cells. This considerably slows the spread of the parasite and renders the heterozygous person effectively resistant to malaria. This is an advantage in regions of the world, such as tropical Africa, that are afflicted with the malaria parasite. In tropical Africa, therefore, people who are homozygous for the mutant hemoglobin often die from SCA, people who are homozygous for the normal hemoglobin often die from malaria, while heterozygous people survive malaria and pay for it with mild anemia. The damage caused by the mutant hemoglobin is balanced by the advantage it provides. A similar situation explains the prevalence of the blood disorder *thalassemia* in some Mediterranean and Asian populations.

Another possible example of balanced polymorphism involves cystic fibrosis. People who are homozygous for this mutation have a mutant CTFR protein in their cell membranes. The mutant protein disrupts salt balance between the cell and the body. As a result, in people who are homozygous for the mutation, a buildup of mucus occurs in the respiratory passages and digestive tract. The mucus buildup encourages bacterial infections. Certain bacteria that are harmless to most people, such as *Pseudomonas aeruginosa* and *Burkholderia cepacia,* thrive in the mucus that results from cystic fibrosis. People who are carriers of the mutant protein have a slightly greater chance of infection and impaired salt balance, but not enough to produce noticeable health consequences.

Scientists have puzzled over why the mutation that causes cystic fibrosis is so common. In some populations, up to one in 25 people are carriers of this mutation. *Salmonella typhi* is the bacterium that causes typhoid fever, a disease that was common in Europe before the middle of the 20th century. This bacterium binds to the normal form of the protein. These bacteria cannot bind to the mutant form of the protein. Therefore, in regions that had typhoid fever, people homozygous for the mutant protein often died of cystic fibrosis and people homozygous for the normal protein often died of typhoid fever, while heterozygous people were less susceptible to typhoid fever and had only mild salt balance problems. Today, with the control of typhoid fever by sanitation and antibiotics, the mutant gene provides no advantage. The cystic fibrosis gene is particularly common in populations of European origin. Scientists have also suggested that the gene that promotes osteoporosis (calcium loss from bones in older adults) may confer resistance to the bacteria that cause tuberculosis.

The possibility has also been raised that blood type proteins have been selected by disease resistance. The human ABO blood groups are determined by a gene on chromosome 9. The gene codes for proteins that are in the membranes of red blood cells. Allele A differs from allele B by seven nucleotides. Three of these nucleotide differences are neutral in their effect, while the other four result in amino acid differences in the blood protein. This is enough to allow the immune system to distinguish between blood protein A and blood protein B. Allele O results from a deletion, which prevents the production of the blood protein. Each person has two alleles for this group of blood proteins:

- A person who carries two A alleles, or an A and an O allele, will have blood type A, since their red blood cells carry protein A.
- A person who carries two B alleles, or a B and an O allele, will have blood type B, since their red blood cells carry protein B.
- A person who carries one A and one B allele will have blood type AB, since their blood cells have both proteins.
- Blood type O results from the absence of these blood proteins.

This has long been considered an example of neutral genetic variation since the differences in blood type proteins are not known to have any adverse health effects, and since no benefit of one blood type over another has been proven. Populations in most parts of the world consist of a mixture of blood types. The only major pattern seems to be that blood type O is relatively more common in populations of Native Americans than in other populations. Research has suggested that people with blood type AB are resistant to the bacterial disease cholera, while people with blood type O are susceptible to it. Although the blood proteins are found mainly on red blood cell membranes, they are also secreted in saliva, where they might function in preventing cholera. On the other hand, people with blood type O are more resistant to malaria and syphilis than those with the A and/or B proteins. If this is the case, scientists would expect that natural selection would make blood type AB very common in populations that frequently encounter cholera. This has not proven to be the case, perhaps because cholera is a disease of relatively recent origin. It did not reach Europe until the 18th century, and it may not have existed in Asia (its place of origin) very many centuries before that. Scientists would also expect blood type O to be more common in regions with malaria and syphilis. However, Native Americans did not encounter malaria until it was brought from the Old World in the 16th century, and the hypothesis of the New World origin of syphilis is controversial.

What looks like neutral variation in blood group proteins may very well be left over from the past history of natural selection for disease resistance. As geneticist and science writer Matt Ridley wrote, "In a sense the genome is a written record of our pathological past, a medical scripture for each people and race." Whether or not this turns out to be the case with blood group proteins, it has certainly happened with sickle-cell anemia, which evolutionary scientists still consider the best example of a balanced polymorphism.

Further Reading

Dean, A. M. "The molecular anatomy of an ancient adaptive event." *American Scientist* 86 (1998): 26–37.

Pier, G. B., et al. "*Salmonella typhi* uses CTFR to enter intestinal epithelial cells." *Nature* 393 (1998): 79–82.

Ridley, Matt. *Genome: The Autobiography of a Species in 23 Chapters*. New York: HarperCollins, 1999.

Bates, Henry Walter (1825–1892) British *Naturalist* Born on February 8, 1825, Henry Walter Bates was a British naturalist who specialized in collecting and studying insects. Bates befriended another British naturalist (see WALLACE, ALFRED RUSSEL). In 1848 Bates and Wallace traveled together to the Amazon rain forest of South America, where both of them extensively collected insects and other animals. Wallace returned to England and then traveled to Indonesia, but Bates remained in the Amazon rain forest for 14 years.

While studying Amazonian insects, Bates discovered that certain nonpoisonous species of heliconid butterflies had wing coloration patterns that resembled those of poisonous heliconid butterfly species. Since Bates's initial discovery, numerous other examples of MIMICRY have been found. This type of mimicry is now called *Batesian mimicry*. Bates's discovery of mimicry contributed significantly to Charles Darwin's work on evolution. Darwin cited Bates's discovery in later editions of *Origin of Species* (see DARWIN, CHARLES; *ORIGIN OF SPECIES* [book]). Bates and Darwin both maintained that such mimicry could not have resulted from the use and disuse of the structures involved (see LAMARCKISM) and must have resulted from NATURAL SELECTION.

Bates returned to England in 1862. He published an account of his work in the rain forest, *The Naturalist on the River Amazons,* in 1863. Bates continued scientific research and was elected a fellow of the Royal Society in 1881. His studies contributed to an understanding of BIODIVERSITY and of evolution, particularly of COEVOLUTION. He died February 16, 1892.

Further Reading

Beddall, Barbara G. *Wallace and Bates in the Tropics: An Introduction to the Theory of Natural Selection, Based on the Writings of Alfred Russel Wallace and Henry Walter Bates*. London: Macmillan, 1969.

***Beagle*, HMS** *See* DARWIN, CHARLES.

behavior, evolution of Behavior is the activity that results from the voluntary nervous activity of animals. Activities that are caused by purely metabolic or hormonal events are not considered behavior. Plants grow toward light. They do so because a hormone diffuses from the growing tip, primarily down the dark side of the stem, causing the cells on that side to grow more. Animals respond to low blood sugar by feeling hungry, which is a hormonal response. Animals respond to hot temperatures by increasing blood circulation to the skin, thereby allowing body heat to diffuse into the air. This is an involuntary nervous response in which the muscles around the animal's arteries relax. None of these actions are behavior. If the animal searches for food in response to hunger, or stretches its limbs to expose more surface to the air, thus allowing more heat loss, these actions would be considered behavior.

Animals raised or kept in captivity often display unusual actions that do not reflect their natural behavior. The study of animal behavior under natural conditions is called *ethology*.

The study of behavior can occur at many different levels. This is illustrated by the question, "Why does a bird sing?" Several different answers could be given to this question. First, a scientist could investigate the short-term causes of the singing of a particular bird. What is it about the environment, the genes, or the previous experience of this bird that makes it sing at the time, at the place, and in the manner that it does? These causes are called *proximate causes,* because they are near or proximate to the organism. Second, the scientist could investigate the evolutionary origin of the genes that make or allow the bird to sing. Since these causes reach far back in time, they are called *ultimate causes.*

Proximate Causes of Behavior

Proximate causes of animal behavior include stimuli, genes, and learning.

1. *Stimuli.* A bird sings in response to stimuli (singular stimulus) from its environment. The warmer temperatures and increasing day length of spring act as *external stimuli* to mating behavior in birds. An internal biological clock acts as an *internal stimulus* to the bird's activities. The external stimuli (in this case, day length) continually reset the internal stimulus (the biological clock).

Many stimuli come from the nonliving environment. Sowbugs, for example, respond to the presence of bright light by crawling toward the shade. Other stimuli come from animals that did not intend to produce the stimulus; sharks, for example, swim toward blood. Many stimuli are deliberately produced as communication signals by other animals. Dominant animals in populations use threat displays to subdue other members of the population. Some animals use startle displays or MIMICRY to avoid predators or to catch prey. Mating behavior in many mammals occurs in response to the timing of the female's period of reproductive readiness (see REPRODUCTIVE SYSTEMS). For the female, it is an internal stimulus; for the male, an external one; for both, it causes physiological changes and behavior that precede or accompany copulation.

A bird can respond to environmental stimuli only because it has sensory structures that intercept, and a nervous system that can interpret, the stimulus. The sensory structures and the nervous system are the products of gene expression. In this way, all animal behavior has a genetic basis.

2. *Genes.* A bird sings because singing is a behavior encoded in its genes and is therefore instinctive. Instinct allows an animal to perform behaviors that it has never seen. Instinctive behavior patterns have a genetic basis, in two ways. First, instinctive behavior patterns may be specifically coded into the animal's brain, the structure of which is encoded in the animal's DNA. The genetic basis of many animal behavior patterns, including a number of human behavioral disorders, has been determined by the study of inheritance patterns. Second, instinctive behavior patterns may be controlled by hormones. Genes regulate and are regulated by hormones, and hormones can influence behavior. During the winter, the days are short; this is the environmental stimulus. In response to the short days, the pineal glands of birds produce high levels of the hormone melatonin. High levels of blood melatonin inhibit birds from singing. When the days become longer in the spring, the pineal gland produces less melatonin, and the birds sing.

Once they have been stimulated, instinctive behaviors continue to completion. For this reason, instinctive behaviors are called *fixed action patterns.* Fixed action patterns may be very complex but are still performed without thinking. Examples include the egg-laying behavior of the female digger wasp, the foraging behavior of honeybees, and the parental behavior of female blue-footed boobies.

Consider the example of the digger wasp. The female digger wasp stings her prey (usually another insect) enough to paralyze but not to kill it. Then she brings the prey to the mouth of a burrow that she has previously dug in the ground. After dragging the prey into the burrow, the wasp lays eggs on it. When the eggs hatch the larvae eat fresh food, as the prey is still alive. Before dragging the prey into the burrow, the wasp enters the burrow to inspect it. In one case, a scientist was watching; while the wasp was inside the burrow, the scientist moved the prey to a nearby location. When the wasp emerged, she noticed that the prey was missing and quickly located it. She dragged the prey back to the mouth of the burrow. Only a few seconds had passed since she last inspected the burrow. It would seem to a human observer that the wasp should now drag the prey into the burrow and finish laying her eggs. However, the wasp dutifully left the prey outside the burrow and went inside for another inspection. The scientist again moved the prey. When the wasp emerged again, she relocated the prey, dragged it back, then inspected the burrow. The scientist repeated this activity many times.

The digger wasp's behavior involved both elements of response to stimulus and of instinct, but not of reasoning. The wasp responded to environmental stimuli in order to locate the prey and the burrow. The timing of the wasp's reproductive activity, and of all of the stages of its life, occurred at the appropriate season of the year because of responses to environmental stimuli such as temperature and day length. The sequence of events that occurred during egg-laying was instinctive, similar to the operation of a machine. When the scientist moved the prey, the wasp's egg-laying program was reset to an earlier stage. The wasp was like a washing machine that had been set back to an earlier stage in its program; the washing machine does not know that it has already washed the same load of clothes before.

In another species of digger wasp, the female visits her various burrows each day. The burrows contain larvae of different ages, therefore of different sizes. The female provides various sizes and numbers of paralyzed insects to them, depending on the amount of food that they need. This appears at first to be reasoning, but it actually is not. When a scientist experimentally exchanged the larva in one burrow for the larva in another, the female did not notice, but continued for a while to provide food to the burrow proportional to the size of its original, not its current, inhabitant. The female wasp is not reasoning but is following a fixed action pattern. Eventually the pattern is reset as the wasp gathers updated information about the sizes of the larvae.

Another example is the honeybee. A colony of honeybees *(Apis mellifera)* can contain up to 80,000 bees, all sisters or half sisters, the offspring of a single queen. Many of the bees collect food within a radius of several kilometers.

Near the entrance, worker bees stand and buzz their wings, creating an air current that cools the nest. Other worker bees, the foragers, bring home loads of nectar and pollen that they have gathered from flowers. The foragers do not visit flowers indiscriminately, but only those flowers that are structurally suited to bee pollination. A bee drinks nectar with its proboscis and stores it in a special sac; a valve prevents the nectar from entering the stomach. While still visiting the flowers, the foragers rake the pollen off of their bristles and cram it into pollen baskets on the hind legs.

The foragers crawl into the darkness of the nest. The nest contains thousands of perfectly hexagonal chambers, which not only serve as honeycombs but also as brood chambers. The chambers are composed of beeswax secreted by workers' wax glands. When the nectar-laden forager finds an empty chamber, she regurgitates the nectar from the nectar sac. She turns perishable nectar into well-preserved honey by regurgitating an enzyme that breaks the sucrose into simpler sugars and an acid which helps prevent bacterial contamination. Numerous other workers inside the nest fan their wings, helping to evaporate water out of the honey.

A forager called a scout arrives, having located a new patch of flowers. She places herself in the midst of a crowd of her sisters and begins to dance. If the nectar source is nearby, she performs a round dance which merely excites the other foragers to rush out and search in the vicinity of the hive. If the patch of flowers is distant, she performs a complex waggle dance. The timing and direction of her movements communicate information about the distance and the direction of the patch of flowers. In the darkness of the hive, this information is communicated by touch and sound. A faster dance communicates that the patch of flowers is farther away. The scout performs the dance in a figure eight, with a straight run in the middle during which the scout waggles her abdomen back and forth rapidly. The direction of the straight run communicates the direction of the food. It does not point directly to the food, because the honeycomb surface on which the scout performs the dance is vertical. The other foragers must interpret the direction of the food as a human would interpret a line drawn on a map on the wall. If the scout had flown a detour in order to find the food, her dance would have communicated not the distance and direction of the detour but the direction of a straight flight to the food!

The other foragers use the direction of the Sun to interpret the waggle dance information. What do they do if the Sun is hidden behind clouds? Bees are able to see the direction of the polarization of light, therefore they can determine where the Sun is located behind the clouds. If the straight run of the scout's waggle dance is vertical, this communicates to the other foragers that the food is in the same direction as the Sun. A straight downward run indicates that the food is in the opposite direction. Intermediate directions of the straight run indicate the approximate direction to the left or right of the Sun.

The Earth turns and the Sun appears to move across the sky. The bees compensate for this movement with their internal biological clocks which provide them with a mental map of the Sun's location in the sky at various times of day. If a scout is caught and imprisoned for several hours in a dark box, then released directly into the dark hive, she communicates the direction of the nectar source relative to where the Sun currently is located, not where the Sun was located when she was captured. She rotates the direction of the straight run of the dance as if it were the hour hand of a 24-hour clock. Bees raised in the Southern Hemisphere also rotate their dance direction, but in reverse. Other workers then leave for the new nectar source, flying in the right direction and carrying only enough food to supply them for the distance that the scout communicated to them.

Although these actions are complex, they are instinctive fixed action patterns. Each bee has relatively little intelligence, but their interactions can result in a hive that displays a collective intelligence that far surpasses that of any constituent individual (see EMERGENCE).

Another example is the blue-footed booby. The blue-footed booby of the GALÁPAGOS ISLANDS defines its nesting territory by dropping a ring of guano, as shown in the figure on page 43. It treats any nestling booby that is within the circle as its own offspring and totally ignores any nestling that is outside the circle, even when its cries of starvation can be plainly heard. The parents' caretaking behavior is a fixed action pattern.

Humans also have fixed action patterns. Many facial expressions are instinctive and are more complex than typical muscular reflexes. People who have been blind from birth, and who have never seen anyone smile, still smile.

3. *Learning.* A bird sings because it learns how to sing. Fixed action patterns can be modified by learning. In all three of the above examples, the actual form of the fixed action pattern is modified by learning. A bird learns many of its songs by watching and listening to other birds, and to birds of other species. Mockingbirds are famous for their diverse patterns of singing. Due to learning, no two mockingbirds have exactly the same repertoire of songs. Some birds, such as parrots, have a very highly developed ability to learn new vocalizations. The domestication and training of many mammals has demonstrated that they are capable of learning a greater variety of behavior than they typically display

under natural conditions. A learned response may not be any more conscious or reasoned than are the instinctive responses within the animal's behavioral repertoire. Many insects are capable of learned behavior. For example, other experiments with digger wasps have demonstrated that they memorize the landmarks surrounding the entrance of their burrows.

In some cases, animals acquire information through learning very rapidly soon after they are born and do not change their minds when they are older. This behavior is called *imprinting*. The most famous example of imprinting occurred when behavioral scientist Konrad Lorenz studied birds that had just hatched. The hatchling imprints upon the first animal that it sees as its mother. The hatchlings, when old enough to leave the nest, followed Lorenz wherever he went.

Animals with brains that are large relative to their body size have a greater ability to learn (see ALLOMETRY). Mammals have brains larger than other vertebrates of their size, primates have brains larger than other mammals of their size, and humans have brains larger than other primates of their size. Evolutionary scientists are uncertain regarding at what point in the history of the human species the brain became large enough to allow reasoning ability and consciousness to emerge (see INTELLIGENCE, EVOLUTION OF). Learning requires neither consciousness nor intelligence, but intelligence greatly enhances learning ability, because the animal can think about environmental stimuli, is not limited to fixed action patterns, and need not learn every behavioral response separately.

Ultimate Causes of Behavior

The behavioral characteristics of animals, just like all other characteristics, are the product of natural selection. It should therefore be possible to investigate the intermediate steps in the evolution of these behaviors, and the fitness advantage that individuals within populations gain from various kinds of behavior: from fixed action patterns, from the ability to learn, and from the ability to reason.

1. *Intermediate steps in the evolution of behavior*. Intermediate stages by which complex behavior has evolved have, in numerous instances, been demonstrated. Consider the example of bee foraging behavior, described above.

The most primitive form of bee foraging behavior, in which a scout can direct other foragers to the food source, would be for the other foragers to follow the scout as it returns to the new food source for another load of nectar. In fact, this behavior frequently occurs in honeybee colonies.

The next intermediate step in the evolution of the bee dance would be for the scout to perform a dance in which the straight run pointed directly at the food source. The honeybee *Apis mellifera* is a temperate zone insect that nests in hollow trees and in caves and builds vertical combs. One of its closest relatives is the tropical dwarf bee *Apis florea* that can

This blue-footed booby in the Galápagos Islands has created a nesting area by depositing a ring of guano. The nesting behavior is a fixed action pattern, not reasoning: The mother bird considers any nestling inside the ring to be her own, and any nestling outside the ring to not be her own. ***(Photograph by Stanley A. Rice)***

make nests with horizontal combs on tree boughs or rocks. It is likely that *A. mellifera* evolved from bees with habits similar to those still found in *A. florea*. When the dwarf bee scout returns from a new food source, she performs a dance on the horizontal surface of the comb in which the straight run points directly toward the food. This intermediate evolutionary step in the evolution of the bee dance has in fact been observed.

Another necessary step in the evolution of the bee dance would be for gravity (on the vertical honeycomb in the dark hive) to substitute for light (on the horizontal honeycomb on the rock or branch). A scientist performed an experiment in which a light bulb was placed in a honeybee hive. When a scout was dancing, the scientist turned on the light. The scout then altered the direction of her waggle dance to align to the light, rather than to gravity. Apparently the honeybee retains a primitive instinct to align its dance to the Sun, an instinct it has not used since it evolved the habit of making vertical honeycombs in the dark.

The evolution of complex behavior patterns is slow and gradual and has not been observed to occur over short periods of time. However, as in the case of bee foraging behavior, some behavior patterns that scientists believe to have been present in the ancestors of honeybees can still be found, in honeybees or their close relatives.

2. *Fitness advantages of specific behaviors.* The study of the fitness advantages conferred by natural selection upon behavior patterns is called SOCIOBIOLOGY. Sociobiology has been particularly successful at explaining the fitness advantages of ALTRUISM, especially in social insects. Sociobiology has often explained the evolution of fixed action patterns, and of learning abilities. A central idea of sociobiology is that an understanding of the evolutionary basis of behavior in one animal species should provide insights into the behavior of other species, including humans. As behavioral scientist Tim Friend says, "No matter where you look, just about every creature is obsessed with sex, real estate, who's the boss, and what's for dinner."

Sociobiology has proven controversial among scientists particularly when sociobiologists have attempted to explain the fitness advantage of specific human behavior patterns. Examples include religion and the fear of strangers (see RELIGION, EVOLUTION OF). Sociobiologists point to the universality of these behaviors, and the fitness advantages that would result from them. This implies, sociobiologists claim, that the specific behavior patterns are genetically based. Sociobiologists do not deny that learning and volition can modify or override these genetically based behaviors. Critics of sociobiology claim that natural selection has caused the evolution of a human brain that is so large and flexible that its ability to learn and modify human behavior far outweighs any residual genetic influence. To the critics, human behavior results from a genetically-based brain, but specific human behavior patterns have no genetic basis (see essay, "How Much Do Genes Control Human Behavior?").

Each species evolves its own set of behavioral patterns. Some behavioral patterns have evolved in more than one species because these patterns allow communication that is beneficial to more than one species. An example is Müllerian mimicry (see MIMICRY) in which dangerous prey all share a common set of warning coloration patterns (usually black alternating with white, yellow, orange, or red). Warning coloration alerts predators to leave these dangerous prey animals alone. More than one animal species can also share the ability to produce and recognize warning calls. Mammals have been seen to respond to bird calls that indicate the presence of a predator.

Behavior patterns may not be only a response to environmental conditions, but can contribute to those conditions. If some animals within a group learn a new behavior, it can give them a fitness advantage; the behavior can spread through the group, or the entire species, by a nongenetic learning process. This newly acquired behavior creates a situation in which there is a fitness advantage for any genetic variation that may make that behavior easier or more effective. A bird may learn to eat a new type of food, which is not an evolutionary modification. However, any mutant birds that had a genetically based preference for that new type of food, or beaks that allowed them to eat it more efficiently, would be favored by natural selection. This is GENE-CULTURE COEVOLUTION.

One of the greatest difficulties in studying the evolution of animal behavior is that humans tend to impute consciousness, volition, even intelligence, to practically every action, not only to the behavior of animals but even to storms and earthquakes. Even though scientists have demonstrated that many complex animal behaviors result from fixed action patterns modified by learning, human observers cannot help but imagine that these behaviors are conscious and purposeful. Evolutionary philosopher Daniel Dennett has focused attention on the conditions that would favor the evolution of true consciousness, as opposed to behavioral patterns that simply appear conscious to human observers.

Further Reading

Alcock, John. *Animal Behavior: An Evolutionary Approach.* Sunderland, Mass.: Sinauer Associates, 2001.

Dennett, Daniel. *Kinds of Minds: Towards an Understanding of Consciousness.* New York: Basic Books, 1997.

Friend, Tim. *Animal Talk: Breaking the Codes of Animal Language.* New York: Free Press, 2004.

Lorenz, Konrad. *King Solomon's Ring: New Light on Animal Ways.* 1952. Reprint. New York: Routledge, 2002.

big bang *See* UNIVERSE, ORIGIN OF.

biodiversity Biodiversity is the total number of species in the world, or within any limited place. Biodiversity is the product of evolution. Evolutionary scientists are interested in how the patterns of biodiversity differ from place to place, and how they have changed over the history of the Earth, because this gives them insight into how the evolutionary process works.

The most general pattern of biodiversity is that the places on the Earth, and the times in Earth history, that have the most species are those with the highest temperature and greatest amount of moisture. Today this is represented by the

How Much Do Genes Control Human Behavior?

It is obvious that genes control human anatomy and physiology, and that genes produce the structure and chemistry of the human brain. It is unclear—and controversial—to what extent genes determine, or influence, the specific moods, behavior patterns, and actions that the brain causes a human to feel or perform. If a brain is a computer, the genes build it; but how much do they program it? This is an important question for understanding the effects of evolutionary history upon human society today.

Geneticist Dean Hamer discovered the Xq28 MARKER, the so-called gay gene. He found that homosexual men had a certain form of Xq28 more often than would be expected on the basis of random patterns of inheritance. This marker is on the X chromosome, which males inherit only from their mothers (see MENDELIAN GENETICS). This would explain the significant pattern that male homosexuals frequently have homosexual uncles. He also found that lesbianism is not associated with any known genetic inheritance pattern. Hamer's peers did not doubt that he had found a significant correlation between this marker and homosexual behavior, but many were skeptical about how important this association is. Hamer's results were also noticed by the press and the general public. They were greeted with consternation by both conservatives (who insist that homosexuality is sin) and liberals (who insist that sexual orientation, and most other behavior, is determined by environment and individual choice), but greeted with elation by some homosexuals (whose T-shirts proclaimed, "Xq28—thanks for the genes, Mom!". One would expect that, if anybody would say that genes control human behavior, it would be Hamer. But he does not. In his book, *Living with Our Genes: Why They Matter More Than You Think,* Hamer points out that genetic influence is not the same as genetic causation. He says regarding the response to Xq28, "Rarely before have so many reacted so loudly to so little."

Hamer believes what all scientists believe: physical characteristics are the product of both genes and environment interacting with one another. While the environment affects the kind of person an individual human becomes, genes may influence which environments a person seeks. Thus genes have some control over environment, just as environment can influence gene expression. Hamer points out that at least 300 different brain chemicals can influence human feelings. The production of these chemicals is controlled by genes. Surely they must affect behavior in some way. Hamer's analogy is that genes are like musical instruments: They determine the range and the sound but not the tune. Science writer and geneticist Matt Ridley uses a similar analogy: Genes create an appetite, not an aptitude.

How does one determine that a gene influences behavior? Hamer cited a correlation between a certain form of Xq28 and the incidence of homosexuality, but he is very careful to point out that a correlation of unrelated individuals could be spurious. Hamer calls it the "chopstick effect." People who have straight dark hair are better at using chopsticks than people with light hair. This is a very significant and true correlation, but no one would claim that black hair enhances the ability to use chopsticks. While this is a humorous example, similar spurious reasoning has been used in much more serious contexts. Some observers note a correlation between black ancestry and incidence of crime. However, all competent scientists know that the correlation disappears once you take such factors as unemployment and poverty, and individual differences, into account (see EUGENICS).

One way to avoid spurious correlations is to use different measurements for each factor. In another study, Hamer found a correlation of the brain chemical serotonin with anxiety. Was this a spurious correlation? He checked for other factors, and the correlation of serotonin with anxiety held true even when he included male vs. female, gay vs. straight, age, ethnicity, education, and income in the analysis. He also measured anxiety levels using two separate personality tests, to make sure the results were not a mere artifact of the test.

Hamer explains that spurious correlations are less likely to happen when found within a genetically limited cultural group or a family, since culture is less of a variable. True, just because something looks like it "runs in families" proves nothing; for example, religions and recipes run in families but are not genetic. But when there are differences among individuals within families, and these differences are correlated with genes, the differences are less likely to be cultural or environmental, and more likely to be genetic.

The influence of genes on behavior is frequently studied by comparison of identical twins separated at birth and raised in different households. Critics such as evolutionary geneticist Richard Lewontin (see LEWONTIN, RICHARD) point out that this procedure is quite imperfect. Imperfect as it is, though, it may be the best way to detect a genetic basis for behavioral characteristics. Researchers analyze data in which each point comes from a pair of twins, the x value for one twin and the y value for the other. Then they determine a correlation coefficient for the trait. Another approach is to compare monozygotic (identical) with dizygotic (fraternal) twins. Identical twins come from a single fertilized egg and are genetically identical. Fraternal twins come from two separate eggs fertilized by two separate sperm and are no more similar to one another than any pair of siblings.

Geneticist Thomas J. Bouchard, Jr., of the University of Minnesota, heads the Minnesota Study of Twins. He tells the story of Jim Lewis and Jim Springer, identical twins separated at birth, who were raised separately and met for the first time at age 39. Both were six feet (two m) tall and weighed 180 pounds (about 90 kg). Both had been married twice, first to a Linda and then to a Betty. One had a son named James Alan, the other a son named James Allen. As boys they both named their dog Toy. They smoked Salem cigarettes and drank Miller Lite. This story is astonishing, but one must ask: Can there be a gene for marrying Betty? Naming a dog? Smoking Salems? No one claims that this is the case. The evidence seems strong, however, that there is a genetic basis for certain preferences that led to the choices that Jim and Jim made—perhaps a brain chemical that influenced their inclinations, attitudes, moods—which in the same culture (they both grew up in America) would lead to similar choices. Many other twins in the study, separated at birth, were not nearly as similar as Jim and Jim.

Environmental factors that influence behavior include not only the environment in which a child is raised, but also the nutritional and hormonal environment when the fetus is in the womb. A correlation between birth order and sexual orientation, for example,

(continues)

How Much Do Genes Control Human Behavior? *(continued)*

suggests that the birth of older siblings influences the hormonal environment that a fetus experiences. Identical twins, separated at birth, shared the same prenatal environment.

As an example of the interaction of genes and environment, Hamer cites a Swedish study of adoptions and petty crimes. The researchers assumed "bad genes" when both biological parents were criminals, and "bad home" when both adoptive parents were criminals. Here were the effects of genes and homes on the incidence of petty crimes in the youth from these homes:

Good genes, good home: 3 percent petty crime rate
Good genes, bad home: 7 percent petty crime rate
Bad genes, good home: 12 percent petty crime rate
Bad genes, bad home: 40 percent petty crime rate

Good genes, good home (3 percent crime rate) was similar to population at large. The influence of genes seems to exceed that of environment, but the difference between 7 percent and 12 percent may not be biologically significant. Hamer says that home environment has most of its influence on the behavior of kids, but genes have more of an effect on behavior in adulthood. Hamer therefore suggests that bad environments put kids in juvenile hall but bad genes put adults in jail.

Consider these other examples of genetic influence (not genetic determination) on human behavior:

Appetite. Another example that Hamer cites is obesity. Body weight is mostly an inherited trait. It takes hard work to change one's body weight. For some people it is easy to be thin—their bodies burn the calories—while for others it is difficult to be thin, because their bodies store calories. Is there a genetic basis for appetite?

Brain structure. In studies with mice, it has been found that the hunger center of the brain is in the hypothalamus. If the hunger center of the hypothalamus is removed, the mice starve even in the presence of food; and if the hunger center is stimulated, the mice overeat.

Brain chemistry. A mutation that causes obesity in mice was found to influence the production of the protein leptin ("thin" in Greek). Other mutations had the same effect—it did not matter which mutation it was, so long as it influenced leptin production. Humans also have the leptin gene, and when the human version of the gene is inserted into mice, it works the same way as the mouse gene. The mutation appears to increase appetite and lower metabolism. Although humans have a leptin gene, the mutant leptin gene was found in only a few grossly obese humans. Therefore the genetic basis of human obesity must be mostly in other genes. Mutations in many different metabolic genes—in any genes that normally cause fat to burn faster—could influence appetite and cause obesity. A mutation affecting the production of a certain brain neuropeptide makes rats eat even materials that taste bad or while receiving a shock. The human counterpart might be that the brain chemical serotonin is known to be associated with anxiety, depression, and a craving for carbohydrates. Hormones influence appetites in a complex fashion: Separate hormonal mechanisms are involved in hunger for carbohydrates, for fats, and for proteins, in a manner not understood.

Population variation. In twin studies, body weight had a 70 percent correlation. There is also a genetic effect on type and location of body fat. A correlation was found between people with potbellies and a gene that causes blood vessels to constrict. This study was conducted in a homogeneous population of Hutterites, in order to restrict other sources of variability. Of course, environment is also important. Many people of the Pima tribe in Arizona are obese, but this is not the case with the Pima in Mexico. The two populations separated just 1,000 years ago.

Evolutionary advantage. How does natural selection fit in to these considerations? In the prehistoric past, individuals who burned calories rather than saving them were less likely to survive famines, unless they compensated for it by violence. The ability to store calories, which evolutionary geneticists call the "thrifty genotype," was beneficial in the human evolutionary past. Today, it can lead to obesity (see EVOLUTIONARY MEDICINE).

Stimulation-seeking behavior. Another example of a genetic influence on behavior is the desire for stimulation. This is an important behavioral factor. People with an inclination to seek stimulation are more likely to divorce; and one of the best predictors for divorce is when one partner is inclined to seek stimulation and the other is not.

tropical rain forests. Two acres (a hectare) of tropical rain forest can contain as many tree or bird species as the entire United States. The rain forests, though covering less than 5 percent of the Earth's surface, may contain half of its species. The equable climatic conditions of the tropics allow them to be a museum of old species. It has so many species interactions that it is also a prime nursery of new species.

Biodiversity has been repeatedly reduced by global catastrophes, most notably the five MASS EXTINCTIONS in Earth history. Biodiversity has increased after each mass extinction event. Sometimes it takes a long time for biodiversity to recover. After the PERMIAN EXTINCTION, more than a hundred million years was required to restore the pre-disaster diversity levels (see figure on page 48).

Many scientists estimate that a sixth mass extinction, caused by human activity, is now under way. A combination of human activities is causing these extinctions. The most important is habitat destruction: Humans convert natural habitats into farmlands and cities that are unsuitable for wild species. Recovery will occur from this extinction event also. New species will eventually evolve suitable to the new, disturbed conditions created by humans. Science fiction writers

Brain structure. The brain influences the desire for stimulation. The nucleus accumbens is the pleasure center of the mammalian brain. When rats are allowed to self-stimulate this pleasure center (through an electrode), they do so, even to the extent of neglecting food.

Brain chemistry. Dopamine stimulates animals to seek pleasure and rewards them when it is found. Genetically altered mice with enhanced activity of the enzyme that makes dopamine explore their environments more; genetically altered mice lacking the enzyme sit and starve. In humans, the D4 dopamine receptor on chromosome 11 has a noncoding region of 48 bases that can be repeated two to 11 times. The longer sequences of this region result in less binding of dopamine—that is, it takes more dopamine to get the same effect. Longer sequences are also associated with people wanting more adventure.

Population variation. Overall, the heritability (genetic component of variability among humans; see POPULATION GENETICS) is 40 percent for the human tendency to seek stimulation. Although the correlation of the form of D4 receptor with stimulation seeking was statistically significant, it explained only 4 percent of the variation. If this gene explains only 4 percent out of the 40 percent total genetic effect, there might be 10 genes involved in dopamine regulation or other aspects of seeking stimulation. Therefore, one cannot claim to have found "the gene" that explains this part of human behavior. There must be many genes, and taken together they influence less than half of the variation in this behavior pattern among humans.

Evolutionary advantage. What evolutionary advantage might one or the other form of the D4 gene have had? Both forms of the gene can confer advantages. Individuals with the long form of the gene would pursue many sexual partners; individuals with the short form would tend to care more for their offspring. Both forms of the gene may enhance fitness, in different ways, depending on the circumstances.

Mating behavior in mammals is also influenced by hormone levels, which have a genetic basis. Variation in the noncoding DNA associated with vasopressin receptors appears to explain the differences in mating behavior between species of rodents. Prairie voles *(Microtus ochrogaster)* are usually monogamous, while the closely related meadow vole *(Microtus pensylvanicus)* is promiscuous. Prairie voles have more vasopressin receptors, perhaps due to the longer noncoding region associated with the vasopressin receptor gene, than meadow voles. Some prairie voles are more monogamous than others and more attentive to their offspring. The more monogamous and attentive prairie vole parents also have longer noncoding DNA regions near the vasopressin receptor gene. In 1999 a researcher inserted the prairie vole vasopressin receptor gene, together with its associated noncoding DNA, into mouse chromosomes. The resulting mice were more monogamous and more attentive to their offspring than normal mice.

The overall message is that genes do not determine, but can strongly influence, human behavior, which therefore has a strong evolutionary component. First, even where there is a genetic basis, many genes can be involved. The effects of the genes can be complex and indirect. For example, Prozac influences serotonin levels, but not in a direct and simple way. It takes several weeks to work, because it must influence the whole brain system. If its effect was only a straightforward effect on serotonin levels, it would start to work right away. Second, there is often evolutionary selection for different alleles of the same gene—for example, both for seeking stimulation and for not seeking it, for storing weight and for not storing weight. Both the pathway of causation and the effects of natural selection can be very complex, but very real.

Further Reading

Hamer, Dean, and Peter Copeland. *Living with Our Genes: Why They Matter More Than You Think.* New York: Bantam, 1998.

Hammock, Elizabeth A. D., and Larry J. Young. "Microsatellite instability generates diversity in brain and sociobehavioral traits." *Science* 308 (2005): 1,630–1,634. Summarized by Pennisi, Elizabeth. "In voles, a little extra DNA makes for faithful mates." *Science* 308 (2005): 1,533.

Lewontin, Richard. *It Ain't Necessarily So: The Dream of the Human Genome and Other Illusions.* New York: New York Review of Books, 2000.

Pennisi, Elizabeth. "A genomic view of animal behavior." *Science* 307 (2005): 30–32.

Ridley, Matt. *Genome: The Autobiography of a Species in 23 Chapters.* New York: HarperCollins, 1999.

———. *The Agile Gene: How Nature Turns on Nurture.* New York: HarperPerennial, 2004.

invent wild scenarios, but even some serious scientists speculate that super rats and tree-sized weeds may evolve. This will occur on such a long timescale that during one's lifetime a human will see virtually nothing but extinctions.

Biodiversity seems like such an easy thing to measure: Just go out and look for species, write them down, and count them up. Sir Joseph Banks, when he was the botanist on Captain Cook's voyages, increased the number of plant species known to Western science by a quarter by this method. This process of discovery has nearly been completed for mammals and birds, which are easy for human explorers to see, though a few new species are occasionally found. A new species of deer was discovered recently in Vietnam. There are 30,000 okapis (relatives of giraffes) in the jungles of Zaire, but they were unknown to Western science until the 20th century. Riwoche horses in Tibet were known only from cave drawings and were assumed to be extinct, until some explorers who got lost found them in a remote valley in 1995.

Biologists are still discovering many hundreds of species of microbes, plants, and terrestrial arthropods such as insects (see BACTERIA, EVOLUTION OF; INVERTEBRATES, EVOLUTION OF). The photosynthetic bacterium *Prochlorococcus,* the most

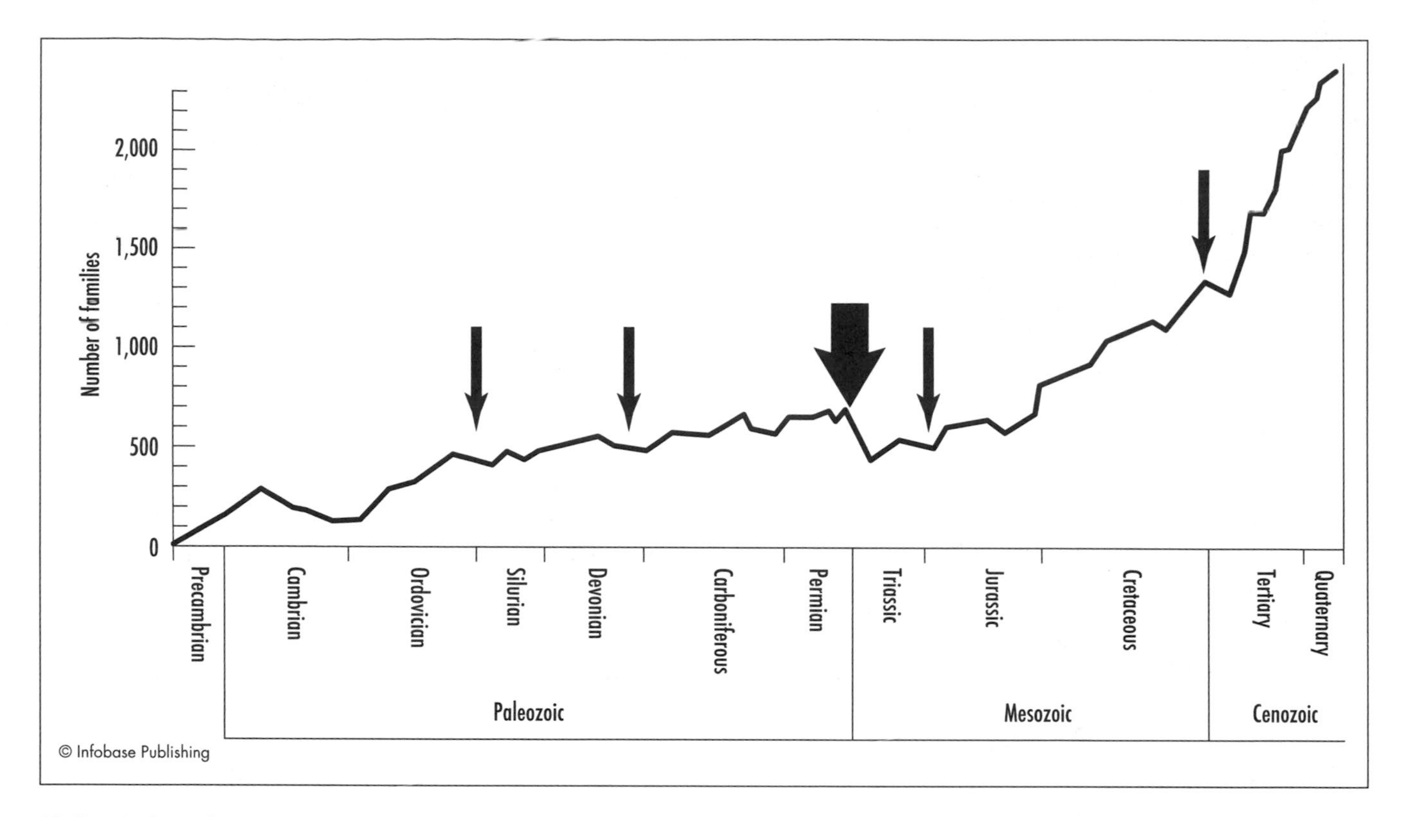

Biodiversity (as indicated by the number of families of fossilized marine and terrestrial organisms) has increased through geological time. After each of the five MASS EXTINCTIONS in earth history (five arrows), biodiversity decreased (especially after the PERMIAN EXTINCTION, largest arrow) but within a few million years began to recover and continue its increase. ***(Redrawn from Benton)***

abundant photosynthetic organism in the oceans, was not discovered until 1988. There is no end in sight for discovering new insect species. Partly this is because there are so many of them (350,000 species of beetles and counting!) but also because they are so hard to see.

Only an expert can recognize many of the distinctions between closely related species. Flies are often classified using the arcane science of *chaetotaxy,* which distinguishes them on the basis of the arrangements of their bristles. Clearly, one limiting factor to the discovery of new species is the availability of experts who can recognize that they are new.

About a million and a half species have been named. Some biologists estimate that there may be as many as 30 million to 100 million species in the world. Ecologist Terry Erwin calculated this estimate. He began by using pesticides to kill all the insects in certain tropical trees and collecting the insects that fell. By estimating how many of these insects were unique to certain species of trees, and extrapolating to the number of tree species, he was able to estimate the number of tropical insect species. From there he could estimate the number of insect species in the world, and then of all other species, using existing proportions of each taxonomic category. Although his estimate is probably high, it is certain that there are many more species than those already discovered. Robert May estimates just under seven million (see table at right).

Estimate of Global Species Diversity (Excluding Prokaryotes)

Taxon	Estimated number of species
Protist kingdom	400,000
Plant kingdom	320,000
Fungus kingdom	500,000
Animal kingdom	
Nematodes	500,000
Molluscs	120,000
Arthropods	
Crustaceans	150,000
Arachnids	500,000
Insects	4,000,000
Vertebrates	
Fishes	27,000
Amphibians	4,000
Reptiles	7,150
Birds	9,800
Mammals	4,800
Others	150,000
Total	6,790,000

The number of species that can exist in any habitat is large but not unlimited. The resources that a species uses, its spatial limits, and the times at which it uses them, is called the *niche* of a species. Two species cannot, in theory, have exactly the same niche. They either use slightly different resources, or in different places, or at different times. There is no general rule for how different the niches of two species must be (the *limiting similarity* of the species) in order for them to coexist. Tropical species can apparently divide up their environment into finer spatial niches, which may contribute to the greater biodiversity of the tropics. Species that live in seasonal climates must tolerate those seasonal changes if they are to survive, or else they migrate great distances in order to avoid these seasonal changes. Tropical species, in contrast, can remain in and specialize on one location.

Keeping track even of the species that are discovered is a difficult task. Sometimes new species turn out to have been already described. Although duplicate descriptions occur about 20 percent of the time, the other 80 percent of the descriptions are of genuinely new species. Usually new species are within previously known phyla, families, and genera (see LINNAEAN SYSTEM). There are occasional surprises. Whole new phyla of animals have been found in recent decades in places such as the deep ocean sediments.

If there are indeed 100 million species, it would take at current rates 14,000 years for biologists to catalog them. Some biologists and information scientists are designing streamlined methods for cataloguing biodiversity that bypass the slow process of academic publication and even allow researchers in distant jungles to send their information via satellite to databases, and which make information available online for identifying known species. Examples include the All Species Foundation, started in 2001 by *Wired* Magazine cofounder Kevin Kelly, and the Encyclopedia of Life project (see WILSON, EDWARD O.). It is also important to train researchers native to each country to catalog their countries' biodiversity. If there are not and will not be enough scientists, then this research must rely on trained amateurs.

The simple count of species (called *species richness*) is not always a sufficient measure of diversity. For example, the 74 species of terrestrial vertebrates from the Karoo Basin in the late PERMIAN PERIOD, contrasted with the 28 species that lived in that location in the early TRIASSIC PERIOD, indicate that species richness decreased by about two-thirds. (The extinction rate was far greater, since almost all of the 28 Triassic species evolved after the Permian extinction event.) The 74 Permian species were more or less equal in abundance, while in the early Triassic, one genus *(Lystrosaurus)* was overwhelmingly common, not only in the Karoo but worldwide. The post-extinction world had one-third the number of terrestrial vertebrate species but was far less than one-third as diverse in terms of ecological interactions, since most of the higher animals were all members of one generalist species.

In ecological and evolutionary terms, scientists distinguish between an ecosystem that contains, say, 50 equally abundant species and one that contains 50 species, 49 of which are rare. Ecologists have, accordingly, devised different *diversity indices* (symbolized by H) rather than to rely simply on species richness. In each of these indices, there are s species, and the proportional importance of the ith species is represented by p_i. If one adds up all of these proportions, the result is 1. The indices are calculated by adding up mathematical derivatives of the proportions. One of these indices, *Simpson's diversity index,* emphasizes the dominant species because it adds up the squares of the proportions:

$$H = \sum_{i=1}^{s} p_i^2$$

The other, the *Shannon-Wiener diversity index,* which is based upon information theory developed by mathematicians Claude Shannon and Norbert Wiener, emphasizes the rare species because it multiplies each proportion by its natural logarithm:

$$H = \sum_{i=1}^{s} p_i \ln p_i$$

Besides diversity, scientists can also quantify *equitability.* In a community of species with high equitability, the species are all about equally important; in a community with low equitability, there are a few dominant species, and the rest are rare. Equitability (E) is the diversity divided by the natural logarithm of the number of species:

$$E = H/\ln s$$

Within the United States, Appalachian cove forests not only have many plant species but these species are equitable. In contrast, a boreal forest has few species and low equitability.

Much of the Earth remains unexplored. The deep oceans, for example, are nearly inaccessible. Until the 1970s, no one even suspected the existence of entire complex communities of species living at the deep-sea vents where ocean floor volcanic eruptions occur. These deep-sea vents may have conditions that resemble those of early life on Earth (see ORIGIN OF LIFE). Not all unknown species hide in remote regions. New species of plants are continually discovered, some of them very close to scientific research centers. A new species of centipede was discovered recently in New York City's Central Park!

The diversity of microbes, especially bacteria, is particularly difficult to quantify. Biologists typically grow bacteria in media in order to estimate the number of species present in a sample; but most media have been designed for medical research. What about the bacterial species (especially the ARCHAEBACTERIA) that require hot, acid, or salty conditions not represented by standard bacterial growth protocols? Many bacteria live off of minerals, and they may do so up to several kilometers deep into the crust of the Earth (see BACTERIA, EVOLUTION OF). This indicates that there is a much greater number of bacterial species than previously suspected. *Bergey's Manual,* the closest thing scientists have to a list of bacterial species, includes about 4,000 species. But in 1990, Norwegian biologists Jostein Goksøyr and Vigdis Torsvik took a random gram of soil and determined that it contained four to five thousand species of bacteria. In another sample, they found a largely different set of 4,000 to 5,000 species. Bacterial species are now often identified by their DNA sequences (see DNA [EVIDENCE FOR EVOLUTION]). Biotechnologist Craig Venter used DNA sequencing techniques to detect thousands of previously unidentified microbial species in the Sargasso Sea in 2004.

Another problem with estimating biodiversity is the definition of species. Traditionally, biologists have classified organisms into the same species if they look the same. Most biologists now use the *biological species concept,* which defines species as populations that are reproductively isolated, that is, they cannot interbreed, even if they should be brought in contact with one another. Sometimes such biologically defined species may look the same to human observers, but they do not act the same in response to one another. The biological species concept is the preferred method because it allows the organisms themselves to indicate their distinctions. Each species represents a group of genes that are well-adapted to work together; hybrids between two species are frequently inferior in their fitness, since the two sets of genes do not work well together. Natural selection favors the isolation of species from one another (see SPECIATION). Reproductive isolation is not perfect. Interspecific hybrids are common; for example, many oak species are capable of interbreeding. Intergeneric hybrids are not unknown (for example, between mustards of the genus *Brassica* and radishes of the genus *Raphanus,* forming the hybrid genus *Raphanobrassica*) (see HYBRIDIZATION). However, hybrids are not as common as the species that produced them. Species distinctions, while imperfect, remain recognizable.

Regardless of how species are defined, and how many species there are, human activity is rapidly destroying many thousands of them. These species may, or may not, be valuable to their ecological communities or to the human economy, and humans destroy them before finding out. The current rate of destruction far exceeds the ability of evolution to replace them. Humans destroy what they do not know and what they do not even know how to know.

Further Reading

Bascompte, Jordi, et al. "Asymmetric coevolutionary networks facilitate biodiversity maintenance." *Science* 312 (2006): 431–433.

Benton, Michael J. "Diversity, extinction and mass extinction." Chap. 6 in *When Life Nearly Died: The Greatest Mass Extinction of All Time.* London: Thames and Hudson, 2003.

Ertter, Barbara. "Floristic surprises in North America north of Mexico." *Annals of the Missouri Botanical Garden* 87 (2000): 81–109. Available online. URL: http://ucjeps.berkeley.edu/floristic_surprises.html. Accessed 23 March 2005.

Holt, John G., ed. *Bergey's Manual of Systematic Bacteriology.* Four volumes. Baltimore, Md.: Williams and Wilkins, 1984–1989.

Jaramillo, Carlos, et al. "Cenozoic plant diversity in the neotropics." *Science* 311 (2006): 1,893–1,896.

Llamas, Hugo. "All Species Foundation." Available online. URL: http://www.all-species.org. Accessed 23 March 2005.

May, Robert M. "How many species?" *Philosophical Transactions of the Royal Society of London Series B* 330 (1990): 293–301; 345 (1994): 13–20.

———. "The dimensions of life on earth." In *Nature and Human Society.* Washington D.C.: National Academy of Sciences, 1998.

Torsvik, V., J. Goksøyr, and F. L. Daae. "High diversity in DNA of soil bacteria." *Applied Environmental Microbiology* 56 (1990): 782–787.

Ward, Peter, and Alexis Rockman. *Future Evolution: An Illuminated History of Life to Come.* New York: Henry Holt, 2001.

Wilmé, Lucienne, Steven M. Goodman, and Jörg U. Ganzhorn. "Biogeographic evolution of Madagascar's microendemic biota." *Science* 312 (2006): 1,063–1,065.

Wilson, Edward O. *The Future of Life.* New York: Vintage, 2003.

———. "The encyclopedia of life." *Trends in Ecology and Evolution* 18: 77–80.

biogeography Biogeography is the study of diversity through time and space; the study of where organisms live and how they got there. An understanding of biogeography is inseparable from evolutionary science. In fact, biogeography allowed some initial insights into the fact that evolution had occurred. In *Origin of Species,* Darwin noted that major groups of animals lived on certain continents, and not on others that had similar climates, because they had evolved on those continents (see DARWIN, CHARLES; *ORIGIN OF SPECIES* [book]). This insight from biogeography was also valuable to Wallace, who also discovered natural selection (see WALLACE, ALFRED RUSSEL). Wallace's discovery that the (primarily placental) mammals of Asia were very different from the (primarily marsupial) mammals of New Guinea and Australia is still one of the best examples of evolutionary biogeography (see MAMMALS, EVOLUTION OF).

Biogeography generally does not deal with the effects of global climatic alterations (for example, the buildup of oxygen in the atmosphere during PRECAMBRIAN TIME), however important they have been in evolution. Instead, it deals with events and forces that have created geographic patterns. Probably the biggest process that has affected biogeography over the history of life on Earth has been CONTINENTAL DRIFT. Continents have moved, coalesced into supercontinents, then split apart in different clusters. Continental coalescence brought different species into contact for the first time, allowing COEVOLUTION to produce new species; and subsequent splitting isolated them, allowing them to pursue separate evolutionary directions (see SPECIATION). Continental movements are the major explanation for the *biogeographic realms,* which have largely separate sets of species. Using the classification system of biogeographer E. C. Pielou, these realms are (see figure on page 51).

- Nearctic (North America)
- Neotropical (Central and South America)
- Palaearctic (Europe and western Asia including Mediterranean region)
- Ethiopian (Africa south of Mediterranean region)
- Oriental (eastern Asia)
- Australasian (Australia and nearby islands)
- Oceanian (Pacific islands)
- Antarctic (Antarctica)

Not only have the continents moved, but mountains have arisen and subsequently eroded away. Climatic changes have also created barriers such as deserts between populations that were once in contact with one another. Mountains and deserts can separate a population into two or more populations as surely as can an ocean. When populations are separated, one species can become many (see ADAPTIVE RADIATION), and different species can evolve similar adaptations independently of one another in each isolated region (see CONVERGENCE).

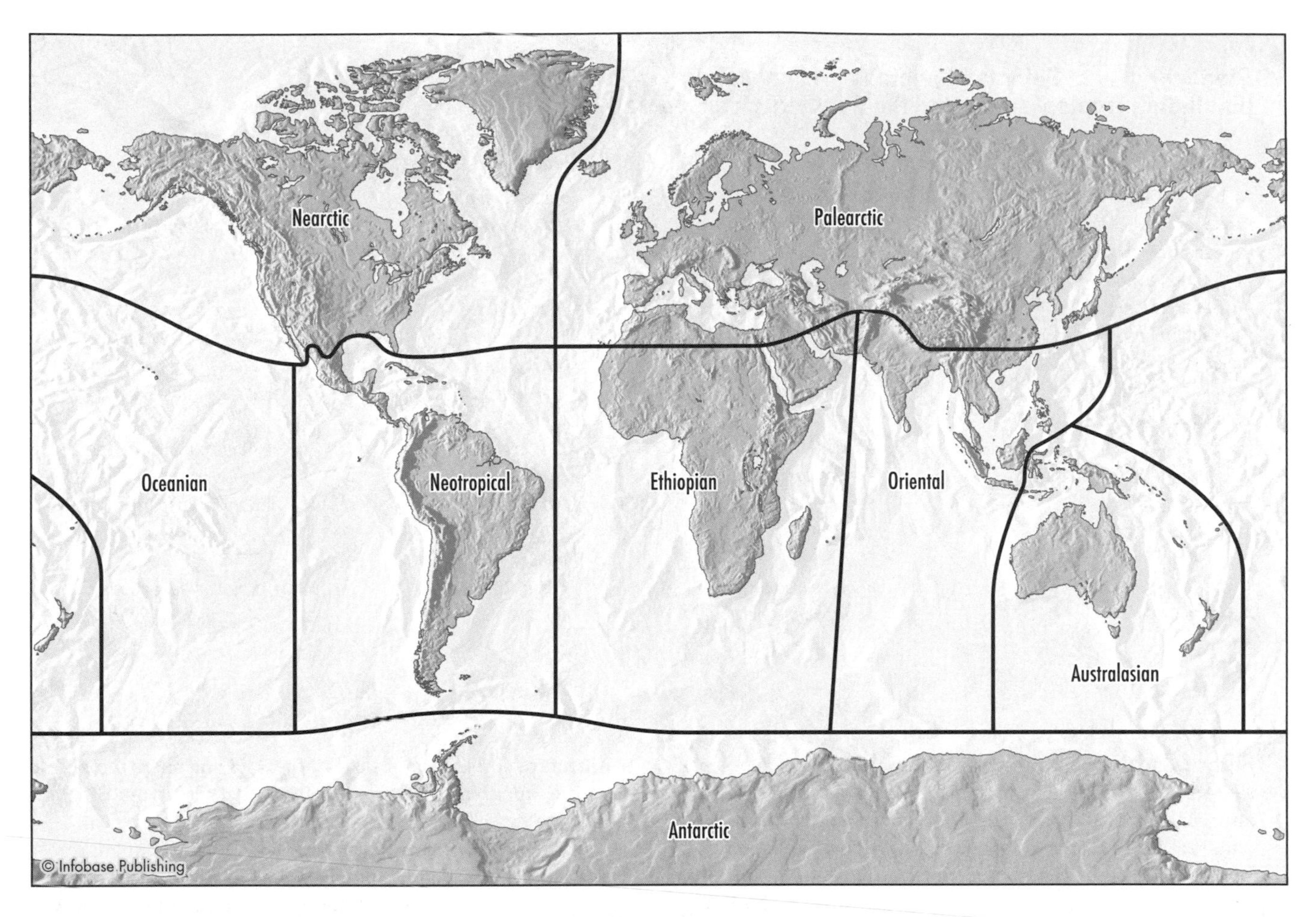

A map showing the approximate extents of the eight world biogeographical realms

Global Patterns of Vicariance and Dispersal

When geographical barriers split a previously unified population or species, allowing speciation, the resulting geographical pattern is due to *vicariance*. Alternatively, individuals of a species can migrate to a new location. The resulting geographical pattern is due to *dispersal*. Since each dispersal event is independent, it is unlikely that broad-scale biogeographic patterns are due to dispersal.

Often, biogeography on a higher taxonomic level (such as class or family; see LINNAEAN SYSTEM) may need to invoke more ancient continental movements and climatic changes, while biogeography on a lower taxonomic level (such as genus or species) may be attributed to more recent continental movements and climatic changes. Consider the example of the oak family, Fagaceae:

The oak family Fagaceae apparently evolved soon after the breakup of the supercontinent Pangaea at the end of the MESOZOIC ERA. Since the northern and southern continents separated, they have remained largely separate, except for the connection of Africa to Eurasia, the collision of the Indian subcontinent into Eurasia, and the recent formation of the isthmus of Panama. Because the family Fagaceae was already in existence before the Northern and Southern Hemispheres had separated very far, both hemispheres contain members of this plant family.

The modern genera within the Fagaceae family evolved after the separation of northern from southern continents. Therefore oaks (genus *Quercus*), chestnuts (genus *Castanea*), and beeches (genus *Fagus*) are restricted to the northern continents, where they evolved (see table), while South America and Australia have the genus *Nothofagus*, where it evolved.

Eighty million years ago, North American and European forests were not widely separated from one another. Thereafter, climates began to change and the northern continents drifted apart. Asian, North American, and European oaks were separated from one another. Within North America, the development of the Rocky Mountains and the southwestern desert separated the western oaks (especially in California) from the oaks of the eastern deciduous forest.

Biogeography of Selected Species of Angiosperms (Excluding Tropical Areas and the Southern Hemisphere)

Plant group	North America	Europe and Western Asia	Eastern Asia
Fagaceae: *Fagus*	*Fagus grandiflora*	*Fagus sylvatica*	*Fagus crenata*
Fagaceae: *Castanea*	*Castanea dentata*	*Castanea sativa*	*Castanea crenata*
Fagaceae: *Quercus*			
Section *Lobatae*	*Quercus rubra*		
Section *Quercus*	*Quercus alba*	*Quercus ilex*	*Quercus lanata*
Section *Cerris*		*Quercus suber*	*Quercus acutissima*
Ulmaceae: *Ulmus*	*Ulmus americana*	*Ulmus glabra*	*Ulmus parvifolia*
Betulaceae: *Betula*	*Betula nigra*	*Betula pendula*	*Betula utilis*
Betulaceae: *Alnus*			
Section *Alnus*	*Alnus serrulata*	*Alnus cordata*	*Alnus japonica*
Section *Clethropsis*	*Alnus maritima*		*Alnus nitida*
Salicaceae: *Salix*	*Salix nigra*	*Salix alba*	*Salix tetrasperma*
Salicaceae: *Populus*			
Section *Aegiros*	*Populus deltoides*	*Populus nigra*	
Section *Populus*	*Populus tremuloides*	*Populus tremula*	*Populus adenopoda*
Aceraceae: *Acer*	*Acer saccharum*	*Acer platanoides*	*Acer palmatum*

Europe and North America had apparently already separated by the time the red oak subgenus (section *Lobatae*) and cork oak subgenus (section *Cerris*) had evolved; red oaks are found only in America, and cork oaks only in Eurasia. Separate species within the white oak subgenus (section *Quercus*) are found in America, Europe, and eastern Asia (see table).

Further speciation occurred after the separation of North America and Europe. California, separated by mountains and deserts, has a completely separate set of oak species from eastern North America.

Many plant families are largely tropical, and they have remained in the tropical areas of the world as the continents have moved. Some of them have a *pantropical* distribution, in tropical forests all over the world, while others are found only in the Old World or in the New World.

There are numerous examples of families, genera, or subgenera of plants that are found only in the Northern Hemisphere, for example, the oaks, beeches, and chestnuts; elms (genus *Ulmus*); birches (genus *Betula*); cottonwoods (genus *Populus* section *Aegiros*); aspens (genus *Populus* section *Populus*); and maples (genus *Acer*) (see table). There are fewer examples of families, genera, or subgenera restricted to the temperate regions of the Southern Hemisphere in the manner of *Nothofagus*. This may be because the cold climates of the Southern Hemisphere are not as cold as those of the Northern Hemisphere. There is almost no land surface in the Southern Hemisphere climatic zone that corresponds to the tundra of the Northern Hemisphere, for example. Furthermore, the Southern Hemisphere is mostly ocean, and water does not change in temperature as readily as does land. The gentler temperature conditions of the Southern Hemisphere temperate zone has apparently allowed many otherwise tropical plant families to extend throughout South America and Africa.

In the above examples, vicariance seems to explain most of the patterns of plant biogeography. In some cases, dispersal appears to have also played a role. For example, most willows (genus *Salix*) are found in northern continents (see table) but a few species are found in the southern continents: *Salix humboldtiana* in South America and *Salix cinerea* in Australia. These may be the evolutionary descendants of willows that dispersed southward.

Dispersal is also the major explanation for the similarity of Eurasian and North American mammal genera. The major modern genera of mammals (such as bears, raccoons, deer, horses, bison, etc.) evolved after North America and Eurasia had separated. These animals dispersed across the Bering land bridge (Beringia). This dispersal occurred more readily in periods between ICE AGES. This explains why North America and Eurasia share many genera but not many species of mammals. Dispersal may explain the biogeography of groups of animals more readily than of plants. Birds, some of which migrate thousands of miles, disperse best of all.

Some groups of organisms are cosmopolitan or worldwide in their distribution. This may be the result of both vicariance and dispersal. The major examples appear to be groups of organisms that disperse very well or live in disturbed habitats: rodents among mammals, and plants in the mustard (Brassicaceae), composite (Asteraceae), bean (Fagaceae), and grass (Poaceae) families.

Placental mammals proliferated in the northern continents, while marsupial mammals proliferated in the southern continents, and they remained largely separate from one another. By about a million years ago, the isthmus of Panama formed a land bridge over which placental mammals dispersed from the Nearctic realm into South America, and marsupial mammals in the opposite direction. Plants of temperate climates, not adapted to tropical conditions and less mobile than animals, did not disperse across this land bridge. As it turned out, placental mammals frequently drove marsupial mammals toward extinction by competition. The placental mammals of South America, now dominant, are the descendants of invaders from North America. Most of the marsupials that moved north into North America became extinct, except for the opossum. Meanwhile, Australia and New Guinea remained separate and the realm of marsupials. Few placental mammals made the journey into the Australasian realm. One exception was humans. *HOMO ERECTUS* ventured into the islands of Indonesia, becoming Java man, but there is no clear evidence that they traveled to Australia. Some of them became isolated on the island of Flores (see FLORES ISLAND PEOPLE). Modern humans dispersed to Australia at least 50,000 years ago and apparently brought dogs with them. For thousands of years, humans and dogs were the main placental mammals of Australia. Australian marsupials included many species that were similar to placental species, but which had evolved separately (see CONVERGENCE). Europeans brought placental mammals, such as rodents and livestock, with them. Once again the placental mammals proved to be the superior competitors. Most of the Australian marsupials, except various species of kangaroo, have become rare or extinct, largely as a result of competition with the placental invaders.

When a continental fragment separates, it may sometimes retain a collection of species that closely resemble their ancestors, because they are not exposed to many of the newly evolved continental species. This appears to be what happened with the island of New Caledonia. The forests consist largely of conifers more similar to those that dominated Mesozoic forests than those that dominate modern forests. This is why film crews who want a realistic backdrop to dinosaur movies choose New Caledonia for their film site. The species of flowering plant that most closely resembles the presumed ancestor of all flowering plants, *Amborella trichopoda,* is found only on New Caledonia (see ANGIOSPERMS, EVOLUTION OF).

Some distribution patterns remain mysterious. One example is the plant family Empetraceae (crowberries), which is found in the cool regions of North America and Eurasia, but also the cool tip of South America, and in the warm climate of Florida. This is difficult to explain in terms of continental movements or dispersal to another habitat with similar climatic conditions.

Islands

Islands, starting with Darwin's experience in the GALÁPAGOS ISLANDS, have proven to be evolutionary showcases. When a new island forms, it receives only a genetic subset of organisms from the nearby mainland. In particular:

- In any species some of whose members disperse to a new island, only a small portion of the mainland genetic variability may be represented on the island. This FOUNDER EFFECT may be followed by genetic drift if the island population remains small.
- The newly arrived species may have fewer parasites, and perhaps no predators or competitors. This may, if the species happens to be well suited to the physical environment of the island, allow a population explosion, leading to evolutionary change and diversification.
- Many small mammals remain small partly because their small size allows rapid population growth, giving them an advantage in competition; but on the island, with fewer or no competitors, the small mammals may evolve into larger forms. The same thing appears to happen in plants. Many small, weedy plants have an advantage due to rapid growth and early reproduction. In this way, they avoid competition with large trees. When the plants disperse to an island and there are no trees, the weedy plants may evolve into a tree form. The plant family Asteraceae consists mostly of weedy herbaceous plants, except on some islands (such as the Channel Islands of California and St. Helena in the Atlantic) where they have evolved into small trees. The nettle family Urticaceae consists mostly of weedy herbaceous plants, but on Hawaii there is a bush nettle.
- Many large mammals remain large partly because their size allows them to resist the attacks of predators; but on the island, with no predators, the large mammals may evolve into smaller forms. Mammoths on Wrangel Island, northeast of Siberia, and on Santa Rosa Island off the coast of California evolved into smaller forms. Another example is the Flores Island people, *Homo floresiensis.* Some large mammals evolved into smaller forms on this island at the same time that the Flores Island folk lived there; the predator release explanation, however, is unlikely to be true for the small size of these people. The explanation of the Flores Island people remains a mystery.
- Many island species lose their defenses. Since defenses (spines or chemicals in plants, behaviors in animals) can be expensive, natural selection favors their loss in conditions where they are not useful. Many island plants have lost their defenses. For example, not only is the Hawaiian nettle unusual in its family for being a bush but also for its lack of stinging hairs. Numerous island animal species (reptiles, birds, and mammals) are notorious for their lack of fear of humans, with whom they have had little contact until recently.
- In order to get to the island in the first place, plants and animals needed to have superior dispersal abilities. Once on the island many of them have lost their dispersal abilities. One extreme example is the flightless cormorant of the Galápagos Islands. On some islands, plant families with small seeds have produced species with seeds larger than is normal for the family.

Consider new islands, such as the volcanic islands of Hawaii, as opposed to islands like New Caledonia that are fragments of continents. The new islands accumulate all of their species through dispersal and subsequent evolution. Larger islands have a greater variety of microhabitats. Moreover, on larger islands, populations within a species are more likely to avoid contact, resulting in greater speciation. For both of these reasons, a greater number of species evolves on large islands than on smaller islands. This is known as the species-area relationship. The source of the colonist species is the closest mainland. Islands that are closer to the mainland receive more immigrant species, which allows the evolution of more species native to the island. The first analysis of the balances among immigration, evolution, and extinction on islands, in which near versus far and large versus small islands were contrasted, was the *theory of island biogeography* developed by ecologist Robert H. MacArthur and evolutionary biologist E. O. Wilson in 1967 (see WILSON, EDWARD O.).

Ice Ages

Genera and even species of plants are shared between the arctic and alpine tundras, enough to allow these two regions both to be called tundra. Arctic tundra is found around the Arctic Ocean, while alpine tundra is on mountaintops. This biogeographical pattern is explained mostly by the ice ages. When glaciers were at their maximum extent, the arctic tundra formed a band across what is now the northern United States. The Rocky Mountains, Sierra Nevada, and Cascades received tundra species from contact with this band of tundra, and some mountains that were not in direct contact with the tundra (e.g., the San Francisco Mountains of Arizona) were close enough to receive tundra species by dispersal. When the glaciers retreated northward, the tundra plants retreated into the remaining zones of cold climate, either northward, where they are found today in the arctic tundra, or up the mountains, where they are today found on alpine peaks. The alpine tundra is now stranded, sometimes in very small patches (as in the San Francisco Mountains). This is not the entire explanation for tundra plant species. The tundra of the Sierra Nevada contains not only species affiliated with the arctic, but also the evolutionary descendants of desert species (such as the buckwheat *Eriogonum*). The tundra of the Sierra Nevada not only has drier soil than other tundras but also is close to the desert. Adaptations to cold and to drought are often similar, and this similarity allowed some desert plants to adapt to tundra conditions in California. This pattern is also evident in some animals: The caribou of the cold climates of North America is the same species as the reindeer of the cold climates of Eurasia.

Most of the North American continent south of the tundra was boreal forest (mainly spruce) and pine forests. Nearly the entire area that is now the Great Plains was covered with white spruce *(Picea glauca)*. There was a little bit of grassland in what is now Texas, but there was no hot desert. It would have been an alien world to modern eyes. There were many modern mammals, but also many mammals such as mammoths and mastodons that have become extinct (see PLEISTOCENE EXTINCTION). Most of the trees and other plants would have been familiar to modern observers, but they were in combinations that no longer exist; for example, in some places there were spruce trees scattered in grassland, an arrangement uncommon today.

As the ice sheets retreated, each surviving species moved to new locations independently, some dispersing more rapidly than others. The glacial retreats, and the movements of species, were not uniform; glaciers sometimes temporarily advanced again. The forests are still moving, though too slowly for human observers to notice without consulting long-term records. The tallgrass prairie did not exist before or during the ice ages. The tall, deep-rooted, warm-weather grasses of the prairie were established during the period of maximum warmth (the hypsithermal period, about 7,000 years ago). The prairies persisted even when conditions became cooler and wetter (until American agriculture and civilization destroyed most of them). The prairies persisted because periodic fires killed any forest trees that began to encroach. In North America, where most mountain ranges are north-south, there were few barriers to the primarily northward movement of plant and animal species. In Europe, however, the Mediterranean Sea and mountain ranges such as the Alps were barriers to northward movement. This helps to explain why British forests have fewer species of trees and spring wildflowers than do many American forests. The explanation for current patterns of distribution and diversity are rooted in the past, rather than in modern climatic conditions. The GREENHOUSE EFFECT may be altering the patterns of species movement that have been occurring for the past few millennia.

Humans have had a tremendous effect on biogeographical patterns. First, humans have carried species of plants and animals from one place to another. In some cases these plants and animals have become INVASIVE SPECIES that have displaced native species and greatly altered the original environment. Humans have allowed species to disperse, especially by ship and airplane, more effectively than almost any species could previously have dispersed. Second, humans have created a great deal of disturbed habitat (such as farming, road building, and construction). Species that specialize in disturbed areas (especially weeds) have found a worldwide network of suitable habitats, courtesy of humankind. Some of the most widespread species, such as dandelions and barn swallows, live in areas altered by human activity.

Further Reading

Bonnicksen, Thomas M. *America's Ancient Forests: From the Ice Age to the Age of Discovery.* New York: John Wiley and Sons, 2000.

Gillespie, Rosemary G. "The ecology and evolution of Hawaiian spider communities." *American Scientist* 93 (2005): 122–131.

Grant, Peter R. *Evolution on Islands.* Oxford: Oxford University Press, 1998.

MacArthur, Robert H., and Edward O. Wilson. *The Theory of Island Biogeography.* Princeton, N.J.: Princeton University Press, 1967.

MacDonald, Glen M. *Biogeography: Space, Time, and Life.* New York: John Wiley and Sons, 2003.

Whitfield, John. "Biogeography: Is everything everywhere?" *Science* 310 (2005): 960–961.

bioinformatics Bioinformatics is the computer-based analysis of the information content of biological molecules. This includes the amino acid sequences of proteins and the nucleotide sequences of nucleic acids such as RNA and DNA (see DNA [RAW MATERIAL OF EVOLUTION]). Bioinformatics has many applications for the study of molecular biology and medicine. Its particular application to evolutionary science is that it allows the comparison of amino acid sequences in the corresponding proteins of different species, or the nucleotide sequences in the corresponding genes or NONCODING DNA of different species.

Determining the nucleotide sequence of a DNA molecule used to be a lengthy and expensive process. The process developed by molecular biologist Frederick Sanger considerably shortened the work, which has now been entirely automated. The process uses nucleotides that cause DNA replication to stop, and that also have components that produce different colors of fluorescent light. When DNA fragments from this process are separated out according to size, they form a series of colored bands that look like a supermarket bar code. The sequence of red, green, blue, and yellow bands directly represents the sequence of the four nucleotides in the original DNA molecule. Thousands of DNA sequences have now been determined in laboratories all around the world. Whenever scientists publish papers that include reference to these sequences, the journal requires that the sequence be submitted to an electronic database that is accessible to scientists (or to anyone else) throughout the world. Two major databases are at the National Center for Biotechnology Information (NCBI), sponsored by the U.S. National Institutes of Health, and the Expert Protein Analysis System (ExPASy) in Switzerland. NCBI has databased DNA from more than 165,000 organisms.

Comparing nucleotide sequences also used to be a lengthy process, in which the nucleotides of each DNA molecule were compared to the corresponding nucleotides of another, one by one. Automated programs now line up the DNA molecules, identify the nucleotides that differ among the molecules, and calculate the percent similarities among them. Perhaps the most commonly used system is the Basic Local Alignment Search Tool (BLAST), also available at the NCBI web site. The single-nucleotide differences among DNA molecules, known as single nucleotide polymorphisms (SNPs), are often used as the data in the construction of phylogenetic trees (see CLADISTICS).

The construction of phylogenetic trees is a mainstay of modern evolutionary science. Phylogenetic trees allow the reconstruction of the evolutionary history of everything from single strains of virus (see AIDS, EVOLUTION OF) to families (see LINNAEAN SYSTEM) to the entire living world (see TREE OF LIFE). Modern phylogenetic studies would be unthinkable without the databases and analyses used in bioinformatics.

Further Reading

National Institutes of Health. "National Center for Biotechnology Information." Available online. URL: http://www.ncbi.nih.gov. Accessed April 17, 2006.

Swiss Institute of Bioinformatics. "ExPASy Proteomics Server." Available online. URL: http://us.expasy.org. Accessed July 12, 2005.

Selected Fields of Study Within Biology

Level	Name	Study of:
Molecules	Molecular biology	Biological molecules in cells
	Molecular genetics	DNA and its use in cells
Cells	Cytology	Cell structure and function
Tissues	Histology	Tissue structure and function
Organs and individuals	Anatomy	Structures of organs
	Physiology	Functions of cells, tissues, and organs
	Genetics	Inheritance patterns
Species and groups of species	Virology	Viruses
	Microbiology	Microorganisms
	Bacteriology	Bacteria
	Mycology	Fungi
	Botany	Plants
	Zoology:	Animals
	Malacology	Molluscs
	Entomology	Insects
	Ichthyology	Fishes
	Herpetology	Reptiles and amphibians
	Ornithology	Birds
	Mammalogy	Mammals
	Physical anthropology	Humans
Interactions	Ethology	Animal behavior
	Ecology	Interactions among organisms
	Parasitology	Interactions between parasites and hosts
	Epidemiology	Spread of parasites
	Population biology	Populations
Evolution	Population genetics	Genetic variation in populations
	Taxonomy	Classification
	Systematics	Evolutionary basis of classification
	Evolutionary biology	Evolutionary processes

biology Biology is the scientific study of life. Other disciplines study life, and life-forms, from other viewpoints, but biology employs the SCIENTIFIC METHOD. Subdisciplines within biology are presented in the table.

Biology has been transformed by the emergence of evolutionary science. Many of the characteristics of organisms make little sense in terms of operational design (see INTELLIGENT

DESIGN) or adaptation to their environments and can only be understood in terms of evolutionary ancestry. This applies to some of their structural and functional characteristics (see VESTIGIAL CHARACTERISTICS) and particularly to much of the DNA (see NONCODING DNA). This is what evolutionary geneticist Theodosius Dobzhansky meant by his famous statement, "Nothing in biology makes sense except in the light of evolution" (see DOBZHANSKY, THEODOSIUS). Furthermore, evolution has given an organizing principle to biology, which might otherwise have continued to be a cataloguing of types of organisms and their structures and functions.

Although biologists study life, they cannot precisely define it. Ernst Mayr (see MAYR, ERNST), perhaps the most prominent biologist of modern times, indicated that a life-form must have the following capacities:

- For metabolism, in which energy is bound and released, for example, to digest food molecules
- For self-regulation, whereby the chemical reactions of metabolism are kept under control and in homeostasis, for example, to maintain relatively constant internal conditions of temperature, or moisture, or chemical composition
- To respond to environmental stimuli, for example, moving toward or away from light
- To store genetic information that determines the chemical reactions that occur in the organism, for example, DNA, and to use this information to bring about changes in the organism
- For growth
- For differentiation, for example to develop from an embryo into a juvenile into an adult
- For reproduction
- To undergo genetic change which, in a population, allows evolution to occur

While most of these characteristics may not be much in dispute, it is impossible for scientists to imagine all the possible forms that they could take. In addition, no life-form carries out all of these activities all of the time or under all conditions. These considerations become important in two respects. First, would it be possible to recognize a life-form on another planet? Scientists may not be able to witness putative life-forms carrying out metabolic and other activities (see MARS, LIFE ON). Second, at what point might scientists be able to construct a mechanical life-form? Although computerized robots cannot grow or reproduce themselves, they can construct new components and whole new robots. Many computer algorithms already utilize natural selection to generate improvements in structure and function (see EVOLUTIONARY ALGORITHMS). Should robots, or even computer programs, be considered life-forms?

Fundamental assumptions of biology include the following:

- The physical and chemical components and processes of organisms are the same as those of the nonliving world. That is, organisms are constructed of the same kinds of atoms (though not necessarily in the same relative amounts) as the nonliving world; and the laws of physics and chemistry are the same inside an organism as outside. The alternative to this view, *vitalism,* claimed that organisms were made out of material that is different from the nonliving world. This view was widespread along with many biblical views of the natural world until the 19th century, despite the fact that the Bible says "Dust thou art and to dust thou shalt return." The German chemist Friedrich Wöhler put an end to vitalism in 1815 when he synthesized urea, a biological molecule, from ammonia, an inorganic molecule. Some people continue to believe, or hope, that there is some further essence within organisms, at least within humans, that is not shared with the nonliving environment, but no evidence of such an essence has been found. Many of the processes that occur in organisms are more complex than those in the nonliving world; in addition, when complex molecules interact, they can produce emergent properties (see EMERGENCE) that could not have been predicted from a study of the atoms themselves. But this is no different from what happens in the nonliving world. One cannot explain water in terms of the properties of hydrogen and oxygen; the properties of water are emergent; yet nobody claims that water molecules violate the laws of physics and chemistry, or that a nonmaterial essence is needed to make water what it is.
- Explanations of biological phenomena can be made at two levels. Consider, for example, why a mockingbird sings. First, scientists can explain the immediate physical and chemical causes of the singing. The pineal gland in the mockingbird's head senses the increase in the length of the day; this indicates that spring is coming. In response to this information, the mockingbird's brain produces enhanced levels of the hormone melatonin, which activates the nerve pathways that cause the production of sound. The brain uses both instinctive and learned information to determine which songs the mockingbird sings. The brain also stimulates the bird to leap and dance. This complex network of immediate causation, which involves sunlight, a gland, the brain, a hormone, the voice box, and muscles, is the *proximate* causation of the mockingbird's singing. Second, scientists can explain the advantage that the ancestors of the mockingbird obtained from undertaking such behavior. Singing and dancing was a territorial display that allowed dominant male mockingbirds to keep other male mockingbirds away and to attract female mockingbirds. Natural selection in the past favored these activities and selected the genes that now determine this behavior. The evolutionary causation is the *ultimate* causation of the mockingbird's singing (see BEHAVIOR, EVOLUTION OF).
- Structure and function are interconnected in organisms. That is, the structures do what they look like, and they look like what they do. Both the xylem cells of plants and the blood vessels of animals conduct fluid, and they have a long, cylindrical structure. They act, and look, like pipes. A student may memorize anatomical structures, or physiological processes, but will understand biology only when bringing the two together.
- Large organisms have less external surface area relative to their volume than do small organisms (see ALLOMETRY).

Since organisms must bring molecules in from, and eject molecules out into, their environments through surfaces large organisms must have additional surface area to compensate for this relatively lesser surface area. Small animals can absorb the oxygen that they need through their external surfaces, but large animals need additional surface areas (either gills on the outside, or lungs on the inside) that absorb oxygen. The surface area of human lungs, convoluted into millions of tiny sacs, is as great as that of a tennis court. In small organisms, molecules can diffuse and flow everywhere that is necessary, since they do not need to go very far; larger organisms need circulatory systems.

The following is a brief outline of some of the areas of study within biology:

I. *Autecological context.* This is the ecological interaction of an individual organism with its nonliving environment (*aut-* comes from the Greek for self). Organisms must interact with the energy and matter of their environments. This results in the flow of energy and the cycling of matter.

A. Energy flows from the Sun, through the systems of the Earth, and then is lost in outer space. Some of this energy empowers climate and weather and keeps organisms warm. Photosynthetic organisms (see PHOTOSYNTHESIS, EVOLUTION OF) absorb a small amount of the energy and store it in sugar and other complex organic molecules. Other organisms obtain energy by eating photosynthetic organisms, or one another. In this way, energy passes through the food web of organisms. Eventually all organisms die, and the decomposers release the energy into the environment, where it eventually goes into outer space.

B. Matter cycles over and over on the Earth. Photosynthetic organisms obtain small molecules, such as carbon dioxide from the air and nitrates from the soil, from which they make complex organic molecules. Other organisms obtain molecules by eating photosynthetic organisms, or one another. In this way, atoms pass through the food web of organisms. Eventually all organisms die and the decomposers release the atoms into the environment, where they are used again by photosynthetic organisms.

C. Energy can flow, and matter can cycle, on a dead planet, but on the living Earth, these processes are almost completely different than they would be on a lifeless planet. Two examples are oxygen and water. Photosynthesis produces oxygen gas, which is highly reactive. An atmosphere contains oxygen gas only if it is continually replenished. Therefore the presence of oxygen gas in a planetary atmosphere is evidence that there is life on the planet. Water cycles endlessly, through evaporation from oceans, condensation in clouds, and precipitation onto the ground. Forests slow down the rain and allow it to percolate into the soil. In this way, forests prevent the floods and mudslides that would occur on a bare hillside and recharge the groundwater. Trees release water vapor into the air, creating more clouds than would form over a lifeless landscape.

II. *Chemistry of life.* Organisms consist largely of carbon, hydrogen, oxygen, nitrogen, phosphorus, and a few other kinds of atoms. Most of the other chemical elements play no part in organisms except as contaminants. Carbon, hydrogen, oxygen, nitrogen, and phosphorus are the principal components of biological molecules, which include: carbohydrates and fats, which often store chemical energy; nucleic acids, which store genetic information; and proteins, which often control the chemical reactions of organisms. Proteins release genetic information from nucleic acids, allowing that information to determine the chemical reactions (metabolism) of the organism.

III. *Cells and tissues.* All life processes occur within cells. Tissues are groups of similar cells. Cells can replicate their genetic information, which allows one cell to become two (cell division). Cell division allows three things:

A. Old or damaged cells can be replaced by new ones (maintenance).

B. Cells or organisms can produce new cells or organisms (reproduction).

C. A single cell can grow into an embryo, which grows into a new organism (development). The use of genetic information changes during development, which allows a small mass of similar cells (the embryo) to develop into a large mass of many different kinds of cells (the juvenile and adult).

IV. *Organs.* Large organisms (mostly plants and animals) need organs to carry out basic processes necessary to their survival. Plants grow by continually adding new organs (new leaves, stems, and roots) and shedding some old organs (dead leaves). Plants can lose organs and keep on living. Animal growth, however, involves the growth of each organ, the loss of any of which may be fatal to the animal.

A. *Exchange.* Both plants and animals have surfaces through which food molecules enter and waste molecules leave the organism. In plants, most of these surfaces are external (thin leaves, fine roots), while in animals they are internal (in the intestines, lungs, and kidneys).

B. *Internal movement.* Both plants and animals have internal passageways that allow molecules to move from one place to another within the body. In plants this movement is mostly one direction at a time (water from the roots to the leaves, sugar from the leaves to the roots), while large animals have internal circulation of blood.

C. *Internal coordination.* Both plants and animals have structures that support them and functions that allow their organs to work together. Hormones carry messages from one part of a plant to another, allowing it to coordinate its growth. The responses to hormones are mostly on the cellular level in plants. Hormones also carry messages from one part of an animal to another, but in addition animals have nerves that allow internal coordination, for example, maintaining homeostasis of body temperature and balance during movement. The responses to nerves and to hormones in animals can be on the cellular level or involve the movements of muscles and bones.

D. *Response to environment.* Both plants and animals have mechanisms that detect environmental stimuli, and processes that allow a response to them. In plants stimuli are usually detected by molecules such as pigments, and hormones allow the responses, such as bending toward light. Animals have sensory organs and a central nervous system that figures out the appropriate response to the stimuli.

V. *Sexual reproduction.* Nearly all organisms have sexual life cycles. Certain organs produce cells with only half the genetic information that normal cells contain (see MEIOSIS). In many plants and animals, these reproductive cells come in two sizes: The large female ones are either megaspores or eggs, while the small male ones are either microspores or sperm. The male cells of one individual may fuse with the female cells of another, a process called fertilization. Fertilization produces a zygote, which may be sheltered inside of a seed, an eggshell, or a womb (see SEX, EVOLUTION OF; LIFE HISTORY, EVOLUTION OF). Sexual reproduction is not necessary for survival but generates genetic diversity.

VI. *Inheritance patterns.* Because the traits of different parents can be shuffled into new combinations by sexual reproduction, new combinations of traits can occur in each generation (see MENDELIAN GENETICS).

VII. *Populations and evolution.* All the interacting organisms of one species in one location constitute a POPULATION. Populations can grow rapidly, but limited resources prevent them from doing so forever. Because some individuals in a population reproduce more successfully than others, natural selection occurs and the population evolves (see NATURAL SELECTION). If the population separates into two populations, they can evolve into two species (see SPECIATION).

VIII. *Diversity.* Evolution has produced millions of species. Biologists have attempted to classify organisms on the basis of their evolutionary diversification rather than merely on the differences in their appearance. One such classification is:

A. *Domain Archaea:* bacteria-like cells that, today, survive in extreme environments (see ARCHAEBACTERIA).

B. *Domain Eubacteria:* bacteria-like cells that may use sunlight (photosynthetic bacteria) or inorganic chemicals (chemosynthetic bacteria) as energy sources for producing organic molecules or may obtain energy from organic molecules in living or dead organisms (heterotrophic bacteria) (see BACTERIA, EVOLUTION OF). Early in evolutionary history, some heterotrophic bacteria invaded larger cells, and today they are the mitochondria that release energy from sugar in nearly all larger cells. Early in evolutionary history, some photosynthetic bacteria invaded larger cells, and today they are the chloroplasts that carry out photosynthesis in some larger cells (see SYMBIOGENESIS).

C. *Domain Eukarya:* organisms composed of complex cells with DNA in nuclei (see EUKARYOTES, EVOLUTION OF). The evolutionary linages of the eukaryotes are still being worked out. From within these lineages evolved:

1. The plant kingdom. Land plants are descendants of green algae (see SEEDLESS PLANTS, EVOLUTION OF; GYMNOSPERMS, EVOLUTION OF; ANGIOSPERMS, EVOLUTION OF).

2. The fungus kingdom consists of decomposers and pathogens that absorb food molecules.

3. The animal kingdom. Animals, descendants of one of the lineages of protozoa (see INVERTEBRATES, EVOLUTION OF; FISHES, EVOLUTION OF; AMPHIBIANS, EVOLUTION OF; REPTILES, EVOLUTION OF; BIRDS, EVOLUTION OF; MAMMALS, EVOLUTION OF).

IX. *Synecological context.* This is the ecological interaction of organisms with one another (*syn-* comes from the Greek for together).

A. *General ecological interactions.* Evolution has refined the interactions that broad groups of organisms have with one another (see COEVOLUTION): for example, herbivores that eat plants and plants that defend themselves from herbivores; predators that eat prey, and prey that defend themselves from predators; pollinators and flowering plants. Many complex animal behavior patterns have evolved, particularly in response to SEXUAL SELECTION.

B. *Symbiotic ecological interactions.* Symbiosis results from the very close interaction of two species, in which at least one of the species depends upon the other. These interactions have resulted from coevolution.

1. Parasitism occurs when a parasite harms the host.

2. Commensalism occurs when a commensal has no effect on the host.

3. Mutualism occurs when both species benefit. Mutualism is so widespread that, in some cases, different species of organisms have actually fused together and formed new kinds of organisms.

4. Natural selection favors the evolution of hosts that resist parasites and sometimes favors the evolution of parasites that have only mild effects on their hosts. Natural selection can favor the evolution of parasitism into commensalism, and commensalism into mutualism.

C. *Ecological communities* are all of the interacting species in a location. They generally form clusters, based upon temperature and moisture conditions in different parts of the world: for example, tundra, forests, grasslands, deserts, lakes, shallow seas, deep oceans. Each community contains microhabitats with smaller communities within them.

D. Ecological communities continually undergo change. On a large scale, continents drift (see CONTINENTAL DRIFT), MASS EXTINCTIONS occur, and climates fluctuate (see ICE AGES; SNOWBALL EARTH). On a small scale, disturbances such as fires are followed by stages of regrowth called ecological succession. Billions of years of these changes have produced an entire evolutionary history, as described throughout this book, and the world as humans know it today.

The Earth is filled with living organisms. But is the Earth itself alive? It does not match all of the characteristics listed above to qualify as an organism, but it does have some processes that produce a semblance of homeostasis, which is not readily understandable by the operations of the organisms on its surface (see GAIA HYPOTHESIS). At present, most biologists consider the earth to be the home of biology, rather than a biological being.

Further Reading

Keller, Evelyn Fox. *Making Sense of Life: Explaining Biological Development with Models, Metaphors, and Machines.* Cambridge, Mass.: Harvard University Press, 2002.

Mayr, Ernst. *This Is Biology: The Science of the Living World.* Cambridge, Mass.: Harvard University Press, 1997.

———. *What Makes Biology Unique?: Considerations on the Autonomy of a Scientific Discipline.* New York: Cambridge University Press, 2004.

Vandermeer, John. *Reconstructing Biology: Genetics and Ecology in the New World Order.* New York: John Wiley and Sons, 1996.

biophilia Biophilia, a term invented by evolutionary biologist E. O. Wilson (see WILSON, EDWARD O.), refers to a universal love (-philia) of nature and life (bio-). It may have a genetic component, as well as a learned component that is acquired by children during exposure to the outdoors.

The feeling of biophilia, contends Wilson, was as much the product of NATURAL SELECTION as any other aspect of SOCIOBIOLOGY such as religion or the fear of strangers. Biophilia provided a fitness advantage for humans and their evolutionary ancestors. The enjoyment of birds, trees, and mammals encouraged learning about them. By carefully watching the other species, primitive humans could learn to make more effective use of them as natural resources. The people who enjoyed listening to birds would be more likely to recognize a bird alarm call that might indicate the presence of a dangerous predator. While a feeling of biophilia was not necessary to these beneficial actions, it made them easier and more effective. The extensive use of animal motifs in such paintings as those in the caves of Lascaux and Altamira (see CRO-MAGNON) suggests that biophilia may have been closely connected to religion, which also provided important advantages to primitive human societies (see RELIGION, EVOLUTION OF). An appreciation of the beauty of the landscape, and a love for its living community of species, may have kept some primitive humans from giving up in the frequently harsh circumstances of life, and these were the humans who were the ancestors of modern people. Modern humans are, according to this concept, happiest in environments that most closely resemble the savanna in which the human genus first evolved two million years ago.

Psychologists and urban planners have independently arrived at a concept similar to biophilia. Psychologists have found that humans are happier in, and choose environments that contain, a balance of trees and open areas. Urban planners have been careful to include trees and landscaping with buildings, and parks between them. Biophilia may also explain the frequent success of horticultural therapy in the recovery of patients from physical and mental diseases. Programs to help guide children away from social problems often include an outdoor component. Even though many people can be as happy in a big city as if it were canyons, most people seek occasional escape into the natural world. Biophilia may be essential to the successful preservation of natural areas simply because most people cannot conceive of living in a world in which natural areas have been destroyed.

In modern society, the learning component of biophilia has lost a great deal of its inclusion of BIODIVERSITY. Many people like trees but do not particularly care which ones. While Central American native tribal children can identify many plant species, children in the United States can identify very few. Moreover, when modern humans visit the outdoors, they usually take a great deal of their manufactured environment with them. While modern humans insist that they be close to parks and natural areas, they pay very little attention to the protection of endangered species. Wilson and many others believe that, in order to rescue biodiversity from what appears to be the sixth of the MASS EXTINCTIONS that is now resulting from human activity, humans must rediscover the love not just of nature but of species. In this regard, evolution-based biological and environmental education may be essential for the protection of the natural world.

Further Reading

Lewis, Charles A. *Green Nature Human Nature: The Meaning of Plants in Our Lives.* Urbana: University of Illinois Press, 1996.

Wilson, Edward O. *Biophilia.* Cambridge, Mass.: Harvard University Press, 1986.

bipedalism Animals that walk habitually upon two legs are bipeds, and this ability is called bipedalism. As any dog owner or anyone who has visited a zoo knows, many quadrupedal animals (animals that habitually walk on four legs) can stand on two legs for a brief time, and perhaps even walk on two legs. Some apes, such as chimpanzees, are partially bipedal. Their form of locomotion, called knuckle-walking, involves long arms and the use of their knuckles to help them maintain their posture. Only humans and birds are fully bipedal among modern animal species.

Adaptations required for human bipedalism include:

- *Strong gluteus maximus muscle.* The largest muscle in the human body is the gluteus maximus, which connects the backbone and the femur (thighbone or upper leg bone). By contracting, this muscle pulls the backbone into an upright position. This is one reason why humans have big butts, compared to other primates.
- *Foramen magnum underneath the skull.* The foramen magnum is the opening in the skull through which the spinal cord connects to the brain. In quadrupedal animals, this opening is in the rear of the skull, but in humans it is at the base of the skull.
- *The pelvis.* The pelvis (hip bone) of humans must support most of the weight of the body and has a shorter, broader shape than is found in quadrupeds.
- *The knees.* The knee joint must be able to extend to make the femur straight in line with the tibia and fibula (lower leg bones).
- *Angle of femur.* The femur of humans angles inward, while in chimpanzees it does not.

- *The foot.* The human foot has an arch, and a big toe that is in line with the others. In contrast, chimps do not have arches, and their big toes are at an angle resembling that of a thumb. The thumb-like big toe helps chimps to grasp branches while climbing trees, something not necessary to a bipedal animal walking on the ground.

The origin of human bipedalism was gradual. Several other primates are partly bipedal. Gibbons and orangutans, for example, are often upright as they climb and swing through trees. This represents a division of labor between arms (for climbing, gathering fruits, etc.) and legs. The knuckle-walking of chimpanzees was mentioned earlier. The common ancestor of chimpanzees and hominins was already partly bipedal. The earliest members of the human lineage were mostly bipedal, as indicated by the Laetoli footprints, but had a projecting big toe that allowed these individuals to retain at least a vestige of tree-climbing ability (see AUSTRALOPITHECINES). The human lineage was fully bipedal by the origin of the genus *Homo* (see *HOMO HABILIS*).

Bipedalism is much slower and less efficient than quadrupedalism. Even relatively slow quadrupeds, such as bears, can easily outrun even the most athletic humans. The evolutionary advantages of bipedalism, therefore, must have been tremendous and long-term. The problem is that nobody is sure what the advantage or advantages might have been. Many theories have been advanced. An early theory was that, with hands no longer used for walking, early humans could make tools. However, even the earliest hominins, such as *Sahelanthropus tchadensis,* were bipedal, more than three million years before the first stone tools were manufactured.

The origin of bipedalism seems to have coincided with the gradual drying of the climate of Africa, so that continuous rain forest was replaced by savannas and scattered forests. Hominins would have had to disperse between clusters of trees in grasslands. This still does not explain bipedalism, for quadrupedal animals would be able to do this.

Some scientists have suggested that upright posture allowed hominins to look out over the grass and see predators coming. This suggestion can hardly be taken seriously, as the full suite of bipedal adaptations is clearly not necessary for simply standing up once in a while. Others have suggested that it allowed human ancestors to throw rocks at threatening carnivores. Again, this cannot be taken seriously, since lions could hardly be scared off by rocks; how much better it would be to run away like an antelope, an option not open to bipedal hominins.

The most likely advantage came from something that the hominins were doing with their hands. But what were these hominins doing with their hands that was so important that it made up for the loss of speed, out in the open, and a reduced tree-climbing ability? Some scientists suggest that they may have been carrying food. If early humans were scavengers of carcasses killed by lions and left by hyenas, or stealers of leopard kills stashed in trees, they had to run while carrying a hunk of meat. However, baboons are not fully bipedal yet they can run and carry food at the same time. Furthermore, the evidence that humans ate meat is unclear until about two million years ago. Others suggest that it allowed mothers (and possibly fathers) to carry their infants and children. In other primates, infants cling to the fur of the mother. Humans, being without significant fur except on the head, would have to carry their kids ("Let go of my hair! Okay, I'll carry you!"). This explanation is believable mostly because of the failure of all the others. A more recent proposal combines the two previous explanations: that bipedal characteristics were selected not so much for walking as for running long distances while carrying resources or children.

Three sisters and a brother living in Turkey are human quadrupeds. Although they have fully human anatomy, they prefer to walk on their feet and palms. This behavior has been traced to a mutation on chromosome 17. Therefore the evolution of bipedalism not only required anatomical changes but also changes in the brain.

Bipedalism was apparently the first characteristic that distinguished the hominin line from earlier primates. Its origin remains unexplained.

Further Reading

Chen, Ingfei. "Born to run." *Discover,* May 2006, 62–67.

Lieberman, Daniel, and Dennis Bramble. "Endurance running and the evolution of *Homo.*" *Nature* 432 (2004): 345–352.

Summers, Adam. "Born to run." *Natural History,* April 2005, 34–35.

birds, evolution of Birds are vertebrates that have feathers, warm blood, and lay eggs externally. Birds are one of the lineages of reptiles (see REPTILES, EVOLUTION OF); in fact, they probably form a lineage from within the DINOSAURS. Because "reptile" is not a coherent group if mammals and birds are excluded, and because "dinosaur" is not a coherent group if birds are excluded, many scientists include birds and mammals with the reptiles and include birds with the dinosaurs (see CLADISTICS). There are almost twice as many bird species (more than 9,000) as mammal species.

The single most recognizable feature of birds is feathers. Feathers are complex and lightweight structures consisting of a central shaft with a vane consisting of barbs and hooks. Flight feathers are very specialized, with the shaft off-center in a manner that permits aerodynamic efficiency. Other kinds of feathers are less complex. Down feathers, for example, are small and function mostly in holding in body heat. Since simpler feather structures are ineffective for flight, the evolution of feathers probably began with simple feathers that held in body heat, from which more complex flight feathers evolved (see ADAPTATION). Feathers also repel water, especially when birds preen them, applying water-repellent materials from glands.

Birds have many other features that adapt them to flight. In fact, the entire body of the bird appears adapted to flight, especially by the reduction of weight:

- *Skeletal features.* The bones are partially hollow, which retains most of their strength while greatly reducing their weight. Because flight requires enormous muscular energy, bird flight muscles are enormous relative to the rest of the body and require a large breastbone *(sternum)* with a keel for muscle attachment. Modern birds do not have bony tails; instead, the tail vertebrae are fused into a short *pygostyle.*

- *Miniaturized organs.* Bird organs, in many cases, work very efficiently for their size. Their small lungs allow very efficient oxygen uptake relative to their size, which is essential for flight. Reproductive organs remain atrophied except during the reproductive season. Birds excrete nitrogenous wastes in the form of dry uric acid rather than urea dissolved in water; therefore they need to carry very little water, which is heavy. They excrete white uric acid and dark intestinal contents through a single opening, the *cloaca.* Bird brains are relatively smaller than the brains of most mammals, since brains (and thickened skulls to protect them) are heavy.
- *High metabolic rate.* Birds have high body temperatures (several degrees higher than those of mammals), which encourage the rapid metabolism necessary for supplying energy to flight. The small size of their cells and the small amount of DNA in each also allows rapid cell growth. This means that birds have to eat a lot, in some cases half of their body weight each day, or more! To "eat like a bird," a human would have to eat at least 50 pounds of food daily.

The evolutionary origin of birds from among reptiles is evident from modern characteristics and from fossil evidence.

Modern characteristics. Birds and reptiles share many skeletal and muscular features. Some reptile skull features are present in embryonic but not adult birds. Birds, like reptiles and amphibians, have nucleated red blood cells. Also, feathers are just scales that develop greater complexity. Birds have scales on their feet. Birds, like reptiles, excrete uric acid and have nasal salt glands; both of these adaptations allow them to excrete wastes and maintain salt balance without excessive secretion of water. Bird claws resemble reptile claws; one bird, the hoatzin, retains claws on the wings in the juveniles. Birds, like reptiles, have a cloaca, a common opening for the expulsion of intestinal and kidney wastes. In many birds, as in many reptiles, the cloaca also releases sperm, and mating occurs when males and females press their cloacae together. In some reptiles, and some birds (ratites such as ostriches, cassowaries, and kiwis; and the geese, ducks, and swans), males have penes; bird penes resemble reptile, not mammal, penes.

Fossils. Protoavis, from the TRIASSIC PERIOD, is the earliest fossil that may have been a bird, but its identity as a bird is uncertain. The oldest undisputed bird fossil is *ARCHAEOPTERYX,* which lived 150 million years ago during the JURASSIC PERIOD, at the same time as the dinosaurs. The skeleton of *Archaeopteryx* was mostly reptilian, but this species had feathers. *Archaeopteryx* was intermediate between earlier dinosaurs and modern birds.

Fossils of birds and other feathered dinosaurs reveal an entire series of characteristics intermediate between Mesozoic dinosaurs and modern birds (see figure on page 62). *Sinosauropteryx* had down feathers that would not have been suitable for flight. *Caudipteryx* had flight feathers on its forelimbs and tail, but these feathers were symmetrical. *Confuciusornis* had down feathers on its trunk and asymmetrical flight feathers on its wings. The skeleton of *Confuciusornis* was mostly reptilian but had reduced tail vertebrae and a toothless beak instead of a toothed jaw. These birds were apparently part of one of the lineages of coelurosaur dinosaurs.

The feathered dinosaur *Archaeoraptor lianingensis,* the discovery of which was announced in 1999, turned out to be a hoax. It was not an evolutionary hoax, since there was no need for one. Several other feathered dinosaurs, including *Sinornithosaurus* and *Beipiaosaurus,* had already been legitimately described from the same fossil bed in China (see figures on page 63). The hoax was probably intended for monetary gain, since fossils can bring a lot of money to a Chinese peasant.

During the Mesozoic era, feathered coelurosaurian dinosaurs with a whole range of intermediate characteristics ran around all over the place. And it was not just the coelurosaurs. In 2004 scientists announced the discovery of a 130-million-year-old tyrannosaur fossil, named *Dilong paradoxus,* that had primitive featherlike structures. Scientists are uncertain which of these feathered dinosaurs may have been the ancestor of modern birds, precisely because fossils of so many animals intermediate between reptiles and birds have been found.

Numerous birds during the CRETACEOUS PERIOD also had characteristics intermediate between those of reptiles and of modern birds. At least 12 genera are known. Most had toothed jaws rather than toothless beaks, and a sternum not enlarged to the size of that of modern birds. *Sinornis* had teeth and an enlarged sternum, but the sternum was not keeled as in modern birds. *Hesperornis* in North America had very small wings, large feet, and was probably a flightless aquatic diving bird. Other Cretaceous birds resembled modern birds in their small size and ability to perch. Birds diversified into numerous lineages, most of which became extinct.

Before the end of the Cretaceous period, three modern lineages of birds may have evolved and survived the CRETACEOUS EXTINCTION. These are the lineages that today include waterfowl (represented by the Cretaceous *Presbyornis*), loons (represented by *Neogaornis*), and gulls (represented by *Graculavus*). The albatross lineage may also have been differentiated by the Cretaceous. Although fossil evidence is lacking, cladistic analysis suggests that the earliest divergence in the history of modern birds was between the *paleognaths* (large flightless birds such as ostriches) and all other birds, the *neognaths.* Certainly after the Cretaceous Extinction, the evolutionary radiation of birds was spectacular. Modern bird groups include:

- waterfowl (ducks, geese, swans)
- swifts and hummingbirds
- wading birds
- herons, storks, and New World vultures

- doves and pigeons
- raptors (eagles, hawks, and Old World vultures)
- chickens
- pelicans
- woodpeckers
- penguins
- owls
- ratites (large flightless birds)
- passerines (perching birds)

Birds have sex chromosomes (see MENDELIAN GENETICS). Male birds have two Z chromosomes, while female birds have a Z and a W chromosome. In paleognaths, Z and W chromosomes are about the same size, while in neognaths the W chromosome is smaller.

Birds represent very interesting examples of CONVERGENCE. Vultures, for example, are a polyphyletic category. New World vultures have a separate evolutionary origin (from among the herons and storks) from Old World vultures

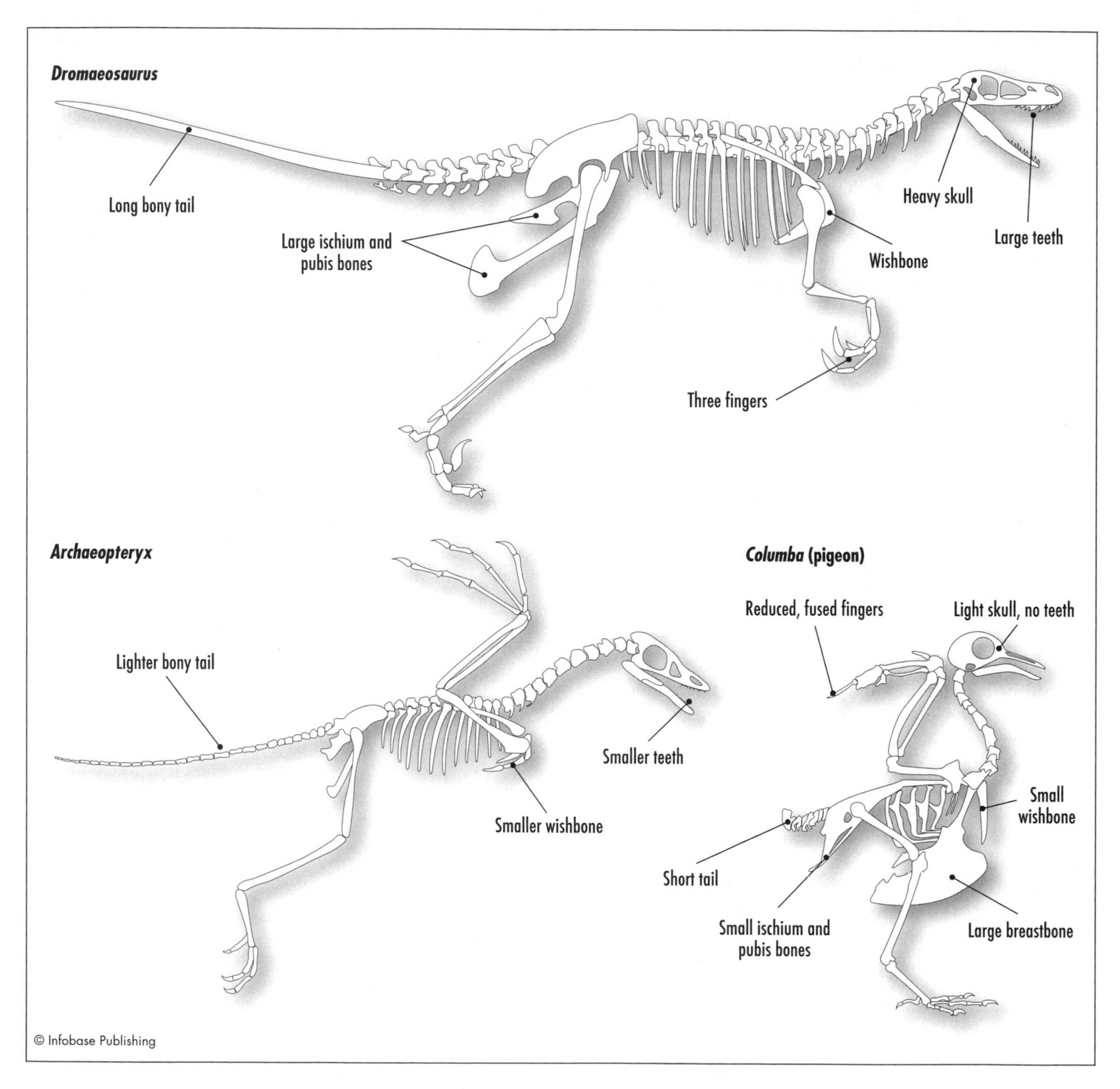

The skeleton of the Jurassic bird *Archaeopteryx* is intermediate between the dinosaur *Dromaeosaurus* and the modern pigeon *(Columba)*. During the evolution of birds, the tail shortened, the wishbone got smaller and the breastbone got larger, and teeth were lost. *(Redrawn from Benton)*

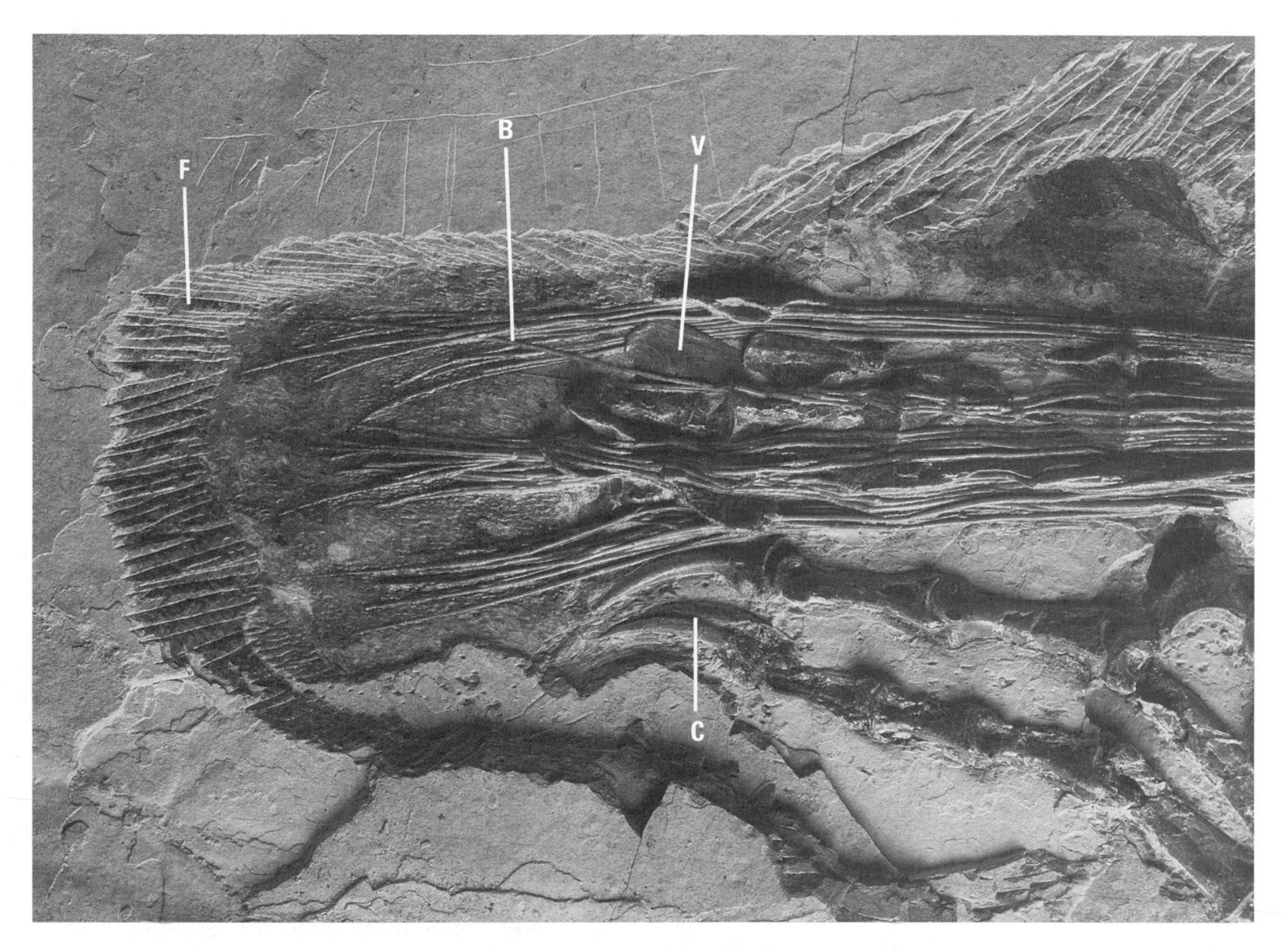

A close-up of a fossil of *Sinornithosaurus millennii,* found in China: C, claws; V, tail vertebrae; B, bony rods; F, feathers. ***(Courtesy of O. Louis Mazzatenta, National Geographic Society)***

(from among the raptors), which converged upon characteristics (such as bald heads) suitable for eating carrion. Another example is a group of Australian birds that radiated into many species resembling the warblers, flycatchers, creepers, and thrushes of other parts of the world. This example parallels the convergence of Australian marsupial mammals upon the placental mammals of other parts of the world (see MAMMALS, EVOLUTION OF).

Further Reading

American Museum of Natural History. "Newly discovered primitive tyrannosaur found to be feathered." Available online. URL: http://www.amnh.org/science/papers/feathered_tyrannosaur.php. Accessed March 23, 2005.

Benton, Michael. "Dinosaur summer." Chap. 4 in Gould, Stephen Jay, ed., *The Book of Life: An Illustrated History of the Evolution of Life on Earth*. New York: Norton, 1993.

Chiappe, Luis M., and Lawrence M. Wittmer, eds. *Mesozoic Birds: Above the Heads of Dinosaurs*. Berkeley: University of California, 2002.

——— and Gareth J. Dyke. "The Mesozoic radiation of birds." *Annual Review of Ecology and Systematics* 33 (2002): 91–124.

Gould, Stephen Jay. "Tales of a feathered tail." Chap. 23 in *I Have Landed: The End of a Beginning in Natural History*. New York: Harmony, 2002.

Norell, Mark. "The dragons of Liaoning." *Discover* (June 2005): 58–63.

Padian, Kevin, and Luis M. Chiappe. "The origin of birds and their flight." *Scientific American* 278 (1998): 38–47.

Paul, Gregory S. *Dinosaurs of the Air: The Evolution and Loss of Flight in Dinosaurs and Birds*. Baltimore, Md.: Johns Hopkins University, 2002.

Short, Lester L. *The Lives of Birds: Birds of the World and Their Behavior*. New York: Henry Holt, 1993.

Bryan, William Jennings *See* SCOPES TRIAL; EUGENICS.

Buffon, Georges (1707–1788) French *Naturalist* Georges-Louis Leclerc de Buffon (later Comte de Buffon) was born September 7, 1707. He studied mathematics and physics in his early career. By heating metal spheres and allowing them to cool, Buffon made one of the earliest estimates of the age of the Earth (see AGE OF EARTH): it was, he said, between 75,000 and 168,000 years old. He was one of the French scientists who, early in the 18th century, championed a mechanical vision of the world in which matter, operated on by forces such as gravity, could explain all that occurs in the natural world. Buffon's beliefs conflicted with those of church authorities, but by carefully worded recantations he was able to avoid disaster. Buffon also believed that scientific research should provide some practical benefits to society, which is why he performed research on such topics as forestry.

In 1739, Louis XV appointed Buffon the director of the Jardin du Roi (Royal Gardens and Natural History Collections) in Paris. Thousands of plant, animal, and mineral specimens had accumulated in this collection from around the world. Rather than simply cataloging the specimens contained in the natural history collections, Buffon used his position as an opportunity to write a natural history of all plants, animals, and minerals. Beginning in 1749, Buffon and collaborators published the first three volumes of *Histoire Naturelle* (Natural History), which eventually comprised 36 volumes published over a half century. This work was not yet completed at Buffon's death. For each animal species in the Natural History, Buffon described internal and external anatomy, life stages, breeding habits and behavior, geographical distribution and variation, economic value, and a summary of the work of previous naturalists.

Buffon also included essays that were precursors of the concepts of evolution and of how to define species. He argued that a species consisted of animals that can interbreed and produce offspring, which is very similar to the modern concept of species (see SPECIATION). He performed breeding experiments, for example, between horses and donkeys, and between dogs and wolves. Since they produced hybrid offspring (see HYBRIDIZATION), Buffon concluded that horses and donkeys, and dogs and wolves, shared a common ancestor. He classified animals into categories based on their anatomy, and claimed that "molding forces" of the Earth had modified primitive stocks of animals into the diversity of animal species that exists today. These were early evolutionary concepts. However, he believed that the ability of animals to change was limited to change within the primitive lineages. Even though Buffon's evolutionary concepts were limited, they were among the earliest serious scholarly proposals that eventually led to Darwin's evolutionary theory (see DARWIN, CHARLES).

The *Histoire Naturelle* was not entirely an objective work of scientific description. In this book Buffon claimed that North America was a land of stagnant water and unproductive soil, whose animals were small and without vigor because they were weakened by noxious vapors from dark forests and rotting swamps. He claimed that the lack of beards and body hair on the Native Americans indicated a lack of virility and ardor for their females. When Thomas Jefferson was American Ambassador to France, he wanted to prove that Buffon was wrong. Jefferson asked a friend in the American military to shoot a moose and send it to France. It was not a particularly good moose specimen, and Buffon was not impressed.

Buffon believed that it was useless to attempt to classify organisms until they had been cataloged and studied completely. This put him in conflict with the Swedish scientist Karl von Linné, who invented the modern taxonomic system (see LINNAEUS, CAROLUS; LINNAEAN SYSTEM). Linné, moreover, rejected any evolutionary transition. Legend has it that Linné disliked Buffon enough to name the bullfrog (genus *Bufo*) after him. This appears to be untrue, as *bufo* is the Latin word for toad, but the coincidence must have pleased Linné.

Buffon's two principal contributions to evolutionary science were: First, he established the importance of thoroughly assembling information about the natural world before proposing theories about it, a process that is still under way (see BIODIVERSITY); second, he brought up the possibility that life had, to at least a limited extent, evolved. He died April 16, 1788.

Further Reading

Farber, Paul Lawrence. *Finding Order in Nature: The Naturalist Tradition from Linnaeus to E. O. Wilson.* Baltimore, Md.: Johns Hopkins University Press, 2000.

Roger, Jacques, Sarah Lucille Bonnefoi, and L. Pearce Williams. *Buffon: A Life in Natural History.* Ithaca, N.Y.: Cornell University Press, 1997.

Burgess shale The Burgess shale is a geological deposit in Canada, from the CAMBRIAN PERIOD, in which a large number of soft-bodied organisms are preserved. Today, the Burgess shale is high in the Rocky Mountains of British Columbia, Canada, but during the Cambrian period it was a relatively shallow ocean floor. About 510 million years ago, mudslides suddenly buried many of these organisms. In part because of the rapid burial in fine sediments, and because of the lack of oxygen, even the soft parts of the organisms (and organisms that had no hard parts) were exquisitely preserved. It therefore provides a window into the biodiversity of the Cambrian period. Other deposits, with specimens similarly preserved, include the Chengjiang deposits in China and the Sirius Passet deposit in northern Greenland. During the Cambrian period, these locations were all tropical, but they were as distant from one another then as they are now. Many of the organisms are similar in all three deposits. Therefore the Burgess shale represents a set of species with nearly worldwide distribution. The Cambrian period was the first period of the Phanerozoic era, when multicellular organisms became common. Prior to the Cambrian period, there were few multicellular organisms aside from the EDIACARAN ORGANISMS and some seaweeds. The relatively rapid origin of a great diversity of multicellular organisms near the beginning of the Cambrian period has been called the CAMBRIAN EXPLOSION.

Charles Doolittle Walcott was secretary of the Smithsonian Institution in Washington, D.C. He was a geologist who had thoroughly explored the high mountains of western North America for many years. He was nearing the end of his career when, in late August of 1909, he discovered the

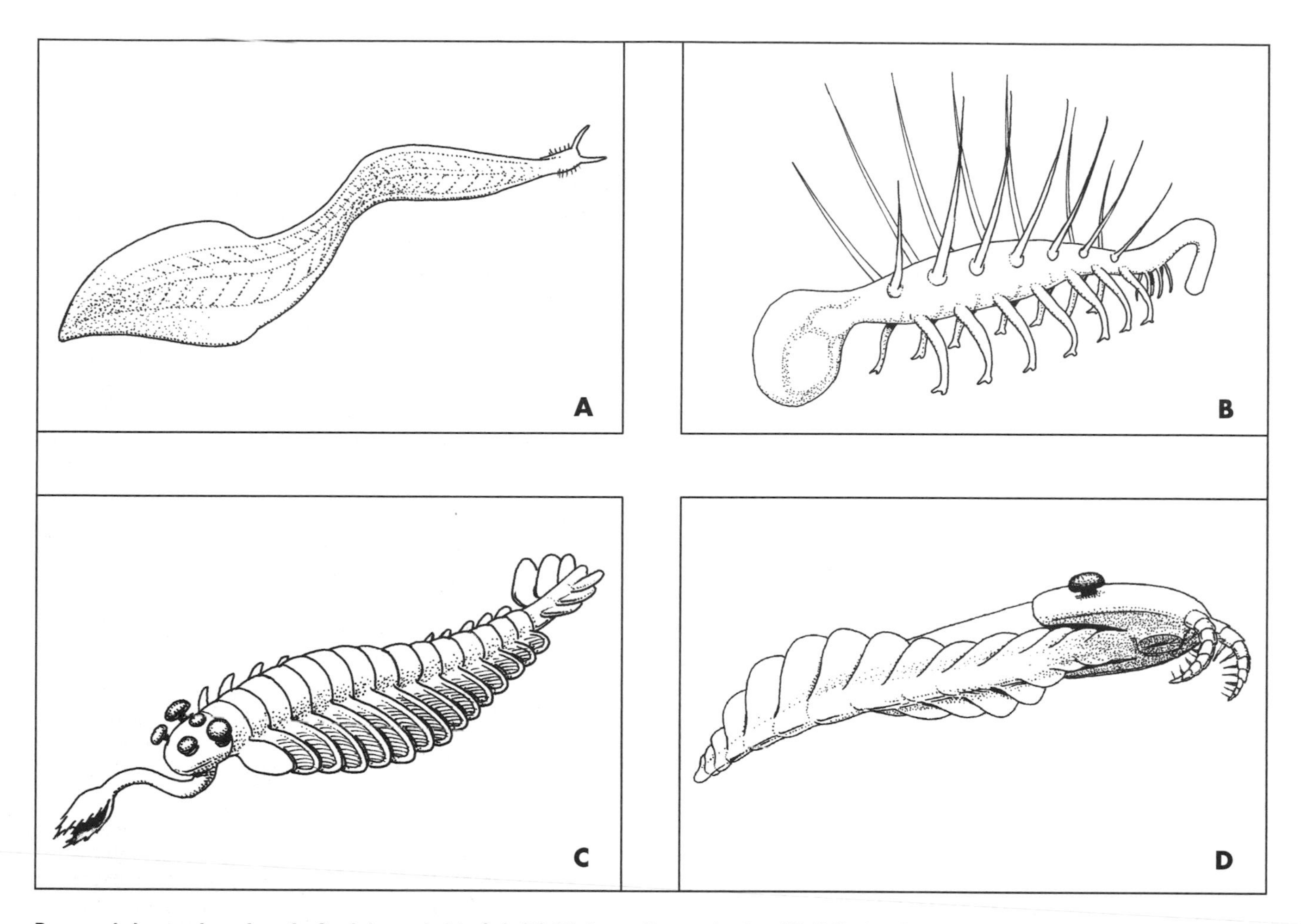

Burgess shale organisms, from the Cambrian period, include (A) *Pikaia gracilens,* a chordate; (B) *Hallucigenia sparsa,* probably an onychophoran worm; (C) *Opabinia regalis,* an arthropod, and (D) *Anomalocaris canadensis,* another arthropod.

Burgess shale deposit. He hastily gathered a few specimens, then returned in later summers for more specimens. He took specimens back to the Smithsonian and studied them. He recognized their strangeness but classified them into existing groups of animals. After his death, the Burgess shale specimens lay in obscurity for many years in the National Museum of Natural History at the Smithsonian Institution in Washington, D.C., until they were again examined and reanalyzed by three paleontologists from Cambridge University in England: Harry Whittington, Derek Briggs, and Simon Conway Morris.

Many of the organisms preserved in the Burgess shale represent groups of invertebrates that are recognizable today, or are well known from other parts of the fossil record. This includes: sponges; segmented worms similar to those in the oceans today; onychophoran worms that are today considered primitive arthropods; and TRILOBITES, which looked like cockroaches on the Paleozoic seafloor and which declined during the late Paleozoic era and vanished in the PERMIAN EXTINCTION. There are possible cnidarians in the Burgess shale, but none of them are jellyfish (see INVERTEBRATES, EVOLUTION OF).

The affinities of many of the Burgess shale organisms were not immediately apparent. Examples include (see figure on page 65):

- *Anomalocaris canadensis.* The name *Anomalocaris* means "weird shrimp" because researchers found numerous fossils of what appeared to be headless shrimp. Other fossils were found that resembled a pineapple slice and were thought to be jellyfish. Other fossils had what appeared to be swimming flaps. It was not until Whittington recognized a fossil in which two of the above fossils were attached to one another that he realized all three fossils really came from one animal: What had been considered shrimp were actually the front appendanges of the animal, what had been considered a jellyfish was the animal's round mouth, both attached to a body that had flaps. The animal grabbed trilobites with its big front appendages, and jammed them into the round mouth. A mechanical model of a reconstructed *Anomalocaris* mouth was able to produce a bite mark on a model of a trilobite that resembled actual damage known from Cambrian trilobite fossils. Because of the jointed legs, *Anomalocaris* is now considered an arthropod. At more than a meter in length, it was one of the largest arthropods ever to live, though not as large as some modern crabs.

- *Hallucigenia sparsa.* As the name might suggest, when the fossil was first studied it seemed as strange as a hallucination. At first, Conway Morris thought that it walked upon strange stilt-like legs and had tubes coming out of its back. No such organism exists anywhere today, so *Hallucigenia* might have represented an extinct phylum. New discoveries in China revealed that the tubes were feet, and the stilts were actually defensive spines. *Hallucigenia* appears to have been related to onychophorans.
- *Opabinia regalis.* This animal also had swimming flaps but had a long jointed snout with which it apparently ate, and a cluster of five eyes on its head. Because of its general similarity to *Anomalocaris, Opabinia* is also now considered an arthropod.
- *Wiwaxia corrugata.* It resembled a flat snail with an armor-plated body that had sharp spines sticking up from it. It may have had a radula, which is the structure with which mollusks scrape single-celled algae or other cells from surfaces and eat them. Because of its similarities to annelids and to mollusks, it may have been a relative of the common ancestor of these two groups.
- *Odoraia alata* was almost completely surrounded by a large shell, from which two stalked eyes and a tail with three flukes emerged. It was probably also an arthropod.

The Burgess shale fossil *Pikaia gracilens* resembled a modern lancelet, which is an invertebrate chordate. Because true fishes have been found from the Chengjiang deposits (see FISHES, EVOLUTION OF), it appears that *Pikaia* was not part of the group ancestral to modern vertebrates.

Besides being a window into the tremendous diversity of marine animal life in the earliest period of multicellular life on the planet, the Burgess shale fossils also became the focal point of controversy about the general pattern of evolution. Most of the scientifically literate public had not heard of the Burgess shale until paleontologist Stephen Jay Gould published *Wonderful Life,* in which he presented numerous details about it. He interpreted most of the animals that were difficult to classify as representing phyla of animals that are now extinct. Life, he said, was experimenting during the Cambrian period with a great diversity of different structures, like a bush with profuse branches; what happened in the Ordovician period and afterward was the extinction of many of them, and domination by a few of the more familiar types of animals. Because life during the Cambrian had such wild diversity, there was no way to tell which of the many body plans might have survived. If *Anomalocaris* or *Opabinia* had survived, and vertebrates had not, what would animal life look like on the Earth today? There would probably be no vertebrates, and therefore no humans. Life could have evolved an almost unlimited number of ways, very few of which would have produced a world similar to the one that actually exists.

One of the members of the Whittington group, Simon Conway Morris, initially thought along similar lines but ultimately took a very different view not only of the Burgess shale but of the direction of evolution in particular. As noted above, he and others at Cambridge have now concluded that many of the weird and wonderful animals of the Burgess shale were actually members of animal phyla familiar to scientists today, such as arthropods. Walcott had not been too far off, it turns out, for classifying Burgess shale animals into modern groups. If primitive vertebrates had become extinct, the terrestrial vertebrates (such as birds and mammals) would not have evolved, but Conway Morris maintains that something similar would have evolved from a different ancestor. Only a limited number of animal structures will work successfully, and the diversity of animal life on Earth seems to converge upon them (see CONVERGENCE).

The Burgess shale animals also give an idea of what might have caused the Cambrian Explosion. The affinities of the Ediacaran organisms are unclear, but all researchers agree that they appeared quite defenseless. The near or complete extinction of the Ediacaran organisms may have been caused by the evolution of the first predators. Many of the Burgess shale animals appear almost outlandishly defended. In a world without predators, a low diversity of unprotected organisms may be possible; the advent of predators sparked an arms race of COEVOLUTION between ever more efficient predators and ever more cleverly defended prey.

If scientists had only the animals with hard parts preserved from the Cambrian period, they would be able to see less than one-fifth of the animal diversity that is now known to have existed. This might have led to some serious errors in reconstructing evolutionary history. The importance of the Burgess shale, the Chengjiang deposits, and Sirius Passet deposits can scarcely be exaggerated.

Further Reading

Conway Morris, Simon. *The Crucible of Creation: The Burgess Shale and the Rise of Animals.* New York: Oxford University Press, 1998.

Gon, Sam. "The Anomalocaris homepage." Available online. URL: http://www.trilobites.info/anohome.html. Accessed March 23, 2005.

Gould, Stephen Jay. *Wonderful Life: The Burgess Shale and the Nature of History.* New York: Norton, 1989.

Cambrian explosion The CAMBRIAN PERIOD began about 540 million years ago. Life had existed for about three billion years before the Cambrian period (see ORIGIN OF LIFE; PRECAMBRIAN TIME). The fundamental processes of life had evolved before the Cambrian period: metabolism, prokaryotic cells (see ARCHAEBACTERIA; BACTERIA, EVOLUTION OF), eukaryotic cells (see EUKARYOTES, EVOLUTION OF), and multicellularity. Protists, including seaweeds, existed before the Cambrian. Fossilized animal embryos and burrows indicated the existence of at least simple animal life before the Cambrian. Immediately prior to the Cambrian period, the EDIACARAN ORGANISMS dominated fossil deposits worldwide. But at the beginning of the Cambrian, a rapid diversification of complex animals, particularly animals with hard skeletons, occurred in the oceans of the world. Because much of this evolution occurred right at the beginning of the Cambrian period, it has been called the Cambrian explosion. Paleontologist Andrew Knoll wrote that the Cambrian explosion looked like a "brief moment ... when all things were possible in animal evolution." The appearance of new animal forms was rapid, but not sudden; it occurred over a period of about five million years.

Some scientists explain the Cambrian explosion as the appearance of animals that had previously existed but had not been fossilized. According to this view, complex animals without hard parts had previously existed although complex animals with hard skeletons had not. Hard skeletons are much more readily preserved in the fossil record (see FOSSILS AND FOSSILIZATION). However, organisms without hard parts have often been preserved in the fossil record, including the Precambrian Ediacaran organisms and animal embryos mentioned previously. The Cambrian explosion, then, must be explained as the rapid evolution of new animals, not just the evolution of hard skeletons. Some observers have classified Ediacaran organisms as relatives of modern animal groups such as cnidarians (see INVERTEBRATES, EVOLUTION OF). But even if this is true, the Cambrian explosion represents spectacular diversification within those invertebrate groups, as well as the origin of new animal groups such as the arthropods.

A number of worldwide environmental changes occurred in the late Precambrian that were necessary for the Cambrian explosion but did not precipitate it.

- Oxygen gas accumulated in the atmosphere from the photosynthesis of bacteria (see BACTERIA, EVOLUTION OF) and eukaryotes. Metabolism that uses oxygen is essential for complex animals, therefore the explosive evolution of animals could not have occurred before the atmosphere and oceans contained abundant oxygen. Recent evidence suggests that oxygen gas accumulated in the atmosphere, then in the oceans, not long before the Cambrian period.
- At least two major worldwide glaciations occurred during the Precambrian (see SNOWBALL EARTH). The most recent of them may not have ended until about 600 million years ago. The explosive evolution of animals was unlikely to occur under such cold conditions. The melting of the ice also made more mineral nutrients available to the seaweeds and microscopic protists that were the basis of the marine food chain. However, glaciation does not explain the 60-million-year delay before the Cambrian explosion. Late Precambrian conditions were certainly suitable for the Ediacarans; why not complex animals?

The leading explanation for the Cambrian explosion takes its cue from the rapid appearance of animals with hard parts. This explanation suggests that the Cambrian explosion began with the evolution of predation. The first predators were animals with a feeding apparatus that allowed them to chew their food. Previously, large organisms such as the Ediacarans were defenseless but safe. Now any soft-bodied animal was helpless before the predators. The Cambrian explosion occurred when animals began to evolve different kinds of defenses as predators or against predators:

- Natural selection favored many different structures that permitted predators to search for, catch, and chew their prey.
- Some evolved hard external skeletons. In many cases, this required the evolution of new body configurations. For example, the bivalve body form (as in mussels and oysters) could not function without hard shells. Hard shell defenses also evolved in a wide variety of forms. Even single-celled photosynthetic eukaryotes evolved hard and complex external coverings.
- Some prey, then as now, have soft bodies but hide from predators by burrowing into the mud. The complexity and depth of animal burrows increased rapidly during the Cambrian explosion.

As a result of these processes, the Cambrian world was filled with diverse and complex animals, as exemplified by faunas such as the BURGESS SHALE and others. Animals with internal skeletons did not evolve until the later part of the Cambrian period (see FISHES, EVOLUTION OF).

Further Reading

Canfield, Don E., Simon W. Poulton, and Guy M. Narbonne. "Late-neoproterozoic deep-ocean oxygenation and the rise of animal life." *Science* 315 (2007): 92–95.

Conway Morris, Simon. *The Crucible of Creation: The Burgess Shale and the Rise of Animals.* New York: Oxford University Press, 1998.

Knoll, Andrew H. *Life on a Young Planet: The First Three Billion Years of Evolution on Earth.* Princeton, N.J.: Princeton University Press, 2003.

Rokas, Antonis, Dirk Krüger, and Sean B. Carroll. "Animal evolution and the molecular signature of radiation compressed in time." *Science* 310 (2005): 1,933–1,938.

Cambrian period The Cambrian period (540 million to 510 million years ago) was the first period of the Phanerozoic Eon and the PALEOZOIC ERA (see GEOLOGICAL TIME SCALE). During the previous eons, major evolutionary innovations had occurred, such as the origin of prokaryotic cells, of eukaryotic cells, and of multicellular organisms. At the beginning of the Cambrian period, complex animal forms rapidly evolved (see CAMBRIAN EXPLOSION). This is why the Cambrian period is considered the beginning of the Phanerozoic Eon, or eon of "visible life."

Climate. All Cambrian life lived underwater. Temperatures are much more stable underwater than on land.

Continents. Because there were no continents located near the poles, ocean currents were able to circulate freely. As a result there was no significant formation of ice sheets.

Marine life. Most and perhaps all of the major animal groups evolved before or during the Cambrian period. Fossil deposits such as the BURGESS SHALE preserve fossils of many animal phyla. The groups of animals within these phyla were different from the groups within those phyla today. For example, the arthropods (see INVERTEBRATES, EVOLUTION OF) were represented by TRILOBITES and animals such as *Anomalocaris,* a large predator with a round mouth, both of which have long been extinct. The first vertebrates, jawless fishes, evolved during the Cambrian period. The 530-million-year-old Chengjiang deposit in China contains two jawless vertebrate fossils, *Myllokunmingia* and *Haikouichthys* (see FISHES, EVOLUTION OF). Conodonts existed as far back as 540 million years. These fishlike animals had long bodies, large eyes, and conelike teeth, but no jaws. Single-celled and multicellular photosynthetic protists (such as seaweeds) were the basis of the marine food chain.

Life on land. There is no clear evidence that either animals or plants existed on land during the Cambrian period. The continents consisted of bare rock, sand, silt, and clay.

Extinctions. Although the Cambrian period is not considered a time of MASS EXTINCTIONS, the EDIACARAN ORGANISMS and other Precambrian forms apparently became extinct during this period. There were far fewer species in the Cambrian period than in the following ORDOVICIAN PERIOD but, on a percentage basis, Cambrian extinctions were considerable.

Further Reading

White, Toby, Renato Santos, et al. "The Cambrian." Available online. URL: http://www.palaeos.com/Paleozoic/Cambrian/Cambrian.htm. Accessed March 23, 2005.

Carboniferous period The Carboniferous period (360 million to 290 million years ago) was the fifth period of the PALEOZOIC ERA (see GEOLOGICAL TIME SCALE). In North America, the earlier Mississippian time, with extensive marine limestone deposits, is distinguished from the later Pennsylvanian time, with extensive coal deposits derived from terrestrial forests, within the Carboniferous period.

Climate. In land that is now in the northern continents, climatic conditions were very warm and wet, with shallow lakes and seas. These conditions were perfect for the growth of extensive swamps, from which coal deposits formed. This is why most of the coal and oil deposits are found today in the Northern Hemisphere. Much of the land that is now in the southern continents had cold polar conditions. Glaciations on the large southern continent caused intermittent reductions in sea level during the Carboniferous period.

Continents. The land that is today the northern continents (such as Europe and North America) formed the continent of Laurasia, which was near the equator, creating warm climates. The land that is today the southern continents (such as Africa and South America) formed the continent of Gondwanaland, much of which was over the South Pole. During the later part of the Carboniferous, Laurasia collided with Gondwanaland (see CONTINENTAL DRIFT). The resulting geological forces produced mountain ranges even in the middle of tectonic plates, such as the Appalachian Mountains of North America (see PLATE TECTONICS). The land that is now Siberia collided with eastern Europe, creating what is now the Ural Mountains. The worldwide ocean (Panthalassic Ocean) was not divided by continents as today.

Marine life. All modern groups of marine organisms existed during the Carboniferous, except aquatic mammals. Many older lineages of fishes, such as the armored fishes of the preceding DEVONIAN PERIOD, became extinct, while the fish lineages that are dominant today proliferated (see FISHES, EVOLUTION OF).

Life on land. Extensive swamps filled the shallow seas and lakes of the Carboniferous period in the tropical areas. Plant and animal life included:

- *Plants.* Large trees had been rare in the Devonian period, but trees up to 100 feet (30 m) in height became abundant in the Carboniferous. Most of these were tree-sized versions of modern club mosses and horsetails (see SEEDLESS PLANTS, EVOLUTION OF). These trees had systems of water transport and sexual reproduction that would not work very well for tree-sized plants on the Earth today but functioned well in the very wet conditions of the Carboniferous. The first seed plants evolved, possibly in drier upland regions, during this time (see GYMNOSPERMS, EVOLUTION OF). The seed plants became more common, and the seedless trees began their slide toward extinction, during a period of drier weather during the Carboniferous. Some seed plants, such as the seed ferns, were in evolutionary lineages that are now extinct. There were no flowering plants.
- *Animals.* Insect groups such as dragonflies and beetles were abundant during this period. Many of them, particularly dragonflies, were larger than any insects now alive on the Earth. Insects breathe through openings along their abdomens, which is an efficient method of gas exchange only in small animals or under conditions of elevated oxygen concentration. The large size of some Carboniferous insects has suggested to some paleontologists that the amount of oxygen in the atmosphere greatly exceeded that of today. The insects were predators or ate plant spores; herbivores, that ate leaf material, were apparently very rare. The shortage of herbivores may have contributed to the great accumulation of plant material into what became coal. Amphibians had evolved during the preceding Devonian period (see AMPHIBIANS, EVOLUTION OF) and proliferated during the Carboniferous into many forms, some large. The first vertebrates with hard-shelled eggs, which would today be classified as reptiles, evolved during the Carboniferous period (see REPTILES, EVOLUTION OF). Although some reptile lineages are not known until the Permian period, all of the major reptile lineages had probably separated before the end of the Carboniferous period. There were not yet any dinosaurs, mammals, or birds.

Extinctions. The Carboniferous period is not recognized to contain one of the major extinction events. A worldwide period of warmer, drier climate near the end of the Carboniferous created conditions that favored the spread and diversification of seed plants and reptiles.

Further Reading

White, Toby, Renato Santos, et al. "The Carboniferous." Available online. URL: http://www.palaeos.com/Paleozoic/Carboniferous/Carboniferous.htm. Accessed March 23, 2005.

catastrophism Catastrophism was a set of geological theories that claimed that Earth history was dominated by a series of worldwide catastrophes. This approach dominated geology through the early 19th century, before UNIFORMITARIANISM gained prominence (see HUTTON, JAMES; LYELL, CHARLES).

Catastrophists, among whom Georges Cuvier in France and William Buckland in England figured prominently (see CUVIER, GEORGES), believed that the Earth's geological history was divided into a series of ages, separated by spectacular worldwide catastrophes. They believed that the Flood of Noah was the most recent of these catastrophes. Some modern creationists have called themselves catastrophists. They attribute all of the geological deposits to the Flood of Noah, on an Earth recently created. However, this position was not held among catastrophist geologists of the 19th century. It is primarily a product of 20th-century fundamentalism (see CREATIONISM).

In studying the geological deposits of the Paris Basin, Cuvier determined that the Earth had experienced a long history, with many ages previous to this one. He found skeletons of mammals of kinds that no longer exist, from terrestrial deposits, with oceanic deposits over them; to him this meant that a catastrophic flood had destroyed the world in which those mammals lived. Each previous age of the Earth was destroyed by a flood, then repopulated with plants and animals created anew by the Creator. Each time, the world became more and more suitable for human habitation, which Cuvier thought was the Creator's ultimate goal. The last of the floods, the Flood of Noah, was so sudden as to have left mammoths, hair and all, frozen in ice. As a result of the research of Louis Agassiz (see AGASSIZ, LOUIS), geologists had to admit that this most recent supposed flood was actually an ice age, in which glaciers, not floodwaters, had piled up moraines of rubble (see ICE AGES).

Prior to the 19th century division between catastrophists and uniformitarians, the 18th century had been dominated by the Plutonists and the Neptunists. The Plutonists claimed that the geological history of the Earth had been dominated by volcanic eruptions and earthquakes. The Neptunists, in contrast, emphasized the role of worldwide floods. The Plutonists demanded to know, where did the water go after worldwide floods? The Neptunists demanded to know, without worldwide floods, how did fossilized seashells end up on mountaintops? Nineteenth-century catastrophists accepted both volcanic and flood processes, while uniformitarians rejected the worldwide effects of both processes. None of them suspected that the continents might actually move (see CONTINENTAL DRIFT).

The teleological and supernaturalistic approach of catastrophism had to give way to Lyell's scientific approach in order for progress to continue in geological sciences. Cuvier was not totally wrong nor Lyell totally right. The catastrophists believed in an Earth in which biological progress occurred, while Lyell believed in uniformity of state; the former approach opens the way to evolutionary science. Catastrophists, furthermore, realized that there were abrupt transitions between one geological age and another, while Lyell believed in uniformity of rate; the former approach opens the way to an understanding of MASS EXTINCTIONS, followed by rapid evolution, that punctuate the history of life on Earth (see PUNCTUATED EQUILIBRIA). Modern geology, by presenting a background of uniformitarian processes punctuated by catastrophes such as asteroid impacts (see PERMIAN EXTINCTION; CRETACEOUS EXTINCTION), has preserved the best of both 19th-century approaches: uniformitarianism and catastrophism.

Further Reading

Benton, Michael. "The death of catastrophism." Chap. 3 in *When Life Nearly Died: The Greatest Mass Extinction of All Time.* London: Thames and Hudson, 2003.

Gould, Stephen Jay. "Uniformity and catastrophe." Chap. 18 in *Ever Since Darwin: Reflections in Natural History.* New York: Norton, 1977.

Cenozoic era The Cenozoic era (the era of "recent life") is the third era of the Phanerozoic Eon, or period of visible multicellular life, which followed the PRECAMBRIAN TIME in Earth history (see GEOLOGICAL TIME SCALE). The Cenozoic era, which is the current era of Earth history, began with the mass extinction event that occurred at the end of the CRETACEOUS PERIOD of the MESOZOIC ERA (see MASS EXTINCTIONS; CRETACEOUS EXTINCTION). The Cenozoic era is traditionally divided into two geological periods, the TERTIARY PERIOD and the QUATERNARY PERIOD.

The Cretaceous extinction left a world in which many organisms had died and much space and many resources were available for growth. The DINOSAURS had become extinct, as well as numerous lineages within the birds, reptiles, and mammals (see BIRDS, EVOLUTION OF; REPTILES, EVOLUTION OF; MAMMALS, EVOLUTION OF). The conifers that had dominated the early Mesozoic forests came to dominate only the forests of cold or nutrient-poor mountainous regions in the Cenozoic (see GYMNOSPERMS, EVOLUTION OF). The flowering plants evolved in the Cretaceous period but proliferated during the Cenozoic era, into the forest trees that dominate the temperate and tropical regions, and many shrubs and herbaceous species (see ANGIOSPERMS, EVOLUTION OF). The explosive speciation of flowering plants paralleled that of insect groups such as bees, butterflies, and flies, a pattern most scientists attribute to COEVOLUTION.

The Cenozoic world was cooler and drier than most of previous Earth history. Climatic conditions became cooler and drier during the late Cenozoic than they had been early in the Cenozoic. During the middle of the Tertiary period, dry conditions allowed the adaptation of grasses and other plants to aridity, and the spread of grasslands and deserts. Grazing animals evolved from browsing ancestors, taking advantage of the grass food base (see HORSES, EVOLUTION OF). Five million years ago, the Mediterranean dried up into a salt flat. The coolest and driest conditions began with the Quaternary period. About every hundred thousand years, glaciers build up around the Arctic Ocean and push southward over the northern continents (see ICE AGES). Meanwhile, lower sea levels and reduced evaporation result in drier conditions in the equatorial regions.

Many evolutionary scientists and geologists now divide the Cenozoic era into the Paleogene (Paleocene, Eocene, Oligocene) and the Neogene (Miocene, Pliocene, Pleistocene, Holocene) rather than the traditional Tertiary and Quaternary periods.

Chambers, Robert (1802–1871) British *Publisher* Born July 10, 1802, Robert Chambers was a British publisher, whose role in the development of evolutionary theory was not widely known until after his death. Long before Darwin (see DARWIN, CHARLES) wrote his famous book (see *ORIGIN OF SPECIES* [book]), evolutionary thought was in the air in Europe. The naturalist Buffon (see BUFFON, GEORGES) had presented a limited evolutionary theory in France in the 18th century. Charles Darwin's grandfather (see DARWIN, ERASMUS) had speculated about the possibility, and the French scientist Lamarck had proposed a scientific evolutionary theory (see LAMARCKISM). In October of 1844, an anonymous British book, *Vestiges of the Natural History of Creation,* presented an evolutionary history of the Earth, from the formation of the solar system and plant and animal life, including even the origins of humankind. More than 20,000 copies sold in a decade. Political leaders like American president Abraham Lincoln, British Queen Victoria, and British statesmen William Gladstone and Benjamin Disraeli read it. Poets, such as Alfred Tennyson, and philosophers like John Stuart Mill did also. Many scientists also read it (see HUXLEY, THOMAS HENRY). The purpose of the book, said its author, was to provoke scientific and popular discussion about evolution.

It certainly succeeded in this objective. Responses ranged from enthusiasm to condemnation. A British medical journal, *Lancet,* described *Vestiges* as "a breath of fresh air." Physicist Sir David Brewster wrote that *Vestiges* raised the risk of "poisoning the fountains of science, and sapping the foundations of religion." Scottish geologist Hugh Miller published an entire book, *Foot-Prints of the Creator,* as a rebuttal to *Vestiges.* Charles Darwin called it a "strange, unphilosophical, but capitally-written book," and noted that some people had suspected him of being the author. Huxley recognized it as the work of an amateur whose author could "indulge in science at second-hand and dispense totally with logic." A professional scientist would have dismissed the fraudulent claim of the amateur scientist, Mr. W. H. Weekes, who claimed to have created living mites by passing electric currents through a solution of potassium ferrocyanate, a claim that the *Vestiges* author was credulous enough to believe.

It was not until 1884, after Darwin's death and evolutionary science had become respectable, that the author was officially revealed to be Robert Chambers, one of Britain's most successful publishers. Chambers had chosen to remain anonymous because he feared the reputation of *Vestiges* would hurt his business. (His company published, among many other things, Bibles.) Because of Chambers's interest in science, some people, including Darwin, had already guessed the identity of the notorious "Mr. Vestiges."

Chambers and his older brother had begun their publishing business by selling cheap Bibles and schoolbooks from a book stall on a street Edinburgh in 1818. In little over a decade, W. & R. Chambers had become one of the most successful publishing houses in Britain. They published science books for the general public and a weekly magazine.

Little of Chambers's book was of lasting scientific value. In particular, it provided no explanation for what caused evolution to occur. Without a mechanism, no evolutionary theory could be credible to scientists. It was a mechanism, NATURAL SELECTION, that Darwin provided and Mr. Vestiges did not. The one lasting effect of *Vestiges* is that it vividly demonstrated to Darwin, who was secretly working on his theory, that publishing an ill-formed and unsupported evolutionary theory would be a disaster. It was perhaps because of Chambers that Darwin waited until 1858 before presenting even a summary of his ideas, and Darwin would not have done so even then had it not been for a letter from a young British naturalist who had come up with the same idea as Darwin (see WALLACE, ALFRED RUSSEL). Chambers died March 17, 1871.

Further Reading

Eiseley, Loren. *Darwin's Century: Evolution and the Men Who Discovered It.* New York: Anchor, 1961.

Secord, James A. "Behind the veil: Robert Chambers and Vestiges." In J. R. Moore, ed., *History, Humanity and Evolution.* Cambridge, U.K.: Cambridge University Press, 1989.

character displacement Character displacement is an evolutionary change that occurs within a trait when species compete for the same resource. It is called character displacement because, with respect to that character or trait, the species evolve to become less similar to one another, thus displacing the character from its average state. Charles Darwin, in *Origin of Species* (see DARWIN, CHARLES; *ORIGIN OF SPECIES* [book]) stated that NATURAL SELECTION should favor divergence between closely related species that had similar diet and habitat. Competition is strongest when the species are most similar in that trait. Each species benefits from evolving versions of the trait that minimize competition.

Extensive observational evidence to support this concept was not available until the 20th century. Ornithologist David Lack studied DARWIN'S FINCHES on the GALÁPAGOS ISLANDS. The beaks of the small ground finch *Geospiza fuliginosa* were smaller, on the average, than those of the medium ground finch *G. fortis*. There was considerable overlap between the two species. The amount of overlap in beak sizes between the species was much greater when populations were compared from islands on which only one of the species occurred. There was much less overlap in beak sizes on islands on which the two species competed for seeds. Evolutionary biologists William Brown and E. O. Wilson (see WILSON, EDWARD O.) presented other examples in the 1956 paper in which they presented the term *character displacement*. The problem with observational evidence is that evolutionary processes other than character displacement may be able to account for them. In some cases, what appears to be character displacement results from plasticity rather than from true evolutionary ADAPTATION. Many proposed examples of character displacement, including that of Darwin's finches, have withstood the test of alternative explanations.

Experiments have begun to confirm observations of character displacement. Evolutionary biologist Dolph Schluter studied three-spined sticklebacks, which are freshwater fish in coastal lakes of British Columbia. When two species of sticklebacks are present in the same lake, one of them (the limnetic form) is small, slender, and feeds mainly on small organisms throughout the depths of the lake, while the other (the benthic form) is larger and feeds mainly on animals that live at the bottom of the lake. The limnetic and benthic forms have recognizable differences in the structures of their mouths. When a stickleback species occurs alone in a lake, its characteristics are more intermediate between the limnetic and benthic extremes. In one experiment, Schluter placed individuals of an intermediate species (the target species) on both sides of a divided experimental pond, then added individuals of a limnetic species to one side. Because of competition, the individuals of the target species that fed in the limnetic zone had lower fitness than the individuals that fed on the bottom, where there were no competitors. In another experiment, Schluter placed individuals of the target species on both sides of the pond, then added a limnetic species to one side and a benthic species to the other. The individuals of the target species that had highest fitness were those that were different from the species that were added to their half of the pond. Schluter had demonstrated that natural selection favored displacement.

For over 30 years, Peter and Rosemary Grant have studied evolutionary changes in DARWIN'S FINCHES on the GALÁPAGOS ISLANDS. During a prolonged drought beginning in 1977, average beak size in the medium ground finch (*Geospiza fortis*) on the island of Daphne Major increased, an example of directional selection (see NATURAL SELECTION). This shift allowed the birds to consume larger seeds. When the rains returned, natural selection favored a return to smaller beak sizes in this species. At the time, there was no other species of bird that ate seeds on the ground on this island. Then, in 1982, *G. magnirostris*, a species of finch with larger beaks, invaded Daphne Major. When another drought struck in 2003, the beaks of *G. fortis* evolved to be smaller, not larger. This occurred because in 2003, unlike 1977, G. fortis had to compete with *G. magnirostris*, which could more effectively consume the large seeds. The evolutionary shift in 1977 was due directly to drought, while the 2003 evolutionary shift was due to competition during drought—making it perhaps the best example of yet documented of character displacement in the wild.

Further Reading

Brown, William W., Jr., and E. O. Wilson. "Character displacement." *Systematic Zoology* 5 (1956): 49–64.

Grant, Peter R. "The classic case of character displacement." *Evolutionary Biology* 8 (1975): 237.

Grant, Peter R., and B. Rosemary Grant. "Evolution of character displacement in Darwin's finches." *Science* 313 (2006): 224–226. Summary by Pennisi, Elizabeth. "Competition drives big beaks out of business." *Science* 313 (2006): 156.

Pritchard, J., and Dolph Schluter. "Declining competition during character displacement: Summoning the ghost of competition past." *Evolutionary Ecology Research* 3 (2000): 209–220.

Rundle, H. D., S. M. Vamosi, and Dolph Schluter. "Experimental test of predation's effect on divergent selection during character displacement in sticklebacks." *Proceedings of the National Academy of Sciences (USA)* 100 (2003): 14,943–14,948.

———. "Experimental evidence that competition promotes divergence in adaptive radiation." *Science* 266 (1994): 798–801.

———. "Ecological character displacement in adaptive radiation." *American Naturalist,* supplement, 156 (2000): S4–S16.

———. "Frequency dependent natural selection during character displacement in sticklebacks." *Evolution* 57 (2003): 1,142–1,150.

Vamosi, S. M., and Dolph Schluter. "Character shifts in defensive armor of sympatric sticklebacks." *Evolution* 58 (2004): 376–385.

Chicxulub *See* ASTEROIDS AND COMETS.

chordates *See* INVERTEBRATES, EVOLUTION OF.

cladistics Cladistics is a technique that classifies organisms on the basis of shared derived characters, which are similarities that they have inherited from a common ancestor. This method assumes that two species with a greater number of shared derived characters are more closely related than two species that have fewer shared derived characters. The common ancestor of two species represents a branch point (Greek *clados* means branch) in the evolutionary histories of these species. Cladistic analysis can be applied to any set of evolutionary lineages, not just to species. Cladistic analysis is also called *phylogenetic* analysis. This method was developed in 1950 by German entomologist Willi Hennig. A similar method, which was developed by American botanist Warren Wagner, contributed methodologies that are now part of cladistic analysis.

Cladistic analysis generates clusters of branches entirely on the basis of the characters of species. The diagram that represents the branching pattern is called a *cladogram.* The older systems of classification, in contrast, attempt to reconstruct the evolutionary history of the organisms in question. By making no a priori evolutionary assumptions, cladistic analysis generates a set of possible evolutionary relationships and allows the investigator to choose the one that most closely matches the data of modern organisms or, if known, the data from the fossil record.

Grouping the species objectively on the basis of similarities has two advantages over older systems of classification.

- All of the species are placed at the tips of the branches, rather than at branch points. Modern green algae and modern flowering plants are descendants of a common ancestor, but the lineage leading to green algae has undergone less evolutionary modification than the lineage leading to flowering plants. Green algae and flowering plants are at separate branch tips of the cladogram. The ancestral population of organisms from which both green algae and flowering plants are descended no longer exists.
- Cladistics allows a more objective approach to classification than the traditional systems. The cladogram can be generated by mathematical rules. This is often done by a computer. The investigator's preferences do not influence the results.

As a result of this approach, cladists will not say that one modern group evolved from another modern group. Consider the example of green algae and flowering plants. If one could actually have looked at their common ancestor, it would have looked like green algae. Cladists will not generally say that flowering plants "evolved from" green algae. They will say that green algae and flowering plants evolved from a common ancestor.

Choosing the traits

The investigator must choose which traits are to be used in the analysis. Some traits are ancestral *(plesiomorphies)* and some are derived *(apomorphies).* There are four categories of traits:

1. *Shared ancestral traits (symplesiomorphy).* Ancestral traits are shared by almost all of the species in the analysis; ancestral traits cannot be used to distinguish among the species. In a cladistic analysis of flowering plants, for example, chlorophyll would not be a useful trait, since almost all flowering plants have chlorophyll in their leaves, a trait they inherited from their common ancestor (see ANGIOSPERMS, EVOLUTION OF). In human evolution, most HOMININ characteristics are symplesiomorphies.

2. *Unique traits (autapomorphy).* Characteristics that are totally unique to one species also cannot be used to distinguish among the species. Consider a cladistic analysis of flowering plants in which one of the species is a parasitic plant, lacking chlorophyll in its leaves. The lack of chlorophyll is not a useful trait for this analysis, since only one of the species has this trait. An example from human evolution would be the unique skull characteristics, such as the projecting face and strong bite, of the NEANDERTALS.

3. *Convergences (homoplasy).* In many cases, a trait may evolve more than once (see CONVERGENCE). This could happen in either of two ways. First, the trait may have evolved into a more advanced form, then reverted back to the primitive form, during the course of evolution. This would cause a species to resemble a distantly related species that had retained the ancestral form all along. Second, two species with separate origins may have evolved the same trait. Evolutionary biologist Caro-Beth Stewart calls homoplasy the "ultimate trickster" of cladistics. Consider examples of homoplasy from plants, from nonhuman animals, and from humans:

- A cladistic analysis of flowering plants may include two parasitic plant species that lack chlorophyll, one of them a parasitic relative of heath plants (such as genus *Pterospora*), the other a parasitic relative of morning glories (genus *Cuscuta*). The common ancestor of these two plants had chlorophyll. The chlorophyll was lost in two separate evolutionary events. It would be incorrect to classify *Pterospora* and *Cuscuta* together into one group, with the other heath and morning glory plants into another group.

- A cladistic analysis of vertebrates would include both mammals and reptiles. All mammals except monotremes (see MAMMALS, EVOLUTION OF) have live birth. Most reptiles lay eggs, but some snakes have live birth (see REPTILES, EVOLUTION OF). No biologist would classify live-bearing snakes together with the mammals, and the monotremes together with the reptiles. Live birth evolved more than once: in the ancestors of certain snakes, and in the ancestors of most mammals.
- Increased brain size evolved independently in *HOMO ERECTUS*, Neandertals, and in the lineage that led to modern humans.

It is generally easier for convergence to occur in the loss of an old adaptation (e.g., the loss of chlorophyll) than in the acquisition of a new one. Traits that result from convergence are called *analogous* to each other. In contrast, traits that are inherited from a common ancestor are considered to be *homologous* to each other. Convergence represents a problem for cladistics, just as it does for older methods of evolutionary classification.

4. *Shared derived traits (synapomorphy)*. These are the traits that evolutionary scientists try to use in cladistic analysis. A cladistic analysis requires the choice of traits that distinguish different lineages: The species within each lineage share the traits with one another, but not with species in other lineages. Examples of synapomorphies include hair (which all mammals, and only mammals, possess) and feathers (which all birds, and only birds, possess). It is generally easy to eliminate ancestral and unique traits from a phylogenetic analysis. The recognition and elimination of homoplasy can sometimes be difficult.

Evolutionary scientists use more than one trait in a cladistic analysis. Not only must the traits be shared derived traits, but they must be independent of one another. To consider leaf length and leaf width as two traits would be incorrect, since longer leaves are usually wider as well. Rather than two traits, these may be just two ways of measuring leaf size.

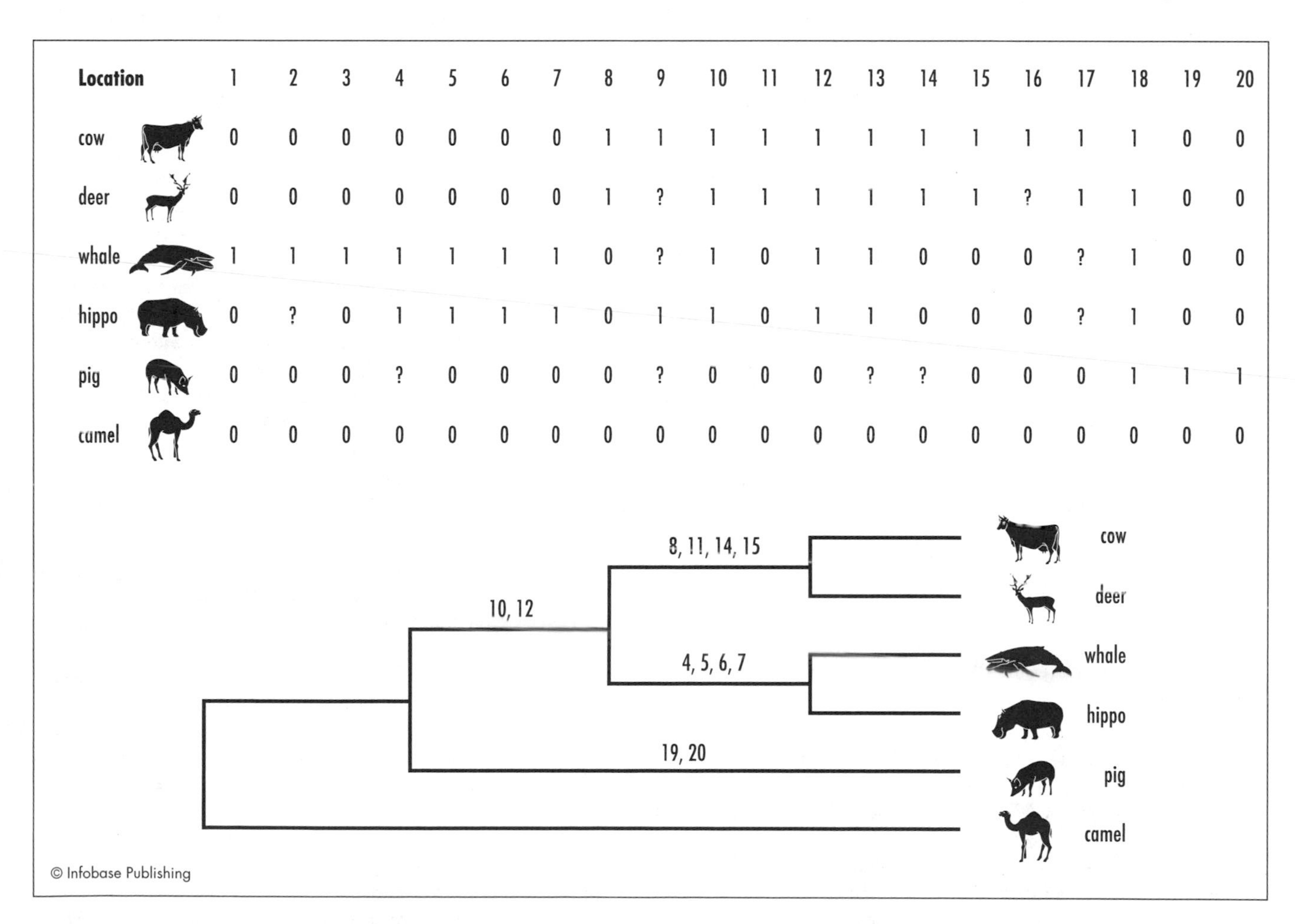

Location	1	2	3	4	5	6	7	8	9	10	11	12	13	14	15	16	17	18	19	20
cow	0	0	0	0	0	0	0	1	1	1	1	1	1	1	1	1	1	1	0	0
deer	0	0	0	0	0	0	0	1	?	1	1	1	1	1	1	?	1	1	0	0
whale	1	1	1	1	1	1	1	0	?	1	0	1	1	0	0	0	?	1	0	0
hippo	0	?	0	1	1	1	1	0	1	1	0	1	1	0	0	0	?	1	0	0
pig	0	0	0	?	0	0	0	0	?	0	0	0	?	?	0	0	0	1	1	1
camel	0	0	0	0	0	0	0	0	0	0	0	0	0	0	0	0	0	0	0	0

This cladogram is based upon 20 noncoding sequences ("locations") in the genomes of six mammal species—"1" refers to the presence, and "0" to the absence, of the sequence. The camel (the outgroup) has none of the 20 sequences and is less related to any of the others than they are to one another. Pigs form a clade separate from the remaining four species; they have sequences 19 and 20, while the others do not. The others all have sequences 10 and 12. Cows and deer share four sequences that none of the others have; whales and hippos share four sequences that none of the others have. This cladogram is meant to show patterns, not degrees, of relationship, therefore the horizontal line lengths are arbitrary. ***(Based on Nikaido, et al.)***

The above examples used easily visible characteristics to classify organisms into groups. The characteristics need not be visible. In fact, many cladistic analyses use DNA sequence information (see BIOINFORMATICS; DNA [EVIDENCE FOR EVOLUTION]). Suppose the scientist has segments of DNA from three species. If the segment of DNA is 100 nucleotides in length, this allows 100 characters to be used to test each cladogram. Nearly every phylogenetic study includes at least one analysis that uses DNA nucleotides as traits. It is no easier to use DNA than to use visible characters in cladistic analysis. The reason is that one cannot compare just any old piece of DNA of one species with any old piece of DNA in another, any more than one can make comparisons of just any visible trait. Just as visible traits must be homologous, so the DNA segments must be homologous. The investigator must compare the same gene in all of the species in the analysis. Further complications arise when one considers that some MUTATIONS exchange one nucleotide for another, while other mutations add or delete nucleotides, or reverse segments, or even move segments to different places. Usually, researchers use certain well-studied DNA sequences rather than trying to figure a new one out for themselves. In comparisons among plants, matK (the gene for an enzyme that acts upon the amino acid tyrosine) and rbcL (the gene for the large subunit of the enzyme rubisco) are frequently used.

Some analyses use NONCODING DNA for phylogenetic analysis. Researchers may use transposable elements (see HORIZONTAL GENE TRANSFER) because homoplasy is very unlikely to occur with them. It is very unlikely that the same transposable element would evolve twice: It would neither insert itself in the same location in the chromosomes of two different species, nor transfer to a new location without gaining or losing a few nucleotides in the process. Other analyses use short or long interspersed sequences in the noncoding DNA. The figure on page 73 uses the presence or absence of 20 segments of noncoding DNA to construct a cladogram of mammals. Cows and deer are clustered together, because they share all of the segments for which information is available. Whales and hippos are clustered together, because they share 15 of the 17 segments for which information is available. Whales and cows are not clustered closely together, because they share only nine of the 17 segments for which information is available. Camels are not closely related to cows, as they have none of the 11 DNA segments that cows possess.

Choosing the best cladogram

A computer can generate many possible cladograms for any set of species that are being compared. Which of these cladograms is the correct one? The correct one is the one that most closely matches what actually happened during evolutionary history. Nobody can know exactly what happened during evolutionary history; cladists must make an assumption about which cladogram is best. They often make the assumption of parsimony (see SCIENTIFIC METHOD), that is, they choose the simplest explanation. To do this, they choose the cladogram that requires the fewest number of character changes. Consider this simplified example in the classification of dogs, cats, and lions. Since each group has two branches, one could classify them in three ways:

1. [Dogs] [Cats, Lions]
2. [Dogs, Cats] [Lions]
3. [Dogs, Lions] [Cats]

Cats and lions both have retractable claws; dogs do not. Cladograms 2 and 3 require that retractable claws should have evolved twice, separately in cats and lions. Therefore cladogram 1 is the most parsimonious. It would not be safe to draw a conclusion based upon just one characteristic (in this case, retractable claws). Cladograms based on any of a number of other characteristics would lead to the same conclusion: Cats and lions form the feline group, while dogs are part of a separate (canine) group. From this one would conclude that cats and lions shared a common ancestor that lived much more recently than the common ancestor of canines and felines.

It is easy enough to compare the three cladograms that are possible with three species. However, as more species are compared, the analysis becomes exponentially more difficult. When comparing even a slightly larger number of species, the calculations become so complex that they can only be performed by a computer. As the number of species in the analysis increases, even a computer cannot sort through all of the possibilities. For 10 species, there are more than two million possible cladograms; for 50 species, there are 3×10^{76} possible cladograms. To provide some idea of how large a number this is, consider that the universe has existed for less than 10^{18} seconds! The computer must randomly generate a subset of several hundred possible cladograms and make comparisons among them.

Even when a cladistic analysis compares only a few hundred of the possible cladograms, there may be a large number of almost equally parsimonious ones. The computer must then generate a *consensus tree* that preserves the patterns that the best cladograms share in common with one another.

Applying cladistics to classification

A cladistic approach to classification would group organisms together on the basis of the branch points in the cladogram. A *clade* (whether it corresponds to a genus, a family, an order, a class, a phylum, or a kingdom; see LINNAEAN SYSTEM) must consist of all species that share a common ancestor, and only those species. Such a clade is called *monophyletic* (mono- means one). Cladists will not accept a genus, family, order, class, phylum, or kingdom that includes only some of the species that have evolved from a common ancestor. Such a group is called *paraphyletic* (para- means beside). Cladists most vigorously reject taxonomic groups that both omit some species that share a common ancestor and include some species that do not, a group referred to as *polyphyletic* (poly- means many).

A cladistic approach can lead to a different system of classification than what is traditionally recognized, as in these examples:

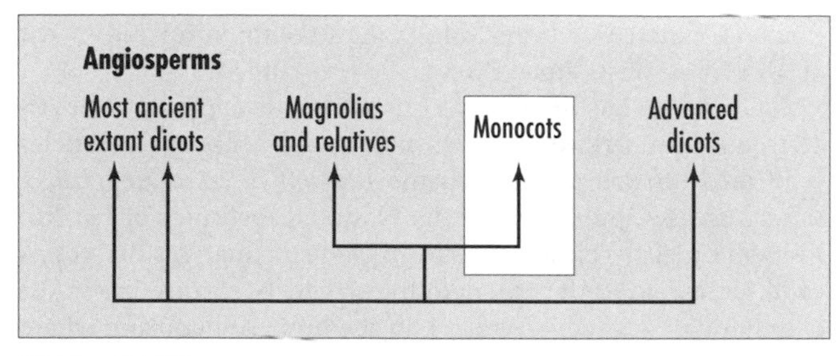

The flowering plants (angiosperms) are a monophyletic group that consists of all of the species descended from a common ancestor; within the angiosperms, the monocots are a monophyletic group.

- "Monkey" is a paraphyletic category. The new world monkeys diverged from the old world lineage that includes old world monkeys and all of the apes (see PRIMATES). To a cladist, either a human is a monkey or else there is no such thing as a monkey.
- "Reptile" is a paraphyletic category. Terrestrial vertebrates are traditionally classified into reptiles, birds, and mammals. Reptiles are cold-blooded and have scales; birds are warm-blooded and have feathers; mammals are warm-blooded and have hair and milk. This sounds like a straightforward classification based upon characteristics: Classes Reptilia, Aves, and Mammalia. Cladists point out that several lineages of reptiles branched off from one another, producing separate lineages such as turtles, lizards, tuataras, more than one branch of DINOSAURS, and therapsids. Mammals evolved from therapsids, and birds from dinosaurs. Reptiles, birds, and mammals are not separate clades, or branches. "Reptile" is therefore a paraphyletic group that contains some, but not all, of the species descended from the group's common ancestor. Strict cladists insist that either the term *reptile* should not be used at all, or else the term *reptile* should include birds and mammals. To a cladist, the term *dinosaur* includes birds, and birds are dinosaurs. One can always tell a cladist because he or she will refer to the big lumbering beasts of the MESOZOIC ERA as "nonavian dinosaurs" rather than simply as dinosaurs. To a cladist, either a human is a reptile, or else there is no such thing as a reptile.
- "Fish" is a paraphyletic category. Vertebrates include fishes, amphibians, reptiles, mammals, and birds (see FISHES, EVOLUTION OF; AMPHIBIANS, EVOLUTION OF). The reptile clade (which includes birds and mammals) and the amphibian clade share a common ancestor. But the amphibian-reptile clade branched off from the lobe-finned fishes, which is just one branch of several branches that are usually called "fishes"; the others are the ray-finned fishes, cartilaginous fishes, hagfishes, and lampreys. "Fishes" is therefore also a paraphyletic group. Cladistically speaking, then, there is no such thing as a fish, unless by "fish" you include amphibians and reptiles (which includes birds and mammals). To a cladist, either a human is a fish, or else there is no such thing as a fish.

Some cladists, in fact, insist that scientists should not even bother with the Linnaean system. Instead of kingdoms, phyla, classes, orders, families, and genera scientists should refer only to clades. This approach is called the *PhyloCode*. Most people, from common experience and also from the biology classes they took in school, tend to think of the old-fashioned classifications such as fish, amphibian, reptile, bird, and mammal. The new classifications are unfamiliar and confusing to most laypeople. What many science teachers do is accept the evolutionary insights of cladistic analysis but do not attempt cladistic consistency in terminology.

Cladistics confirms evolution

The grouping patterns produced by one cladistic analysis often closely resemble the grouping patterns produced by another. For example, a cladistic analysis using visible characteristics usually comes out almost the same as a cladistic analysis using DNA. This is highly unlikely to happen by chance and almost certainly means that the resulting classification represents what really happened in the evolutionary history of the organisms.

Moreover, the grouping patterns produced by cladistic analyses usually bear a close resemblance to the grouping patterns that had been proposed previous to the development of cladistics. One example is within the walnuts (genus *Juglans*). For at least a century, botanists have recognized four major groups of trees in this genus, based upon leaf and fruit characteristics, as well as upon the geography of where the trees grew:

- the black walnuts of North and South America, section *Rhysocaryon* (16 species);
- the Asian butternuts, section *Cardiocaryon* (3 species);
- the butternut of eastern North America, section *Trachycaryon* (1 species);
- the Eurasian walnut, section *Dioscaryon* (1 species).

Botanist Alice Stanford and colleagues performed cladistic analyses upon the walnuts, using three different regions of DNA: the matK gene, an ITS (internal transcribed spacer, a type of noncoding DNA), and RFLP (restriction fragment length polymorphism DNA). They did not base their results on just one region of DNA but on a consensus of three regions. The result was the same four groupings of walnut species. Cladistics, therefore, often confirms that the evolutionary classifications that had been previously produced were not the mere product of evolutionists' imaginations. Computers, which have no imagination, produce the same groupings.

This result is similar to what happened when Charles Darwin published *Origin of Species* (see *ORIGIN OF SPECIES* [book]). Before Darwin's theory, in fact before the acceptance of any form of evolution, scientists had been classifying organisms by the Linnaean system, based on what seemed to them to be the most "natural" groupings. Darwin said that these "natural" groupings were, in fact, the product of common descent, and that scientists had been studying evolution all along without knowing it.

Cladists pay close attention to the occasional differences between cladistic and traditional classification systems. One significant example of these differences is the reclassification of the flowering plants. Traditionally, all flowering plants were classified as either monocots or dicots. More recent analysis indicates that the monocots are a monophyletic group but the dicots are not. The monocots are one branch from within the dicots; therefore "dicot" is a paraphyletic group (see figure on page 76). Cladistics has shown that the monocot/dicot dichotomy that every college biology student learns is an oversimplification.

Perhaps the main advantage of cladistics in evolutionary science is that it allows investigations to proceed without having detailed knowledge of which ancient species, represented today by fossils, were or were not ancestral to which modern species. For example, the existence of *HOMO HABILIS* about two million years ago is well known. However, scientists cannot know if *H. habilis* represented the actual ancestral population from which the modern human species evolved, or whether modern humans evolved from another population similar to *H. habilis*. With cladistics, research can continue despite this uncertainty.

The cladistic method has received some experimental confirmation. Evolutionary microbiologist Daniel Hillis grew cultures of viruses in his laboratory, allowing them to undergo mutations and evolve into different lineages. He knew precisely what the pattern of evolutionary branching was for these viruses. He then performed cladistic analyses, using DNA base sequences. The resulting cladogram closely matched the evolutionary pattern that had actually occurred.

Some creationists (see CREATIONISM) have pointed to cladistics as an example of evolutionists starting to lose faith in evolutionary science. When a cladistic approach was used in classifying organisms in the British Museum of Natural History in 1981, some creationists hailed it as the first step in the death of evolutionary assumptions, and some evolutionary scientists also worried that this was what was happening. Cladists are quite certain that all of life represents one immense pattern of branching from a common ancestor, just as all evolutionists since Darwin have assumed.

Cladistics has also proven its worth in the study of EVOLUTIONARY MEDICINE. Cladograms regularly appear in articles published in such journals as the *Journal of Emerging Infectious Diseases*, published by the National Institutes of Health. Epidemiologists routinely perform cladistic analyses on strains of diseases, in order to determine which strains are most closely related to which others, in the hope of deducing where new strains of diseases originated. This knowledge helps in the production of vaccines, to control the spread of the disease, and to predict the emergence of new diseases. It was cladistic analysis that allowed epidemiologists to identify that the strain of anthrax that was used in bioterrorist attacks in the United States in 2001 came from within the United States. Cladistics has also revealed the origin of HIV (see AIDS, EVOLUTION OF). Cladistics has proven not only a revolutionary technique within evolutionary science but has allowed evolutionary science to produce practical benefits to humankind.

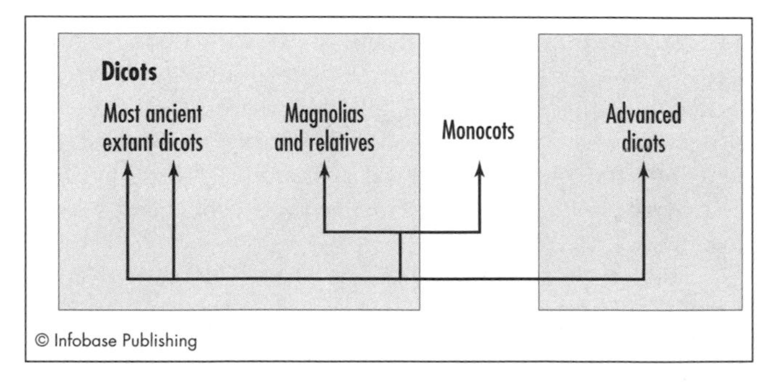

The dicots are a paraphyletic group of flowering plants because the monocot lineage is excluded from it. The term *dicot* is used descriptively but is not a phylogenetic classification.

Further Reading

Baum, David A., et al. "The tree-thinking challenge." *Science* 310 (2005): 979–980.

Cracraft, Joel, and Michael J. Donoghue. *Assembling the Tree of Life.* New York: Oxford University Press, 2004.

Foer, Joshua. "Pushing PhyloCode." *Discover,* April 2005, 46–51.

Freeman, Scott, and Jon C. Herron. "Reconstructing evolutionary trees." Chap. 14 in *Evolutionary Analysis,* 3rd ed. Upper Saddle River, N.J.: Pearson Prentice Hall, 2004.

Hillis, Daniel M., et al. "Experimental phylogenies: Generation of a known phylogeny." *Science* 255 (1992): 589–592.

Nikaido, M., A. P. Rooney, and N. Okada. "Phylogenetic relationships among cetartiodactyls based on insertions of short and long interspersed elements: Hippopotamuses are the closest extant relatives of whales." *Proceedings of the National Academy of Sciences USA* 96 (1999): 10,261–10,266.

Stanford, Alice M., Rachel Harden, and Clifford R. Parks. "Phylogeny and biogeography of *Juglans* (Juglandaceae) based on matK and ITS sequence data." *American Journal of Botany* 87 (2000): 872–882.

Stewart, Caro-Beth. "The powers and pitfalls of parsimony." *Nature* 361 (1993): 603–607.

coevolution Coevolution occurs when the evolution of one species influences, and is influenced by, the evolution of another species. Probably every species has experienced coevolution, since every species exists in a web of ecological relationships. In the general sense, every species coevolves with every other species. In the specific sense, the term *coevolution* is usually restricted to important and specific relationships between species that influence the evolution of each of them, rather than the diffuse background of ecological relationships. For example, many plants have evolved the ability to compensate for the loss of tissue to herbivores, and herbivores have evolved the ability to digest plants, but this interaction is usually considered too diffuse to qualify as coevolution. Coevolution occurred when plants evolved the ability to produce toxins, and herbivores evolved the ability to detoxify them, because the toxins and the ability to detoxify them are both often specific to certain groups of plants and herbivores.

In coevolution, natural selection favors those individuals within a population that most effectively resist, or cooperate beneficially with, individuals of another species. In many cases, coevolution has produced *symbiotic* relationships between species *(symbiosis),* in which the life processes of individuals within one species are closely linked to those of individuals of another species (see table). The following are examples, rather than a thorough analysis, of coevolution.

Types of Symbiotic Interactions

Type of interaction	Effect on host	Effect on other species
Parasitism	Harmful	Beneficial
Commensalism	Unaffected	Beneficial
Mutualism	Beneficial	Beneficial

Coevolution of Plants and Herbivores

Around the 1950s, ecologists such as Gottfried Fraenkel began to ask, why is the world green? Insects are capable of massive population growth, and by now they should have eaten every bit of green leaf on the planet. Part of the reason that this has not happened is that many insects die during the winter, and their populations must grow again the following spring. But even the winterless tropical forests are green. Why? The answer lies with coevolution.

Plants produce a great variety of *secondary compounds* in their tissues, especially their leaves. Secondary compounds have no primary metabolic role in organisms; they usually serve a defensive role instead. Many secondary plant compounds are famous for the uses to which humans put them. The active ingredients of spices, of many pharmaceuticals, as well as many miscellaneous compounds such as caffeine are all secondary compounds of plants. Some of the chemicals, such as tannins, are digestion inhibitors, while many others are mildly to strongly poisonous. In some cases, plant secondary compounds can work in ways other than as toxins. Compounds in the needles of some conifers function as inhibitors of insect growth, so that insects that eat them cannot reach maturity. Certain members of the nightshade family of plants produce compounds that aphids mistake for predator alarm signals; this scares the aphids away. Most plants have evolved chemical defenses against herbivores. Because of coevolution, the world is not a big salad bowl.

Plants usually produce the greatest amounts of secondary compounds in their leaves, as the stems are tougher (and often defended by thorns or spines), while roots are underground where herbivorous animals are fewer. Leaves need to be relatively free of defensive spines, and to be somewhat soft, because they have to release water and oxygen into, and absorb carbon dioxide and light from, their environment.

Some plants have relatively little toxic defense against herbivores. These plants are usually the ones that can grow quickly after being eaten. Grasses, for example, have few toxic compounds in their leaves, but they grow back quickly from underground structures after being grazed. Grass leaves have high levels of silicon dioxide, which makes them tough. Only specialized grazing animals can chew and digest them. Other plants get by without secondary compounds because they live for only a short time and vanish from the scene after reproducing.

Chemical defenses are costly. Some plants may produce fewer chemical defenses in years of abundant rainfall, when they can more easily afford to grow new leaves to replace those lost to herbivores. In some cases, plants wait until the herbivores begin to eat them, then they undergo a rapid response and produce toxins. In some cases, plants respond to drought by producing chemicals that make them more toxic. They may respond to drought and to herbivory by some of the same reactions. In all these and other ways, herbivores have influenced plant evolution.

Herbivores have evolved ways of eating plant tissue despite these defenses. Some animals consume massive quantities of herbage, get what little nutrition they can from it, and evacuate the rest. These animals (such as caterpillars, koalas, and sloths), deriving little energy from their food, are generally lethargic. Other animals can tolerate the toxins of specific plants, and even specialize upon them. The caterpillar of the monarch butterfly *(Danaus plexippus)* eats milkweeds (genus *Asclepias*), which have chemicals that interfere with heartbeats, but they sequester the poisons in special sacs. They even use the chemical for their own defense: Having stored up the milkweed poison, they are now themselves poisonous to their predators. Caterpillars of *Heliconius* butterflies similarly use toxic compounds from the passionflower vines (genus *Passiflora*) upon which they feed. Humans eat tissues of plants as spices. The fact that spices contain potent chemicals prevents humans from eating very much plant tissue. The human attraction to spices may have evolved because some spices have antimicrobial activity. In all these and other ways, plants have influenced herbivore evolution.

Plants and herbivores are participating in what has been called an arms race, in which increased plant defense selects for increased ability of the herbivore to tolerate or to avoid the plant defenses; and increased herbivore success selects for increased plant defense. Since plant-herbivore coevolution occurs over a long time period, the only thing similar to experimental confirmation is that on some islands plants do not have strong defenses. On Hawaii, which has no native browsing mammals, there is a stingless bush nettle.

Evolution of defense against herbivores has occurred separately in every evolutionary lineage of plants. Therefore hardly any two plant defenses are the same. So vast is the diversity of plant chemical defenses that these chemicals are sometimes used to assist in the evolutionary classification of plants.

Coevolution of Organisms and Predators

Predators are animals that kill entire organisms for food. Usually their prey is other animals, but animals that eat seeds are usually called seed predators because a seed contains an entire plant.

The evolution of predators has been overwhelmingly influenced by their search for prey. They tend to have greater intelligence, sensory acuteness, and agility than prey. Some predators have evolved pack behavior. The evolution of the prey has been just as strongly influenced by the need to avoid or resist predators. In some cases, this involves sensory acuteness, agility, and herd behavior just as it does with predators. Some prey form large herds in which the members can watch out for one another; in some flocks of birds and schools of fish, the coordinated and sudden movements of the members can disorient and confuse the predators. In some cases, the prey animals have evolved to simply become too big for most predators to be able to handle. The evolution of predators constantly influences the evolution of prey, and vice versa, because individuals that are slower, less intelligent, or less able to function as part of the group are more likely to be eaten or to not eat.

Seed predators have evolved the ability to find, and sometimes store and retrieve, seeds. The fact that seed predators, such as squirrels, fail to retrieve all the seeds they store is what makes them effective dispersers of seeds to new locations; as such they often constitute a net benefit to the plant some of whose seeds they eat. Seeds are a rich source of food, for predators as well as for the embryonic plants inside of them. Seed plants have evolved, in turn, to resist seed predators, at least the ones like weevils that offer them no benefit. Many seeds are mildly or strongly toxic. For example, while the apple fruit is delicious and nutritious, the seeds contain cyanide. This encourages the evolution of animals that will eat the apple but pass the seeds unharmed through their digestive tracts, obtaining nutrition from the fruit but not from the seeds. Many fruits have a laxative effect on animals, which encourages the animals to evacuate the seeds before they die in the intestines. Perhaps the most interesting way in which plants have evolved in response to seed predators is *mast seeding*. Many trees produce large seed crops every few years, and few seeds at other times. In this way, the seed predator populations do not build up; when the trees produce a huge (mast) crop, the predator populations cannot consume them all. Not only has coevolution occurred between plants and the animals that both eat and disperse their seeds, but it has produced a radiation of species within each. Many pines have seeds with wings that blow in the wind, but some have wingless seeds that are stored, and eaten, by crows and jays. A proliferation of both pine and jay species occurred in the Miocene epoch (see TERTIARY PERIOD), which some evolutionary scientists attribute to coevolution between them.

Coevolution of Flowering Plants and Pollinators

Pollination is necessary for sexual reproduction in seed plants (see SEX, EVOLUTION OF). The pollen must be carried from the male organs (male cones of conifers, or stamens of flowers) to the female organs (female cones of conifers, or the pistils of flowers) (see GYMNOSPERMS, EVOLUTION OF; ANGIOSPERMS, EVOLUTION OF). The pollen lands upon the surface of the female cone, or the stigma of the flower's pistil, and grows a tube down to the immature seed.

In the ancestors of flowering plants, pollen was carried from one plant to another primarily by the wind. Modern conifers continue to rely on wind pollination. The earliest flowering plants relied upon animals (usually insects) to carry pollen from one plant to another. Very soon after the evolution of these early insect-pollinated flowers, some flowering plants reverted to wind pollination, and many flowering plants today rely on wind pollination. The evolution of wind pollination is not coevolution, since the wind is not an organism. Flowers that rely on wind pollination cannot control where their pollen goes, and they have characteristics such as the following:

- Their petals are reduced or absent, as petals would only get in the way of the transportation of pollen by the wind from stamens of one flower to pistils of other flowers.
- They produce massive amounts of pollen, since most of the pollen misses its target.
- The stigmatic surfaces are large, to increase the chance of the pollen hitting its target.
- The flowers of many wind-pollinated trees, such as oaks, open in the early spring, before leaves of deciduous plants emerge; otherwise, leaves would slow down the wind, and pollen might stick, uselessly, to the leaves.

In contrast, flowers that are pollinated by animals have characteristics such as the following:

- Their petals and nectar attract the animals.
- They produce less pollen, since animals can carry the pollen directly to another flower.
- The stigmatic surfaces are not usually large.
- Their flowers may open at a later time during the growing season.

Flowers often attract pollinators with a variety of rewards: nectar is high in calories; pollen is a high-protein food; in some cases, flowers in cold climates offer a warm spot (by concentrating sunlight and blocking wind) for the pollinators to get in from the cold. In some cases, flowers produce extra pollen, which is sterile and intended specifically to feed the pollinators.

In many cases, flower characteristics attract and offer rewards to specific kinds of pollinators. Red tubular flowers (such as those of trumpet creepers, *Campsis radicans*) attract hummingbirds, whose long beaks and tongues fit into the tube, while they exclude animals that are too large to crawl into the tube and do not have long tongues (for example, bumblebees). Hummingbirds may have evolved a preference for red because it helped them find flowers, and the flowers may have evolved red pigment production because it brought pollinators to them. In contrast, red flowers do not strongly attract bees, which cannot see red. Moths and butterflies have long tongues and often visit tubular flowers. Some flowers have closed petals and a landing platform; only large bees can push their way into the flower. In these and many other examples, coevolution has produced characteristics that benefit both species: the pollinators have evolved the ability to find and exploit these flowers as food sources, and the flowers benefit from being pollinated.

In other cases, flowers attract pollinators by deception. Many species of flowers have dark reddish petals and produce the putrid scent of rotten meat, which attracts flies as pollinators. A few species of orchids attract male wasps by imitating the appearance and chemical characteristics of female wasps; the male wasps attempt to mate with the flower, but succeed only in either getting pollen stuck to it from stamens, or rubbing pollen off onto the stigma. In these examples, the flowers benefit, but the insects do not. It is questionable whether these adaptations were produced by coevolution. Natural selection favored flowers that most closely imitated dead meat or female wasps; however, the evolution of the flies and wasps was primarily influenced by the search for food and mates, not by the flowers.

Whether or not the pollinator benefits from the relationships, the pollination relationships just mentioned are specific. Only certain wasp species will respond to the orchid that fakes the female wasp. While the specificity is not always absolute (hummingbirds sometimes drink nectar from flowers that are not red, and bees sometimes visit red flowers), it does assure that, in most cases, the animal carries pollen from one flower to another flower of the same species. An indiscriminate animal, carrying pollen from or to a flower of a different species, would not benefit either of the plants, in fact it would not even be a pollinator. Coevolution that leads to specific pollination relationships, rather than indiscriminate ones, helps the plants transfer their pollen more efficiently and may help the pollinators find a more reliable food source.

In other cases flowers are open to a wider variety of pollinators. Many flowers are shallow, their stamens and pistils open to any insect that can fly or crawl in. In these cases, the plant population may be locally dense, so that pollen is seldom carried to the wrong species, or the pollinators are relatively faithful to one species even when not forced to be by the flower structure.

Once coevolution had begun between flowering plants and insects, 120 million years ago (see CRETACEOUS PERIOD), it resulted in an astonishing diversification of flowering plants. A floral innovation that allowed even a slight change in which insects pollinated it would cause an almost immediate isolation of the new form from the rest of the population (see ISOLATING MECHANISMS), starting it on the road to becoming a new species of flowering plant. This is probably the reason that each flowering plant family usually contains a considerable variety of growth forms and environmental adaptations, but all of its species have similar flowers: The family itself began its evolution as a partner in a new coevolutionary pollination relationship. Today there are nearly a quarter million species of flowering plants. Hymenopterans (bees and wasps) and lepidopterans (butterflies and moths) had existed since the Paleozoic era, and their diversification was not stimulated as much by flowers as the diversification of flowers was stimulated by them.

Coevolution Leading from Parasitism to Commensalism

The preceding discussion focused on how coevolution influenced general relationships among groups of species. Coevolution has also produced symbiotic species relationships in which one species depends upon another.

Parasitism occurs when individuals of one species (the parasite) obtains its living from individuals of another species (the host). Some parasites are occasional visitors, such as mosquitoes; some, like lice, may live on the host a long time; some, like intestinal worms, live inside the host; microbial parasites live among, even in, the host cells. Parasites have a negative impact on the host, either by consuming it, consuming some of its food (as when intestinal worms harm the host's nutrition), by producing toxic waste products (as do many bacterial and viral parasites such as cholera and bubonic plague), or by causing the host's own immune system to have harmful effects on the host. The evolution of parasites has been strongly influenced by the hosts. For example, the life cycles of fleas that live on rabbits are coordinated with hormone levels in the rabbits' blood, which ensures that the young fleas hatch at the same time as the baby rabbits are born. The evolution of the hosts has been influenced by parasites; the very existence of immune systems may be considered proof of this.

In contrast, a *commensalistic* relationship is one in which the commensal obtains its living from the host but the host is indifferent to its presence. (This is why, according to some biologists, commensalism is not true symbiosis but represents an intermediate condition between parasitism and mutualism.) The human body contains trillions of commensalistic microorganisms, primarily on the skin and in the digestive tract; in fact, there are 10 times as many bacteria as there are human cells in the human body. The commensals consume materials that the body does not need; the worst thing they do is to produce disagreeable odors. They are commensalistic not out of goodwill toward humans, but because the body's defenses keep them in check. People with compromised immune systems (such as from HIV) may be infected, and killed, by microbes that would otherwise have been commensalistic (see AIDS, EVOLUTION OF).

Humans are surrounded by commensals. A mattress can have two million microscopic mites eating the oils and skin flakes produced by their human hosts. They are so small they can go right through pillowcases and sheets. One-tenth of the weight of a six-year-old pillow may consist of skin flakes, living mites, dead mites, and mite dung. As shown in the photo on page 82, mites are found in the hair follicles of nearly every human being, no matter how thoroughly they wash.

Coevolution may select parasites that are more effective, and hosts that have adaptations that will limit, resist, or eliminate the parasite. If the host species' evolution is successful in causing the parasite to become harmless to it, the relationship is no longer parasitism but commensalism. Similarly, in many but not all cases, the parasites actually benefit

What Are the "Ghosts of Evolution"?

Many evolutionary adaptations can only be understood as evolution of one species in response to another species, a process known as COEVOLUTION. Coevolution can modify general interactions such as herbivory or predation, or it can result in very close symbiotic relationships between two species. In coevolution, the evolution of each species is influenced not by the mere presence, but by the evolution, of the other species. To understand an adaptation that results from coevolution between two partners, one has to at least know the identity of the other partner. But what happens if one of the partners has become extinct? The other species may continue manifesting its adaptations, perhaps for thousands of years, even though the adaptations are now meaningless. The adaptations of the surviving species now become puzzling, because the other species has become one of the "ghosts of evolution."

In order for symbiotic adaptations to continue being expressed, even when the other partner is a ghost, the adaptations must not be detrimental to the organism, otherwise its cost would be so great that natural selection would get rid of the adaptations or the species that has them. Furthermore, the extinction of one species should have been relatively recent, otherwise natural selection, operating over long periods of time, would presumably eliminate the adaptations.

One of the types of symbiotic interaction is parasitism, in which the parasite benefits at the expense of the host. Coevolution favors hosts that resist parasites. If the parasite becomes extinct, the host may continue defending against it. Some observers maintain that some human blood proteins are examples of defenses against bacterial parasites that are now rare. A mutated form of the CCR5 protein, which is on the surfaces of some human white blood cells, may have conferred resistance to bubonic and pneumonic plague, which would explain why it is most prevalent (even though it is still less common than the normal CCR5 protein) in northern European countries. Calculations of the rate of evolution of this mutant protein suggest that it originated at about the time of the Black Death of 1347–50, and it may help to explain why subsequent outbreaks of the plague were less severe than the Black Death. The plague bacillus, *Yersinia pestis,* is not actually a ghost; it still exists. However, it is sufficiently rare—mainly because it is spread by rat fleas, and public health measures now keep rats and humans from as close contact as occurred in the Middle Ages—that it is almost a ghost. The CCR5 protein has recently become a subject of intense interest, as it appears to be one of the proteins that HIV uses to gain entry into certain white blood cells (see AIDS, EVOLUTION OF). It has also been suggested that the mutant form of the cell membrane chloride transport protein, a mutation that causes cystic fibrosis, was once favored by natural selection because it conferred resistance to diseases. This would explain why the mutation is so common: one in 25 Americans of European descent carry the mutation. Since the adaptations carry little cost, and the parasites have only recently become uncommon, the adaptations persist.

Another type of symbiotic interaction is mutualism, in which both species benefit. One major category of examples of ghosts of evolution is certain types of fruits. The function of a fruit is to get the seeds within it dispersed to a new location. Some fruits have parachute-like structures or wings that allow the wind to disperse the seeds to new locations. Other fruits use animal dispersers. Spiny fruits cling to the fur of mammals. These fruits have probably not coevolved with specific mammalian species; any furry mammal can carry a cocklebur fruit and scatter its seeds. But coevolution is likely to occur between animals and plants with fruits that are soft, sweet, fragrant, and colorful. All four of these adaptations are costly to the plant that produces them, and specific to a relatively small group of animals that eat the fruits. A fruit may appeal to one kind of animal, but its particular characteristics, especially flavor, may be uninteresting or even disgusting to other animals. Therefore if a species of animal that eats fruits becomes extinct, the seeds in those particular fruits may no longer be dispersed to new locations.

If this should happen, the species of plant will not necessarily become extinct, although it will probably suffer a reduced population because fewer seeds are dispersed to suitable new locations. The fruits will simply fall to the ground near the parent and grow there. Some seeds will not germinate right away unless they have passed through an animal intestine; however, these seeds often germinate eventually even without this treatment. The result will be clumps of unhealthy, competing plants, but at least they will not immediately die. Another species of animal may become attracted to the fruit, but they are unlikely to be as effective as the original species of animal with which the plant species coevolved. Consider an example, in which a large animal consumes the fruits of a species of tree. The animal chews and digests the fruits but does not chew the seeds. The seeds, with hard coats, pass through the digestive tract intact and can germinate. This animal is an effective dispersal agent. If this large animal species becomes extinct, a smaller animal species may consume the fruits. However, the smaller animal may not swallow the whole fruit and may either pick out the seeds or actually crush and eat them. In either case, the smaller animal is not acting as an effective dispersal agent, the way the large animal did. In some extreme cases, the fruits pile up on the ground and rot.

North America has an impressive number of plant species that produce fruits that seem to have no animals that disperse them, often because they are too large for any extant animal species to eat. Their dispersers would appear to be ghosts of evolution. North America had many large mammal species, until the end of the last Ice Age, when two-thirds of the genera of large animals became extinct. This included mastodons, mammoths, horses, and giant sloths. It is unclear to what extent this was caused by the climate changes that were occurring at that time, or by overhunting by the newly arrived humans (see PLEISTOCENE EXTINCTION). These animals are probably the ghosts. South America also suffered a wave of extinctions, about the same time as North America. North America has many more ghosts of evolution than Eurasia, and it also suffered a far larger number of large mammal Pleistocene extinctions than Eurasia. Ecologists Daniel Janzen and Paul Martin first suggested that this phenomenon is widespread in North and South America.

Scientists cannot identify ghost dispersers with certainty, because nobody knows whether any of these animals actually would have eaten the fruits. If scientists hypothesize that mastodons dispersed certain fruits, the best that they can do is to see whether elephants eat any of these fruits. This tells the scientists little, because elephants are similar to but not the same as mastodons, and because animals eat not just what their species can eat but what they have learned to eat from their parents. Even if investigators could train elephants to eat some of these fruits, this would not prove that, in the wild, mammoths or mastodons ate them.

Among the examples of North American ghosts of evolution are:

- *Bois-d'arcs.* These are small trees, noted for their very strong wood, that have been planted widely across North America as fencerow windbreaks. Like other trees in the mulberry family, *Maclura pomifera* has separate male and female trees; and the females produce composite fruits that consist of a cluster of berries. While mulberries are small, sweet, and edible by many species of birds and mammals (including humans), the fruits of the bois-d'arc are large (about seven inches, or 15 cm, in diameter), green, hard, and sticky. (These trees are also called Osage oranges, because the fruits look a little bit like oranges and grow wild in the tribal lands of the Osage tribe; and hedge-apples, because the fruits look a little bit like apples and the trees were planted as hedges.) While not highly poisonous, the fruits are definitely repulsive and difficult to digest. There appears to be no animal that eats the fruits. Squirrels consume the seeds, but this is not dispersal; it kills the seeds rather than dispersing them. The fruits commonly fall to the ground and rot; several of the fruit's dozens of seeds grow, resulting in a clump of trunks that have grown together. In some cases the trunks have fused to form what appears to be a single trunk. It is rare to see a single-trunked bois-d'arc tree in the wild. (Single-trunked bois-d'arc trees are common in the countryside, but these were planted in fields that were subsequently abandoned and are not really wild trees.) Another consequence of the absence of a disperser is that wild bois-d'arcs are found only in river valleys of Oklahoma, Texas, Arkansas, and nearby areas. Without dispersers, the only direction for such large fruits to roll is downhill, into river valleys. Had they not been planted widely across eastern North America, they might eventually have become extinct as the last fruit rolled into the sea.
- *Avocados.* Wild avocados *(Persea americana)* are not nearly as large as the ones in supermarkets, but they are still large enough that no extant species of animal swallows them whole. Many animals can nibble at the nutritious fruit, but the seeds are simply left near the parent in most cases. Which large animal might have eaten these fruits in the past?
- *Honey locusts. Gleditsia triacanthos,* a tree in the bean family, produces very long, spiral seed pods. No native species of mammal regularly eats these pods, even though they are juicy and sweet. Once cattle were introduced to North America, many of them would eat these pods when available. Like bois-d'arcs, honey locust trees have a recent native range confined largely to river valleys. Today they have spread again, because they have been widely planted as urban trees and have escaped again into the wild. There appears to be a similar lack of dispersal in the mesquite, a bush in the bean family that also produces sweet pods, though much smaller than those of honey locust.

Other examples that have been suggested include papayas, mangos, melons, gourds, and watermelons.

The idea that North and South America have many fruits that are ghosts of evolution is corroborated by African and Asian analogs. There are numerous fruits in Africa and Asia that are consumed, and the seeds dispersed, by animals (such as elephants) that are similar to the ones that became extinct in the Americas. Some of the ghost fruits are inedible to most animals, but the now-extinct specialist dispersers may have had, as African and Asian fruit dispersers do today, intestinal bacteria that break down otherwise toxic products. Elephants eat clay, which adsorbs toxins from the digestive system. Might mammoths and mastodons have had similar adaptations that allowed them to consume fruits such as bois-d'arc?

Fruits are not the only examples of the ghosts of evolution. Other examples include flowers and stems.

- *Flowers.* Pawpaws *(Asimina triloba),* small wild fruit trees of the eastern deciduous forest of North America, persist primarily by the spread and resprouting of roots and underground stems. No native pollinator appears to be reliable. Although there has been no recent wave of pollinator extinctions as great as that of the Pleistocene extinction of fruit dispersers, apparently the pollinator of this tree has become extinct. Recently, native pollinator populations have precipitously declined, which threatens many native plant species with ultimate extinction. This may especially be true of large cacti in the deserts of southwestern North America, which rely on bats to pollinate their flowers. Thus very soon the bats that pollinate these cacti might become ghosts of evolution.
- *Stems.* The bois-d'arc also has large thorns, widely spaced on the branches, which appear totally ineffective at defending the leaves against the animals that today eat them, such as deer. Perhaps the thorns as well as the fruits are the leftover responses to the now-extinct ghosts of evolution. Wild honey locust trees have three-pointed thorns (hence the name *tri-acanthos*) bristling from their trunks. Hawthorns (genus *Crataegus*) have thorns even longer, and more widely spaced, then those of bois-d'arc. Evolutionary biologists estimate that 10 percent of the woody plant species in New Zealand have a particularly tangled method of branching that may have protected them at one time from browsing by moas, giant birds that were driven into extinction by the first Maori inhabitants. Other revolutionary biologists, such as C. J. Howell, disagree.

Thirteen thousand years is not long enough for the trees to evolve some other mechanism of dispersal. One would expect that, eventually, the unnecessarily large fruits would no longer be produced, as mutant forms of the trees produced fruits that could be easily dispersed by wind or water. This may have happened in the case of the honey locust. An apparently new species of locust, the swamp locust *Gleditsia aquatica* has evolved in southern North America, perhaps from a *Gleditsia triacanthos* ancestor. It produces much smaller pods that float easily in water.

The main point of this essay is that species are "designed" by evolution not to live in their current environments but in those of their immediate ancestors—and if that environment changes, there can be a considerable lag time before evolution designs a response. The result is, as ecologist Paul Martin wrote, "We live on a continent of ghosts, their prehistoric presence hinted at by sweet-tasting

(continues)

What Are the "Ghosts of Evolution"? *(continued)*

pods of mesquite [and] honey locust ..." As the destruction of wild habitats by humans continues at a rapid pace, humans may be inaugurating the sixth of the MASS EXTINCTIONS in the history of the world. Sometimes, when two species are partners, human activities cause the extinction of one partner but not the other. Starting with the 19th century, scientists began to understand that the present is the key to understanding the past. To understand what happened in the past, look at what is happening today in the natural world. However, it is just as true that the past is the key to understanding the present. This is just one of the many meanings that emerge from the brilliant statement made by one of the founders of modern evolutionary science (see DOBZHANSKY, THEODOSIUS): "Nothing makes sense except in the light of evolution."

Further Reading

Barlow, Connie. *The Ghosts of Evolution: Nonsensical Fruit, Missing Partners, and Other Ecological Anachronisms*. New York: Basic Books, 2000.

Buchmann, Stephen L., and Gary Paul Nabhan. *The Forgotten Pollinators*. Washington, D.C.: Island Press, 1996.

Howell, C. J., et al. "Moa ghosts exorcised? New Zealand's divaricate shrubs avoid photoinhibition." *Functional Ecology* 16 (2002): 232–240.

Janzen, Daniel H., and Paul S. Martin. "Neotropical anachronisms: the fruits the gomphotheres ate." *Science* 215 (1982): 19–27.

from *not* harming the host; if the parasite kills the host, it has killed its habitat, and must find another. Natural selection often favors the evolution of milder and milder parasites. This does not happen with parasites such as cholera bacteria, that disperse to new hosts rapidly or impersonally (in the case of cholera, through sewage). But when a parasite must disperse to a new host by personal contact among hosts (as with smallpox), the parasite benefits when the host remains well enough to walk around and infect other people. The Ebola virus is one of the deadliest diseases known to humankind, but it is so deadly that it never has had a chance to spread through more than a small group of people at any one time.

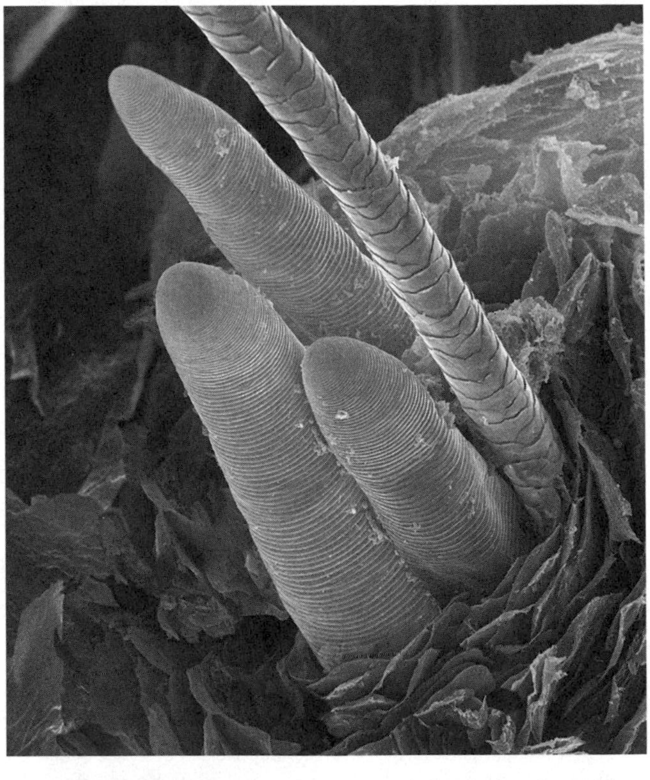

Follicle mites. In this scanning electron micrograph (SEM), the tails of three follicle or eyelash mites *(Demodex folliculorum)* emerge beside a hair in a human follicle. These harmless organisms live in hair follicles around the eyelids, nose, and in the ear canals of humans. One follicle may contain up to 25 growing mites. They feed on oils secreted from sebaceous glands, as well as on dead skin cells. Magnification: ×278 at 6×7 cm size. *(Courtesy of Andrew Syred/Photo Researchers, Inc.)*

Coevolution, therefore, can cause host species to become more resistant, and parasite populations to become milder. Many diseases, such as smallpox, were evolving toward milder forms even before the introduction of modern medical practices. Some diseases have, in fact, disappeared, perhaps because the parasitic relationship evolved into a completely commensalistic one. This process, also called *balanced pathogenicity,* has been directly measured during the course of a disease outbreak in rabbits. Commensalism can be produced by coevolution, but once it has been attained, does not involve coevolution any longer. Natural selection may prevent the relationship from slipping back into parasitism.

Coevolution Leading to Mutualism

Another kind of symbiosis is *mutualism,* in which both species benefit. In some cases, the activities of commensals can evolve into mutualism. Most human intestinal bacteria are commensals. Some of them are mutualists because they provide a benefit to humans. The presence of some kinds of intestinal bacteria may prevent the growth of parasitic bacteria. Some of the mutualistic bacteria may even produce vitamin B which is to them a waste product. Bacteria on the skin and in body orifices may prevent infection by parasitic bacteria and fungi. Nonhuman examples of beneficial internal microbes abound. Cows cannot digest grass; it is the bacteria in their stomachs that digest it (and get food for themselves in the process). Many termites cannot digest wood; it is the microbes in their intestines that digest it. The microbes in cow and termite guts may well have begun as parasites, or commensals, and evolved a mutualistic function.

Insectivorous birds may benefit from associating with grazing animals in a pasture; the large animal stirs up insects which the birds eat. The bird commensals benefit and the large animal hosts are unaffected. Some birds have evolved an extra step. They eat parasites from the skin of the large host animals, which turns the relationship into mutualism.

Scientists may have observed an example of coevolution in the process of turning parasitism into mutualism. A fungus infects, and reduces the growth and reproduction of, a grass. This sounds like parasitism. The infected grass smells bad to grazing animals such as cows, which avoid it, thus benefiting the grass. The cows that fail to avoid this grass suffer numerous physical ailments and may become lame from limb damage. What began as a parasitic infection of the grass may be evolving into a mutualistic protection for the grass.

Mutualism need not evolve through the stages of parasitism and commensalism. It may evolve from antagonistic, or casual, associations, as in the following examples:

- Big fish usually eat little fish. However, little fish can eat parasites from the big fish. If the big fish evolve behavior patterns in which they allow the little fish to come close to them, even to search around inside their mouths, and if the little fish evolve behavior patterns to trust them, a mutualism may result: The big fish are relieved of their parasite load and the little fish get food. Such cleaning mutualisms are common, especially in coral reef ecosystems.
- Ants crawl on many kinds of plants, looking for prey. This may have the effect of helping the plant by killings its herbivores. The plant may, in turn, benefit by rewarding the ants, encouraging them to stay. Acacia trees of tropical dry forests are protected by ants, which kill herbivores and even vines that may threaten the acacias. The acacias, in turn, feed the ants with nectar and special morsels of protein and fat. In some cases, the mutualism has evolved so far that the ants and trees depend upon one another for survival.
- Leaf-cutter ants in tropical forests cannot eat leaves, but they can eat fungi that decompose the leaves. What once may have been a casual association, in which the ants preferred molded leaves, has evolved into an intricate mutualism in which the ants gather leaves, chew them up, and raise underground fungus gardens on the compost. As with the acacias and ants, the leaf-cutter ants and fungi have evolved a mutual dependence: Neither can survive without the other.
- Human agriculture may have begun as a casual association between gatherers and wild plants and evolved into the biggest mutualism in the world (see AGRICULTURE, EVOLUTION OF).

Some mutualists are so intimate that they have begun the process of melding into single organisms. Zooxanthellae are single-celled algae that live in the skins of aquatic invertebrates. Nitrogen-fixing bacteria and mycorrhizal fungi live in the roots of plants, most notably many leguminous plants (beans, peas, clover, vetch, alfalfa, locust), as well as alders, cycads, and casuarinas. *Riftia pachyptila,* a vestimentiferan worm that lives in deep-sea volcanic vents, is more than a meter in length, but it does not eat: It has no mouth, gut, or anus. It has specialized compartments in which symbiotic sulfur bacteria live off of the hydrogen sulfide from the vents and then become food for the worm.

Some organisms have gone all the way, melding mutualistically into unified creatures. The cells of all photosynthetic eukaryotes contain chloroplasts, and almost all eukaryotic cells contain mitochondria, the evolutionary descendants of mutualistic prokaryotes (see BACTERIA, EVOLUTION OF). In fact, the evolution of mutualism may have made the evolutionary advancement of life beyond the bacterial stage possible. Evolutionary biologist Lynn Margulis points out that the formation of mutualisms has been a major process in the evolution of biodiversity (see MARGULIS, LYNN; SYMBIOGENESIS). The evolution of adaptations to the physical environment might have produced only a few species: Just how many ways are there for a plant or an animal to adapt to desert conditions? But the greatest number of evolutionary innovations, from the huge array of plant chemicals to the astonishing diversity of flowers, and even the symbioses that keep human cells alive, have resulted from coevolution of organisms in response to one another.

Further Reading

Bäckhed, Fredrik, et al. "Host-bacterial mutualism in the human intestine." *Science* 307 (2005): 1,915–1,920.

Birkle, L. M., et al. "Microbial genotype and insect fitness in an aphid-bacterial symbiosis." *Functional Ecology* 18 (2004): 598–604.

Bronstein, Judith L. "Our current understanding of mutualism." *Quarterly Review of Biology* 69 (1994): 31–51.

Combes, Claude. *The Art of Being a Parasite.* Transl. by Daniel Simberloff. University of Chicago Press, 2005.

Currie, Cameron R., et al. "Coevolved crypts and exocrine glands support mutualistic bacteria in fungus-growing ants." *Science* 311 (2006): 81–83.

Janzen, Daniel H. "Co-evolution of mutualism between ants and acacias in Central America." *Evolution* 20 (1966): 249–275.

Pichersky, Eran. "Plant scents." *American Scientist* 92 (2004): 514–521.

Proctor, Michael, Peter Yeo, and Andrew Lack. *The Natural History of Pollination.* Portland, Ore.: Timber Press, 1996.

Ruby, Edward, Brian Henderson, and Margaret McFall-Ngai. "We get by with a little help from our (little) friends." *Science* 303 (2004): 1,305–1,307.

Scarborough, Claire L., Julia Ferrari, and H. C. J. Godfray. "Aphid protected from pathogen by endosymbiont." *Science* 310 (2005): 1,781.

Sessions, Laura A., and Steven D. Johnson. "The flower and the fly." *Natural History,* March 2005, 58–63.

Sherman, Paul W., and J. Billing. "Darwinian gastronomy: Why we use spices." *BioScience* 49 (1999): 453–463.

Thompson, John N. *The Coevolutionary Process.* Chicago: University of Chicago Press, 1994.

Zimmer, Carl. *Parasite Rex.* New York: Arrow, 2003.

Wilson, Michael. *Microbial Inhabitants of Humans: Their Ecology and Role in Health and Disease.* New York: Cambridge University Press, 2005.

commensalism *See* COEVOLUTION.

continental drift Continental drift is the theory, now universally accepted, that the continents have moved, or drifted, across the face of the Earth throughout its history, and that they continue to do so. This theory also explains which continents were at which locations and when, with consequences for both global climate and the evolution of life.

Geographers, and even students of geography, have long noticed that the coastlines of Africa and South America have corresponding shapes, as if they were once a single continent that broke apart. This was suggested as long ago as 1858 by the French geographer Antonio Snider-Pellegrini.

In the 19th century, scientists discovered that the coal deposits of India, South America, Australia, and South Africa all contained a relatively uniform set of fossil plants, which is unlikely to have evolved separately on each continent. The plants are referred to as the Glossopteris flora because it was dominated by the now-extinct seed fern *Glossopteris*. The Glossopteris flora is also found in Antarctica. Twenty of the 27 species of land plants in the Glossopteris flora of Antarctica can also be found in India!

Evidence such as this led scientists to propose that the continents had been connected in the past. In 1908 American geologist Frank B. Taylor proposed that continents had split and drifted apart, suggesting that these movements not only explained the shapes of the continents but also explained the mid-oceanic ridges. Most scientists preferred the claim that land bridges had connected the continents. Animals and plants had migrated across the land bridges. Geographer Eduard Suess proposed that land bridges had connected India, South America, Australia, and South Africa into a complex he called Gondwanaland. Other land bridges connected the northern continents. Then these land bridges conveniently sank into the ocean, without leaving a trace. While such land bridges seem highly imaginative today, there was at the time no reasonable alternative, except for Taylor's wandering continents. But sometimes even the proposal of land bridges did not help; there is a TRILOBITE fossil species found on only one coast of Newfoundland as well as across the Atlantic in Europe. Trilobites could not have used land bridges to achieve such a strange distribution pattern.

Alfred Wegener, a 20th-century German meteorologist and polar scientist, claimed not only that some of the continents had been connected, as Taylor and Seuss had suggested, but that all of them had been connected. Wegener claimed that the northern continents (later called Laurasia) and the southern Gondwanan continents had been joined into a gigantic continent, Pangaea (Greek for "the whole Earth"), in the past. Wegener, and geologist Alexander du Toit in South Africa, further claimed that the connecting and splitting were due not to the emergence and disappearance of land bridges but to continental drift. One principal line of evidence that Wegener cited was that coal is formed only under tropical conditions; if so, why are there coal deposits in Spitsbergen, 400 miles north of Norway? Wegener said this could only be explained if the northern continents had, in fact, once been near the equator. Nor was this the only problem facing conventional geology. If the continents have been in their current locations, and eroding, for billions of years, where did all the sediment go?

Most geologists were not prepared for the theories of Wegener and du Toit. American geologists, in particular, did not want to accept a theory proposed by a meteorologist from a country with which they were at war. Wegener, moreover, made some claims that could be readily discounted. He claimed that Greenland was moving west at between 30 and 100 feet (9 to 32 m) per year, which was incorrect. His mechanism of what caused the continents to move was also wrong: He said their movement was caused by the tidal forces of the Moon and the centrifugal force of the rotating Earth. By the time Wegener died (he froze to death on his 50th birthday during an expedition into Greenland in 1930), his theory was still considered outlandish by most scholars.

After Wegener's death, du Toit continued to assemble fossil evidence for Pangaea. He noted, for example, that the now-extinct reptile *Mesosaurus* is found near the Carboniferous-Permian boundary in both Brazil and South Africa (see CARBONIFEROUS PERIOD; PERMIAN PERIOD). Evidence of plants and insects accompanying *Mesosaurus* suggest that it may have lived in fresh water. How could *Mesosaurus* have dispersed across an ocean to be in both places? When du Toit looked more closely, he found a close correspondence between the sequence of fossils in Antarctica, South Africa, South America, and India. In all these places, carboniferous glacial tillites (an assortment of stones dropped from melting glaciers) have the Glossopteris flora and *Mesosaurus* above them. Above this, deposits from the JURASSIC PERIOD of both South America and South Africa contain sand dune deposits. Du Toit also found that late Paleozoic deposits in South America, South Africa, India, and Australia (see PALEOZOIC ERA) contained grooves produced by the movements of glaciers. The grooves radiated out from a common central location, which could best be explained if the glaciers had moved from a central location when these continents were Pangaea in the late Paleozoic era, and if this central location had been near the South Pole.

Scientists have corroborated du Toit's evidence of the Glossopteris flora, and *Mesosaurus,* with the fossils of other species. In 1969, well into the modern era of continental drift theory, fossils of the reptile *Lystrosaurus* were discovered from early Mesozoic deposits of Antarctica, Africa, and southeast Asia (see MESOZOIC ERA). This pattern corroborated what du Toit had found with *Mesosaurus. Lystrosaurus*

(Opposite page) **Several fossil plant and animal species, such as the seed fern *Glossopteris* and the reptiles *Lystrosaurus* and *Mesosaurus*, have a distribution pattern that makes sense only if they lived in the same place that was subsequently separated by continental drift into South America (A), Africa (B), India (C), Madagascar (D), Antarctica (E), and Australia (F). Arrows and shading indicate directions of ice movement.**

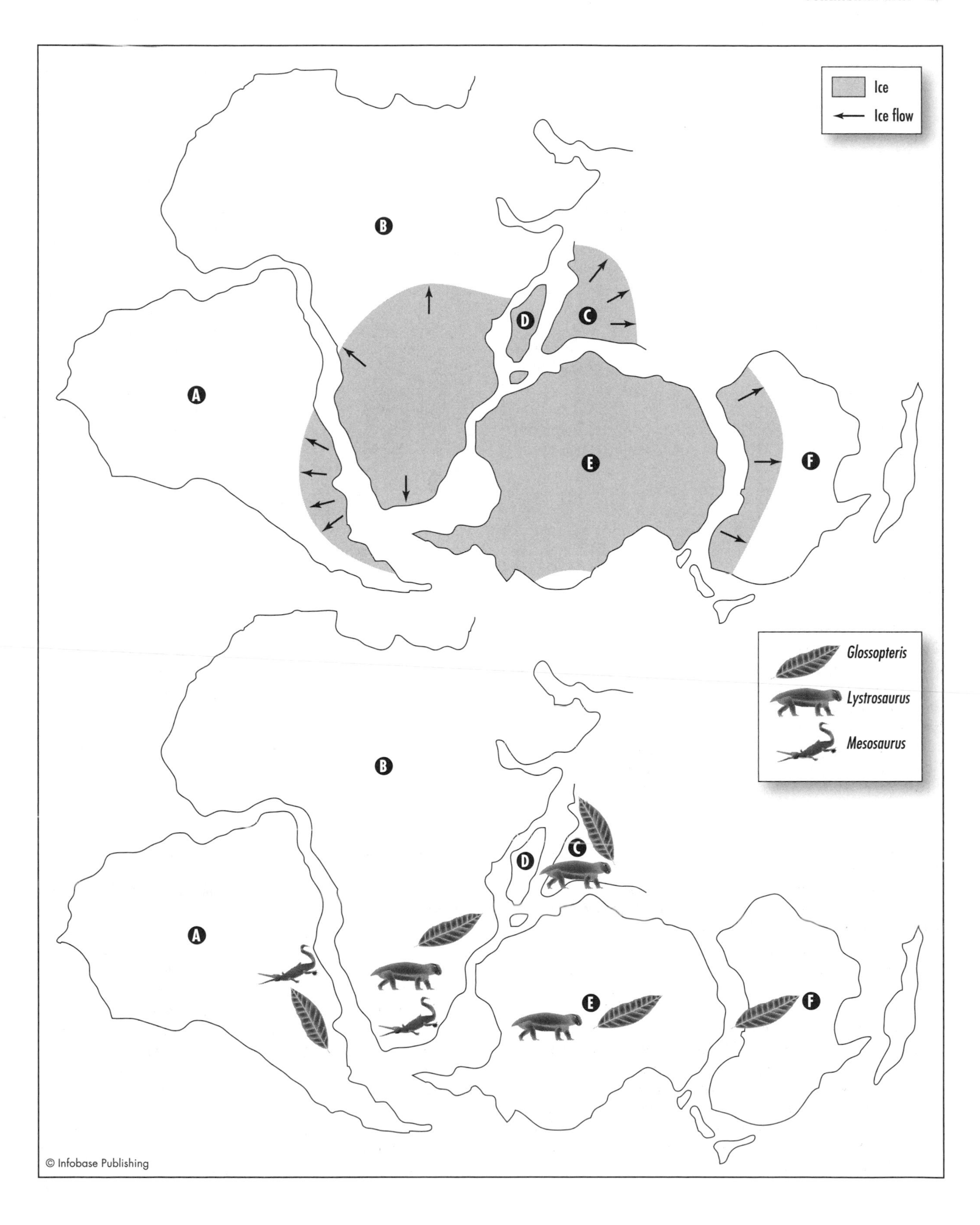
Ice
Ice flow
B
D
C
A
E
F
Glossopteris
Lystrosaurus
Mesosaurus
© Infobase Publishing

was an animal that spread widely through the world continents after the PERMIAN EXTINCTION (see figure).

There were two major reasons that the proposal of Wegener and du Toit met resistance and criticism, sometimes even hostility. First, Wegener had set out to prove, rather than to test, his proposal; he was preaching it, rather than investigating it, according to his critics. Second, Wegener could not explain how continents could drift. What would move them? And how could they plow through the ocean floor without leaving huge piles of rubble? Scientists now understand that it is not the continents themselves that move, but the heavy rock of the crustal plates of the Earth, upon which the lighter rock of the continents rests (see PLATE TECTONICS). Plate tectonics provides the explanation for how continental drift occurs.

A major breakthrough that enhanced the credibility of continental drift was the discovery of the Mid-Atlantic Ridge, a mountain range that runs the entire length of the bottom of the Atlantic Ocean. The existence of this ridge had been known ever since it was encountered while the transatlantic cable was being laid in the late 19th century. Princeton mineralogist Harry Hess designed a depth-sounder during World War II that used sound waves to produce images of underwater objects. After the war he used this technique to study the Mid-Atlantic Ridge. By the 1950s, oceanographers had found ridges through all the world's oceans, with a total length of 46,600 miles. All of these ridges looked like fractures from which volcanic material emerged, and which could produce new ocean floor, the driving force for plate tectonics.

Another advancement in scientific understanding that paved the way to the acceptance of continental drift was the discovery of PALEOMAGNETISM. Molten lava contains small pieces of iron that align with the Earth's magnetic field. When the lava hardens, the pieces of iron are locked into place. If the rock subsequently moves, the pieces of iron will no longer line up with the Earth's magnetic field. Also, the Earth's magnetic field changes polarity every few hundred thousand years. Volcanic rocks that hardened during an earlier period of Earth history will have pieces of iron that align with the direction that the Earth's magnetic field had at that time, not what it is today. Geologist Fred Vine, along with Harry Hess, discovered bands of alternation of paleomagnetism in the rocks of the Atlantic Ocean floor (see figure on page 87). These bands were symmetrical on either side of the Mid-Atlantic Ridge. This evidence strongly indicated that the Mid-Atlantic Ridge had, for millions of years, been producing new seafloor rock, both to the east and to the west. This demonstrated the mechanism of plate tectonics and made continental drift a compelling theory. The newly developed techniques of RADIOMETRIC DATING, furthermore, revealed that no part of the ocean floor was more than 175 million years old, while many continental rocks were more than three billion years old. This can now be explained because the continents have rafted along on top of the continually recycling rock of the ocean floors.

Acceptance of continental drift theory represented a swift revolution in geological science. Surveys indicated that by 1950 about half of geologists accepted continental drift. Almost no scientists had accepted it in 1930, and by the 1960s nearly all geologists and evolutionary biologists accepted it. One exception was George Gaylord Simpson, who resisted it for many years (see SIMPSON, GEORGE GAYLORD).

The modern reconstruction of continental drift reveals a complex and surprising history of the continents. What is now Kazakhstan was once connected to Norway and to New England. One corner of what is now Staten Island is European, as is a corner of what is now Newfoundland. What is now Massachusetts is part of the North American continent, but part of its coast is of European origin. The Scottish Highlands and Scandinavia contain much rock of American origin. The future movement of continents is no less surprising. Coastal California will move northwest, pulling Los Angeles past Oakland and, later, Seattle. Africa will push into Europe, destroying the Mediterranean and creating a mountain chain all the way to India. Australia will run into Asia. This will occur at about the pace that a fingernail grows.

The further back in time one looks, the less evidence is available for reconstructing the patterns of continental drift. Most rocks over a billion years old have been destroyed by crustal movements. Geologist J. J. W. Rogers summarized available evidence and proposed that four continents merged together about a billion years ago to form a single continent, Rodinia. Others have proposed that the effect of this world-continent on oceanic circulation patterns caused the Earth to freeze, forming SNOWBALL EARTH. By 700 million years ago, when the Snowball was beginning to thaw, Rodinia split into separate continents, which became Laurasia and Gondwana during the Paleozoic era.

In the middle of the Paleozoic era, most of the landmass of the Earth was in the Southern Hemisphere. Toward the end of the Paleozoic, the continents crashed back together, forming once again a single world-continent, Pangaea (see figure on page 88). The loss of shallow oceans, and changes in world climate, that accompanied the formation of Pangaea contributed to the PERMIAN EXTINCTION.

Near the beginning of the Mesozoic era, Pangaea began to split again, into continents roughly corresponding to modern Asia, Europe, and North America; and a cluster that was later to become South America, Antarctica, Australia, Africa, and India (see figure on page 88). By about 50 million years ago, North America and Europe were separated, as well as Africa and South America; Antarctica separated from South America and Australia. India separated from the other southern continents and moved northward, crashing into Asia and forming the Himalayan mountains. Only about a million years ago, the isthmus of Panama formed, connecting North and South America for the first time since Pangaea (see figure on page 89).

Continental drift also helps to explain otherwise puzzling features about modern distributions of organisms. For example, the animals of Madagascar resembled those of India, several thousand miles away, rather than nearby Africa. This can be explained by the previous unification of these continents in Gondwanaland. Many types of trees (for example, oaks) are found in Eurasia and North America but not in the Southern Hemisphere. Apparently the first oaks evolved at a time when the northern continents were united together. After the continents separated, and mountain ranges arose, the oaks evolved into separate species in the different locations (see ADAPTIVE RADIATION; BIOGEOGRAPHY).

The same evidence that confirmed continental drift also confirms the evolutionary interpretation of the history of the Earth. The patterns of the fossils, such as the Glossopteris flora and *Lystrosaurus,* could not have been produced

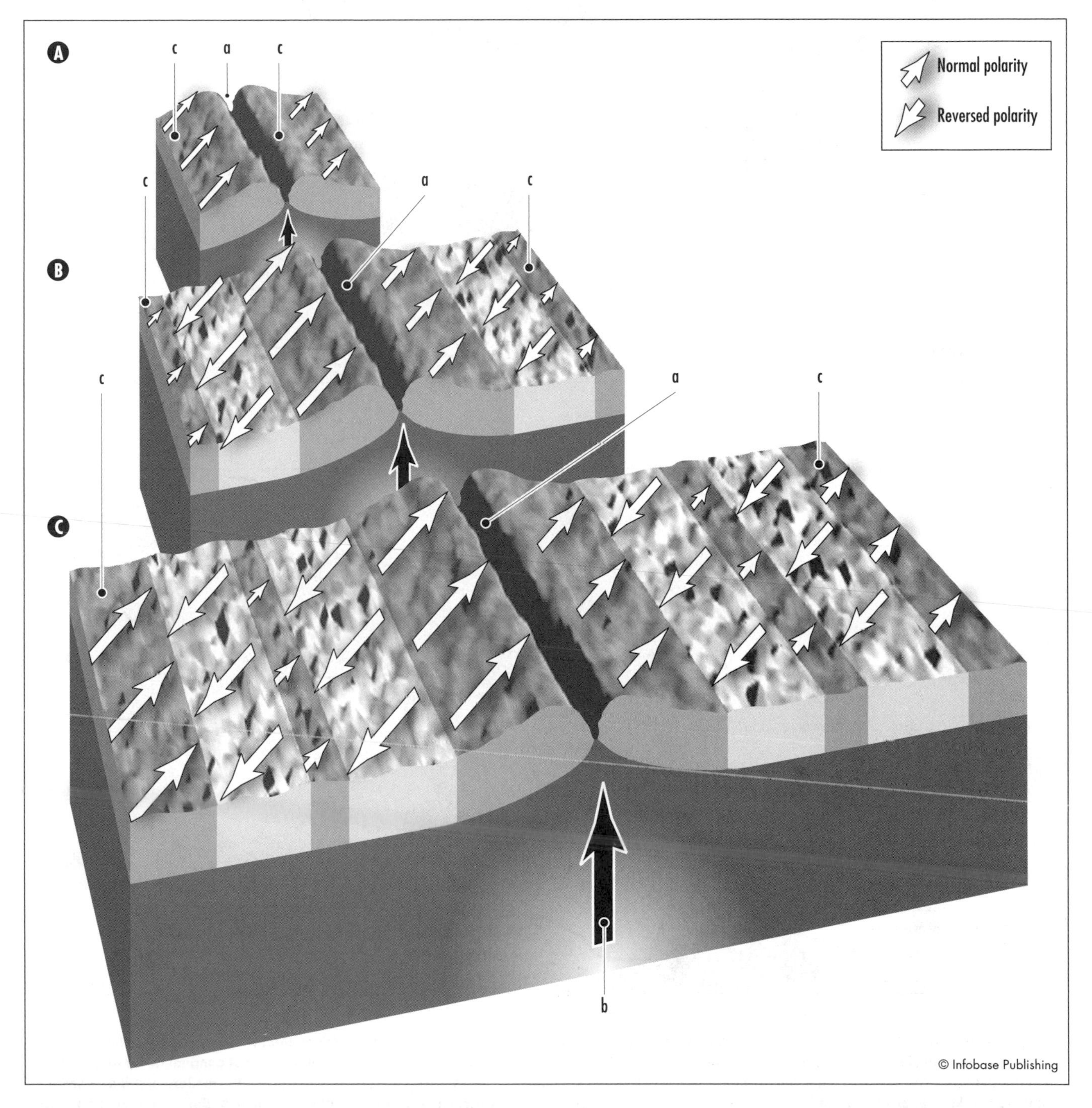

Paleomagnetism confirms that new ocean floor was formed along the Mid-Atlantic Ridge, pushing North America and Europe apart. In this hypothetical drawing, new ocean floor forms at a mid-ocean ridge (a) by volcanic material rising (b). Each of the stripes (c) represents a change in the polarity (represented by directions of arrows) of magnetism; the time axis is from A to C.

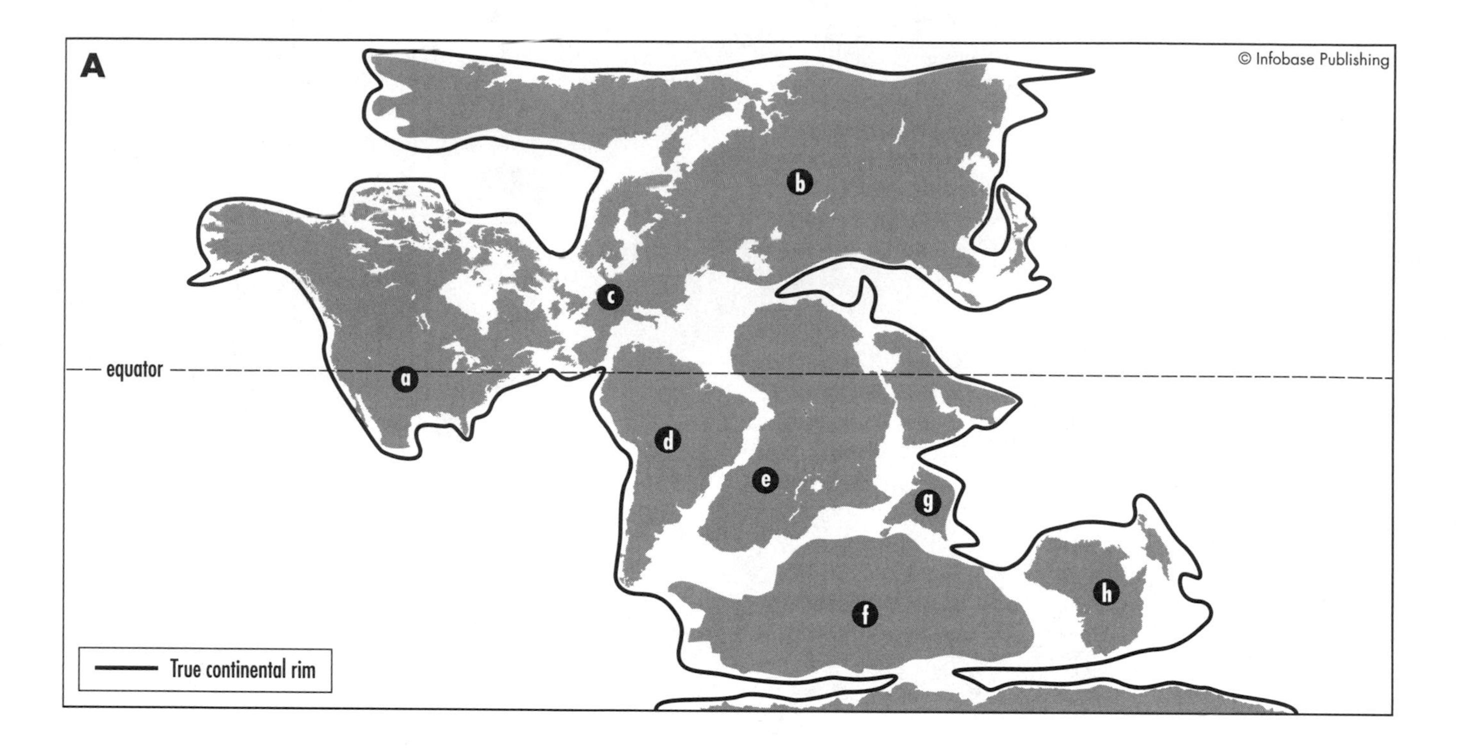

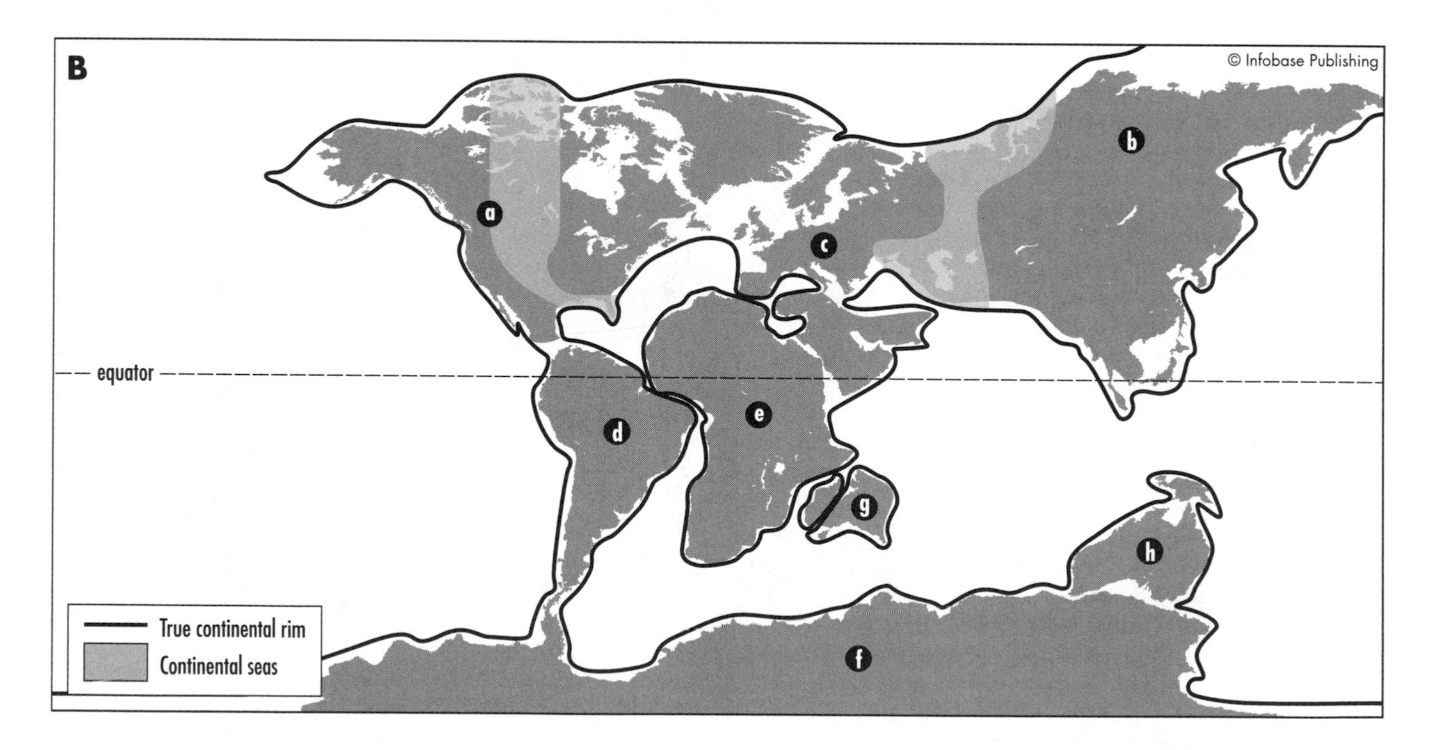

During geological time, the continents have moved and rearranged. In these diagrams, the shapes of the modern continents are indicated by dark shading, the true rim of each continent by heavy lines. Lightly shaded areas indicate where shallow seas covered areas of continents. (a) At the end of the Permian period, the continents had mostly coalesced into a single continent, Pangaea. (b) By the Cretaceous period, the modern continents were beginning to take shape. South America (d), Africa (e), Antarctica (f), India (g), and Australia (h) were still connected or very close together and shared many plants and animals. North America (a) and Eurasia (b, c) were not yet widely separated and shared many plants and animals. (c) The continents today. ***(Figures continue on opposite page.)***

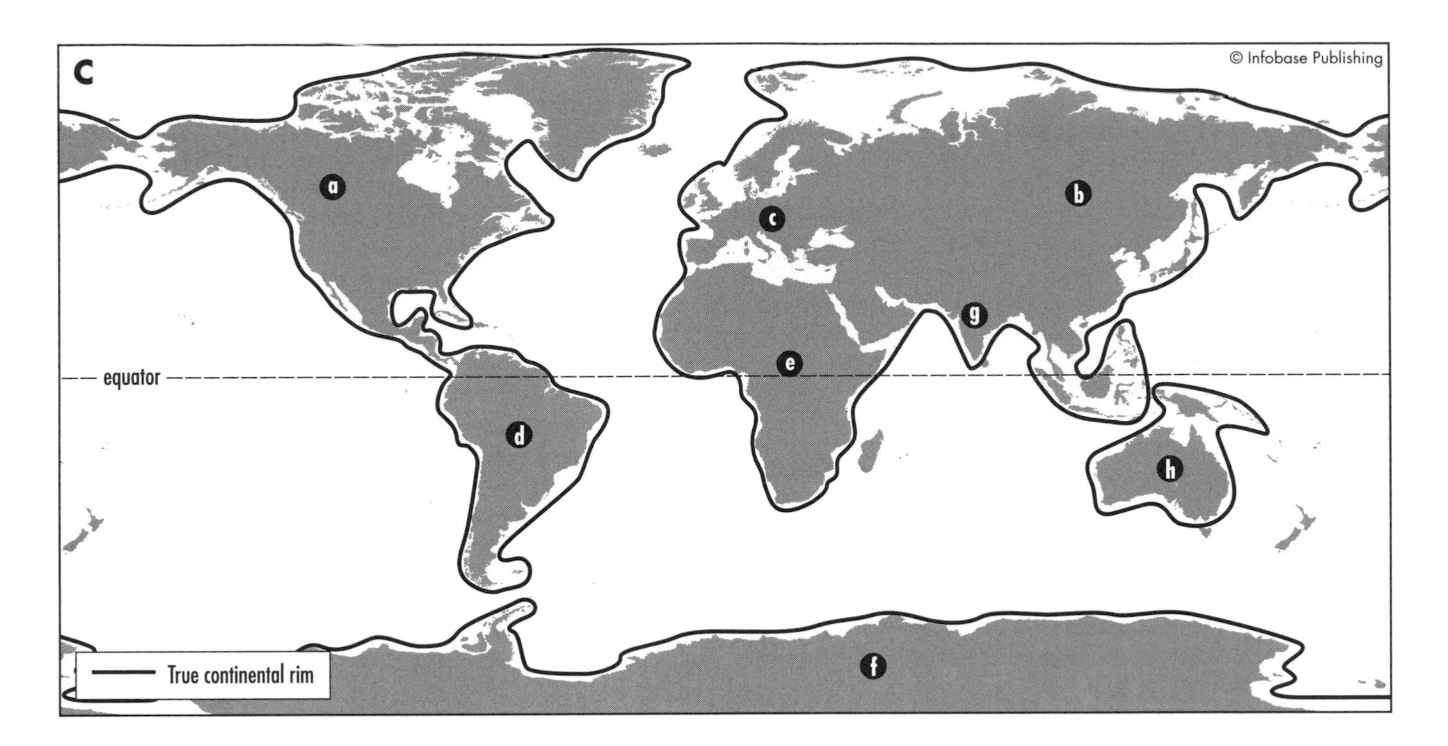

as deposits of a single gigantic flood, as has been claimed by creationists (see CREATIONISM).

Further Reading

Eldredge, Niles. "Evolution of the Earth." Chap. 4 in *The Pattern of Evolution.* New York: Freeman, 1999.

Murphy, J. Brendan, and R. Damian Nance. "How do supercontinents collide?" *American Scientist* 92 (2004): 324–333.

convergence Convergence occurs when similar adaptations evolve in separate lineages of organisms. Convergent adaptations have evolved separately (by *convergent evolution*), rather than being inherited from a common ancestor. Convergence can often be recognized because the organisms evolve the adaptation in different ways. This has been experimentally demonstrated in bacteria. Scientists grew some populations of the gut bacterium *Escherichia coli* in the absence of the sugar maltose, and the bacteria lost the genes that would allow them to use maltose as a food source. Then they were exposed to maltose again. The populations produced bacteria that had MUTATIONS that allowed them to metabolize the maltose. Each of these populations evolved a different set of mutations. The reinvention of the ability to use maltose was a convergence in the separate bacterial populations; but each population achieved this result in a different way.

The following are just a few of the many thousands of possible examples, from plants and from animals.

Plant Convergences

Plant adaptations to climatic conditions. Plants have evolved convergent adaptations to stressful environmental conditions.

- Bushes that live in dry climates have small leaves. Small leaves dissipate heat into the air more efficiently than large leaves, which is important in hot, dry climates where plants must restrict the amount of water that they use in evaporative cooling. Small-leaved bushes have evolved within many different plant families, all of which have species with larger leaves. Small leaves are therefore a new adaptation in many different plant families. In some environments, these bushes are also adapted to sprouting or seed germination after periodic fires. Habitats that are dominated by bushes with small leaves, and that experience periodic fires, have evolved separately in California *(chaparral),* Chile *(matorral),* France *(maquis* and *garrigue),* Australia *(mallee),* and South Africa *(fynbos).*
- The succulent growth form, which allows water storage and CAM (see PHOTOSYNTHESIS, EVOLUTION OF), has evolved in several plant families, for example the cactus and lily families in North American deserts and the spurge family in South African deserts.
- The C_4 form of photosynthesis may have evolved more than 30 different times.
- Several lineages of plants have evolved the ability to grow in soil that is toxic due to high concentrations of heavy metal ions.

Plant adaptations to pollination methods. Plants have evolved convergent adaptations to the same pollination mechanisms.

- Flowers that attract hummingbirds as pollinators are usually tubular and red. The tubular red form of flower has evolved in many different plant families. In one instance, a single mutation began the process that changed the yellow, bee-pollinated flowers into red, hummingbird-pollinated flowers and resulted in an immediate reproductive isolation of two populations, resulting in the evolution of two separate species (see ISOLATING MECHANISMS; QUANTITATIVE TRAIT LOCI).

- Different lineages of plants have evolved adaptations for wind pollination, in which the petals are reduced in size or lost, pollen production is increased, and the pollen-capturing surfaces of the flowers increase in size.

Plant defenses. Plants have evolved convergent adaptations as defenses against herbivores.

- Spines and thorns have evolved many times in separate lineages of plants. Some plants have seeds with sharp spines that entangle animal fur, causing the animal to disperse the seeds to a new location. Strikingly similar structures for animal fur dispersal have evolved in *Tribulus terrestris,* a dicot, and genus *Cenchrus,* a grass (see photo at left). Grasses and dicots evolved from a common ancestor, without spiny seeds, that lived over a hundred million years ago.
- The use of latex (a sticky, milky sap) as a defense against herbivores and as a way of sealing damaged areas has also evolved separately in several plant families. As a result, rubber latex is produced by a spurge *(Hevea brasiliensis),* a fig *(Ficus elastica),* and the dita bark tree *Alstonia* in the dogbane family.
- Chemical defense in plants has evolved thousands of times (see COEVOLUTION). Usually, each lineage of plant uses a different set of chemicals, which is why the plant kingdom produces such an astounding variety of pharmaceutical and spice substances. Some plant chemicals such as limonene reappear over and over, producing the lemon scent separately in lemons (a citrus), lemon basil and lemon balm (mints), lemon grass, and the conifer *Thuja standishii.* Onion scent has evolved separately in onions *(Allium cepa),* which are monocots, and in the salmwood tree, a dicot *(Cordia alliadora).*

Monocot and dicot angiosperms have had separate lineages of evolution for about 100 million years. A monocot (the sandbur grass, genus *Cenchrus*) (labeled C) and a dicot (the puncture vine, *Tribulus terrestris*) (labeled T) have independently evolved spines around their seeds that puncture animal skin and entangle animal fur, allowing the seeds to be carried to new locations. ***(Photograph by Stanley A. Rice)***

Other examples of convergent evolution in plants include the following:

- Parasitism has evolved separately in several lineages of flowering plants. These plants have lost their chlorophyll and obtain food by growing on other plants. Examples include plants in the heath family (genera *Pterospora* and *Monotropa*), relatives of snapdragons (genus *Epifagus*), and plants in the orchid family (genus *Corallorhiza*).
- Within the genus *Calochortus,* the same flower shapes appear to have evolved in several different lineages: the cat's ear form, which is hairy and has open petals; the mariposa form, which is cuplike; the fairy lantern form, which hangs down, with the petals together; and the star tulip form, with smooth spreading petals.
- At least three lineages of plants have separately evolved mutualisms with bacteria in their roots that convert nitrogen gas into nitrogen fertilizer: the legume family Fabaceae, alders (in the birch family Betulaceae), and *Vitex keniensis* (in the verbena family Verbenaceae). The bacteria involved in these mutualisms are different: Legumes have coevolved with bacteria such as *Rhizobium,* while alders have coevolved with *Frankia* bacteria.
- Strikingly similar leaf forms, so similar as to be almost indistinguishable except by experts, have evolved in *Jacaranda mimosifolia* (family Bignoniaceae) and the mimosa (*Albizzia julibrissin* of the Mimosoideae).

Animal Convergences

Eyes. Eyes have evolved several times in the animal kingdom.

- Among the earliest eyes were the crystalline eyes of TRILOBITES, which evolved separately from the compound eyes found in other arthropods and the calcitic eyes of certain echinoderms.
- Most arthropods have compound eyes, but some spiders (the jumping spiders that rely on acute vision) have a lens

eye similar to that of vertebrates. Some shrimp have both compound and lens-type eyes.
- Eyes have evolved in three lineages of gastropod mollusks (snails and their relatives), separately from the eyes of cephalopods such as squids and octopus.
- Some marine worms have also evolved eyes.
- One group of coelenterates has evolved eyes.
- Vertebrates have also evolved eyes.

Each of these groups has eyes that receives visual information, but in a different way. For example, the eyes of the squid are nearly as complex as vertebrate eyes. In vertebrates, the nerve cells are in front of the retina, while in the squid they are behind the retina. Humans, other apes, and Old World monkeys have three-color vision; this type of vision has evolved independently in howler monkeys of the New World. All of the above examples are true eyes, with lenses and retinas, as opposed to simple eyespots that detect the presence of light.

Animal societies. Social evolution has converged in different lineages of animals.

- The "army ant" lifestyle may have evolved separately in Africa and South America, including the behavior of forming a living nest out of their own bodies that has a limited ability to regulate the internal temperature. In turn, seven different lineages of staphylinid beetles have evolved into army ant mimics, which can live among the ants without being recognized as intruders.
- The reproductive system in which one dominant female reproduces and the other females do not, is well known in the insects (ants, bees, and wasps). Three different lineages of coral reef shrimp have evolved a similar system. It is also found in one group of mammals, the naked mole rats. The insect and mammal lineages achieved this system differently, however. In ants, bees, and wasps, the males have only half the chromosomes of the females, and the difference between queens and workers results from being fed different foods as they developed. In mole rats, the males have the same number of chromosomes as females, and the dominant female inhibits reproduction in the other females by behavioral means.

Animals finding their way in the dark. Different animal lineages have converged on alternatives to vision.

- *Echolocation* has evolved several times independently. Bats emit sound waves and use the reflection of those sounds with which they may create mental maps of their surroundings, including the locations of prey insects. Some kinds of fishes, and many cetaceans such as dolphins, use a similar system in underwater locations that are too murky for vision to be of much use. Oilbirds in South America, and swiftlets in Asia, use echolocation also. In turn, some of the prey animals sought by these echolocators have evolved a defense mechanism: the production of rapid clicks to confuse the predator. This defense has evolved separately in the moths hunted by bats, and the shad hunted by cetaceans.
- *Electrolocation* has also evolved independently, at least six times in different lineages of fishes. These animals use electric currents generated by muscles to create a mental map of their surroundings, but the electric currents are different and come from different sets of muscles in the six lineages. They also use electrical signals for communication. Electrolocation can involve an increase in brain size, just as did the development of vision; in electrolocating fishes, it is the cerebellum that has increased in size, not the cerebrum as in vertebrates that rely on vision.

Convergences in Marsupial and Placental Mammals

Adaptation	Marsupial example	Placental example
Burrowing form with poor eyesight	Marsupial mole	Mole
Burrowing form with good eyesight	Wombat	Badger
Long snout, eating ants	Numbat	Anteater
Small, eating seeds	Marsupial mouse	Mouse
Climbs trees	Cuscus	Lemur
Glides from trees	Flying phalanger	Flying squirrel
Small, hopping	Bandicoot	Rabbit
Large herbivore	Kangaroo	Deer
Feline form	Tasmanian cat	Bobcat
Canine form	Tasmanian wolf	Wolf

Animal body form and physiological adaptations. Different lineages of animals have converged upon similar body forms.

- Marsupial mammals in Australia and placental mammals in the northern continents (see MAMMALS, EVOLUTION OF) have undergone separate ADAPTIVE RADIATION but have converged upon some striking similarities of form (see table). Some of the convergences produce forms that are scarcely distinguishable (as in placental and marsupial mice), while others are convergences of function more than of appearance (as in deer and kangaroos).
- A similar convergence has occurred in the placental mammals that evolved in Africa and the placental mammals that evolved in the northern continents (North America and Eurasia) that separated near the beginning of the CENOZOIC ERA. This convergence has occurred in separate lineages of grazing mammals, of otters, of insectivores, of burrowers, and of anteaters.
- Flight has evolved separately in insects, reptiles, birds, and mammals (see INVERTEBRATES, EVOLUTION OF; REPTILES, EVOLUTION OF; BIRDS, EVOLUTION OF; MAMMALS, EVOLUTION OF). Insect wings are outgrowths of the external skeleton. The flying reptiles such as the pterosaurs of the MESOZOIC ERA had wings supported by one long finger. Birds have wings without fingers. Bats have wings with several fingers (see figure on page 92). The ability to glide evolved separately in the now-extinct kuelineosaurs, in which extended ribs produced a gliding surface; flying

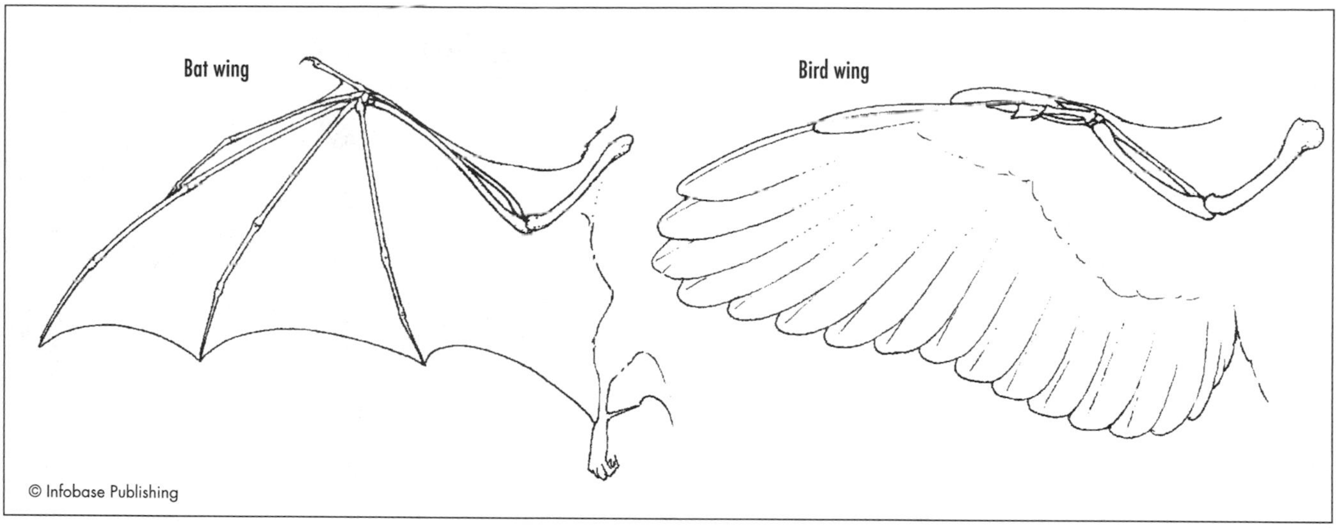

Both birds and bats have wings, but their structure is very different. Bird wings do not have fingers; bat wings have three. Wings evolved from forelegs, separately in birds and bats, an example of convergent evolution.

squirrels, in which flaps of skin between the legs produces a gliding surface; and even flying snakes, in which the entire flattened body produces a gliding surface. There is even a species of gliding ant.

- The fishlike body form has evolved several times: in fishes; in aquatic reptiles of the Mesozoic era; and in dolphins and other cetaceans (see WHALES, EVOLUTION OF).
- Several different families of birds, on different continents and islands, have evolved the hummingbird body form and way of living: small bodies, long bills, and rapid fluttering of wings. This includes the hummingbirds of North and South America (family Trochilidae), Australian honeyeaters (family Meliphagidae), Hawaiian honeycreepers (family Drepanididae), and African sunbirds (family Nectariniidae). This adaptation has also evolved in sphinx moths, which many people mistake for hummingbirds when they first see them.
- Mammals in 11 families in three different orders have evolved the mole-like lifestyle of digging through the soil and eating underground plant parts, with big claws and reduced eyes.
- Earthworms (which are invertebrates) and caecilians (which are vertebrate amphibians; see AMPHIBIANS, EVOLUTION OF) have separately evolved long, flexible bodies without limbs, and the use of a hydrostatic skeleton (muscles pressing against water compartments) to move through the soil.
- One type of barnacle produces a shell made of calcium phosphate and protein, like the bones of vertebrates, although most barnacles produce shells of calcium carbonate.
- Insect species with a grasping foreleg (the praying-mantis body form) have evolved in three separate insect orders.
- Both bombardier beetles and certain kinds of mites spray caustic liquids against animals that threaten them; however, it is not the same chemicals in the two groups.
- Both dipterans and strepsipteran flies, as well as African beetles, have separately evolved structures that help them to maintain balance during flight. In dipterans (such as the housefly) it is the hind wings that become halteres; in strepsipterans, it is the forewings; in some African beetles, it is the wing cover.
- Cephalopod mollusks such as the octopus have evolved a closed circulatory system, in which the blood cells remain inside of the vessels, independently of vertebrates.
- Live birth has evolved separately in mammals, some reptiles, some bony fishes, and some sharks.
- Warm-bloodedness has evolved separately in birds, in mammals, and in tuna.
- Arctic and Antarctic fishes have separately evolved a very similar set of antifreeze proteins.
- Vultures, which are birds that have featherless heads and eat carrion, have evolved separately in the New World (the family Cathartidae is related to storks) and the Old World (the family Accipitridae is a type of hawk).

Coevolution between animal species and other species. Similar types of coevolution have occurred in different sets of organisms.

- At least five different species of ants have evolved the ability to grow underground gardens of fungus. These ant gardens may represent an evolution of agriculture separate from humans (see AGRICULTURE, EVOLUTION OF).
- In 70 different lineages, insects and spiders have evolved characteristics that allow them to parasitize ant nests without being recognized as intruders.
- Bioluminescence, in which living tissues produce light, has evolved separately in insects and in vertebrates. In vertebrates, the light is usually produced by mutualistic bacteria that live in the vertebrate tissues. Bioluminescence may have evolved 30 separate times.
- In some vertebrates that eat plant material, bacteria in the digestive system digest the cellulose. The vertebrate then

uses a mutant form of lysozyme to digest the bacteria. This adaptation has evolved separately in cows, langur monkeys, and hoatzin birds.

In many cases, convergent evolution does not require a total reinvention of an adaptation.

- Limonene has evolved separately in several lineages of plants (see above), but most flowering plants have terpene synthases, the enzymes that produce the immediate chemical precursors of limonene.
- Ornithologist N. I. Mundy explains that the evolution of similar color patterns in separate bird lineages has resulted from the modification of an underlying genetic pattern that they inherited from their common ancestor.
- Eyes have evolved perhaps 40 separate times in the evolutionary history of animals (see above). All of these animals share an organizer gene, called *pax,* that stimulates the development of eyes. In each of these lineages, the actual structure of the eye has evolved separately. There is a marine worm that has compound eyes, like those of other invertebrates, but also has brain lobes that have photosensitive chemicals similar to those of vertebrate eyes. Apparently vertebrate eyes evolved from invertebrate brain lobes rather than from invertebrate eyes.

In modern organisms, scientists can recognize convergence, because despite the external similarities, the organisms are internally different and have achieved the convergence in different ways. In looking at fossils, scientists cannot always tell whether organisms shared a characteristic because they inherited it from a common ancestor, or because they evolved it separately by convergence. This sometimes makes reconstructing fossil history difficult. Sometimes in the fossil record a species appears to have become extinct then reappears. Is it because the species was rare for a long time, then reappeared in greater abundance (what paleontologist David Jablonski calls *Lazarus taxa,* because they appeared to rise from the dead), or because another lineage evolved similar features by convergence, and it just looked like a reappearance (what paleontologist Douglas Erwin calls *Elvis taxa*)?

Every evolutionary scientist understands that there are many examples of convergent evolution. They differ in how much importance they give convergence in the overall pattern of the evolution of life. Some evolutionary scientists claim that evolution has so many different possible directions that the actual course taken by evolution is a matter of historical contingency, a viewpoint expressed by evolutionary biologist Stephen Jay Gould (see GOULD, STEPHEN JAY; PROGRESS, CONCEPT OF). Gould said that evolution produced a bush of different forms, not a ladder leading upward to human perfection. These scientists consider convergence to be details that have occurred at the tips of the branches of the bush of life. Others, like paleontologist Simon Conway Morris, insist that convergence is so frequent that it constitutes one of the major features of evolution. They point out that even though there may be an almost infinite number of possible biological adaptations, there are only a limited number of strategies that work, and evolution keeps finding these strategies over and over. Conway Morris agrees with Gould that evolution is a bush, not a ladder; but in this bush, he says, many of the branches are parallel. Conway Morris cites the parallel evolution of large brains in cetaceans (such as dolphins) and in humans. In the human lineage, there was a separate increase in brain size in the lineage that led to *HOMO SAPIENS,* the lineage that led to NEANDERTALS, perhaps also in the lineages that led to Asian *HOMO ERECTUS;* and in the robust AUSTRALOPITHECINES. Agriculture and civilization in humans evolved separately in East Asia, the Middle East, and America. Intelligence does not always offer an evolutionary advantage (see INTELLIGENCE, EVOLUTION OF), but when it does, one would expect that the large processing center for the nerves would be near the front end of the animal, where most of the sensory organs are located (a big brain in a head with eyes and nose). Conway Morris says that the evolution of intelligence is a rare event, but that if humans encounter an intelligent life-form somewhere else in the universe it would probably be something that they would recognize as a person, perhaps even as a human. This is why he refers to "inevitable humans in a lonely universe" (see essay, "Are Humans Alone in the Universe?").

Further Reading

Conway Morris, Simon. *Life's Solution: Inevitable Humans in a Lonely Universe.* Cambridge University Press, 2003.

Mundy, N. I., et al. "Conserved genetic basis of a quantitative plumage trait involved in mate choice." *Science* 303 (2004): 1,870–1,873. Summarized by Hoekstra, Hopi E., and Trevor Price. "Parallel evolution is in the genes." *Science* 303 (2004): 1,779–1,781.

Patterson, Thomas B., and Thomas J. Givnish. "Geographic cohesion, chromosomal evolution, parallel adaptive radiations, and consequent floral adaptations in *Calochortus* (Calochortaceae): Evidence from a cpDNA phylogeny." *New Phytologist* 161 (2003): 253–264.

Pichersky, Eran. "Plant scents." *American Scientist* 92 (2004): 514–521.

Shadwick, Robert E. "How tunas and lamnid sharks swim: An evolutionary convergence." *American Scientist* 93 (2005): 524–531.

Weinrech, Daniel M., et al. "Darwinian evolution can follow only very few mutational paths to fitter proteins." *Science* 312 (2006): 111–114.

Copernicus, Nicolaus (1473–1543) Polish *Astronomer* Mikołaj Koppernigk (who Latinized his name to Nicolaus Copernicus) was born February 19, 1473. He attended university in Krakow, Poland, and appeared to be headed for a church career. But he also studied astronomy, geography, and mathematics and was particularly interested in planetary theory and patterns of eclipses. He was offered a church administrative position but delayed taking it so that he could continue scientific studies in astronomy and medicine, which he undertook in Italy. He then returned to Poland and continued studying astronomy while he performed church duties.

At the time, astronomers considered the Earth to be the center of the universe. The Sun and the planets circled the Earth. This belief made it difficult to explain the apparent movements of the planets in the sky, especially when the planets appeared to reverse their motion temporarily. Astronomers explained these observations by appealing to epicycles, which are smaller revolutions within the larger revolution of the planets around the Earth. Copernicus realized that the calculations were much simpler if one assumed that the planets were circling the Sun, not the Earth. (Interestingly, Copernicus also had to appeal to epicycles, since he assumed planetary motions were circular. Physicist and astronomer Sir Isaac Newton was later to demonstrate that planetary orbits were elliptical.)

Copernicus feared that such a view might be considered heretical. When he published his views in 1514, it was in a handwritten treatise *(Little Commentary)* circulated only among his friends. He stated basic axioms which became the basis of his major book, *De Revolutionibus Orbium Coelestium* (On the revolutions of the heavenly orbs), which he began writing the following year. Among the axioms were that the Earth is not the center of the universe; the distance from the Earth to the Sun is miniscule compared with its distance to the stars; the rotation of the Earth accounts for the apparent movement of stars across the sky each night; the revolution of the Earth accounts for the apparent movements of the Sun and stars over the course of a year; and the apparent backward motion of the planets was caused by the movement of the observer, on the Earth. All of these things are understood by educated people today, but in Copernicus's time the very concept that "the Earth moves" was revolutionary.

Copernicus's progress on his book was slow, partly because he hesitated to announce his theories, and partly because he spent a lot of time as a government administrator, attempting to prevent war between Poland and the Germans and then dealing with the consequences of war. The book was not published until immediately before his death. Copernicus left his manuscript with a Lutheran theologian, who without the author's knowledge removed the preface and substituted his own, in which he claimed the work was not intended as literal truth but just as a simpler way of doing astronomical calculations. Although many scholars were and are appalled at this substitution, it may have prevented Copernicus's work from being condemned outright. Had Copernicus lived to see what happened to other astronomers such as Galileo Galilei, who defended the view that the Earth moved around the Sun, he might have been even more hesitant about publishing his work.

Copernicus's revelation that the Earth revolved around the Sun is considered by many historians to be the decisive work that broke the unscientific medieval view of the world and opened the way for scientific research. Copernicus's dedication to observation over theology made all subsequent scientific work, including the development of evolutionary science, possible. He died on May 24, 1543.

Further Reading

Copernicus, Nicolaus. *On the Revolutions of the Heavenly Spheres.* Trans. by Charles Glenn Wallis. Amherst, N.Y.: Prometheus, 1995.

Gingerich, Owen. *The Book Nobody Read: Chasing the Revolutions of Nicolaus Copernicus.* New York: Walker and Company, 2004.

O'Connor, J. J., and E. F. Robertson. "Nicolaus Copernicus." Available online. URL: http://www-groups.dcs.st-and.ac.uk/~history/Mathematicians/Copernicus.html. Accessed March 23, 2005.

creationism Creationism is the belief that evolutionary science is wrong and that the history of the Earth and life can be explained only by numerous miracles. There is a range of beliefs ranging from the most extreme forms of creationism to evolutionary science:

- *Young-Earth creationists* insist that the universe and all its components were created recently by God and the entire fossil record was produced by a single flood, the Deluge of Noah.
- *Old-Earth creationists* believe that the Earth is old but that life-forms were created miraculously throughout the history of the Earth.
- *Theistic evolutionists* accept the general outline of the evolutionary history of life on Earth but believe that God intervened miraculously, perhaps subtly, in the evolutionary process.
- *Deists* believe that God set the natural laws in motion at the beginning and has not physically acted upon the universe since that time. Deism, in terms of scientific evidence, cannot be distinguished from evolutionary science, except perhaps in cosmology (see UNIVERSE, ORIGIN OF).

In the review that follows, the young-Earth definition is used. Creationism consists of a creationist theology, in which the Christian Bible is interpreted literalistically, and "creation science," in which scientific information is presented to support those literalistic Bible interpretations.

Although many evolutionary scientists believe in God (see essay, "Can an Evolutionary Scientist Be Religious?"), evolutionary science operates under the assumption that God, if God exists, has not influenced the processes that are being studied.

The fundamental assumption of creationists is that creationism is the only alternative to a completely nontheistic evolutionary philosophy; most consider creationism to be the only alternative to atheism. For this reason, creationists spend a great deal of time attacking aspects, sometimes even details, of evolutionary science. They imply, but seldom directly state, that if evolutionary science has even one flaw, then any reasonable person must accept all of the tenets of biblical literalism, from the six days of Genesis to the Flood of Noah. Most religious people, not only in mainstream but also in conservative Christian churches, accept at least some aspects of evolutionary science and have done so for more than a century (see "American Scientific Affiliation" and the book by Matsumura in Further Reading). Many prominent evolutionists have been religious (see GRAY, ASA; TEILHARD DE CHARDIN, PIERRE; DOBZHANSKY, THEODOSIUS). Despite this, creationists attempt to increase their support base by presenting themselves as the only alternative to atheism.

There is a limited amount of diversity in creationist beliefs. For example, many creationists do not believe that God created each species, as defined by modern science, separately; they believe instead that God created separate "kinds," an undefined category that may refer to species, genus, family, or other level of taxonomic classification (see LINNAEAN SYSTEM). Creationists have even invented a new taxonomic system called "baraminology" (from the Hebrew words for "created kinds") to express these ideas.

Creationism is rare outside of the Judeo-Christian-Islamic tradition, partly because polytheistic religions closely identify their gods with natural forces (however, see the Hindu references in Further Reading). Even though most of the biblical foundation of creationism comes from the Old Testament, which is also scripture for Jews and Muslims, very few Jews or Muslims publicly proclaim opposition to evolutionary science in Western countries (for exceptions, see Further Reading). Throughout its history, creationism has been associated with Christianity, and, beginning in the 20th century, with Christian fundamentalism, although conservative Muslims also accept creationism. Creationism is not simply belief in the existence of a creator; thus when Theodosius Dobzhansky described himself as a creationist, he used the term in a highly unusual manner.

Before the rise of modern science, deriving science from the Bible was the only available explanation for the origin of the world and life in Western countries. Medieval church leaders insisted that the Earth was the center of the universe, because, they believed, the Bible said so; Copernicus (see COPERNICUS, NICOLAUS) and Galileo Galilei got in trouble for presenting scientific evidence against this. As people began to realize that the world was far larger and far older than they had imagined, Western thinkers began to more easily accommodate concepts that did not fit with biblical literalism. As far back as the Scottish Presbyterian scholars Thomas Chalmers in the late 18th century and Hugh Miller in the early 19th century, the creation chapter of the Christian Bible (Genesis 1) was being reinterpreted. Chalmers believed there was a big gap between Genesis 1:1 and 1:2, into which countless eons of Earth history could fit; and Miller believed that each day in the Genesis creation week represented a long period of time. When theologian William Paley presented the clearest and most famous defense of the idea that God had designed the heavens and all of Earth's organisms (see NATURAL THEOLOGY), the only competing theory was that all of these things had simply been produced by chance. Even though Paley and others like him were not biblical literalists, their viewpoints were similar to those of modern creationists, because no credible theory of evolution had been presented. The evolutionary theories of Lamarck and Chambers (see LAMARCKISM; CHAMBERS, ROBERT) were not convincing to most scholars.

All that changed in 1859 when Darwin published *Origin of Species* (see DARWIN, CHARLES; *ORIGIN OF SPECIES* [book]). Here, at last, was a credible alternative to supernatural design. If God did not supernaturally create the world, then what reason was there to believe in God? Some people became atheists; others, like Darwin and Huxley (see HUXLEY, THOMAS), believed that the existence of God could not be known, a belief for which Huxley invented the term *agnosticism*. Some religious people viewed Darwinism as a major threat to their entire system of beliefs. But many Christians, particularly those who were prominent in the study of theology and science, did not. Some, like botanist Asa Gray, blended evolutionary theory into their religious views. Others, such as the prominent theologian B. B. Warfield, defended evolutionary science even while championing the divine inspiration, and the inerrancy, of the Bible. Some evangelical Christians even found alternatives to a literal interpretation of Adam and Eve. Geologist Alexander Winchell, for example, believed that the Genesis genealogies referred only to the peoples of the Mediterranean area, not the whole world; he, and many others, accepted the possibility that Adam and Eve were neither the first nor the only human ancestors.

Even those prominent scholars who opposed Darwinian evolution in the late 19th century seldom did so from the standpoint of biblical literalism that characterizes modern creationists. One of the prominent antievolutionists of the 19th century, Louis Agassiz, was a Unitarian (see AGASSIZ, LOUIS). Paleontologist and anatomist Sir Richard Owen studied the comparative anatomy of vertebrates and saw God's signature in the ideal themes upon which animals were designed, rather than upon the details of supernatural creation (see OWEN, RICHARD). Swiss geographer Arnold Henry Guyot rejected evolution but continued Hugh Miller's tradition of saying that the creation days of Genesis 1 may have been long periods of time. Guyot did not believe in human evolution but admitted nevertheless that the ape was a perfectly good intermediate form linking humans to animals, conceptually if not by origin. Geologist and theologian John William Dawson preferred Paley's natural theology to Darwin's evolution. He admitted that these designs did not flash into existence and may have developed over long periods of time. Historian and theologian William G. T. Shedd insisted that God created different "kinds" of animals, but that these kinds were categories such as birds, fishes, and vertebrates, rather than species. Within each of these kinds, he believed, transmutation was possible.

For the first 70 years after Darwin's book was published, there was very little opposition to evolution that resembled modern creationism. This began to change between 1910 and 1915 when a series of books, *The Fundamentals,* was issued. These books proclaimed biblical inerrancy; it is from these books that the term *fundamentalist* is derived. Several prominent scholars contributed essays to these books. Some of the essays were antievolutionary, but two contributors (theologians George Frederick Wright and James Orr) accepted some components of evolutionary science. Therefore even the book from which fundamentalism gets its name did not have a unified antievolutionary stance.

The earliest creationist writers of the modern variety were George McCready Price, a teacher and handyman from the Seventh-Day Adventist tradition, and Harry Rimmer, a Presbyterian preacher. Supported by themselves and a small number of donors, they published books and preached creationism and Flood Geology in the 1920s.

Creationism got a boost to nationwide notice by the support of William Jennings Bryan, three-time candidate for president. His opposition to Darwinism was based primarily upon the abuses to which he thought evolutionary philosophy was being put, particularly with "survival of the fittest" being used to justify militarism and the oppression of the common man (see EUGENICS). Theologically, he was not a fundamentalist. He debated the famous lawyer Clarence Darrow in the 1925 SCOPES TRIAL in Dayton, Tennessee. Bryan's support was far from the only reason that creationism increased in popularity. The political and social optimism of the Teddy Roosevelt era in America and the Edwardian era in England had been punctured by World War I and the influenza pandemic that followed it. Unprecedented social changes swept America during the 1920s, and traditional Bible Christians scrambled to oppose this tide of modernism. The tension between fundamentalist Christianity and social modernism continues, and along with it, creationism has persisted unto this day. In addition to this, much of the fuel of creationism has come from southern Christianity, which had long been a bastion of Bible belief, particularly after it suffered post-Civil-War social upheaval. For the most part, then, creationism was a product of the 20th century, rather than being a holdout of old-time religion.

Creationism is based upon a number of assumptions that are disputed by most Christian theologians as well as by scientists, including:

- The Creator operates only through miracles. Even though this Creator presumably brought the natural laws of physics and chemistry into existence, creationists do not believe that the operation of these laws counts as God's action. If it evolved, then God did not make it, they think.
- The only kind of truth is literal truth. Thus, they claim, that when Genesis 1 says six days, it means 144 hours, and that Adam was no symbol of humankind but was a historical figure.

Creationists are very selective about the biblical passages to which they apply a literalistic interpretation. They exempt obvious poetry, as in the Psalms, and prophetic imagery, from such interpretation. They also tend to exempt every biblical passage that deals with any kind of science other than evolution. Thus when Job referred to "storehouses of the wind," creationists do not build a creationist version of meteorology upon the belief there are actually big rooms where God keeps the wind locked up, nor that God opens up literal windows for rain as is written in Genesis 6. When the second book of Samuel says that a plague was caused by the Angel of Death, with whom King David actually had a face-to-face conversation, creationists do not reject the germ theory of disease and champion a creationist version of medical science. Finally, First Kings 1:40 describes King Solomon's inaugural parade by saying that "the earth was split by their noise." Creationists do not claim that this literally happened. Generally, creationists insist that even though humans can employ figures of speech, God cannot.

Creationism is based partly on Genesis 1, and upon the genealogies in later chapters of Genesis, from which most creationists obtain their belief in a young Earth. Though few of them accept that creation occurred exactly in 4004 B.C.E. as Irish archbishop James Ussher insisted (see AGE OF EARTH), creationists usually limit the age of the planet to less than 10,000 years. To do so they must reject all of the RADIOMETRIC DATING techniques.

Creationism is also based partly upon Genesis 6–9, the account of the Flood of Noah. Creationists believe the whole Earth was covered by a flood, which killed almost everything, except Noah and his family, terrestrial animals on the ark, aquatic animals, and floating mats of plant material. It is no surprise that the late 20th century revival of creationism resulted from a book about the Flood of Noah: *The Genesis Flood,* by Henry Morris, an engineer, and John C. Whitcomb, a preacher. This flood could not have been tranquil, they said, but would have produced the entire fossil record (see FOSSILS AND FOSSILIZATION). Fossils do not represent a progression in time over the history of the Earth, according to creationists; instead, they were produced by sedimentation during this single flood. In their literal acceptance of Noah's flood, the creationists take a much more extreme position than the catastrophists and diluvialists of earlier centuries (see CATASTROPHISM). Many geologists of the early 19th century believed that Noah's flood produced only the uppermost layer of fossils, and that previous floods had produced the earlier deposits. To them, the fossil layers really did represent a record of Earth history. Moreover, most of them became convinced by Agassiz that the uppermost layer was the product of an ice age, not a flood. But to modern creationists, the fossil layers represent deposits of plants and animals that were alive all at once on the Earth on the day Noah walked into the ark.

The sheer number of fossils makes this belief at once impossible. If all the coal and oil in the world has been derived from swamps that were in existence all at one time, the Earth could scarcely have held that many trees! The lowest aquatic fossil layers contain many marine organisms but no fishes; the lowest terrestrial fossil layers contain relatives of ferns, club mosses, and horsetails, but no flowering plants. Somehow Noah's flood must have sorted out the primitive organisms from the complex and buried the primitive ones on the bottom. Somehow both marine and freshwater organisms survived the mixture of freshwater and saltwater.

Perhaps the biggest problem with Flood Geology is mud. Many geological deposits have been tilted into strange angles, some almost vertically. This had to happen after they were rock. In many cases, further sediments were deposited on top of the tilted layers (see UNCOMFORMITY). Creationists have to believe that the tilting occurred while the layers were still mud. These layers would have collapsed rather than retaining their shape. The sedimentary fossil deposits are often interlayered with volcanic deposits. How could hundreds of temporally separate volcanic eruptions have spread ash and cinders during a 371-day Flood? Some geological deposits, such as layers of salt that were produced by evaporation, simply could not have been produced under such conditions.

Creationist Flood Geology also fails to explain what would have happened after the Flood. All of the rotting vegetation (and dead sinners) would have contributed a massive amount of carbon dioxide to the atmosphere, causing a devastating GREENHOUSE EFFECT. The "two of every kind" of animal would have held insufficient genetic variability to reestablish populations (see FOUNDER EFFECT). If all the animals of the world are descended from the ones that emerged from Noah's Ark, why are some groups of animals found only in certain localized regions of the world, such as marsupials found almost exclusively in Australia? Flood Geology can be rejected by known facts of science without making even the slightest reference to evolution!

One of the most popular creationist tactics is to claim that the fossil record documents the sudden appearances of species, rather than their gradual evolutionary origin through transitional forms. In fact, in many cases, the transitional forms are well known from the fossils, including those of humans, and from living organisms (see MISSING LINKS). Many scientists interpret the relatively rapid origin of species to the PUNCTUATED EQUILIBRIA model, which works on strictly Darwinian evolutionary principles. But the most important point is that the creationists themselves do not believe that the fossil record represents the passage of time at all! They believe it represents deposition that occurred all at once during Noah's Flood, therefore it cannot be for them a record of the sudden appearance of species.

The "argument from design" is still popular among creationists, in a form essentially unchanged since William Paley's natural theology. Its most recent version (INTELLIGENT DESIGN or ID) does not necessarily incorporate all creationist beliefs here described. Proponents of ID describe instances of complex design, usually from the world of plants and animals, and claim that random evolution could not have produced it. What they mean is that they cannot imagine a way in which evolution could have produced it. One evolutionary scientist (see DAWKINS, RICHARD) calls this the "argument from personal incredulity." Creationists furthermore claim the Bible to be without error, but they assume that when they read the Bible, they cannot possibly have interpreted it incorrectly. They therefore also use what the author of this volume calls the "argument from personal inerrancy." With both of these arguments, creationists attribute to themselves a truly superhuman mental capacity. Whatever else creationists may be, they are most certainly human.

The "argument from design" also neglects to explain the countless instances of what appear to most observers to be evil design. From the malaria plasmodium to the intricate life cycle of tapeworms, the biological sources of suffering in the world do not lead most observers to a belief in a loving God. This was perhaps the major reason that Darwin became an agnostic. In a letter to his Christian friend and evolutionist colleague Asa Gray, he said that he could not believe in a God that created parasitic insects and predatory mammals.

In order to concoct what they call "creation science," creationists have had to highly develop their skills of creative twisting of both Bible and science. It is not surprising that they have violated their own publicly stated beliefs regarding both.

- *The Bible.* Creationists claim to believe the Bible, all of it and only it, as God's inerrant communication. But to make their Bible interpretations fit the known facts of the world, they have had to invent numerous fanciful theories. For example, in order to explain where all the water came from in Noah's Flood, they had to invent the theory of a "vapor canopy" that floated over the atmosphere of the Earth in antediluvian times. The Bible mentions no such canopy; there is no scientific evidence for it, nor even a scientific explanation of how it would be possible. The Genesis story of the Fall of Man tells about the origin of human sin and suffering and includes God's "curse upon the ground." Many creationists claim that this was the point in time in which all predators, parasites, disease, and decay began; some even claim that the second law of THERMODYNAMICS began at that time! Needless to say, Genesis contains no such statements. In their zeal to defend the Bible, creationists have had to write a little scripture themselves to add to it.
- *Science.* It is easy to see why a creationist would be unable to undertake evolutionary research. But a creationist ought to be able to conduct scientific investigations that involve current operations of scientific processes. In other words, creationist plant physiology or bacteriology or molecular genetics ought not to differ from that of scientists who accept evolution, and yet some creationists have published many examples of ordinary scientific research that was conducted sloppily.

It is difficult to overestimate the degree to which creationism is antagonistic to modern science and science education, from the creationist viewpoint as well as from that of scientists. Henry Morris, one of the leading creationists of modern times, has written that the theory of evolution was invented during a conference between Satan and Nimrod at the top of the tower of Babel! He considers all evolutionary science, not just the atheistic personal views that many people derive from it, to be literally Satanic in origin. To creationists, even Asa Gray would be an infidel.

Throughout the 20th century and into the 21st century, creationists have attempted to introduce their version of science into public school curricula. They have done so in three phases:

1. In the first phase, during the early to mid 20th century, creationists attempted to directly introduce biblical teachings into school curricula (see SCOPES TRIAL). The Supreme Court found this approach unconstitutional in *Epperson v. Arkansas* decision in 1968.

2. In the second phase, during the late 20th century, creationists attempted to gain equal time in science classrooms for creation science, without making reference to religious beliefs. Federal courts have consistently interpreted creationism as religion, even if it makes no references to a specific religion. In 1982 federal judge William Overton overturned an Arkansas law requiring equal time for creation. In 1987 the Supreme Court overturned a Louisiana creationism law in the *Edwards v. Aguillard* decision.

3. In the third phase, creationists have attempted to introduce nontheistic alternatives to evolution.

- In the early 21st century, creationists in several states attempted to require stickers, containing an antievolution statement, to be affixed to science textbooks. In all cases, these attempts have failed or were reversed upon legal challenge (as of 2006). In one case, the stickers had to be manually removed.
- Also in the early 21st century, creationists in several states attempted to require the teaching of Intelligent Design, either at the state or local levels. As of 2006, all of these attempts have failed. The most famous case was in Dover, Pennsylvania, in which a 2005 lawsuit *(Kitzmiller v. Dover Area School District)* not only ended the attempt of the school board to require the teaching of Intelligent Design but also resulted in a published opinion, from judge John Jones, that became instantly famous in discrediting Intelligent Design. Many state and local governing bodies took this federal decision as clear evidence that any attempt to introduce Intelligent Design would result in costly litigation that they would almost certainly lose. The school board that lost the lawsuit in Dover was quickly replaced, in the next election, by another that did not support the required teaching of Intelligent Design.

In some cases, the creationist controversy is almost entirely political, rather than religious or scientific. Creationism is overwhelmingly a component of Republican or other conservative agendas. Votes concerning a 2003 proposal in the Oklahoma House of Representatives to require antievolution stickers were almost precisely down party lines: Republicans for it, Democrats against it. The Oklahoma representative who proposed the antievolution disclaimer in 2003 also created controversy by announcing that Oklahoma had too many Hispanic people. Former Congressman Tom Delay (R-TX) blamed the Columbine High School shootings on children being taught that humans are "...nothing but glorified apes who are evolutionized out of some primordial soup of mud..." Republican commentator Ann Coulter extensively attacks evolution as being part of a "godless society." This situation has persisted for a long time. Rousas Rushdoony, who defended creationism in the 20th century, started an organization that continues to promote the institution of an Old Testament style of government in the United States. There are exceptions to this pattern. Judge John Jones, who issued the Kitzmiller decision, is a Republican appointee of President George W. Bush, and the new Dover school board, united in its opposition to Intelligent Design, consisted equally of Republicans and Democrats. According to research by Jon D. Miller, about half of Americans are uncertain about evolution. Nevertheless, the close association between creationism and a conservative political agenda led Miller and colleagues to say that there was an era when science could avoid open partisanship. They concluded, however, "That era appears to have closed."

Scientists generally oppose any movement the primary motivation of which is political. For example, most scientists strongly defend environmental science and an environmental political agenda, but not fringe environmentalism that has no scientific basis.

Today, as in William Jennings Bryan's day, many creationists are motivated not so much by a pure zeal of scientific research as by the concern that the decay of religion is leading to the breakdown of society. They make the assumption that the evolutionary process cannot produce minds or behavior patterns other than utter and ruthless selfishness and sexual chaos. However, evolutionary scientists have explained how the evolutionary process could produce, and has produced, many instances of cooperation (see ALTRUISM) and even monogamy (see REPRODUCTIVE SYSTEMS), in many animal species and, almost certainly, in the human species as well (see EVOLUTIONARY ETHICS; SOCIOBIOLOGY).

Some creationists have claimed not so much that creationism is a science as that evolution is a religion, the teaching of which is not legal under the Establishment Clause of the U.S. Constitution. One tract, which was popular in the 1970s and is still available, used cartoons to present the idea that evolution, as taught in college classrooms, was nothing but a religion. The professor asked the students if they "believed in" evolution. Most of the students, who looked like hippies, waved their arms and said "We do!" But one student, who had noticeably Aryan features, stood up and said he did not, and stated his beliefs calmly and respectfully. The professor proceeded to yell at him with the kind of vehemence historically associated with bishops denouncing heretics. The author of this encyclopedia, through numerous years of contacts with educators across the country, has never heard of an instance in which a professor displayed this level of anger and abuse toward creationist students. It is safe to label this tract as propaganda.

Many of the scientists who oppose creationism are Christians, and they oppose it for several reasons. First, as described above, creationism is not, strictly speaking, Christianity, but is something of a cult religion based upon highly idiosyncratic Bible interpretations and even wild imagination. Second, creationism (as here described) is so extreme that it brings embarrassment upon all of Christianity. The second-century C.E. Christian theologian St. Augustine wrote, "It is a disgraceful and dangerous thing for an infidel to hear a Christian ... talking nonsense on these topics ... we should take all means to prevent such an embarrassing situation, in which people show up vast ignorance in a Christian and laugh it to scorn." Thomas Burnet, a 17th-century theologian, wrote, "'Tis a dangerous thing to engage the authority of scripture in disputes about the natural world, in opposition to reason; lest time, which brings all things to light, should discover that to be evidently false which we had made scripture assert." Third, the Christian scientific opponents of creationism believe that scientific inquiry, if it is to be of any use to the human race, must be independent of ideology and religion. If it is merely another version of religion, then why even have it? Either let science be science, they say, or else do not bother with it (see SCIENTIFIC METHOD).

Creationism, then, can be summarized as: (1) a product of the 20th century rather than a holdout of pre-Darwinian Christianity; (2) based upon highly imaginative Bible interpretation; (3) supported by bad science; and (4) often politically motivated. Creationism contradicts nearly every article in this Encyclopedia; refer to any or all of them for further information.

Further Reading

American Scientific Affiliation. "Science in Christian Perspective." Available online. URL: http://www.asa3.org. Accessed March 23, 2005.

Barbour, Ian G. *When Science Meets Religion.* San Francisco, Calif.: HarperSanFrancisco, 2000.

Bhaktivedanta Book Trust International. "Science." Available online. URL: http://www.krishna.com/main.php?id=72. Accessed July 14, 2005.

Chick, Jack. "Big Daddy?" Ontario, Calif.: Chick Publications. Available online. URL: http://www.chick.com/reading/tracts/ 0055/0055_01.asp. Accessed August 29, 2005.

Cremo, Michael A. *Human Devolution: A Vedic alternative to Darwin's theory.* Los Angeles: Bhaktivedanta Book Publishing, 2003.

Dawkins, Richard. *A Devil's Chaplain: Reflections on Hope, Lies, Science, and Love.* New York: Houghton Mifflin, 2003.

Dobzhansky, Theodosius. "Nothing in biology makes sense except in the light of evolution." *American Biology Teacher* 35 (1973): 125–129.

Duncan, Otis Dudley. "The creationists: How many, who, where?" *National Center for Science Education Reports,* September–October 2004, 26–32.

Elsberry, Wesley R. "The Panda's Thumb." URL: http://www.pandasthumb.org/archives/2004/03/welcome_to_the.html. Accessed April 17, 2006.

Fischer, Dick. "Young-earth creationism: A literal mistake." *Perspectives on Science and Christian Faith* 55 (2003): 222–231.

Forrest, Barbara Carroll, and Paul R. Gross. *Creationism's Trojan Horse: The Wedge of Intelligent Design.* New York: Oxford University Press, 2003.

Gould, Stephen Jay. *Rocks of Ages: Science and Religion in the Fullness of Life.* New York: Ballantine, 1999.

Jones, John E. Memorandum Opinion, *Tammy Kitzmiller, et al., vs. Dover Area School District.* Available online. URL: http://coop.www.uscourts.gov/pamd/kitzmiller_342.pdf. Accessed April 17, 2006.

Livingstone, David. *Darwin's Forgotten Defenders: The Encounter between Evangelical Theology and Evolutionary Thought.* Grand Rapids, Mich.: Eerdmans; Edinburgh, UK: Scottish Academic Press, 1987.

Matsumura, Molleen. *Voices for Evolution.* Berkeley, Calif.: National Center for Science Education, 1995.

Miller, Jon D., Eugenie C. Scott, and Shinji Okamoto. "Public acceptance of evolution." *Science* 313 (2006): 765–766.

Miller, Keith B., ed. *Perspectives on an Evolving Creation.* Grand Rapids, Mich.: Eerdmans, 2003.

Moore, John A. *From Genesis to Genetics: The Case of Evolution and Creationism.* Berkeley, Calif.: University of California Press, 2002.

Morris, Henry M. *The Troubled Waters of Evolution.* San Diego, Calif.: Master Books, 1982.

Overton, William R. "Creationism in Schools: The Decision in *McLean v. Arkansas Board of Education.*" *Science* 215 (1982): 934–943.

National Center for Science Education. "National Center for Science Education: Defending the Teaching of Evolution in the Public Schools." Available online. URL: http://www.ncseweb.org. Accessed March 23, 2005.

Pigliucci, Massimo. *Denying Evolution: Creationism, Scientism, and the Nature of Science.* Sunderland, Mass.: Sinauer, 2002.

Ramm, Bernard. *The Christian View of Science and Scripture.* Grand Rapids, Mich.: Eerdmans, 1954.

Rice, Stanley A. "Scientific Creationism: Adding Imagination to Scripture." *Creation/Evolution* 24 (1988): 25–36.

———. "Faithful in the Little Things: Creationists and 'Operations Science.'" *Creation/Evolution* 25 (1989): 8–14.

Ruse, Michael. *The Evolution-Creation Struggle.* Cambridge, Mass.: Harvard University Press, 2005.

Rushdoony, Rousas J. "The necessity for creationism." Available online. URL: http://www.creationism.org/csshs/v03n1p05.htm. Accessed June 5, 2005.

Scott, Eugenie C. *Evolution vs. Creationism: An Introduction.* Westport, Conn.: Greenwood Press, 2004.

Seely, Paul H. "The GISP2 ice core: Ultimate proof that Noah's flood was not global." *Perspectives on Science and Christian Faith* 55 (2003): 252–260.

Shurkin, Joel. "Evolution and the zoo rabbi." Available online. URL: http://cabbageskings.blogspot.com/2005/02/evolution-and-zoo-rabbi.html. Accessed May 6, 2005.

Whitcomb, John C., and Henry M. Morris. *The Genesis Flood.* Philadelphia, Pa.: Presbyterian and Reformed Publishing, 1962.

Working Group on Teaching Evolution, National Academy of Sciences. *Teaching about Evolution and the Nature of Science.* Washington, D.C.: National Academies Press, 1998.

Yahya, Harun. "The Evolution Deceit." Available online. URL: http://www.evolutiondeceit.com. Accessed July 12, 2006.

Cretaceous extinction Sixty-five million years ago, at the end of the CRETACEOUS PERIOD, one of the five MASS EXTINCTIONS of Earth history occurred. It was not the greatest of the mass extinctions (the PERMIAN EXTINCTION resulted in the disappearance of more species), but has long been the subject of intense speculation and research because it was the most recent of the mass extinctions, and it involved the disappearance of the DINOSAURS, the prehistoric animals that have attracted perhaps the most attention from scientists and the general public.

Part of the problem in explaining the Cretaceous extinction, as with the other mass extinctions, was that not all groups of organisms suffered the same rate of disappearance. The largest land animals were dinosaurs, all of which suffered extinction, and the large marine reptiles also became extinct. However, not all dinosaurs were large; even the small ones became extinct. Most groups of mammals, amphibians, crocodiles, turtles, insects, and land plants survived the Cretaceous extinction. Among the marine mollusks, the ammonoids became extinct but the nautiloids, which appeared to have very similar ecological roles, did not. Birds survived, but only a few lineages of them. Some plankton, the floating microscopic organisms of the oceans, experienced a dramatic set of extinctions (for example, the foraminiferans), while others, such as diatoms, did not. Extinction rates were much higher on land (about 90 percent of species) than in freshwater (about 10 percent of species).

There was also the question of whether the Cretaceous extinction was a single catastrophic event or the outcome of gradual declines toward extinction. Several lineages of dinosaurs were already in decline during the Cretaceous period;

however, others were not. And although land plants survived the extinction, their populations were apparently set back by some catastrophe. At the end of the Cretaceous, ferns and fern spores were very numerous. Ferns often lived in areas from which the larger vegetation had been removed by some disturbance. The Cretaceous extinction, then, appeared to be an event in need of an explanation.

Some investigators noticed that a catastrophic collapse of the food chain could have resulted in the pattern of extinctions. The large reptiles (which may have had relatively rapid metabolism) would need continual food, often in vast quantities. In contrast, many amphibians, reptiles with slow metabolism such as crocodiles and turtles, and insects can hibernate until food resources again became available. The Cretaceous mammals were small and, despite their high metabolism, might also have been able to hibernate, as many do today. Plants could have survived as rootstocks or seeds. Some researchers speculated that plankton may have been less capable of surviving an interruption of conditions suitable for photosynthesis, although most plankton have resting phases that can resist very cold and dry conditions. Furthermore, there was no agreement on what might have interrupted the food supply. Volcanic eruptions can have a massive impact on organisms around the world. The Indonesian volcano Tambora erupted in 1815 and was followed by what has been called the "year without a summer" because the volcanic dust, propelled into the stratosphere where it could not be cleared by rain, blocked sunlight. It was the worst agricultural year on record, with crop failures; the resulting famine killed 65,000 people in Ireland alone. All this happened, even though the global temperature decreased by only a couple of degrees F (one-and-a-half degrees C). A "supervolcano" eruption occurred 74,000 years ago (Toba, in Sumatra). Evidence from Greenland ice cores, which preserve a record of global climate, suggests that this volcanic eruption was followed by six cold years. Clearly, massive volcanic eruptions could have interrupted the food chain at the end of the Cretaceous. However, the evidence that such eruptions (or, alternatively, changes in sea level) were severe and abrupt at the end of the Cretaceous was slim.

In a 1980 paper, physicist Luis Alvarez and his son, geologist Walter Alvarez, and their associates made a dramatic and risky proposal: that the Cretaceous extinction was caused by an extraterrestrial impact (see ASTEROIDS AND COMETS). Such an impact would in fact explain global extinctions. The impact itself would project dust high into the atmosphere and ignite fires that would add great clouds of smoke, both of which would darken the Earth, plunging it into wintry conditions disastrous for the food chain. Massive acid rain events might be possible also, as a result of the fires, and if the rock underlying the impact contained sulfur. The shock to the crust could have set off volcanic eruptions and tsunamis (tidal waves), which would have contributed further to local and worldwide extinctions. Some researchers estimate that such an impact would have influenced climate around the world for up to 10 thousand years afterward.

Because the prevalence of UNIFORMITARIANISM in geological science made paleontologists extremely hesitant to invoke processes that are not normally occurring on the earth as explanations for events in the history of the Earth, the Alvarez suggestion was resisted at first. But their initial evidence was good and more evidence later reinforced his proposal.

The Alvarez team had not begun their research with the intention of explaining the Cretaceous extinction. They were more interested in determining accurate dates for the strata that spanned the Cretaceous-Tertiary interval. They expected that the rare element iridium might be useful in studying these strata. They found an astonishing "iridium anomaly" in rocks over a wide area, right at the Cretaceous-Tertiary boundary: a transient and very high concentration of this element. Iridium is much more common in interplanetary particles and objects than on Earth. The iridium spike in the sedimentary layers at the end of the Cretaceous period strongly suggested an asteroid impact, but scientists sought corroboration. They found it.

- Not only was iridium common in the last Cretaceous layers, but also microtektites, which are small glass spheres characteristic of the impact of an asteroid with terrestrial rocks that contain silicon. Shocked quartz grains, characteristic of powerful impacts, were also found.
- Soot-like carbon particles were also unusually common in the last Cretaceous deposits.
- Sediments in the Caribbean from the end of the Cretaceous suggest that a very large tsunami had struck coastal areas.
- Perhaps most important, scientists wanted to find the crater caused by the asteroid. If the impact had occurred in an area with abundant rainfall and erosion, or in the ocean, it might have eroded away but still left some evidence. Just such evidence was found on the Yucatan Peninsula. The Chicxulub Crater has mostly filled in, but evidence of sedimentary layers shows that there was once a massive impact crater.

More than 20 years after the Alvarez proposal, evidence was presented that the Permian extinction may also have been caused by an asteroid impact, which created the remnants of the Bedout Crater off the coast of Australia. The possibility is emerging that the background rate of extinction is caused by normal climate changes and biological interactions (see RED QUEEN HYPOTHESIS), but that the five mass extinctions in Earth history may have resulted from rare and spectacular events such as asteroid impacts.

Further Reading

Alvarez, Luis W., Walter Alvarez, Frank Asaro, and Helen V. Michel. "Extraterrestrial cause for the Cretaceous-Tertiary extinction." *Science* 208 (1980): 1,095–1,107.

Alvarez, Walter. *T. Rex and the Crater of Doom.* Princeton, N.J.: Princeton University Press, 1997.

Cretaceous period The Cretaceous period (140 million to 65 million years ago) was the third and final period of the

MESOZOIC ERA (see GEOLOGICAL TIME SCALE), following the JURASSIC PERIOD. The Mesozoic era is also known as the Age of Dinosaurs, because DINOSAURS were the largest land animals during that time. Dinosaurs became extinct during the CRETACEOUS EXTINCTION, which is one of the two most important MASS EXTINCTIONS in the history of the Earth. The Cretaceous period was named for the deposits of chalk (Latin *creta*) that were abundant in European strata.

Climate. The mild climate of the Jurassic period continued into the Cretaceous period. Swamps and shallow seas covered widespread areas. As continents continued to drift apart, the climate over much of the land became cooler and drier during the Cretaceous. Occasional glaciations in the Antarctic regions caused sea levels to fall, then rise again when the glaciers melted.

Continents. The massive continent of Pangaea continued to break apart, mainly into what would become the northern continents and the southern continents of today (see CONTINENTAL DRIFT). Africa and South America split apart, allowing the southern part of the Atlantic Ocean to form. The Alps formed in Europe. India broke free from Gondwanaland and became a separate continent. The continents began to assume their modern shapes and locations. The geographical isolation of plants and animals on the newly separated continents allowed the evolution of new species in the separate areas.

Marine life. All modern groups of marine organisms existed during the Cretaceous, except aquatic mammals. Very large aquatic reptiles, some like the elasmosaurs with very long necks, lived in the oceans. Ichthyosaurs, which were reptiles that looked like porpoises, became less common, perhaps because of competition from modern forms of fishes that were diversifying during the Cretaceous (see FISHES, EVOLUTION OF).

Life on land. Terrestrial forests were transformed by the appearance of the flowering plants at or just before the beginning of the Cretaceous (see ANGIOSPERMS, EVOLUTION OF). The replacement of coniferous forests by angiosperm trees strongly affected the dominant animals. Examples of plants and animals of the Cretaceous period include:

- *Plants.* Gymnosperms had dominated all Mesozoic forests (see GYMNOSPERMS, EVOLUTION OF). Angiosperm trees evolved in tropical regions. During the Cretaceous, these trees displaced most of the gymnosperms toward the polar regions. The forests of tropical and temperate regions began to look like they do today. By the end of the Cretaceous, temperate forests were dominated by angiosperms such as oaks, maples, beeches, birches, magnolias, and walnuts. The gymnosperm forests consisted mostly of conifers such as pines and spruces, in mountainous and northern regions where they are found today.
- *Animals.* Dinosaurs such as hadrosaurs, which may have specialized upon eating gymnosperms, were abundant worldwide at the beginning of the Cretaceous. As angiosperm forests spread, new kinds of dinosaurs such as ceratopsians evolved that specialized upon them. The gymnosperms and the dinosaurs that ate them were displaced into the polar regions, which were cold but not nearly as cold as today. Primitive birds may have driven flying reptiles such as pterosaurs toward extinction during the Cretaceous (see BIRDS, EVOLUTION OF). Mammals were small and represented by few species, compared to the dinosaurs (see MAMMALS, EVOLUTION OF). Certain groups of insects, such as butterflies and bees that pollinated flowers, stimulated the diversification of angiosperms (see COEVOLUTION).

Extinctions. No major extinction event separated the Jurassic from the Cretaceous period. Many dinosaur lineages went into decline during the Cretaceous, while new groups of dinosaurs evolved. Some but not all dinosaurs would probably have become extinct if the Cretaceous extinction had not occurred. The large marine and flying reptiles also became extinct at the end of the Cretaceous. The extinction of the dinosaurs allowed the mammals to diversify into the forms that dominated the CENOZOIC ERA.

Further Reading

Morell, Virginia. "When sea monsters ruled the deep." *National Geographic,* December 2005, 58–87.

Museum of Paleontology, University of California, Berkeley. "Cretaceous period." Available online. URL: http://www.ucmp.berkeley.edu/mesozoic/cretaceous/cretaceous.html. Accessed March 23, 2005.

Skelton, Peter, et al. *The Cretaceous World.* New York: Cambridge University Press, 2003.

White, Toby, Renato Santos, et al. "Cretaceous period." Available online. URL: http://www.palaeos.com/Mesozoic/Cretaceous/Cretaceous. htm. Accessed March 23, 2005.

Cro-Magnon French geologist Louis Lartet found the first remains of humans who lived a very long time ago but were fully modern in anatomy in 1868. In the Cro-Magnon Cave near the town of Les Eyzies in the Dordogne region of France, Lartet found six skeletons, stone tools, pierced shells, ivory pendants, and carved reindeer antlers. The term *Cro-Magnon* now refers to all of the *HOMO SAPIENS* in Europe who lived prior to the establishment of villages. Although the Cro-Magnon remains provided no evidence of the evolution of the human body, they demonstrated that Europeans had a primitive past just like other peoples that modern Europeans encountered around the world. This implied that the entire human race had a long prehistory. The Cro-Magnon people used stone tools, had no agriculture, and usually lived in caves rather than building structures to live in, at least in the winter. Since the discovery of the Cro-Magnon cave, many other caves have been found throughout Europe that contain Cro-Magnon remains.

Other caves, also throughout Europe, contained remains of humans who were not identical to modern people (see NEANDERTALS). The Neandertals also used only stone tools, had no agriculture, and built no structures to live in. The Neandertals were different enough in appearance that they provided some credible evidence that human evolution had occurred in Europe. The Cro-Magnon and Neandertals did not appear to have lived in the same caves at the same time. Until the development of RADIOMETRIC DATING, the ages of the Cro-Magnon and Neandertal caves could not be determined. After dates became available, researchers discovered that the Neandertal remains were much older than most of the Cro-Magnon remains. The Neandertals lived between 200,000 and 30,000 years ago; most Cro-Magnon remains were between about 35,000 and 15,000 years old. This could mean either that the Cro-Magnon had evolved from Neandertal ancestors, or that Cro-Magnon people had migrated from somewhere else and replaced the formerly dominant Neandertals. Most scholars preferred the former interpretation. They believed that PILTDOWN MAN demonstrated the origin of humankind in Europe. Piltdown man had evolved into Neandertals, and Neandertals into modern people. Piltdown man turned out to be a hoax, but even after the hoax was revealed the belief persisted that Neandertals were the ancestors of the Cro-Magnon people.

The evolutionary advance of Cro-Magnon over Neandertal was also believable because the Cro-Magnons had a great capacity for art and religion, possibly equal to that of modern humans. This was revealed by the discovery of the following:

- *Cave paintings.* Dozens of caves have now been discovered across southern Europe that are filled with paintings, some of astounding beauty and complexity. Most of the paintings are of animals and some are quite large. The artists used natural pigments primarily for black, red, and yellow colors, which are strikingly well preserved in the caves, and used artistic techniques that were even better than those of many medieval European artists (see figure on page 102). Artist Pablo Picasso said, regarding the Altamira Cave paintings, "None of us could paint like that." In addition, there are many smaller paintings, sketches, and complex lines, some with geometric shapes. There are many silhouettes of hands. Modern reenactments have revealed that the artists applied much of the paint by spitting the pigments from their mouths. Cosquer Cave has entire surfaces covered with lines drawn by the fingers of the artists. When these caves were first discovered, their age was not known. They had to be very ancient, however, for they depicted animals such as the mammoth and the hippopotamus that had been extinct in Europe for thousands of years. The most famous of these caves are the Lascaux, Salon Noir, Cosquer, Le Portel, Pech Merle, Les Trois Frères, and Chauvet caves in France, and Altamira in Spain.
- *Carvings.* Numerous portable carvings, often in bone and ivory, have been discovered among Cro-Magnon remains.

This cave painting, from Lascaux in the Dordogne of France, beautifully depicts a bull and a horse. ***(Courtesy of Art Resource, New York)***

The Venus of Willendorf, a limestone figure carved about 25,000 years ago, was found in Germany. ***(Courtesy of Erich Lessing, Art Resource, New York)***

Many are of animals, and some are of humans. The carvings are artistically advanced. Some carvings were very realistic. Some were very abstract, such as the numerous Venus figurines (see figure on page 103). Abstraction included the melding of human and animal motifs. Most of the portable carvings are from what is now Germany and areas of Eastern Europe (in particular Dolní Věstonice in what is now the Czech Republic) rather than from the regions of France and Spain where most cave paintings are found.

- *Bodily ornaments.* Cro-Magnon people made many thousands of beads, and carved holes in many thousands of shells, for use in necklaces.
- *Ceramics.* Many clay figurines were produced and baked into ceramics at Dolní Věstonice.
- *Musical instruments.* Some Cro-Magnon people made flutes from limb bones of animals, into which they carved finger holes. Modern researchers have produced musical scales from replicas of these instruments. Bull-roarers are flutelike objects which Cro-Magnon musicians attached to a rope and whirled around in a circle.

An explorer named Ruben de la Vialle entered the cave now known as Salon Noir in 1660. He discovered many great paintings in this cave but assumed they were of recent origin. He inscribed his name and date on the wall. The earliest discovery of a painted cave whose antiquity was recognized occurred in 1879. Amateur archaeologist Marcelino Sanz de Sautuola explored a cave on his property at Altamira in Spain. He looked down and collected artifacts, but his young daughter looked up and saw the paintings on the ceiling. He publicized the discovery, and even King Alfonso XII visited the cave. Sautuola was convinced that the paintings were very old but could not prove it. Interest in this discovery waned. The Lascaux Cave was discovered in 1940 by three schoolboys when Marcel Ravidat's dog Robot fell down a sinkhole. After they rescued the dog, the schoolboys decided to explore the hole. When they found large chambers filled with animal images, they told their schoolmaster, Léon Laval, who contacted the anthropologist and clergyman Abbé Henri Breuil.

At first, some scholars resisted the idea that cave people could produce great art. The French archaeologist Emile Cartailhac claimed that caves such as Altamira had been produced by modern artists. Another French archaeologist, Édouard Harlé, even dismissed the Altamira paintings as a forgery Sautuola had made himself. Long after Sautuola's death, Harlé changed his mind and admitted in a 1902 paper that Sautuola had been right after all.

All of these discoveries revealed a rich cultural and spiritual life among the Cro-Magnon (see RELIGION, EVOLUTION OF). The community spiritual function of the cave paintings was further suggested by the fact that (as in the cave at Lascaux in France) the big, colorful paintings were in large chambers near the mouth of the cave, while smaller, overlapping pieces of art were found deeper in the cave, as if the former represented public religious gatherings and the latter represented individual spiritual quests. In addition, many Cro-Magnon artifacts were made of materials that had to be transported long distances, for example seashells in inland caves. Some Cro-Magnon burials contained such a large number of artifacts that they could not have been produced by a single family or small band. A Cro-Magnon burial site of two children at Sungir in Russia contained 10,000 beads, each of which would have taken at least an hour to produce. All of these discoveries revealed, in addition to cultural and spiritual awareness, a complex social structure at least at the tribal level and perhaps trading networks that spanned great distances. In contrast, the Neandertal remains were virtually devoid of any artifacts other than stone tools and provided no evidence of complex social structure or trading networks. All of this seemed to suggest that primitive Neandertals, who thought only of getting food and shelter for survival, evolved into Cro-Magnons, to whom culture was an essential part of survival.

Another major difference between Cro-Magnons and Neandertals was that each Cro-Magnon tribe appeared to be culturally different from each other tribe. There were recognizable regional differences in the stone tools that they produced. From place to place, Cro-Magnon artistry also took on different forms: animal tooth pendants in some areas, perforated shells in others, sculptures in others, limestone engravings in yet others. Different caves even bore the mark of individual artists. The oldest known cave, Chauvet in France, has about 260 paintings, many of them of rhinoceroses. The rhinoceros is rare among the paintings in other caves. The Chauvet rhinoceros paintings have a distinctive type of horn and ears, which anthropologists interpret as the work of a single artist. By comparison, the culture of the Neandertals, revealed only by their stone tools, was more geographically uniform.

A closer examination of the evidence called into question the idea that Cro-Magnons had evolved from Neandertals.

- There were no clear examples of humans intermediate between Neandertals and Cro-Magnons. Some traits of some remains appeared intermediate, but the vast majority of specimens of Neandertals and Cro-Magnon were clearly one or the other.
- There appeared to be no evidence for the gradual evolution of culture among the Cro-Magnon. The oldest of the Cro-Magnon caves, Chauvet in France, was painted 33,000 years ago and exhibited as advanced an artistic capacity as any of the later caves. There was no evidence of gradual advancement from the Mousterian stone tool industry used by the Neandertals to the more advanced stone tools made by the Cro-Magnon (see TECHNOLOGY). Some Neandertal sites in southern France adopted a more advanced stone tool technology, called the Châtelperronian. The fact that the only Neandertals to do so were those who lived at the same time and place as the Cro-Magnon implies that they adopted the technological advance from the Cro-Magnon, or perhaps even stole the tools from them.
- It appeared that there was a general trend for the older caves to be in eastern Europe, and the newer caves in western Europe (see figure on page 104).

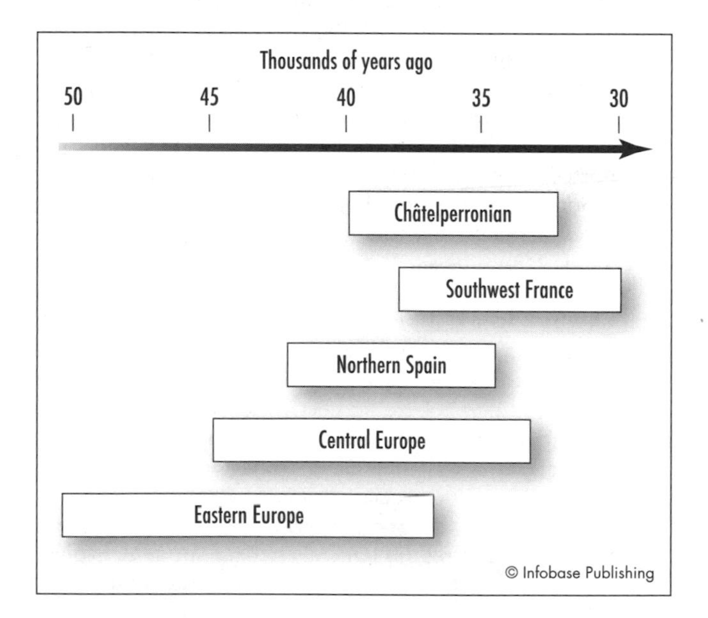

Bars represent approximate periods of time in which Cro-Magnon sites were occupied in each region. There is a general pattern of migration from Eastern Europe and northern Spain to southwest France. The Neanderthal Châtelperronian culture existed at the same time and place (southern France) as the Aurignacian Cro-Magnon culture. *(Adapted from Lewis-Williams)*

The lack of intermediate fossils, the sudden emergence of art, and the progress of culture westward across Europe, all suggested that the Cro-Magnons were descendants of immigrants rather than of Neandertals. The immigrants, presumably *H. sapiens* from Africa, would have encountered the Neandertals and displaced them by competition. This hypothesis is part of the Out of Africa theory that claims all modern humans, including Europeans, Orientals, Australian aborigines, and Native Americans, dispersed from Africa less than 100,000 years ago. It is supported by DNA evidence from both *H. sapiens* and from Neandertals, which most evolutionary scientists now consider to have been a species of humans, *H. neanderthalensis,* that evolved separately from an African ancestor similar to *HOMO ERGASTER* (see DNA [EVIDENCE FOR EVOLUTION]; NEANDERTALS).

The Out of Africa hypothesis would explain not only the lack of anatomical and cultural intermediates between Neandertals and Cro-Magnon but also the rapid development of art and religion among the Cro-Magnon. The immigrants would have brought with them the capacity for art and religion. The immediate ancestors of the immigrants, the *Homo sapiens* of the Middle East and Africa, did not produce such an astounding array of artistic works. Artifacts such as the intricate ivory harpoon from Katanda in Africa, believed to be 90,000 years old, showed that *H. sapiens* had the capacity to produce the kind of art that the Cro-Magnon had (see *HOMO SAPIENS*). The earlier *H. sapiens* had the capacity but had not experienced the stimulus to produce great art. The encounter with the Neandertals, who were just similar enough to them to be a threat, may have stimulated the Cro-Magnon to develop cultural complexity, which provided them with superior survival skills as well as integrating the tribes with cultural identity.

The cultural explosion of the Cro-Magnon continued after the last Neandertals became extinct about 30,000 years ago. Anthropologists usually attribute this to the most recent of the ICE AGES which was at its climax between 30,000 and 15,000 years ago. Northern Europe was covered with glaciers, and this forced some Cro-Magnon southward into regions already inhabited by other Cro-Magnon. The resulting conflict created conditions that favored the development of cultural identity and cohesion within each tribe.

No one knows why the Cro-Magnon people, who became the European tribes, stopped painting caves and producing as many other cultural artifacts about 10,000 years ago. The end of the Cro-Magnon high culture coincided with the end of the ice age. After that time, migrations of tribes from the east displaced or mixed with many of the descendants of the Cro-Magnon. The new immigrants, revealed by both DNA (see MARKERS) and language, were primarily the speakers of Indo-European languages. They may have driven some tribes into extinction, while other tribes adopted the languages of the immigrants. One of the few surviving languages of people who lived in Europe before the arrival of the Indo-Europeans is Basque, spoken by natives of the Pyrenees Mountains that separate France and Spain. A later wave of immigrants brought agriculture across Europe from the Middle East (see AGRICULTURE, EVOLUTION OF). Among the earliest agricultural cities in the Middle East are Jericho (Israel) and Çatal Höyük (Turkey). The transformation of Cro-Magnon into modern civilization involved an increasing rapidity of technological, cultural, and artistic innovation and increasing geographical variation.

Further Reading

Hitchcock, Don. "Cave Paintings and Sculptures." Available online. URL: http://donsmaps.com/cavepaintings.html. Accessed March 23, 2005.

Lewis-Williams, David. *The Mind in the Cave.* London: Thames and Hudson, 2002.

O'Neil, Dennis. "Early Modern Human Culture." Available online. URL: http://anthro.palomar.edu/homo2/sapiens_culture.htm. Accessed March 23, 2005.

cultural evolution *See* EVOLUTION.

Cuvier, Georges (1769–1832) French *Biologist, Paleontologist* Georges Cuvier was born on August 23, 1769. He made contributions to the understanding of animal anatomy and the history of the Earth that allowed the later development of evolutionary science. Cuvier's main contributions were to establish the science of comparative anatomy and the fact that extinctions have occurred during the history of the Earth.

In 1795 French biologist Étienne Geoffroy Saint-Hilaire invited Cuvier, who was then a naturalist and tutor, to come to Paris. While Geoffroy started Cuvier's career, the two men were later to split over fundamental scientific issues. Soon Cuvier became a professor of animal anatomy at the Musée National d'Histoire Naturelle (National Museum of Natural History), established by the Revolutionary government. Not only did Cuvier stay at his post when Napoleon came to power, but Napoleon appointed him to several government positions, including Inspector-General of Public Education. Cuvier held government positions under three French kings thereafter. Cuvier may have been the only public figure to have held French government leadership positions under Revolutionary, Napoleonic, and monarchical regimes.

Cuvier's careful studies of the anatomy of invertebrate and vertebrate animals allowed him to develop the science of comparative anatomy. The function of one organ within an animal could only be understood as it related to the other organs; all organs fit and functioned together perfectly. Because of this, Cuvier had the famous ability to reconstruct organisms from the most fragmentary of fossil remains. When more complete fossils were later found, Cuvier's reconstructions turned out to be amazingly accurate. A corollary of this principle was that no part of an animal could change without destroying its ability to interact with all the other parts. This is one reason that Cuvier did not accept any evolutionary theories, such as that proposed by his fellow French scientist Lamarck (see LAMARCKISM). Another reason is that when Cuvier studied the mummified cats and ibises that had been brought to France from Napoleon's conquest of Egypt, he found that they were no different from modern cats and ibises. From this Cuvier concluded that animals did not undergo evolutionary change.

Cuvier was one of the first to recognize that groups of animals had fundamental structural differences. He classified animal life into four branches or *embranchements:* the vertebrates, the articulates (arthropods and segmented worms), mollusks, and the radiates (cnidarians and echinoderms) (see INVERTEBRATES, EVOLUTION OF). These embranchements correspond roughly to modern evolutionary classifications, although echinoderms have bilaterally symmetrical larvae even if adult starfish have external radial symmetry. Because all organs must work together, the organs of animals in one embranchement would not work with organs of animals in another; therefore Cuvier considered evolutionary transformations among the embranchements to be impossible.

Because he recognized the separate embranchements of animal life, Cuvier opposed the theories of contemporaries such as Buffon (see BUFFON, GEORGES), Lamarck, and Geoffroy, all of whom suggested or championed some form of evolutionary transformation. Cuvier pointed out that vertebrates have a central nerve cord along the back (dorsal) surface of the body, while articulates have a nerve cord along the front (ventral) surface, and that no transformation is possible between them. Geoffroy's response, which was at the time unsupported by evidence, was that vertebrates might have developed as upside-down articulates. Geoffroy's view was not vindicated until the discovery of Hox genes (see DEVELOPMENTAL EVOLUTION). One of these genes, found both in flies and frogs, affected the ventral surface of fly embryos but the dorsal surface of frog embryos. Furthermore, Buffon and Geoffroy claimed that VESTIGIAL CHARACTERISTICS were evolutionary leftovers. Cuvier claimed that all organs were perfectly designed, and a vestigial organ was one whose function had not yet been discovered.

Cuvier was not the first to believe that extinction had occurred, but his studies removed all reasonable doubt. Fossilized mammoths had been found in Italy and the United States. Some scientists claimed that the Italian bones were merely those of elephants that had died during Hannibal's invasion of Rome. American President Thomas Jefferson believed that mammoths were still alive somewhere in the American wilderness; one of the purposes of the Lewis and Clark Expedition was to find them. Cuvier's careful study of living elephants and of mammoth and mastodon bones proved that Indian and African elephants were different species, and that mammoths and mastodons were not elephants. He also studied the skeletons of the Irish elk, a kind of deer with huge antlers, and demonstrated that they were unlike any existing animal. Cuvier explained extinction as having resulted from "revolutions" during Earth history. He avoided the word *catastrophe* because of its supernatural overtones, but his view was similar to that of the supporters of CATASTROPHISM.

Cuvier held strong views about the superiority of some races and regions over others. He proclaimed in 1817, regarding the "Negro race," that "... with its small cranium, flattened nose, its protruding jaw, and its large lips, this race clearly resembles the monkeys." Within two decades anatomist Friedrich Tiedemann had demonstrated the fundamental equality of the races (see EUGENICS; *HOMO SAPIENS*), but by then Cuvier was dead.

Although Cuvier rejected evolution, his approach to the study of anatomy and his proof of extinction were breakthroughs that made evolutionary science possible. He died on May 13, 1832.

Dart, Raymond (1893–1988) South African *Anatomist, Anthropologist* Raymond Dart made some of the most famous early discoveries of fossils of extinct human ancestors.

Born on February 4, 1893, Dart grew up in Queensland, Australia, where his family raised cattle. His excellent studies at the University of Queensland in Brisbane allowed him to begin medical studies in Sydney. He went to England to join a medical corps in World War I. When the war ended, one of the world's leading neuroanatomists, Grafton Elliot Smith, accepted Dart into graduate study at the University of Manchester. In 1922 Dart joined the faculty of the University of Witwatersrand in Johannesburg, South Africa, as a professor of anatomy. Dart displayed brilliance but also challenged many orthodox opinions of his peers. He worked hard to build up the university facilities for the study of anatomy.

Dart's interest in anatomy extended to fossils. In 1924, a fossil baboon skull was found at a limestone quarry at Taung. Dart examined a specimen from this quarry, and spent a month removing the minerals from the specimen. It turned out to be the fossil face and jaw of a young primate, now known as the Taung Child (see AUSTRALOPITHECINES). Dart immediately recognized that the fossil may represent an animal intermediate between apes and humans. He quickly wrote a paper for *Nature* which described this species and named it *Australopithecus africanus* ("Southern ape from Africa"). Scientists in Britain did not believe that this species was anything other than an ape. This was partly because the leading anthropologists of the day (including Grafton Elliot Smith) considered the English PILTDOWN MAN to be the ancestor of humans and therefore concluded that humans evolved in Europe, not Africa. In fact, the Taung specimen, having an apelike skull and teeth that resembled those of modern humans, seemed difficult to reconcile with the human skull and apelike jaw of Piltdown man. Almost the only scholar who supported Dart was the Scottish doctor and paleontologist Robert Broom. When Dart finally traveled to London in 1930 to try to convince anthropologists of the legitimacy of the Taung skull, he was overshadowed by the recent discovery of Peking man (see *HOMO ERECTUS*). Giving up the fossil hunting for many years, Dart concentrated on teaching and administration at the university.

Broom discovered many more australopithecine fossils in South Africa. As more fossils accumulated, many scientists finally accepted australopithecines as possible human ancestors. Encouraged, Dart resumed fossil hunting. At Makapansgat, Dart found fossils which he named *Australopithecus prometheus,* in the erroneous belief that their blackened state indicated the use of fire (Prometheus was the Greek mythological figure who gave humans fire). These specimens are now considered to be *A. africanus.*

Dart is perhaps most famous for his "killer ape" theory. He claimed that, even though they did not make stone tools or weapons, the australopithecines used bones, teeth, and horn as weapons and were bloodthirsty killers. If modern humans are their descendants, then *Homo sapiens* must have inherited these violent tendencies. This idea was popularized by writers such as Robert Ardrey in his 1961 book *African Genesis.* The Dart theory is the inspiration behind the opening scene of the 1968 movie version of Arthur C. Clarke's *2001: A Space Odyssey.* Subsequent fossil discoveries, notably by anthropologist C. K. Brain, showed that the australopithecines were not the hunters but the prey, and that the killer ape theory was wrong.

Even though he erred about the killer apes, Dart made a vital contribution to the study of human evolution: He demonstrated that Darwin had been right in speculating that Africa was the place where humans evolved (see DARWIN, CHARLES; *DESCENT OF MAN* [book]). The genus that Dart named is still considered almost certainly to be the ancestor of modern humankind, even though the species that Dart discovered was most likely a side-branch of human evolution. Dart died November 22, 1988.

Darwin Awards "The Darwin Awards" is a Web site and series of books started in 1993 by Wendy Northcutt, a molecular biology graduate at the University of California in Berkeley. She began collecting stories of human stupidity, then opened a Web site in which people from all over the world can submit similar stories. Although the books and Web site claim to offer examples of human evolution in action, it is highly unlikely that any of the stories actually represent the process of NATURAL SELECTION. They are, however, an interesting and entertaining way to examine the difference between the popular and scientific conceptions of natural selection.

The people in the Darwin Awards stories commit outlandish acts of stupidity that either kill them or prevent them from reproducing. Either way, their genes will not be represented in the next generation. They have removed themselves from the human gene pool in what the author calls "a spectacularly stupid manner," which will contribute to the eventual evolutionary improvement of the entire species. This is the fundamental premise of the Darwin Awards.

Examples include: A woman burned herself to death because she was smoking a cigarette while dousing anthills with gasoline; a man took offense at a rattlesnake sticking its tongue out at him, and when he stuck his tongue out at the snake, the snake bit it; a college student jumped down what he thought was a library laundry chute, but libraries do not have laundry chutes, and the chute led to a trash compactor; a man electrocuted himself trying to electroshock worms out of the ground for fish bait; and a man tried to stop a car that was rolling down a hill by stepping in front of it. These are all stories that Northcutt attempted to verify before publishing them, and they are probably true.

Despite the tremendous value of these stories as entertainment, it is unlikely that they actually represent evolution in action. The principal reason for this is that the death of these individuals does not necessarily represent the removal of what could be called judgment impairment genes from the population. The lack of intelligence demonstrated by some of these people may not arise from any consistent genetic differences between them and the rest of humankind. Even in cases where a specific genetic basis for intelligence has been sought, it has not been found; in the Darwin Awards database, it has not been sought. If the stupidity of the people arose from habits of thought they acquired from upbringing and society, then their deaths will have contributed nothing to natural selection.

These stories may not represent a lack of intelligence so much as excessive impulsiveness. The proper order for intelligent action is (1) think, (2) act; to reverse these two steps is not necessarily as stupid as it is impulsive. The question then becomes, are there conditions in which impulsiveness confers a possible evolutionary benefit? He who hesitates long enough to think will sometimes live longer, but he will also sometimes miss an opportunity to acquire status, resources, or reproductive opportunities. Modern humans are sometimes intelligent, sometimes impulsive, and modern humans have inherited both of these behavioral patterns from prehistoric ancestors (see INTELLIGENCE, EVOLUTION OF; SOCIOBIOLOGY).

Most scientists believe that human genetic evolution has not recently progressed in any particular direction (although random genetic changes have continued to occur). Rather than allowing natural selection to eliminate inferior versions of genes, humans have invented technological responses to everything from the need for more food, to responding to the weather, to the control of diseases. Most scientists who study human evolution will insist that no measurable increase in brain size has occurred in the last 100,000 years, and no increase in brain quality in about 50,000 years.

Further Reading

Northcutt, Wendy. *The Darwin Awards: Evolution in Action.* New York: Dutton, 2000.

———. *The Darwin Awards II: Unnatural Selection.* New York: Dutton, 2001.

———. *The Darwin Awards III: Survival of the Fittest.* New York: Dutton, 2003.

———. "The Darwin Awards." Available online. URL: http://www.DarwinAwards.com. Accessed March 23, 2005.

Darwin, Charles (1809–1882) British *Evolutionary scientist* Charles Darwin (see figure on page 109) was the scientist whose perseverance and wisdom changed the way professional scientists, amateurs, and almost everyone else viewed the world. He presented a convincing case that all life-forms had evolved from a common ancestor, and for a mechanism, NATURAL SELECTION, by which EVOLUTION occurs. Earlier scholars had proposed evolutionary theories (see BUFFON, GEORGES; CHAMBERS, ROBERT; DARWIN, ERASMUS; LAMARCKISM), but, until Charles Darwin, no proposals of evolution were credible.

Charles Robert Darwin was born February 12, 1809, in Shrewsbury, England, the same day that Abraham Lincoln was born in the state of Kentucky. Both men, in different ways, led the human mind to consider new possibilities and freedoms without which future progress would not have been possible. Darwin, like many other aristocratic Englishmen, was born into a rich and inbred family. His father, Robert Darwin, was a prominent physician, and his mother, Susannah Wedgwood Darwin, was an heir of the Wedgwood pottery fortune. Robert Darwin and Susannah Wedgwood were, in addition, cousins. Unlike other rich inbred Englishmen, Charles Darwin was to put his inherited wealth to better use than perhaps anyone ever has.

Darwin had three older sisters, an older brother, and a younger sister. Scholars do not know what kind of impact the death of Darwin's mother had upon him. Her death was painful and prolonged, and Darwin was only eight years old. Whether this event might have inclined him to think more of evolutionary contingency, rather than divine benevolence, as the ruling order of the world, scholars can only speculate.

The Darwin family was respectably but not passionately religious. Robert Darwin was secretly a freethinker but remained associated with the Anglican Church. Susan-

Charles Darwin at his home, circa 1880 ***(Courtesy of HIP/Art Resource, New York)***

nah Darwin continued the Wedgwood family tradition of Unitarianism. The family had a history of creative thinking that went beyond the bounds of common opinion. Robert Darwin's father, Erasmus, also a physician, doubled as a poet and wrote books in which he openly speculated on evolution—even to the extent of suggesting that life began in a warm little pond from nonliving chemicals. Darwin grew up, however, with a negative view of evolution. The only major evolutionary theory that had been considered by scientists was Lamarckism, which to the British elite was associated with the chaos of the French Revolution. While young Darwin was not repressed into religious narrowness, he explored the path of evolutionary theory hesitantly and on his own.

There was plenty of opportunity for Darwin to develop scientific interests. His family kept a dovecote, and Darwin collected insects and rocks and performed chemical experiments (earning the nickname "Gas" among his friends). Darwin got none of this important training from the schools he attended. He learned adequately in school, but nothing except direct contact with the world of nature inspired him. Being a naturalist, however, was not a career. When it came time for him to go to college, he followed his older brother to the University of Edinburgh to study medicine.

Darwin stayed in Edinburgh about a year. First, he could not keep his mind on the study of medicine. He spent more time attending lectures in geology and zoology, and he was more interested in the reproduction of marine invertebrates than in surgery. During his brief stay, he was exposed to some of the radical anticlerical views of fellow students and even of one professor, the zoologist Robert Grant, who openly defended Lamarckian evolution. Edinburgh University, being free from Anglican theology, was a haven for scientific and political debate. While Darwin experienced the bracing atmosphere of free inquiry, he also saw what could happen if one went too far. One student presented a paper at a local scientific society, claiming that the human mind consisted only of the brain, without any soul; his remarks were stricken from the record. During his year and a half in Edinburgh, working with Grant, Darwin learned how to conduct scientific research, and he presented his first paper there, in 1827, on a discovery he had made about the life cycle of the cnidarian *Flustra*. Second, Darwin could not stand the sight of blood. When he had to witness operations administered without anesthesia, he decided to abandon plans for a medical career.

Darwin chose to attend Cambridge and study theology. He planned to become a country parson. At the time, when the Church of England was securely supported by the government, a country parson needed only to perform his round of duties, then he would have plenty of time for studying birds and flowers and rocks. Some of the best natural history studies, such as the *Natural History of Selborne* written by parson-naturalist Gilbert White, were the work of country ministers. Even at Cambridge, Darwin spent far more time attending lectures in science classes in which he was not actually enrolled than in studying theology. He befriended the botany professor, John Stevens Henslow, and frequently went on field trips with the botany students. He was particularly passionate about collecting beetles. Robert Darwin was openly skeptical about the success prospects of his son, who, he said, cared "for nothing but shooting, dogs, and rat-catching," adding, "You will be a disgrace to yourself and all of your family." At the last minute, before his comprehensive examinations, Darwin studied enough to earn a respectable grade in theology.

There was one aspect of his theological studies that Charles was eager to pursue. He was required to read Reverend William Paley's book (see NATURAL THEOLOGY) which presented the Design Argument. This argument maintained that the evidences of geology and biology strongly indicated that there was a Creator. At the time, the only alternative to Natural Theology was some unspecified and mysterious process of self-formation, of which most scientists, and of course Darwin, were understandably skeptical. Little did Darwin, who nearly memorized the book, suspect that he would be the one to present a credible, non-miraculous mechanism by which nature could design itself.

For several years, Darwin had shown an interest in the daughter of a family that was close to the Darwins: Fanny Owen. His future seemed clear to him: They would marry, he would assume a parsonage, and they would settle down to live in the country, where he could collect beetles forever. It would not pay much, but Charles was in line to inherit at least part of a small fortune. Then, at the last minute, an unexpected opportunity arose. The captain of the HMS *Beagle* (see FITZROY, ROBERT) needed a companion for a voyage around the world. It was a scientific expedition, dedicated mainly to surveying the major ports of the world (into which the British assumed they had unlimited access), but also to every other aspect of scientific exploration. The ship already had a naturalist, the ship's surgeon, but the captain wanted a fellow aristocrat with whom to share meals and conversation. FitzRoy contacted Henslow, who told him the perfect person for the job would be Charles Darwin. It was hardly a job—Darwin had to pay his own way—but it was an opportunity to see the wonders of creation around the world in a way that, as it turned out, Darwin would never have again. FitzRoy interviewed Darwin and offered him the post. This voyage changed Darwin and changed the world. FitzRoy chose Darwin despite misgivings about the shape of Darwin's nose. It is possible that never has such an important event in history been determined by something so trivial. Robert Darwin at first refused to pay for his son's trip, but his brother-in-law Josiah Wedgwood convinced Robert to support Charles. When Charles accepted the appointment, it came at considerable personal cost: while he was on the voyage, he found that Fanny was not willing to wait for him to return, and she married another man.

During the 1831–36 voyage of HMS *Beagle,* Darwin was constantly putting his knowledge of natural history to work and almost never used his college degree in theology. The *Beagle* went down the east coast of South America, to the west coast, then across to Australia, the Indian Ocean, around Africa, and back to England. There were long periods in which the ship remained in harbor for surveying work, and during which Darwin was able to go ashore and travel far inland. This was fortunate for him, for he suffered constantly from seasickness. Darwin studied the geology of islands and coastlines; he collected fossils, as well as plants and animals. He took careful notes on everything and periodically sent his specimens back to England. He visited tropical rain forests and the heights of the Andes Mountains. The rain forest was, for him, a particularly religious experience; one could not look at the beauties of such a forest, he said, without believing that there was more to man than merely the breath of his body.

The voyage had more than spiritual inspiration for him. He read the first volume of geologist Sir Charles Lyell's famous and controversial *Principles of Geology* (see LYELL, CHARLES; UNIFORMITARIANISM). While on the voyage he received the other volumes and read them as well. He was convinced that Lyell was correct about the long ages of Earth history and their gradual changes. And many of the things that Darwin saw made him question some aspects of his belief in creation. He collected fossils and found that animal fossils most nearly resembled the animals that were currently alive in the region. For example, South America is where llamas, capybaras, sloths, and armadillos now live; this is also where what were then believed to be the fossils of extinct species of llamas, capybaras, sloths, and armadillos are found. Australia was the place to find most of the marsupials, and most of the marsupial fossils, of the world. It suggested strongly to him that the fossils represented ancestors of the animals now present. And when he saw the animals of the GALÁPAGOS ISLANDS, he began to wonder if at least some species had come into existence on the islands where they are now found (see DARWIN'S FINCHES).

Darwin also had experiences with people on this voyage that would influence his later thought. In particular, one of the objectives of the *Beagle* voyage was to return three Fuegian Native Americans to their tribes as missionaries. Two men and one woman had been captured from a Fuegian tribe and taught Christianity and English life. Now they were going back to enlighten their tribe. Instead, they converted back to their tribal ways. This made Captain FitzRoy upset but made Darwin think about why the people whom Europeans called "primitive" had levels of intelligence nearly, or completely, equal to theirs. He also saw slaves in Brazil (he did not visit the United States) and developed a deep hatred of slavery and of violent and brutal European colonialism. In particular, he was bothered by the way slave owners used biblical passages to justify slavery. Perhaps these experiences laid the groundwork for his later rejection of Christianity.

Darwin returned to England, as he had left, still believing in creationism, but his mind was prepared for rapid changes of viewpoint. Upon his return, he found that he had become a little famous among the scientific community, because of his correspondence and his specimens. Lyell, instead of just being a hero who wrote a book, became a friend. Darwin spent a couple of years in London cataloging his specimens and writing his travel notes, which became a popular book, *The Voyage of HMS* Beagle.

While in London, living with his brother Erasmus, Charles Darwin began to think about marriage. Since it appeared that he would inherit much of a sizable fortune, there was no need for him to marry, or to pursue any career. Now that Darwin was moderately famous, his father no longer considered him a disgrace and was willing to support him in his scientific studies. Darwin's life was to be characterized by taking a scientific approach to everything, and marriage was no exception. He listed his reasons for finding a wife (such as companionship), and his reasons for remaining single (such as solitude), and eventually reached a "Q.E.D.—Marry, marry, marry!" He was attracted to Emma Wedgwood, who also happened to be his cousin, and she liked him as well, and they were married. They had six sons and four daughters. One son later became a famous botanist, another a famous astronomer, and another a member of Parliament.

Emma was very firmly religious. Even before marriage Darwin had already begun to entertain doubts about the religious view of the world. In later years, as Darwin developed his evolutionary theories, he had an ever more difficult time reconciling them with Christianity, even though some of his friends (especially the Harvard botanist Asa Gray; see GRAY, ASA) managed to retain conventional Christian faith while embracing evolutionary science. By the end of his life Charles Darwin was an agnostic. Historians have considered that the death of Darwin's beloved daughter Annie at age 10 may have been the precipitating factor in making him abandon belief in, at least, a providential God. All this time, Darwin was in great pain that his loss of faith was distressing to Emma, which indeed it was. When they wrote notes to one another about it, she remained his loyal partner even though she feared he was losing his soul, and he responded, "You do not know how many times I have wept over this." Darwin's religious agnosticism, therefore, was not easy or arrogant but came at the expense of the pain that he caused to his dearest friend.

Charles and Emma bought a country house in Down, just far enough from London to be able to visit it but not be bothered by it. As it turned out, it would not have mattered how far away from London he lived, because Darwin became ill at this time and remained so for the rest of his life and hardly ever traveled. He suffered recurring symptoms of headaches, heart flutters, indigestion, boils, muscle spasms, and eczema, on which he kept scientific notes but for which neither he nor any physician had a solution. Several possibilities have been proposed for the cause of Darwin's illness. Some scholars have suggested that he had contracted Chagas' disease, caused by a parasite from a bug bite, while in South America. This is unlikely, because Darwin was developing some of his symptoms shortly before the voyage began. Other scholars have suggested that his illness was genetic in origin, resulting from the inbreeding of the Darwin and Wedgwood families. Emma's brother Josiah and his wife, who were also first cousins, lost a sickly baby. It has also been suggested that recessive mutations from inbreeding caused the death of Darwin's daughter Annie.

Emma was the perfect companion and nurse. Her care and friendship enabled Darwin to continue his research and writing despite his illness, and despite the criticisms he received throughout his life. She would play the piano for him, and they would read together in the evenings. The conclusion seems inescapable that Charles Darwin could not have accomplished what he did had it not been for Emma.

Darwin's illness kept him home, and his wealth meant that he did not have to work. He was, accordingly, able to devote full time to the study of science—that is, the few hours of the day in which he was not suffering the symptoms of his malady. He read every scientific book and article he could find and corresponded widely with fellow scientists all over the world. The solitude, especially during his strolls on the Sandwalk around his property, allowed him to think. He put the time to good use, not only proposing the modern theory of evolution but also almost single-handedly inventing the sciences of ecology, pollination biology, and plant physiology. Right to the very end of his life, he noticed things that hurried and shallow people did not: His last book was about the slow but cumulative effects of earthworms on transforming the landscape and creating the soil.

It was not long after the Darwins moved to Down House that Charles had to deal with a crisis of thought. By 1837 he had worked out the theory of natural selection. He thought of natural selection, first, because he was very familiar with individual differences within populations, whether of pigeons or of people; second, because he realized that populations could grow rapidly. He got this latter idea from reading Malthus's *Essay on Populations* (see MALTHUS, THOMAS). Malthus was a clergyman who used his essay to argue against the doctrine of Divine Providence: Because populations grow by doubling but resources increase, if at all, only in a linear fashion, then people were condemned into an eternal cycle of overpopulation and famine. Darwin, his Christian faith already slipping, undoubtedly agreed; but he also realized that the whole world of plants and animals was similarly condemned to eternal overpopulation. Out of this crisis came an insight that changed the world. Darwin realized that overpopulation could be a creative force as well as a destructive one, as the superior individuals in a population would be the ones to survive, and this would cause the population to evolve. One of the consequences of Malthus's doctrine was the political opinion that any attempts to help the poor, through public works and welfare, would only cause their populations to grow larger and to increase the total sum of their misery. Throughout Darwin's adult life, there were riots and political movements in which the poor and their populist leaders threatened the entrenched power of lords, priests, and the leaders of industry. A new set of Poor Laws in England specifically invoked Malthus as the justification for forcing the poor to leave England, work for lower wages, or to be content with poverty. The political theory of the Malthusians was that ruthless competition among companies guaranteed the progress of England and of humankind. It was exactly the political viewpoint that philosopher Herbert Spencer would later defend, inventing the phrase "survival of the fittest" (see SPENCER, HERBERT). Darwin feared that by incorporating Malthus into his theory of natural selection, he might be seen to side with those who oppressed the poor, an association he did not desire. Even today, Darwinian natural selection is often associated with ruthless competition. Novelist Kurt Vonnegut, for example, said that Darwin "taught that those who die are meant to die, that corpses are improvements." During the 20th century, evolutionary scientists proposed ways in which natural selection could lead to cooperation rather than just to competition (see ALTRUISM).

Darwin began to keep a notebook on the subject of natural selection, but told almost nobody about it for many years. The book *Vestiges of Creation* had demonstrated what would happen if someone came up with a half-baked theory of evolution and rushed it into print, and Darwin did not want his theory to suffer such deserved ridicule. He was afraid to admit that he accepted evolution (called at that time *transmutation*) partly because it was the reigning doctrine of political revolutionaries. He told his friend, the botanist Sir Joseph Hooker, that admitting that he believed in transmutation was "like confessing a murder." Darwin wanted to make sure that he had all of the scientific facts in order, to certify every aspect of his theory, before he published it. This was a tall order. Darwin had to demonstrate that populations of plants and animals did have a tremendous amount of variability, that plants and animals did experience overpopulation and other natural stresses, that selection can produce significant changes in populations, and that an evolutionary explanation fit the facts of geology and of the distribution patterns of plants and animals. He gathered information for 21 years, showing that pigeon breeders could use ARTIFICIAL SELECTION to produce astonishing new varieties starting only with the variation present in wild pigeon populations, that plants and animals did suffer the same population stresses that humans do, and that an evolutionary pattern would explain the living and extinct species of barnacles. Darwin spent eight years studying every kind of barnacle in the world, living or extinct. It became quite a burden to him; no man, he said, hated a barnacle quite as much as he did. His son George grew up seeing him study barnacles, and he assumed that is what grownups did. George asked one of his friends, "Where does your father study *his* barnacles?" Darwin's evidence piled up, nicely matching his theory of natural selection; but after 21 years, he had hardly begun, and he might have worked until he died without completing the "big book" in which he planned to explain everything.

In 1858 Darwin received a letter and manuscript from a young English naturalist in Indonesia (see WALLACE, ALFRED RUSSEL). Wallace had figured out all the same aspects of natural selection that Darwin had, only he was ready to rush into print with it. Fortunately, Darwin had written a brief summary of his theory in 1842 and had shown it to Lyell, Hooker, and Emma. Lyell could therefore vouch that Darwin had thought of natural selection first. Natural selection was presented as a joint theory of Darwin and Wallace in what he considered to be only a brief abstract of his ideas, which became a 400-page book that changed the world (see *ORIGIN OF SPECIES* [book]). A summary of this book is included as an Appendix to this Encyclopedia. The editor of the *Quarterly Review* read an advance copy of the book. He thought it was good but that nobody would be interested in it. He suggested that Darwin write a book about pigeons instead; he pointed out that everybody likes pigeons.

The first printing, 1,250 copies, of *Origin of Species* sold out the first day. Perhaps the buyers were expecting to find something about human evolution in this book, but Darwin stayed away from that subject, saying only that "Light will be thrown on the origin of man and his history," a statement that remained unchanged in later editions. In the book, Darwin argued convincingly that all life had a common evolutionary origin and history, and that evolution was caused by natural selection. When Darwin's friend Huxley (see HUXLEY, THOMAS HENRY) read it, he is reported to have said, "How stupid not to have thought of this before!" It was primarily to the fact of evolution rather than to the mechanism of natural selection that Huxley referred. Natural selection remained a theory seldom embraced, until the MODERN SYNTHESIS of the 20th century. The codiscoverer of natural selection, Alfred Russel Wallace, defended both evolution and natural selection but argued from it (quite differently from Darwin) that the human mind was a product of miraculous creation. Darwinian evolution (soon called Darwinism) survived numerous attacks, for example from biologist St. George Jackson Mivart. Some of the criticisms, such as those of engineer Fleeming Jenkin, stemmed not so much from problems with evolution as from the fact that biologists did not as yet understand how genetic traits were inherited (see MENDELIAN GENETICS). One of his fiercest critics was a man who had once been his friend, the anatomist Sir Richard Owen (see OWEN, RICHARD), who accepted some aspects of transmutation but vehemently rejected Darwin's nontheistic mechanism of it. After this brief flurry of controversy, within a decade most scientists had accepted Darwin's proposal of the fact of evolution. (CREATIONISM is primarily a product of the 20th century.)

Darwin, being ill, did not defend his views in public. He left this primarily to his "bulldog" friend Huxley. Meanwhile in America, it was the botanist (and traditional Christian) Asa Gray who publicly defended Darwinian evolution.

The world did not have to wait long for Darwin to make very clear statements about human evolution. In 1871 he published a book (see *DESCENT OF MAN* [book]), stating that humans evolved from African apes. Since the size of the human brain, and the diversity of human racial characteristics, seem to be almost outlandish in proportion to what survival would require and natural selection would produce, Darwin proposed another mechanism of evolution in this same book: SEXUAL SELECTION, by which characteristics evolved in one sex because they were pleasing to the other sex, allowing the individuals that had these characteristics to produce more offspring. In case anyone thought Darwin meant that only the human body had evolved, and not the human spirit, they could read another of his books, *The Expression of Emotions in Men and Animals*.

Although Charles Darwin remains a popular target for antievolutionists, and for evolutionists who have their own new theories of evolution, the carefulness of Darwin's work and breadth of his imagination have assured that his theory has survived far longer than he might have imagined. His personal integrity and humility have also made him one of the most admired scientific figures in history.

Charles Darwin died on April 19, 1882, while Emma was cradling his head in her arms. Darwin's friends (primarily Huxley) successfully arranged to have him buried in Westminster Abbey, right next to Sir Isaac Newton.

Further Reading

Adler, Jerry. "Charles Darwin: Evolution of a Scientist." *Newsweek*, November 28, 2005, 50–58.

American Museum of Natural History. "Darwin Digital Library of Evolution." Available online. URL: http://darwinlibrary.amnh.org. Accessed July 8, 2006.

Armstrong, Patrick. *Darwin's Other Islands*. London: Continuum, 2004.

Browne, Janet. *Charles Darwin: Voyaging*. Princeton, N.J.: Princeton University Press, 1995.

———. *Charles Darwin: The Power of Place*. New York: Knopf, 2002.

Darwin, Charles. *Autobiography of Charles Darwin 1809–1882*. Edited by Nora Barlow. New York: Norton, 1993.

Desmond, Adrian, and James Moore. *Darwin: The Life of a Tormented Evolutionist*. New York: Warner Books, 1991.

Eiseley, Loren. *Darwin's Century: Evolution and the Men Who Discovered It.* New York: Doubleday, 1961.

Eldredge, Niles. *Darwin: Discovering the Tree of Life.* New York: Norton, 2005.

Healey, Edna. *Emma Darwin: The Inspirational Wife of a Genius.* London: Headline, 2001.

Keynes, Richard Darwin. *Fossils, Finches, and Fuegians: Darwin's Adventures and Discoveries on the Beagle.* New York: Oxford University Press, 2003.

Lamoureux, Denis O. "Theological insights from Charles Darwin." *Perspectives on Science and Christian Faith* 56 (2004): 2–12.

Leff, David. "AboutDarwin.com: Dedicated to the Life and Times of Charles Darwin." Available online. URL: http://www.aboutdarwin.com/index.html. Accessed May 3, 2005.

Moore, James. *The Darwin Legend.* Grand Rapids, Mich.: Baker Book House, 1994.

Quammen, David. *The Reluctant Mr. Darwin: An Intimate Portrait of Charles Darwin and the Making of His Theory of Evolution.* New York: Norton, 2006.

Van Wyhe, John, ed. "The writings of Charles Darwin on the Web." Available online. URL: http://pages.britishlibrary.net/charles.darwin. Accessed March 23, 2005.

Darwin, Erasmus (1731–1802) British *Physician, Naturalist* Born December 12, 1731, Erasmus Darwin was a prominent 18th-century intellectual. As a physician with wealthy patrons, he became wealthy, which allowed him the means to pursue a wide array of studies. He investigated many aspects of natural history, including botany. He also wrote works of philosophy and poetry.

In 1794 Erasmus Darwin published *Zoonomia, or, The Laws of Organic Life.* This was one of the first formal proposals of evolution, almost a decade earlier than the French biologist Lamarck (see LAMARCKISM). Erasmus Darwin also expressed his evolutionary ideas in the posthumously published poem *The Temple of Nature,* in which he wrote about the spontaneous ORIGIN OF LIFE in the oceans, and its development from microscopic forms into plants and animals:

Nurs'd by warm sun-beams in primeval caves
Organic life began beneath the waves;
First forms minute, unseen by spheric glass,
Move on the mud, or pierce the watery mass;
These, as successive generations bloom,
New powers acquire and larger limbs assume;
Whence countless groups of vegetation spring,
And breathing realms of fin and feet and wing ...
Hence without parent by spontaneous birth
Rise the first specks of animated earth.

Erasmus Darwin struggled with the question of how one species could evolve into another. His explanation was similar to that of Lamarck, but Erasmus Darwin also wrote about what we now call competition and SEXUAL SELECTION: "The final course of this contest among males seems to be, that the strongest and most active animal should propogate the species which should thus be improved." To formulate his theories, Erasmus Darwin drew upon his extensive observations of domesticated animals, the behavior of wildlife, and his knowledge of paleontology, biogeography, classification, embryology, and anatomy.

Erasmus Darwin died April 18, 1802, before his grandson Charles Darwin was born (see DARWIN, CHARLES). Erasmus Darwin's contribution to Charles Darwin's intellectual accomplishments may have been crucial. Charles Darwin, like his grandfather, integrated knowledge from many fields to reach his conclusions. Erasmus Darwin's fearlessness in announcing an evolutionary theory may have been the stimulus that made the development of his grandson's theory possible. Erasmus Darwin's fortune contributed significantly to allowing Charles Darwin to devote all of his time to scientific research.

Further Reading

Browne, Janet. "Botany for gentlemen: Erasmus Darwin and *The Loves of the Plants.*" *Isis* 80 (1989): 593–621.

Desmond, Adrian, and James Moore. "Catch a falling Christian." Chap. 1 in *Darwin: The Life of a Tormented Evolutionist.* New York: Warner Books, 1991.

Darwin's finches Darwin's finches are a group of 14 species of finches that inhabit the GALÁPAGOS ISLANDS off the coast of Ecuador. The finches, and the islands, are famous because Charles Darwin (see DARWIN, CHARLES) visited them in September 1835 during an around-the-world trip on HMS *Beagle.* Because evolution is occurring rapidly in these finches, they represent an almost unique opportunity for studying evolution in action. Because the species are still differentiating, some experts classify the finches into a slightly different set of species than do other experts.

As had every previous visitor, Darwin noticed the giant tortoises that are unique to the islands. He also observed that the islands seemed to have similar geography and climate yet were inhabited by slightly different groups of animals. He collected specimens of many of the small birds he found. Because their beaks were all so different from each other, Darwin did not recognize that the birds were all closely related finches. When Darwin returned to London, ornithologist John Gould identified them as different species of finches. Darwin realized that each island might have had its own species of finch, but he had not taken careful notes regarding from which island he had obtained which finch specimen. Fortunately, other visitors to the islands, including the captain of the *Beagle* (see FITZROY, ROBERT), had kept more careful notes. When Darwin published the account of his voyage on HMS *Beagle,* he had not yet begun to fully develop an evolutionary theory. A famous passage indicates that he had already begun thinking about evolution: "Seeing this gradation and diversity of structure in one small, intimately related group of birds, one might really fancy that from an original paucity of birds in this archipelago, one species had been taken and modified for different ends." These finches, then, could be considered the spark that started the fire of evolutionary theory.

About three million years ago, a group of finches, probably genus *Tiaria,* migrated to the Galápagos Islands from the mainland. From this original founding group, the 14 species of finches present today evolved (see table). Most of the finch species live on more than one of the islands, but none live on all of the islands. No island in the Galápagos archipelago has its own unique species of finch. Some species live on or near the ground and eat seeds; others live on or in cactuses and eat seeds, pollen, and nectar; the others live in trees. The tree finches eat fruits, leaves, seeds, and insects; one of them strips the outer bark and eats the rich inner bark. One group of finches, though primarily consuming insects, occasionally drinks the blood or eats the eggs of seabirds such as blue-footed boobies. Some finches occasionally perch on iguanas and eat the parasites from them. The finches have undergone ADAPTIVE RADIATION in two ways: first, by evolving on different islands; second, by specializing on different food resources.

One of the earliest extensive studies of these finches, which brought them to the attention of the world as an excellent example of evolution in action, was the 1947 study by David Lack. He visited for only four months, not long enough to get extensive data regarding the ongoing processes of evolution in these finch populations. He built an aviary, in which he was able to confirm that the different kinds of finches did not readily interbreed, therefore they were different species (see SPECIATION). He showed that evolution had occurred, but his evidence did not show that evolution was still occurring.

In order to obtain convincing evidence about NATURAL SELECTION, it is important to closely monitor entire isolated populations of organisms. If this is not done, the researcher cannot determine whether the disappearance of individuals is due to natural selection (the individuals die or fail to reproduce) or due to migration (the individuals leave). This was one of the problems with what had once been considered a classic study of natural selection, zoologist H. B. D. Kettlewell's studies on the PEPPERED MOTHS. Darwin's finches provided a perhaps unique opportunity to study natural selection and other evolutionary processes, since they are isolated on different islands, to and from which they seldom migrate. A prodigious amount of work is required to keep track of each individual finch, how many offspring it produces, how it lives, and its physical characteristics, even with relatively small populations on small islands. But that is precisely what Peter and Rosemary Grant, evolutionary biologists at Princeton University, and their associates, have been doing on some of the islands since the 1970s. Having monitored the lives of more than 20,000 individual finches, they have been able to present convincing evidence of natural selection and other evolutionary processes at work. They can recognize the individual finches, and know the genealogies of each, the kinds and sizes of seeds that these individuals eat, and their individual songs, as well as the kinds of seeds available on the islands and in what amounts.

The Grants studied the beak sizes of *Geospiza fortis* ground finches, and the relationship between beak size and the kinds of seeds that these finches ate. The ground finches eat seeds. Some plants (for example, grasses) produce small seeds that are relatively easy to crack. These plants grow abundantly during rainy seasons. During dry seasons, large seeds (such as those of the puncture vine) are the most abundant. The large seeds also have very hard coats. Finches with larger beaks can crack larger seeds. The finches with smaller beaks can more easily handle the smaller seeds. A difference of less than a millimeter in beak size can make a tremendous difference on the food resources available to a finch. A finch with a 0.44 inch (11 mm) beak can crack open a puncture vine seed, while an individual with a 0.42 inch (10.5 mm) beak may not be able to do so. The individuals with smaller beaks may starve. Or they may eat seeds from the *Chamaesyce* bush, which produces a sticky, milky sap. The sap gets on the heads of the birds. The birds use their heads to push gravel around as they search for seeds. When they do this, the birds with sticky heads rub the feathers off of the tops of their heads, leaving the skin exposed, which can lead to overheating of the brain, followed by death. As Darwin wrote, "the smallest grain in the balance, in the long run, must tell on which death shall fall, and which shall survive."

The Grants have gathered data about natural selection, CHARACTER DISPLACEMENT, and incipient speciation among Darwin's finches. The finches also provide evidence about the limits to adaptive radiation.

Natural Selection in Finch Populations

In order to show that natural selection is occurring, the Grants had to establish that the beak size differences among the individual *Geospiza fortis* birds are genetically based. To do this, the Grants switched eggs among nests and found that individual birds had beak sizes similar to those of their biological, not their adoptive, parents (see figure on page 115).

Next, the Grants had to measure changes that occurred in genetically based characteristics such as beak size. They measured beak sizes of *Geospiza fortis* during a long drought when vegetation was sparse, and most of the seeds came from

Darwin's Finches

Ground finches	
Small ground finch	*Geospiza fuliginosa*
Medium ground finch	*Geospiza fortis*
Large ground finch	*Geospiza magnirostris*
Cactus finch	*Geospiza scandens*
Large cactus finch	*Geospiza conirostris*
Sharp-beaked ground finch	*Geospiza difficilis*
Tree finches	
Small tree finch	*Camarhynchus parvulus*
Medium tree finch	*Camarhynchus pauper*
Large tree finch	*Camarhynchus psittacula*
Woodpecker finch	*Camarhynchus pallidus*
Vegetarian finch	*Platyspiza crassirostris*
Gray warbler finch	*Certhidea fusca*
Olive warbler finch	*Certhidea olivacea*
Cocos finch	*Pinaroloxias inornata*

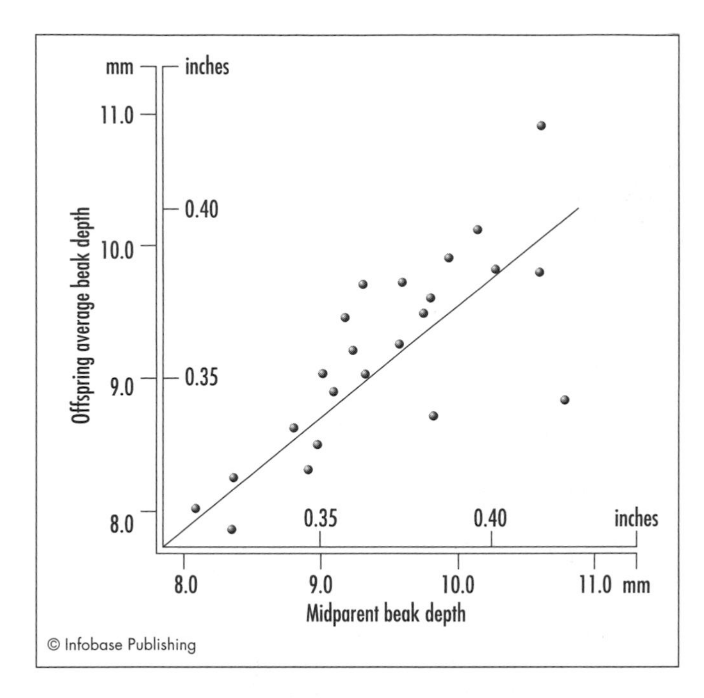

Beak depth (from the top of the beak to the bottom) is one of the measurements of beak size in Darwin's finches *(Geospiza fortis)* used by Peter and Rosemary Grant to study natural selection. Across a wide range of beak depths, from about 0.30 to 0.45 inch (8 to 11 mm), the offspring beak sizes were correlated with the parental beak sizes in 1976. The horizontal axis is the average of the parental beak sizes; the vertical axis is the average of the offspring beak sizes. The slanted line is the statistical regression line. Results for 1978 were similar.

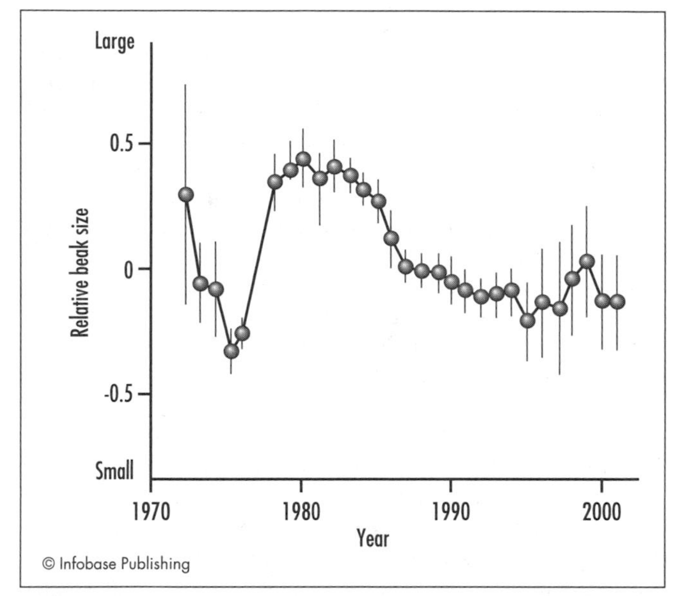

Beak sizes in Darwin's finches *(Geospiza fortis)* on Isla Daphne Major from 1972 to 2001. In response to the severe drought that began in 1977, the relative beak size in the population increased dramatically. This occurred because larger birds, with larger beaks, could eat the large, hard seeds of plants that survived the drought. After the drought ended, starting about 1982, the average beak size decreased to its previous levels. Natural selection caused beak sizes to fluctuate as climatic conditions changed. The vertical axis is a mathematical combination of various measurements of beak size. Small vertical lines indicate error ranges of the data.

the puncture vine. Then, during an El Niño Southern Oscillation event, a tremendous amount of rain fell, and the island was overgrown with plants that produced smaller seeds. Almost immediately, the average sizes of the birds, and the sizes of their beaks, began to decrease (see figure on page 115). When the drought returned, and the plants that produced small seeds became progressively less abundant, the average body and beak size began to increase in the *G. fortis* population. This is an example of natural selection in action, first one direction, then the other.

Character Displacement between Finch Species

Not only does natural selection occur in response to weather patterns but also in response to contact between species of finches. On the island of Genovesa, the large cactus finch *Geospiza conirostris* and the large ground finch *G. magnirostris* (which has a larger average beak size) live together. During rainy seasons, when there are plenty of seeds, both species exhibit a wide range of beak sizes, and there is overlap between the beak sizes of the largest *G. conirostris* and the smallest *G. magnirostris.* The finches in the zone of overlap would compete, but they do not because there are plenty of seeds for both. During droughts, when seeds become scarce, the two species specialize: Natural selection favors the *G. magnirostris* with the largest beaks but favors the *G. conirostris* with the smallest beaks. This is an example of character displacement between two species that minimizes competition between them. Character displacement is expected to occur only when two species are in contact, and this is precisely what we see with these finches. On the island of Española, the large ground finch *G. magnirostris* is absent. On this island, without competition from the large ground finch, the cactus finch *G. conirostris* has a much greater range of beak sizes regardless of the weather conditions. Competition between species has not only caused character displacement but may have caused some local extinctions as well. Generally speaking, finches that are very similar to one another in size and the type of food they eat are not found on the same island. An even more striking example of character displacement, induced in *G. fortis* by *G. magnirostris,* was announced in late 2006 (see CHARACTER DISPLACEMENT).

Incipient Speciation

The Grants have also observed some possible examples of speciation beginning to occur. During a dry year, they identified two different groups of cactus finch *Geospiza conirostris:* One group had shorter beaks than the other group. The differences were only slight: The two groups differed by only 15 percent. This was more than twice as much difference in beak size as there was among individuals within each group. The birds with the longer beaks extracted seeds from cactus fruits, while those with shorter beaks frequently tore open cactus pads and ate the pulp. Moreover, the two groups tended to breed separately; each group had its own type of mating song. This had the elements—genetic variation and reproductive isolation—needed to begin the formation of separate species. When the rains returned, the two groups began to blend back together.

Computer simulation can retrace the pathway of evolution for the seed-eating finches of the Galápagos. A program designed by evolutionary biologist Dolph Schluter begins with the ranges of beak sizes within each species, the range of seed sizes available on each island, and the presence or absence of competition, and calculates the expected ranges of beak sizes on each of the islands. The result of Schluter's analysis matches the real pattern of beak sizes very closely.

Limits to Adaptive Radiation

When the finches first arrived on the Galápagos Islands, there were no other similar kinds of birds with which to compete. The islands represented a new world open to them for adaptive radiation. They were able to exploit new ways of making a living, without having to compete with other birds species that already possessed superior adaptations. The woodpecker finch *(Camarhynchus pallidus),* for example, would not have been able to evolve on an island that already had woodpeckers. Some of the cactus finches eat nectar from the flowers of *Waltheria ovata,* and smaller finches are able to obtain and use nectar more efficiently. (Most birds that consume nectar, such as hummingbirds, are small.) On islands that do not have native bees, some of the finches specialize upon nectar, although not exclusively: Nectar comprises 20 percent of their diet, and they are smaller than the finches that do not eat nectar. On islands that already have populations of the native bee *Xylocopa darwini,* nectar makes up only 5 percent of the finch diet, and the finches do not differ in size from those that do not eat nectar.

HYBRIDIZATION is rare between species of Galápagos finches, but when it does occur, the Grants often make note of it. In more than one case, hybrids have proven fertile for several more generations, during wet seasons; but under dry seasons, when the successful individuals are the ones that specialize, the hybrids (with intermediate beak sizes) have inferior fitness. It appears that it is the dry seasons in which natural selection enforces postzygotic reproductive isolation (see ISOLATING MECHANISMS) and keeps the species apart.

In the Pacific Ocean, the island of Cocos is the home of one of the same species of finches that is found in the Galápagos: the Cocos Island finch, *Pinaroloxias inornata.* Unlike the Galápagos Islands, the island of Cocos is not large enough to promote reproductive isolation among populations of the finches. Speciation is not occurring in these finches. Rather than specializing on food resources, the Cocos finches have remained generalists, exploiting many different food resources, and learning how to do so by watching other finches.

Further Reading

Grant, Peter R. *Ecology and Evolution of Darwin's Finches.* Princeton, N.J.: Princeton University Press, 1999.

Sato, A., et al. "On the origin of Darwin's finches." *Molecular Biology and Evolution* 18 (2001): 299–311.

Weiner, Jonathan. *The Beak of the Finch: A Story of Evolution in Our Time.* New York: Knopf, 1994.

Dawkins, Richard (1941–) British *Evolutionary scientist* Richard Dawkins is one of the leading spokespersons in the world today for the understanding of evolution by the general public. He has generated numerous original ideas that have proven controversial and productive.

Dawkins was born in Kenya on March 21, 1941. He left when he was too young to see many of the wild animals of Africa but developed an interest in animal behavior. When he became an undergraduate at Oxford University in 1959, he studied with the eminent ethologist Niko Tinbergen, who helped to establish the modern study of behavior (see BEHAVIOR, EVOLUTION OF). Tinbergen said that the study of behavior required an interdisciplinary approach, involving psychology, physiology, ecology, sociology, taxonomy, and evolution. This was the kind of approach Dawkins was later to take in all of his work. At this time, the repercussions of the discovery of DNA as the basis of the inheritance of all organisms were being felt throughout all fields of biology (see DNA [RAW MATERIAL OF EVOLUTION]). As Dawkins studied bird behavior, learned how to use computers, and read about DNA, he began to put all three perspectives together and arrived at an understanding that genes, like animals, have behavior. Dawkins remained with Tinbergen for his doctoral research. Dawkins spent two years at the University of California at Berkeley before returning to Oxford University, where he has been on the faculty ever since. Since 1995 he has held the Charles Simonyi Chair of Public Understanding of Science at Oxford.

Two of Dawkins's ideas have had a particular impact on evolutionary science. One is that NATURAL SELECTION operates primarily on the genes; genes use bodies as their machines for recognition, survival, and propagation. Dawkins presented this idea in his first book, *The Selfish Gene,* in 1976. Since that time, many researchers have found examples of SELFISH GENETIC ELEMENTS that pervade the natural world at the molecular level. Molecular biologists consider the phenotype of an organism to be the chemicals that make up its cells, which are produced under the direction of DNA. In his book *The Extended Phenotype,* Dawkins expanded the phenotype concept. He argued that bodies are not the only machines that genes use for their benefit, but the immediate environment that the animal manipulates, such as the nests of birds, the dams of beavers, and the effects of animal signals on other animals, even of other species.

A second and related major idea is that genes are not the only replicators. Ideas and cultural innovations can spread through the minds of an animal population just as genes can spread through any biological population (see EVOLUTION). Dawkins called these replicating ideas *memes* and speculated that they have been crucial in the astonishing evolutionary development of humans. Others have gone further in suggesting that memes are part of the extended phenotype of humans. Humans create technology but must also adapt to it; anyone who does not adapt to technological innovation is at an evolutionary disadvantage. This idea is very similar to the GENE-CULTURE COEVOLUTION proposed by evolutionary biologist E. O. Wilson (see WILSON, EDWARD O).

An idea that Dawkins did not invent but that he has been a leading world figure in promoting is that natural selection can produce complexity by gradual steps. There is no need for a supernatural Designer to produce complexity; for any apparently complex structure or function, one can find simpler ones. It is largely through his works such as *The Blind Watchmaker* and *Climbing Mount Improbable* that many readers today have encountered the argument first made by Darwin (see DARWIN, CHARLES; *ORIGIN OF SPECIES* [book]). *The Blind Watchmaker* takes direct aim at the metaphor used by the 18th-century English theologian William Paley to present perhaps the most famous defense of the Design Argument (see NATURAL THEOLOGY). Because Dawkins has promoted the concept that complexity can arise out of the interactions of simple components, he has taken an interest in computer simulations of the origins of complexity (see EMERGENCE).

Dawkins is famous for proclaiming (see especially *A Devil's Chaplain* for his personal reflections) that religion, even if it may have played an essential role in human evolution (see RELIGION, EVOLUTION OF), is a source of irritation, grief, and destructiveness in modern human society.

Further Reading

Catalano, John. "The World of Richard Dawkins." Available online. URL: http://www.simonyi.ox.ac.uk/dawkins/WorldOfDawkins-archive/index.shtml Accessed March 23, 2005.

Dawkins, Richard. *The Ancestor's Tale: A Pilgrimage to the Dawn of Evolution.* New York: Houghton Mifflin, 2004.

———. *The Blind Watchmaker: Why the Evidence of Evolution Reveals a Universe Without Design.* New York: Norton, 1986.

———. *Climbing Mount Improbable.* New York: Norton, 1997.

———. *A Devil's Chaplain: Reflections on Hope, Lies, Science, and Love.* New York: Houghton Mifflin, 2003.

———. *The Extended Phenotype: The Long Reach of the Gene.* New York: Oxford University Press, 1982.

———. *The God Delusion.* New York: Houghton Mifflin, 2006.

———. *River out of Eden: A Darwinian View of Life.* New York: BasicBooks, 1995.

———. *The Selfish Gene.* New York: Oxford University Press, 1976.

———. *Unweaving the Rainbow: Science, Delusion, and the Appetite for Wonder.* New York: Houghton Mifflin, 2000.

Grafen, Alan, and Mark Ridley. *Richard Dawkins: How a Scientist Changed the Way We Think.* Oxford, U.K.: Oxford University Press, 2006.

***Descent of Man* (book)** When Darwin (see DARWIN, CHARLES) wrote his famous book (see *ORIGIN OF SPECIES* [book]) in 1859, he gathered together evidence from all areas of scientific research to establish the common evolutionary ancestry of all species, and to present NATURAL SELECTION as the mechanism of evolution. It was clear that Darwin did not exclude humans from an evolutionary origin, but he said very little about it: "Much light will be thrown on the origin of man and his history." Soon after the publication of *Origin of Species,* other scientists did not hesitate to present the evidence of human evolution. *Man's Place in Nature,* published in 1863 by the British zoologist Huxley (see HUXLEY, THOMAS HENRY), presented the fossil evidence that showed humans as one of the apes. *Natural History of Creation,* published in 1866 by the German zoologist Haeckel (see HAECKEL, ERNST) left no doubt that evolutionary theory includes human origins.

It was not until 1871 that Charles Darwin published a book that clearly presented the evolutionary origin of the human species, entitled *The Descent of Man, and Selection in Relation to Sex.* Part I of the book presents the evidence for the descent of humans from earlier animal forms. Part II of the book presents Darwin's theory of SEXUAL SELECTION, which he applies to humans in Part III. Rather than being three books in one, *The Descent of Man* uses sexual selection as the explanation for the origin of unique human characteristics and of human diversity. The diversity of human bodies and, even more clearly, the diversity of human cultures require more than just adaptation to different environmental conditions. The only way Darwin could explain human diversity was through sexual selection. Sexual selection remains today an important part of evolutionary theory and is still considered by evolutionary scientists to be the main process that produced unique human characters and human diversity.

Part I. The first chapter presents evidence of human evolutionary origin, including the homology of human organs with those of apes and VESTIGIAL CHARACTERISTICS in humans left over from previous evolutionary stages. In the second chapter, Darwin speculates about how the transformation from lower animal into human might have occurred. He proposed that the evolution of humans began when human ancestors started to rely on tools rather than teeth for processing food, and that the increase in brain size came later, allowing the use of ever more complex tools. Evolutionary scientists still recognize that brain size began to increase and tool use started at about the same time in human evolution. Darwin also proposed that many human characteristics could be understood as examples of what is now called NEOTENY, or the retention of juvenile characteristics into adulthood. The third chapter explains that, although humans have vastly greater mental powers than the other apes, there are many mental characteristics (such as emotions) that humans share with other animals. Chapters 4 and 5 extend this argument to the evolution of human moral instincts (see EVOLUTIONARY ETHICS), including their advancement through the barbarous period that preceded human civilization. Chapter 6, like Huxley's book, places humankind in the animal kingdom. In Chapter 7 Darwin speculates about the origins of human races. While he explains that races differ in their adaptations to different climates, he concludes that, "We have thus far been baffled in all our attempts to account for the differences between the races of man; but there remains one important agency, namely Sexual Selection, which appears to have acted powerfully on man ..." An appendix presents Huxley's explanations of the similarities between the brains of humans and other animals.

Part II. In Chapter 8 Darwin develops his general theory of sexual selection, in which he points out that males are often brightly colored and ornamented as well as "pugnacious," and, just as important, females often choose the males that are not just the most pugnacious but the most highly ornamented. He maintains that sexual selection also explains why there are more males than are necessary in animal populations. Chapter 9 applies these principles to invertebrate groups such as annelids and crustaceans; Chapters 10 and 11 apply these principles to insects, especially to butterflies and moths. Chapter 12 applies the principles of sexual selection to fishes, amphibians, and reptiles; Chapters 13 through 16, to birds; and Chapters 17 and 18, to mammals.

Part III. In Chapters 19 and 20, Darwin applies the principles of sexual selection to humans. He points out not only that males tend to fight and to choose females on the basis of their ideas of beauty, but that females in tribal societies frequently choose males. Darwin explains that sexual selection is essential for explaining both the universal and the particular traits of humans. One universal trait is human hairlessness. "No one supposes that the nakedness of the skin is any direct advantage to man; his body therefore cannot have been divested of hair through natural selection." Since the absence of hair is a secondary sexual characteristic, it must have arisen through sexual selection. Chapter 21 presents a general summary and conclusion.

Because of the clarity of writing, and the organization of evidence, *The Descent of Man* is still considered one of the great works of scientific investigation. Darwin treated races other than his own with respect. In so doing, and by his application of the principles of sexual selection, Darwin's explanation of human evolution agrees more closely with modern views than do the racist and speculative works of many of the anthropologists in the late 19th and early 20th centuries. Recently, sexual selection has been used as the explanation even of human intelligence (see INTELLIGENCE, EVOLUTION OF). With this, evolutionary scientist Richard Dawkins (see DAWKINS, RICHARD) has written, "Sexual selection, Darwin's 'other' theory, has finally come in from the cold ..."

Further Reading

Darwin, Charles. *The Descent of Man, and Selection in Relation to Sex,* 2nd ed. London: John Murray, 1871. Reprinted with introduction by James Moore and Adrian Desmond. New York: Penguin Books, 2004.

developmental evolution Developmental evolution (evo-devo) provides a new understanding of how evolutionary complexity, and evolutionary differences, result from the modification of genes that act very early in the development of a multicellular organism. These genes, called *homeotic* genes, establish the spatial arrangement of cells in the developing embryo or tissue by turning other genes on or off.

In animals, some of the homeotic genes are called *Hox genes.* They begin to work very early in embryonic development and are responsible for the overall body plan of the animal. All animals have at least a few Hox genes, which they inherited from their earliest multicellular common ancestor during PRECAMBRIAN TIME. Sponges and cnidarians, which do not have bilateral (left and right) symmetry, have only these fundamental Hox genes. All bilaterally symmetrical animals appear to have a set of 10 Hox genes that they inherited from their common ancestor. Plants also have genes, called *MADS genes,* that control their structure and symmetry. Animal Hox and plant MADS genes are not homologous; they evolved independently of one another.

All Hox genes contain a sequence of 180 nucleotides, called the *homeobox,* where transcription factors bind (see DNA [RAW MATERIAL OF EVOLUTION]). The homeobox has remained virtually unchanged during the course of evolution because changes in the homeobox would be very likely to disrupt the fundamental processes of embryonic development.

Gene duplication, followed by modification, has been important in the evolution of Hox genes. One of the Hox genes in the common ancestor of all animals was duplicated to produce the three Hox genes *scr, dfd,* and *pb,* and another Hox gene in the common ancestor produced the three Hox genes *Antp, abdA,* and *Ubx.* The five *AbdB* Hox genes found in mammals were produced by duplication from an ancestral *AbdB* gene; fruit flies, for example, have only one *AbdB* gene. Each time new Hox genes appeared in animal evolution, they allowed an increase in anatomical complexity. As a result, mammals have one each of eight Hox genes and five of the *AbdB* gene, a total of 13 (see figure on page 119). The largest gene duplication event in the history of Hox genes occurred in the ancestor of vertebrates. Vertebrates have four sets of Hox genes. If the ancestral 13 Hox genes duplicated twice, this would have produced 52 Hox genes, from which some were lost; mice and humans today have 39 Hox genes.

The Hox genes are arranged in the same order on the chromosomes as the parts of the body whose development they control. The *lab* gene is first, and it plays a role in the development of the head, which is the most anterior body region. Further down the chromosome are the *scr* and *Antp* genes, which play a role in the development of the middle body region. They activate the development of the legs and wings of flies, as well as the front limbs of vertebrates. Although flies and vertebrates separately evolved the ability to fly (that is, their wings are analogous; see CONVERGENCE), their front limbs share a common evolutionary origin (they are homologous). Finally come the *Ubx*, *abdA*, and *AbdB* genes, which play a role in the development of the posterior body regions. An animal may not use all of the Hox genes that it inherited from its ancestors. The five *AbdB* genes (Hox genes 9–13) code for tail segments. In humans, as in all apes (see PRIMATES), these genes are present but are not used. They constitute evidence that the ancestors of apes were animals with tails (see DNA [EVIDENCE FOR EVOLUTION]). Not only are the Hox genes in the same order as the portions of the body, but they are also in the same order as embryonic development (from head to tail) and as evolution (the last five Hox genes evolved most recently).

The fact that Hox genes all came from a common set of ancestral genes has been confirmed experimentally. Mice, from which one of the Hox genes was experimentally removed, were able to develop normally when the corresponding Hox

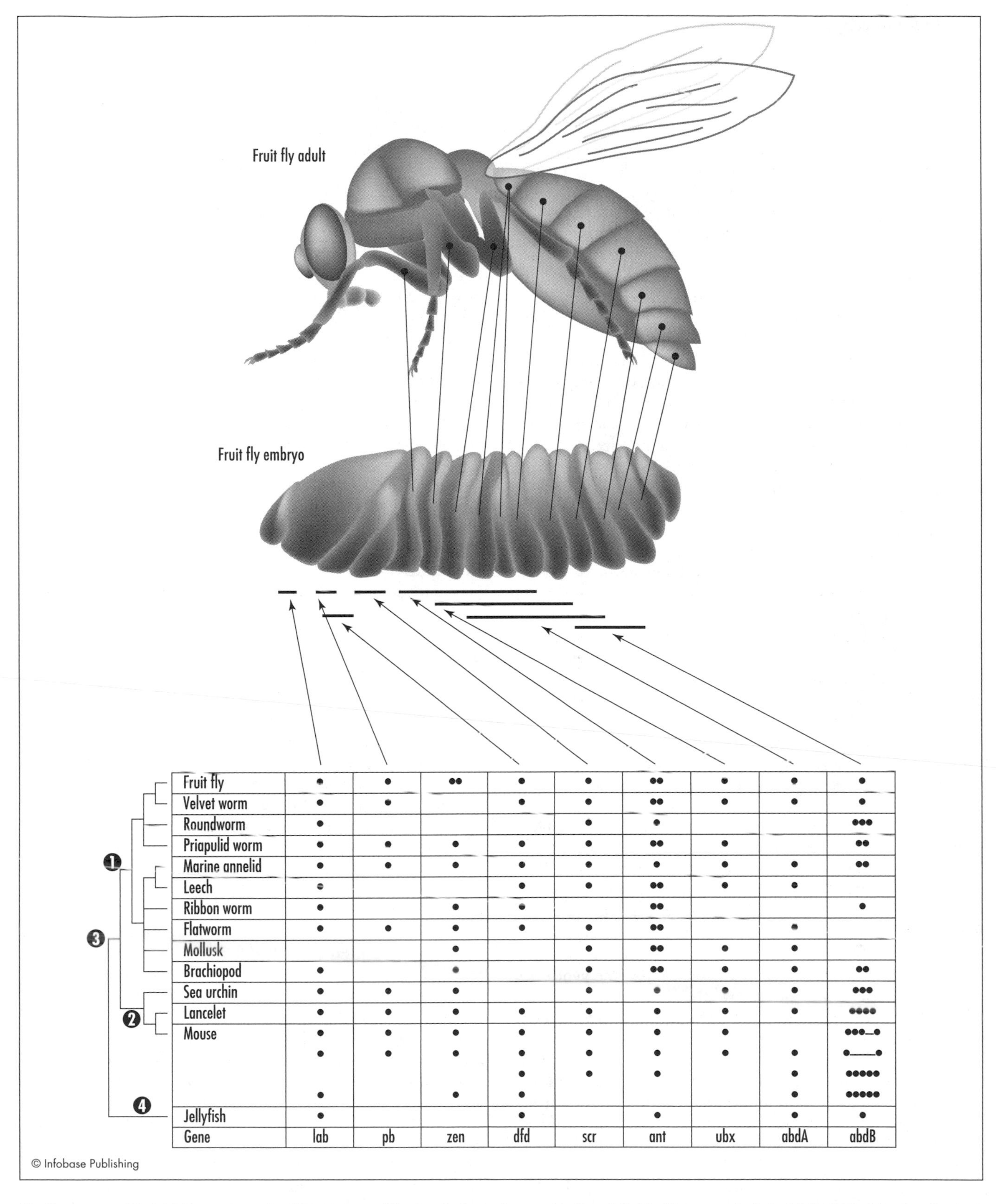

The Hox genes of the fruit fly correspond to the segments of the developing embryo and in the adult. The Hox genes include *lab* (labial), *pb* (proboscipedea), *dfd* (deformed), *ant* (antennapedia), *ubx* (ultrabithorax), *abdA* and *abdD* (abdominal A and B), which are often named after the deformities caused by mutations in these genes. The Hox genes are found throughout the animal kingdom, and their similarity reflects their pattern of evolutionary diversification. On the cladogram, 1 represents protostomes, 2 represents deuterostomes, 3 represents bilateral symmetry, 4 represents radial symmetry. Vertebrates such as the mouse have one of each gene except *AbdB*, of which they have five. Gene duplication has resulted in four sets of mammalian Hox genes. *(Modified from two figures in Freeman and Herron.)*

gene from an invertebrate chordate amphioxus was inserted into its DNA.

Hox genes do not control the structure of the body parts but makes them develop in their proper places. One example of this is the Hox gene *Antp*. This gene normally controls the development of legs in the thorax of flies. A mutation in this gene can cause legs to grow out of the fly's head, where the antennae would have been, a mutation called antennapedia (*antenna,* plus Greek *pedia* for foot). A missing Hox gene product causes a structure, such as a leg, to misunderstand its location in the developing animal. The normal gene, like the mutation, does not affect the structure of the legs but their location.

Homeotic genes other than Hox genes are essential to the order of development in animal bodies. For example, the *sog* gene in flies promotes the development of a nerve cord along the bottom or front (ventral) surface, while the chordin *(chd)* gene in vertebrates promotes the development of a nerve cord along the top or back (dorsal) surface. Molecular biologists E. M. DeRobertis and Y. Sasai reported in 1996 that these two genes were derived from a common ancestor. Experiments have confirmed this. If RNA transcripts from either a fly *sog* gene or a frog *chd* gene are injected into an embryo, they induce the normal development of the nerve cord: on the ventral surface in flies, on the dorsal surface in frogs. Thus, it appears, the embryonic development of arthropods is inverted with respect to that of vertebrates. This was the first scientific confirmation of the theory presented by Geoffroy St.-Hilaire in the early 19th century that vertebrates developed as upside-down arthropods (see INVERTEBRATES, EVOLUTION OF; CUVIER, GEORGES).

Another homeotic gene (not a Hox gene) in flies is the hedgehog gene. This gene got its name from a mutant form, in which the fly was covered with prickles. The normal function of the hedgehog gene is to control the front-to-back order of development within each of the segments defined by the Hox genes. When this gene was also discovered to be present in vertebrates, it was given the name sonic hedgehog *(shh)*. The *shh* gene is involved in the development of vertebrate limbs. It is expressed for a brief period in the development of fish fins; its prolonged expression allows the development of legs and feet in four-legged vertebrates (see AMPHIBIANS, EVOLUTION OF). Within each of the vertebrate limbs, the genes are expressed in order.

Other homeotic genes (not Hox genes) found widely among animal genomes are the *pax* and *tinman* genes. The first gene stimulates the development of eyes. The structure of the eye is very different in flies and vertebrates, but both are stimulated by versions of the same *pax* gene inherited from an ancestor that may have had simple eye spots. The *tinman* gene specifies the development of a heart, although the form of the heart is different in each animal.

Most of the studies of MADS genes in plants have involved the development of flowers from buds. Plants retain embryonic tissues in their buds, which are free to develop in different ways, like the stem cells of animals. A bud resembles an embryo in its developmental possibilities. Different MADS genes influence the development of different flower parts: the sepals on the outside, the petals inside of them, then the male stamens, and the female carpels in the middle (see ANGIOSPERMS, EVOLUTION OF). Mutations in the MADS genes can cause unusual developmental patterns, such as the development of sepals and petals where stamens and carpels would normally develop, or vice versa. A single mutation of a homeotic gene in a flowering plant can change the entire shape of the flower, which will change its relationship with pollinators (see COEVOLUTION). For example, a single mutation changes snapdragon-like flowers *(Linaria vulgaris),* which are bilaterally symmetrical, into the Peloria version, which is radially symmetrical. Genes similar to the MADS genes can be found in plants related to the ancestors of the angiosperms (see GYMNOSPERMS, EVOLUTION OF), even though these plants do not have flowers.

Some biologists, such as the botanist Sonia Sultan, have pointed out that it is not only necessary to understand the genetic processes of development but also to understand the ecological consequences of development (such as phenotypic plasticity; see ADAPTATION), in order to understand the process of evolution. This approach has been called "eco-devo."

Further Reading

Amundson, Ron. *The Changing Role of the Embryo in Evolutionary Thought: The Roots of Evo-Devo.* Cambridge, U.K.: Cambridge University Press, 2005.

Carroll, Sean B. *Endless Forms Most Beautiful: The New Science of Evo Devo and the Making of the Animal Kingdom.* New York: Norton, 2005.

Carroll, Sean B. *The Making of the Fittest: DNA and the Ultimate Forensic Record of Evolution.* New York: Norton, 2006.

Coen, Enrico. *The Art of Genes: How Organisms Make Themselves.* New York: Oxford University Press, 1999.

Davidson, Eric H., and Douglas H. Erwin. "Gene regulatory networks and the evolution of animal body plans." *Science* 311 (2006): 796–800.

DeRobertis, E. M., and Y. Sasai. "A common plan for dorsiventral patterning in bilateria." *Nature* 380 (1996): 37–40.

Freeman, Scott, and John C. Herron. "Development and evolution." Chap. 18 in *Evolutionary Analysis.* 3rd ed. Upper Saddle River, N.J.: Pearson Prentice Hall, 2004.

Maynard Smith, John. *Shaping Life: Genes, Embryos, and Evolution.* New Haven, Conn.: Yale University Press, 1998.

Shubin, Neal, C. Tabin, and Sean B. Carroll. "Fossils, genes, and the evolution of animal limbs." *Nature* 388 (1997): 639–648.

Zimmer, Carl. "A fin is a limb is a wing: How evolution fashioned its masterworks." *National Geographic,* November 2006, 110–135.

Devonian period The Devonian period (410 to 360 million years ago) was the fourth period of the PALEOZOIC ERA (see GEOLOGICAL TIME SCALE). The first significant advances of life onto land had occurred during the preceding SILURIAN PERIOD. During the Devonian period, plants filled the wet areas of the Earth, making large areas of the Earth's surface green for the first time.

Climate. The climate in the equatorial regions was warm and wet. Portions of the southern continent, over the South

Pole, were cold. Some researchers have suggested that the oxygen content of the atmosphere of the Earth began to rise during the Devonian period, reaching 35 percent by the subsequent CARBONIFEROUS PERIOD, in contrast to today's 21 percent. This would have resulted from the extensive growth of plants, which produced oxygen (see PHOTOSYNTHESIS, EVOLUTION OF), and which accumulated into thick deposits of peat (later to become coal) rather than decomposing. The high concentration of oxygen would explain how some animals, particularly insects, could grow so large during the Devonian and Carboniferous periods. Other researchers disagree, saying that such high levels of atmospheric oxygen would have promoted spontaneous combustion of the plant materials, producing extensive fires for which there is no clear evidence.

Continents. During the Devonian, there were three major continental areas. What is now North America and Europe formed a single landmass near the equator, much of it covered by shallow oceans. A portion of what is now Siberia lay to the north. Gondwanaland, a continent composed of what is now South America, Africa, Antarctica, India, and Australia, formed a large mass in the Southern Hemisphere.

Marine life. Many kinds of invertebrate animals, such as TRILOBITES, were abundant in the oceans of the Devonian period (see INVERTEBRATES, EVOLUTION OF). In addition, many kinds of fishes, in groups that had evolved earlier, proliferated into many species (see FISHES, EVOLUTION OF). So great was their diversity that the Devonian period has been called the Age of Fishes, even though there were also many species of plants and animals on land. These included jawless agnathan fishes; the heavily armored placederms ("plateskins"), a group now extinct; sharks; ray-finned fishes, which are the most abundant group of fishes today; and the lobe-finned fishes, which had finger-like bony reinforcements in their fins. One group of lobe-finned fishes was the lungfishes. A few lungfishes still live in tropical rivers today. Their lungs allow them to survive long periods in mud when ponds become seasonally dry.

Life on land. At the beginning of the Devonian period, there were only small plants (about 3 feet [1 m] in height), and a few arthropods, in terrestrial wetlands. By the end of the Devonian period, the first trees and terrestrial vertebrates had evolved.

- *Plants.* Early Devonian *Rhynia* was not much bigger than the late Silurian *Cooksonia.* These plants had xylem cells, which carry water from the ground up to the tips of the stems. Xylem allowed the evolutionary descendants of these plants to grow to the size of trees. By the end of the Devonian, some trees were about 30 feet (10 m) tall. Their descendants became much larger trees during the subsequent Carboniferous period. Silurian plants evolved into Devonian club mosses, horsetails, and ferns, which reproduced by spores rather than by seeds (see SEEDLESS PLANTS, EVOLUTION OF). Silurian plants had only simple branching stems; but during the Devonian, the first leaves evolved: Some of the branches were united by layers of green tissue, which became the veins and blades of the leaves. "Progymnosperms" such as *Archaeopteris* had primitive forms of seeds (see GYMNOSPERMS, EVOLUTION OF).
- *Animals.* Living among the small plants in the early Devonian mud were the first land animals, relatives of modern arthropods such as scorpions, mites, and spiders. Shortly thereafter, the ancestors of insects, centipedes, and millipedes were living among the small plants of early Devonian wetlands. Also during the Devonian period, amphibians evolved from ancestral lobe-finned fishes. These amphibians still closely resembled their fish ancestors. The Devonian fish *Eusthenopteron* had bones in its paired fins that corresponded to the leg bones of terrestrial vertebrates (see AMPHIBIANS, EVOLUTION OF). The Devonian amphibian *Ichthyostega* retained many of the features of its fish ancestors. The shape of the body and head, and the tail fin, made *Ichthyostega* look very much like a walking fish.

Extinctions. One of the mass extinction events in Earth history occurred at the end of the Devonian, at which time an estimated 22 percent of families, representing 83 percent of species, became extinct (see MASS EXTINCTION). The causes of this extinction event are unclear.

Further Reading

Dudley, Robert. "Atmospheric oxygen, giant Paleozoic insects and the evolution of aerial locomotor performance." *Journal of Experimental Biology* 201 (1998): 1,043–1,050.

Museum of Paleontology, University of California, Berkeley. "The Devonian." Available online. URL: http://www.ucmp.berkeley.edu/devonian/devonian.html. Accessed March 23, 2005.

DeVries, Hugo (1848–1935) Dutch *Botanist* Hugo DeVries, born February 16, 1848, studied the genetics of plants in the late 19th and early 20th centuries. In 1889 DeVries suggested that the heritable characteristics of organisms were transmitted from one generation to another by "pangenes," now called genes. He also argued that new species were produced by sudden large changes or mutations in the pangenes of the ancestral species, producing the new species instantly. This is known as the mutationist theory. DeVries maintained that environmental stress could produce these changes simultaneously within multiple individuals within the ancestral species. While he like most other scientists of his day accepted Darwin's proof that evolution had occurred (see DARWIN, CHARLES; *ORIGIN OF SPECIES* [book]), he did not accept Darwin's proposal of NATURAL SELECTION as the mechanism by which evolution worked. He believed that natural selection operated within species, to produce local varieties, and selected among the species that had been produced by sudden mutation. Today, geneticists believe that most MUTATIONS are small and that the same mutation does not occur simultaneously in more than one individual.

DeVries had experimental evidence that he believed confirmed the mutationist theory. He did crossbreeding experiments with the evening primrose, *Oenothera lamarckiana.* He found that strikingly new varieties arose within a single generation. Later geneticists showed that what DeVries thought were mutations were actually recombinations of existing genes.

DeVries was one of three biologists (along with Erich von Tschermak-Seysenegg and Carl Correns) who rediscovered the work of Gregor Mendel (see MENDEL, GREGOR; MENDELIAN GENETICS). Mendel had demonstrated that traits were passed on from one generation to another in a particulate fashion rather than by blending, and DeVries believed it was pangenes that did this. Largely because of DeVries, who influenced later geneticists such as Henry Bateson and Thomas Hunt Morgan, Mendel's mutations were considered inconsistent with Darwinian evolution. It was not until the MODERN SYNTHESIS of the late 1930s that Darwinian evolution and Mendelian genetics were brought together in a way that showed DeVries to have been wrong. Interestingly, this might not have happened had DeVries not helped to rediscover and publicize the work of Gregor Mendel. The rediscovery of Mendel's work remains DeVries's major impact on evolutionary science. He died May 21, 1935.

dinosaurs The term *dinosaur* usually refers to reptiles, many of them large, that dominated the Earth during the MESOZOIC ERA. The earliest dinosaurs lived during the TRIASSIC PERIOD. Dinosaurs diversified into numerous forms during the JURASSIC PERIOD and CRETACEOUS PERIOD, perhaps because one of the MASS EXTINCTIONS, which occurred at the end of the Triassic period, removed competition from other vertebrate groups. The dinosaurs themselves became extinct at the end of the Cretaceous period (see CRETACEOUS EXTINCTION).

Modern phylogeny (see CLADISTICS) classifies organisms based on evolutionary divergence. Since birds and mammals diverged from the group of vertebrates usually called reptiles, there is no such thing as reptiles unless one includes birds and mammals in the group (see REPTILES, EVOLUTION OF). Dinosaurs had many characteristics of skin, teeth, and bones that would remind observers of the reptiles with which they are familiar. In fact, since birds evolved from dinosaurs, the cladistic approach would indicate that the dinosaurs still exist: They are birds. Therefore the dinosaurs that most people think of are more properly called *non-avian dinosaurs*. Although dinosaurs are truly ancient, the therapsid (mammal-like) reptile lineage had diverged before the dinosaur lineage, and true mammals already existed by the time the first dinosaurs roamed the Earth.

The first dinosaurs to be studied were large, which is why Sir Richard Owen (see OWEN, RICHARD) gave them the name dinosaurs (Greek for "terrible lizards") in 1842. Many famous dinosaur discoveries came from the northern plains of the United States. In the late 19th century, almost 150 new dinosaur species were revealed by the excavations and reconstructions of American paleontologists Edward Drinker Cope and Othniel Charles Marsh. They began as collaborators but ended up in a three-decade intense rivalry. Marsh did not perform very much fieldwork. One account claims that he visited a spot where dinosaur bones were lying about like logs but did not recognize them. Marsh was, however, rich enough to buy fossils and to hire field workers. His uncle, George Peabody, endowed a museum at Yale specifically so that Marsh could study dinosaurs. Cope, who was also wealthy, spent much productive time in the field. He worked in Montana at the same time that General Custer's army was being annihilated nearby. When suspicious Natives checked out his camp, he entertained them by taking out his false teeth until they went away. A schoolteacher who found some dinosaur bones genially informed both Marsh and Cope about them. Cope sent the schoolteacher money and told him to ask Marsh to send all the specimens back to him. The rivalry grew intense enough that the excavators of the Marsh and Cope camps threw rocks at one another. Cope lost his money, but not his ego, in financial speculation, dying poor. He willed his body to a museum, hoping to have his skeleton used as the "type specimen" to represent the human species. When the skeleton was prepared, it was found to have syphilitic lesions and was unsuitable as a type specimen.

Most of what scientists know about dinosaurs comes from the study of fossilized bones, in the tradition of Cope and Marsh. Recently this has included the microscopic study of the spaces left by blood vessels in the bones. Other kinds of fossil evidence have also allowed many insights into the lives of dinosaurs:

- *Gastroliths.* These are "stomach stones" that helped to grind coarse plant material in the stomachs of some dinosaurs that ate branches and leaves. Many modern birds swallow stones that help them to grind their food.
- *Eggs.* Fossilized eggs with embryos inside reveal some of the process of dinosaur development. In addition, the discovery of caches of eggs reveals that dinosaurs cared for nests of eggs in a manner similar to that of most modern birds.
- *Soft material.* In 2005 paleontologist Mary Higby Schweitzer discovered preserved soft tissue, including what may be blood vessels and cells, in a dinosaur bone.

Since the time of Owen, Marsh, and Cope, dinosaurs have entered the popular imagination as emblematic both of huge size and of evolutionary failure. In both ways, the image is unfair. First, many dinosaurs were small or medium sized. Many of the smaller dinosaurs were sleek and rapid, and some had feathers (see BIRDS, EVOLUTION OF). Second, there were many kinds of dinosaurs, all over the world; they were, for more than a hundred million years, an astounding evolutionary success.

Some dinosaurs were extremely large. *Sauroposeidon proteles,* which lived 110 million years ago in what is now Oklahoma, had a body 100 feet (30 m) in length, with a neck more than 30 feet (10 m) long. The size estimate of the neck of this dinosaur is based upon the fact that most vertebrates have the same number of neck vertebrae, and the longest vertebrae found for this dinosaur were about five feet (140 cm) long. The sauropod *Supersaurus* weighed nearly a hundred tons, the largest animal ever to live on land (see figure on page 123). They were also the largest four-legged animals that would possibly have lived; if they had grown to 140 tons, the size of the legs necessary to support the weight would have become so great that the legs would not be able to move (see ALLOMETRY).

Sauropods were herbivorous dinosaurs, whose large size and long necks allowed them to eat leaves even in the tallest trees. Instead of having large teeth to grind the branches and leaves, they had small sharp teeth that could tear leaves and branches to be swallowed whole. Grinding occurred in the stomach with the aid of gastroliths. If dinosaurs such as *Apatosaurus* (formerly *Brontosaurus*) held their heads and necks upright, they would have needed seven times the blood pressure of a human, and the huge heart would have required the energy equivalent to about half of the food intake of the dinosaur. Some paleontologists say that huge, long-necked dinosaurs did not raise their heads. Perhaps they lifted their heads just long enough to eat. In contrast, the carnivorous dinosaurs, even the famous *Tyrannosaurus rex,* were much smaller (about five tons) because meat was much less available than vegetation.

The huge sauropod dinosaurs were probably also stupid. The huge, long-necked dinosaurs had small brains (and small teeth) because the force of leverage that would have been necessary to raise a large head at the end of a long neck would have been too great. Large dinosaurs were also stupid because a single, large brain in the head would have been too far away from the tail to allow quick control over muscle movement. The large dinosaurs had a brain, but also had a secondary ganglion of nerves near the base of the tail, which aided in control of posterior muscles.

Among the oldest known dinosaur fossils are *Herrerasaurus* and *Eoraptor,* from the late Triassic period about 230 million years ago. Because these fossils already had dinosaur specializations, the ancestral dinosaur must have lived even earlier. These earliest dinosaurs were about the size of a human and had narrow heads, long pointed snouts, mouths that extended from ear to ear, many small sharp teeth, and large eyes. Unlike nearly all other reptiles, these early dinosaurs walked on their two hind legs, which allowed their hands to evolve the ability to grasp prey. From animals similar to *Herrerasaurus,* two lineages of reptiles evolved: the bird-hipped dinosaurs *(ornithischians)* and the lizard-hipped dinosaurs *(saurischians),* which evolved into numerous forms during the ensuing Jurassic and Cretaceous periods.

Ornithischians had a pelvis that reminded early paleontologists of birds. Their teeth and jaws allowed the tearing and grinding of plant materials, and the enlarged ribs and pelvis allowed the digestion of large quantities of plant material. Despite their name, the ornithischians were not the ancestors of birds; they all met extinction at the end of the Cretaceous period. As time went on, ornithischians diverged into different lineages, all of which became much larger than their *Herrerasaurus* ancestors. They included:

- *Pachycephalosaurs* or "thick-headed" dinosaurs had skulls with thick domes that may have served as battering rams during fights. Recent research has failed to find evidence of tiny fractures that would have resulted from such conflicts.
- *Ceratopsians,* or horned dinosaurs, reverted to walking on all fours. By the late Paleozoic, some were quite large and had large horns, such as *Triceratops.*

Supersaurus **was one of the largest land animals ever to live. At 100 tons, it was nearly as large as an animal could be and still be able to support its weight on four legs. A single leg bone was larger than a human being.**

- *Thyreophores* or "shield-bearing" dinosaurs evolved into large animals with armored plates and, in the stegosaurs, a paired row of large plates along the back.
- *Ornithopods* also evolved into large forms but remained primarily bipedal.

Saurischian dinosaurs had long necks, and hands with long index fingers, a basic pattern now found in many birds. They diverged into the sauropods and the theropods.

- *Sauropods* became very large. An example is the *Supersaurus* mentioned above.
- *Theropods* had long, flexible necks, sharp teeth and claws, and their hands (with three curved claws) were good at grasping. Their bones were thin and light and hollow like those of modern birds. Their collarbones were fused into a wishbone or furcula, a feature found only in birds today. These features made the later development of flight possible.

From large theropod ancestors, coelourosaurians evolved. Some of them, such as *Tyrannosaurus,* were quite large, with big heads but very small arms. Others were much smaller, with larger brains and eyes like modern birds, resembling an ostrich in body form even to the point of having a toothless beak. All but one of these birdlike lineages became extinct. The lineage that survived to eventually become the birds was the maniraptorans, which were dinosaurs with feathers.

The maniraptoran dinosaurs apparently had hollow bones, a wishbone, and feathers before they had the ability to fly. Evolutionary biologists conclude from this that the original function of feathers was not for flight but to hold in body heat. This may have been the first step in warm-bloodedness, which is important in providing the energy needed for flight. Specimens of one of them, *Oviraptor,* have been found brooding eggs in the fashion of modern birds, another indication of warm-bloodedness.

One evolutionary branch of maniraptorans not only had feathers but could fly. The earliest known of these was *ARCHAEOPTERYX* from 150 million years ago, during the Jurassic period. Although skeletally similar to other maniraptorans, *Archaeopteryx* resembled modern birds in having long arms that formed a wing with flight feathers, reduced teeth, and large brain. Because maniraptorans were runners, evolutionary biologists believe that flight originated from rapid running "from the ground up" rather than from gliding "from the trees down." The evolutionary lineage of *Archaeopteryx* became extinct, but related lineages continued to evolve more birdlike characteristics, such as a larger breastbone that allowed massive flight muscles to anchor. By the early Cretaceous period, before the extinction of non-avian dinosaurs, there were finch-sized dinosaur birds capable of a modern type of flight. Other lineages adapted to an aquatic existence, resembling modern waterfowl, some of them even losing the ability to fly. All but one lineage, the birds, became extinct by the end of the Cretaceous period.

Dinosaurs diversified into many forms, from huge consumers of branches and leaves to small, sleek predators, mainly during the Jurassic and Cretaceous periods. When flowering plants evolved at the beginning of the Cretaceous (see ANGIOSPERMS, EVOLUTION OF), some dinosaurs evolved the ability to consume their leaves, but there was apparently no COEVOLUTION between non-avian dinosaurs and flowering plants for pollination of flowers or dispersal of fruits. Insects (and some birds) evolved many pollinating forms, and mammals were particularly good dispersers of fruits and seeds. The increasing dominance of flowering plants in the landscape may have been one reason that dinosaurs were in decline during the late Cretaceous. Massive volcanic eruptions changed the climate and further hastened the decline of dinosaurs. It was apparently the asteroid that hit the Earth 65 million years ago that sent all remaining dinosaurs, except the birds, into extinction. Until almost the last moment, there were at least a hundred species of dinosaurs.

Further Reading

Achenbach, Joel. "Dinosaurs: Cracking the mystery of how they lived." *National Geographic,* March 2003, 2–33.

Bakker, Robert T. *The Dinosaur Heresies.* New York: William Morrow, 1986.

Carrano, Matthew T., and Patrick M. O'Connor. "Bird's-eye view." *Natural History,* May 2005, 42–47.

Currie, Philip E., and Kevin Padian. *Encyclopedia of Dinosaurs.* New York: Academic Press, 1997.

Fiorillo, Anthony R. "The dinosaurs of arctic Alaska." *Scientific American,* December 2004, 84–91.

Fipper, Steve. *Tyrannosaurus Sue: The Extraordinary Saga of the Largest, Most Fought Over T. Rex Ever Found.* New York: Freeman, 2000.

Horner, John R., Kevin Padian, and Armand de Ricqlès. "How dinosaurs grew so large—and so small." *Scientific American,* July 2005, 56–63.

Levin, Eric. "Dinosaur family values." *Discover,* June 2003, 34–41.

Norell, Mark. *Unearthing the Dragon: The Great Feathered Dinosaur Discovery.* New York: Pi Press, 2005.

Parker, Steve. *Dinosaurus: The Complete Guide to Dinosaurs.* Richmond Hill, Ontario: Firefly Books, 2003.

Sander, P. Martin, and Nicole Klein. "Developmental plasticity in the life history of a prosauropod dinosaur." *Science* 310 (2005): 1,800–1,802.

Schweitzer, Mary H., Jennifer L. Wittmeyer, and John R. Horner. "Gender-specific reproductive tissue in ratites and *Tyrannosaurus rex.*" *Science* 308 (2005): 1,456–1,460.

Weishampel, David B., Peter Dodson, and Halszka Osmólska. *The Dinosauria.* 2nd ed. Berkeley: University of California Press, 2004.

Yeoman, Barry. "Schweitzer's dangerous discovery." *Discover,* April 2006, 36–41, 77.

diseases, evolution of *See* EVOLUTIONARY MEDICINE.

disjunct species Disjunct species contain or consist of populations (disjunct populations) that are distant from one another. The distances that separate the populations of disjunct species are greater than those that would allow frequent, or even occasional, gene flow between them. Gene flow occurs when pollen or seeds (of plants), or animals, travel from one population, or part of a population, to another, and this generally cannot occur in disjunct species. If the populations of a disjunct species are genetically different, the populations can often be described as subspecies. If enough time passes, it is practically inevitable that the disjunct populations will evolve into separate species, unless they become extinct (see SPECIATION).

Two ways are generally recognized in which populations within a species could have become disjunct. The first is that the populations are *relictual,* that is, they are relicts of a common, ancestral population. Populations of the species that were formerly present between the disjunct populations have disappeared, leaving the relictual populations isolated from one another. If, before or after the disappearance of the intervening populations, the two relictual populations diverge (perhaps becoming distinct subspecies), they are considered the product of *vicariance* (see BIOGEOGRAPHY). The other way is that one of the populations dispersed (as seeds or as a small flock) from one population and settled down in a distant location, an explanation called *dispersal.*

How can scientists determine which explanation, the relictual versus the dispersal explanation, is correct? The best evidence is if fossil or other evidence of the formerly intervening populations can actually be discovered. In some cases, evidence for populations of that particular species cannot be found, but if it is found for other species with similar disjunct distributions, then it can be inferred for the species in question as well. For example, there are many species of wildflowers in the alpine tundra of the high elevations of the Sierra Nevada, Cascade, Olympic, Rocky, and Sangre de Cristo Mountains of western North America that are very similar, in some cases nearly identical, to those of the arctic tundra of northern Alaska and Canada. Hundreds or thousands of miles separate the disjunct arctic and alpine populations. These distributions were, in fact, somewhat difficult to explain prior to Agassiz and Darwin (see AGASSIZ, LOUIS; DARWIN, CHARLES). Agassiz explained the ICE AGES: At various times, much of North America was, in fact, in the arctic tundra zone. As the glaciers retreated, some tundra plants migrated north, where they are now found as arctic tundra; others migrated up the mountains, where they are now stranded as alpine tundra. Darwin explained the subsequent vicariance of these populations by evolution occurring separately in each. Researchers have not found fossil evidence to confirm this for each of the wildflower species. The evidence for the glaciation and the overall patterns of vegetation change is overwhelming, and scientists can infer that it is true for each of these disjunct species. This is also the explanation of how disjunct populations of tundra species such as the sedge *Eriophorum vaginatum,* and boreal species such as the larch tree, came to be stranded in the bogs of northern states in the United States such as Michigan.

The next best evidence comes from genetic comparisons. If the disjunct populations can be shown to have several distinct mutations within their DNA, mutations not shared with one another, then they may be relictual (see DNA [EVIDENCE FOR EVOLUTION]). If one of the populations has DNA that is a subset of another population, then the first population clearly dispersed from the second. The only problem with this approach is that even if one population dispersed from another, the new population can evolve its own distinct mutations, which would make it appear as relictual. Genetic studies are most useful to detect recently dispersed disjunct populations.

Numerous examples of disjunct populations have been found. One example is the seaside alder, *Alnus maritima.* Many alders are large bushes that grow in swamps or next to streams and rivers with rocky beds. The hazel alder, *Alnus serrulata,* grows in this kind of streamside habitat throughout much of the eastern United States. In contrast, the seaside alder is a disjunct species, consisting entirely of three small disjunct populations. One population lives on the Delmarva Peninsula east of Chesapeake Bay. Another lives almost entirely within Johnston County in central Oklahoma. The third lives in a single swamp in Bartow County in northwest Georgia. Why is the seaside alder so rare, despite its apparent similarity to the hazel alder, which grows throughout eastern North America?

First, consider the vicariance explanation. Alders reproduce in two ways. Their seeds, produced in the autumn, grow best in sunny locations with moist mineral soil. Along many rivers, locations that have moist mineral soil are often shaded and not good for the growth of alder seedlings. In a second reproductive mode, once a bush is established, it can keep resprouting from the same roots even if floods wash many of its branches away, or if it is subsequently shaded by the growth of larger trees. Botanists Stanley Rice and J. Phil Gibson have demonstrated that both species of alders prefer sunny over shady conditions, but the hazel alders are found in the shade more frequently than seaside alders.

In the past (perhaps after the most recent Ice Age), both species of alders may have established themselves abundantly in the new landscape of sunny, moist, mineral soil. Subsequent regrowth of forests shaded the alder habitats, and the hazel alders have tolerated these shaded conditions at least a little better than the seaside alders. Hazel alders have therefore persisted throughout eastern North America, whereas the seaside alders have persisted in only three places. The seaside alder survived in these three places by chance, not because the three places were the most suitable habitat. The climatic conditions in Oklahoma, Georgia, and Delaware are quite different, and without doubt there are many places between these three locations that are better for seaside alder growth but in which the seaside alder by chance became extinct. Botanists James Schrader and William Graves have found that separate mutations have occurred in the three populations, enough to cause noticeable differences in leaf shape. These three populations are now classified as three distinct subspecies. There is no direct evidence that earlier populations of the seaside alder lived in places between Oklahoma, Georgia, and Delaware. Fossilized leaves that are identical to modern seaside alder leaves have been found in northwestern North America, which suggests that this species once grew across the entire continent. The species of alder that are most closely related to the seaside alder, in fact, grow in eastern Asia. This is the vicariance explanation for the disjunct distribution of the seaside alder.

Now consider the dispersal explanation. The seaside alder might have originated in one location, perhaps Delmarva, and subsequently dispersed to the other locations. How could this have happened? Something, or someone, might have carried the seeds from Delmarva to the other locations. Neither wind nor watershed patterns seem to suggest how this could have occurred. Native Americans use alder bark as a medicine and may have carried seeds or cuttings with them. The Georgia population is within the former Cherokee Nation territory, and the Cherokees are known to have moved south from the Iroquoian lands of what is now the northeastern United States. Subsequently the Cherokees were relocated by the U.S. government to Oklahoma in 1838. Could Cherokees have carried the plants with them, from the Northeast to Georgia to Oklahoma, connecting all three of the existing populations? This explanation is unlikely to be true, because the three populations (subspecies) are different enough that they must have been separated for a long time, much longer than the Cherokee migrations of ca. 1000 and 1838 C.E.

In the case of the seaside alder, and in many other cases as well, the disjunct populations are threatened with EXTINCTION. Not only do small populations have a greater risk of extinction, but disjunct populations are frequently not in the habitats most suitable for them. Oklahoma seaside alders have been planted, and are surviving well, in Iowa, while they are barely surviving in Oklahoma, where they can withstand the summer heat only if the soil is continually wet. This is because the climatic changes that isolated the disjunct populations in the first place are now threatening them with extinction. The Georgia and Delmarva populations are similarly threatened.

In other cases, a species may have a large, central population and several disjunct populations at some distance away from the central population. The least tern, for example, is a shorebird that is abundant in many areas, but the interior subspecies of the least tern (which migrated up the Mississippi and Arkansas Rivers into Oklahoma) is rare. The white spruce *(Picea glauca)* is abundant in the subalpine forests of the mountains of central North America and the boreal forests of Canada. The limber pine *(Pinus flexilis)* is fairly common in the Rocky Mountains. In the Black Hills of South Dakota, relictual populations of both the white spruce and the limber pine survive in conditions that are not cool enough for their optimal growth. They became stranded in the Black Hills after the most recent Ice Age and have declined in abundance since then. The spruce remains abundant in the Black Hills because it grows well in cool, moist, shaded areas of the northern hills. In contrast, the limber pine has nearly become extinct in the Black Hills, because it requires cool, dry, sunny conditions, which are now rare in the Black Hills. The few remaining limber pines are being crowded and shaded out by the extremely abundant ponderosa pines (see figure on page 126).

This is one of the last limber pines in the Black Hills of South Dakota. *Pinus flexilis* (labeled L) is a widespread pine species of high mountain areas in North America, but this population in South Dakota, a remnant from the most recent Ice Age, is disjunct, small, and heading toward extinction. Behind the pine is a spruce, *Picea glauca* (labeled S), which is also a remnant of the most recent Ice Age, and which is part of a relatively abundant population. A ponderosa pine is labeled P. *(Photograph by Stanley A. Rice)*

Disjunct populations are frequently the failed dead ends of evolution. They can also be wellsprings of evolutionary novelty. Because they are disjunct, these populations are reproductively isolated from the larger population, which allows them to evolve in a different direction. If they have dispersed from a large, central population, they may also by chance be genetically different from the original population (see FOUNDER EFFECT). Evolution can occur more rapidly in small populations (see NATURAL SELECTION). If the disjunct population evolves rapidly, it can produce a new species which will appear on the scene in what amounts to a geological instant. Small peripheral (sometimes disjunct) populations may be largely responsible for the speciation events in the PUNCTUATED EQUILIBRIA model.

Further Reading

Schrader, James A., and William R. Graves. "Infraspecific systematics of *Alnus maritima* (Betulaceae) from three widely disjunct provenances." *Castanea* 67 (2002): 380–401.

———, Stanley A. Rice, and J. Phil Gibson. "Differences in shade tolerance help explain varying success in two sympatric *Alnus* species." *International Journal of Plant Sciences,* 167 (2006) 979– 989.

DNA (evidence for evolution) DNA is not only the molecule that stores information about the physical characteristics of individuals (see DNA [RAW MATERIAL OF EVOLUTION]), but also the molecule that stores a record of evolution in past generations. From generation to generation, MUTATIONS accumulate in the DNA. If these mutations occur within genes, NATURAL SELECTION may act upon the organism that has the mutation. Natural selection may reduce the frequency of a harmful mutation in a population, although it may never eliminate it (see POPULATION GENETICS), or it may increase the frequency of a beneficial mutation. Natural selection does not affect the frequency of a mutation that has no effect on the function of the organism (a neutral mutation). Mutations that occur in the NONCODING DNA will also be unaffected by natural selection. Neutral genetic mutations, and mutations in the noncoding DNA, accumulate in the DNA as it is passed from one generation to another (see MARKERS). A comparison between the coding and noncoding regions of DNA indicates that there are more differences between humans and mice in their noncoding DNA than in their genetic DNA (see figure).

Organisms that are part of a single population interbreed and share genes. Although each organism in a population is genetically unique, all are genetically similar because they draw their genes from a single *gene pool.* But when a population is separated into two populations, new and different mutations occur in each of the populations (see SPECIATION). The new mutations present in one population are not found in the other. Over time, the DNA of the two gene pools becomes different. The longer the populations are separated, the greater the differences in their DNA. If the populations diverge enough to become separate species, the DNA is even more different; DNA of two different genera are even more different; the DNA of two families even more different; and so on (see LINNAEAN SYSTEM). By quantifying how different the DNA is between two organisms, scientists can determine how much evolutionary divergence has occurred since their separation. If scientists assume that mutations occur at a constant rate, these mutations constitute a MOLECULAR CLOCK. The DNA molecules have been keeping a cumulative record of the passage of time that helps scientists to reconstruct evolutionary history.

Because neutral mutations and mutations in the noncoding DNA are not related to the structure and function of organisms, comparisons of DNA and comparisons of anatomy provide independent data for the reconstruction of evolutionary relationships. DNA nucleotides, just like the physical characteristics of organisms, can be used as data in phylogenetic analysis (see CLADISTICS). Usually, phylogenetic reconstructions of evolutionary history based upon DNA correspond closely to phylogenetic reconstructions based upon physical traits, thus providing independent confirmation of the evolutionary patterns. This could scarcely have happened unless evolution really occurred. The DNA confirmation of evolution is one of the best sources of evidence against CREATIONISM. A creationist could explain the evolutionary patterns of neutral and noncoding DNA only by claiming that a Supreme Being created mutations that had an evolutionary pattern but were otherwise useless.

Evolutionary hypotheses can be tested not only by comparing DNA sequences but also the presence and absence of segments of noncoding DNA. The DNA of an organism contains the DNA from genes used by its ancestors, even though the organism no longer uses these genes. The functionless ancestral gene is now called a *pseudogene.* The presence of a pseudogene can be evidence of evolutionary ancestry. It can also be used to quantify evolutionary relatedness. Two species that share a greater number of the same pseudogenes share a more recent common ancestor (are more closely related to one another) than two species that share fewer pseudogenes. This is particularly true for the older pseudo-

Typical coding region:
Human nucleotides
ATGGTTTGATGTcCTCCAGAAAGTGTCTaCCCAgTTGAAGACaAACCTcACgAGtGTCACAAAGAACCGTCGAGATAA
gTGGTTTGATGTaCTCCAGAAAGTGTCTgCCCAaTTGAAGACgAACCTaACaAGcGTCACAAAGAACCGTCGAGATAA

Mouse nucleotides
70 of 78 nucleotides match = 90% similarity

Typical intron:
Human nucleotides
TaAATGGtgCcgttTGtgGCatGtGAactCAgGCGtGtcAgtgCTaGaGaGGAAACtgGAGCTgAGACTTTcC-AG
TgAATGGcaC----TGcaGCtaGaGAtgaCAtGCT-GatAtcaCTgGgGtGGAAAC-aGAGCTcAGACTTTtCtAG

Mouse nucleotides
46 of 73 nucleotides match = 63% similarity

This is a comparison of human and mouse nucleotide sequences for the same genetic region. Capital letters indicate nucleotides that have not changed since humans and mice evolved from a common ancestor; lowercase letters indicate changes that have occurred since the common ancestor of humans and mice. Dashes indicate loss of nucleotides in one of the species. There are more differences in the noncoding (intron) region than in the coding region of DNA. *(Modified from Watson)*

genes, which lost their genetic function further back in the past. The age of a pseudogene can be determined by molecular clock techniques.

DNA is found in the nucleus of each cell of complex organisms (see EUKARYOTES, EVOLUTION OF). It is also found in the mitochondria of eukaryotic cells, and the chloroplasts of photosynthetic eukaryotes, because mitochondria and chloroplasts used to be free-living bacteria (see SYMBIOGENESIS). Mitochondrial DNA (mtDNA) is inherited only maternally, that is, through egg cells. Phylogenetic analyses based upon mtDNA reconstruct only the matrilineal history of a species. This can be a convenient simplification in the study of the evolution of sexual species, in which nuclear DNA is reshuffled each generation. In mammals, the Y chromosome is inherited only paternally (see MENDELIAN GENETICS). Studies of mutations in Y chromosomes have been used to reconstruct the patrilineal evolutionary history of humans.

Mitochondrial DNA mutates 20 times faster than nuclear DNA. More mitochondrial mutations accumulate per million years than nuclear mutations. Noncoding DNA (which makes up a large proportion of the Y chromosome) experiences mutations at the same rate as genetic DNA, but the mutations accumulate. Therefore mtDNA and the Y chromosome are often used in evolutionary studies of the modern human species, which has existed only for about 100,000 years (see *HOMO SAPIENS*).

Because DNA usually decomposes quickly, scientists perform most DNA comparisons on specimens from living or recently dead organisms. In some cases, they can obtain enough DNA from ancient specimens to allow comparisons. It is usually DNA from mitochondria or chloroplasts that is sufficiently abundant in ancient specimens to allow such comparisons. Mitochondrial DNA from Neandertal bones more than 30,000 years old (see below) and chloroplast DNA from leaves almost 20 million years old have been recovered and studied.

How do evolutionary scientists compare different samples of DNA?

- The actual base sequence (of A, C, T, and G bases) of each sample can be determined. Then the base sequence of one species is lined up with the base sequence of the other species, for a specified part of a chromosome. The number of similarities as a proportion of the total number of comparisons serves as a measure of evolutionary relatedness. Once a laborious process, DNA sequencing is now automated. However, it remains expensive. For this process to be used, the DNA must be amplified, that is, copied over and over in a test tube. This can only be done if the correct *primers* are used. Primers are needed because the enzyme that copies the DNA cannot create a whole new DNA strand but can only lengthen a strand that already exists; the primer is that short strand that the enzyme lengthens. It usually takes a lot of work to find, or design, the correct primers; once this has been done, it is generally easy (at major research sites with the right equipment) to determine base sequences. Once the base sequences are determined, they are submitted to a worldwide data bank. Researchers can make computer-based comparisons among any of the millions of sequences in the database (see BIOINFORMATICS).
- The DNA can be broken up into fragments of various lengths. The fragments can be separated out from one another on the basis of their size, on a gel subjected to an electric current. The process of *gel electrophoresis* that is used to separate DNA fragments from a mixture is very similar to the process that separates proteins from a mixture. This produces a *DNA fingerprint* that looks something like a supermarket bar code. DNA fingerprints are usually used to identify criminals; if the suspect's DNA fingerprint matches that of the DNA sample at the crime scene, after more than one independent test, it is virtually certain that the DNA came from this individual. DNA fingerprints of different species can be compared by this method. Two organisms with similar fingerprints are closely related, while two organisms with very different fingerprints are more distantly related, by evolutionary descent.
- Two samples of DNA can also be compared by reannealing. Each DNA molecule in an organism consists of two strands that match perfectly or almost perfectly. These strands can separate if an investigator heats them in the laboratory. The investigator can mix the DNA from two different organisms, heat it, and allow the strands to come back together, or reanneal, into pairs. Sometimes the DNA molecules come back together in their original pairs. But sometimes the DNA strands from the two organisms form pairs. When the two DNA strands from the different species match closely, they bind tightly, and a higher temperature is required to separate them. When the two DNA strands do not match closely, they bind loosely, and a lower temperature is required to separate them. There is roughly a 1 degree Celsius difference in reannealing temperature for each 1 percent difference in the base sequences of the two DNA strands. The original evolutionary comparisons of DNA were conducted before the invention of biotechnology techniques and relied on reannealing data, but the method is much less common in evolutionary studies today.

DNA comparisons among species of organisms have revealed some fascinating insights into evolutionary history. The following are examples of research that has used the techniques described above.

Evolutionary Patterns of Life

Origin of mitochondria and chloroplasts. The laboratory of Carl Woese (see WOESE, CARL R.) compared DNA sequences of many species of bacteria, as well as the DNA from mitochondria and chloroplasts. His results show that mitochondria are related to aerobic bacteria and chloroplasts are related to cyanobacteria (see BACTERIA, EVOLUTION OF). Woese's results, and similar results of other researchers, confirm the symbiogenetic theory (see MARGULIS, LYNN) of the origin of chloroplasts and mitochondria.

Tree of life. Woese's laboratory also found, by comparing DNA sequences from many species of bacteria and eukaryotes, that the diversity of life falls into three major groups or domains: the archaea (see ARCHAEBACTERIA), the

Amino Acid Similarities among Vertebrates for the β Chain of Hemoglobin

Animal	Number of amino acid differences, out of 146, compared to humans	Percentage of amino acid similarities to humans
Human	0	100%
Gorilla	1	99%
Gibbon	2	98%
Rhesus monkey	8	94%
Dog	15	90%
Horse and cow	25	83%
Mouse	27	82%
Gray kangaroo	38	74%
Chicken	45	69%
Frog	67	54%

true bacteria, and eukaryotic nuclear DNA. Even though archaebacteria and true bacteria look alike to most human observers, they have numerous differences in their chemical makeup and diverged from one another very early in the history of life. Woese has accordingly classified all organisms into three domains (Archaea, Eubacteria, and the domain Eucarya that contains the protist, fungus, plant, and animal kingdoms)—all on the basis of DNA comparisons (see TREE OF LIFE).

Origin of multicellular animals. The common ancestor of multicellular animals is not known. By the time animals show up in the fossil record, they have already differentiated into many forms (see BURGESS SHALE; CAMBRIAN EXPLOSION; EDIACARAN ORGANISMS). Comparisons of DNA of modern animals suggest that a wormlike organism lived in the oceans a billion years ago. Dubbed the RFW or "roundish flat worm," it is the supposed ancestor of all bilaterally symmetrical animals (see INVERTEBRATES, EVOLUTION OF). If the fossil of such an animal is ever found, it will not come as a complete surprise to scientists, for the existence and even the general appearance of this animal has been reconstructed from DNA comparisons.

Evolutionary Patterns of Vertebrates

Evolution of hemoglobin. All vertebrates (except the hagfishes, the most ancient branch) have hemoglobin (the red protein that carries oxygen in blood) that consists of four protein chains (two α and two β chains) surrounding a heme group that has an iron atom in the middle. Over time, since the origin of vertebrates, mutations have accumulated in the gene for the β chain. As a result, different species of vertebrates have different sequences of amino acids in the β chain. Each vertebrate has a β chain amino acid sequence that most closely resembles the sequence in the animals from which its lineage diverged most recently. The further back in time the lineages diverged, the less similar the amino acid sequences are. The hemoglobin data confirm the evolutionary pattern (see first table). The human β chain sequence is almost identical to that of gorillas, a little less similar to that of the rhesus monkey, even less similar to that of the dog, even less similar to that of the kangaroo, even less to that of the chicken, and least of all to that of the frog. Some invertebrates and even plants also have hemoglobin, but it appears to have evolved independently from simpler heme proteins, for which this kind of comparison would not be meaningful.

Evolution of birds. Fifty million years ago birds had teeth, but modern birds do not (see BIRDS, EVOLUTION OF). Modern birds still have the gene for making a portion of their teeth. This gene has lost its promoter and cannot be used. The gene can be experimentally stimulated in chickens, resulting in chickens with rudimentary teeth (see PROMOTER).

Evolution of horses. The ancestors of modern horses had three toes, unlike modern horses that have just one; occasionally, mutations in DNA control sequences stimulate the production of extra toes in modern horses (see HORSES, EVOLUTION OF).

Evolutionary Patterns of Primates

Humans and other primates. As noted above, species that share a more recent common ancestor should share more pseudogenes than species that have a common ancestor that lived further back in the past. Evolutionary scientists Felix Friedberg and Allen Rhoads have confirmed this pattern among PRIMATES. They studied six human pseudogenes, for which ages could be estimated. They then looked for these pseudogenes in six primate species, and in the hamster (see

Pseudogenes in Six Species of Primates (from Friedberg and Rhoads)

Pseudogene	Estimated age (millions of years)	Human	Chimp	Gorilla	Orangutan	Rhesus monkey	Capuchin monkey
α-Enolase Ψ_1	11	•	•	•			
AS Ψ_7	16	•	•		•		
CALM II Ψ_2	19	•	•	•	•		
AS Ψ_1	21	•	•	•	•	•	
AS Ψ_3	25	•	•	•	•	•	
CALM II Ψ_3	36	•	•	•	•	•	•

second table). Humans and chimps shared all of them; gorillas and orangutans shared five of them with humans; the rhesus monkey, an Old World monkey, shared three of them with humans; the capuchin monkey, a New World monkey, shared one; the hamster had none of the six pseudogenes. This pattern corresponded perfectly to the primate evolutionary lineage.

Humans and chimpanzees. Two species of chimpanzees are the living animals most closely related to *Homo sapiens.* Humans and chimpanzees have a 99 percent similarity of nucleotide base sequences. The genetic differences between humans and chimpanzees have major effects but are few in number. The molecular clock also suggests that the evolutionary lineages leading to humans and to chimps may have diverged as recently as five million years ago (see AUSTRALOPITHECINES).

Modern human diversity. Most anthropologists are now convinced that African HOMININS (similar to *HOMO ERGASTER*) spread throughout Europe (as *HOMO HEIDELBERGENSIS*) and Asia (evolving into *HOMO ERECTUS*). *H. erectus* became extinct. *H. heidelbergensis* evolved into NEANDERTALS and became extinct. The only survivors were the *H. ergaster* individuals that remained in Africa and evolved into *H. sapiens* in Africa between 100,000 and 200,000 years ago. This is consistent with the fossil evidence, which shows *H. sapiens* appearing only in Africa, and then spreading around the world. This has been called the Out of Africa hypothesis.

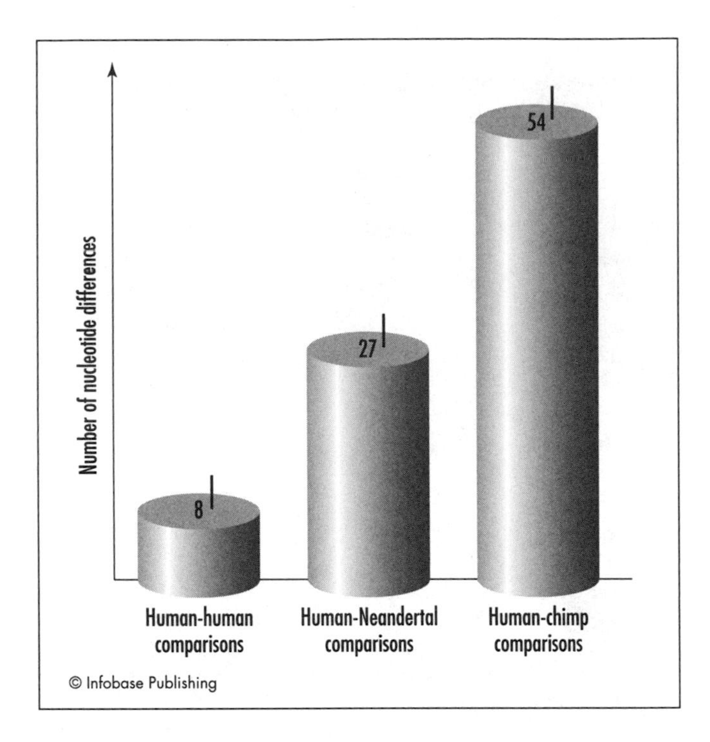

Investigators compared nucleotide sequence differences between different humans; between humans and Neandertals; and between humans and chimpanzees. The comparisons were made between homologous (corresponding) locations on chromosomes. The narrow vertical lines represent approximate standard deviations of the data. Comparisons among humans averaged eight nucleotide differences; comparisons between humans and Neandertals averaged 27, over three times as great; and comparisons between humans and chimpanzees averaged about 54. Neandertals differed from humans about half as much as chimpanzees differ from humans.

If the Out of Africa hypothesis is correct, then there should be very little DNA variability among modern humans. University of California geneticists Allan Wilson, Mark Stoneking, and Rebecca Cann analyzed mtDNA from 150 women from Africa, Asia, Europe, Australia, and New Guinea. By using computers that can make thousands of comparisons among the nucleotide sequences of DNA, the researchers were able to reconstruct a human family tree. Their results showed that the evolutionary divergence of human races was relatively recent, having occurred only 100,000 to 200,000 years ago. Because mtDNA is inherited only through mothers, the conclusion of the Wilson group has been popularized as the "mitochondrial Eve" or "African Eve." Their results also showed that DNA variability in the human species is very low, compared to the variability found in populations of most other species. The greatest degree of variability was from the African samples, which implies that the evolution of *H. sapiens* has been going on in Africa longer than anywhere else.

More than a decade of study since the original report of the Wilson group has failed to find human genetic variation outside of Africa that cannot be found within African populations. More recent studies on nuclear genes confirm the pattern that was revealed by mtDNA. Studies on the Y chromosome suggest that the pattern inherited through males is the same as that inherited through females. Not surprisingly, the Y chromosome studies have been popularized as the "African Adam." These results indicate that all modern humans share a recent African origin.

These results also indicate that all non-African human populations were originally emigrants from Africa. Ancestors of aborigines left Africa and arrived in Australia over 60,000 years ago. Their burial customs, and the fact that they would have to build boats to reach Australia, strongly suggest that the people who colonized Australia were fully modern humans. Bones of some of these people, found in Mungo by anthropologist Alan Thorne, appear to be 62,000 years old. Mitochondrial DNA extracted from the Mungo bones is outside the modern range of human genetic variation, which to some researchers suggests that they came from a population of humans that no longer exists. However, the Mungo DNA is only slightly outside the range of modern humans, and most researchers do not consider it to discredit the Out of Africa hypothesis.

Humans and Neandertals. Another test of the hypothesis that all modern humans are descended from African *H. sapiens* would be to compare modern humans with Neandertals. Comparisons of bone structure have suggested that Neandertals were a separate lineage of evolution from *H. ergaster.* However, a DNA comparison would clinch the case. It would seem impossible to find Neandertal DNA—they have been extinct for almost 30,000 years—but the laboratory of Svante Pääbo was able to obtain a little bit of mtDNA from some of the youngest Neandertal bones. Because mtDNA is so much more abundant than nuclear DNA, it was more likely to be preserved in these ancient bones. The Neandertal DNA turned out to have a base sequence very different from that of any modern human (see NEANDERTALS). Since 1997,

DNA comparisons between humans and Neandertals have been performed three times, with the same results (see figure on page 130). James Noonan and others working with Pääbo at the Max Planck Institute have begun an analysis of the nuclear DNA of Neandertals, and hope to complete a rough Neandertal genome by about 2008. Because nuclear DNA evolves more slowly than mtDNA, there is a much greater similarity (over 99%) of human and Neandertal nuclear DNA than of mtDNA. Geneticist Edward Rubin is pursuing similar research.

Native American origins. The mitochondrial DNA of tribes native to North and South America clusters into three groups. The first group is the Amerindians, which constitute most of the Native American tribes, and which may have been the first group of immigrants from Siberia into the New World after the most recent of the ICE AGES. The second group is the Inuit, also called Eskimos, whose ancestors arrived later, and remained in northern North America. The third group is the Na-Dene, which include the Apache and Navaho tribes. There is no clear archaeological evidence that the ancestors of the Apache and Navaho arrived in North America later than the other Native American tribes, although Navaho oral tradition indicates that when they arrived, the cliff dwellings of earlier people had already been abandoned. Their DNA reveals this important part of their history.

Human linguistic diversity. Geneticist Luigi Luca Cavalli-Sforza has compared genetic variability (first of proteins, later of DNA) and patterns of linguistic diversity among human groups. The major genetic lineages of the human species correspond roughly to the major language groups in the world (see LANGUAGE, EVOLUTION OF).

Other Uses of DNA Evidence

Genetic variability of endangered populations. The quantification of DNA variability is also useful to assess how much genetic variability is present in a population. Populations need genetic variability in order for natural selection to occur, and to avoid inbreeding depression. Endangered species and subspecies often suffer from inbreeding depression (see EXTINCTION). Also, when habitats are fragmented into small bits by human activity, such as when one big forest becomes many small woodlots surrounded by civilization, the small populations in the fragments may suffer the loss of genetic variability. DNA studies are now performed to determine the amount of genetic variability within populations of endangered species or in fragmented habitats. Without genetic variation, the evolutionary future, and therefore the future, of these populations and species is likely to be hopeless. In order to qualify for governmental protection, a population must be shown to be genetically different from other populations in order to qualify as a distinct species or subspecies.

Genes and archaeology. DNA from human remains can help to identify the origins of archaeological artifacts and help reconstruct recent human prehistory. The result is a new branch of research called "bio-archaeology." Bio-archaeology uses the analysis of DNA traces to determine what people ate, what people hunted, how they cooked, and how long ago humans domesticated various animals.

Noncoding DNA, though of limited and possibly of no use to the organism, turns out to be very useful in evolutionary studies.

Further Reading

Carroll, Sean B. *The Making of the Fittest: DNA and the Ultimate Forensic Record of Evolution.* New York: Norton, 2006.

Cavalli-Sforza, Luigi Luca. "Genes, peoples, and languages." *Proceedings of the National Academy of Sciences USA* 94 (1997): 7,719–7,724.

Freeman, Scott, and Jon C. Herron. "The evidence for evolution." Chap. 2 in *Evolutionary Analysis,* 3rd ed. Upper Saddle River, N.J.: Pearson Prentice Hall, 2004.

Friedberg, Felix, and Allen Rhoads. "Calculation and verification of the ages of retroprocessed pseudogenes." *Molecular Phylogenetics and Evolution* 16 (2000): 127–130.

Golenberg, E. M., et al. "Chloroplast DNA sequence from a Miocene *Magnolia* species." *Nature* 344 (1990): 656–658.

Gould, Stephen Jay. "Hen's teeth and horse's toes." Chap. 14 in *Hen's Teeth and Horse's Toes: Further Reflections in Natural History.* New York: Norton, 1983.

Jones, Martin. *The Molecule Hunt: Archaeology and the Search for Ancient DNA.* New York: Arcade Publishing, 2002.

Krings, Matthias, H. Geisert, Ralf Schmitz, Heike Krainitzki, Mark Stoneking, and Svante Pääbo. "DNA sequence of the mitochondrial hypervariable region II from the Neanderthal type specimen." *Proceedings of the National Academy of Sciences USA* 96 (1999): 5,581–5,585.

———, Anne Stone, Ralf Schmitz, Heike Krainitzki, Mark Stoneking, and Svante Pääbo. "Neanderthal DNA sequences and the origin of modern humans." *Cell* 90 (1997): 19–30.

Lindsay, Jeff. "Does DNA evidence refute the book of Mormon?" Available online. URL: http://www.jefflindsay.com/LDSFAQ/DNA.shtml. Accessed April 18, 2006.

Marks, Jonathan. *What It Means to Be 98 Percent Chimpanzee: Apes, People, and Their Genes.* Berkeley: University of California Press, 2002.

Murphy, William J., et al. "Dynamics of mammalian chromosome evolution inferred from multispecies comparative maps." *Science* 309 (2005): 613–617.

Noonan, James P. et al. "Sequencing and analysis of Neanderthal genomic DNA." *Science* 314 (2006): 1113–1118.

Pennisi, Elizabeth. "The dawn of stone age genomics." *Science* 314 (2006): 1068–1071.

Soltis, Pamela S., Douglas E. Soltis, and C. J. Smiley. "An rbcL sequence from a Miocene *Taxodium* (bald cypress)." *Proceedings of the National Academy of Sciences USA* 89 (1992): 449–451.

Wallace, Douglas C., and A. Torroni. "American Indian prehistory as written in the mitochondrial DNA: A review. *Human Biology* 64 (1992): 403–416.

Watson, James D., with Andrew Berry. *DNA: The Secret of Life.* New York: Knopf, 2003.

Wells, Spencer. *The Journey of Man: A Genetic Odyssey.* Princeton, N.J.: Princeton University Press, 2003.

Wilson, Allan C., and Rebecca L. Cann. "The recent African genesis of humans." *Scientific American* 266 (1992): 68–73.

DNA (raw material of evolution) DNA is the molecule by which information passes from one generation to the next.

DNA is the basis of inheritance and of evolution. DNA, or deoxyribonucleic acid, is important to evolutionary science in four ways:

- Because of DNA, traits are heritable.
- Because of DNA, traits are mutable.
- The study of DNA allows comparisons of evolutionary divergence to be made among individuals (see DNA [EVIDENCE FOR EVOLUTION]).
- The study of DNA allows the genetic variability of populations to be assessed (see POPULATION GENETICS).

In order for natural selection to work on traits within a population, those traits must be heritable (see NATURAL SELECTION). Characteristics that are induced by environmental conditions, though sometimes called adaptations, are not heritable (see ADAPTATION). Scientists, and most other people, have long known that traits are passed on from one generation to another, that organisms reproduce not only "after their own kind" but that the offspring resemble their parents more than they resemble the other members of the population. Some scientists in the 17th century believed that tiny versions of organisms were contained within the reproductive cells, such as homonculi within sperm; that is, entire structures were passed on from one generation to another, a theory called *preformation*. Other scientists believed that structures formed spontaneously from formless material, a theory called *epigenesis*. They were both partly right and partly wrong. In the 20th century scientists discovered that the instructions for making the structure, rather than the structure itself, were passed from one generation to another through the reproductive cells.

In the 19th century, many scientists believed that characteristics that an organism acquired during its lifetime could be passed on to later generations. The scientist most remembered for this theory is Jean Baptiste de Lamarck (see LAMARCKISM). The *inheritance of acquired characteristics* was not his special theory, but rather the common assumption of scientists in his day. Until Gregor Mendel (see MENDEL, GREGOR; MENDELIAN GENETICS), nobody had performed experiments that would adequately test these assumptions. Charles Darwin (see DARWIN, CHARLES) was so perplexed and frustrated by the general lack of scientific understanding of inheritance that he invented his own theory: *pangenesis*, in which "gemmules" from the body's cells worked their way to the reproductive organs and lodged there, carrying acquired genetic information with them. His theory was essentially the same as Lamarck's, and equally wrong. Darwin either had never heard of Mendel's work or overlooked its significance.

How DNA Stores Genetic Information

By the early 20th century, data had accumulated that a chemical transmitted genetic information from one generation to the next, and that the chemical was DNA. In 1928 microbiologist Frederick Griffith performed an experiment in which a harmless strain of bacteria was transformed into a deadly strain of bacteria by exposing the harmless bacteria to dead bacteria of the deadly strain. Some chemical from the dead bacteria had transformed the live harmless bacteria permanently into generation after generation of deadly bacteria. Research in 1944 by geneticist Oswald Avery and associates established that it was the DNA, not proteins, that caused the transformation that Griffith had observed. Research in 1953 by geneticists Alfred Hershey and Martha Chase established that it was the DNA, not the proteins, of viruses that allowed them to reproduce: The DNA from inside the old viruses produced new viruses, while the protein coats were merely shed and lost.

Many scientists doubted that DNA could be the basis of inheritance and suspected that proteins might carry the genetic information. DNA was a minor component of cells, compared to the abundance of protein. Furthermore, DNA was a structurally simple molecule, compared to proteins, and was thus considered unlikely to carry enough genetic information. It was not until the structure of DNA was explained by chemists James Watson and Sir Francis Crick in 1953, based upon their data and data from colleagues such as chemist Rosalind Franklin, that DNA became a truly believable molecule for the transmission of genetic information.

DNA is an enormously long molecule made up of smaller nucleotides (see figure on page 133). Each nucleotide consists of a sugar, a phosphate, and a nitrogenous base. In DNA, the sugar is always deoxyribose. The nitrogenous base in a DNA nucleotide is always one of the following: adenine, guanine, cytosine, or thymine. The general public has been well exposed to the abbreviations of these bases (A, G, C, and T). The nucleotides are arranged in two parallel strands, like a rope ladder. The parallel sides of the ladder consist of alternating molecules of phosphate and sugar. The rungs consist of the nitrogenous bases meeting together in the center. A large base is always opposite a small base, and the correct bonds must form; therefore, A is always opposite T, and C is always opposite G. Because of this, both strands of DNA contain mirror-images of the same information: if one strand is ACCTGAGGT, the other strand must be TGGACTCCA. DNA not only stores information but stores it in a stable fashion: All of the information in one strand is mirrored in the other strand. If MUTATIONS occur in one strand, the base sequence in the other strand can be used to correct them. Mutations in DNA are usually but not always corrected. All cells use DNA to store genetic information. Therefore the mutations that occur in one strand are frequently corrected by an enzyme that consults the other strand.

Mutations occur relatively infrequently, and evolution proceeds slowly, in all species of organisms. Some viruses (which are not true organisms) use a related molecule, RNA (ribonucleic acid), to store genetic information. RNA is single-stranded, and its mutations cannot be corrected. Because of this, RNA viruses evolve much more rapidly than DNA viruses. RNA viruses such as colds and influenza evolve so rapidly that the human immune system cannot keep up with them. This is why a new flu vaccine is needed every year. Last year's flu vaccine is effective only against last year's viruses. In contrast, DNA viruses such as the ones that cause poliomyelitis (polio) evolve slowly enough that old forms of the vaccine are still effective. The human immunodeficiency virus is an RNA virus and evolves rapidly (see AIDS, EVOLUTION OF).

DNA is capable of *replication*. If the two strands separate, new nucleotides can line up and form new strands that exactly mirror the exposed strands. In this way, one DNA

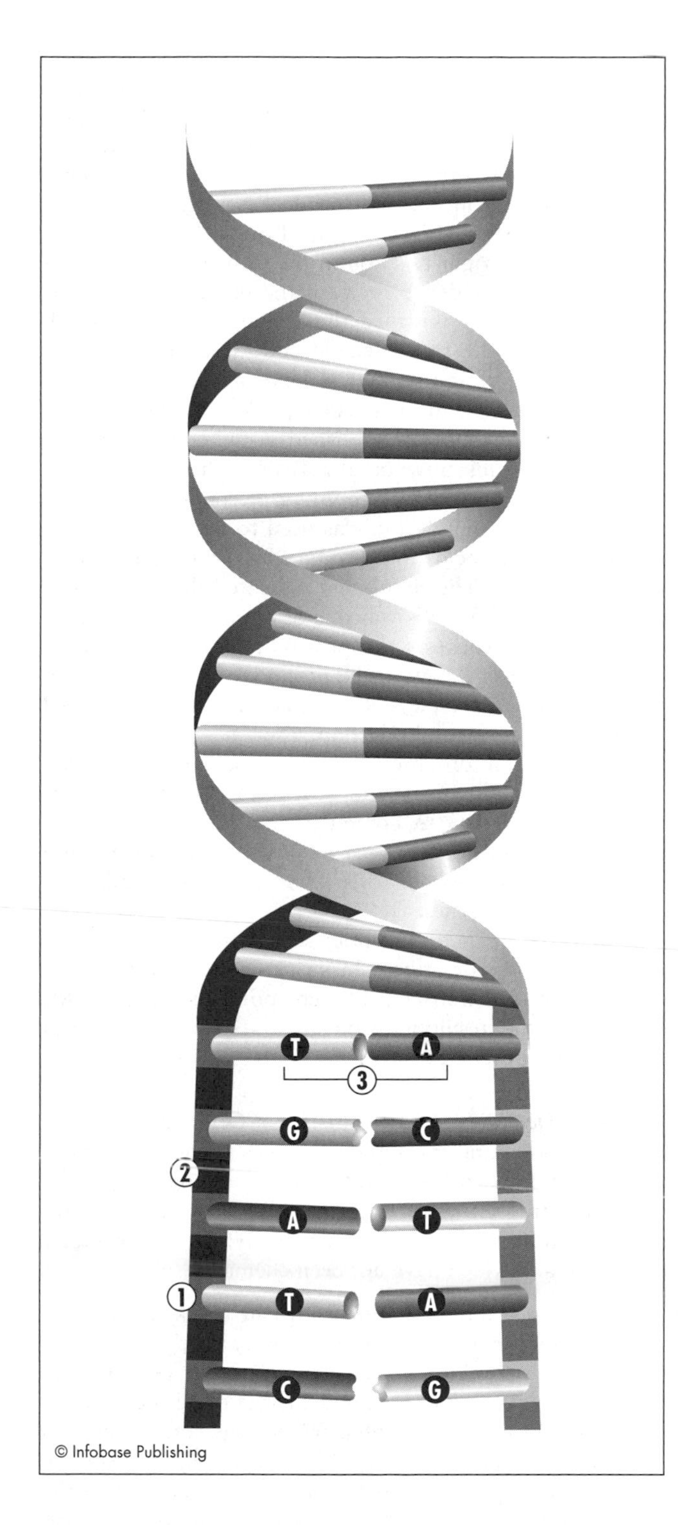

DNA is an extremely long molecule that consists of two strands that form a double helix (spiral). Each strand contains sugars, phosphates, and nitrogenous bases. Within each strand, sugars (1) and phosphates (2) form a backbone. Nitrogenous bases A, C, T, and G meet in the middle of the molecule between the two strands and hold them together by weak bonds. A is always opposite T, and C is always opposite G. This illustration does not show the arrangements of atoms.

molecule can become two, two can become four, and so on. This occurs only when the appropriate enzymes and raw materials are present to control the process.

Cells use DNA information in three ways:

- A single cell, such as a fertilized egg cell, can develop into a multicellular organism. Most of the 70 trillion cells in a human body contain nearly an exact copy of the DNA that was in the fertilized egg from which the person developed.
- During the daily operation of each cell, the genetic instructions in the DNA determine which proteins are made, and what the cell does.
- DNA can be passed on from one generation to the next. Usually this involves separating the information into two sets and then recombining them in sexual reproduction (see MEIOSIS; SEX, EVOLUTION OF). DNA is a potentially immortal molecule: Each person's DNA molecules are the descendants of an unbroken line of replication going back to the last universal common ancestor of all life on Earth (see ORIGIN OF LIFE).

In eukaryotic cells, DNA is organized into chromosomes (see EUKARYOTES, EVOLUTION OF). The evolutionary origin of chromosomes has not been explained. One hypothesis, proposed by John Maynard Smith (see MAYNARD SMITH, JOHN), is that within a chromosome all of the DNA must replicate at the same time. This system prevents some segments of DNA from replicating more often than others, and prevents any of them from getting lost. If each segment of DNA replicated on its own, some of them might replicate faster than others and lead to the death of the cell (see SELFISH GENETIC ELEMENTS). A nucleus can simultaneously replicate all of its chromosomes, of which most nuclei have fewer than 50, more easily than it could simultaneously replicate many separate DNA segments.

Two important facts result from the fact that genetic information is passed on by DNA:

- Acquired characteristics cannot be inherited. Lamarck, and Darwin, were wrong about this. Although various regulatory molecules can pass from one generation to another through egg cells, the only characteristics that are transmitted from one generation to the next are the ones coded by DNA. One 20th-century fundamentalist preacher said, "If evolution is true, why are Hebrew babies, after thousands of years of circumcision, still born uncircumcised?" It was not only evolution but DNA that this preacher misunderstood. His question, though ridiculous to modern scientists, might have bothered Lamarck.
- Traits that seem to have disappeared can reappear in a later generation. Traits do not blend together like different colors of paint, as scientists once believed; instead, the traits are discrete units. A trait may be hidden by other traits, but its DNA is still there and can reappear under the right conditions. This was also a problem with which Darwin wrestled, and which has been solved by an understanding of DNA and of genetics.

DNA encodes genetic information, allowing it to be passed on from one generation to the next in almost perfect form, but with just enough imperfection to allow mutations,

The Genetic Code

First base	Second base: A	G	T	C	Third base
A	AAA phe	AGA ser	ATA tyr	ACA cys	A
	AAG phe	AGG ser	ATG tyr	ACG cys	G
	AAT leu	AGT ser	ATT stop	ACT stop	T
	AAC leu	AGC ser	ATC stop	ACC trp	C
G	GAA leu	GGA pro	GTA his	GCA arg	A
	GAG leu	GGG pro	GTG his	GCG arg	G
	GAT leu	GGT pro	GTT gln	GCT arg	T
	GAC leu	GGC pro	GTC gln	GCC arg	C
T	TAA ile	TGA thr	TTA asn	TCA ser	A
	TAG ile	TGG thr	TTG asn	TCG ser	G
	TAT ile	TGT thr	TTT lys	TCT arg	T
	TAC met*	TGC thr	TTC lys	TCC arg	C
C	CAA val	CGA ala	CTA asp	CCA gly	A
	CAG val	CGG ala	CTG asp	CCG gly	G
	CAT val	CGT ala	CTT glu	CCT gly	T
	CAC val	CGC ala	CTC glu	CCC gly	C

There are 64 codons. The three-letter abbreviations are for the 20 kinds of amino acids found in cells. Three codons cause translation to stop. *The met (methionine) codon also marks the place where translation begins.

the raw material of the genetic variability of populations, and of natural selection.

How Genes Determine Proteins

DNA stores information as a four-letter alphabet (A, C, T, G) forming three-letter words called *codons*. Each codon of three nucleotides specifies one amino acid. Proteins are large molecules made up of smaller amino acids; therefore 3,000 DNA nucleotide pairs specify the order of amino acids in a protein that is made of 1,000 amino acids. The DNA that specifies the structure of one protein or group of related proteins is called a *gene*. The proteins do all of the work of the cell. A human is different from a snail largely because many of their proteins differ. An organism's characteristics result from the work of its proteins; and its proteins are specified by its DNA.

DNA remains in the nucleus of the cell. Proteins are manufactured out in the cytoplasm of the cell. Enzymes copy or *transcribe* genetic information from the DNA into *messenger RNA;* it is the messenger RNA that travels from the nucleus out to the cytoplasm. Each gene has a group of nucleotides (see PROMOTER) which identifies it and indicates where the gene begins. Different groups of genes have different kinds of promoters. A cell transcribes only the genes that have promoters that are appropriate for that cell's functions.

Once the messenger RNA molecule arrives in the cytoplasm, structures called *ribosomes* produce proteins whose amino acid sequence matches the RNA nucleotide sequence. Small molecules called *transfer RNA* attach to amino acids and bring them to the ribosomes. Each transfer RNA molecule recognizes and attaches only to its particular kind of amino acid. Each transfer RNA molecule recognizes only particular codons on the messenger RNA molecule. This is how the transfer RNA molecule brings the appropriate amino acid to the right position in the growing protein molecule. The correspondence between the nucleic acid codons and the amino acids is called the *genetic code*. Nearly all cells use exactly the same genetic code (see table at left). Mitochondria, and some ciliates, have a slightly different genetic code. Evolutionary scientists take this as evidence that the genetic code was established in the common ancestor of all life-forms now on the Earth. The genetic code seems not to be an arbitrary coupling of codons and amino acids. A computer simulation was used to randomly link up codons and amino acids and produced over a million alternate genetic codes. The simulation indicated that the genetic code actually used by cells was one of the most efficient possible codes. This suggests the possibility that the common ancestor of all cells was itself the product of a long period of evolution, during which less efficient genetic codes were tried and eliminated by natural selection.

DNA stores information digitally, just like a computer. A computer uses the bits 0 and 1, while DNA uses the four bases A, C, T, and G. A computer has bits organized into bytes, which specify letters, just as bases are organized into codons that specify amino acids. Bytes make up words, just as codons make up genes.

The processes of transcription and translation are more complex than here described. In particular, cell components can transcribe and translate different portions of the DNA, then modify the resulting protein, so that one gene can encode several different proteins.

How Genes Determine Characteristics of Organisms

The transcription and translation of genes produces proteins, which form many structures and do nearly all the work in the cell. No cell transcribes or translates all of its genes. It transcribes and translates only the genes for which the promoter site is open, and which have not been chemically altered:

- In some cases, inhibitor molecules can block a promoter site. The inhibitor molecule may consist partly of the end product of the series of reactions that the gene begins. When the end product is abundant, the end product itself helps to block the promoter. When the end product is scarce, the promoter is open. This process helps to keep the amount of the gene product more or less constant in the cell. Usually, the interactions of control molecules, most of them proteins, is very complex, especially in eukaryotic cells.
- In some cases, the genes can be altered by a process called *methylation*. The nucleic acid sequence of the methylated gene is intact, but the gene cannot be transcribed. In some cases an entire chromosome can be inactivated, as with one of the two X chromosomes in female mammals.

Methylation can have lifelong effects on an organism. In mice, when a mother neglects her offspring, the gene for the glucocortoid receptor is inactivated, resulting in lifelong anxiety in the offspring. Interestingly, it is not just the genes that can be inherited from one generation to another, but also the status of their activation. A gene can be passed on into later generations in an inactivated form. Examples include coat color in mice and flower form in toadflax. *Genetics* is the study of the inheritance of genes; *epigenetics* is the study of the inheritance of gene activation patterns. Sometimes, small RNA molecules (such as the interfering RNA molecules, RNAi) can chemically interfere with the transcription of a gene.

Once expressed, genes interact in complex ways. A cascade of gene expression occurs during embryonic development, in which homeotic genes activate other genes, which in turn activate still other genes (see DEVELOPMENTAL EVOLUTION). Even after development of the organism, one gene can affect another, a phenomenon called *epistasis.* Because of epistasis, two mutations that are individually deleterious can work together to produce a beneficial change in the organism; and two mutations that are individually beneficial can work together to produce a harmful effect on the organism. Because of development and epistasis, genes do not usually have individually recognizable effects on an organism. Therefore NATURAL SELECTION does not generally act upon a single gene.

Amount of DNA in Cells

Only a small part of the DNA of eukaryotic cells consists of genes. About 90 percent of the DNA of many eukaryotic species does not encode genetic information. Some of this NONCODING DNA plays an essential role in controlling the expression of the genes. For example, some noncoding DNA may play essential roles in coiling and uncoiling the DNA during cell division, and other noncoding DNA produces RNAi (see above). A great deal of the DNA appears to have no present function in the cells and has been called "junk DNA." Because scientists cannot be sure that it is useless, they prefer to call it noncoding DNA. What, if anything, most of the noncoding DNA does is one of the major unanswered questions of modern genetics.

Most eukaryotic cells have two chromosomes of each type. That is, most eukaryotic organisms consist mostly of *diploid* cells. In the cells of some species, extra copies of chromosomes can be present. Triploid cells have three chromosomes of each type, tetraploid cells have four, and pentaploid cells have five. Collectively, cells with more than two chromosomes of each type are called *polyploid* cells. In many plants, polyploid cells can function normally. Polyploidy allows hybrid plants to reproduce (see HYBRIDIZATION). Most animals cannot tolerate polyploidy. This is the reason that some plants have a huge amount of DNA in the nuclei of their cells.

Largely as a result of noncoding DNA and polyploidy, cells of different species differ enormously in the amount of DNA that they contain, and there is no relationship between the complexity of an organism and the amount of DNA that its cells contain. When only the genes are taken into account, a general pattern emerges in which more complex organisms have more genes (see table).

The number of genes is not a complete explanation for the complexity of an organism. Humans have about 35,000 genes, which may be the most of any organism. Mice have almost as many genes as humans do. Even the nematode worm *Caenorhabditis elegans,* which has only 959 cells in its entire body, has 20,000 genes. The complexity of an organism, therefore, cannot be explained simply by the number of its genes. The EMERGENCE of complexity in humans is largely due to the pathways of interaction among the genes.

Where New Genes Come From

Sometimes, a gene can be copied more than once, and the copy is inserted into another location among the chromosomes. This process is called *gene duplication.* The cell now has two copies of the gene instead of one. The two copies are now free to evolve in different directions and encode two different proteins rather than just one. One example of this is the globin gene, which encodes a protein that binds oxygen gas. This gene is found widely among animal species. This

DNA Comparison among Species

Organism	Genus	Million base pair total	% of DNA that is coding	Million base pairs of coding DNA
Bacterium	Escherichia	4	100	4
Yeast	Saccharomyces	9	70	6
Nematode	Caenorhabditis	90	25	22
Fruit fly	Drosophila	180	33	60
Cress plant	Arabidopsis	200	31	60
Human	Homo	3,500	9–27	320–950
Newt	Triturus	19,000	1.5–4.5	290–860
Lily	Fritillaria	130,000	0.02	30
Lungfish	Protopterus	140,000	0.4–1.2	560–1680

Human globins

ε	93	122	222	850	126
Aγ	93	122	222	866	126
Gγ	93	122	222	886	126
δ	93	128	222	886	126
β	93	130	222	850	126
α	93	113	204	141	129

Mouse globins

α	93	116	222	646	126
β	96	122	204	135	126

Coding sequence

Intron

Mice have two genes (more properly, transcription units) for globin proteins, which have resulted from a duplication of a single gene. Each of the genes has three very similar coding regions, and two introns in the same locations between the coding regions; in the β globin gene, one of the introns is much longer than the corresponding intron in the α globin gene. Humans also have α and β globin genes very similar to those of mice and in addition have genes for δ, Gγ, Aγ, and ε globins that are nearly identical to the β globin gene and resulted from duplication. Numbers inside the boxes represent the number of nucleotides in the coding region or the intron. ***(Redrawn from Freeman and Herron)***

gene has duplicated several times during animal evolution, producing a whole family of globin genes (see figure on page 136). One of them is myoglobin, which is found in muscles; two others are the components of hemoglobin, the protein that carries oxygen gas in the red blood cells. The gene duplication that produced the α and β genes occurred early in vertebrate evolution; all vertebrates have both of these genes, except hagfishes, which are descendants of one of the earliest divergences in vertebrate evolution. Fetal hemoglobin, slightly different from adult hemoglobin, is encoded by yet another gene in this complex. If the duplicated gene has no promoter, it becomes part of the noncoding DNA.

DNA in Evolution

Not only are the physical characteristics of organisms, such as humans, encoded within the DNA but also many behavioral characteristics. Individual human behavior is strongly influenced by the levels of brain chemicals such as neurotransmitters and neuropeptides. Individual humans differ in the levels of these chemicals, which causes some people to eat more than others, or to seek adventure more than others. In part, the levels of these chemicals are determined by things that happen during the individual's lifetime—environmental factors, education, reinforcement of behavior by habit. But they are also strongly influenced by control sequences of DNA that regulate the transcription of the genes and the manufacture of the proteins. Many aspects of human behavior, including sexual orientation, have a genetic as well as an environmental component. While genes are not known to ever force an individual to do something (to eat more or less, or to seek more or less stimulation), they definitely influence human inclinations (see essay, "How Much Do Genes Control Human Behavior?")

DNA provides raw material for evolution by reliably passing genetic traits to the next generation and by allowing limited mutations. Gene duplication also contributes raw material to the evolutionary process. Even noncoding DNA can contribute to evolution if it alters the transcription of genes or is reactivated and becomes a gene. All of this contributes to the stability and variability of genetic information that must be present in populations for evolution to occur. Geneticists have analyzed DNA, and proteins, and traits, and found that wild populations contain a great deal of variability. This confirms what Darwin established in his landmark 1859 book (see ORIGIN OF SPECIES [book]), but the basis of which he did not understand. Modern evolutionary science therefore represents a convergence of Darwinian natural selection, Mendelian genetics (see MODERN SYNTHESIS), and the chemistry of DNA.

Further Reading

Ast, Gil. "The alternative genome." *Scientific American,* April 2005, 58–65.

Chiu, Lisa Seachrist. *When a Gene Makes You Smell Like a Fish... And Other Amazing Tales About the Genes in Your Body.* New York: Oxford University Press, 2006.

Freeman, Scott, and Jon C. Herron. "Mutation and Genetic Variation." Chap. 4 in *Evolutionary Analysis,* 3rd ed. Upper Saddle River, N.J.: Pearson Prentice Hall, 2004.

Lynch, Michael, and John S. Conery. "The evolutionary fate and consequences of duplicate genes." *Science* 290 (2000): 1,151–1,155.

Maynard Smith, John, and Eörs Szathmáry. *The Origins of Life: From the Birth of Life to the Origins of Language.* New York: Oxford University Press, 1999.

Rice, Stanley, and John McArthur. "Computer analogies: Teaching molecular biology and ecology." *Journal of College Science Teaching* 32 (2002): 176–181.

Russo, V. E. A., et al., eds. *Development: Genetics, Epigenetics and Environmental Regulation.* New York: Springer, 1999.

Watson, James D., with Andrew Berry. *DNA: The Secret of Life.* New York: Knopf, 2003.

Dobzhansky, Theodosius (1900–1975) Russian-American *Evolutionary geneticist* Theodosius Dobzhansky was one of the principal architects of the MODERN SYNTHESIS in which MENDELIAN GENETICS was united with Darwinian NATURAL SELECTION (see ORIGIN OF SPECIES [book]) to form the modern theory of evolution.

Born January 25, 1900, Dobzhansky was a teenager during the tumultuous time of the Russian Revolution. Immediately afterward, he attended the University of Kiev and graduated with a degree in biology. He taught genetics at the Polytechnic Institute in Kiev until 1924 when he joined the genetics department at the University of Leningrad (now St. Petersburg). Dobzhansky's wife, Natalia Sivertzev, was also a geneticist, and they conducted some joint research. Dobzhansky was familiar with the work of Thomas Hunt Morgan at Columbia University in New York. Like Morgan, Dobzhansky decided to study the effects of genes and their interaction in the fruit fly, which is still one of the major experimental organisms in genetic studies (see DNA [RAW MATERIAL OF EVOLUTION]). In 1927 Dobzhansky had the opportunity to travel to New York and work in Morgan's lab. He went with Morgan to the California Institute of Technology and received a faculty appointment there. This proved fortunate for him, as it allowed him to escape from the upheaval that was occurring in Russian genetics. Mendelian genetics, which was universally accepted by modern scientists, was being forcibly replaced by a version of Lamarckism that was favored by the Soviet government (see LYSENKOISM). Dobzhansky remained in California until 1940, when he returned to New York for a faculty appointment at Columbia University. In 1962 he joined the faculty of the Rockefeller Institute (now Rockefeller University), from which he retired in 1970. Until his death in 1975, Dobzhansky taught part-time at the University of California at Davis.

Dobzhansky did not confine his biological interests to the laboratory. He loved to hike, and this brought him in contact with the populations of natural species. This allowed him to recognize the tremendous genetic variation that occurs not just in laboratory populations of fruit flies but in every natural population. He understood that mutations, which produced genetic diversity, were usually numerous and had small effects, and that these mutations were acted upon by natural selection. Natural selection among these small mutations was what produced new species (see SPECIATION). In this, he differed from the mutationist view, which prevailed when he started his career. The mutationist view, championed by prominent geneticists (see DEVRIES, HUGO), claimed that mutations were large and that they produced new species. Other architects of the modern synthesis were also biologists who had observed the extensive variation of individuals within wild populations (see MAYR, ERNST; STEBBINS, G. LEDYARD). Dobzhansky's book, *Genetics and the Origin of Species,* published in 1937, was one of the most important books in the modern synthesis.

Dobzhansky also applied his genetic insights to an understanding of human evolution. Appalled by the racism that had gripped Europe during the 20th century, Dobzhansky defended the idea that there was more genetic variation within races than between them, which meant that racial genetic differences were not important in human evolution or in modern society. He expressed these views in influential books such as *The Biological Basis of Human Freedom* (1954), *Mankind Evolving: The Evolution of the Human Species* (1962), and *Genetic Diversity and Human Equality* (1973).

Dobzhansky was also a religious man, who maintained his participation in the Russian Orthodox Church. He also believed that the evolutionary process had allowed humans to transcend mere physical existence, entering into the realm of spiritual awareness. He expressed this view in yet another important book, *The Biology of Ultimate Concern* (1967).

One of Dobzhansky's most important contributions to modern education is his insight that "nothing in biology makes sense except in the light of evolution." This is one of the most often-quoted lines in modern biological education. Everything from the components of the cell, to the structure of genes, to the anatomy of the human body, to the function of ecosystems, seems confusing when looked at from the viewpoint of design by a Supreme Being (see INTELLIGENT DESIGN) but makes sense only as the product of an evolutionary process. Had Dobzhansky lived to see biology at the beginning of the 21st century, he might have been astounded at how correct his statement had proved to be (see DNA [EVIDENCE FOR EVOLUTION]; ECOLOGY; EUKARYOTES, EVOLUTION OF; SELFISH GENETIC ELEMENTS; NONCODING DNA; SYMBIOGENESIS). He died December 18, 1975.

Further Reading

Adams, M. B., ed. *The Evolution of Theodosius Dobzhansky: Essays on His Life and Thought in Russia and America.* Princeton, N.J.: Princeton University Press, 1994.

Ayala, Francisco J. "Theodosius Dobzhansky: A man for all seasons." *Resonance,* October 2000, 48–60.

Dobzhansky, Theodosius. "Nothing in biology makes sense except in the light of evolution." *American Biology Teacher* 35 (1973): 125–129.

Dubois, Eugène (1858–1940) Dutch *Physician, Anthropologist* Marie Eugène François Thomas Dubois was the discoverer of the original Java man specimens (see *HOMO ERECTUS*). He was the first man to actually go looking for fossils of human ancestors. Earlier discoveries of human ancestors had been accidental (see NEANDERTALS).

Born January 28, 1858, Dubois was a Dutch physician who became passionately interested in the study of human evolution. The only ancient human specimens that had been found were Neandertals, and these were similar enough to modern humans as to be classified in the same genus, *Homo.* Dubois decided to find some more ancient ones. In 1887 he set out for Sumatra. Early in the 19th century Buffon (see BUFFON, GEORGES) had speculated that Asia was where humans originated, even though his theory was not evolutionary. Sumatra was part of the Dutch territory in Indonesia, so Dubois could travel there without difficulty.

Dubois joined a hospital staff and searched for fossils in his spare time. His first two years produced no results, but in 1889 he heard that a human-like fossil had been found on the nearby island of Java. Excited, Dubois went to investigate, only to find that the specimen was of a modern human. He decided to remain and look for fossils himself on Java. He established a camp near Trinil. In 1890 he found a molar and a skullcap that had features that appeared to be intermediate between apes and humans. In particular, the skullcap was low and flat on top, having housed a small brain, and had prominent brow ridges that were characteristic of apes but not of humans (except for Neandertals). The estimated size of the brain was larger than that of any ape and smaller than the range of modern humans.

Two years later, he found a femur (upper leg bone). The fact that the femur was found only 15 meters away from the tooth and skullcap, and from the same layer, suggested to Dubois that they were from the same species, perhaps the same individual. The femur bent inward, which is a characteristic of animals that walk upright. The length of the femur, moreover, suggested that this species had relatively long legs and short arms, which characterizes humans, not apes. Dubois was convinced that he had found the remains of a species that was intermediate between humans and apes and walked erect: In the article in which he announced his discoveries, he called it *Anthropopithecus erectus,* the human-ape that walked upright. Later he changed the genus name to *Pithecanthropus,* which was the name that the German evolutionist Haeckel (see HAECKEL, ERNST) had used to describe the hypothetical ape-man that he was sure had to have existed. For the next several decades, scientific articles and books referred to Java man as *Pithecanthropus erectus.* After anthropologists determined that Java man was very similar to Peking man and that it belonged in the human genus, both were subsumed into the species *Homo erectus.*

While not every scientist accepted Dubois's conclusions, many did. Dubois himself began to doubt that the specimens really were an intermediate between apes and humans. The more he studied the Southeast Asian gibbon, the more he came to believe that Java man was just a large gibbon. In his later years, Dubois kept the Java man specimens hidden underneath the floorboards of his house. This reduced, but did not eliminate, the credibility of his discoveries. To this day creationists claim that *Homo erectus* fossils can be ignored because their discoverer lost confidence in them (see CREATIONISM). Many other specimens of *Homo erectus,* and of related species such as *HOMO ERGASTER,* have been found that leave no doubt about either the existence or the characteristics of this hominin species. Dubois died December 16, 1940.

Further Reading

Lewin, Roger, Carl Swisher, and Garniss H. Curtis. *Java Man: How Two Geologists' Discoveries Changed Our Understanding of the Evolutionary Path to Modern Humans.* New York: Scribners, 2000.

Shipman, Pat. *The Man Who Found the Missing Link: Eugene Dubois and His Lifelong Quest to Prove Darwin Right.* Cambridge, Mass.: Harvard University Press, 2002.

duplication, gene *See* DNA (RAW MATERIAL OF EVOLUTION).

ecology Ecology is the study of the relationships between organisms and their living and nonliving environments. As such, it is the context within which evolutionary adaptations operate and evolutionary changes occur. Ecology can be divided broadly into two areas: the relationship between an organism and its nonliving environment (*autecology,* from Greek *auto-,* meaning "self") and the relationship between an organism and other organisms (*synecology,* from Greek *syn-,* meaning "with").

Autecology

I. *Energy balance of the Earth and organisms.* Every organism must maintain the proper balance of heat and chemical energy. An organism must remain within a relatively narrow range of temperature if it is to survive and must obtain energy for its metabolism.

A. *Energy balance of the Earth.* The surface of the Earth obtains almost all of its energy in the form of photons from the Sun. The clouds, ice, and light-colored rocks of the Earth reflect some of the sunlight, called *albedo,* back into outer space. Albedo does not increase the temperature of the Earth. The darker rocks, water, and organisms of the Earth absorb sunshine. The sunshine increases the temperature of the Earth and all its surface components, which results from increased movement (kinetic energy) of its molecules. The Earth cannot conduct kinetic energy into outer space. The only way in which the Earth can cool off is to radiate low-energy infrared photons into outer space. The Sun shines visible photons on one side of the Earth; invisible photons radiate outward in all directions from the Earth.

Over time, the input and output of photons to and from the Earth is in balance. The average temperature of the Earth can temporarily fluctuate. Some of the infrared photons are absorbed by greenhouse gases such as carbon dioxide, methane, and water vapor, with the result that high concentrations of these gases can cause the atmosphere to become warmer. This is the source of the GREENHOUSE EFFECT. Methane reacts with oxygen in the atmosphere, producing carbon dioxide. Plants remove carbon dioxide from the air (see PHOTOSYNTHESIS, EVOLUTION OF). In this way, photosynthesis prevents an excessive greenhouse effect from occurring. In fact, excessive photosynthesis may have been a contributing factor to worldwide glaciations that occurred during PRECAMBRIAN TIME (see SNOWBALL EARTH).

B. *Chemical energy in organisms.* All organisms need chemical energy, to build the complex molecules that make up their bodies and to run their metabolic reactions which they use for the uptake of molecules, for growth, for movement, and for reproduction. Plants absorb photons of sunlight and transfer them to the potential energy of chemical bonds in sugar molecules (see PHOTOSYNTHESIS, EVOLUTION OF). From these sugar molecules, plants make all of their other molecules. Animals obtain chemical energy by eating plants or by eating other animals that have eaten plants. When plants and animals die, decomposers obtain chemical energy from them. The result is the food chain or food web. The evolution of every organism has been strongly influenced by its place in the food web. Since only 10 percent of the food energy is transferred from one step in the food chain to the next, the total weight of photosynthetic organisms is usually 10 times as great as the total weight of herbivorous animals, which is 10 times as great as the total weight of carnivorous animals. Since 100 calories of sugar in plants can support only one calorie of carnivore, carnivores are often rare and have evolved adaptations that allow them to search for and pursue prey over a wide area.

C. *Energy balance of organisms.* Organisms obtain some of their warmth from sunshine and the warm air around them, and the rest from their food. Organisms must also get rid of energy, thus maintaining an energy balance. The energy that originally entered the organisms as photons or as food can be lost in three ways. First, organisms can radiate infrared photons. Warm-blooded animals radiate more infrared energy than do cold-blooded animals, plants, and their surroundings.

Second, organisms can conduct kinetic energy into the air or water around them. Third, plants and animals on dry land can evaporate water. A great deal of energy is required to evaporate water, therefore evaporation (transpiration from plants, perspiration from animals) is a very efficient cooling mechanism.

The evolution of each species has been influenced by the necessity of maintaining energy balance. For example, desert bushes have evolved small leaves that allow them to disperse more of their heat load in the form of kinetic energy rather than the evaporation of water, which is rare in their habitats. At the same time, these bushes have often evolved deep roots, which maximize the amount of water they can obtain. Plants and animals from separate evolutionary lineages have often evolved similar adaptations to these climatic conditions (see CONVERGENCE). Animals need to increase their energy loss in hot environments and restrict it in cold environments. Warm-blooded animals have evolved thick layers of hair or fat to insulate them in cold environments. Hibernation is an evolutionary adaptation that allows animals to avoid excessive energy loss during winter.

The temperature and rainfall patterns of different parts of the Earth are influenced by the movements of the Earth, as well as the arrangement of the continents. Before the Pleistocene epoch (see QUATERNARY PERIOD), oceanic currents carried warm water into North Polar regions, making them much warmer than they are today. The movement of continents subsequently closed off some of this circulation, causing the North Polar region to become permanently cold and beginning a cycle of ICE AGES (see CONTINENTAL DRIFT).

II. *Obtaining matter.* Organisms not only need energy but also need matter. Photosynthesis removes carbon dioxide gas from the air or water and makes sugar from it. Animals and decomposers obtain matter from the food chain at the same time and from the same sources that they obtain energy. All organisms, including plants, release carbon dioxide gas back into the air or water. Organisms also need nitrogen and minerals such as phosphorus, potassium, and calcium. Plants obtain them mostly from the soil by absorbing inorganic chemicals with their roots. Plants store some of these minerals and use others in the construction of their large organic molecules. Animals and decomposers obtain minerals from their food.

Synecology

I. *Interactions within populations.* A POPULATION is all of the individuals of a species that interact. Populations can grow exponentially because of the reproduction of the individuals that make it up. Species have evolved different REPRODUCTIVE SYSTEMS that allow individuals to successfully produce offspring and disperse them into new locations. Populations also contain genetic variability (see POPULATION GENETICS), which is essential for NATURAL SELECTION. Natural selection occurs within populations, and the reproductive success (fitness) of an individual is relative to the other members of the population. Individuals compete with one another not only for resources such as food and territory but also for opportunities to mate (see SEXUAL SELECTION). Different populations in a species can become new species (see SPECIATION). Practically every aspect of evolution is influenced by the synecology of populations.

II. *Interactions between species.* Species have evolved to use one another in many different ways. The food chain has already been mentioned. Since evolution has produced millions of species, each of which makes its living in a slightly different way, evolution has produced many complex interactions among species. Plant species have evolved different ways of protecting themselves from herbivores, including thousands of kinds of toxic chemicals, and herbivores have evolved ways of getting around the plant defenses. Prey animals have evolved many ways of protecting themselves from predators, and predators have evolved many ways of finding and eating the prey. Many animals obtain their food by pollinating the plants that have made use of those animals to transport their pollen. Many species have evolved very close relationships, called *symbioses,* in which the other species is the most important environmental factor. In many cases, symbioses have evolved toward the mutual benefit of both species (mutualism). Many ecological interactions have resulted from COEVOLUTION. In some cases, one species has evolved such a dependence on another that its individuality has been lost and the two have fused into one (see SYMBIOGENESIS).

III. *Ecological communities.* The total of all the interspecific interactions in a location is the ecological community. This is the context within which all evolutionary change occurs. Evolution has produced a great diversity of species within each community (see BIODIVERSITY). Biodiversity results from a balance between speciation and EXTINCTION. The movements of continents not only influence the climate, as explained above, but also separate species and bring species together, allowing new interactions to occur and new biodiversity to evolve. When new islands emerge from the ocean, new species quickly evolve (see BIOGEOGRAPHY). Disturbances such as fire and storm occur within ecological communities, which are disastrous to some individuals, but create openings that can be colonized by others. Species diversity has been much enhanced by the evolution of species that specialize on disturbed areas. Many human activities create disturbances. The disturbances created by humans, however, may be occurring more rapidly than ecological communities or the evolutionary process can handle, with the result that human activity may now be causing the sixth of the Earth's great MASS EXTINCTIONS.

IV. *Ecosystems.* An ecosystem is an ecological system that incorporates the ecological community with the flow of energy and cycling of matter. Ecosystem ecology ties together autecology and synecology by considering both organisms and the nonliving environment as components of an interacting system.

At every level, from the energy balance of an individual organism to the interactions of all the species in the world, evolution occurs within an ecological context and has made and continues making that ecological context. The study of evolution is inseparable from the study of ecology.

Further Reading

Bazzaz, Fakhri A. *Plants in Changing Environments: Linking Physiological, Population, and Community Ecology.* New York: Cambridge University Press, 1996.

Hagen, Joel B. *An Entangled Bank: The Origins of Ecosystem Ecology.* New Brunswick, N.J.: Rutgers University Press, 1992.

Mayhew, Peter J. *Discovering Evolutionary Ecology: Bringing Together Ecology and Evolution.* New York: Oxford University Press, 2006.

Ediacaran organisms The Ediacaran organisms lived immediately before the CAMBRIAN PERIOD, which began about 540 million years ago. They are among the few multicellular organisms in the PRECAMBRIAN TIME. The abundance of multicellular forms after the CAMBRIAN EXPLOSION contrasts sharply with the relatively few multicellular forms represented by the Ediacarans.

The earliest Ediacarans lived about 580 million years ago. These marine organisms may have had a worldwide distribution, as their fossils are found in 24 locations on six continents, including Australia (especially the Ediacara Hills after which they are named), Newfoundland, and Namibia.

Reginald Sprigg, an Australian geologist, was investigating mines in the Ediacara Hills in 1946. While eating lunch, he turned over a rock. He was not expecting to find any fossils, since the rocks were Precambrian in age, and at the time no multicellular organisms were known from before the Cambrian explosion. But he saw many leaflike impressions in the rock. He reported his findings at meetings, but they were rejected for publication. Then in 1957 John Mason, a student, found a fossil in England, also of Precambrian age, that closely resembled Sprigg's Australian fossil. Now that these fossils were being found more than once and in such diverse places, they were taken seriously. Sprigg's priority as discoverer has been recognized by calling them Ediacaran fossils.

Dickinsonia costata **is an example of an Ediacaran organism. Ediacaran organisms are of unknown affinities and lived about 580 million years ago.** ***(Courtesy of Ken Lucas/Visuals Unlimited)***

Nobody is entirely sure what these organisms were. Some of them bear at least a superficial resemblance to existing kinds of marine invertebrates (see INVERTEBRATES, EVOLUTION OF). For example, *Tribrachidium, Arkarua,* and *Albumares* resemble modern echinoderm "sand dollars"; *Mawsonites* resembles a modern cnidarian sea anemone; *Spriggina* and *Bomakellia* resemble the arthropod trilobites that were abundant in oceans until about 250 million years ago; and *Charniodiscus* and *Pteridinium* resemble modern sea pens, all of them invertebrate animals known from the Cambrian period (see figure on page 141). These resemblances are often pointed out because it is reasonable to expect the immediate ancestors of Cambrian animals to have lived during the Ediacaran period.

There are numerous puzzling differences between the Ediacaran organisms and the invertebrate animals they resemble. In particular, the Ediacarans are not known to have had legs, or mouths. While many invertebrate animals lack legs, none lack mouths. If Ediacarans were animals, how and what did they eat?

Another difficulty in identifying Ediacarans as animals related to modern invertebrates is the question of whether the modern phyla of animals were yet in existence. Molecular studies (see DNA [EVIDENCE FOR EVOLUTION]) suggest that the diversification of the modern groups of invertebrates had already begun. Precambrian fossils of what are clearly animal embryos and what were probably bilaterally symmetrical animals have been found. These observations suggest that some modern groups of invertebrates were in existence during Ediacaran times. Although animal burrows appear in the Precambrian fossil record, they do so only right before the Cambrian explosion.

Some researchers claim that the Ediacarans had to be animals because they moved. In several examples, Ediacaran fossils are found in groups of four: three depression fossils, and one with positive relief. The interpretation given in this case is that the Ediacaran organism flopped around four times, leaving three prints in the bottom of the sea where they lived; the fourth, positive structure is the fossil of the organism itself. However, this does not prove that they moved themselves; they may have been moved by ocean currents.

Other researchers believe that the Ediacarans were organisms that are completely unrelated to modern animals; in fact, they may not have been animals at all. Consider, for example, the *Spriggina* fossil. To human observers it looks like a worm or trilobite with a head, but if an observer turns it upside down he or she may see a seaweed-like structure with a holdfast to anchor it to the bottom of the ocean. Researchers generally accept that animals were present at that time. Fossilized burrows and other trace fossils (evidence of things the organisms did, rather than the organisms themselves) have been found. But that does not prove that the Ediacarans were the animals that left the traces.

Although red algae are known from 1.2 billion years ago, long before the Ediacaran organisms, the Ediacarans possess no structures that identify them as seaweeds, just as they possess no structures clearly identifying them as animals. Researchers such as paleontologists Dolf Seilacher and Mark McMenamin have suggested that the Ediacarans were actually symbiotic combinations of protist cells, some of which were photosynthetic (see EUKARYOTES, EVOLUTION OF; SYMBIOGENESIS). Many symbiotic combinations of plantlike and animal-like protist cells are known from the modern world, such as *Ophrydium,* a protist colony that lives underwater. The symbiosis would explain how the Ediacarans obtained food without having mouths: The plantlike cells made food from sunlight. It might even explain why many of these fossils were sometimes found to be filled with sand: They contained sand, as ballast, while alive, which would have been inconvenient for an animal but perfectly workable for symbiotic bags of cells. Alternatively, these organisms could have lived by absorbing dissolved organic materials. They may even have been large single-celled protists, similar to the modern oceanic xenophyophores. In 1992 Dolf Seilacher proposed that the Ediacaran organisms were a separate kingdom of life, which he called Vendobionta.

It is possible that some of the Ediacarans were members of modern invertebrate phyla, while some were not. After all, the resemblance of *Charniodiscus* to the Cambrian cnidarian *Thaumaptilon* (see BURGESS SHALE) and to modern sea pens is very close. Sea pens, like many other cnidarians, form symbiotic associations the way many researchers claim the Ediacarans did. Therefore some Ediacarans may have been cnidarians but also had endosymbiotic algae living in their tissues, just as many coelenterates such as jellyfish and corals do today.

One of the major features of the Ediacarans is that they had no hard shells, spines, or other defenses against predators. The implication is that predatory animals had not yet evolved. Once predators evolved, the Ediacarans may have been driven to extinction by them; only the hard-shelled animals or chemically defended seaweeds could survive. It has been suggested that the appearance of the first predators was the stimulus for the Cambrian explosion.

Ediacaran organisms were the first to become prominent after a prolonged period of global cooling that has been called SNOWBALL EARTH. A long period of Earth history, in which (aside from some seaweeds) most organisms were colonies of single cells (such as stromatolites), came to an end with the global spread of ice on land and over most of the oceans as well. Organisms survived only underneath a few openings in the ice. It was in the rapid and possibly severe period of global warming that followed Snowball Earth that multicellular organisms, first the Ediacarans, then the Cambrian organisms, evolved.

Further Reading

Conway Morris, Simon. *The Crucible of Creation: The Burgess Shale and the Rise of Animals.* New York: Oxford University Press, 1998.

McMenamin, Mark A. S. *The Garden of Ediacara: Discovering the First Complex Life.* New York: Columbia University Press, 1998.

Shu, D.-G., et al. "Lower Cambrian vendobionts from China and early diplobiont evolution." *Science* 312 (2006): 731–734.

Eldredge, Niles (1943–) American *Paleontologist* Niles Eldredge is Curator of Invertebrates at the American Museum of Natural History in New York City. He is most famous for two things: first, he and evolutionary biologist Stephen Jay Gould (see GOULD, STEPHEN JAY) proposed the theory of PUNCTUATED EQUILIBRIA in 1972; second, Eldredge has written 23 books that explain the evidence for evolution, how evolution works, and what it means for the protection of BIODIVERSITY on the Earth.

Niles Eldredge was born August 25, 1943. When he was an undergraduate, his principal interest was in languages but soon turned to anthropology. This led him to an interest in fossils, and he obtained a Ph.D. in paleontology from Columbia University in New York. In 1969 he began employment at the American Museum of Natural History, where he had done both undergraduate and graduate research. The focus of his research for three decades has been TRILOBITES, arthropods that were abundant in the oceans during the PALEOZOIC ERA.

In order to understand how evolution works, it is necessary to have a good fossil record over a long period of time. Trilobites provide such a record. Because of their hard shells and abundance, trilobites are well represented in the fossil record, over almost the entire Paleozoic era. As Gould and Eldredge studied fossils, they noticed two patterns: First, species came into existence in a relatively short period of time (a punctuation), and second, once a species came into existence, it persisted for a long time virtually unchanged (an equilibrium). Most evolutionary scientists had attributed this to the incompleteness of the fossil record, but Gould and Eldredge claimed that scientists should believe the overall reliability of the paleontological data, particularly a data set as rich as that of the trilobites. From this they proposed the theory of punctuated equilibria. Eldredge has presented this argument for the general public in books such as *Reinventing Darwin* and *The Pattern of Evolution.*

After Gould and Eldredge had explained the pattern of evolutionary history, they investigated the processes that may have produced it. While they completely accepted Darwinian NATURAL SELECTION, they rejected the idea that the process of natural selection acting gradually in populations over a long period of time was adequate to explain the pattern of punctuated equilibrium. Evolution, they claimed, does not operate only on the level of populations but also selects among species. In particular, Eldredge has rejected the gene-level selection process proposed by Richard Dawkins (see DAWKINS, RICHARD), in his recent book *Why We Do It.*

Eldredge has explored the effects of the evolution of human intelligence and culture (see INTELLIGENCE, EVOLUTION OF) on the evolution of other species. The way that humans view their role upon the Earth is crucial to the future of life on Earth, a viewpoint Eldredge defended in his books *Dominion* and *Life in the Balance.*

Eldredge has also written and spoken extensively about the evidence against CREATIONISM, which he summarized in his book *The Triumph of Evolution and the Failure of Creationism.*

Further Reading

Eldredge, Niles, and Stephen Jay Gould. "Punctuated equilibria: An alternative to phyletic gradualism." In *Models in Paleobiology,* edited by T. J. M. Schopf, 82–115. San Francisco, Calif.: Freeman, Cooper and Co., 1972.

———. *Darwin: Discovering the Tree of Life.* New York: Norton, 2005.

———. *Dominion.* Berkeley, Calif.: University of California Press, 1997.

———. *Life in the Balance: Humanity and the Biodiversity Crisis.* Princeton, N.J.: Princeton University Press, 1998.

———. *The Pattern of Evolution.* New York: Freeman, 1999.

———. *Reinventing Darwin: The Great Debate at the High Table of Evolutionary Theory.* New York: John Wiley, 1995.

———. *The Triumph of Evolution and the Failure of Creationism.* New York: Freeman, 2000.

———. *Why We Do It: Rethinking Sex and the Selfish Gene.* New York: Norton, 2004.

emergence Emergent properties are complex properties that result from the interaction of simpler components. Emergent properties are essential for understanding evolution because they reveal how complexity can result without a complex control system to produce and maintain it. The information that is contained within an emergent system does not have to be stored in DNA or imposed by an intelligent designer (see INTELLIGENT DESIGN). The simpler components of an emergent system do not themselves contain the information for the structure of the complex system of which they are a part. One important reason for emergent properties is that the components of the system interact with one another and modify their behavior in response to one another, a process which could be called learning. The following are examples, rather than a thorough explanation of the topic.

Emergent properties of atoms and molecules. Atoms are made of particles, but atoms have properties that the particles do not. Particles interact with one another when they make up atoms; electrons in atoms do not act like individual electrons. Molecules are made of atoms, but molecules have properties that the atoms do not have. Neither hydrogen nor oxygen has the properties of water. The atoms interact with one another when they make up molecules; oxygen atoms in water do not behave like individual oxygen atoms but share their electrons with the hydrogen atoms.

Emergent properties of cells in an organism. As an organism develops from a single egg cell, the cells respond to one another. Development occurs as a cascade of reactions, beginning with instructions from some homeotic genes, which cause other gene expressions, which cause yet other gene expressions (see DEVELOPMENTAL EVOLUTION). While each cell contains a complete set of instructions in its DNA regarding what kinds of proteins to make (see DNA [RAW MATERIAL OF EVOLUTION]), no cell consults all of that information. Each cell has its own particular job that it performs in response to the instructions that it originally received. The complexity of a body results from the very limited range of tasks performed by each kind of cell. The genes code not for the structure but for the instructions to make the structure emerge.

In complex animals, nervous systems coordinate movements and responses. Sponges are multicellular animals but do not have nervous systems (see INVERTEBRATES, EVOLUTION OF). Choanocytes are cells that line the internal passages of a sponge. The movements of their whip-like flagella create currents of water that flow in through pores on the side, and out through a chimney-like structure at the top, of the animal. The choanocytes beat their flagella in rhythm, but there is no nervous system that coordinates this rhythm. The chaonocytes develop the rhythm in response to one another, which makes it an emergent property.

Emergent systems of organisms. A colony of ants is a very complex structure. The queen lays eggs, and workers carry out numerous tasks. As a result, many jobs get done by the ant colony. Workers explore their environments, find food, and bring it back to the colony. Yet no individual, and no set of instructions, is in control of the process. The queen is an egg-laying machine, rather than a ruler. None of the individual worker ants have information about the entire colony or about all of the surrounding environment. Each worker explores more or less at random, and if she finds a food source, she deposits pheromones, which are chemicals that signal to the other ants that this is the direction to go in order to find food. The crucial step is that each individual learns a little bit from other individuals. This limited amount of learning on the part of the individuals allows the colony as a whole to learn complex behavior. The ant colony displays what looks like intelligence to a human observer, but none of the individual ants have intelligence. The complex behavior of the colony is an emergent property.

A similar process of chemical communication causes individual slime mold cells to coalesce into a single reproductive structure. Slime mold cells live underneath leaf litter, each one obtaining its own food individually under warm, moist conditions. If conditions become cooler or drier, the behavior of the individual cells changes, and they begin to come together. They form a stalk, in a capsule at the top of which some of the individuals produce spores that blow away in the wind. None of the slime mold cells is in charge of organizing this activity. But as each cell responds to and learns from a few others, they assemble themselves into a complex structure. The reproductive behavior of slime molds is an emergent property.

Emergent properties of brains. A brain, especially the human brain, can store an immense amount of information. Yet each individual neuron stores very little information. The brain as a whole is able to learn and store information, because each neuron responds to and learns from the other neurons with which it is in contact. A complex structure of information storage emerges. While parts of the brain specialize on different functions, most brain functions are shared by several parts of the brain. PET studies of the brain (positron-emission tomography, which reveals the parts of the brain that are active during different functions such as talking, grieving, or experiencing inspiration) have shown that nearly every activity involves more than one part of the brain, and individuals differ at least a little bit in which parts of their brains are active during certain functions.

Brainlike properties can emerge from a computer. When computer subroutines or algorithms interact with one another, and learn from each other, complex outputs can be produced, even though each algorithm is simple and has no information about what the rest of the computer is doing (see EVOLUTIONARY ALGORITHMS). This is particularly true of neural net computing, which is designed to imitate the way neurons interact within a brain.

Emergent properties of human interactions. Groups of interacting humans, who learn from one another and alter their behavior accordingly, show emergent properties. This is why cities and whole societies take on spontaneous structures even when no individual or government agency imposes the order. The emergent properties are unplanned and frequently surprising. One would expect the World Wide Web to produce emergent properties. It usually does not do so, however, because most of the interactions are one-way: Users access information, but the Web sites that they access remain unchanged by the users. Some Web sites gather information about the users or solicit feedback from the users and apply this information to modify themselves; in these cases, emergent properties begin to appear. This is how Web sites can recommend other Web sites to users, without any human being supervising the process.

Some scientists study evolution not by studying the natural world or designing laboratory experiments but by using computer models that explore the kinds of complexity that can emerge from the seeming chaos of simpler interacting components. The Santa Fe Institute in New Mexico attracts scientists who wish to explore the evolutionary implications of what they call *complexity theory.*

Emergence explains how complexity can arise from simplicity rather than having to be imposed by a complex information structure. No cell is in charge of embryonic development, no ant is in charge of the colony, no cluster of neurons is in charge of the brain, and no person is in charge of a human society. Yet complexity emerges from all of them. NATURAL SELECTION acts upon the results of the complexity. This is one reason that natural selection usually cannot act upon a single gene: No gene acts in isolation, but as part of a network of interactions.

Further Reading

Barabási, Albert-László. *Linked: How Everything Is Connected to Everything Else and What It Means for Business, Science, and Everyday Life.* New York: Penguin, 2003.

Buhl, J., et al. "From disorder to order in marching locusts." *Science* 312 (2006): 1,402–1,406.

Franks, N. R. "Army ants: A collective intelligence." *American Scientist* 77 (1989): 138–145.

Johnson, Steven. *Emergence: The Connected Lives of Ants, Brains, Cities, and Software.* New York: Scribner, 2001.

Kaufmann, Stuart A. *The Origins of Order: Self-Organization and Selection in Evolution.* New York: Oxford University Press, 1993.

Prigogine, Ilya. *Order out of Chaos: Man's New Dialogue with Nature.* New York: Bantam, 1984.

Santa Fe Institute. "Santa Fe Institute: Celebrating 20 Years of Complexity Science." Available online. URL: http://www.santafe.edu. Accessed March 28, 2005.

eugenics Eugenics was a pseudoscience that attempted to use genetics to justify delusions of racial superiority or to promote genetic improvements in human populations. Eugenics (from the Greek for "good breeding") grew together with the young science of genetics (see MENDELIAN GENETICS). But while genetics has grown into a major science, the underpinning of evolutionary biology, eugenics has fallen into disrepute in the popular mind as well as among scientists, both for its injustice and its erroneous science.

Theories of "good breeding" predated both genetics and evolutionary theory. For many centuries, some societies viewed "purity of blood" as an important virtue, especially among its elite classes, who eschewed interbreeding with people of lower estate. Royal and aristocratic families bred only among themselves, and sometimes cousin marriages were common, although few went as far as the sibling marriages of the Pharaohs. Scientists now understand that interbreeding of close relatives results in inbreeding depression, the expression of numerous detrimental genes that would have otherwise remained hidden. Some scholars suggest that the illnesses both of Charles Darwin (see DARWIN, CHARLES) and his daughter Annie resulted from the close inbreeding between the Darwin and Wedgwood families (Charles Darwin's father and mother were cousins, and he married his cousin).

Not everyone in previous centuries shared the opinion that royal and aristocratic blood needed to be kept pure. The embarrassing problems of royal families were an open secret in late medieval and early modern Europe. The infamous "Hapsburg lip" of one of Europe's royal families was so striking that even flattering portraits could not hide it, and it passed through 23 generations. The Hapsburg Charles II of Spain could not even chew his own food, because of it, and was in addition a complete invalid and incapable of having children. Several British kings were insane for reasons now known to be genetic. George III (who lost the American colonies) had periodic bouts of madness variously attributed to genetic diseases porphyria or alkaptonuria. Thomas Jefferson, when visiting Europe, became convinced that something was biologically wrong with the European royals, which reinforced his belief that democracy was necessary not only for justice but even for physical health. As Jefferson wrote in 1810 in a letter to Governor John Langdon:

> ... take any race of animals, confine them in idleness and inaction, whether in a sty, a stable, or a state-room, pamper them with high diet, gratify all their sexual appetites, immerse them in sensualities ... and banish whatever might lead them to think, and in a few generations they become all body and no mind ... and this too ... by that very law by which we are in constant practice of changing the characters and propensities of the animals that we

raise for our own purposes. Such is the regimen in raising kings, and in this way they have gone on for centuries ... Louis the XVI was a fool ... The King of Spain was a fool, and of Naples the same ... the King of Sardinia was a fool. All these were Bourbons. The Queen of Portugal, a Braganza, was an idiot by nature. And so was the King of Denmark ... The King of Prussia, successor to the great Frederick, was a mere hog in body as well as in mind. Gustavus of Sweden, and Joseph of Austria, were really crazy, and George of England, you know, was in a strait waistcoat ... These animals had become without mind and powerless; and so will every hereditary monarch be after a few generations ... And so endeth the book of Kings, from all of whom the Lord deliver us ...

Positive Eugenics

When evolutionary theory came along, theories of good breeding were formalized into eugenics by Charles Darwin's cousin Sir Francis Galton (see GALTON, FRANCIS). Galton pioneered the use of twin studies in the estimates of heritability (see POPULATION GENETICS). Galton also developed techniques by which measurements of organisms (such as body proportions) could be analyzed. Galton accepted Darwin's theory that natural selection caused an improvement in the heritable qualities of a population. He became alarmed as he saw what he considered to be heritable degeneration in the British population: first, that people with what he considered superior qualities produced few children, and second, that people with what he considered inferior qualities (mainly people of other ethnic groups) produced many children. He proposed what has come to be called *positive eugenics:* that governments should encourage reproduction in families they believed to be genetically superior. Galton believed that upper-class, educated British should be encouraged by government incentives to have more children. Galton's motivation was actually to help people. He thought that positive eugenics could accomplish the same ends as natural selection against inferior types, "more rapidly and with less distress." Galton called eugenics "participatory evolution." Eugenics was also championed by Charles Darwin's son Leonard. Galton influenced other prominent British intellectuals, notably geneticist Sir Ronald Aylmer Fisher (see FISHER, R. A.), one of the architects of modern evolutionary science. Geneticist H. J. Muller, who discovered how genetic load can result from mutations, also espoused positive eugenics. In his tract *Out of the Night,* Muller advocated mass artificial insemination of women with sperm of men superior in intellect and character.

Charles Davenport was the director of the Cold Springs Harbor Laboratory on Long Island in the early 20th century. He and his research associates gathered genetic information, mostly pedigrees, about characteristics, ranging from epilepsy to criminality, that he believed had a genetic basis. Many of his researchers were women, whom Davenport considered to have superior observational and social skills necessary for collecting eugenic data. Consistent with his belief in eugenics, he would not employ these women for more than three years, since he wanted these genetically superior women to go home and have kids. Davenport would analyze pedigrees of traits and determine whether they were dominant or recessive. He correctly identified albinism as a recessive trait, and Huntington's disease as a dominant trait, each caused by a single mutation. But many other "traits" had either a complex genetic basis or were clearly caused by upbringing and environment. For example, Davenport claimed in all seriousness that there was a gene for the ability to build boats.

Eugenics was embraced with particular enthusiasm by people who would today be classified as the political left, such as the Fabian socialists. George Bernard Shaw, the critic of free market imperialism, said that "nothing but a eugenic religion can save our civilisation." Politicians who championed 20th-century progressivism, from President Teddy Roosevelt to Prime Minister Winston Churchill, admired eugenics. Margaret Sanger, founder of Planned Parenthood, said in 1919, "More children from the fit, less from the unfit—this is the chief issue of birth control." As eugenics became a big and important science in the United States, local chapters of the Eugenics Society sponsored Fitter Families contests at state fairs across the nation.

Negative Eugenics

Positive eugenics encouraged the reproduction of supposedly superior people; *negative eugenics* discouraged or even prevented the reproduction of supposedly inferior people, by forced sterilization or detention if necessary.

Some eugenicists studied human and other primate skulls and claimed to show that the skulls of people of the dark races had a more apelike shape than the skulls of Europeans. This is sometimes the case and sometimes not, so the eugenicists had to look until they found a measurement that confirmed the racial superiority of Europeans. They found one: facial angle, which is the degree to which the face slopes forward. They claimed that apes had a greater facial angle than humans, and therefore humans with a greater facial angle were more apelike. There were major problems with this approach. First, the facial angle of nonhuman apes results from a sloping of the entire face, while the facial angle of humans results from the jaw. Second, eugenicists carefully selected data to prove their point. They conveniently bypassed the fact that the Inuit have the smallest facial angles. By carefully selecting the data, eugenicists could demonstrate that Europeans had bigger brains than members of other races. Today scientists know their claims to be both trivial and wrong.

The claims were trivial and wrong, but not harmless. Phrenology (the study of skulls) arose as a branch of eugenics. Practitioners such as eugenicist Cesare Lombroso went so far as to claim that one could tell "criminal types" by the details of skull structure. People with supposedly more apelike skulls (which were later called "throwbacks") were more likely to be criminals. Therefore, reasoned Lombroso, if a criminal could be identified by skull shape when still a child, he could be institutionalized before he had a chance to do any harm. Hundreds of people in Europe were unjustly detained because government officials thought they looked like they would turn into criminals.

With the advent of Mendelian genetics, eugenics took on a new form, involving the dominant and recessive inheritance patterns of racial traits. For example, eugenicists invented the term *feeblemindedness* and considered it a discrete Mendelian trait. Supposedly scientific studies described criminal and abnormal mental traits in extended families. Eugenicist Richard Dugdale wrote an account in 1875 (before the acceptance of Mendelian genetics) about the Juke clan of upstate New York. Eugenicist Henry Goddard's 1912 study (after the acceptance of Mendelian genetics) of the "Kallikak family" (a fictitious name) introduced the term *moron* into the language. Goddard described two family lines from one male ancestor: the legitimate family line, which produced a normal family, and an illegitimate line that sprang from the man's affair with a "feebleminded wench" in a tavern. Goddard considered this a natural experiment that demonstrated good vs. bad genes. It did not occur to these researchers that families not only share genes but also upbringing and other environmental conditions.

One of the main problems that eugenicists wanted to solve was what to do about people who were institutionalized. If, as seemed apparent to eugenicists, the institutionalized people were insane because of bad genes, the obvious solution was to prevent them from having offspring. As Julian Huxley (see Huxley, Julian S.) wrote in 1926, "... we have no more right to allow [people with hereditary defects] to reproduce than to allow a child with scarlet fever to be visited by all his school friends."

An early event in negative eugenics occurred in 1899 when a doctor named Harry Sharp performed a vasectomy on a young man named Clawson. Sharp claimed to have cured Clawson of his "obsession" with masturbation, which was taken as an indicator of mental degeneracy. The idea caught on: If criminals could be rendered safe for release into society by vasectomy, the taxpayers would save a lot of money. Preventing supposedly inferior women from having babies would be a similar benefit to society. Indiana became the first state, in 1907, to pass a compulsory sterilization law authorizing the state to sterilize "criminals, idiots, rapists, and imbeciles." Eventually 30 states passed similar laws, and by 1941 60,000 Americans had been sterilized, half of them in California. The Virginia law was challenged in the Supreme Court in 1927 in the famous case of institutional resident Carrie Buck. Chief Justice Oliver Wendell Holmes wrote the decision:

> It is better for all the world if, instead of waiting to execute degenerate offspring for crime, or to let them starve for their imbecility, society can prevent those who are manifestly unfit from continuing their kind ... Three generations of imbeciles are enough.

Another form of negative eugenics was to restrict immigration. In the early 20th century, educational administrator Henry Goddard introduced the IQ (intelligence quotient) test into the United States that had been developed by educator Alfred Binet in Europe. The IQ test had originally been developed to determine the likelihood that a child would be successful in the educational system, and the test worked well for that purpose. In Europe and soon thereafter in America it was used as a measure of basic intellectual capacity, a purpose for which it was not appropriate. Goddard administered the test to people whose families had been in America for generations (mostly of English descent) and to immigrants who spoke no English, and the data clearly showed lower scores on the part of the latter. This was hardly surprising, since the test was administered in English. Goddard, like other eugenicists, overlooked the interpretation that is obvious to scientists today and assumed that it constituted evidence of genetic inferiority of the new immigrants. Eugenicist Madison Grant's 1916 book *The Passing of the Great Race* used these data to supposedly prove that Nordic Europeans were a superior race, even over other Europeans, and that non-Nordic immigration should be restricted.

Not only did eugenicists make rash assumptions about the genetic basis of the traits they studied, but they also classified their data in whatever way was necessary to produce the conclusion they had already decided. Harry Laughlin, assistant to Charles Davenport, gathered IQ data from immigrants and analyzed the results. To his embarrassment, he found that Jewish immigrants from Europe had higher scores than native-born Americans. Rather than accept this conclusion, he averaged the Jewish scores in with those of immigrants from whatever European country the particular Jews had come from, which lowered their scores. He presented his evidence before Congress, and it was perhaps the most important factor behind the 1924 Johnson-Reed Immigration Act, which set quotas to immigration similar to what had occurred prior to the influx of southern Europeans in the late 19th century. One effect this law had was to turn away many Jews who might otherwise have been saved from the Nazi Holocaust.

Some 19th-century evolutionary scientists drew the ultimate cruel interpretation from eugenics. Ernst Haeckel (see Haeckel, Ernst) claimed in 1868 that Christian religious sensibilities of "loving your enemies" was absurd, while German scientist Friedrich von Hellwald claimed in 1875 that evolution leads to superior people "striding over the corpses of the vaniquished." Some leaders of western society in the early 20th century also believed astonishingly ugly versions of negative eugenics. The science and science fiction writer Herbert George Wells described his vision of a future society, called the New Republic. He wrote,

> And how will the New Republic treat the inferior races? How will it deal with the black? ... the yellow man? ... the Jew? ... those swarms of black, and brown, and dirty-white, and yellow people, who do not come into the new needs of efficiency? Well, the world is a world, and not a charitable institution, and I take it they will have to go ... And the ethical system of these men of the New Republic, the ethical system which will dominate the world state, will be shaped primarily to favour the procreation of what is fine and efficient and beautiful in human-

> ity—beautiful and strong bodies, clear and powerful minds ... And the method that nature has followed hitherto in the shaping of the world, whereby weakness was prevented from propagating weakness ... is death ... The men of the New Republic ... will have an ideal that will make the killing worth the while.

The worst political and historical fruit of eugenics was National Socialism in Germany. Soon after coming to power in 1933, the Nazis began both positive and negative eugenics. They encouraged "Aryans," through incentive and propaganda, to produce children, and they passed mandatory sterilization laws. By 1936 they had sterilized 225,000 people. Harry Laughlin's work was admired in Germany, and he received an honorary doctorate from Heidelberg University in 1936. He translated Germany's sterilization law into English and publicized it in the United States. (Laughlin's health later deteriorated. He suffered from late-onset epilepsy, a disease he had claimed was evidence of genetic degeneracy.) The Nazis also prohibited marriage or extramarital intercourse between Germans and Jews. By 1939 the Nazis began to reason that it was too much trouble to sterilize inmates in institutions—why waste food? So they began to euthanize (kill) those whose lives they considered to be "not worth living." The world is only too familiar with what happened next, as the Nazis expanded eugenics and euthanasia to encompass entire races such as the Jews, Poles, Slavs, and Gypsies. Hitler used religious arguments against Jews. He once forced a clergyman to preach to Jews, "You are being killed because you killed the Christ," prior to their execution. But Hitler also used evolution as propaganda. One of his films equated Jews with rats and called for the survival of the fittest, by which he meant himself and others who had the guns.

Opposition to Eugenics

Some scientists stood up against eugenics. Perhaps the most prominent British scientist of the late 19th century (see WALLACE, ALFRED RUSSEL) denounced eugenics as "the meddlesome interference of an arrogant, scientific priestcraft." Geneticist Reginald Punnett, who invented the Punnett square method of quantifying the inheritance of genetic traits, showed that eugenics would not work. Punnett explained that even with the strongest negative eugenics carried out against individuals who were homozygous for the inferior gene, it would take 22 generations (almost 500 years) to reduce the incidence of feeblemindedness from 1 percent to 0.1 percent in a population, according to his calculations. R. A. Fisher pointed out, in response, that this calculation made the assumption of random mating, which would not be correct in the case of feeblemindedness. Fisher concluded that, in order to effect positive eugenics, it would be necessary to identify not only those who were homozygous but those who were carriers as well.

Darwin's closest ally (see HUXLEY, THOMAS HENRY), wrote,

> Of the more thoroughgoing of the multitudinous attempts to apply the principles of cosmic evolution ... to social and political problems ... a considerable proportion appear to me to be based upon the notion that human society is competent to furnish, from its own resources, an administrator ... The pigeons, in short, are to be their own [breeder]. A despotic government ... is to be endowed with the preternatural intelligence and ... preternatural ruthlessness required for the purpose of carrying out the principle of improvement by selection ... a collective despotism, a mob got to believe in its own divine right by demagogic missionaries, would be capable of more thorough work in this [eugenic] direction than any single tyrant, puffed up with the same illusion, has ever achieved.

Huxley may have foreseen that eugenics could turn into a nightmare, which is what Hitler did. One of the founders of modern evolutionary theory (see DOBZHANSKY, THEODOSIUS), wrote,

> The eugenical Jeremiahs keep constantly before our eyes the nightmare of human populations accumulating recessive genes that produce pathological effects ... These prophets of doom seem to be unaware of the fact that wild species in the state of nature fare in this respect no better than man does with all the artificiality of his surroundings, and yet life has not come to an end on this planet. The eschatological cries proclaiming the failure of natural selection to operate in human populations have more to do with political beliefs than with scientific findings.

Politicians and writers, in both England and the United States, stood up against eugenics. In England the writer G. K. Chesterton wrote a book against eugenics. Chesterton wrote that "eugenicists had discovered how to combine the hardening of the heart with the softening of the head." In America, William Jennings Bryan, the antievolutionary prosecutor at the SCOPES TRIAL, is often dismissed as a hopelessly outdated anti-scientist. Popular portrayals of Bryan, such as in the movie *Inherit The Wind,* usually fail to point out that one of his principal reasons for opposing evolution was that it was associated (by many scientists, and in popular social movements) with eugenics. Bryan spent his life battling for the rights of the "common man," and he saw eugenics as being the doorway to despotism and oppression.

After World War II, eugenics fell into extreme disrepute. It is safe to say that no competent scientist espouses it today. As a matter of fact, it is used by some critics as a label for scientists who accept any manner of genetic influence on human behavior. Even though SOCIOBIOLOGY is a scientific theory very different from eugenics, its proponents (such as WILSON, EDWARD O.) have been unfairly denounced as eugenicists who wish to direct the world back to the days of scientifically sponsored racism. It is also safe to say that no competent scientist espouses complete genetic determinism of the sort that led Charles Davenport to posit the existence of a boat-building gene. All scientists recognize an important role of environmental causation. Sociobiologists do not want to manipulate human breeding, either through positive or negative means; they want people to recognize the effects of genes on human behavior so that destructive behavior might be avoided. Eugenics is safely dead in evolutionary science. It lives on only in a few scattered groups of racial supremacists. Although eugenics has been entirely discredited, scientists still universally use the methods of statistical analysis that were invented by the eugenicists of the late 19th and early 20th centuries: Mathematician and eugenicist Karl Pearson invented the correlation coefficient and the chi-square test, and Fisher invented the variance ratio.

Some scientists fear that positive eugenics may make a comeback. In the early 1980s, the government of Singapore instituted tax incentives for rich people to have more children, a renascence of positive eugenics. Negative eugenics may also make a comeback. Genetic screening of unborn fetuses may soon be widely practiced. A disproportionate number of girls are aborted in India, and a disproportionate number of ethnic minorities in China. It appears that modern technologies may be applied under the guidance of old prejudices. It is therefore essential that scientists and policy makers keep the failure of past eugenic theories in mind.

Further Reading

Gould, Stephen J. *The Mismeasure of Man.* New York: Norton, 1981.

Hubbard, Ruth, and Elijah Wald. *Exploding the Gene Myth: How Genetic Information Is Produced and Manipulated by Scientists, Physicians, Employers, Insurance Companies, Educators, and Law Enforcers.* Boston, Mass.: Beacon Press, 1999.

Jefferson, Thomas. "Letter from Thomas Jefferson to Gov. John Langdon, March 5, 1810." Pages 1,218–1,222 in Peterson, Merrill D., ed. *Thomas Jefferson: Writings.* New York: Library of America, 1984.

Landler, Mark. "Results of secret Nazi breeding program: Ordinary folks." *New York Times,* November 7, 2006.

Lewontin, Richard. *It Ain't Necessarily So: The Dream of the Human Genome and Other Illusions.* New York: New York Review of Books, 2000.

United States Holocaust Memorial Museum. "Deadly Medicine: Creating the Master Race." Available online. URL: http://www.ushmm.org/museum/exhibit/online/deadlymedicine. Accessed July 9, 2006.

Watson, James D., with Andrew Berry. *DNA: The Secret of Life.* New York: Knopf, 2003.

Weikart, Richard. *From Darwin to Hitler: Evolutionary Ethics, Eugenics, and Racism in Germany.* London: Palgrave Macmillan, 2004.

eukaryotes, evolution of Eukaryotes are organisms that consist of cells with nuclei. Every cell has a cell membrane, which is a thin double layer of lipids that maintains a chemical difference between the inside of a cell and the outside; cytoplasm, which is the liquid contents of a cell with its suspended components; ribosomes, which make proteins; and DNA, which stores genetic information (see DNA [RAW MATERIAL OF EVOLUTION]). The simplest cells (though by no means simple) are the *prokaryotic* cells which have very little more than this. Most bacteria (see ARCHAEBACTERIA; BACTERIA, EVOLUTION OF) have, in addition, a cell wall and some bacteria have internal membranes.

The cells of protists, fungi, plants, and animals are *eukaryotic* cells, which are often about 10 times larger than prokaryotic cells and much more complex. Each eukaryotic cell has a nucleus, which contains the DNA in the form of chromosomes inside of a membrane; an internal membrane system involved in making large molecules inside the cell; and organelles such as chloroplasts and mitochondria. "Eukaryotic" refers to the nucleus (Greek for "true kernel"). The DNA molecules in the nucleus are associated with proteins called histones, and the membrane surrounding the nucleus has pores, which are not merely holes but are passageways controlled by proteins. These pores allow genetic instructions and materials, in the form of RNA, to go out into the cytoplasm and allow hormone messages to enter the nucleus. Part of the reason bacteria have rapid metabolism is that as soon as the genetic information of DNA is transcribed into RNA, the ribosomes start using the information to make proteins. In eukaryotic cells, the RNA must travel out to the cytoplasm first, a process that is slower but may allow a greater degree of control over the use of genetic information.

The events involved in the origin of eukaryotes have not all been resolved. It is known that chloroplasts and mitochondria are the degenerate evolutionary descendants of prokaryotes that invaded and persisted in ancestral eukaryotic cells. The bacterial ancestors of chloroplasts and mitochondria had enough DNA to control their own activities, but chloroplasts and mitochondria do not now have enough genes to survive on their own. Many genes transferred from the primitive chloroplasts and mitochondria to the nucleus of the host eukaryote (see HORIZONTAL GENE TRANSFER). In some cases, part of a complex gene transferred to the nucleus, and part remained behind in the organelle. An important enzyme in photosynthesis, the rbc gene for the rubisco enzyme, is constructed from genetic instructions in both the chloroplast (the large component, rbcL) and the nucleus (the small component). Mitochondrial genes occasionally act in a manner that some scientists consider selfish (for example, cytoplasmic male sterility factors in plants; see SELFISH GENETIC ELEMENTS). Smaller cells moving into and living inside a larger cell is an example of symbiosis (see COEVOLUTION), and when it leads to the origin of a new species, it is called SYMBIOGENESIS.

The evolution of the nucleus has proven much more difficult to explain. Three theories have been proposed, each with provocative evidence. The three theories for the origin of the nucleus are symbiogenesis, the evolution of the nucleus from structures in bacterial ancestors, and that the nucleus evolved from a virus.

The nucleus originated through symbiogenesis. Some evolutionary scientists propose that a primordial archaebacterium entered into a larger eubacterial host and became its nucleus. Subsequently, horizontal gene transfer merged the eubacterial and the archaebacterial DNA. According to this theory, the nucleus was originally archaebacterial, to which eubacterial genes were added.

It is not clear what kind of eubacterium might have been the original host. It may have resembled today's myxobacteria, which unlike other bacteria use molecules for intercellular communication and can form multicellular structures, just like eukaryotes.

It is also not clear where the membrane that surrounds the nucleus came from. Lynn Margulis, who convinced most evolutionary scientists of the symbiogenetic origin of chloroplasts and mitochondria (see MARGULIS, LYNN), proposes that the nuclear membrane began as a fusion of the membranes of eubacterial and archaebacterial attachment structures.

The nucleus evolved from structures in bacterial ancestors. Most biologists, biology teachers, and biology students assume that eukaryotes have fully formed nuclei, and bacteria have nothing of the sort. It turns out that an obscure group of bacteria called planctomycetes have their DNA almost completely enclosed in a membrane that looks like a nuclear membrane. Another group of recently discovered bacteria, which live only inside of sponges, also appear to have structures that look like nuclei. Some researchers even claim that these bacteria have nuclei, and that the eukaryotic nucleus may have evolved from this bacterial nucleus. Critics point out that, despite appearances, the eukaryotic nuclear membrane and the membrane inside these bacteria are structurally very different.

The nucleus evolved from a virus. A few researchers point out that nuclei, like viruses, are genetic material surrounded by a protein coat. They maintain that an ancient virus invaded an archaebacterium, and instead of killing the host, the virus maintained a multigenerational mild infection inside of it. It could have been a virus similar to modern mimiviruses, which have up to a hundred times as many genes as the smallest viruses. Gradually, horizontal gene transfer moved bacterial genes into the virus, which became the nucleus. Their evidence is that nuclei, like viruses, depend on the cell around them for metabolism, and while bacterial DNA molecules are circular, the DNA of viruses is linear, like that of eukaryotic nuclei. Some viruses not only have a protein coat but have an envelope, which they make by stealing a chunk of the host cell membrane, just like the eukaryotic cell uses internal membranes to make the nuclear membrane. Some large viruses even have telomeres, nucleotide sequences at the ends of chromosomes which, in eukaryotes, help keep the process of DNA replication itself from breaking off chunks of essential genes.

These three possibilities are not mere guesses but can be framed as scientific hypotheses (see SCIENTIFIC METHOD). If the first eukaryotic cells were mergers between archaebacteria and eubacteria, then some eukaryotic genes should more closely resemble those of archaebacteria, while others should more closely resemble those of eubacteria (see DNA [EVIDENCE FOR EVOLUTION]). This appears to be the case. DNA sequences for the genes in eukaryotes that are involved in the use of genetic information resemble archaebacterial genes. This includes the production of histone proteins, known only from eukaryotic and archaebacterial cells. DNA sequences for the eukaryotic genes involved in metabolism (energy use and construction of molecules) resemble eubacterial genes. Cell biologists are pursuing the information necessary to test other hypotheses. If the nucleus began as an extension of an undulipodium, which began as a spirochete, there should be significant similarities between the proteins that move the chromosomes during cell division, the proteins of undulipodia, and the proteins of spirochetes. If the first eukaryotic cells evolved from planctomycete bacteria, these bacteria should have a closer genetic resemblance to eukaryotes than do other kinds of bacteria.

According to these views, the first nucleated cell was a *chimera,* a fusion of two different life-forms. In Greek mythology, a chimera was a combination of two different animals. This would mean that the origin of many structures found in the eukaryotic cell occurred through symbiogenesis. If this is the case, the reason that scientists have not found a gradual series of transitional forms in the evolution of eukaryotic cell structures is that they never existed.

The TREE OF LIFE constructed from DNA analysis shows that the three main branches of life are the Archaea (archaebacteria), Eubacteria, and Eukaryotes. The branches within the eukaryotes cannot be clearly distinguished from one another. Most of the diversity of eukaryotes exists among the primarily single-celled organisms usually called *protists* (formerly "algae" and "protozoans"). Being usually small, of relatively simple structure, and living in places humans seldom look (such as oceans, ponds, soil, and inside of animal guts), protists have been less intensively studied than plants and animals. Not only is less known about protists than about plants and animals, but new groups of protists, including some as small as bacteria, continue to be discovered. Each such discovery has the possibility of changing scientific understanding of eukaryotic evolutionary relationships. One recent reconstruction of these relationships, given by biologist S. L. Baldauf, divides eukaryotes into the following groups:

- amoebalike organisms mostly without hard outer coverings
- amoebalike organisms that usually have hard outer coverings
- ciliates, dinoflagellates, and their relatives
- euglenas and their relatives
- anaerobic single-celled organisms that lost their mitochondria
- brown algae, diatoms, and their relatives
- red algae, green algae, and plants
- fungi, animals, and their relatives

Many eukaryotes are multicellular. Multicellularity evolved several times independently: in brown algae, in some slime molds, in fungi, in plants, and in animals. The lineages that contain multicellular organisms also contain single-celled organisms. The classification above also contains some surprises. One surprise is that the evolution of chloroplasts from cyanobacteria occurred only in the lineage leading to red algae, green algae, and plants. There are other photosynthetic eukaryotes, such as the brown algae and diatoms in one lineage, some euglenas in another lineage, and dinoflagellates in yet another linage. The chloroplasts in these last three lineages did not evolve directly from cyanobacteria. Instead, they are the evolutionary descendants of eukaryotic algae (perhaps green algae). This means that eukaryotes such as brown algae and dinoflagellates do not consist of cells within cells, but rather cells within cells within cells! Another surprise is that fungi and animals are more closely related to each other than to the other eukaryotes.

Disagreement remains on the status of the eukaryotes that do not have mitochondria. For example, parabasalids such as *Trichomonas vaginalis* are single-celled protists that have no mitochondria. While they may have evolved from ancestors that did not yet have mitochondria, it is more likely that they evolved from ancestors that did have mitochondria and then lost them. Trichomonads have hydrogenosomes, which appear to be the degenerate descendants of mitochondria that lost their genes to the nucleus through horizontal gene transfer.

Within the LINNAEAN SYSTEM of classification, the Kingdom Protista consists of the eukaryotic organisms that are either single-celled or consist of multicellular bodies that have no clearly defined tissues. They are usually classified separately from Kingdom Fungi, Kingdom Plantae, and Kingdom Animalia. If the classification presented above is correct, the fungi, plants, and animals occur within branches of the protist kingdom. This may require a rethinking of the overall scheme of eukaryotic classification.

The origin of the complex eukaryotic cell allowed an explosion of evolutionary diversification and novelty. Although some bacteria can form multicellular structures, it appears that only eukaryotes can form complex multicellular organisms. The early evolution of eukaryotes occurred during PRECAMBRIAN TIME and left few fossils. Today scientists attempt to reconstruct these evolutionary events, and to do so they must study the organisms that most closely resemble the earliest eukaryotes: the protists.

Further Reading

Baldauf, S. L. "The deep roots of eukaryotes." *Science* 300 (2003): 1,703–1,706.

Kurland, G. C., L. J. Collins, and D. Penny. "Genomics and the irreducible nature of eukaryote cells." *Science* 312 (2006): 1,011–1,014.

Lane, Nick. *Power, Sex, and Suicide: Mitochondria and the Meaning of Life.* New York: Oxford University Press, 2005.

Margulis, Lynn, Michael F. Dolan, and Ricardo Guerrero. "The chimeric eukaryote: Origin of the nucleus from the karyomastigont in amitochondriate protists." *Proceedings of the National Academy of Sciences USA* 97 (2000): 6,954–6,959.

———,———, and Jessica H. Whiteside. "'Imperfections and oddities' in the origin of the nucleus." *Paleobiology* 31 (2005): 175–191.

Pennisi, Elizabeth. "The birth of the nucleus." *Science* 305 (2004): 766–768.

Eve, mitochondrial *See* DNA (EVIDENCE FOR EVOLUTION).

evolution *Evolution* is a term that has been used, in the broad sense, to denote many different kinds of change, usually gradual change. Stars evolve, life evolved from nonliving molecules, one form of life evolves into another, societies evolve, ideas evolve. This broad-sense meaning, referring even to the evolution of London, is used in the book by evolutionary scientist A. C. Fabian. The term has been used in so many ways that scientists have imposed a narrow-sense meaning to prevent misunderstandings. A similar problem occurs with the concept of ADAPTATION. In the narrow sense, evolution refers to genetic changes over time in populations of organisms. In this narrow sense, neither stars nor ideas evolve; the process of evolution did not exist until life-forms existed; and individual organisms can change but cannot evolve.

One fundamental distinction between evolution and other processes of change involves directionality, or *teleology*—that is, movement toward a goal. The original word *evolvere* in Latin refers to an unfolding of a predetermined set of events. However, most scientists use the word *evolution,* even in the broad sense, to refer to changes that have no predetermined direction or destination. Many natural processes are largely predetermined: For example, the development of an embryo from a fertilized egg is determined by genetic instructions, and the development of a star, from nebula to nova, follows the laws of physics. Even in these processes there is a random element: Two identical eggs can develop in different ways, depending on events and environmental conditions. Their overall predetermined directionality prevents these processes from being legitimately described as evolution. In contrast, the evolution of humans was not inevitable. Had different mutations occurred in ancestral populations, whether of microbes or the earliest vertebrates or of primates; or had environmental conditions, events, or opportunities been different, something other than *Homo sapiens* in the modern form would have evolved. Some evolutionary scientists such as Simon Conway Morris point out that natural selection would inevitably have produced something similar to human beings (see CONVERGENCE). This would have resulted from natural selection happening at each step, and not from change predetermined within ancestral DNA. Therefore humans evolved, but embryos and stars do not.

Another fundamental distinction involves typological versus population thinking. Typological thinking dates back to Platonist idealism, which claims that somewhere there is a perfect world, of which our world is but a shadow. Typological thinking in biology would imply that somewhere there is a perfect redwood tree, a perfect mockingbird, a perfect monarch butterfly—each is the type of its species, from which all other individuals in the species are deviations. One of Darwin's great realizations was that populations contain variation; the species consists of these variations. There is no perfect mockingbird; there are just mockingbirds. As evolutionary biologist Ernst Mayr (see MAYR, ERNST) has said, the variation is real; the type is an illusion. Darwin became convinced of this mainly through his work with barnacles, in which each individual of each species was unique. Therefore, in evolution, just as there is no goal that the species must attain or has attained, there is also no perfect specimen of the species.

Another fundamental distinction involves the inheritance of acquired characteristics. Things that happen accumulate over time, whether scars on human skin or mutations in human chromosomes. However, in organisms, most of these accumulated things do not persist. When a person dies, the scars are lost forever; and the mutations in most of the chromosomes die when the cells die. The only changes that persist beyond death are some of the mutations that occurred in the germ line cells (the cells that produce eggs or sperm), and then only if the person has children, and then only the mutations that occurred before the person had children. In the narrow sense, evolution acts only upon these inherited (genetic) changes. If some individuals reproduce more than others, the relative abundances of the genes of these individuals will be different in the next generation (see NATURAL SELECTION). This can happen only in populations. The alternative, that the scars and experiences from a person's entire body or entire life can be passed on to future generations, is named after one of its major 19th-century proponents, Jean Baptiste de Lamarck (see LAMARCKISM). Even though Charles Darwin himself considered this possibility, calling it pangenesis, biologists now know that this process cannot occur in organisms.

Lamarckian evolution can, however, occur in social interactions, which is perhaps the major reason that the narrow sense of evolution is not applied to them. Once the biological basis of language had evolved, all further changes (such as the development of specific languages by different tribes, the learning and modification of languages by individuals) could be passed on from one person, society, or generation to another (see LANGUAGE, EVOLUTION OF). Biological evolution usually occurs slowly, because it has to wait for the right mutations to come along in the germ line—and they may never do so. Social evolution can occur rapidly, because as soon as a good new idea comes along, everybody can adopt it. This is as true with the invention of tool use by macaque monkeys, or the ability of birds to open milk bottles, as it is with the rapid development of new technologies in human society (see GENE-CULTURE COEVOLUTION). Lamarckian social evolution has some parallels with biological evolution, an idea explored by evolutionary biologist Richard Dawkins (see DAWKINS, RICHARD) with the concept of *memes,* which are units of social evolution that correspond to the genes of natural selection.

As Ernst Mayr has explained, evolution as presented by Darwin has five tenets:

- Species are not constant; they change over time.
- All organisms had a common ancestral population.
- Evolutionary change occurs gradually.
- One species can diversify into more than one.
- Evolution occurs by natural selection.

Hardly anybody accepted all five tenets when Darwin first presented them (see *ORIGIN OF SPECIES* [book]). Darwin's evidence convinced most readers of the first two tenets. The third, gradualism, is still debated (see PUNCTUATED EQUILIBRIA). The fourth tenet is accepted, but scientists still disagree about how it occurs in many cases (see SPECIATION). The triumph of the fifth tenet had to wait until 60 years after Darwin's death (see MODERN SYNTHESIS).

Social evolution, though not evolution in the narrow sense, may resemble biological evolution in a macroscale pattern. When, for example, a new religion forms (as when Christianity diverged from Judaism about two millennia ago), a period of crisis occurs while the small, new religion seeks patterns of belief and practice that make it recognizably distinct from the large, old religion. Therefore a period of rapid evolution can be said to occur in which most of the new beliefs and practices take form, followed by a longer period of either stasis or of much slower change. These changes allow the new religion to be isolated from the old religion, permitting it to evolve in its own direction. Isolation is necessary for the evolution of new species and also appears to be important in social evolution. In this manner, the speciation of a new religion resembles the punctuated equilibria model of the evolution of species. The resemblance, however, is only an analogy.

Finally, a distinction must be made between the process of evolution (how it occurred) and the products of evolution (what happened). It is one thing to describe an outline of what happened in human evolution, and quite another to explain why it happened the way it did. Evolution is both a process and a product.

Of the many possible uses of the word *evolution,* scientists usually restrict the use of the term to genetic changes within populations, and the production of new species.

Further Reading

Blackmore, Susan. *The Meme Machine.* Oxford, U.K.: Oxford University Press, 1999.

Fabian, A. C., ed. *Evolution: Science, Society, and the Universe.* Cambridge, U.K.: Cambridge University Press, 1998.

Gould, Stephen Jay. "What does the dreaded 'E' word mean anyway?" Chap. 18 in *I Have Landed: The End of a Beginning in Natural History.* New York: Harmony, 2002.

Mayr, Ernst. *What Evolution Is.* New York: Basic Books, 2001.

Ormerod, Paul. *Why Most Things Fail: Evolution, Extinction, and Economics.* New York: Pantheon, 2006.

evolutionary algorithms Also known as genetic algorithms, evolutionary algorithms are computer program subunits that use natural selection to achieve optimal solutions. The concept of a genetic algorithm was originally proposed by John Holland, the first person to receive a Ph.D. in computer science. The evolutionary algorithm begins with a relatively simple process, generates random mutations in that process, then selects the best of the mutations. The choice among mutations is made on the basis of the efficiency of the process at producing a desired outcome. After numerous iterations, an algorithm results that is typically much superior to what a human programmer could have produced.

The similarity of evolutionary algorithms to natural selection is evident. First, there must be a genetic process. In the computer, it is the computer itself, both the hardware and the original software with which it is provided. In organisms, it is the genetic software (biochemical genetic machinery [see DNA (RAW MATERIAL OF EVOLUTION)], which stores and expresses genetic information, and which allows mutations to occur) and hardware (the cells, which provide metabolism). Second, the mutations are generated at random, not by a programmer or designer. Third, there is a clear definition of what constitutes success or fitness. In the computer, it is the ability to produce the desired effect. In the organism, it is reproductive success.

Millions of people have seen the effects of evolutionary algorithms without necessarily knowing it, in movies. In early motion pictures, a large battle scene would require a cast of thousands. In recent movies such as the *Lord of the Rings* trilogy and *Troy,* the characters in battlefields were generated by computers. Rather than a designer programming the movement of each character, evolutionary algorithms selected the movements of each of these characters relative to one another to produce lifelike simulations of movements, trajectories, and conflicts. The real power of evolutionary algorithms is that, given enough computing capacity, they can produce a whole battlefield of movements just as easily as the movement of a single entity. A computer can produce what would take a designer a very long time, or would take a lot of designers, to accomplish.

Besides producing complex movie scenes, evolutionary algorithms have also been used to design:

- plane wings
- jet engine components
- antennae
- computer chips
- schedule networks
- drugs (which must fit into protein or nucleic acid binding sites)
- medical diagnosis

Both evolutionary algorithms and biological evolution appear wasteful. But with billions of bits of random access memory, or billions of bacteria, evolution can afford to be wasteful. Far from being the hopelessly random process that critics claim (see CREATIONISM; INTELLIGENT DESIGN), the process of natural selection is both sufficiently productive and efficient that modern industry has adopted it, in many instances, as a replacement for intelligent design. Using evolution allows many companies to make money. As evolutionary scientist Robert Pennock says, "Evolution got its credibility the old fashioned way: it earned it."

Further Reading

Coley, David A. *An Introduction to Genetic Algorithms for Scientists and Engineers.* Hackensack, N.J.: World Scientific Publishing, 1997.

Zimmer, Carl. "Testing Darwin." *Discover,* February 2005, 28–35.

evolutionary ethics Evolutionary ethics is the investigation of human evolution to determine ethical principles. Evolutionary science investigates and explains what has happened in human evolution, which in large measure helps scientists to understand what is happening with the human species today. Evolutionary science can, in turn, be used as a basis for ethics, although this is not part of evolutionary science itself. One of the earliest claims that ethics can be derived from an evolutionary basis was by Charles Darwin (see DARWIN, CHARLES) in one of his notebooks in the 1830s, long before he published his theory (see *ORIGIN OF SPECIES* [book]): "He who understands baboon would do more towards metaphysics than Locke."

Human behavior patterns emerge from the interaction of genes and environment, or "nature and nurture," to use the dichotomy first proposed in the 19th century (see GALTON, FRANCIS). Behavior patterns, however, hardly ever result from single genes. For example, a gene that contributes to impulsive behavior (being violent, taking risks) has been identified (see essay, "How Much Do Genes Control Human Behavior?"), but it is only one of possibly many genes that influence this aspect of our behavior. Usually, genes interact with one another, one gene switching others on in a complex cascade (see DEVELOPMENTAL EVOLUTION). Environmental effects on behavior range from chemical conditions and events during gestation, to childhood upbringing, to social conditions. All these influences contribute to a *proximate* explanation of human behavior. As with the behavior of all other animals, there is also an *ultimate* explanation that elucidates the adaptive advantage, over evolutionary time, that those behavior patterns have conferred on humans (see BEHAVIOR, EVOLUTION OF). Humans behave as they do because of genes and environment; genes are as they are because of evolution. SOCIOBIOLOGY is an example of a scientific explanation of the genetic and evolutionary basis of human behavior patterns.

Just because humans do behave in certain ways does not mean that they should behave in these ways. The *naturalistic fallacy* recognized by ethicists is that "is" is not the same as "ought." It is in addressing the questions about how humans should behave, rather than just how they do, that evolutionary ethics is different from sociobiology.

Two general approaches to ethics have been proposed. *Transcendentalism* claims that standards of right and wrong, good and evil, come from beyond humans and transcend them. The source of ethics may be God, gods, or an overarching universal spiritualism. *Empiricism* claims that humans have invented moral standards, and that what is right in one society might be wrong in another. Transcendentalists point out that premeditated murder is always wrong. Empiricists point out that, while murder may be universally wrong, polygamy is not: Some societies (including Old Testament Israelite society to which many modern religious ethicists turn for guidance) say it is good, while others say it is evil. The extreme form of empiricism is that each person makes up his or her own standards of good and evil and can change them at will.

Michael Shermer, an American philosopher and skeptic, claims that both approaches are deficient. Morality comes neither from Heaven nor from the whims of individuals or society. Morality, he claims, evolved. It provided a fitness advantage to those who possessed it (see NATURAL SELECTION). Shermer's evidence is that human moral sense, and the sense of shame that humans feel (and exhibit by physiological characteristics such as blushing and the changes of electrical conductivity in the skin that polygraphs can measure), are deep and genetic. People from all societies, for example, feel shame if they are dishonest. Upbringing can enhance or reduce this feeling, and there will always be pathologically dishonest people. For morality, as for any other trait, there is genetic variability within populations.

The human species has existed for at least 100,000 years (see *HOMO SAPIENS*), and the sense of morality may have existed in the immediate ancestors of modern humans (such as *Homo ergaster; Homo habilis*). The moral sense evolved, along with all other human behaviors, feelings, and thought patterns, throughout that time. No individual or society can come along and invent a new kind of morality and expect it to work. Ethics, Shermer insists, must be based upon the behavior patterns that have proven successful over evolutionary time in the human species. Rather than doing what one thinks God told him or her to do, or whatever one wants, one should do what has worked for thousands of years in the human species.

It has been popular among religious communities to claim that evolutionary beliefs lead to total immorality. Without God, they claim, humans behave like animals. Shermer disagrees. For one thing (as famously pointed out by author Mark Twain), most animals do not "behave like animals." Despite the recent discovery of warfare among chimpanzees, humans still stand out as the species that is perhaps most cruel to its fellow members. Nonhuman animals may, indeed, "behave better" than humans.

For another thing, natural selection has favored some types of morality. This is how it happened, according to Shermer. Some behavior patterns worked, and the individuals (and the societies in which they lived) prospered as a result. An example is honesty. A person might gain a temporary advantage by being dishonest, but eventually other members of society will detect a repeated pattern of dishonesty. People who are generally honest will prosper by the goodwill of others within a social network (see ALTRUISM). Natural selection also favors the ability to detect even the most subtle hints of dishonesty in other people. Therefore, natural selection has not only favored honest behavior but also a feeling of satisfaction that accompanies honest behavior. People who not only were honest, but also wanted to be honest, were most successful at gaining status, resources, and mates, allowing the production of children, who inherited their genes and learned honest behavior patterns from honest parents. (This discussion omits situations in which ethical dilemmas arise.) The same argument would apply to stealing, murder, and adultery as to dishonesty.

What does one get when one adds all of this together? The result is the evolution of conscience. The result is what traditional religion calls The Golden Rule (Do unto others as you would have them do unto you) and philosophers call the Categorical Imperative (Do those things that, if everybody did them, would result in the kind of world you would like to live in). This explains why it feels good to be good; to be moral, Shermer says, is as human as to be hungry or horny. This would also explain the negative side of the issue: why humans feel shame for immorality. Although the specific behaviors that elicit feelings of shame and guilt may differ from one society to another, the feelings themselves are universal. Evolutionary ethics explains why, as Shermer says, most people are good most of the time.

Likewise, an evolutionary approach explains why human morality is different from the behavior patterns of other animal species. There is no selective advantage of honesty among birds, for example, because they are not smart enough to recognize and remember which individual birds are honest and which are not. The ability to recognize and remember which individual is a friend and which is not may be the principal force behind the evolution of our higher mental characteristics (see INTELLIGENCE, EVOLUTION OF; LANGUAGE, EVOLUTION OF) and even the human moral sense. As the product of evolution and environment, the human moral sense has both universal truths and situational variations.

One major difference between evolutionary ethics and traditional religious ethics, says Shermer, is that traditional ethics identifies some things as 100 percent right, others as 100 percent wrong. Shermer claims that it is necessary to replace this absolutist logic with what mathematicians call fuzzy logic. A behavior pattern may be wrong most of the time, but beneficial under rare circumstances; it might, therefore, be 10 percent right. Another behavior pattern might be good most of the time but detrimental under rare circumstances and be 90 percent right. The human legal system and common sense already recognize this (for instance, by distinguishing murder from manslaughter), and it is time that human ethical systems incorporate it also. Ethical principles can be 100 percent right or wrong only if they are transcendental, which, Shermer insists, they are not.

Many scholars believe that evolutionary ethics cannot lead to a completely satisfactory society or personal life. Although many of the behavior patterns that worked during prehistoric times, during which time almost all human evolution occurred, will work today, many will not. Throughout human prehistory, natural selection may have favored morality within societies, but did not favor friendliness between competing societies. Perhaps the human tendency to go to war has a genetic basis, but even if it does not, human antipathy toward outsiders definitely does seem ineradicably genetic. Nearly everybody agrees that the time has long passed for humans to use war as a solution to problems. In light of this, it is difficult to surpass the judgment that a close friend of Charles Darwin (see HUXLEY, THOMAS HENRY) placed upon evolutionary ethics. Instead of deriving morality from evolution, said Huxley, a human's duty consists of resisting natural impulses: "Let us understand, once for all, that the ethical process of society depends, not on imitating the cosmic process, still less in running away from it, but in combating it." Huxley primarily intended his statement to be against the simplistic political application of "survival of the fittest" (see EUGENICS), but it could also apply today at least in part to evolutionary ethics.

Other philosophers and evolutionary scientists have emphasized the fact that evolution need not produce violent competition. Evolution can lead to mutualism as readily as to antagonism (see COEVOLUTION; SYMBIOGENESIS). Thomas Henry Huxley said that evolutionary ethics consisted of resisting the violence of evolution; his grandson (see HUXLEY, JULIAN S.) said that evolutionary ethics consisted of embracing the cooperation that evolution produces. Russian philosopher Petr Kropotkin derived evolutionary ethics from mutualism in his 1902 book *Mutual Aid.*

Many scientists, including evolutionary scientists, are religious (see essay, "Can an Evolutionary Scientist be Religious?"). To these scientists, the empirical evolutionary basis of ethics is not enough. They admit the evolutionary origin of ethics, morality, altruism, and the behavior patterns of religion. Their desire to believe in transcendental truths, they admit, is beyond scientific proof or disproof.

Can an Evolutionary Scientist Be Religious?

Yes. As philosopher Michael Ruse points out, one might conclude that it is impossible for an evolutionary scientist to be religious, but one should not begin with this as an assumption. The compatibility of belief in God and acceptance of evolutionary science in human origins appears to be, according to a 2004 Gallup poll, the view of 38 percent of the American population. About 13 percent of Americans accept only evolution, while 45 percent accept only creationism.

Almost all scientists answer the question posed in the title in the affirmative, for both personal and scientific reasons. The personal reason is that many scientists are religious individuals, although fewer scientists are religious (about 40 percent believe in a personal God) compared to the general American population (over 90 percent). For example, some of the most famous evolutionary scientists have been openly religious (see FISHER, R. A.; DOBZHANSKY, THEODOSIUS). The scientific reason is that the SCIENTIFIC METHOD does not require an assumption that a supreme being (God) does not exist. What science does require is that, for the phenomena being investigated, this supreme being, or other supernatural beings, have not caused the results that are observed. Geneticist Richard Lewontin (see LEWONTIN, RICHARD) points out that science cannot coexist with belief in a deity that might intervene at any moment. A biologist must assume that no angel, demon, or God makes biological phenomena occur as they do today; and an evolutionist must assume that no supernatural being brought biological phenomena into existence in the first place. If it should happen that supernatural processes or beings did or do influence the operation of natural phenomena, then the scientific explanation will fail. It is not necessary to assume that supernaturally caused events never occur. A scientist need not deny miracles, but just not try to study them, or to use scientific credentials to try to get other people to believe them. Scientific investigation excludes a consideration of miracles, whether or not they occur. Therefore, science in general, and evolution in particular, do not prohibit religious belief.

Just because religious belief is permissible does not mean that it is a good idea. While evolutionary science and religious belief can mix, there may be so many points of contradiction in the minds of individuals that this mixture is unreasonable. This criticism comes from two directions. Adherents of CREATIONISM insist that evolutionary science contradicts what they consider to be true and correct religion; the kind of religion that is compatible with evolutionary science is, to them, a waste of time or worse. Many atheists insist that any scientist who maintains religious beliefs is not really thinking rigorously through the issues. Some (see DAWKINS, RICHARD) believe that scientists who have religious beliefs are retaining mutually contradictory thoughts for reasons of personal satisfaction. For instance, many scientists who are Christians believe that "God is love" but that God has used natural selection, which is a painfully unfair process (see below), as God's method of creation. Dawkins considers this a "delicious irony." From this viewpoint, the capacity for religion may be a product of evolution and may have played an important role during human prehistory (see RELIGION, EVOLUTION OF), but religion is something that humans should now get rid of. Especially in the post–September 11 world, Dawkins says, one should see that religion (such as fundamentalist terrorism, of which all major religions have provided examples) does more harm than good. Flexible human minds should learn how to get along without it as much as possible, according to this viewpoint.

In particular, one problem that creates difficulty for the peaceful coexistence of science and religion (in particular, evolutionary science and monotheistic religion) is the problem of evil. This is not a problem that was noticed just since the beginning of evolutionary science. A whole field of theological inquiry, called *theodicy,* addresses how a good God can allow evil to occur. Some of the earliest Christian theologians, such as Augustine, wrote extensively about theodicy, as did non-Christian philosophers such as Plotinus. Jewish and Christian theologians have found elements of this issue as far back as the story of Abraham, who is said to have confronted God himself and questioned God about God's plans to destroy Sodom and Gomorrah. The Old Testament prophets, New Testament apostles, and theologians of Judaism, Christianity, and Islam have given many different answers to why an all-powerful God permits evil events to occur, but none has gained widespread acceptance. Among the speculations of theodicy are:

- Suffering is punishment for sin. The fact that suffering afflicts so many innocent people (especially infants and children) discredits this explanation.
- Suffering strengthens the righteousness of character. The fact that suffering vastly exceeds what is necessary for promoting the growth of character discredits this explanation. A victim of genocide, along with his or her millions of fellow victims, is dead, without opportunity for subsequent growth of character.
- Everything that happens is part of the ultimate good. This was the viewpoint of Dr. Pangloss that the writer Voltaire lampooned in *Candide.*
- God permits random suffering but shares it vicariously, through God's spiritual presence in each human. This is an unprovable statement of faith even if it is true.
- God cannot create a world in which there is no possibility of suffering. To prevent pain, God would have to make fire no longer burn; but then it would not be fire. Although this sounds like a denial that God is all-powerful, it actually is an admission of inescapable logic. One of the most famous Christian writers, Clive Staples Lewis, noted that one cannot make nonsense into sense by sticking the words "God can" in front of it.
- Suffering is caused by demons, but as God permits these demons to run about, the problem of theodicy remains.

Deists, who believe in a God who is disconnected from the world (and perhaps not even a person), have little problem with the abundance of suffering; but how could the God of Love, to whom Jesus and Christian writers so abundantly testified, permit it? How can a supreme being who has all power permit such incredible sufferings as have bloodied the pages of history, and which seem totally unconnected to whether the victims were good or evil? Among those who despaired of finding an answer to theodicy was the writer of the biblical book of Ecclesiastes.

Evolutionary science has only sharpened the problem of evil. The process of NATURAL SELECTION practically ensures the suffering of great masses of organisms. Malthus (see MALTHUS, THOMAS) first noted a principle that Darwin (see DARWIN, CHARLES) applied to evolution: Populations always surpass their resources, with the inevitable result of violent competition and starvation. Intelligence should allow humans to foresee this result and prevent it by restraining reproduction, but as Malthus noted when he surveyed the condition of Europe, humans seldom do so.

Darwin's friend, the botanist, evolutionist, and Christian Asa Gray (see GRAY, ASA) maintained a belief in the ultimate purposes of a good God. In a letter of May 22, 1860, Darwin wrote to Gray:

> *With respect to the theological view of the question. This is always painful to me. I am bewildered. I had no intention to write atheistically. But I own that I cannot see as plainly as others do, and as I should wish to do, evidence of design and beneficence on all sides of us. There seems to me to be too much misery in the world. I cannot persuade myself that a beneficent and omnipotent God would have designedly created the Ichneumonidae with the express intention of their feeding within the living bodies of caterpillars, or that a cat should play with mice.*

Darwin noted a couple of silver linings in this cloud. First, his principal contribution was to point out that the victims of natural selection were primarily those that had inferior adaptations, with the result that the destructive process of death produced improvements in ADAPTATION. At least something good—in fact, a whole world of BIODIVERSITY—comes from it. The author of this encyclopedia, in younger and more naïve days, published this view in an unsuccessful attempt at Christian theodicy. Second, Darwin assured his readers that most animals were incapable of feeling pain, and even for those that could, "... we may console ourselves with the full belief, that the war of nature is not incessant, that no fear is felt, that death is generally prompt, and that the vigorous, the healthy, and the happy survive and multiply."

One reason that many in the general population, and even many scientists, had a hard time accepting natural selection was that it seemed so unfair. Neither the individuals with superior nor those with inferior characteristics deserved them; they were born with them, and most have paid the price for it. Genetic variation comes from MUTATIONS (see POPULATION GENETICS). For every good mutation that benefits its possessor there are numerous deleterious ones that cause their bearers to suffer. LAMARCKISM, the inheritance of acquired characteristics, seemed much more fair: An individual that worked hard could pass on to its descendants the progress that it had made. While perhaps all evolutionary scientists wish that Lamarckism were true, it simply is not. Furthermore, the fossil record is littered with millions of extinct species. The unfair and painful process of natural selection has therefore occurred everywhere for billions of years. If there is a supreme God, then this unfair and painful process was the method God used to create the living world. Without mutation, there is no variation, and EXTINCTION almost inevitably results; yet these mutations cause much suffering.

Darwin's contemporaries struggled with these issues. Theologian Henry L. Mansell published a Christian theodicy the same year and through the same publisher as Darwin's *Origin of Species* (see *ORIGIN OF SPECIES* [book]). Alfred Lord Tennyson was a literary friend of Darwin and Huxley, and one of his most famous poems *(In Memoriam)* captured the essence of this problem. The poem was published before the *Origin of Species* but reflected much of the thinking prevalent among his scientific acquaintances:

> *Who trusted God was love indeed*
> *And love Creation's final law—*
> *Tho' Nature, red in tooth and claw ...*

And while there may be cruelty at present, at least, Tennyson wondered, would not a benevolent God at least preserve species from extinction? But no:

> *"So careful of the type?" but no.*
> *From scarped cliff and quarried stone*
> *She cries, "A thousand types are gone:*
> *I care for nothing, all shall go."*

Evolutionary biologist Stephen Jay Gould (see GOULD, STEPHEN JAY) claimed that religion and science were compatible because they had non-overlapping realms of competence (which he called *non-overlapping magisteria*). Science explained how the world works, and its physical history; religion focuses on what is morally right and wrong. The distinction is between what happens or has happened, and what should happen. He considered both science and religion to be components of our "coat of many colors called wisdom."

Earlier scientists held beliefs that appear to match this approach. Galileo, who was punished for an astronomical theory that contradicted what Church authorities claimed was biblical teaching, said the Bible was about "how to go to Heaven, not how the heavens go." And Huxley (see HUXLEY, THOMAS HENRY), Darwin's contemporary and defender, said that even though natural selection is a violent and unfair process, human responsibility was to resist acting in a violent and unfair manner (see EVOLUTIONARY ETHICS). Natural selection produced humans, but humans should not practice "survival of the fittest" in society or between nations. In this, Huxley directly opposed the social Darwinism of people like Spencer (see SPENCER, HERBERT).

Therefore, Christianity and evolutionary science are, or can at least be forced to be, compatible. But many thinkers have wanted to go beyond mere compatibility. They have aspired to bring religion and science together. One of the most famous attempts to do so was William Paley's NATURAL THEOLOGY. Natural theology claimed that the existence and attributes of God could be discerned by a study of God's creation. As Richard Dawkins has written, "Paley's argument is made with passionate sincerity and is informed by the best biological scholarship of his day, but it was wrong, gloriously and utterly wrong." Natural theology, which today exists in the form of INTELLIGENT DESIGN theory, never seems to go away.

Still other scientists attempt a union of science and religion without embracing natural theology. Some of them detect evidence that supports their faith within the ANTHROPIC PRINCIPLE. Others claim that natural law itself, uniform throughout the universe, shows that there is a universal and constant God. Perhaps the minimalist version of the union of science and religion is the statement, of uncertain origin, that God is the answer to the question of why anything exists rather than nothing. Because the presence of God may be

(continues)

Can an Evolutionary Scientist Be Religious? *(continued)*

easier to detect, or imagine, in broad universal terms rather than in biological particulars, most of the scholars who attempt this union have been physicists such as Erwin Schrödinger and John Polkinghorne. The Templeton Foundation, started by a rich philanthropist, gives awards for investigations that bring science and Christianity together. These awards have a greater cash value than a Nobel Prize.

Of course, it also works the other way. Richard Dawkins has pointed out that, while atheism has always been possible, it was the Darwinian revolution that allowed people like him to be honest and intellectually fulfilled atheists.

Evolutionary science has eroded the credibility of many specific religious beliefs that have prevailed for centuries. This includes not only the claims of the creationists about the age of the Earth, the origin of humans, and the biblical flood, but also some more general tenets of Christian theology. Most people who identify themselves today as Christians would not insist on a literal interpretation of the biblical account of Adam and Eve as the first humans. They identify Adam and Eve as symbols of prehistoric humans. But Christian theology has long maintained that the sinful nature of humans entered into a previously sinless human race through the sin of Adam and Eve. According to this view, dominant even today, "original sin" entered humans through "the Fall" of Adam and Eve. Evolutionary science, however, claims that human nature is the product of evolution (see SOCIOBIOLOGY), as are all other human mental attributes (see INTELLIGENCE, EVOLUTION OF). As more and more human mental processes are explained by brain structures and functions (for example, stimulation of the right temporal lobe produces experiences that appear identical to religious visions), the role of a separate spirit becomes more and more difficult to believe, unless the human spirit is an exact replica of the human brain. While some religious people claim to have had revelations that could not have been generated by their brains, other people are understandably skeptical of these claims, especially when prominent religious leaders claim that "God told them" which American political party God preferred.

Most religious scientists are quite silent about their beliefs. One prominent evolutionary scientist who has taken a stand, but done so very cautiously, is paleontologist Simon Conway Morris. He wrote, "We do indeed have a choice, and we can exercise our free will. We might be a product of the biosphere, but it is one with which we are charged to exercise stewardship. We might do better to accept our intelligence as a gift, and it may be a mistake to imagine that we shall not be called to account."

If science and religion, especially evolutionary science and Christian religion, are to be compatible, a rethinking of both science and religion must occur. Science already undertakes a constant process of rethinking (see SCIENTIFIC METHOD). It is religion that must take the unaccustomed step of questioning ancient assumptions.

Further Reading

Banerjee, Neela, and Anne Berryman. "At churches nationwide, good words for evolution." *New York Times,* February 13, 2006.

Conway Morris, Simon. *The Crucible of Creation.* New York: Oxford University Press, 1998.

Darwin, Charles. *On The Origin of Species By Means of Natural Selection,* 1st ed. London: John Murray, 1859.

Dawkins, Richard. *A Devil's Chaplain: Reflections on Hope, Lies, Science and Love.* New York: Houghton Mifflin, 2003.

Dean, Cornelia. "Scientists speak up on mix of God and science." *New York Times,* August 23, 2005. Available online. URL: http://www.nytimes.com/2005/08/23/national/23believers.html.

Easterbrook, Greg. "The new convergence." *Wired,* December 2002, 162–185.

Gould, Stephen J. *Rocks of Ages: Science and Religion in the Fullness of Life.* New York: Ballantine Books, 2002.

———. "The narthex of San Marco and the pangenetic paradigm." Chap. 20 in *I Have Landed: The End of a Beginning in Natural History.* New York: Harmony, 2002.

Harris, Sam. *The End of Faith: Religion, Terror, and the Future of Reason.* New York: Norton, 2004.

Haught, John F. *God after Darwin: A Theology of Evolution.* New York: Westview Press, 2001.

Heilbron, J. L. *The Sun in the Church: Cathedrals as Solar Observatories.* Cambridge, Mass.: Harvard University Press, 2001.

Mansell, Henry L. *The Limits of Religious Thought,* 4th ed. London: John Murray, 1859.

McGrath, Alister. *Dawkins's God: Genes, Memes, and the Meaning of Life.* Oxford, U.K.: Blackwell Publications, 2004.

Pigliucci, Massimo. *Tales of the Rational: Skeptical Essays about Nature and Science.* Smyrna, Ga.: Freethought Press, 2000.

Rice, Stanley A. "Bringing blessings out of adversity: God's activity in the world of nature." *Perspectives on Science and Christian Faith* 41 (1989): 2–9.

———. "On the problem of apparent evil in the natural world." *Perspectives on Science and Christian Faith* 39 (1987): 150–157.

Ruse, Michael. *Can a Darwinian Be a Christian? The Relationship between Science and Religion.* New York: Cambridge University Press, 2000.

Thomson, Keith. *Before Darwin: Reconciling God and Nature.* New Haven, Conn.: Yale University Press, 2005.

Further Reading

Hauser, Marc D. *Born to Be Good: How Nature Designed Our Universal Sense of Right and Wrong.* New York: HarperCollins, 2006.

Shermer, Michael. *The Science of Good and Evil: Why People Cheat, Gossip, Care, Share, and Follow the Golden Rule.* New York: Henry Holt, 2004.

Zimmer, Carl. "Whose life would you save?" *Discover,* April 2004, 60–65.

evolutionary medicine Evolution is a subject seldom studied in detail by premedical undergraduates or by students in medical school. All of modern medicine would benefit from evolutionary insights, according to physician and medical educator Randolph Nesse and evolutionary biologist George C. Williams, for reasons that include the following. First, some symptoms of disease may actually be responses that have been favored by NATURAL SELECTION. Second, some genetic diseases may have evolved as responses to infectious diseases. Third, evolution occurs in populations of pathogens and of hosts. Fourth, some medical conditions involved with pregnancy may best be understood as an evolutionary conflict of interest between the fetus and the mother. Fifth, an understanding of the evolution of the immune system may be essential to an understanding of hypersensitivities. In some of these cases, an evolutionary approach will actually change the type of treatment a physician may choose to administer to a patient and how the population as a whole views its approach to health.

Symptoms as evolutionary responses. Most people consider fever to be a bad thing. But medical researchers are now beginning to understand fever as an evolutionary adaptation. The higher temperature of the body slows down the reproduction of the pathogen, perhaps long enough for the patient's immune system to successfully eliminate the pathogen. This is demonstrated by the behavior of desert iguanas. Being cold-blooded, iguanas do not have fever, but when they are infected by pathogens they lie in the sun, which raises their body temperature beyond the limits of their normal preference. The direct application of this evolutionary understanding of fever is that using medicines to reduce fever might prolong the infection; drugs should be used to reduce fever only in cases where the fever itself may endanger the patient.

Fever is a host defense against pathogen infection, but it does not always work. The *Plasmodium falciparum* pathogen that causes malaria produces heat-shock proteins that defend it from the effects of human fever and even uses the fever as a stimulus for the stages of its life cycle. This explains the cycles of fever in victims of malaria.

Evolution of genetic diseases. Many genetic diseases result from MUTATIONS that interrupt the normal function of proteins in the human body. While such mutations are usually rare, some are widespread, as with sickle-cell anemia, cystic fibrosis, Tay-Sachs disease, and G6PD deficiency. Evolution explains why these mutations are so widespread. The sickle-cell gene and the deficiency of the G6PD enzyme confer resistance to malaria; the cystic fibrosis gene confers resistance to typhoid fever. Tay-Sachs disease is particularly common among people of European Jewish ancestry. The mutation may have allowed people, living in crowded ghettos, to resist tuberculosis. The greater health of the heterozygous carriers of these mutations is balanced by the poor health of those people who are homozygous for the mutation (see BALANCED POLYMORPHISM).

Even diabetes and obesity may have evolutionary explanations:

- *Diabetes.* Diabetes results when cells fail to respond to blood sugar, a situation called glucose resistance. Could glucose resistance ever be a beneficial process? In prehistoric times, human populations frequently experienced periods of famine, and under those conditions, glucose resistance may have actually helped some individuals to survive by limiting the amount of body growth that occurs in response to blood sugar.
- *Obesity.* The protein leptin reduces appetite; genetically based leptin deficiency can lead to obesity. Leptin deficiency might have allowed humans to consume, and store up, large quantities of food energy during the times in which it was briefly available. There are even mutations that cause mitochondria to use fatty acids less efficiently, which can cause fat buildup in the liver. This mutation also might have allowed humans to use their food reserves more slowly during times of famine.

These genetic conditions contribute to what has been called a "thrifty genotype," which was advantageous under prehistoric conditions but which under modern conditions of food abundance produce obesity and diabetes. The "thrifty genotype" hypothesis may, however, be an oversimplified explanation. Rather than having an unalterable "thrifty genotype," an individual may develop a "thrifty phenotype" before birth, if the mother experiences malnutrition, and develop diabetes and obesity when he or she grows up.

Elevated blood cholesterol contributes to heart disease. The ApoE gene has several alleles (see MENDELIAN GENETICS), some of which contribute to cholesterol buildup in arteries. Could high blood cholesterol ever be beneficial? Cholesterol is a steroid derivative, chemically related to steroid hormones. Cholesterol is also an important component of animal cell membranes. The evolution of high blood cholesterol may have resulted from selection for the availability of steroids as hormones and of cholesterol for cell membranes. The role of cholesterol in cell membranes is particularly important in the neural cells of the brain. Especially during the phase of rapid brain size increase in human evolution, it may have been important for the human body to have abundant cholesterol resources. The advantages conferred by high blood cholesterol for their role in steroid hormone production may also have been favored by natural selection. Natural selection may thus have favored high cholesterol levels in the past. Today, in modern society where fatty foods are continuously available, high cholesterol is a problem. Humans have lived in modern society for only a small portion of their evolutionary history. Moreover, cholesterol buildup in arteries is a problem mostly in older individuals. Through most of human evolution, individuals did not usually live long enough for natural selection to operate against cholesterol buildup.

Evolution of pathogens and hosts. Evolutionary biologist Walter Fitch has investigated the evolution of the strains of the influenza virus. The strains of influenza virus are identified on the basis of two proteins found in their coats: H (hemagglutinin), which is the protein that allows the virus to invade human cells, and N (neuraminidase). The H protein evolves at the rate of 6.7 mutations per nucleotide per millennium, producing new strains of the influenza virus. The most successful new strains are those with mutations in the portion of the H protein to which the immunoglobulins bind. Public health investigators now take the evolution of the influenza viruses seriously. The strains of influenza virus that are most variable in the binding region of the H protein are the ones that the investigators predict will be next year's big flu, and these are the strains against which the flu vaccine is produced. This is a clear application of evolution to medicine.

Evolutionary analyses, such as CLADISTICS, have helped researchers to understand the origin of the strains of diseases such as influenza. If the phylogeny of influenza viruses is constructed using the nucleoprotein (not the same as neuraminidase), the branches clearly correspond to those of the major hosts of the virus (horses; humans; swine; ducks and fowl). If the phylogeny of the influenza viruses is constructed using the H protein, each major branch is a mixture of viruses of different hosts, except for the horse influenza viruses, which have remained distinct. The meaning of this result is that the influenza viruses exchange genes for the H protein (see HORIZONTAL GENE TRANSFER). In particular, human viruses and pig viruses can both infect ducks, and while inside the ducks the viruses exchange segments of nucleic acid that code for H genes, but not those that code for nucleoprotein genes. The cladistic reconstruction of the evolution of influenza viruses has allowed public health investigators to discover where new strains of influenza come from: from regions of high human population density where people are in contact with both pigs and ducks. This is the reason that most new strains of influenza evolve in China. This evolutionary insight has allowed public health researchers to know where to look for new strains of influenza. This is another clear application of evolution to medicine.

The major role of evolution in pathogens has been the evolution of bacteria that resist antibiotics (see RESISTANCE, EVOLUTION OF). Without an evolutionary viewpoint, physicians thought that the best thing to do was to use antibiotics as much as possible. The result of the overuse of antibiotics was the evolution of bacteria that resisted them. Most physicians and public health researchers now understand the lesson of evolution: Antibiotics must be used only when necessary, to prevent the evolution of resistant strains. This is yet another clear application of evolution to medicine.

Recently, zoologist Bryan Grenfell and colleagues compared the phylogenies and the infection (epidemiological) patterns of pathogens. They found a correspondence between the patterns by which pathogens evolve and the patterns by which they spread:

- Some pathogens cause brief infections but there is cross-immunity between the strains. For example, infection by one strain of measles makes the host immune to other strains as well. In such pathogens, many strains can coexist in the host population.
- Some pathogens cause brief infections but there is little cross-immunity between the strains. For example, infection by one strain of influenza does not make the host immune to other strains. In such pathogens, there is a rapid turnover of strains, with few strains coexisting at any one time.
- Some pathogens cause the host immune system to respond in a way that weakens the host such that the host can be infected by other strains. For example, infection by one strain of dengue fever puts the host at greater risk of infection by another strain. In such pathogens, a small number of major strains coexist.
- Some pathogens such as HIV cause a persistent infection. In such pathogens, many strains evolve within the individual host (see AIDS, EVOLUTION OF).

Therefore evolution is essential not only to the practice of medicine against infectious diseases, but also to the explanation of infection patterns.

Pathogens and hosts coevolve (see COEVOLUTION). A pathogen can be successful either by spreading rapidly from one host to another, or by letting the host live. In the first case, particularly common in diseases such as cholera that are spread without direct personal contact, pathogens remain virulent. In the second case, particularly common in diseases such as smallpox that spread directly from one person to another, pathogens evolve into milder forms. A host is most successful when it resists the pathogen. Evolutionary biologist Paul Ewald pointed out that humans can tip the competitive balance away from virulent diseases by restricting the transmission of virulent strains, for example through the control of water pollution. He called this the "domestication" of the pathogens.

Relationship between fetus and mother. The placenta allows the mother to nourish the fetus (see MAMMALS, EVOLUTION OF). The placenta allows food and oxygen from the mother's blood to enter the fetus, and wastes such as carbon dioxide from the fetus to enter the mother's blood, without a direct contact of the blood. Direct contact would allow the mother's immune system to attack the parasitic fetus. In Rhesus (Rh) factor incompatibility, this happens anyway.

Some aspects of pregnancy are not fully explained by the assumption of a complete harmony between the interests of the mother and of the fetus. The placenta (which contains much tissue of embryonic rather than maternal origin) produces Igf2 (insulin-like growth factor) that raises the blood pressure and blood sugar levels of the mother. This benefits the fetus, but may endanger the health of the mother, in particular her ability to have further offspring. Having more children will enhance the evolutionary fitness of the mother but may not be in the interests of the fetus. This has also been interpreted as a conflict of interest between mother and father, in which the father's fitness is promoted by the mother providing so many resources to the fetus that the mother's own future reproduction or even survival is impaired (see SELFISH GENETIC ELEMENTS). Although the paternal alleles may work against the interests of the mother, they do not usually produce immediate harm upon her. In the case of gestational trophoblast disease, however, this is not the case. The enhanced expression of some paternal alleles in the embryo causes the embryo to become cancerous; this can endanger the mother's life.

Most of the time, the relationship between fetus and mother is mutualistic rather than a conflict of interest. But even here, some symptoms that are often interpreted as disease may actually arise from evolution. Morning sickness can be interpreted as an excessive aversion to any foods that could possibly be spoiled or toxic—a response that benefits both the fetus and the mother.

The immune system. Allergies are also called hypersensitivities because they occur when the immune system (especially the IgE immunoglobulins) respond to some chemicals in the environment as if they are harmful, when they are not really harmful. Examples include the immune response to urushiol, a chemical found in poison ivy, which is harmless except for the immune response that it provokes; hay fever, in which the immune system responds to otherwise harmless pollen; and massive shock that results from beestings in some individuals. Asthma is also interpreted as a response to potentially harmful particles in the environment. The normal function of IgE immunoglobulins is to stimulate the production of histamines which cause gut, skin, and respiratory passages to produce mucus and to massively shed tissue to get rid of viruses and bacteria. Medical researchers have found that, in order to develop normal operation, the immune system needs to be stimulated by exposure to soil mycobacteria and to animals. This is what the immune system evolved to do. Children who are raised in a sterile environment, in the mistaken belief that this is healthier than allowing them to get dirty, are more likely to develop allergies and asthma. Before the last century, all children were exposed to dust and animals. An evolutionary understanding of the immune system leads to the recommendation that children be allowed to have contact with dirt and animals.

In these five ways, evolutionary science has proven valuable in the study and practice of medicine.

Further Reading

Amábile-Cuevas, Carlos F. "New antibiotics and new resistance." *American Scientist* 91 (2003): 138–149.

Ewald, Paul W. *Evolution of Infectious Disease.* New York: Oxford University Press, 1994.

Fitch, Walter M., et al. "Positive Darwinian evolution in human influenza A viruses." *Proceedings of the National Academy of Sciences USA* 88 (1991): 4,270–4,274.

Foster, Kevin R. "Hamiltonian medicine: Why the social lives of pathogens matter." *Science* 308 (2005): 1,269–1,270.

Grenfell, Bryan T., et al. "Unifying the epidemiological and evolutionary dynamics of pathogens." *Science* 303 (2004): 327–332.

Nesse, Randolph M. "DarwinianMedicine.org, EvolutionaryMedicine.org: A resource for information on evolutionary biology as applied to medicine." Available online. URL: http://www.darwinianmedicine.org. Accessed July 19, 2005.

———, and George C. Williams. *Why We Get Sick: The New Science of Darwinian Medicine.* New York: Random House, 1996.

Stearns, Steven C., ed. *Evolution in Health and Disease.* New York: Oxford University Press, 1998.

Trevathan, W. R., E. O. Smith, and J. J. McKenna, eds. *Evolutionary Medicine.* New York: Oxford University Press, 1999.

Wills, Christopher. *Yellow Fever, Black Goddess: The Coevolution of People and Plagues.* Reading, Mass.: Addison-Wesley, 1996.

exaptation *See* ADAPTATION.

exobiology *See* ORIGIN OF LIFE.

extinction Extinction is the disappearance of a species. Extinction usually involves (1) a decrease in population sizes within a species and (2) a decrease in genetic diversity within a species. These two processes are related, since small populations usually have less genetic diversity than large populations (see POPULATION GENETICS).

Extinction usually occurs for a variety of interrelated reasons, and it occurs one species at a time. When a large number of species becomes extinct at the same time, for the same reason, a mass extinction event is said to have occurred (see MASS EXTINCTIONS). Five mass extinctions have occurred during the history of life and have been important in the pattern of evolutionary history. Many biologists consider that a sixth mass extinction of species is now occurring, largely due to human disturbance of the environment, and that the mass extinction now under way may well prove to be the biggest in the history of life on Earth.

Changes in the environment, whether naturally occurring or caused by human activity, may impair the ability of organisms to reproduce; this will cause populations to decline. A species with a small number of populations is in much greater danger of extinction than a species with larger population number and size. For example, a plant species that is limited to a single locality may become extinct if that locality is disturbed by fire, flood, or human activity. The bush *Franklinia alatamaha* was discovered by the American botanists John and William Bartram in the 18th century. It was growing in just one location in the mountains of the colony of Georgia, on the plateau adjacent to the Altamaha River. The Bartrams took cuttings from these bushes and planted them in a garden; descendants of these bushes are still alive in gardens throughout the United States and Europe. The wild population has disappeared, perhaps when the forest that contained it was cleared for cultivation, perhaps even by a single farmer. The tree *Ginkgo biloba* was very nearly extinct in its habitat, the forests of north central China. Buddhist priests planted specimens of this tree in their monasteries. Ginkgo trees are now abundant throughout the world where they have been planted in cities, and more recently on plantations for the production of medicinal compounds. Franklinias and ginkgos survive today largely because a small sample of them was rescued from their very small and vulnerable wild populations.

Loss of genetic diversity also makes a species more vulnerable to extinction. A species with fewer genes is less capable of responding through NATURAL SELECTION to changes in the environment. A species with little diversity of genes that confer resistance to specific diseases runs the risk of being devastated by a disease for which it has no resistance. In addition, the loss of genetic diversity can lead to inbreeding. Within most populations, deadly genetic mutations are hidden in the cells of even apparently healthy individuals. In small populations, the chance that two closely related individuals will mate and produce offspring is greater than in large populations. If two individuals that have the same hidden mutations mate, these mutations may show up in double dose in the offspring, causing the offspring to die before or shortly after birth (see MENDELIAN GENETICS). This can be a problem even in a population that has begun to recover from a previous decline. The number of individuals in the population may grow but the number of genes will not, except over a very long period of time. A loss of genetic diversity in the past (a *genetic bottleneck;* see FOUNDER EFFECT) can condemn a population to reduced survival ability even if circumstances should become favorable to its growth.

An example of a species that is near extinction because of a genetic bottleneck is the nene, or Hawaiian goose *(Nesochen sandiviensis)*. Due to the efforts of conservation workers and public cooperation, populations of the nene (the Hawaiian state bird) are growing, following its near extinction in the first half of the 20th century. DNA studies that compare modern birds with birds preserved in museums indicate that modern populations of the nene have much lower genetic diversity than did populations in the 19th century. Many goslings, whether hatched in captivity or in the wild, have birth defects, and many of them die. Small, isolated human populations also exhibit traits that remain hidden in larger populations, for example extra fingers and toes (polydactyly) and metabolic disorders.

Saving species from extinction can provide a tremendous benefit to science and to human well-being. Plant breeders, for example, wish to introduce genes for drought resistance into maize (*Zea mays;* also called corn) raised in the United States. Drought resistance genes are available in *Tripsacum* grass, but *Tripsacum* cannot crossbreed with maize. A wild species of perennial maize, *Zea diploperennis,* can crossbreed with both *Tripsacum* and *Zea mays* and can thereby act as the bridge for introducing *Tripsacum* genes into agricultural maize populations. *Zea diploperennis* was almost extinct when botanist Hugh Iltis discovered it in the 20th century.

Conservationists frequently say that "extinction is forever." This statement is obviously true for a species—once it is gone it cannot come back—but is also true of the genetic diversity within a species.

Further Reading

Cardillo, Marcel, et al. "Multiple causes of high extinction risk in large mammal species." *Science* 309 (2005): 1,239–1,241.

Fisher, R. A. (1890–1962) British *Mathematician, Geneticist* Sir R. A. Fisher paved the way for the acceptance of natural selection as the mechanism of evolution and helped to make the MODERN SYNTHESIS possible.

Born February 17, 1890, Ronald Aylmer Fisher was a mathematical prodigy. He was accustomed to solitude, both because of his upbringing and because he was extremely nearsighted. Because his doctor had forbidden him to read by electric light, he had to receive his mathematics tutoring at Harrow School without the aid of pencil, paper, and blackboard. He developed the ability to work out complex mathematical solutions in his head.

He began to attend Cambridge University in 1909, the centennial of Darwin's birth and the fiftieth anniversary of the publication of his most famous work (see DARWIN, CHARLES; *ORIGIN OF SPECIES* [book]). Despite the ceremonies that attended the anniversary, Darwin's theories did not meet with widespread acceptance at that time. Nevertheless, young Fisher was strongly influenced by them.

Fisher was even more strongly influenced by the recent rediscovery of MENDELIAN GENETICS, which were being investigated by William Bateson, geneticist at Cambridge. Both from his studies of Mendelian inheritance patterns, and from his association with the statistician Karl Pearson, Fisher became well-grounded in the study of statistics. His 1927 book *Statistical Methods for Research Workers* was a landmark in the history of scientific research methods, and he eventually became one of the giants of statistical theory. He developed the variance ratio, one of the most widely used statistical parameters in the world today. This ratio in effect expressed the variation in a sample that could be explained relative to the proportion that could not be explained. He developed a probability table that showed how likely the results were to be significant, depending on the sample size and the number of categories. For example, a variance ratio of 3 (three times as much variation explained as unexplained) might be credible for a large sample size but not for a small one. As a result of Fisher's table, scientists no longer need to say that "the results appear to not be due to chance" but can say something like "the probability that the results occurred by chance are less than one percent." Modern scientific research, and its application in fields as diverse as space travel, industrial quality control, and the control of diseases, would scarcely be possible without the statistical tools pioneered by Fisher. The variance ratio is called the F ratio, after Fisher.

Fisher was one of the founding fathers of POPULATION GENETICS (see also HALDANE, J. B. S.; WRIGHT, SEWALL). Fisher demonstrated mathematically how Mendelian genetics, which had previously been considered anti-Darwinian, fit together with Darwinian natural selection, as proclaimed in the title of his 1930 book *The Genetical Theory of Natural Selection.* The concept of "fitness," which had previously been vague, was given a precise mathematical definition. His equations made some oversimplifications—for example, he assumed that each gene had an independent effect on the phenotype, rather than interacting with other genes—but were more advanced than anything that had gone before. His equations also showed that even the smallest fitness advantages, over enough time, allowed natural selection to produce big effects. At the same time he demonstrated that it was small mutations, not large ones, that caused evolutionary change in populations.

Fisher's interest in genetics spilled over into a passion for EUGENICS, especially from his association with Charles Darwin's son, Major Leonard Darwin. Fisher came to believe that natural selection was weakened as a result of the comforts of civilization, that the genes of Englishmen were becoming weaker, which he believed was the reason that the British Empire was in decline. One of the founders of eugenics was Darwin's cousin (see GALTON, FRANCIS), who was also an expert in statistics, which made it appeal even more to Fisher. Fisher noted with alarm that the upper class in British society had fewer offspring than the lower classes. He put his eugenic theories (with himself at the top, despite his obvious visual defect) into practice: he championed government subsidies for rich families to have more children; got married and had eight children; and left academia to start a dairy farm. He justified this last move by saying that farming was the only occupation in which having many children was an advantage. The farm was disastrous and Leonard Darwin had to rescue him financially.

Modern evolutionary science continues to benefit from Fisher's insights, both through population genetics and by using statistical methods of which Fisher was the pioneer. Fisher died July 29, 1962.

Further Reading

Fisher, R. A. *The Genetical Theory of Natural Selection.* Oxford, U.K.: Oxford University Press, 1930.

fishes, evolution of Fishes were the first vertebrates. Fishes are the only group of vertebrates that is almost exclusively aquatic, which was the primitive condition of all life. When referring to more than one individual of the same species, "fish" is the correct plural; for more than one species, "fishes" is correct.

All vertebrates are *chordates* and have evolved from invertebrate ancestors that resembled modern lancelets (see INVERTEBRATES, EVOLUTION OF). Lancelets have a cartilaginous rod along the back (a *notochord*) in association with the main nerve cord and gill slits. All vertebrates have these features, although in most adult vertebrates the notochord is replaced by a backbone of vertebrae and may possess the gill slits only during the embryonic stage. Lancelets do not have jaws. An animal of the CAMBRIAN PERIOD known as *Pikaia* closely resembled a lancelet (see BURGESS SHALE).

The earliest known fishes lived during the Cambrian period. One example is *Myllokunmingia,* found from fossil deposits in China. Conodont animals also appeared in the Cambrian period and were probably also classifiable as fishes. Neither they nor the fishes of the early ORDOVICIAN PERIOD had jaws. Two classes of modern *agnathan* (jawless) fishes survive: lampreys and hagfishes, which today live by sucking blood from larger fishes.

One lineage of fishes evolved jaws during the Ordovician period. This allowed a major evolutionary advancement in the efficiency of predation. Many of the earliest jawed fishes, including the *placoderms,* were covered with bony armor, as the predatory arms race became severe. Placoderms apparently became extinct without descendants. Other lineages of jawed fishes survived to become the Chondrichthyes (cartilaginous fishes) and the Osteichthyes (bony fishes).

Cartilaginous fishes include the sharks and rays. Their skeletons consist only of cartilage. Teeth are not bones; shark teeth are structurally similar to their scales and are often the only part of the shark to be preserved as fossils. Many kinds of sharks diversified throughout the PALEOZOIC ERA. Today, some of them have very specialized features, such as the ability to navigate by electrolocation.

Bony fishes evolved in freshwater conditions and were restricted to freshwater for the first 160 million years of their existence. Evolutionary scientists speculate that their bones were more important as a way of storing calcium, a mineral that could be scarce in freshwater, than as a skeletal reinforcement. Bony fishes diverged into two major lineages:

- the *ray-finned fishes,* which have fins reinforced with bony rays that do not correspond to the fingers or toes of other vertebrates. The major lineage of ray-finned fishes is the *teleosts,* which includes most modern fish species.
- the *flesh-finned fishes.* These fishes diverged into two lineages: the lungfishes and the crossopterygians. Some lungfishes survive today in shallow tropical ponds. When the ponds have water, the oxygen levels are low and the lungfishes gulp air. When the ponds dry up, the lungfishes continue to breathe while estivating in the mud. The crossopterygian fishes have bones at the bases of their fins that correspond to the one upper and two lower limb bones of tetrapods (four-legged animals). Two branches of crossopterygian fishes became the coelacanths and the tetrapods.

Coelacanths were thought to have been extinct since the CRETACEOUS EXTINCTION 65 million years ago. However, in 1938, Captain Hendrick Goosen of the trawler *Nerine* brought a fish that had been caught deep in the Indian Ocean to Marjorie Courtenay-Latimer, curator of a museum in East London, South Africa. She identified the fish, now named *Latimeria chalumnae,* as a crossopterygian. Subsequent searches revealed that native fishermen of the Comoros Islands (in the Indian Ocean off the coast of Africa) reported that they had caught these fish for years and had thrown them back as inedible. In 1998 another species, *Latimeria menadoensis,* was found 10,000 kilometers away, in Indonesia, by a scientist on his honeymoon.

The lineage that became tetrapods included fishes such as *Eusthenopteron,* which had skeletal characteristics intermediate between those of earlier fishes and later amphibians. Later species in this lineage had legs but still retained many fish skeletal characteristics (see AMPHIBIANS, EVOLUTION OF). As is the case with any so-called MISSING LINKS, there is no clear line of division between ancestor and descendant.

Lungs are a primitive condition for bony fishes. Lungs began as expansions of the upper digestive tract that were open to the pharynx, allowing the fish to gulp air. Gars and bowfins can still do this. In most bony fishes today, the lung has evolved into the air bladder, a pocket not connected to the pharynx, which the fish uses for buoyancy.

Teleost fishes represent tremendous evolutionary diversity. Some groups of fishes, such as the cichlids of African lakes, are some of the best examples of rapid evolution (see SPECIATION).

Further Reading

Barlow, George. *The Cichlid Fishes: Nature's Grand Experiment in Evolution.* New York: Perseus, 2002.

Forey, P. L. "Golden jubilee for coelacanth *Latimeria chalumnae.*" *Nature* 336 (1988): 727–732.

Long, John A. *The Rise of Fishes: 500 Million Years of Evolution.* Johns Hopkins University Press, 1996.

fitness *See* NATURAL SELECTION.

FitzRoy, Robert (1805–1865) British *Naval Officer* Robert FitzRoy was the captain of HMS *Beagle,* the ship on which a young Charles Darwin (see DARWIN, CHARLES), only a few years younger than himself, sailed around the world. FitzRoy, a staunch believer in the Bible, was very upset to discover that it was the trip on board his vessel that opened Darwin's eyes to the science of evolution.

Born July 5, 1805, FitzRoy attended the Royal Naval College at Portsmouth, where he studied with distinction. He held positions of responsibility on two ships. At the age of 23 he was appointed captain of HMS *Beagle.* The crew of the *Beagle* surveyed the coasts and ports of South America. Also on this voyage, they brought three Fuegian Native Americans from southern South America to England for religious and cultural instruction. A second voyage was required, for further surveying and to take the Fuegians back as missionaries to their tribe. The second voyage departed England in 1831 with Charles Darwin on board as companion to Captain FitzRoy. FitzRoy and Darwin got along reasonably well most of the time, but Darwin had a difficult time adjusting to FitzRoy's explosive temper. The crew referred to their captain as "Hot Coffee" because he was always boiling over. The *Beagle* returned to England in 1836. In 1839 a three-volume narrative of the Beagle voyages was published; FitzRoy was the editor, and the author of the first two volumes. Charles Darwin wrote the third volume, which became a popular book. FitzRoy's volumes demonstrated that he was not only an excellent navigator and surveyor but an observant man of science as well. In some ways, his scientific technique exceeded Darwin's: In his collection of what are now called DARWIN'S FINCHES, FitzRoy indicated which island each finch had come from, while Darwin did not.

When he returned to England, FitzRoy married and began a family. He also briefly served in Parliament. In 1843 he was appointed the governor of New Zealand. He was dismissed from this position in 1846. Some historians say that it was because of his explosive temper, but others have pointed out that he treated native Maori land claims as equally valid to those of the white settlers, which was against the imperialistic interests of the Crown.

In 1854 FitzRoy was appointed the head of the British Meteorological Department. He developed some of the meteorological and forecasting techniques that modern people take for granted, for example the printing of weather forecasts in daily newspapers and a system of storm warnings. FitzRoy also invented a cheap, useful new kind of barometer.

FitzRoy spent much time gathering information that he believed supported a literalistic interpretation of the Bible. He was present at the famous Oxford debate that included an exchange between Huxley (see HUXLEY, THOMAS HENRY) and Bishop Samuel Wilberforce. FitzRoy walked around waving a Bible in the air, proclaiming, "The Book! The Book!"

Suicides had occurred in FitzRoy's family, and it is possible that he had an inherited mental instability that finally overcame him. He shot himself on April 30, 1865.

Further Reading

Nichols, Peter. *Evolution's Captain: The Dark Fate of the Man Who Sailed Charles Darwin Around the World.* New York: HarperCollins, 2003.

Flores Island people A dwarf species of early humans *(Homo floresiensis)* may have evolved in isolation on Flores Island in Indonesia less than 30,000 years ago. A team led by anthropologist Michael Morwood discovered a largely complete adult female skeleton and fragments of as many as seven others in a limestone cave in 2003 (see photo on page 164).

For more than a century scientists have known that early humans known as Java man lived in Indonesia (see DUBOIS, EUGÈNE; *HOMO ERECTUS*). These early humans, dating back to almost a million years ago, have been considered the same species as Peking man in northeast Asia. They are considered to be descendants of an African HOMININ species, perhaps *HOMO ERGASTER,* that dispersed into Asia. Once in Asia, these hominins evolved some distinctive skeletal features, perhaps slightly larger brains, but experienced a regression in tool-making abilities, from the Acheulean phase of *H. ergaster* back to the Oldowan phase characteristic of *HOMO HABILIS* (see TECHNOLOGY). Some radiometric dates (see RADIOMETRIC DATING) suggested that *H. erectus* might have persisted until as recently as 26,000 years ago. By that time, modern *H. sapiens* had arrived in a separate wave of migration from Africa and undoubtedly saw the smaller-brained, shorter humans who were already there. Like *H. ergaster, H. erectus* used fire. *H. erectus* must have known how to make rafts, because even when ocean levels were at their lowest (during each of the ICE AGES), there were several miles of ocean to cross to reach the various islands on which *H. erectus* lived. This included Flores Island.

Scientists had studied what they assumed were *H. erectus* specimens on Flores Island. Like other *H. erectus* populations, these people made stone tools and used fire. Strangely, the tools seemed to be what one researcher described as "toy-sized" versions of typical *H. erectus* tools. The one puzzling aspect was that they seemed to find only the skeletons of children, about a meter in height. Since it was unlikely that Flores Island was a prehistoric Boy Scout camp, where were the adults? The newly discovered skull, with a cranial capacity of about 25 cubic inches (400 cc), was assumed to be that of a child.

Then someone took a close look at the teeth and the joints between the skull bones. The degree of wear upon the teeth could only have been produced by decades of use, and the cranial bones were fused like those of an adult. Estimating age from tooth wear is standard procedure for all hominin species. Therefore the researchers concluded that the Flores Island hominins were a species of dwarf hominins. Presumably, *H. erectus* individuals colonized Flores Island, lost either the ability or desire to reconnect with other populations, and over many generations evolved into a miniature form, now called *Homo floresiensis.*

The reason that this was big news among evolutionary scientists (and made headlines around the world) was that

Christopher Stringer, head of Human Origins at the Natural History Museum in London, displays a skull of a *Homo erectus,* left, a cast taken from the skull of the newly discovered *Homo floresiensis,* center, and the cast of a modern *Homo sapiens* skull during a news conference in London, October 27, 2004. *(Courtesy of Richard Lewis/AP)*

this was the only known example of human evolution going in the direction of smaller bodies and brains. The evolutionary myth (myth in the sense of a big story that helps humans understand their place in the world) as understood by most laymen is that evolution produced ever larger and brainier humans. Human evolution had frequently been portrayed as a march upward (see PROGRESS, CONCEPT OF). For people who adhered to the evolution myth rather than evolutionary science, *H. floresiensis* came as a shock.

Why might these hominins have evolved small size? On islands, immigrant animal species with large bodies evolve smaller bodies, and immigrant animal species with small bodies evolve larger bodies. Extremes of body size may be the result of competition: Animals can avoid direct competition with other animals of their species either by being bigger, and overpowering them, or by being smaller, and avoiding competition by specializing on more limited resources. This is why islands may have big rodents and small elephants. Flores Island itself had giant tortoises, giant lizards, elephants as small as ponies, and rats as big as hunting dogs, at the same time that *H. floresiensis* lived there. Isolated from competition with other *H. erectus,* the Flores islanders may have been free to evolve into smaller form. Smaller bodies can survive more easily upon limited food supplies than can larger bodies.

One particularly surprising aspect was the very small size of the brain. The cranial capacity of *H. floresiensis* was in the range of chimpanzees and AUSTRALOPITHECINES. Amazingly, despite such small brains, the Flores Island people continued making stone tools and building fires, neither of which activities had been invented by australopithecines. Presumably it takes less brain power to maintain a technological tradition than to develop one. Although it has been suggested that the smaller brain might have resulted simply from the overall decrease in body size (see ALLOMETRY), in this case the decrease in brain size was disproportionately more than the decrease in body size: *H. floresiensis* bodies were about half the size, but their brains only a third the size, of those of some *H. erectus*. This suggests that there was some evolutionary advantage to the decrease in brain size. Nobody has a clue as to what this might have been.

Other possibilities remain. Might the Flores Island hominins have been descendants not of Java man but of earlier hominin species such as *H. habilis,* or even australopithecines? This is unlikely, since Flores Island is thousands of miles away from the only places in which *H. habilis* and australopithecines are known to have lived, and there is no evidence that these earlier hominin species knew how to make rafts. It is always possible that the small-brained skull came from a pathological individual. Evidence against this interpretation included the following: First, there are no recognizable deformities in the skull that would suggest microcephaly, other than the small size of the brain itself. Anthropologist Dean Falk explains that even though the brain of *H. floresiensis* was small, it had some structural characteristics that resemble the normal brains of larger-brained hominins, rather than those of microcephalics. Second, the individual lived enough years to wear down his or her teeth and was smart enough to make fire and tools. Third, how likely is it that the very first skull one happens to find would be of a deformed individual? Indonesian anthropologist Teuku Jacob insists that the Flores Island people were not deformed but were merely dwarf modern humans. But why would there have been a whole population that consisted only of dwarves?

Another surprise was that the archaeological deposits associated with the Flores Island hominins were produced as recently as 18,000 years ago. The hominins might have persisted even longer than this. Modern humans were certainly in Indonesia by that time and may have caught a glimpse of them. Every culture has legends about little people. Indonesian legend describes "Ebu Gogo" little people. Indonesian little-people legends might have had a basis in fact. The only thing of which scientists are certain is that this species no longer exists on the small and very well explored Flores Island.

Human evolution has produced the biggest brains relative to body size, and among the biggest brains in absolute size, that the world has ever known. However, the Flores Island hominins demonstrate that human evolution can proceed in the other direction, if conditions permit—a vision that science fiction writer Herbert George Wells developed when he invented the Eloi in his novella *The Time Machine* in the early 20th century.

Further Reading

Brown, Peter, et al. "A new small-bodied hominin from the Late Pleistocene of Flores, Indonesia." *Nature* 431 (2004): 1,055–1,061.

Culotta, Elizabeth. "Battle erupts over the 'hobbit' bones." *Science* 307 (2005): 1,179.

Diamond, Jared. "The astonishing micropygmies." *Science* 306 (2004): 2,047–2,048.

Falk, Dean, et al. "The brain of LB1, *Homo floresiensis.*" *Science* 308 (2005): 242–245.

Morwood, Michael, et al. "Archaeology and age of a new hominin from Flores in eastern Indonesia." *Nature* 431 (2004): 1,087–1,091.

———. "The people time forgot." *National Geographic,* April 2005, 2–15.

Wong, Kate. "The littlest human." *Scientific American,* February 2005, 56–65.

fossils and fossilization Fossils are the physical evidence of organisms that have been dead for many years, past the normal duration of decomposition; fossilization is the set of processes by which they are formed. When volcanic or other rocks erode, water carries sediment to the continental shelves at the edges of the oceans, where it accumulates in layers called *strata.* This process is occurring right now. As silt and mud layers are buried, the increased temperature and pressure transforms them into *sedimentary* rocks. The layers in sedimentary rocks are visually distinguishable. Clay particles become shale, while sand particles become sandstone. Fossils are usually found in sedimentary rocks, since dead plants and animals are often buried within the sediment. Intense heat and pressure can transform volcanic or sedimentary rocks into *metamorphic* rocks, in which the fossils are usually destroyed. Fossils are not always destroyed by metamorphosis. Structures that may be fossilized cyanobacteria can be seen in the 3.5 billion-year-old Apex chert of Australia (see ORIGIN OF LIFE).

Fossils have been known for millennia; but until the last three centuries, most people thought fossils were simply peculiar rocks that just happened to look like plants and animals, rather than being the remains of actual plants and animals. Leonardo da Vinci was one of the earliest scholars to recognize fossils for what they really were, and he puzzled about why fossils of seashells were found at the tops of mountains. His solution to the problem reflected more of the medieval than the modern mind. He conceived of the Earth as an organism, therefore it must have a circulatory system, which means that ocean water must circulate underground to the tops of mountains, where it comes out as creeks and rivers—and sometimes brings seashells with it. He was right about the fossils, though wrong about how they got there. Another early scholar to recognize fossils as remnants of formerly living organisms was Danish geologist Niels Stensen (Nicholas Steno).

When an organism dies, it almost always decays. Under some circumstances, decomposition is delayed, such as in the presence of water and the absence of oxygen. As the organism decays slowly, the space inside its body, sometimes even inside of its cells, is filled with mineral deposits from the water (see figure on page 166). The fossil may be completely mineralized; alternatively, *permineralization* occurs when minerals enter the dead organism without completely replacing the original molecules. Under conditions of high temperature and pressure, the sediments surrounding the organism and the organism itself are both transformed into rock. Because of chemical differences between the fossil and its surrounding *matrix,* fossils usually stand out and may cause a zone of weakness in the rock that contains them. Fossils are therefore frequently found by geologists who strike the rocks with hammers; the fracture reveals the fossils. Mineral deposits inside of a coal seam *(coal balls),* rather than the coal itself, often contain fossils of the plants from which the coal was

The grain of the wood is still visible in a petrified trunk of a tree that lived about 300 million years ago, now in Petrified Forest National Park, Arizona. ***(Photograph by Stanley A. Rice)***

derived. The chemical differences between the fossil and its matrix can sometimes be of help in extracting the fossil. For example, shells of invertebrates can sometimes be isolated if the matrix is dissolved in concentrated hydrofluoric acid.

Sediments are the major, but not the only, place to find fossils. Many insects, stumbling into conifer sap, became entangled and engulfed; the sap became amber, and the insects were preserved in exquisite detail. This happened frequently enough that there is a worldwide market for insects preserved in amber.

Ancient wood other than petrified wood, human bodies from a bog, and mummies preserved in Andean caves are not usually considered fossils, nor are bodies preserved by human activity (such as Egyptian mummies). Fossils retain the structure of the original organisms, sometimes with considerable internal detail; therefore coal and oil are not considered fossils, since they have been transformed into uniform organic materials (even though they are called fossil fuels). These remains can be as valuable as fossils in scientific research. Leaves that have been preserved for 20 million years in cold, wet, anoxic conditions in deposits near Clarkia, Idaho, are still green but oxidize into black films upon exposure to air. The leaves were so chemically intact that scientists could extract chloroplast DNA from them (see DNA [EVIDENCE FOR EVOLUTION]).

The hard parts of organisms (such as the calcium carbonate shells of invertebrates, and the calcium phosphate bones of vertebrates) are most easily preserved and constitute the vast majority of fossils. Teeth are particularly popular fossils to study, not only because they are hard (and abundantly preserved) but also because they can reveal what kind of diet the animal had—grinding teeth for grazers, sharp teeth for carnivores. The study of fossil bones can also reveal more than just the anatomical structure of the animal. Study of tissue layers inside of bones reveals the growth rate of the animal and its age at death. This is how scientists have determined that *Tyrannosaurus* dinosaurs had a high metabolic rate and grew rapidly, up to 880 pounds (400 kg) by the age of 20 years (see DINOSAURS).

Under conditions of very slow decomposition and very thorough burial, soft parts of organisms, and entire organisms without hard parts, can be preserved. A famous example is the BURGESS SHALE in Canada. Impressions of soft-bodied animals, many parts of plants, and even single-celled microbes have frequently been preserved. Since fossilization almost always requires burial in sediment, most fossils are of the aquatic organisms that already live in or near the sediments. Aquatic invertebrates therefore constitute the best fossil assemblages. Terrestrial organisms are fossilized when they fall into or are washed into the sediments, as when primitive birds fell into low-oxygen pools that later became the limestone deposits of Solnhofen limestone in Germany (see *ARCHAEOPTERYX*). The odds are very much against the preservation of fossils from terrestrial organisms: It is estimated that only one bone in a billion is preserved as a fossil. If all 290 million Americans, most of whom have 206 bones, were to die, only about 60 bones would be fossilized under normal conditions. The odds are not even good for organisms of shallow oceans, where only 40 percent of the organisms have even the possibility of being fossilized. Rarely, massive graveyards of fossils *(Lagerstätten)* such as the Burgess shale or the Solnhofen limestone have resulted from conditions that were remarkably good for preservation.

An intimate knowledge of biology is necessary in the study of fossils. This allows a great deal of information about the living animal to be reconstructed from the fossils of just its bones. Sites of muscle attachment leave traces on the bones, therefore the musclature of an extinct animal can be reconstructed from bone fossils. Blood vessels inside the living bones may have left traces in the fossilized bones, allowing scientists to determine the extent of blood vessel formation, and hence whether the animal was warm-blooded. Occasional lucky finds reveal even more about the animal's life. Gastroliths are smooth stones found associated with dinosaur skeletons. Like modern birds (which are classified with the dinosaurs; see BIRDS, EVOLUTION OF), many dinosaurs swallowed stones which helped to grind plant food in their stomachs. The grinding action caused the stones to become smooth. Coprolites (fossilized feces) often allow scientists to reconstruct the diet of the animal.

It is usually easy to tell which structures in sedimentary rocks are fossils and which are inorganic in origin, but not always. There was some controversy about whether a "shell" that was found in a Russian fossil deposit was actually a shell or a percussion fracture from the drill bit that extracted the sample.

Unfortunately for researchers, fossils are usually found in fragments and have to be put back together. Examples include:

- Plants often shed their organs separately, or are torn apart during fossilization. Thus it is often impossible to tell which leaf grew in connection with which trunk.
- The same thing can happen to animals. The largest animal of the Cambrian seas, *Anomalocaris,* was formerly described as three separate animal species. The animal's body had been described as a sea-cucumber; its mouth had been described as a jellyfish; and its front legs had been described as headless shrimp. The true identity of these fragments was revealed only when a fossil was found that had these fragments together.
- One of the most abundant kinds of Paleozoic fossils was the *conodont* (cone-tooth) (see FISHES, EVOLUTION OF). Everybody suspected that it was an animal tooth but nobody could find the animal. In 1983, after many decades of research, a conodont was found in association with a soft-bodied vertebrate, and then only from a faint minnow-like depression.
- When fossils are found individually, it is often impossible to determine which of them belonged to the same species. Juvenile dinosaurs have sometimes been classified as a separate species from the parents. When whole herds or populations of dinosaurs are found fossilized together, it may be possible to determine which ones were the juvenile forms.

There are a few huge deposits of fossils, such as the hills of diatomaceous earth near Lompoc, California, and the white hills of Dover in England, which consist of billions of shells of tiny single-celled organisms. A huge graveyard of dinosaur bones, turned to stone, can be seen at Dinosaur National Monument in Colorado. Numerous skeletons of rhinoceroses and horses that died from breathing ash from a volcanic eruption 12 million years ago are preserved, in exquisite detail and still in their original arrangement, at Ashfall Beds State Historic Park in Nebraska (see figure on page 167).

Trace fossils are the preserved evidence of the activities of the organisms. This includes worm burrows and animal footprints. Worm burrows at the beginning of the Cambrian period are the earliest signs of the rapid evolution of animal life in the Phanerozoic Eon (see CAMBRIAN EXPLOSION). Dinosaur footprints have allowed the analysis of how large the dinosaurs were and how fast they ran, which has fueled the suggestion that some of them, at least, were warm-blooded. Perhaps the most convincing evidence that *Australopithecus afarensis* walked on two legs was the discovery, by Mary Leakey, of a trail of remarkably human footprints produced by this species (see LEAKEY, MARY; AUSTRALOPITHECINES).

Fossils can be used to reconstruct the climate in which the organisms lived.

- Plants with large leaves today grow in moist areas, and plants with smaller leaves in drier areas. Serrated leaf margins allow modern plants to cool more efficiently than plants with smooth margins, without the evaporation of as much water. The study of leaf shapes can help reconstruct ancient temperature and moisture conditions.
- The pollen of seed plants has an outer covering that is very resistant to decay, and usually each genus of plant has its own visually distinguishable kind of pollen. Even when all other parts of a plant have decomposed, the pollen (though dead) may still be recognizable. Pollen from mud deposited during the most recent ice age, for example, indicates that spruce trees (now found only far north and on mountaintops) grew as far south as Oklahoma, and that much of what is now Amazonian rain forest was grassland. The study of ancient pollen is called *palynology.*

The *isotope ratios* of fossils may also help determine the climate in which they grew (see ISOTOPES).

These rhinoceroses died 12 million years ago and were buried in volcanic ash in what is now Ashfall Beds State Historic Park in Nebraska. It is rare for such fossils to be found with all the bones still articulated, as in these specimens. ***(Photograph by Stanley A. Rice)***

Further Reading

Fortey, Richard. *Earth: An Intimate History.* New York: Knopf, 2004.

Gould, Stephen Jay. "Magnolias from Moscow." Chap. 31 in *Dinosaur in a Haystack: Reflections in Natural History.* New York: Harmony, 1995.

———. "The upwardly mobile fossils of Leonardo's living earth." Chap 1. in *Leonardo's Mountain of Clams and the Diet of Worms.* New York: Three Rivers, 1998.

———. "The lying stones of Marrakech." Chap. 1 in *The Lying Stones of Marrakech: Penultimate Reflections in Natural History.* New York: Harmony, 2000.

Niklas, Karl J., Robert M. Brown, Jr., and R. Santos. "Ultrastructural states of preservation in Clarkia Angiosperm leaf tissues: Implications on modes of fossilization in late Cenozoic history of the Pacific." Pages 143–159 in Smiley, Charles J., ed., *Late Cenozoic History of the Pacific.* San Francisco: American Association for the Advancement of Science, 1985.

Palmer, Douglas. *Fossil Revolutions: The Finds That Changed Our View of the Past.* New York: Collins, 2004.

Prasad, Vandana, et al. "Dinosaur coprolites and the early evolution of grasses and grazers." *Science* 310 (2005): 1,177–1,180.

founder effect The founder effect may occur when a population is founded (for example, on an island) by a small number of emigrants that disperse from a larger, central population. A small number of founders will not carry all of the genetic diversity that was present in the central population. Consider, for example, a central population that contains individual plants with red and with white flowers. A few seeds disperse to an island, but they happen to be seeds of red-flowered individuals. From the moment that the island population is founded, it already has lower genetic diversity than the population that produced it. It is unlikely that red flowers confer any advantage on the island population; the fact that the island population has no white-flowered plants is due to chance. The founder effect can result in a population that, even if it grows, remains depauperate in genetic diversity.

The founder effect is similar to the phenomenon of *genetic drift.* In genetic drift, a population undergoes a "bottleneck" event in which the population is reduced to a small number, perhaps by a disaster. In a small population, just by chance, some of the genes may be lost. Consider another flower example in which the population, consisting both of red-flowered and white-flowered individuals, experiences a population crash. The red-flowered individuals may be the only ones to survive, but their survival had nothing to do with the color of their flowers. By chance, the genes for white flowers have been lost. Genetic drift can almost permanently reduce the genetic diversity in a population. When the population begins to grow, it remains depauperate in genetic diversity. Both the founder effect and genetic drift reduce genetic diversity in a population: the founder effect, when a new population is formed, and genetic drift, when the old population suffers a bottleneck event. Both the founder effect and genetic drift were investigated by geneticist Sewall Wright (see WRIGHT, SEWALL).

Genetic drift has been experimentally confirmed. In one experiment with fruit flies, geneticist Peter Buri investigated a gene with two alleles (see MENDELIAN GENETICS). At the beginning of each experiment, all of the flies were heterozygous, therefore both of the alleles were equally common. Every generation, Buri chose a small sample of males and females, from which he raised the next generation. Every generation of flies experienced a genetic bottleneck. By the 19th generation, in over half of the populations, one or the other of the two alleles became more and more common until it was the only allele that was present.

Genetic drift can leave its mark on a population for many generations afterward (see POPULATION GENETICS). One in 20 humans who live on Pingelap Atoll in the Pacific Ocean have a mutation that causes color blindness and extreme light sensitivity. The usual population average for this mutation is one in 20,000 people. The reason the mutation is so common on Pingelap Atoll is that it was carried by one of the few survivors of a typhoon and famine that struck in 1776.

Genetic drift can affect entire species. Cheetah populations have a reduced genetic diversity due to a bottleneck event in the past. The Hawaiian goose, or nene, also experienced a genetic bottleneck which greatly reduced the genetic variability in its populations. Both species, principally for this reason, face EXTINCTION. Even when a species consists of a large number of individuals, the fragmentation of the species into many small populations can cause genetic drift to occur in each of the small populations, resulting in the catastrophic loss of genetic diversity. This is occurring in many natural habitats, which are being set aside as only small, disconnected nature preserves.

Humans provide one of the best examples of a genetic bottleneck. The human species has a remarkably low genetic diversity. Geneticists estimate that the entire human species experienced a genetic bottleneck event about 70,000 years ago, at which time the entire human species consisted of only a few thousand individuals.

Further Reading

Buri, Peter. "Gene frequencies in small populations of mutant *Drosophila.*" *Evolution* 10 (1956): 367–402.

Templeton, Alan, et al. "The genetic consequences of habitat fragmentation." *Annals of the Missouri Botanical Garden* 77 (1990): 13–27.

Young, Andrew, et al. "The population genetic consequences of habitat fragmentation for plants." *Trends in Ecology and Evolution* 11 (1996): 413–418.

fungi *See* EUKARYOTES, EVOLUTION OF.

G

Gaia hypothesis The Gaia hypothesis claims that organisms have transformed the entire Earth and maintain it in a state very different than it would have without organisms. "Gaia" is the name of an ancient Greek Earth goddess and is intended by the scientists who defend the hypothesis as an image, not referring to the Earth as a literal organism or person. Biologist Lewis Thomas proposed the concept that the Earth was like a cell in his 1984 book *Lives of a Cell.* However, Thomas's imagery resulted in no program of scientific investigation. The scientific predictions based upon the Gaia hypothesis began with the work of atmospheric scientist James Lovelock. Its principal proponent within the world of biological and evolutionary science is evolutionary biologist Lynn Margulis (see MARGULIS, LYNN).

The Gaia hypothesis is sometimes worded as "The Earth is alive," but this is not correct. The scientists who defend the Gaia hypothesis claim not that the Earth is an organism but that the atmosphere and water of the Earth are a nonliving part of the Earth's living system of interacting organisms the same way that a shell is a nonliving part of a mollusk. The Gaia hypothesis states that organisms transform the Earth in a way that has resulted in a regulation of the temperature and chemical composition of the Earth.

Biological activities of organisms have caused the Earth to be very different from other planets and maintain it in a chemically unstable condition. Thus, even without visiting a distant planet, observers should be able to determine whether life exists on the planet. If all of its atmospheric components (which can be studied by analyzing the light that they reflect) are in equilibrium, the planet is either lifeless or nearly so. Life creates chemical and physical disequilibrium which can be visible from millions of miles away. This was the basis on which James Lovelock predicted that no life would be found on Mars, a prediction that is either entirely or very nearly accurate (see MARS, LIFE ON). Consider the following examples.

Carbon dioxide. The other Earth-like planets (Venus and Mars) have atmospheres with tremendous amounts of carbon dioxide. In contrast, the Earth's atmosphere consists of less than 0.04 percent carbon dioxide. The primordial Earth, during the Hadean era (see PRECAMBRIAN TIME), also had a high concentration of carbon dioxide, although sediments deposited at that time indicate that the Earth's atmosphere never had as much carbon dioxide as Venus and Mars now have. The process of photosynthesis (see PHOTOSYNTHESIS, EVOLUTION OF), carried out by many bacteria during the Archaean era, reduced the atmospheric carbon dioxide concentration by at least two orders of magnitude. Where did the carbon dioxide go? The bacteria made it into organic material, whether living material (the bacteria themselves) or deposits of dead material. Eukaryotic photosynthetic organisms, such as plants, continued this process once they evolved.

Oxygen. Photosynthetic bacteria, from which the chloroplasts of many protists and of green plants have evolved, also filled the Earth's atmosphere with huge amounts of oxygen gas. The buildup of oxygen made the sky turn blue. Through most of the late Precambrian time, the Earth's atmosphere had approximately the level of oxygen gas that it now has. The huge amount of oxygen gas in the air is perhaps the most strikingly unstable result of biological activities. Oxygen atoms attract electrons from many other kinds of atoms and are therefore very reactive. All of the oxygen gas in the air would react with other molecules (such as nitrogen gas, which is the principal component of the atmosphere; the ferrous form of iron; and with organic molecules) and disappear, if it were not continually replenished by photosynthesis.

Other atmospheric components. Methane (CH_4; the principal form of natural gas) is continually produced by organisms, mostly by bacteria that live in anaerobic conditions (for example, mud or animal intestines, away from the presence of oxygen). While there is not much methane in the atmosphere, the little that is there quickly reacts with oxygen to form carbon dioxide. Its continued presence results from bacterial metabolism. The atmosphere also has nitrogen (N_2), nitrogen oxide (NO), ammonia (NH_3), methyl iodide, and hydrogen (H_2) that are produced by microbes.

Temperature. At least two biological processes appear to affect global temperature.

- *Greenhouse gases.* When the Earth first cooled off enough for the oceans to form at the end of the Hadean, the Sun produced much less radiation. If the Earth had its present atmospheric composition, the Earth would have frozen. The Earth apparently did almost completely freeze on more than one occasion, but this occurred later in Precambrian times (see SNOWBALL EARTH). Apparently the reason the Earth did not freeze is that the atmosphere contained large concentrations of greenhouse gases such as carbon dioxide and methane (see GREENHOUSE EFFECT). These gases absorbed infrared radiation that would otherwise have been lost into outer space; the infrared radiation made the atmosphere, thus the whole Earth, warmer than it would have been without them. As the Sun became brighter, photosynthesis of bacteria removed much of the carbon dioxide and produced oxygen that reacted with much of the methane. Thus at the very time that the Sun was becoming brighter, the greenhouse effect was becoming weaker, thus keeping the average temperature of the Earth relatively constant. Throughout the history of the Earth, there has been an approximate balance of photosynthesis (which removes carbon dioxide from and releases oxygen into the air) and respiration (which removes oxygen from and releases carbon dioxide into the air).
- *Aerosols.* Marine algae, cyanobacteria, and some plants produce a compound that, when acted upon by bacteria, becomes dimethyl sulfide gas. Clouds of dimethyl sulfide gas appear to be an important component of the global sulfur cycle. Under warm conditions, the algae have more photosynthesis and bacteria are more active. This results in more dimethyl sulfide, which produces more clouds, which causes conditions to become cooler. The molecule from which dimethyl sulfide is produced serves a protective function in the algae, but the dimethyl sulfide itself may be just a waste product. Airborne biological particles and aerosols (such as hair, skin cells, plant fragments, pollen, spores, bacteria, viruses, and protein crystals) may also affect the climate of the Earth.

At the same time that the Sun became brighter, the biological activities of the Earth altered the atmospheric composition in such a way that the temperature did not become too hot. Throughout the last half billion years, atmospheric oxygen concentration has remained approximately constant. Apparently atmospheric oxygen was much higher than it is today during the DEVONIAN PERIOD and the CARBONIFEROUS PERIOD and very low during the PERMIAN EXTINCTION, which may have contributed to the deaths of large animals in the ocean and on land. Had the oxygen content of the air increased too much, spontaneous combustion would have caused plants all over the world to burn even if they were wet and alive. All it would take would be a few lightning strikes to ignite huge conflagrations.

Minerals. Most of the major mineral elements on the surface of the Earth circulate through the food chain. On a planet without life, these elements (such as calcium, phosphorus, and sulfur) would accumulate in their most stable form and remain in that form.

To a certain extent, the relative stability of atmospheric carbon dioxide and oxygen can be explained by natural selection acting upon the populations of the organisms, in their own interest, rather than a process that operates to maintain the carbon dioxide and oxygen content of the atmosphere. When there is a lot of carbon dioxide in the air, photosynthetic rate increases, and this reduces the amount of carbon dioxide. Similarly, when oxygen is abundant, chemical reactions between oxygen and minerals in rocks increases, respiration in cells increases, and photosynthesis decreases (oxygen is actually a direct chemical inhibitor of photosynthesis). For both carbon dioxide and oxygen, photosynthesis and respiration operate as negative feedback mechanisms. Photosynthesis and respiration help to maintain a happy medium of oxygen, carbon dioxide, and temperature on the Earth. The organisms carry out photosynthesis and respiration to keep themselves alive, not to take care of the Earth.

The activities of organisms, acting in their own interests, help to explain the Gaia hypothesis. Lovelock developed a computer simulation called Daisyworld that makes this point. Consider a planet populated entirely by daisies that are either black or white. When the climate on this planet is cold, the black daisies absorb more sunlight and are warmer, and their populations increase relative to those of the white daisies. When the climate on this planet is warm, the white daisies reflect more sunlight and are cooler, and their populations increase more than those of the black daisies. Under cold conditions, when the black daisies spread, the heat from the daisies may actually cause the climate to become warmer; and under warm conditions, the reflection of light from the white daisies may actually cause the climate to become cooler. This is an example of simple negative feedback: Cold conditions select for the spread of black daisies, which raise the temperature of the environment, just as a cold room trips a thermostat which turns on a heater. Lovelock and Margulis use Daisyworld as a picture of how individual organisms can cause a Gaia-like maintenance of stable global conditions.

The Gaia hypothesis indicates that organisms do not simply live upon the Earth; evolution has not simply caused organisms to adapt to the physical conditions of the Earth. Organisms have radically transformed the Earth and have partly created the very conditions to which evolution has adapted them. This is the largest scale of symbiosis (see COEVOLUTION); Lynn Margulis says that Gaia is just symbiosis seen from outer space. It may therefore be not just certain species interactions but the entire planet that is symbiotic.

Further Reading

Jaenicke, Ruprecht. "Abundance of cellular material and proteins in the atmosphere." *Science* 308 (2005): 73.

Lovelock, James E. *Gaia: A New Look at Life on Earth.* New York: Oxford University Press, 1987.

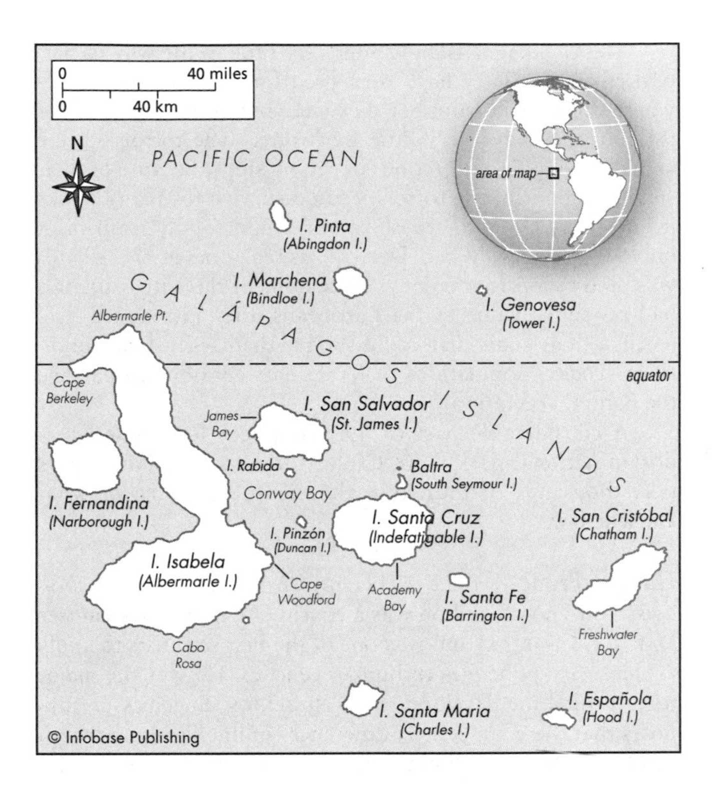

A map of the Galápagos Islands, off the coast of Ecuador

Meskhize, Nicholas, and Athanasios Nenes. "Phytoplankton and cloudiness in the southern ocean." *Science* 314 (2006): 1419–1423.

Malin, G., and G. O. Kirst. "Algal production of dimethyl sulfide and its atmospheric role." *Journal of Phycology* 33 (1997): 889–896.

Margulis, Lynn, and Dorion Sagan. *Slanted Truths: Essays on Gaia, Symbiosis, and Evolution.* New York: Copernicus, 1997.

Galápagos Islands The Galápagos Islands are volcanic islands that have played and continue to play a very important role in evolutionary science. They belong to Ecuador and are about 600 miles (1,000 km) off the coast of that country (see figure above). Much of the volcanic activity that produced them occurred about five million years ago, which is very recent in evolutionary time.

Being directly on the equator, these islands experience almost no variation in sunlight. They are right in the path of oceanic currents that influence temperature and moisture conditions. There is an annual fluctuation of temperature caused by the ocean currents. Because of this, the cactuses can use a temperature cue that allows them to bloom every January. The El Niño Southern Oscillation alternates between dry conditions, when the ocean water near the South American coast is cool, and wet conditions, when the water is warm. During an El Niño year, large amounts of rainfall stimulate plant growth, particularly of grasses, vines, and weeds, but diminish the productivity of the oceans. The food chain of the ocean depends upon the minerals brought up from cool, deep ocean waters; when the water is warm, these minerals are not available. However, periods of heavy rainfall are uncommon, and the lower slopes of the Galápagos Islands are usually hot and dry, with scattered trees and a lot of hot, bare ground. The upland areas of the larger islands are frequently shrouded in mist, which provides enough moisture for forests to grow.

The islands are famous for their unusual plant and animal life. There are no large native grazing animals or predators. The native animals have evolved, until recently, in the absence of predators and human contact. From the fur seals to the ground-nesting birds, they show no fear of humans. Many of the species are unique to these islands, or almost so. This is because they evolved on the islands from small founding populations of mainland species. All over the world, islands are home to unique species (see BIOGEOGRAPHY). The unique species found on islands was one of the observations that led Charles Darwin (see DARWIN, CHARLES) to begin thinking about the process of evolution. During his round-the-world voyage on HMS *Beagle,* he visited many islands, including the Galápagos. His 1835 visit to the Galápagos turned out to be perhaps the single most important event that sparked Darwin's understanding that evolution occurred by means of natural selection.

Among the plants and animals that have undergone ADAPTIVE RADIATION on the Galápagos Islands are:

- *Finches.* These birds have evolved from a single founding population into about 14 species that specialize on different ways of life and types of food (see DARWIN'S FINCHES). Ever since David Lack studied them in 1947, they have

Iguanas bask in the sunlight on the stone marker of Ecuador's Galápagos Islands National Park. These iguanas are just one of many species of organisms unique to these islands. *(Photograph by Stanley A. Rice)*

constituted a premier example of adaptive radiation. Since Peter and Rosemary Grant began their studies in the 1970s, the finches have provided some of the most detailed information in the world about NATURAL SELECTION; ISOLATING MECHANISMS; and HYBRIDIZATION.

- *Mockingbirds.* A single founding population, probably of the Ecuadorian long-tailed mockingbird *Mimus longicaudatus,* has evolved into four species of mockingbirds on the Galápagos: *Nesomimus trifasciatus, N. macdonaldi, N. melanotis,* and *N. parvulus.* They are distinct from one another in coloration and behavior; they differ enough from their mainland ancestors that they have been placed in a different genus, *Nesomimus,* from the genus *Mimus* of which the North and South American mainland mockingbirds are members.
- *Land iguanas, or lava lizards.* Seven species of land iguanas have evolved on the Galápagos Islands (see figure on page 171).
- *Tortoises.* The giant tortoises after which the Galápagos Islands are named are the most famous animals in the islands. Several of the islands have recognizably different kinds of tortoises—a fact that the governor of the islands pointed out to Charles Darwin when he visited. The tortoises move very slowly and live for a long time; a slow metabolism is a major part of their adaptation to these islands. Darwin rode on some when he visited. A few of the tortoises alive today may have been alive when Darwin visited.
- *Daisy trees.* The Galápagos Islands have 13 species of *Scalesia* trees, in the composite (daisy) family Asteraceae. Almost all members of this huge, worldwide family of plants are herbaceous (like the familiar daisies, dandelions, thistles, and ragweeds)—except on islands, such as the Channel Islands of California, St. Helena in the Atlantic, Hawaii, and the Galápagos. These daisy trees are somewhat flimsy and small as trees go and probably could not compete with trees from plant families with a long evolutionary history of producing trees, such as pines and oaks. On these islands, there are relatively few trees, which has allowed natural selection to produce tree-sized daisies.
- *Tree cactuses.* Cactuses in mainland North and South American deserts do not have bark; the six species of Galápagos prickly pear cactuses, however, do. The bark may have evolved as protection against the abundant sea animals that rest in their sparse shade.

There are also species that are unique to the Galápagos Islands, although they have not undergone as much adaptive radiation as can be seen in the animals mentioned above. Two hundred of the 700 species of Galápagos plants are unique, including the Galápagos tomato. Marine iguanas are highly unusual in the reptile world because they swim and eat algae. The Galápagos penguin is the only penguin that lives in a warm environment. There are Galápagos species of sharks, doves, hawks, rats, bees, butterflies, centipedes, martins, and flycatchers.

Other animals such as frigate birds travel widely in the oceans and have not evolved into unique forms confined to the islands. The isolation of the islands makes them a haven for birds and mammals that travel the oceans.

The Galápagos Islands offer very little in the way of natural resources. They have been used for centuries as a place where ships could stop, but they were not even very good for this, for they have very little freshwater. The tortoises were actually a popular attraction for sailing ships. Because of their slow metabolism, the tortoises could survive for many weeks in the cargo holds of the ships and could supply fresh meat on long voyages. When Darwin visited, one of the islands was used as a prison colony. Since none of the native animals had economic value to the Europeans, they introduced livestock such as goats that could survive the harsh island conditions. Today, populations of feral goats continue to damage the natural vegetation.

A considerable part of the Galápagos Islands is Ecuadorian national park land. Cruise ships and airplane flights take thousands of tourists each year to see some of what Darwin saw.

Galton, Francis (1822–1911) British *Evolutionary psychologist* Sir Francis Galton was a cousin of Charles Darwin (see DARWIN, CHARLES) and was one of the first scientists to apply evolutionary principles to human genetics. He was the major force behind the development of EUGENICS. Eugenics presupposes that there are genetic differences in intelligence between races, and that the "lower" races, which have lower intelligence, should have restricted immigration quotas and be discouraged from breeding. Although there are genetic influences upon intelligence, modern geneticists have concluded that these influences do not correspond in any way to races or to cultures. Eugenics has become almost synonymous with racism and is not taken seriously by modern scientists. Galton's championing of eugenics has eclipsed his reputation, causing many modern scholars to overlook his many legitimate and important contributions to science.

Born January 16, 1822, Galton began his studies, as did many other 19th-century British scientists, as a medical student. His lack of interest and his undistinguished performance prevented him from starting a practice. Instead, he traveled in the Middle East and Africa. He gathered many data about the land, people, and climate as he traveled. His data were characterized, perhaps more than anyone else's (even Darwin's) by careful measurements and by statistical sampling methods and relatively large data sets.

Upon his return to England, he used the things he had learned while traveling, and the statistical methods that he had refined, to undertake research in different fields. He created some of the first weather maps and made significant contributions to meteorology. During his studies of human measurements, he noticed that each individual has a unique fingerprint, and he convinced Scotland Yard to adopt a fingerprinting system—the first in the world and now used universally. In order to analyze his eugenics measurements, Galton developed some important statistical techniques, which led to the development of regression. These methods, which analyze the probability that an association between two variables is not simply due to chance, are essential to modern scientific research (see SCIENTIFIC METHOD) even though the eugenics to which Galton applied them has now been discredited. Galton also developed the use of percentile scores and the survey method of psychological research. He was the first to use identical (monozygotic) and fraternal (dizygotic) twins in the study of human inheritance patterns even though the eugenic conclusions he reached are no longer believed (see essay, "How Much Do Genes Control Human Behavior?"). Therefore Galton's techniques, although not his scientific theories, pervade modern genetic and psychological research.

Galton's 1869 book *Hereditary Genius* outlined his conviction that intelligence was primarily inherited. Galton believed that the rapid population growth of the lower classes, and the declining fertility of the upper classes, would lead to a national loss of intelligence. (It is interesting that Galton, who considered himself one of the intelligent elite, had no children.) He even advocated restrictions that would prohibit reproduction by people whom his tests identified as feebleminded. His statistical methods did, in fact, prove that there was a significant correlation between intelligence and heredity. What he seems to have ignored is that it is not just genes that are inherited in human societies. Culture is passed on through families as well. The lower classes may have scored lower on Galton's tests of mental ability not because of inferior genes but because, generation after generation, they had been denied access to economic and educational opportunities. This would have explained his significant correlations just as well as the eugenic explanations Galton offered. Moreover, the very questions that were asked on his surveys presupposed a knowledge of English culture and language; Africans scored as low on his English tests as Englishmen would have on tests conducted in Swahili about African culture. Today, all scientists recognize what Galton apparently did not: Correlation is not the same as causation.

Although Galton's principal scientific accomplishments have been discredited, many of his techniques, from fingerprints to statistics, play an essential role in the world today. He died February 17, 1911.

Further Reading

Allen, G. "The measure of a Victorian polymath: Pulling together the strands of Francis Galton's legacy to modern biology." *Nature* 145 (2002): 19–20.

Bynum, W. F. "The childless father of eugenics." *Science* 296 (2002): 472.

Gillham, Nicholas Wright. *A Life of Sir Francis Galton: From African Exploration to the Birth of Eugenics*. New York: Oxford, 2001.

Tredoux, Gavan. "Galton.org Home Page." Available online. URL: http://www.mugu.com/galton/index.html. Accessed March 28, 2005.

gene *See* DNA (RAW MATERIAL OF EVOLUTION).

gene-culture coevolution Gene-culture coevolution refers to the influence of cultural innovation on genes in a population, and the effects of those genes on culture. The term *coevolution* is usually restricted to the mutual evolutionary effects of one species on another. In this case, one of the participants in coevolution is the culture of a species, and the other is the genes within its populations. Some evolutionary scientists (see WILSON, EDWARD O.; SOCIOBIOLOGY) use gene-culture coevolution as an important mechanism for explaining the evolution of societies and intelligence. Gene-culture coevolution is a more modern form of a concept known as the *Baldwin Effect,* which evolutionary biologist George Gaylord Simpson named after a concept proposed in the late 19th century by evolutionary scientist James Mark Baldwin.

In order to explain this process, consider one famous example of it. In Britain, some individual blue tits (*Parus caeruleus,* a species of bird) figured out how to peck the foil caps off of milk bottles left on doorsteps and drink the milk inside. Other birds watched the ones that could perform this trick and began doing it themselves. The first part of the process, therefore, was cultural: The birds learned a new behavior from one another. The existence of this behavior causes a selective advantage for those birds that have the appropriate physical and behavioral abilities to do this action well: the ones that are best at recognizing the foil, at performing the movements to remove the foil. A new resource, unlocked by behavior, now becomes a selective force that can influence the gene frequencies of the bird population (see NATURAL SELECTION). The second part of the process, then, is genetic. In turn, these new gene frequencies might allow yet more cultural innovations, in which the birds can more effectively steal milk from bottles. It is important to note that the behavior itself is culturally transmitted, rather than instinctual. The new behavior cannot imprint itself on the genes; this would be LAMARCKISM. No one will ever know what further genetic changes might have occurred in the bird populations, because dairies began using bird-proof caps on milk bottles.

Another example, more familiar to modern Americans, is that birds such as grackles are often observed to pick recently killed insects from automobile grilles in parking lots. This behavior is beneficial because the birds can eat insects without using energy to pursue them, do so safely because the automobiles are not moving, and do so efficiently because many automobiles are in one location. This behavior creates a situation in which natural selection favors birds that have an innate preference for parking lots over natural insect habitats. The behavior is now cultural, but in later centuries may become genetic.

Another famous example involves Imo, a very smart Japanese macaque monkey. Imo learned how to wash food items before eating them. This allowed Imo to eat the food more efficiently. Other monkeys began imitating Imo. Now there is a selective advantage for those monkeys that are capable of learning and efficiently executing this behavior. The monkeys that cannot do so might find themselves lower in the social hierarchy, with reduced or nonexistent reproduction as their reward.

On the GALÁPAGOS ISLANDS, DARWIN'S FINCHES have undergone ADAPTIVE RADIATION into numerous species that specialize on different foods. Another Pacific island, Cocos, off the coast of Costa Rica, shares one of these species of finches *(Pinaroloxias inornata)* with the Galápagos. But

Cocos Island is too small for isolating mechanisms to break them into different populations, a necessary step if they are to evolve into different species (see SPECIATION). Instead of different populations specializing on different foods, individual finches specialize on different foods, whether seeds, bark, fruit, or insects, and teach younger finches how to process these foods. Behavioral specialization on foods creates a situation that selects some genes more than others, which in turn will influence behavior. In this case, gene-culture coevolution will probably never go any further than it has, since the genes of all the birds mix together in a single population.

An example of gene-culture coevolution that has influenced human physiology is the evolution of lactose tolerance. Lactose is milk sugar, and juvenile mammals have digestive enzymes that allow them to metabolize lactose. In most adult mammals, including many humans, the genes for these enzymes are not used. Lactose intolerance is the norm for adult mammals. Some human societies use domesticated livestock not just for meat but for milk as an important source of nutrition for adults. In these societies, the cultural choice of milk as a food source created conditions that favored the evolution of lactose tolerance, in which the adults continued to use the genes for the enzymes that digest lactose. Evolutionary analyses have shown that the prevalence of genetically based lactose tolerance in human societies corresponds closely to the importance of livestock herding as an economic activity. For example, lactose tolerance is rare among Oriental peoples but common in Mongolians, and it is rare among Africans but common in herding tribes. The Masai, although they depend heavily upon milk, are mostly lactose intolerant, but they curdle their milk before consuming it, which reduces the lactose content.

The selection that favors the behavioral ability need not always be natural selection. SEXUAL SELECTION, in which one sex favors members of the other sex that have certain behavior patterns, can favor the genetic establishment of these behavioral abilities.

Gene-culture coevolution may have been crucial in the evolution of some important human characteristics. To many evolutionary biologists, language is the defining ability of the human species. Many explanations have been proposed for the origin of language (see LANGUAGE, EVOLUTION OF). One thing that most of them have in common is that the advantages of language had to be social. Language does not help an isolated human being survive or reproduce better. If a rudimentary form of language began as a behavioral innovation, and provided an advantage, natural selection would then favor those individuals with a superior genetic ability to communicate in this fashion. The new genes would then allow yet more innovations in communication. Genes and culture would influence one another, in a positive feedback spiral, until language came into existence. Note again that the languages themselves and the ways in which they are used are not genetically determined; they are culturally transmitted. The brain structures that allow a person to understand and to form language (the Wernicke's area and the Broca's area of the brain, respectively) are genetically determined, and probably evolved because of the cultural advantage they provided to the people who possessed them. In fact, gene-culture coevolution was probably involved in the evolution of all aspects of human intelligence (see INTELLIGENCE, EVOLUTION OF).

Further Reading

Baldwin, James Mark. "A new factor in evolution." *American Naturalist* 30 (1896): 441–451, 536–553.

Cavalli-Sforza, Luigi L., and Marc L. Feldman. *Cultural Transmission and Evolution.* Princeton, N.J.: Princeton University Press, 1981.

Griffiths, Paul E. "Beyond the Baldwin Effect: James Mark Baldwin's 'social heredity,' epigenetic inheritance and niche construction." Available online. URL: http://philsci-archive.pitt.edu/archive/ 00000446/00/Beyond_the_Baldwin_Effect.pdf.

Lumsden, Charles J., and Edward O. Wilson. *Promethean Fire: Reflections on the Origin of Mind.* Cambridge, Mass.: Harvard University Press, 1983.

Simpson, George Gaylord. "The Baldwin Effect." *Evolution* 7 (1953): 110–117.

gene pool *See* POPULATION GENETICS.

genetic code *See* DNA (RAW MATERIAL OF EVOLUTION).

genetic drift *See* FOUNDER EFFECT.

genetics *See* MENDELIAN GENETICS.

geological time scale The geological time scale is the time scale of Earth history. It is based upon the deposits of sedimentary, igneous, and metamorphic rocks throughout Earth history (see FOSSILS AND FOSSILIZATION). No place on Earth contains a complete column of geological deposits. By the process of stratigraphy, which was developed in the early 19th century by geologist William Smith (see SMITH, WILLIAM), geologists can compare the deposits in one location with those in another and, by lining them up, reconstruct a complete geological column. Dates can be assigned to these rocks by RADIOMETRIC DATING of igneous rocks that are found between many of the sedimentary layers.

The major divisions of the geological time scale are eons. Eons are divided into eras. The first three eons are often informally lumped into the PRECAMBRIAN TIME. The Precambrian represents almost 90 percent of Earth history. Human civilization represents an almost unmeasurably small portion of Earth history (see AGE OF EARTH). Eons and eras represent major events in the history of the Earth.

Precambrian time is divided into:

- *Hadean time* ("hellish"; sometimes called *Hadean era*). The Earth formed about 4.5 billion years ago but was so hot that oceans formed only toward the end of the Hadean.
- *Archaean Eon* ("old"; sometimes called *Archaean era*). Organisms similar to modern bacteria (see ARCHAEBACTERIA; BACTERIA, EVOLUTION OF) may have lived in the oceans almost as soon as they formed (see ORIGIN OF LIFE). There was little oxygen gas in the atmosphere.

- *Proterozoic Eon* ("earlier life"; also called *Proterozoic era*). During the Proterozoic, oxygen gradually accumulated in the atmosphere, and complex cells evolved (see EUKARYOTES, EVOLUTION OF). In the middle of the Proterozoic, a few multicellular organisms such as seaweeds evolved. Toward the end of the Proterozoic, during the Ediacaran period, there were a few animals, as evidenced by embryos and burrows. Most of the inhabitants of the Ediacaran period were the enigmatic EDIACARAN ORGANISMS. By the end of the Proterozoic, the atmosphere had its modern concentration of oxygen. At least two worldwide glaciations occurred during the Proterozoic (see SNOWBALL EARTH). The Ediacaran period (defined in 2004) is the first new geological period to be defined in over a century. It lasted from about 600 to about 540 million years ago.

All of Earth history after the Precambrian comprises the *Phanerozoic Eon* ("visible life"). This eon, which consists of three eras, includes the evolution of all complex organisms that were related to modern plant and animal groups:

- *Paleozoic era* ("ancient life"). Animal life proliferated in the oceans (see INVERTEBRATES, EVOLUTION OF; FISHES, EVOLUTION OF). Land plants and animals evolved (see SEEDLESS PLANTS, EVOLUTION OF; GYMNOSPERMS, EVOLUTION OF; AMPHIBIANS, EVOLUTION OF; REPTILES, EVOLUTION OF). The land had been barren at the beginning of the Paleozoic era but by the end was covered with extensive forests, which had modern groups of plants except the flowering plants. The end of the Paleozoic era was marked by the greatest of the MASS EXTINCTIONS, in which over 95 percent of the species died (see PERMIAN EXTINCTION). In most major groups of organisms, a few representatives survived this extinction.
- *Mesozoic era* ("middle life"). Life proliferated in the oceans. On land, flowering plants evolved (see ANGIOSPERMS, EVOLUTION OF), and DINOSAURS were the largest and most diverse of the vertebrates. Mammals and birds also evolved during the Mesozoic era (see BIRDS, EVOLUTION OF; MAMMALS, EVOLUTION OF). The Mesozoic era ended when a gigantic asteroid hit the Earth (see CRETACEOUS EXTINCTION).
- *Cenozoic era* ("recent life"). A cooler, drier Earth was dominated by flowering plants, and the mammals were the largest vertebrates. Toward the end of the Cenozoic era, particularly cool, dry conditions caused the ICE AGES.

Eras (particularly the last three) are divided into periods, most of which also reflect major events in Earth history. All of the periods are divided into epochs, although most scientists are familiar only with the epochs of the Tertiary and Quaternary periods. Because of the intense interest of humans in their own history, the Recent epoch has been defined to begin just 10,000 years ago, coinciding with the end of the most recent ice age and the beginning of villages, agriculture, and then civilization. The Earth is currently in the Phanerozoic Eon, the Cenozoic era, the Quaternary period, and the Recent epoch.

The *Encyclopedia of Evolution* contains entries for each of the eras and periods of the Phanerozoic Eon. The epochs of the Tertiary and Quaternary periods are included with their respective periods. In the accompanying table, as well as in each entry, the time periods have been rounded to the nearest five million years.

Geological Time Scale with Selected Periods and Epochs (to nearest 5 million years)

Eons	Eras	Periods	Epochs	Duration (million years ago)
Hadean				4500–3800
Archaean				3800–2500
Proterozoic				2500–540
		Ediacaran		600–540
Phanerozoic				540–
	Paleozoic			540–250
		Cambrian		540–510
		Ordovician		510–440
		Silurian		440–410
		Devonian		410–360
		Carboniferous		360–290
		Permian		290–250
	Mesozoic			250–65
		Triassic		250–210
		Jurassic		210–140
		Cretaceous		140–65
	Cenozoic			65–
		Tertiary		65–2
			Paleocene	65–55
			Eocene	55–35
			Oligocene	35–25
			Miocene	25–5
			Pliocene	5–2
		Quaternary		2–
			Pleistocene	2–0.01
			Holocene	0.01–

Note: The complete list of periods is presented only for the Phanerozoic Eon.
Note: The complete list of epochs is presented only for the Cenozoic era.
Note: The Holocene epoch is also called the Recent epoch.
Note: The Cenozoic era is now frequently divided into the Paleogene period (Paleocene, Eocene, Oligocene epochs) and the Neogene period (Miocene, Pliocene, Pleistocene, and Holocene epochs).

Further Reading

Museum of Paleontology, University of California, Berkeley. "The paleontology portal." Available online. URL: http://www.paleoportal.org/. Accessed May 3, 2005.

Goodall, Jane (1934–) British *Primatologist* Jane Goodall is a leading world authority on the species that is one of humankind's closest relatives, the chimpanzee (see PRIMATES). In order to understand how humans are different from other animals, and the changes that occurred during the course of human evolution, it is necessary to thoroughly study the behavior of other animals, especially chimpanzees, under natural conditions. This required many years of close observation and the ability to recognize individual chimpanzees. Goodall's close and prolonged observation yielded surprises that other researchers of animal behavior had not seen.

Goodall was born April 3, 1934. In 1960 Goodall began observations in the Gombe Preserve in Tanzania. One persistent problem in the study of animal behavior (see BEHAVIOR, EVOLUTION OF) is that the behavior of the animals, particularly intelligent ones like chimpanzees, is altered by the presence of human observers. Months passed before the chimpanzees allowed Goodall to approach them closely enough for observation. Once they had accepted her, she was able to observe their normal lives. In her first year of observation Goodall made two discoveries. First, she found that chimpanzees hunt and eat meat. Second, she found that chimpanzees use tools. She observed chimps stripping leaves from twigs and using the twigs to fish termites out of a nest. This observation proved that humans were not the only animals that made and used tools (see TECHNOLOGY). Goodall earned her Ph.D. in ethology from Cambridge University in 1965.

Soon thereafter Goodall returned to Tanzania and established the Gombe Stream Research Center. She continued to make observations that showed chimpanzees to be capable of behavior previously associated only with humans. Particularly striking is the complexity of chimpanzee social organization and the advance planning that allows it. She observed a female chimpanzee adopt an unrelated baby chimpanzee. Chimpanzees were also capable of actions that remind observers of the bad side of human behavior. In 1974 a four-year war began between two bands of chimpanzees, resulting in the annihilation of one. Goodall also observed one chimpanzee that cannibalized several infants. Perhaps the most striking observation was in 1970 when Goodall observed chimpanzees apparently dancing for joy in front of a waterfall. All of these observations call into question the vast gulf that most people, including many scientists, assume to exist between humans and other animal species, and demonstrate that individual chimpanzees can differ as much from one another as individual humans.

Goodall is interested in far more than just understanding animal behavior. She began and continues efforts to rescue orphaned chimpanzees and to improve the conditions of chimpanzees that are used in medical research. The president of a medical research company who at first despised Goodall's interference ended up thanking her for her work. In 1977 Goodall founded the Jane Goodall Institute for Wildlife Research, Education, and Conservation, which encourages local people in Africa and all over the world to undertake conservation efforts in their own region. Her work both in the scientific study of behavior and in conservation have earned her many awards and worldwide recognition.

Further Reading

Goodall, Jane, and Phillip Berman. *Reason for Hope: A Spiritual Journey.* New York: Warner Books, 1999.

Jane Goodall Institute. "The Jane Goodall Institute." Available online. URL: http://www.janegoodall.org. Accessed March 28, 2005.

Miller, Peter. "Jane Goodall." *National Geographic,* December 1995, 102–129.

Peterson, Dale. *Jane Goodall: The Woman Who Redefined Man.* New York: Houghton Mifflin, 2006.

Gould, Stephen Jay (1941–2002) American *Evolutionary biologist* Stephen Jay Gould was one of the most creative thinkers in the study of evolution and one of the most famous and effective popularizers of evolution to the general public. His first scientific book, *Ontogeny and Phylogeny,* and his first popular book, *Ever Since Darwin,* were both published in 1977.

Born September 10, 1941, Gould became interested in evolution when he was five years old and saw the *Tyrannosaurus* skeleton at the American Museum of Natural History in New York. He stayed with his youthful decision to become an evolutionary scientist. After graduating from Antioch College, Gould earned his Ph.D. from Columbia University. Until his death he was a professor at Harvard University and a curator at Harvard's Museum of Comparative Zoology.

One of Gould's insights into evolution was that evolution was not entirely a gradual process. Instead, periods of rapid change, associated with the appearance of a new species (see SPECIATION), were followed by long periods of stasis. Together with a colleague at the American Museum of Natural History (see ELDREDGE, NILES), Gould presented the theory of PUNCTUATED EQUILIBRIA in 1972. Gould, Eldredge, and Yale paleontologist Elisabeth Vrba maintained that evolution occurs not just on the level of populations (see NATURAL SELECTION) but on the species level as well. While defending the proof of evolution and the mechanism of natural selection that Darwin had presented (see DARWIN, CHARLES; *ORIGIN OF SPECIES* [book]), Gould maintained that Darwin had unnecessarily constrained the operation of evolution by claiming that it always occurred gradually. Gould summarized a lifetime of insights into the processes of evolution in *The Structure of Evolutionary Theory.* Another major insight was that many features of organisms were not adaptations but were side effects of adaptations, resulting from structural constraints. Gould and paleontologist Elisabeth Vrba called these features exaptations (see ADAPTATION).

A related insight was that chance events have played a major role in the histories of individual humans, of societies, of species, and of the entire Earth. Gould's doctoral research focused on land snails (genus *Cerion*) in Bermuda. He closely studied the individual variations among these snails, the same way that Darwin focused upon barnacles. Gould found that many of the variations could be better explained as the products of chance, or of structural constraint, than of adaptation. Gould also survived cancer for more than two decades. This experience contributed to his thinking about the importance of chance events. Gould's 1989 book *Wonderful Life* made the fossils of the BURGESS SHALE, from a 510-million-year-old marine environment of the CAMBRIAN PERIOD, well known to the general public. He claimed that these fossils demonstrated an early diversification of life, followed by the extinction of many lineages. Had the lineage that led to vertebrates become extinct, something that could have happened by chance, humans might not exist at all. Gould's 1996 book *Full House* explained what appeared to be progress, whether in evolution or in baseball, to results from chance events followed by unequal success, rather than an inherent upward progression (see PROGRESS, CONCEPT OF).

For several decades, Gould contributed columns entitled "This View of Life" to *Natural History* magazine and collected them into popular books. These essays also reveal Gould's great breadth of knowledge into almost every subject, and his ability to connect them. He also made many readers aware (in his book *The Mismeasure of Man*) of how science can be used by some humans to oppress others, as in the pseudoscience of EUGENICS. Gould was also a major contributor to the educational campaign against CREATIONISM, which, as he explained in *Rocks of Ages,* he considered an unnecessary pitting of religion against science. As a result of his popular writings, Gould made almost every aspect of evolution, from the history of the development of the theory, to the evidence that supports it, to the way it works, accessible and interesting to the general public. He died May 20, 2002.

Further Reading

Eldredge, Niles, and Stephen Jay Gould. "Punctuated equilibria: An alternative to phyletic gradualism." In *Models in Paleobiology,* edited by T. J. M Schopf, 82–115. San Francisco, Calif.: Freeman, Cooper and Co., 1972.

Gould, Stephen Jay. *Bully for Brontosaurus.* New York: Norton, 1991.

———. *Crossing Over: Where Art and Science Meet.* New York: Three Rivers Press, 2000.

———. *Dinosaur in a Haystack.* New York: Harmony Books, 1995.

———. *Eight Little Piggies.* New York: Norton, 1993.

———. *The Flamingo's Smile.* New York: Norton, 1985.

———. *Full House: The Spread of Excellence from Plato to Darwin.* New York: Harmony Books, 1996.

———. *Leonardo's Mountain of Clams and the Diet of Worms.* New York: Harmony Books, 1998.

———. *The Lying Stones of Marrakech.* New York: Harmony Books, 2000.

———. *The Hedgehog, the Fox, and the Magister's Pox.* New York: Harmony Books, 2003.

———. *Hen's Teeth and Horse's Toes.* New York: Norton, 1983.

———. *I Have Landed: The End of a Beginning in Natural History.* New York: Harmony Books, 2002.

———. *The Mismeasure of Man.* New York: Norton, 1981.

———. *Ontogeny and Phylogeny.* Cambridge, Mass.: Harvard University Press, 1977.

———. *The Panda's Thumb.* New York: Norton, 1980.

———. *Questioning the Millennium: A Rationalist's Guide to a Precisely Arbitrary Countdown.* New York: Harmony Books, 1997.

———. *Rocks of Ages: Science and Religion in the Fullness of Life.* New York: Ballantine Publications, 1999.

———. *The Structure of Evolutionary Theory.* Cambridge, Mass.: Harvard University Press, 2002.

———. *Time's Arrow, Time's Cycle.* Cambridge, Mass.: Harvard University Press, 1987.

———. *An Urchin in the Storm: Essays about Books and Ideas.* New York: Norton, 1987.

———. *Wonderful Life: The Burgess Shale and the Nature of History.* New York: Norton, 1989.

———, and Richard Lewontin. "The spandrels of San Marco and the Panglossian paradigm: A critique of the adaptationist programme." *Proceedings of the Royal Society of London B* 205 (1979): 581–598.

Grant, Peter *See* DARWIN'S FINCHES.

Grant, Rosemary *See* DARWIN'S FINCHES.

Gray, Asa (1810–1888) American *Botanist, Evolutionary Scientist* Asa Gray was an early defender of Darwinian evolution in the United States. He was representative of a large number of English and American scientists and theologians who accepted evolutionary science while maintaining their traditional Christian beliefs. Antievolutionists, also called creationists (see CREATIONISM), were much more common in the 20th century than in the half century following the publication of Darwin's *Origin of Species* (see DARWIN, CHARLES; *ORIGIN OF SPECIES* [book]). Gray is representative of the Christian scholars who embraced Darwinian science, not only because of Gray's prominence but also because he was a personal correspondent of Charles Darwin even before the publication of the *Origin of Species.*

Born November 18, 1810, Gray was trained as a medical doctor. Even while practicing medicine, his main interest was the study of plants. He learned much from America's leading botanist, John Torrey, and by 1831 Gray was ready to abandon medical practice and pursue botany full-time. For the next five years, he conducted botanical investigations with Torrey, which culminated in their coauthorship of the *Flora of North America.* Gray spent a year as a member of the scientific corps of the American Exploring Expedition in Europe and also studied what was just becoming known about Japanese botany. In 1842 he received a faculty appointment in natural history at Harvard University.

It was not only botany that Gray learned from Torrey. Torrey's firm Presbyterian faith led Gray to consider his own, and in 1835 Gray underwent a Christian conversion experience. Gray found that Christianity led him to investigate science ever more enthusiastically. Gray put his faith into action by church work that included teaching a black Sunday school class at a time when prejudice against African Americans was still standard, even in the North.

At this time, as a devotee of William Paley (see NATURAL THEOLOGY) and opponent of Chambers's *Vestiges of Creation* (see CHAMBERS, ROBERT), Gray resisted the idea of the transmutation of species (today called evolution). In this, he was consistent with most scientists of his time. He did not use the Bible as a source of information about Earth history;

he had long accepted Lyell's geology (see LYELL, CHARLES; UNIFORMITARIANISM) and accommodated his interpretation of the biblical book of Genesis accordingly. He was experiencing difficulty in explaining many facts of botany, in particular the similarity of plant genera in North America and eastern Asia. It looked to him as though these plants had not been separately created but were the descendants of a common flora that had been pushed southward by glaciation (see ICE AGES).

Gray's scientific reputation grew from his insistence upon facts over speculation, and this is what brought him to the attention of Charles Darwin. When Darwin wanted a botanist in whom to confide about his slowly developing evolutionary theory, he turned to Gray, even though on the other side of the Atlantic, as well as to botanist Sir Joseph Hooker (see HOOKER, JOSEPH DALTON). Reading Darwin's manuscript of the *Origin of Species* convinced Gray that transmutation of species had occurred and would solve the botanical mysteries he had been studying. Gray announced this conclusion at a scientific meeting in January 1859, before the appearance of Darwin's book. Once the *Origin of Species* had appeared later that year, Gray was in the perfect position to defend it.

In defending Darwinian evolution, Gray came in public conflict with his Harvard colleague, Louis Agassiz (see AGASSIZ, LOUIS), particularly during a public debate in 1859. This debate was very different from what one might expect to see today: Agassiz, the Unitarian who rejected biblical reliability, attacked Darwinian evolutionary science, while Gray, the Christian who believed in the inspiration of the Bible, defended it. They had many points of conflict, not the least of which was one of the major issues on the minds of people that year: the impending possibility of War Between the States. Agassiz believed that different races were separately created by God, and that God intended some of them (whom he considered inferior) to be dominated by others. Gray believed that all races of humankind had a common origin, and that each individual had equal worth.

Darwin appreciated Gray's support, even though Darwin's agnosticism differed profoundly from Gray's Christianity. Darwin told Gray that he had never intended to write atheistically; but he could not see the world, as Gray did, to be the creation of a beneficent God. There was, Darwin said, too much misery in the world. Darwin was willing to admit that the general laws of nature may have a divine origin, but he could not see evidence of design in the particulars. These admissions on Darwin's part are nearly all that historians have of his religious doubts, as Darwin resisted public debate even about evolutionary science, not to mention religion.

Gray accepted evolutionary science, albeit in a form different from what scientists understand today, but resisted what he considered a fanciful extension of evolutionary science into the kind of social Darwinism that was being promoted by writers such as Spencer (see SPENCER, HERBERT).

Even while accepting Darwinian evolution, however, Gray still admired Paley's NATURAL THEOLOGY. This was the one point at which Gray differed from Darwin: Gray believed that the evolutionary process has a purpose or direction, a concept subsequently called *teleology*. Gray believed that God guided the evolutionary process in directions that would suit God's ultimate purposes. This occurred, he believed, because God created the heritable variation within populations upon which natural selection depended. Since the science of genetics was unknown prior to Mendel (see MENDEL, GREGOR; MENDELIAN GENETICS), there was no proof against this concept. By adding teleology to evolution, Gray thought that he had reconciled Paley's Natural Theology with Darwin's evolution. Even though Darwin knew that this reconciliation was unlikely, he was pleased at the attempt; he reprinted three of Gray's essays at his (Darwin's) own expense, with "Natural Selection Not Inconsistent With Natural Theology" printed at the top. The concept of teleology is rejected by scientists today, as there is no known physical or chemical mechanism for it; it is a religious concept.

Adding teleology to Darwinian evolution was the same accommodation that the other Christian evolutionists of the late 19th century made, and which allowed them to accept a form of evolutionary science. These included geologists George Frederick Wright and James Dwight Dana. Wright, a minister and geologist, later on the faculty of Oberlin College, worked with Gray on some of Gray's last publications about science and religion. Wright insisted that while God was the primary cause of the existence of everything, Darwinian science adequately explained the secondary causes. Wright even saw parallels between evolutionary science and some traditional Calvinist doctrines such as original sin. Dana was a geologist, later at Yale, whose around-the-world journey while he was young transformed his outlook as much as the *Beagle* voyage had influenced Darwin. This was almost the same time that Dana had a religious conversion experience. He, like Gray, had his view of an orderly, created world shaken by reading Darwin's *Origin of Species*. He resisted it at first, but in later works admitted that the history of the Earth had occurred by means of evolution. He pointed out, as do many Christian writers today, that Genesis 1 phrases such as "let the Earth bring forth" allowed, even encouraged, a belief in secondary causes such as evolution. Gray and others believed that God's world was designed to make itself, design was seen not in the organisms themselves but in the overall system of natural laws that had produced them. They believed that the "image of God" referred to in Genesis 1 was not the physical body of humans but the spirit; and that an animal origin of humankind was no more demeaning than what Genesis 1 said: that humans were made of dirt. A number of famous late-19th-century preachers, such as A. H. Strong and Henry Ward Beecher, actually praised Darwinian evolution in part because it removed some of the difficulties that had been faced by the old style Natural Theology. It had previously been difficult to explain why a world created by a good God would contain so much apparent evil; now these ministers, and religious scientists, could claim that these evils such as the death of the less fit were a necessary part of God's evolutionary system of creation. In this, they followed the lead of Asa Gray.

In the 20th century, many creationists proclaimed that evolutionary science is incompatible with belief in God. Asa Gray, and many other scientists and theologians of the 19th century, show that this is not true. Gray died January 30, 1888.

Further Reading

Livingstone, David N. *Darwin's Forgotten Defenders.* Grand Rapids, Mich.: Eerdmans; Edinburgh, U.K.: Scottish Academic Press, 1987.

greenhouse effect The greenhouse effect is the absorption of infrared radiation by certain atmospheric gases, which causes the atmosphere to become warmer. The concept was proposed by Swedish chemist Svante Arrhenius at the end of the 19th century. When sunlight shines upon the Earth, most of the visible light penetrates the atmosphere. Some of the sunlight *(albedo)* reflects off of light-colored surfaces such as ice and snow, directly back into outer space. Reflected sunlight does not contribute to warming the Earth. The rest of the sunlight is absorbed by things like rocks, oceans, plants, and animals and causes them to become warmer. They conduct heat into the air, causing the air to become warmer. This is the reason that air is warmest near the average surface level of the ground (low altitudes and elevations) rather than at high altitudes or high elevations. Rocks, oceans, plants, and animals also glow with invisible photons known as infrared radiation, at wavelengths just beyond the red end of the visible light spectrum. Most of these infrared photons radiate into outer space. Certain atmospheric gases, such as carbon dioxide, methane, and water, absorb these photons on their way out. This causes the gases to become warmer, and they impart their warmth to the other gas molecules in the atmosphere. In this way they act metaphorically like the glass roof of a greenhouse which holds in the heat of the sunlight. This is how the presence of greenhouse gases causes the atmosphere to become warmer. The greenhouse effect is essential to the survival of life on this planet. Without the greenhouse effect, the Earth would have an average temperature similar to that of Mars.

Three atmospheric gases are important in the greenhouse effect:

- Water is the most abundant greenhouse gas. Clouds block sunlight but also absorb infrared radiation; the net effect, according to climatological calculations, is that clouds have a slight cooling effect on the Earth.
- The most potent major naturally occurring greenhouse gas is methane. Methane is produced mainly by bacterial fermentation but becomes carbon dioxide when it reacts with oxygen gas. Although methane is continually produced, it does not accumulate in the atmosphere.
- Most atmospheric scientists consider carbon dioxide to be the most important greenhouse gas because it absorbs more infrared radiation than water and is more stable than methane.

The importance of atmospheric carbon dioxide to the greenhouse effect is demonstrated by studies of ice cores. In Greenland and in Antarctica, annual layers of ice have accumulated for over many thousands of years. For example, ice cores from Vostok, Antarctica, contain ice layers from the recent past to over 400,000 years ago. By counting back from the present, the age of each layer of ice can be determined. The oxygen and deuturium isotope ratios (see ISOTOPES) of each layer of ice is an estimate of the global average temperature. In addition, each layer of ice contains dissolved atmospheric gases. When these gases are released in a laboratory and analyzed, scientists can directly measure the concentration of methane and carbon dioxide that was in the air at that time. There is a close correlation between atmospheric carbon dioxide and global temperature over the past 400,000 years. Both temperature and carbon dioxide have fluctuated dramatically during that time, but always together, at least in the past 400,000 years.

The greenhouse effect is just one influence upon global temperatures. Another major influence is the arrangement of continents. Continents have shifted position drastically during the history of the Earth (see CONTINENTAL DRIFT). Before the Pleistocene epoch, the average Earth temperature was warmer than it is today. During much of the CENOZOIC ERA, forests grew up to the North Pole itself. The trees were deciduous but may have been so in response to the six months of darkness rather than to particularly cold temperatures. The reason that the high northern latitudes were warm was because ocean currents brought warm water from tropical areas. Ocean currents have always been a major factor in distributing the heat on planet Earth. Today, the Gulf Stream brings warmth from the Caribbean to Western Europe, so that England's climate is much milder than that of Labrador, which is at the same latitude. Before the Pleistocene, a global ocean current (a very large version of the Gulf Stream) carried warm water into the north polar regions. When the islands between North and South America coalesced into the isthmus of Panama this global ocean current was interrupted. It was at this time that the cycle of ICE AGES began. There have been periods of cooling in the past history of the Earth; there is evidence for glaciation around the South Pole during the PALEOZOIC ERA. At other times, forests grew at the South Pole.

The principal process by which carbon dioxide is removed from the atmosphere is by photosynthesis (see PHOTOSYNTHESIS, EVOLUTION OF). Cyanobacteria (see BACTERIA, EVOLUTION OF) and the chloroplasts of eukaryotic cells (mostly of plants; see SYMBIOGENESIS) absorb carbon dioxide and make it into carbohydrates, using sunlight energy. Not only is this the source of all the food in the world, but it also removes carbon dioxide from the air. Cellular respiration of organisms releases carbon dioxide into the air. At many times in the past history of the Earth, photosynthesis absorbed more carbon dioxide than respiration released. Especially during the CARBONIFEROUS PERIOD, huge forests of plants were buried and became deposits of coal and oil, with the result that the carbon atoms were stored in the Earth rather than returning to the atmosphere. Also, many marine protists and invertebrates (see EUKARYOTES, EVOLUTION OF; INVERTEBRATES, EVOLUTION OF) consumed single-celled plants, and used the carbon to produce calcium carbonate shells, which have often become fossils (see FOSSILS AND FOSSILIZATION) and in some cases have produced thick deposits of limestone, chalk, or diatomaceous earth. In some regions, such as the tundra, much partially decomposed organic material (humus) remains frozen, which also represents storage of carbon that had previously been atmospheric carbon dioxide. The inorganic production of limestone (calcium carbonate) in the oceans, in which carbon dioxide reacts with calcium silicate to produce limestone and silicon dioxide, can also remove carbon dioxide from the atmosphere.

It is possible that the relatively cool conditions of the modern Earth are partly the result of photosynthesis, followed by the entrapment of carbon in the Earth as fossil fuels, limestone, and humus, in addition to the configuration of the continents. It has been suggested that excessive photosynthesis by oceans full of green scum and stromatolites was a cause of SNOWBALL EARTH, when the planet nearly froze over. Scientists know from studies of oxygen ISOTOPES that the temperature of the Earth has fluctuated over the course of many millions of years. They cannot directly measure atmospheric carbon dioxide levels any further back than several hundred thousand years. They can, however, estimate atmospheric carbon dioxide levels from carbon isotope ratios of fossilized organic molecules. During many of the geological periods of the past, warm temperatures have been associated with worldwide forests, which would suggest that the greenhouse effect was a minor contributor to climatic fluctuations at those times. Scientists are fairly certain of the importance of the greenhouse effect to the climate of the past several hundred thousand years, during which time continental movements have been slight.

Carbon dioxide makes up only 0.035 percent of the atmosphere of the Earth. Yet this tiny amount of carbon dioxide makes life possible on the Earth. Both Mars and Venus have tremendous amounts of carbon dioxide in their atmospheres. This is why Venus has a surface temperature much warmer than Mercury, even though it is farther from the Sun, and why Mars actually has a few days with equatorial temperatures above the freezing point of water. In the past, the Martian atmosphere was much thicker and Mars had water. Today the Martian atmosphere is too thin for the greenhouse effect to help it very much, but four billion years ago the greenhouse effect may have made Mars a warm little pond of bacterialike life-forms (see MARS, LIFE ON).

The greenhouse effect is part of the set of global processes that some scientists believe regulates the climatic and atmospheric conditions of the Earth. According to the GAIA HYPOTHESIS, the activity of organisms affect the atmosphere of the Earth and help to maintain it in a state in which life can continue to function.

Human activities such as industry and transportation are releasing a tremendous amount of carbon dioxide into the atmosphere. At the same time, humans have been destroying forests that absorb carbon dioxide. The net effect is that carbon dioxide levels in the atmosphere have been increasing more rapidly than at any time in the last 400,000 years or perhaps ever. Exactly parallel to this has been a rapid increase in the average temperature of the Earth, as indicated by the studies of ice cores and tree rings. The average temperature of the Earth has vacillated wildly (but, ever since Snowball Earth, not wildly enough to threaten the existence of life), whether as a result of wild changes in carbon dioxide or not. For example, the warming trend at the end of the most recent ice age was interrupted by a period of cooling known as the Younger Dryas period (named after a species of arctic wildflower), which began, and ended, over a period of decades or centuries rather than millennia. The increase in carbon dioxide and in global temperature since about 1850 have been the most rapid changes known to have occurred in the history of the Earth, with the exception of the events following the asteroid that caused the CRETACEOUS EXTINCTION. Even the global warming that occurred at the end of each of the recent ice ages was less rapid than the one that is now occurring as a result of human activity. Human activity is causing the extinctions of many species, which will only increase more as a result of continued global warming. Many scientists estimate that the Earth is entering the sixth of its MASS EXTINCTIONS. The greenhouse effect is highly unlikely to cause massive disasters by itself, but it certainly can alter the climatic patterns of the Earth to an extent that the economic systems of modern human civilization would be unable to cope with them. The greenhouse effect, then, keeps all humans alive, but human activity may be creating too much of a good thing.

Further Reading

Alley, Richard B. "Abrupt climate change." *Scientific American*, November 2004, 62–69.

Barnola, J. M., et al. "Historical carbon dioxide record from the Vostok ice core." In *Trends: A Compendium of Data on Global Change*. Carbon Dioxide Information Analysis Center, Oak Ridge National Laboratory, U.S. Department of Energy, Oak Ridge, Tenn., 2000. Available online. URL: http://cdiac.ornl.gov/trends/co2/vostok.htm. Accessed April 21, 2006.

Bradshaw, William E. and Christina M. Holzapfel. "Evolutionary response to rapid climate change." *Science* 312 (2006): 1477–1478.

Pagani, Mark, et al. "Marked decline in atmospheric carbon dioxide concentrations during the Paleogene." *Science* 309 (2005): 600–603.

Petit, J. R., et al. "Historical isotopic temperature record from the Vostok ice core." In *Trends: A Compendium of Data on Global Change*. Carbon Dioxide Information Analysis Center, Oak Ridge National Laboratory, U.S. Department of Energy, Oak Ridge, Tenn., 2000. Available online. URL: http://cdiac.ornl.gov/trends/temp/vostok/jouz_tem.htm. Accessed April 21, 2006.

Weart, Spencer R. *The Discovery of Global Warming*. Cambridge, Mass.: Harvard University Press, 2003.

group selection The process of NATURAL SELECTION favors characteristics that maximize the fitness of individuals within populations. Most scientists maintain that evolution operates by this process of *individual selection*. Any characteristics, for example behavior patterns that favor an entire population or other group at the expense of the fitness of an individual, could not evolve by individual selection. Such behavior would require individuals to surrender some of their fitness for the common good of the population or group. Any individual that selfishly accepted the benefits of the group, but did not contribute to the common good (evolutionary biologists often call such organisms cheaters), would have a higher fitness, and their descendants, with their selfish behavior patterns, would soon spread and dominate the population or group.

An example of the failure of group selection is optimal population size. Consider an animal population in which an optimal population size can be maintained if every pair of animals produced three offspring. If all the animals in the population had a genetically based tendency to produce only three offspring, the population size would not exceed the available food supplies and other resources. Any animal that had a genetically based tendency to produce four offspring would have a higher fitness than those that produced just three. The genes that caused the tendency to produce four offspring would increase in frequency until they dominated the population. This reproductive rate would lead to population growth that exceeds the available resources and thus would result in a population collapse. Darwin, applying the principles of Malthus to animal and plant populations, noted the universal tendency among organisms for populations to exceed resource availability and made it the basis of his theory of evolution (see DARWIN, CHARLES; MALTHUS, THOMAS; *ORIGIN OF SPECIES* [book]). The population as a whole would benefit if group selection produced a restricted population growth, but individual selection produces populations that grow beyond their resources (see POPULATION).

There are many examples of animal behavior called ALTRUISM that appear to benefit the group rather than the individual. One example occurs when honeybee workers attack animals that threaten their hive. The honeybees leave their stingers in the intruder's skin and die soon afterward. Evolutionary biologists have been quite successful at explaining altruism in terms of individual, not group, selection. Self-sacrificing behavior can benefit the fitness of an individual if the definition of fitness includes the genetic relatives of the individual along with the individual itself *(inclusive fitness)*. Self-sacrificing behavior can also benefit the fitness of an individual if the beneficiary is likely to help the individual, or its offspring, in the future *(reciprocal altruism)*. Altruism can also enhance an animal's social status (see also BEHAVIOR, EVOLUTION OF). In general, natural selection often favors cooperation because cooperation benefits individuals. As one evolutionary scientist (see DAWKINS, RICHARD) has said, "Selection doesn't favor a harmonious whole. Instead, harmonious parts flourish in the presence of each other, and the illusion of a harmonious whole emerges."

Many examples of what may at first appear to be group selection can also be explained because the population or group is an important part of the environment in which an individual lives. Animals that accept a dominance hierarchy, and fight by display rather than mortal combat, benefit from living in a society that is safer. A subordinate animal loses its immediate chances for reproduction but lives for another day in which an opportunity might come along for it to become a dominant animal. Cooperation in animal societies can usually be explained in terms of individual benefits. Animal societies differ widely in the degree of cooperation among their members.

Other examples involve organisms that must cooperate to maintain their common environment. For many parasites, the host organism is the environment. Parasite populations in which some individuals reproduced so much as to kill the host would destroy their own environment. There are many examples of *balanced pathogenicity* in which parasites evolve into milder forms (see COEVOLUTION; EVOLUTIONARY MEDICINE). For the parasite, there is a balance between reproducing too little and reproducing so much as to kill the host. Group selection is not necessary to explain balanced pathogenicity, for all the parasites benefit individually from the survival of their host. Just as animal societies differ widely in how cooperative their members are, parasites differ widely in their tendency to evolve balanced pathogenicity. Germs such as the bacteria that cause cholera, and which spread quickly through water supplies, have not evolved into milder forms, while germs such as the viruses that caused smallpox, and which spread by direct contact among humans, have evolved balanced pathogenicity.

Some evolutionary biologists such as Stephen Jay Gould have used the term *species selection* to refer to the tendency of some species to evolve into a greater number of species than others (see GOULD, STEPHEN JAY). In a species that produces many new species, at least one or a few of the species is likely to survive, while a species that does not produce new lineages is at increased risk of extinction. Paleontologist Niles Eldredge (see ELDREDGE, NILES) prefers to call this process *species sorting* rather than species selection. Natural selection does not actually favor the ability to speciate. Natural selection favors home ranges and interactions that promote individual fitness. Species that happen to have smaller home ranges and more limited interactions are more likely to experience reproductive isolation and evolve ISOLATING MECHANISMS that promote SPECIATION. These are the lineages that persist over time. Since EXTINCTION (particularly MASS EXTINCTIONS) is usually a random event, the lineages that radiate into the most species may dominate the world even if their adaptations are less effective than those of lineages that speciate less (see ADAPTIVE RADIATION).

Most evolutionary scientists would say that group selection does not occur with respect to genetically based adaptations. In intelligent species such as humans, group selection can occur with respect to social characteristics. Consider again the example of population growth. Through most of human history, populations have grown as much as resources permitted. So many humans died that populations remained relatively small. With the invention of agriculture, then again with the invention of industry, the human population began periods of extremely rapid growth, with the result that human overpopulation has threatened many human societies and the world as a whole. In recent decades, however, almost every country in the world has experienced a decline in population growth, as a result of education and economic development rather than evolutionary changes. If the human species avoids a devastating population explosion, it may well be the first species that has ever limited its own population growth by a nongenetic version of group selection.

Further Reading

Eldredge, Niles. *Reinventing Darwin: The Great Debate at the High Table of Evolutionary Theory.* New York: John Wiley, 1995.

Gould, Stephen Jay. *The Structure of Evolutionary Theory.* Cambridge, Mass.: Harvard University Press, 2002.

Sober, Elliott, and David Sloan Wilson. *Unto Others: The Evolution and Psychology of Unselfish Behavior.* Cambridge, Mass.: Harvard University Press, 1999.

Wilson, David Sloan. *Darwin's Cathedral: Evolution, Religion, and the Nature of Society.* Chicago: University of Chicago Press, 2003.

gymnosperms, evolution of Gymnosperms constitute a broad category that includes all extant seed plants that are not flowering plants (see ANGIOSPERMS, EVOLUTION OF), as well as many extinct forms. Seed plants (gymnosperms and angiosperms) have the following characteristics:

- *Vascular tissue.* Vascular tissue consists of xylem and phloem. Xylem tissue conducts water up from roots into stems and leaves. Phloem tissue conducts water with sugar and other organic molecules, usually down from the leaves where the sugar is made (see PHOTOSYNTHESIS, EVOLUTION OF) to the roots. Some seedless plants (see SEEDLESS PLANTS, EVOLUTION OF) such as ferns also have vascular tissue. Roots, stems, and leaves are defined partly by the presence of vascular tissue. Mosses and their relatives do not have vascular tissues and do not have true roots, stems, or leaves. Because of vascular tissue, seed plants are well adapted to life on dry land.
- *Pollen.* Pollen grains contain one or more sperm nuclei inside a hard protein coat that protects the sperm and other male cells while they are transported through the air by the wind or by pollinators (see COEVOLUTION). In gymnosperms, pollen grains usually develop inside of small conelike structures. In seedless plants, sperm must swim through layers of water from one plant to another. Because pollen carries sperm through the air, seed plants are well adapted to life on dry land.
- *Seeds.* Seeds contain an embryonic plant and a food supply, surrounded by a protective coat. Because the embryo is already partly grown, and has a food supply, a plant can germinate from a seed and grow quickly, whereas seedless plants must begin their growth from single spores on the soil. In gymnosperms, seeds usually develop inside of conelike structures. Seeds of gymnosperms (Greek for "naked seed") develop without being surrounded by parental tissue.

The first gymnosperms evolved during the PALEOZOIC ERA. The earliest gymnosperms were *Elkinsia* and *Archaeosperma,* which appeared late in the DEVONIAN PERIOD. Gymnosperms remained a relatively minor component of the forest vegetation during the CARBONIFEROUS PERIOD when seedless plants grew to a very large size. During the PERMIAN PERIOD, as cooler and drier conditions spread, the large seedless trees declined in abundance, and the forests were dominated by gymnosperms. Among these gymnosperms were the trees known as cordaites. Two modern groups of gymnosperms, the conifers and the cycads, evolved during the Permian period. Also common in the late MESOZOIC ERA were seed ferns (pteridosperms) that were not ferns and were probably not related to modern gymnosperms.

In the cooler, drier conditions of the Mesozoic era, gymnosperms (especially conifers and cycads) dominated the forests of the TRIASSIC PERIOD and the JURASSIC PERIOD. Many of the conifers resembled the modern Norfolk Island pine, which is not really a pine. The Wollemi pine, which is also not really a pine, was thought to be extinct until it was discovered in Australia in 1994. True pines and their relatives evolved during the Mesozoic also.

Another group of Mesozoic gymnosperms was the bennettitalean plants. The details of their leaf and wood anatomy, as well as of their reproductive structures, suggest that they may have been the ancestors of the flowering plants. Some bennettitalean plants had female conelike structures surrounded by male reproductive structures inside of bracts, which is very similar to the structure of a flower. When the flowering plants evolved at or before the beginning of the CRETACEOUS PERIOD, they displaced the gymnosperms from the tropical areas.

Modern gymnosperms spread during the CENOZOIC ERA to their present habitats. Modern gymnosperms include these groups:

- *Conifers.* Most conifers have stiff needle-like or scale-like evergreen leaves. Some primitive conifers, such as the podocarps of New Zealand and South America, and the auracarians of Chile, are found in the Southern Hemisphere. Most modern conifers are junipers, cypresses, pines, spruces, firs, hemlocks, and larches. Pines evolved during the Mesozoic era but have spread extensively since, especially in cool regions with poor soils.
- *Cycads.* Cycads look like small palm trees.
- *Ginkgoes.* There is only one modern species of ginkgo (the maidenhair tree, *Ginkgo biloba*) which closely resembles ginkgoes that lived millions of years ago (see LIVING FOSSILS). This species may have been saved from EXTINCTION by Chinese monks who grew them in monasteries. Ginkgoes have leaves with veins that diverge rather than branching out. They produce seeds singly on stalks rather than inside of cones.
- *Gnetales.* This is a loosely defined group of plants with gymnosperm reproductive structures, but some of which have thin, flat leaves that resemble those of flowering

plants. One member of this group is the desert ephedra bush.

Further Reading

McLoughlin, Stephen, and Vivi Vadja. "Wollemi." *American Scientist* 93 (2005): 540–547.

Museum of Paleontology, University of California, Berkeley. "Introduction to the Bennettitales." Available online. URL: http://www.ucmp.berkeley.edu/seedplants/bennettitales.html. Accessed March 29, 2005.

Wollemi Pine International. "The Wollemi Pine." Available online. URL: http://www.wollemipine.com. Accessed March 28, 2005.

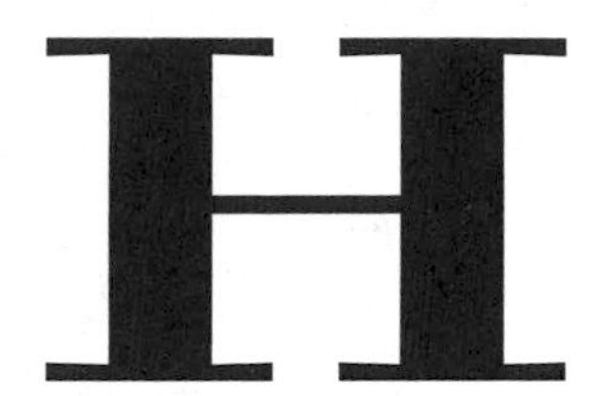

Haeckel, Ernst (1834–1919) German *Evolutionary biologist* Ernst Heinrich Philipp August Haeckel is best known as the principal champion for Darwinian evolution in late 19th-century continental Europe, and for evolutionary concepts that turned out to be wrong in their original form.

Born February 16, 1834, Ernst Haeckel was a physician but in 1859, upon reading Darwin's book (see DARWIN, CHARLES; *ORIGIN OF SPECIES* [book]), he abandoned his practice. He studied zoology at the University of Jena, then became a professor of comparative anatomy at that university in 1862. He extensively studied the embryology of invertebrates (see INVERTEBRATES, EVOLUTION OF) and a group of protists (see EUKARYOTES, EVOLUTION OF) known as radiolarians.

Although Haeckel defended evolution, he like many other prominent scientists did not agree with Darwin that NATURAL SELECTION was the mechanism of evolution. In particular, Haeckel and many other scientists accepted a version of LAMARCKISM. He also believed that evolution had a predetermined direction. This direction was not impelled by spiritual forces—as many other German scholars believed, following the tradition of the scientist and writer Johann Wolfgang von Goethe—but by natural law. This direction could be glimpsed in the stages of development of an animal's embryo. As Haeckel described it in a still famous quote, "Ontogeny recapitulates phylogeny." This means that an animal's embryo, as it develops, passes through the same evolutionary stages by which its species evolved. Thus the human embryo goes through fish, amphibian, reptile, and earlier mammalian stages, according to Haeckel's interpretation. While evolutionary biologists still accept this overall pattern, it is considered an effect rather than a cause, and they do not accept it as a natural law (see RECAPITULATION). The figures that Haeckel published, illustrating stages of embryonic development, contained errors. In particular, Haeckel used dog embryos to fill in some missing stages in human embryonic development. The differences were minor and did not alter the overall pattern but tarnished Haeckel's reputation among evolutionary scientists. He remained popular with the reading public.

Haeckel also championed EUGENICS. He stated that "Politics is applied biology," by which he meant that evolution led upward toward white Europeans and that politics should advance the social and biological interests of white Europeans. After Haeckel's death, the Nazis used this quote and his justifications for racism, nationalism, and SOCIAL DARWINISM to support their cause.

Haeckel applied his scientific concepts broadly, as an amateur scholar in anthropology, psychology, and cosmology. He used evolution as the basis for attacking religious views. He could be considered the German counterpart of T. H. Huxley (see HUXLEY, THOMAS HENRY) but without Huxley's rigorous regard for facts. Haeckel died August 8, 1919.

Further Reading

Gasman, D. *The Scientific Origins of National Socialism: Social Darwinism in Ernst Haeckel and the German Monist League*. London: MacDonald, 1972.

Museum of Paleontology, University of California, Berkeley. "Ernst Haeckel (1834–1919)." Available online. URL: http://www.ucmp.berkeley.edu/history/haeckel.html. Accessed April 28, 2005.

Haldane, J. B. S. (1892–1964) British *Zoologist* John Burdon Sanderson Haldane was an idiosyncratic genius who contributed to the modern understanding of evolution in several important ways. His father, John Scott Haldane, was a professor of physiology at Oxford University, whose diverse studies included parasites, heatstroke, and altitude sickness. J. S. Haldane also studied the intervals that were necessary during an ascent from deep diving in order to avoid the bends. A fearless researcher, J. S. Haldane even took notes on his symptoms while mildly poisoning himself with carbon monoxide.

Born November 5, 1892, J. B. S. Haldane worked alongside his father, J. S. Haldane, and conducted his own experiments, even in childhood. As a teenager, he bred guinea pigs with his sister Naomi and studied their genetics. Here he discovered for himself a phenomenon now known as genetic linkage, in which two different characteristics are passed down together from one generation to another because they are linked on the same chromosome (see MENDELIAN GENETICS). With a father so fearless, and who researched such diverse phenomena, and with his own childhood experience in scientific research, it is little wonder that J. B. S. Haldane made important contributions to more than one field of scholarship.

J. B. S. Haldane fought in World War I. After the war he attended Oxford and majored in classics, and for the rest of his life he would recite long passages of Milton, or of Homer in the original Greek. However, in his spare time, he also studied physiology, mathematics, evolution, and the origin of life. He became so accomplished in these and other scientific fields that they became his principal work when he joined the faculty of Oxford:

- *Physiology:* Following in the footsteps of his father, J. B. S. Haldane studied the reaction of the human body to extreme conditions such as high pressures that would be experienced in submarines. He built a chamber for studying the effects of pressure on human volunteers, including himself. One time, when he was in this chamber, rapid decompression caused the fillings in his teeth to pop out. Another time, he went into seizures from experiencing high oxygen pressure. Due to the importance of this research to understanding the human body under combat conditions, he received Admiralty funding for the work.
- *Mathematics and evolution:* Haldane became interested in Darwinian evolution and in Mendelian genetics. During the 1920s, many scientists (such as DEVRIES, HUGO) considered Darwinism and Mendelism to contradict one another. Haldane's proficiency with mathematics allowed him to make important contributions, along with others (see FISHER, R. A.; WRIGHT, SEWALL), to a theoretical reconciliation of the two. Haldane's *The Causes of Evolution* was published in 1933. The mathematical works of Haldane, Fisher, and Wright were essential precursors to the MODERN SYNTHESIS proposed by biologists (such as DOBZHANSKY, THEODOSIUS; MAYR, ERNST; SIMPSON, GEORGE GAYLORD). He published a paper in 1924 in which he calculated the effects of NATURAL SELECTION on the PEPPERED MOTHS. This paper astonished the scientific community because it showed that natural selection could be an extremely powerful force. He also took the first steps in working out the mathematics of the evolution of ALTRUISM. He pointed out that an individual's genes could be passed on to the next generation by his own offspring, or by the offspring of his relatives. Therefore an individual who sacrificed himself to save two of his brothers would have just as high an evolutionary fitness as if he had his own children, since he was related by a genetic factor of one-half to his brothers. The same would be true of a man who sacrificed himself to save eight cousins. "I would die for two brothers or ten cousins," said Haldane—using ten, not eight, just to be on the safe side.
- *Origin of life:* Haldane was, along with the Russian chemist Aleksandr I. Oparin, the principal defender of the idea that the first cells evolved in a soup of chemicals in the primordial oceans of the Earth, before oxygen became abundant in the atmosphere (see LIFE, ORIGIN OF). This theory is now called the Oparin-Haldane Hypothesis.

Haldane published two dozen books and hundreds of articles. Many of the articles were in a communist newspaper, *The Daily Worker*. He especially enjoyed attacking religious people. He was particularly incensed at his fellow Oxford faculty member and famous defender of Christianity, Clive Staples Lewis. One of his newspaper articles was "Anti-Lewisite," in which he compared C. S. Lewis to the poison gas used in the Great War. In 1924 he presented a paper before the Heretics Society in which he speculated about the possibility of ectogenesis (birth outside the body). Although he was ridiculed for these ideas, it was his intellectual fearlessness that enabled him to make so many important contributions to evolutionary science. Haldane died December 1, 1964.

Hardy-Weinberg equilibrium *See* POPULATION GENETICS.

hominid *See* HOMININ.

hominin Hominin is the term that is preferred by most scientists for the unique evolutionary lineage of human ancestors, since the divergence of the human from the chimpanzee lineages. The previously employed term *hominid* referred to the family Hominidae, as opposed to the chimpanzee family Pongidae. Because of the genetic closeness of humans and chimpanzees, both are now classified in the family Hominidae. The term *hominid*, therefore, now refers to humans and chimpanzees (which includes the pygmy chimpanzee, or bonobo; see PRIMATES). The common ancestor of humans and chimps would look like a chimp to a human observer, because chimps have undergone fewer evolutionary changes of anatomy than have humans since the time of the common ancestor.

HOMO SAPIENS is the only surviving hominin. This simple fact of history has permitted humans to consider themselves entirely separate from "the animals." Human uniqueness is of recent origin; for all but the last 20,000 years, two or more species of hominins have coexisted.

Hominin evolution corresponds closely to, and may have been stimulated by, climatic changes in Africa. When ICE AGES occurred in northern latitudes, droughts as well as cooler temperatures occurred in Africa, transforming forests into savannas. Periods of drier climate occurred about five million, 2.7 million, and 1.8 million years ago, which corresponded to the origin of hominins, of the genus *Homo*, and of advanced *Homo* (the immediate ancestors of *H. sapiens*), respectively. Part of the hominin adaptation to new climates was through TECHNOLOGY. Technology (such as clothes and fire) allowed members of the genus *Homo* to live in cold climates. Many scientists consider this to be the first and only time that a species has migrated to an entirely different climate without evolving into another species.

The initial divergence between hominins and the other apes may have involved an abrupt genetic change. The other great apes have 24 pairs of chromosomes in each cell, while humans have just 23 pairs. The human chromosome 2 is the second largest human chromosome and resulted from the fusion of two ape chromosomes. The hominins that possessed this new chromosomal arrangement could not interbreed with other apes and then evolved in their own separate direction (see ISOLATING MECHANISMS; SPECIATION). There are relatively few differences in the genes that are on the chromosomes. Human and chimp DNA differs by four percent. Just under 3 percent of the difference is due to insertions or deletions (see DNA [RAW MATERIAL OF EVOLUTION]). Twenty-nine percent of the proteins are identical in humans and chimps. BIOINFORMATICS studies indicate that the major differences between humans and chimpanzees are not in the genes that they have but the genes that they use, especially in brain tissue. The human lineage has evolved genetic differences that are important in resistance to certain diseases. Humans and chimps also differ markedly in which portions of the chromosomes experience frequent recombination.

The earliest hominins represented numerous genera and species, which can be lumped unofficially into a single category (see AUSTRALOPITHECINES). The most important feature that distinguished these australopithecines was BIPEDALISM. Australopithecines walked habitually on two legs, but many retained some features of foot anatomy that allowed the continued ability to climb trees. Australopithecines probably could not run for long distances. Australopithecine bipedalism was, however, no unstable transitional form between chimp-like locomotion and modern human walking; it persisted for more than a million years.

The earliest members of the genus *Homo* (two or more species often lumped into *HOMO HABILIS*) possessed several evolutionary innovations:

- Increase in brain size
- Production of simple stone tools (Oldowan technology)
- Reduction in tooth size
- More vertical face

These features were related to the consumption of meat, from small prey or large carcasses. Fats and proteins from meat were important in allowing larger brains to evolve, and larger brains enabled the earliest humans to more effectively find meat. Stone tools allowed the processing of meat, for example to dismember carcasses so that the parts could be carried home. Large teeth were no longer necessary for grinding coarse plant materials, as meat constituted a larger portion of the diet. To a large extent, stone tools made large teeth unnecessary. The earliest *Homo* were taller than australopithecines, but not as tall as modern humans.

The next phase of hominin evolution is represented by several species (*HOMO ERECTUS; HOMO ERGASTER; HOMO HEIDELBERGENSIS*) often lumped into the species *H. erectus*. *H. ergaster* was the African ancestor of Asian *H. erectus* and of *H. heidelbergensis*, some populations of which migrated to Europe. Asian *H. erectus*, as well as the FLORES ISLAND PEOPLE *(H. floresiensis)*, became extinct, some as recently as about 20,000 years ago. These hominins had no major evolutionary innovations, but some of them had further development of the unique features of *Homo:*

- Even larger brains
- More complex stone tools (Acheulean technology)
- More efficient bipedalism, allowing long-distance running
- Longer juvenile period

The nearly complete Nariokotome skeleton indicates that *H. ergaster* was almost identical to modern humans except for skull characteristics, particularly in having a brain that was the largest that a primate had ever had but still much smaller than the brains of modern humans. The rapid juvenile brain growth required that birth occur earlier in the period of gestation, otherwise the large head of the baby would not have been able to emerge. Babies born in a more helpless stage require longer parental care, a situation that would have favored the evolution of a more intricate and close-knit structure of immediate and extended families. This situation, in turn, permitted the retention of juvenile characteristics into adulthood (see NEOTENY). This may have occurred first in *H. ergaster*.

Different *Homo heidelbergensis* populations followed different evolutionary paths. In Europe, some *H. heidelbergensis* populations evolved into NEANDERTALS *(H. neanderthalensis)*. Neandertals had even larger brains (as large as those of modern humans) and more advanced stone tools (Mousterian technology). By about 130,000 years ago, Neandertals were the only surviving European descendants of *H. heidelbergensis*. In Africa some *H. heidelbergensis* populations evolved into *H. sapiens*. *H. sapiens* had large brains (of full modern size by 100,000 years ago) and may have had more advanced stone tools and even art, as evidenced by the Katanda harpoon point and the Blombos Cave ochre stone. When *H. sapiens* coexisted with Neandertals in the Middle East, both had the same Mousterian technology of tools. The explosion of technology and art did not occur until *H. sapiens* began to migrate out of Africa (see CRO-MAGNON).

The evolution of modern human characteristics followed a mosaic pattern, in which different traits evolved at different rates in different populations. In most australopithecines, the face sloped more and the teeth were larger than in the genus *Homo*. However, *Sahelanthropus* and *Kenyanthropus* had relatively vertical faces, even while retaining other australopithecine features. *Kenyanthropus* had smaller teeth than other australopithecines. There was no single path of evolution that all hominins followed.

The smaller teeth of *Homo* than those of earlier hominins were not entirely due to eating more meat. The evolution of larger brains allowed the development of technology (tools and fire) for processing food and allowed the intelligent location and selection of food. This allowed the evolution of smaller teeth. In *H. sapiens*, the gut is smaller than what an average primate of the same size would possess. By intelligently selecting foods, *H. sapiens* had less need for digestive detoxification.

Hominin evolution occurred in numerous lineages, with (except for the last 20,000 years) more than one lineage existing at the same time and often in the same general location. In general, hominin evolution involved bipedalism, the increase in brain size, the reduction in tooth size, and the invention of technology. Art and religion (see RELIGION, EVOLUTION OF) are now and may always have been unique to *H. sapiens,* which is the only surviving hominin species.

Further Reading

Behrensmeyer, Anna K. "Climate change and human evolution." *Science* 311 (2006): 476–478.

Calvin, William H. *A Brain for All Seasons: Human Evolution and Abrupt Climate Change.* Chicago: University of Chicago Press, 2002.

Chimpanzee Sequencing and Analysis Consortium. "Initial sequence of the chimpanzee genome and comparison with the human genome." *Nature* 437 (2005): 69–87.

Gibbons, Ann. *The First Humans: The Race to Discover Our Earliest Ancestors.* New York: Doubleday, 2006.

Hart, Donna L., and Robert W. Sussman. *Man the Hunted: Primates, Predators, and Human Evolution.* New York: Westview, 2005.

Hillier, LaDeana W., et al. "Generation and annotation of the DNA sequences of human chromosomes 2 and 4." *Nature* 434 (2005): 724–731.

Johanson, Donald, Lenora Johanson, and Blake Edgar. *Ancestors: In Search of Human Origins.* New York: Villard, 1994.

Khaitovich, Phillip, et al. "Parallel patterns of evolution in the genomes and transcriptomes of humans and chimpanzees." *Science* 309 (2005): 1,850–1,854.

Leakey, Richard, and Roger Lewin. *Origins Reconsidered: In Search of What Makes Us Human.* New York: Doubleday, 1992.

Sawyer, G. J., et al. *The Last Humans: A Guide to Twenty Species of Extinct Humans.* New Haven, Conn.: Yale University Press, 2006.

Stringer, Chris, and Peter Andrews. *The Complete World of Human Evolution.* London: Thames and Hudson, 2005.

Tattersall, Ian, and Jeffrey Schwartz. *Extinct Humans.* New York: Westview, 2000.

Templeton, Alan R. "Out of Africa again and again." *Nature* 416 (2002): 45–51.

Walker, Alan, and Pat Shipman. *The Wisdom of the Bones: In Search of Human Origins.* New York: Knopf, 1996.

Winckler, Wendy, et al. "Comparison of fine-scale recombination rates in humans and chimpanzees." *Science* 308 (2005): 107–111. Summary by Jorde, Lynn B., *Science* 308 (2005): 60–62.

Homo antecessor *See* HOMO HEIDELBERGENSIS.

Homo erectus In the broad sense, *Homo erectus* refers to the human species intermediate between *HOMO HABILIS* and modern humans. Most modern anthropologists divide these humans into at least three species: *HOMO ERGASTER,* the African "erectus" species that evolved from *H. habilis* and was the ancestor of the other "erectus" species, as well as of modern humans (see *HOMO SAPIENS*); *HOMO HEIDELBERGENSIS,* the "erectus" humans that migrated to Europe and evolved into NEANDERTALS; and the Asian "erectus" humans, including Java man and Peking man, which retain the original name *H. erectus.*

Not long after *H. ergaster* evolved in Africa, some of its populations began to migrate northward and eastward. Before a million years ago, *H. erectus* individuals were probably living in caves in what is now China, as well as on islands in what is now Indonesia. Specimens from the Chinese populations (Peking man) are known from 670,000 to 410,000 years ago. Specimens from the Indonesian populations (Java man) that are 1.8 million years old are known, and Java man may have persisted until as recently as 50,000 years ago. On Flores Island, a population of *H. erectus* apparently evolved into a species of miniature humans, *H. floresiensis,* who survived until perhaps 18,000 years ago (see FLORES ISLAND PEOPLE). None of these populations, Peking man, Java man, or the Flores Island people, evolved into modern humans.

Java man was the first human fossil to be found outside of Europe, and the first fossil that could clearly be interpreted as being more primitive than modern humans. A Dutch physician (see DUBOIS, EUGÈNE) found a skullcap in Trinil, Indonesia, in 1891, and a thighbone in 1892. Dubois at first called this species *Anthropopithecus,* the man-ape, then changed it to *Pithecanthropus erectus,* using the name of a hypothetical human ancestor that had been proposed by a leading German evolutionary scientist (see HAECKEL, ERNST). In the early 20th century, anthropologist Davidson Black excavated fossils from Dragon Bone Hill in China. For many years, local people had gathered the bones (which they called dragon bones) and ground them as medicine. Black found enough bones to reconstruct a species intermediate between apes and humans, which he called Peking man *(Sinanthropus pekinensis).* When World War II started, Europeans evacuated from China ahead of Japanese troops. American Marines were taking the Peking man bones when they were arrested by Japanese soldiers, who may have discarded the bones. Black and anthropologist Franz Weidenreich had made casts of the bones, which have survived. Early in the 20th century, anthropologist Ralph von Koenigswald found bones of Solo man, near the Solo River in Indonesia. He offered 10 cents per bone fragment to local excavators to bring him specimens, only to find that the excavators shattered the bones that they found in order to get more money. Java, Peking, and Solo man are now all considered to be populations of *H. erectus.*

H. erectus evolved some unique skull characteristics that allow it to be classified as a distinct species. First, some populations of *H. erectus* may have evolved a larger brain size, independently of the lineages that led to Neandertals and to modern humans. However, too few reliable estimates of brain size are available to determine whether this was a general trend. Second, *H. erectus* had thick skulls. Anthropologists Noel Boaz and Russell Ciochon point out that the particular pattern of thickness appears to be just right for protecting the brain from blows delivered to the top of the head. *H. erectus* individuals apparently hit each other with clubs often enough and for a long enough time that it influenced the evolution of their skulls.

Although *Homo ergaster,* the presumed ancestors of *H. erectus,* made advanced stone tools of the Acheulean industry, *H. erectus* tools represented only the most primitive stone tool technology, the Oldowan. A diagonal line from England to India, called the Movius line, separates the old and new industries. This pattern could be explained either of two ways. First, *H. erectus* may have evolved from very early *H. ergaster,* or even from populations that could be called *H. habilis,* before the Acheulean tool culture had developed. Second, the *H. erectus* migrants may have lost the ability or desire to produce Acheulean tools and reverted to the simpler Oldowan style.

Homo erectus constitutes evidence that human evolution consisted of several or many parallel lineages, all of which except one ultimately became extinct.

Further Reading

Boaz, Noel T., and Russell L. Ciochon. *Dragon Bone Hill: An Ice-Age Saga of Homo erectus.* New York: Oxford University Press, 2005.

Fischman, Josh. "Family ties." *National Geographic,* April 2005, 16–27.

Tattersall, Ian, and Jeffrey Schwartz. *Fossil Humans.* New York: Westview, 2000.

Homo ergaster *Homo ergaster* ("worker man") refers to the "erectus" humans who lived in Africa from about two million to about one and a half million years ago, and who were the ancestors of modern humans. "Erectus" humans were transitional between *HOMO HABILIS* and later human species (see NEANDERTALS; *HOMO SAPIENS*). They were anatomically nearly modern in all respects except that their brains were smaller than those of Neandertals and modern humans. "Erectus" humans lived in Africa, then spread throughout Europe and Asia. Most anthropologists consider the African, European, and Asian "erectus" specimens to represent distinct species. African "erectus" humans, *H. ergaster,* that remained in Africa evolved into modern humans *(H. sapiens); H. ergaster* that migrated into Europe (see *HOMO HEIDELBERGENSIS*) evolved into Neandertals; *H. ergaster* that migrated into Asia became the Asian "erectus" species (see *HOMO ERECTUS;* FLORES ISLAND PEOPLE). Since the name *H. erectus* was first defined from southeast Asian "Java man" specimens, which are considered to be the same species as "Peking man" specimens, the Asian "erectus" species retains the original name.

Numerous specimens of *H. ergaster* have been found. The most complete is the Nariokotome skeleton, also called the Turkana boy (see figure on page 189), found by Kenyan anthropologist Kamoya Kimeu and associates (see LEAKEY, RICHARD). Some molars were still erupting at the time of death, therefore the Nariokotome boy was less than 10 years of age when he died. He was already five feet three inches (160 cm) tall when he died, and probably would have grown to six feet one inch (185 cm) by adulthood. The tall, skinny body form of this species, like that of modern tribes of the African savanna, appears well adapted to dispersing body heat. The skeleton is complete enough that almost nothing is left to the scientific imagination. *H. ergaster,* as represented by this specimen, had a fully modern anatomy except that the brain was the size of a modern one-year-old child. Because of the fact that there are no obvious structural abnormalities, and that the boy lived to be almost a teenager, the small brain cannot be considered to represent a deformed microcephalic individual. The Nariokotome boy was normal for his species. *H. ergaster* in general, and the Nariokotome skeleton in particular, represent one of the almost perfect MISSING LINKS between earlier hominins and modern humans.

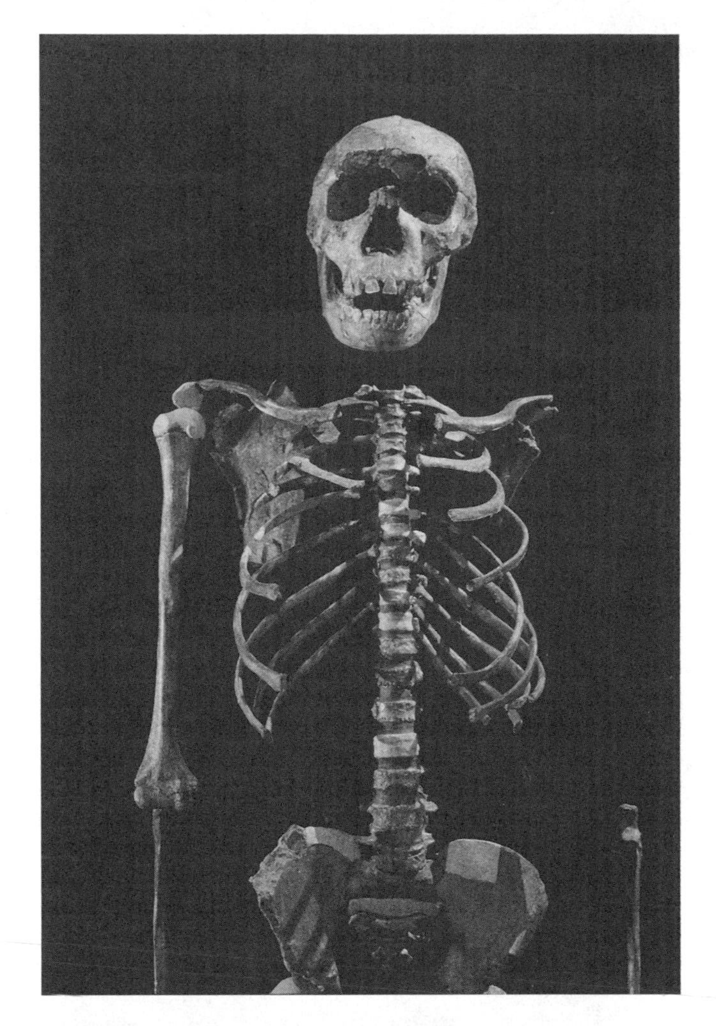

The Nariokotome or "Turkana boy" skeleton, excavated and reconstructed by Alan Walker and Richard Leakey, was about nine years of age when he died. He was over five feet (almost 2 m) tall but had the brain of a modern one-year-old child. He represents an almost perfect "missing link" in human evolution. ***(Courtesy of Kenneth Garrett/National Geographic Society)***

H. ergaster differed from earlier *H. habilis* by making more advanced stone tools, having larger brains and more complex cultures, making fire, and possibly by having the first languages:

Stone tools. Like earlier humans, *H. ergaster* made stone tools. The earliest tools of *H. ergaster* were Oldowan, like those made by *H. habilis.* Later, *H. ergaster* made Acheulean tools, which were more structurally complex (see TECHNOLOGY). Stones were worked into complex teardrop-shaped hand axes by the removal of numerous flakes from stone cores. Both the flakes and the cores were apparently useful as tools, for cutting meat and skin. The most likely use was for cutting pieces of flesh and cracking open bones from animals that had been killed by large predators. *H. ergaster* may have been a scavenger. This interpretation is reinforced by the microscopic analysis of cut marks upon the bones of prey animals from about two million years ago. Cut marks made by stone tools cross over tooth marks from hyenas, indicating that *H. ergaster* cut meat from prey animals after the hyenas had finished with them. A few of the teardrop-shaped axes were quite large, and anthropologists have not figured to what practical use, if any, these cores might have been put. One (see photo below) was almost 20 inches (nearly a half meter) in length and almost a foot (15 cm) wide at its widest point, and weighed 25 pounds (11 kg). These structures may have represented the earliest example of human art.

Ian Tattersall, of the American Museum of Natural History in New York, holds a cast of a stone tool found at Isimila, Tanzania, which resembled the Acheulean tools made by later populations of *Homo ergaster.* Only a few of the tools were this large; too large and heavy to use, these tools served an unknown function. *(Photograph by Stanley A. Rice)*

Larger brains and complex cultures. The brain of *H. ergaster,* at about 50 cubic inches (800 cc), was smaller than that of modern humans (which is more than 75 cubic inches or 1,200 cc), but still large enough that a very human type of birth was necessary. If *H. ergaster* fetuses developed along the same timeline as those of modern apes such as chimpanzees, the head at the time of birth would have been too large to have passed through the birth canal of an upright walking woman. In *H. ergaster,* as in modern humans, the solution was for the infant to be born prematurely and in a condition much more helpless than that of chimpanzee infants. Once born, the infant could continue its brain development. Enhanced brain growth is just one of many juvenile characteristics that modern humans possess even into adulthood, a process known as NEOTENY. This process had already begun with the evolution of *H. ergaster.* The helpless infant would need exceptionally active parental care, and this may have been the starting point of family life and human culture, in at least a rudimentary form in *H. ergaster.*

Another indicator of complex culture is that individuals of this species apparently took care of one another. A skeleton of a woman (specimen KNM-ER 1808) who died of hypervitaminosis A, caused perhaps by eating the liver of a carnivore, has been found in the Turkana region of Africa. She lived for several weeks with her affliction, during which time other humans apparently took care of her.

Use of fire. H. ergaster was apparently also the first human species to make controlled use of fire. It is very difficult to demonstrate evidence for hearths that are more than a million years old.

Language. Although human culture may have begun with *H. ergaster,* modern language apparently did not. The canal inside of the spinal cord of modern humans remains wide until near the coccyx. This allows a great abundance of nerves to receive sensory information from, and control the movements of, all the body parts. The canal inside the spinal cord of *H. ergaster* was restricted even by the time it reached the neck, implying that it not only had restricted manual coordination but also may not have been able to have articulate speech. Uncertainty remains, however, as to whether the brains of *H. ergaster* possessed structures that allow modern humans to produce and understand language (see LANGUAGE, EVOLUTION OF).

Homo ergaster coexisted, at least during its early period, both with the robust australopithecines (genus *Paranthropus*) and with populations of more primitive hominins usually assigned to *H. habilis.* Numerous hominin lineages existed at the same time and very nearly in the same place as *H. ergaster,* but it was the only one that was the ancestor of modern humans.

The larger brains and more complex tools of *H. ergaster* evolved quickly, at a time when cooler and drier conditions were spreading through Africa. After an initial burst of evolution anatomical characteristics remained relatively unchanged. Toolmaking culture also changed little during that time. Although the tools became thinner, they remained recognizable members of the Acheulean industry. When *H. ergaster* began to evolve into modern humans, about a million years ago, another rapid period of physical and cultural evolution began. Not only is *H. ergaster* itself an almost perfect example of a "missing link" between *H. habilis* and modern humans, but also there are numerous examples of intermediate forms between *H. ergaster* and modern humans in Africa.

Further Reading

Leakey, Richard, and Roger Lewin. *Origins Reconsidered: In Search of What Makes Us Human.* New York: Doubleday, 1992.

Potts, R., and Pat Shipman. "Cut marks made by stone tools on bones from Olduvai Gorge, Tanzania." *Nature* 291 (1981): 577–580.

Tattersall, Ian, and Jeffrey Schwartz. *Extinct Humans.* New York: Westview, 2000.

Walker, Alan, and Richard Leakey. *The Nariokotome Homo Erectus Skeleton.* Cambridge, Mass.: Harvard University Press, 1993.

Walker, Allan, and Pat Shipman. *The Wisdom of the Bones: In Search of Human Origins.* New York: Knopf, 1996.

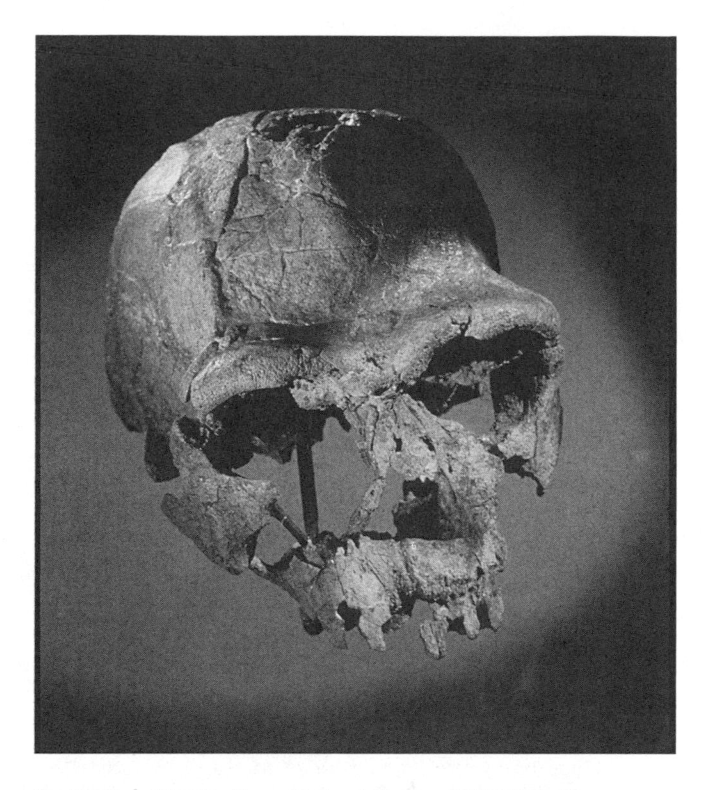

The "1470 skull" at the Kenya National Museum (KNM-ER 1470) represents one of the earliest specimens of increased brain size in human evolution. Found by Richard Leakey and associates, this skull housed a larger brain than most *Homo habilis* and may represent a different species. *(Courtesy of Kenneth Garrett/National Geographic Society)*

Homo habilis *Homo habilis* ("Handy man") is the scientific name usually assigned to the earliest species of the human genus. Louis Leakey (see LEAKEY, LOUIS) assigned this name because this species was the first to make and use stone tools. The tools, which consisted of little more than stones from which sharp-edged flakes were struck, are called Oldowan because they were first found in Olduvai Gorge in Tanzania (see TECHNOLOGY). *H. habilis* lived in eastern Africa between about 2.5 million and about 1.5 million years ago.

It is far from clear that the specimens usually assigned to *Homo habilis* represent a single species of HOMININ. Many anthropologists consider the specimens to represent at least two species: the larger brained *Homo rudolfensis* (named after Lake Rudolf, near which many specimens have been found) and the smaller brained *H. habilis.* Some specimens show variation in characteristics other than brain size. For example, Olduvai Hominid 62 (OH 62) had relatively longer arms and shorter legs than other *H. habilis,* and may therefore represent yet another, unnamed species. However, the variation in brain size within *H. habilis,* broadly defined, is less than the differences between males and females of some species of apes, which raises the possibility that *H. habilis* was a single species. For convenience, the more inclusive definition of *H. habilis* will be used for the remainder of this entry.

It is clear that one of the genetic lineages within *H. habilis* was the ancestor of later humans (see *HOMO ERECTUS; HOMO ERGASTER; HOMO HEIDELBERGENSIS*), but anthropologists do not know which one. Specimens usually assigned to *H. rudolfensis* had more modern skull characteristics, but specimens usually assigned to *H. habilis* had more modern dental characteristics. Some populations of *H. habilis* coexisted with the earliest populations of later *Homo* species, such as *H. ergaster.* It is also clear that *H. habilis* evolved from an earlier ancestor (see AUSTRALOPITHECINES), although which one is also unclear. All of the gracile australopithecines may have been extinct by the time *H. habilis* became common, but the robust australopithecines coexisted with *H. habilis. H. habilis* may therefore represent a state of evolution, transitional between australopithecines and later humans, rather than a single species.

The evolutionary state of *H. habilis* differed from that of the australopithecines in two important respects:

- *Brain size. H. habilis* represents the first significant, and somewhat rapid, advancement in brain size over the range represented by australopithecines and all modern nonhuman apes. The Kenya National Museum specimen 1813 (KNM-ER 1813), usually assigned to *H. habilis,* had a cranial capacity of 32 cubic inches (510 cc), and the KNM-ER 1470 specimen (see figure on page 191), usually assigned to *H. rudolfensis,* had a cranial capacity of 47 cubic inches (750 cc), both significantly exceeding the approximate value of 22 cubic inches (350 cc) that represents australopithecines and modern nonhuman apes.
- *Stone tools.* Australopithecines probably used stick and stone tools, just as modern monkeys and chimpanzees do.

H. habilis was the first species to deliberately fashion stone tools. The simple Oldowan tools (see photo below) do not look much different from natural rocks, but microscopic analysis shows bulbs of percussion where the tools were deliberately struck to form sharp surfaces.

The fact that the oldest Oldowan tools predate the earliest *H. habilis* fossils brings up the possibility that the tools were made by other hominins, such as the *Paranthropus* robust australopithecines. However, even though the hands of *Paranthropus* were probably capable of making stone tools, *Paranthropus* did not make stone tools in other parts of its range, or later in its history. The best explanation is that Oldowan tools were more likely to be preserved than *H. habilis* fossils, and that scientists have simply not yet found fossils from the earliest *H. habilis* populations.

The increase in brain size, the consumption of meat, and the use of stone tools may all have depended upon one another. Larger brains allowed *H. habilis* to locate carcasses by watching vultures and by learning the habits of leopards, which stash their prey in trees. The meat and marrow provided fats and calories that were important to the growth of the brain, which is a physiologically expensive organ. Larger brains allowed the ability to design and produce stone tools, and the tools unlocked new sources of calories for the brain. Once brain size had begun to increase, natural selection favored the ability to find and process carcasses, which in turn created a selective advantage for yet larger brains. The evolutionary increase in brain size may therefore have resulted from a positive feedback process.

Most of the increase in the size of the *H. habilis* brain over that of australopithecine ancestors, and the design of stone tools, occurred early in the evolutionary history of *H. habilis*. After an initial burst of brain growth and creativity in tool production, *H. habilis* may have experienced a million year anatomical and cultural stasis (see PUNCTUATED EQUILIBRIA). According to paleontologists such as Steven Stanley, the initial burst of evolution that produced *H. habilis* may have resulted from rapid climate change that made much of Africa cooler and drier, with the resulting spread of savannas at the expense of forests, about two and a half million years ago. The next stage of human evolution, which produced species such as *H. ergaster*, occurred when another climate change spread across Africa, resulting in yet more extensive savannas.

These Oldowan stone tools, found in East Africa, represent the earliest stone tool technology in the human genus. ***(Courtesy of Katherine Schick and Nicholas Toth, Indiana University)***

Further Reading

Johanson, Donald, and James Shreeve. *Lucy's Child: The Discovery of a Human Ancestor.* New York: William Morrow, 1989.

Leakey, Richard, and Roger Lewin. *Origins Reconsidered: In Search of What Makes Us Human.* New York: Doubleday, 1992.

Stanley, Steven M. *Children of the Ice Age: How a Global Catastrophe Allowed Humans to Evolve.* New York: Harmony Books, 1996.

Tattersall, Ian, and Jeffrey Schwartz. *Extinct Humans.* New York: Westview Press, 2000.

Homo heidelbergensis This species of humans, named after the Heidelberg man fossils found in Germany in 1907, is intermediate between *HOMO ERGASTER* (the African "erectus") and the two large-brained human species. *H. heidelbergensis* evolved in Africa from *H. ergaster* ancestors, principally by an increase in brain size from about 56 to about 75 cubic inches (900 to 1,200 cc). Numerous specimens of this species have been found (see table).

Some *H. heidelbergensis* populations migrated to Europe. *H. heidelbergensis* may not have been the first human species to migrate from Africa to Europe. Acheulean stone flakes found at 'Ubeidiya near the Jordan River are between 1.0 million and 1.4 million years old and may represent northward and eastward migrations of *H. ergaster*. The human remains found at Gran Dolina (the "large depression"), one of the limestone caves of the Sierra de Atapuerca of Spain, may have been descendants of an early, failed migration of *H. ergaster* into Europe. Anthropologist Juan Luis Arsuaga recognizes these populations as a separate species, *H. antecessor* ("explorer"). These specimens represent at least six humans, and numerous other animals, that died about 780,000 years ago. About a quarter of the human bones have chop and cut marks, which suggests cannibalism. (See the table for examples of possible intermediates between *H. ergaster* and *H. heidelbergensis*.)

H. heidelbergensis persisted in Europe until about a quarter million years ago, by which time some of them had evolved into the earliest *H. neanderthalensis* (see NEANDERTALS) and the rest had become extinct. Skeletal remains of individuals intermediate between *H. heidelbergensis* and Neandertals have been found in Europe, such as at Petralona in Greece (see MISSING LINKS) and the Sima de los Huesos (the bone pit) in the Sierra de Atapuerca of Spain. At 400,000 years of age, the Sima de los Huesos remains strongly resemble *H. heidelbergensis*, particularly in brain size, but have some Neandertal characteristics as well: in particular, a projecting face and an oval of porous bone on the rear of the skull. Although the age distribution of these 28 or more individuals is not what would be expected from normal death, they show no evidence of violence or cannibalism. All of the bones are broken, but some researchers believe that whole bodies were originally thrown in and therefore

Selected Examples of *Homo heidelbergensis* and Related Human Sites

Sites	Country	Approximate age (thousand years ago)
Sites possibly intermediate between *Homo ergaster* and *H. heidelbergensis*:		
Dmanisi	Georgian Republic	1,800
'Ubeidiya	Jordan	1,000
Ceprano	Italy	800
Gran Dolina	Spain	780
Homo heidelbergensis sites:		
Boxgrove	England	500
Mauer (Heidelberg)	Germany	500
Terra Amata	France	400
Menez-Dregan	France	400
Bilzingsleben	Germany	400
Terefine	Algeria	400
Vértesszöllös	Hungary	350
Schöningen	Germany	350
Sites possibly intermediate between *H. heidelbergensis* and *H. neanderthalensis*:		
Arago	France	450
Sima de los Huesos	Spain	400
Swanscombe	England	250
Steinheim	Germany	200
Petralona	Greece	200
LaChaise-Suard	France	200
Fontéchevade	France	200
Baude l'Aubesier	France	200
Brache-Saint-Vaast	France	200
Pontnewydd	Wales	200
Ehringsdorf	Germany	200
Altamura	Italy	200
Krapina	Croatia	120
Saccopastore	Italy	120
Sites possibly intermediate between *H. heidelbergensis* and *H. sapiens*:		
Bodo	Ethiopia	600
Sterkfontein	South Africa	250
Berekhat Ram	Syria/Israel	230
Kabwe	Zambia	175

probably represent a primitive form of hygienic disposal. Their descendants, the Neandertals, dug shallow graves for dead individuals. (See the table for examples of possible intermediates between *H. heidelbergensis* and Neandertals.)

Some of the *H. heidelbergensis* populations that remained in Africa evolved into HOMO SAPIENS. Skeletal remains of individuals intermediate between *H. heidelbergensis* and *H. sapiens* have been found in Africa, such as the Kabwe skull (see MISSING LINKS). (See the table for examples of possible intermediates between *H. heidelbergensis* and modern humans.)

Some of the artifacts of *H. heidelbergensis* suggest a technology intermediate between *H. ergaster* and their evolutionary descendants (Neandertals and modern humans):

- The Acheulean stone tools (named after St. Acheul in France, an early site of their discovery) resemble those of late *H. ergaster*.
- At least in Europe, *H. heidelbergensis* made wooden spears. If they were hunters, they must have had cooperation and advanced social structure. At a coal mine near Schöningen in Germany in 1994, anthropologist Hartmut Thieme found a stick, from the *H. heidelbergensis* period, that had been sharpened at both ends. Less than two weeks remained before the coal company planned to use a rotor digger on the site. Thieme's discovery allowed him to get permission to excavate for another year. He found three fire-hardened wooden javelins, each over six feet (2 m) in length, each made from the trunk of a spruce sapling. They were 350,000–400,000 years old. They were too flimsy to be spears used for thrusting, and their aerodynamically advanced shape suggested that they were thrown. Nearby bones suggest that horses were the prey. At a site near Lehringen, also in Germany, a 125,000-year-old spear was found embedded in elephant ribs. Spears have also been found at the Bilzingsleben site.
- The production of art by *H. heidelbergensis* is controversial. Some anthropologists consider a small rock found at Berekhat Ram in the Golan Heights between Syria and Israel to be a female figurine, while others consider it to be just a rock that happens to look like a figure.

H. heidelbergensis also constructed huts and made use of fire. The Terra Amata site in France provides evidence that humans may have created large oval huts by bending saplings over and tying them together at the top. While the use of fire by *H. ergaster* is controversial, there is no doubt that *H. heidelbergensis* had mastered the control of fire. There is evidence of the controlled use of fire 790,000 years ago in what is now Israel, and hearths inside of huts have been found from Terra Amata and Bilzingsleben.

Further Reading

Arsuaga, Juan Luis. *The Neanderthal's Necklace: In Search of the First Thinkers.* New York: Four Walls Eight Windows, 2001.

Goren-Inbar, Naama, et al. "Evidence of hominin control of fire at Gesher Benot Ya'aqov, Israel." *Science* 304 (2004): 725–727.

Klein, Richard G., and Blake Edgar. *The Dawn of Culture: A Bold New Theory on What Sparked the "Big Bang" of Human Consciousness.* New York: John Wiley and Sons, 2002.

Tattersall, Ian, and Jeffrey Schwartz. *Extinct Humans.* New York: Westview, 2000.

This harpoon tip, carved from ivory almost 90,000 years ago, was found in Katanda, Zaire. It demonstrates that *Homo sapiens* had the capability of producing complex craftsmanship at this early date. *(Courtesy of Chip Clark, Smithsonian Institution)*

Homo rudolfensis *See* HOMO HABILIS.

Homo sapiens The modern human species *Homo sapiens* is the only surviving evolutionary lineage of HOMININS. *H. sapiens* evolved in Africa from *HOMO HEIDELBERGENSIS* ancestors. Several examples have been found, such as the Kabwe and Bodo skulls, of human remains that are intermediate between *H. heidelbergensis* and *H. sapiens* (see MISSING LINKS). Humans had reached their modern anatomical form by about 200,000 to 100,000 years ago.

Cultural evolution of *Homo sapiens* in Africa began early and progressed slowly. Remains of modern humans who lived 100,000 years ago at what is now the Klasies River Mouth at the tip of South Africa provide no evidence of artistic culture beyond that of *H. heidelbergensis.* However, a harpoon tip made from bone about 90,000 years ago, found at Katanda in Zaire, exhibits workmanship far in advance of the Achculean technology of *H. heidelbergensis* (see photo at left). A piece of red ochre, found in Blombos Cave in South Africa, was inscribed with intersecting lines 77,000 years ago (see figure on page 195). Ostrich egg shell beads inscribed with patterns date back 50,000 years. Artistic productions in Africa before 10,000 years ago are rare, perhaps because of the decomposition of the media in which they were produced. Shell beads discovered in the Middle East may be 100,000 years old and may represent the oldest art of *H. sapiens*.

Some *Homo sapiens* populations migrated out of Africa. When they did so, they often encountered other human species. By 90,000 years ago, they had already moved northward into what is now the Middle East. They coexisted for almost 60,000 years with NEANDERTALS who had evolved in Europe and moved southward. Although anatomically distinct and apparently not interbreeding, modern humans and Neandertals at that time had similar tool technologies and lifestyles. Pierced shells, suitable for bodily decoration, date back 70,000 years. A 50,000-year-old flint plaque inscribed with concentric circles was found in Israel. The shells and plaque were probably from *H. sapiens* but the possibility that Neanderthals made them cannot be discounted. Modern humans encountered *HOMO ERECTUS* (and perhaps also the FLORES ISLAND PEOPLE) in Indonesia and may have arrived near what is now Mungo in Australia more than 60,000 years ago. Almost immediately, *H. sapiens* began producing rock art in Australia.

When populations of *Homo sapiens* migrated from the Middle East into Europe beginning about 50,000 years ago, their encounter with European Neandertals may have stimulated an explosion of artistic productions such as cave art, sculptures, ceramics, and body adornments (see CRO-MAGNON). Many anthropologists (such as Christopher Stringer and David Lewis-Williams) suggest that this rapid cultural explosion of *H. sapiens* may have resulted first from competition with Neandertals, then, after the extinction of the Neanderthals, from competition among human tribes, to establish their distinctiveness and their superiority over others. Eventually, advanced art, culture, and in many cases agriculture evolved independently in *H. sapiens* populations in every part of the world (see AGRICULTURE, EVOLUTION OF).

Racial Diversity

It has always been difficult to either classify or explain the diversity of human appearance. Before the development of MENDELIAN GENETICS, it was not clear to most observers that racial characteristics passed from one generation to another without environmental modification (see LAMARCKISM). From Greek philosophers such as Aristotle to 18th-century scientists such as the botanist Karl von Linné (see LINNAEUS, CAROLUS) and the anatomist naturalist J. F. Blumenbach, the prevailing idea was that dark skin was caused by the bright sunlight of tropical regions, and that black people would become lighter and white people darker within a few generations of moving to a new climatic region. Furthermore, as reflected in religious beliefs such as the creation account in Genesis chapter 1, the prevailing idea was that all humans shared a common origin. The prescientific and early scientific view was that humans formed a single species and that human physical diversity was labile.

This did not prevent scientists from classifying human races. Linnaeus gave the name *Homo sapiens* (wise man) to the human species in 1758 and divided this single species into four races. From his Enlightenment view of the unity of all knowledge, Linnaeus tied racial diversity into geography, physiology, and behavior. Linnaeus divided the four races on the basis of skin color and posture, the four geographical regions of the world, the four humors of classical physiology, and four behavioral generalizations:

- Americans (Indians) are red and upright; the choleric (angry) humor dominates their personality.
- Europeans are white and muscular; the sanguine (cheerful) humor dominates their personality.
- Asiatics are yellow and stiff; the melancholy (sad) humor dominates their personality.
- Africans are black and relaxed; the phlegmatic (lazy) humor dominates their personality.

Even though the prejudice and racism of these attributes are obvious to modern scientists, Linnaeus did not apparently mean to imply a hierarchy of humanness or superiority. (Linnaeus also included human varieties and even a different human species that were based upon rumors about gorillas, orangutans, and abandoned "wolf children": variety *ferus,* variety *monstrosus,* and *Homo troglodytes.*)

Still less did Linnaeus's disciple J. F. Blumenbach espouse a hierarchy of superiority. He collected an entire library of the writings of black authors, a practice almost unique in his day. He greatly admired the character of current and former slaves who were good people despite the horrors that they endured. Blumenbach held views, however, that would be considered racist today. Like almost everyone else in his day, he considered Europeans to be the pinnacle of beauty and wisdom. Unlike most others, he believed that other races could transform to meet the European standard. Blumenbach then reclassified human races in a way that permitted scientific support for the very racism that he deplored. He defined Europeans as the original and perfect form, from which humans developed into other forms under the influence of habit and climate: through the American Indian form into the Asiatic form, and through the Malay form into the African form. In order to accommodate this model, Blumenbach had to invent a fifth category, the Malay, that Linnaeus had not recognized. In 1795 Blumenbach assigned the name "Caucasian" to the European race because, he believed, the people of the Caucasus Mountains were the most beautiful, and geographically close to the supposed Middle Eastern center of human creation.

Many scholars (see, for example, Cuvier, Georges) used racial distinctions as a basis for declaring darker races

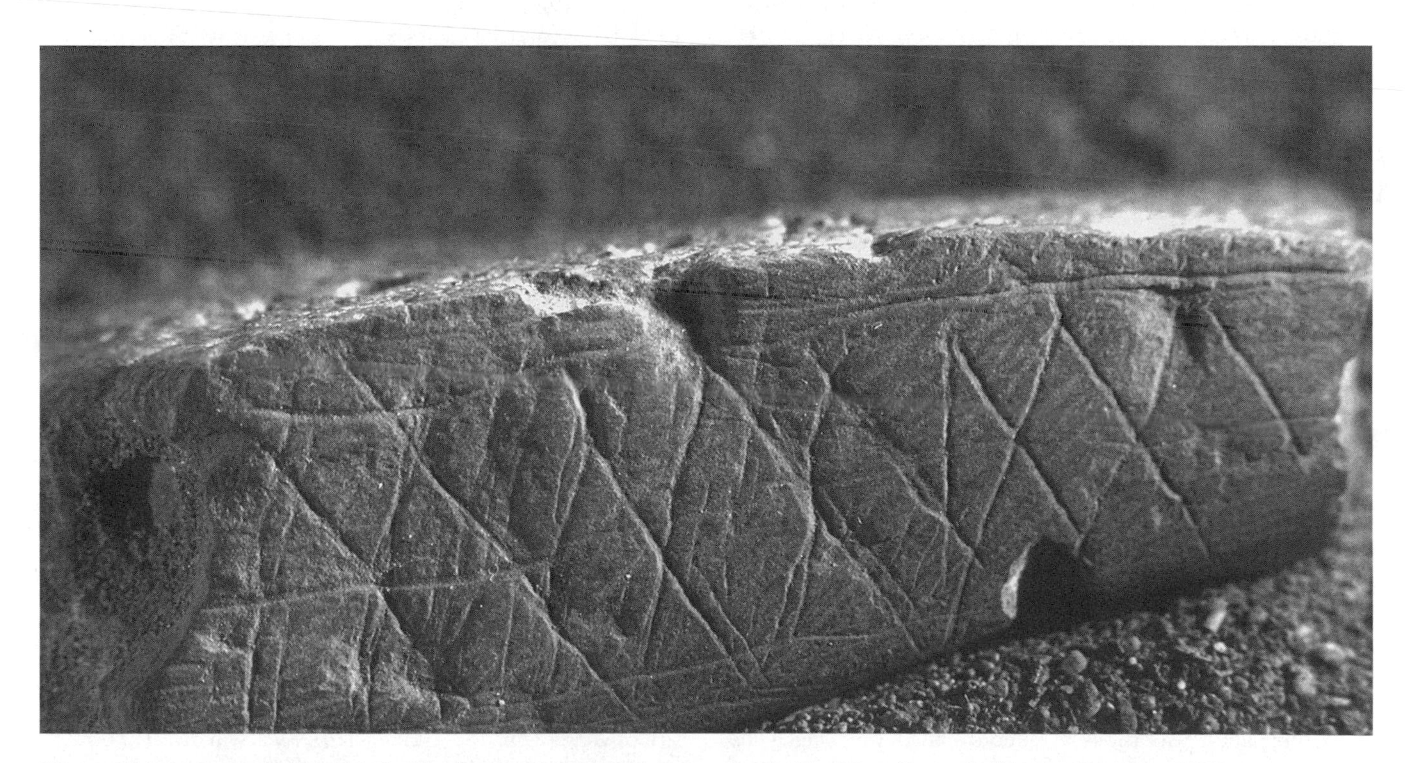

This small block of red ochre, found in the Blombos Cave in South Africa, was carved with abstract figures by *Homo sapiens* about 75,000 years ago. ***(Courtesy of Christopher Henshilwood, African Heritage Research Institute and University of Bergen)***

to be inferior. After the acceptance of an evolutionary origin of humans (see *DESCENT OF MAN* [book]), many scholars simply translated pre-evolutionary racist views into an evolutionary context by claiming either that darker races had a separate evolutionary origin from the European race, or that the darker races represented inferior genetic variation within the human species (see EUGENICS).

Even before widespread evolutionary thought, a few scholars insisted upon the essential equality of all races. German anatomist Friedrich Tiedemann measured the brain sizes from skulls of whites, blacks, and orangutans. Many scholars considered blacks intermediate between whites and baboons, perhaps even more nearly resembling the latter than the former. Tiedemann found that, while there were slight differences in average brain size among races, the range in brain sizes among individuals within each race was an order of magnitude greater than the average differences among the races, which is to say that there is no significant difference in brain size or intelligence among the races. He found that all human races had larger brains than orangutans. Tiedemann claimed that the supposed inferiority of blacks was based upon two scientifically unacceptable procedures. First, some researchers had chosen small black skulls to compare to large white skulls, rather than choosing a large and representative sample from each group. Second, the black people whom scientists had studied were often slaves or former slaves, whose characteristics were the result of abuse and deprivation rather than natural inferiority. Today, scientists are careful to choose samples that are large, representative, and unbiased (see SCIENTIFIC METHOD). Tiedemann published his results in 1836 in English to praise the British government for abolishing slavery.

The *Descent of Man* (see DARWIN, CHARLES) not only presented evolution as an explanation for the origin of human races but also presented a new reason for them. Darwin claimed that some racial differences reflected NATURAL SELECTION, most notably the darker skins of races that live in climates with bright sunshine. He suggested, however, that most racial differences were arbitrary, as each race, in its own separate line of evolution, made different choices of what constituted physical beauty (see SEXUAL SELECTION). Such arbitrariness of racial diversification strongly implied that no race was superior in its evolutionary adaptation to any other.

The development of molecular techniques, allowing comparisons of protein structure and of DNA (see DNA [EVIDENCE FOR EVOLUTION]), created a revolution in thinking about the races of *Homo sapiens*. First, geneticists found that the protein and DNA diversity within the human species was much less than within most animal species. This result implied the recent origin of the entire human species. Even when the relatively small amount of human DNA variation is considered, roughly 85 percent of that variation is within races; all differences between races occur in the remaining 15 percent. Further study of human genetics suggested that the human species had gone through a genetic bottleneck, in which population size and genetic variation was severely reduced and during which the human species came close to EXTINCTION (see FOUNDER EFFECT; POPULATION GENETICS). Geneticists have suggested a date of about 70,000 years ago for this event. All of this evidence suggests that all human racial diversity has evolved in less than 0.1 million years, a miniscule length of evolutionary time. Scientists continue some interesting studies of human racial differences (for example, why some races are more prone to diabetes or heart disease than others), but evolutionary scientists no longer have much interest in racial differences, since these differences are so small and, in most cases, literally only skin-deep.

The fact that *Homo sapiens* has been the only surviving hominin species for the last 20,000 years considerably simplifies EVOLUTIONARY ETHICS and ANIMAL RIGHTS. No one has any idea how a human legal system might deal with Neandertals or Heidelberg people. With the possible exception of cetaceans, no other animal challenges the mental superiority of humans (see INTELLIGENCE, EVOLUTION OF). This was not the case prior to 20,000 years before the present. *H. sapiens* and their evolutionary ancestors may have actively killed off other hominin species that they encountered. The belief of early *H. sapiens* that they were superior created a situation in which they eliminated, directly or indirectly, all challenges to this belief.

Further Reading

Balter, Michael. "Are humans still evolving?" *Science* 309 (2005): 234–237.

Gould, Stephen Jay. "The geometer of race." Chap. 26 in *I Have Landed: The End of a Beginning in Natural History.* New York: Harmony Books, 2002.

———. "The great physiologist of Heidelberg." Chap. 27 in *I Have Landed: The End of a Beginning in Natural History.* New York: Harmony Books, 2002.

Holden, Constance. "Oldest beads suggest early symbolic behavior." *Science* 304 (2004): 369.

Lewis-Williams, David. *The Mind in the Cave.* London: Thames and Hudson, 2002.

Tattersall, Ian. *Becoming Human: Evolution and Human Uniqueness.* New York: Harvest Books, 1999.

Vanhaeren, Marian, et al. "Middle paleolithic shell beads in Israel and Algeria." *Science* 312 (2006): 1785–1788. Summary by Balter, Michael. "First jewelry? Old shell beads suggest early use of symbols." *Science* 312 (2006): 1731.

Hooker, Joseph Dalton (1817–1911) British *Botanist* Sir Joseph Dalton Hooker was a botanist who was one of the closest confidantes of Charles Darwin (see DARWIN, CHARLES) especially as Darwin prepared for the inevitable controversy connected with publishing a theory of evolution. Born on June 30, 1817, Joseph Hooker was the son of the prominent botanist Sir William Hooker. He began attending his father's botany lectures at the University of Glasgow when he was seven years old. He was especially interested in tales of travel and the plants of distant lands, which would grow into the governing passion of his career.

Like many other biologists at the time, Joseph Hooker was a physician, obtaining his medical degree from the University of Glasgow in 1839. Like Charles Darwin and T. H. Huxley (see HUXLEY, THOMAS HENRY), Joseph Hooker began his naturalist's career by taking a long ship voyage. Unlike Darwin, Hooker could not pay his own way. Like Huxley, Hooker was hired as ship surgeon. On board HMS *Erebus,* Hooker sailed in 1839 for the southern seas, exploring islands around Antarctica during the summers and visiting New Zealand and Van Diemen's Land (now called Tasmania) during the winters.

When he returned to Great Britain in 1843, Hooker assisted his father, who had become the director of Kew Gardens. Under the directorship of the elder Hooker, Kew became one of the world's leading centers of botanical research, which it remains. Joseph Hooker had no regular employment but managed to obtain money to pay for the publication of books that documented the numerous plants that he discovered on the islands of the southern seas.

Joseph Hooker obtained a government grant to cover the expenses of an expedition to the Himalayas, from 1847 to 1849. He also visited India from 1850 to 1851. Hooker accompanied a mapmaking expedition. The expedition crossed the border into Tibet, which they were not authorized to do, and Hooker was arrested along with the others. The prisoners were released only when the British government threatened force. While a botanist may not have seemed much of a threat to local Asian governments, Hooker's research actually helped to advance British colonial interests. His father William Hooker had supervised the illegal transfer of *Cinchona* feverbark trees (the source of quinine) from Brazil to British plantations in India. Joseph Hooker supervised a similar illegal transfer from Brazil to India, this time of *Hevea* rubber trees. The new kinds of rhododendron Joseph Hooker brought back from the Himalayas became a profitable commodity for the gardens of England.

Joseph Hooker's botanical research caught the attention of his hero, Charles Darwin. Because of his irregular employment (he was mainly occupied with publishing his discoveries from the Himalayas and India), Joseph Hooker could spend a lot of time with Darwin and answer Darwin's many questions about the distribution of plants. Charles Darwin very much appreciated Hooker's assistance, and Emma Darwin made him feel welcome at their home.

Hooker and Darwin did not always agree. The BIOGEOGRAPHY of southern hemisphere plants was puzzling. Hooker suggested that it could be explained if the separate lands of the southern hemisphere, such as New Zealand, Australia, and Tasmania, had once been connected together in a single continent which had subsequently sunk beneath the sea. Darwin rejected Hooker's speculation, preferring to say that the various species of plants had floated across the seas to get to their present locations. This discussion prompted Darwin to begin a landmark series of experiments regarding how long various kinds of seeds could tolerate soaking in saltwater. Darwin was surprised and pleased to report that many kinds of seeds could germinate quite well after long periods of inundation in brine. Today scientists know that Hooker was right, not Darwin. The southern islands and continents were once part of a southern supercontinent that broke up and moved to its current locations (see CONTINENTAL DRIFT).

Sir Joseph Hooker was one of the few people (along with geologist Sir Charles Lyell and American botanist Asa Gray; see LYELL, CHARLES; GRAY, ASA) with whom Darwin dared to discuss his theory of evolution by means of natural selection. It was to Hooker that Darwin said that his belief in transmutation of species was almost like confessing a murder. In the midst of their discussions, Hooker interviewed for a professorship in Edinburgh. Darwin was relieved when Hooker did not get the job and had to stay around and help him. Joseph Hooker became assistant director of Kew Gardens in 1855. Upon the death of his father in 1865, Hooker became the director. Darwin no longer needed to fear the departure of his friend. Darwin later said that Hooker was "the one living soul from whom I have constantly received sympathy." Darwin and Hooker both experienced the tragic loss of a young daughter.

Joseph Hooker married the daughter of John Stevens Henslow, the botany professor from whom Darwin learned so much. Hooker died December 10, 1911.

Further Reading

Endersby, Jim. *Sir Joseph Dalton Hooker.* Available online. URL: http://www.jdhooker.org.uk. Accessed April 11, 2005.

horizontal gene transfer Most genes are passed vertically from one generation to another by sexual reproduction (see REPRODUCTIVE SYSTEMS), within a species or during HYBRIDIZATION between closely related species. Genetic information can also move horizontally among different species of organisms by means other than sexual reproduction. Though much rarer than vertical gene transmission, horizontal gene transfer has occurred many times. It does not alter the process of NATURAL SELECTION, which acts upon genetic variability in populations, whatever its source, but it does alter the scientific understanding of where genetic variability comes from. Most evolutionary explanations assume that genetic variability is enhanced mostly by MUTATIONS occurring in the genes of individuals in a population, or by immigrants from outside the population but from the same species (see POPULATION GENETICS). With horizontal gene transfer, a whole new source of genetic variation appears: injection of genes from entirely different species.

For several decades scientists have known that bacteria can exchange segments of DNA with one another (see BACTERIA, EVOLUTION OF). In a process called *conjugation,* one bacterium grows a tube that connects with another bacterium, and segments of DNA (usually small circles of genes called *plasmids*) travel through the tube. Bacteria can also absorb intact chunks of DNA from their fluid environments, although usually this requires some sort of temperature or chemical shock to make their cell membranes receptive to the transfer. Both conjugation and the absorption of DNA segments increase under conditions of environmental stress. This allows populations of bacteria to increase their variability of genetic combinations.

The surprise came when microbiologists realized that conjugation and other forms of gene transfer could occur between bacterial species. At first this was considered an interesting but unimportant aberration, but since antibiotic resistance genes are often on plasmids, interspecific conjugation turned out to be very important indeed: Antibiotic-resistant bacteria can make other bacteria, even of a different species, resistant to the antibiotic (see RESISTANCE, EVOLUTION OF). This is how antibiotic resistance can spread so rapidly, even from one species of bacterium to another. Recent research suggests some pathogenic bacteria have obtained resistance genes from bacteria that live in the soil. Horizontal gene transfer has turned out to be one of the central facts in public health research (see EVOLUTIONARY MEDICINE). It is widespread enough that one prominent evolutionary biologist (see MARGULIS, LYNN) has claimed that there is really only one species of bacterium: According to the biological species concept, potentially interbreeding organisms can, if they produce fertile offspring, be considered members of the same species.

When biologists began to determine the nucleotide sequences of different species of bacteria, they were surprised by a further discovery: Horizontal gene transfer has affected a large portion of the DNA of many bacteria. In the gut bacterium *Escherichia coli,* one of the most common bacteria used in education and research, 755 of the 4,288 genes (18 percent) have resulted from 234 horizontal gene transfers during the past 200 million years. This fact was not known until the exact sequence of nucleotides could be determined and compared with those of other bacterial species. Horizontal gene transfer occurs mostly between bacteria that are closely related to one another (see TREE OF LIFE). However, it can even occur between bacteria and ARCHAEBACTERIA. Horizontal gene transfer allows rapid evolutionary response to new environmental conditions. For example, some species of bacteria are able to metabolize and destroy naphthalene, the main ingredient of mothballs, and other species of bacteria cannot. When both kinds of bacteria are mixed and exposed to naphthalene, within 24 hours the first species confers this genetic ability to the second species. It does not need to happen very often; just a single horizontal gene transfer creates a new variant of the second species, the populations of which can then explode.

Viruses can also participate in horizontal gene transfer. Sometimes, when two kinds of virus inhabit the same host cell, they exchange genetic material. This is how human influenza viruses can obtain new genes from the influenza viruses of other species, and why new influenza vaccines are needed each year. Sometimes a virus, when leaving a host chromosome to travel elsewhere, can take some host DNA with it.

Horizontal gene transfer is not limited to viruses and bacteria, although microbes frequently aid in the process. As noted above, viruses can take some of their host DNA with them as they leave, and put it into the chromosome of a new host. Since some viruses (like influenza viruses) have more than one species of host, they may transfer genes from one host species to another. One species of bacterium, *Agrobacterium tumifaciens,* infects plants and causes a swelling known as crown gall on the stems of several kinds of plants. It has a plasmid that inserts into the chromosomes of its host plant. If the plasmid carries the gene from one species of plant host, it can insert this gene into another species of plant host.

There is even some evidence that RNA molecules can travel through the phloem tissue of plants, which contains the vessels that carry sugar solution in plant stems. It is conceivable that RNA molecules from one species of plant can transfer directly to the cells of another species of plant, so long as the two plants are in direct phloem contact. This is the situation with some parasitic plants (see COEVOLUTION). Perhaps the ultimate plant parasite is *Rafflesia,* a Sumatran plant that has lost its leaves, stems, roots, and chlorophyll and lives inside the stems of *Tetrastigma,* a tropical grape vine. *Rafflesia* looks like strands of fungus until it produces a flower: the biggest flower in the world, nearly three feet (a meter) across, and smelling like rotten flesh, which attracts flies as pollinators. For most of its life, *Rafflesia* soaks up whatever organic molecules are in the phloem of its host plant, and this apparently includes RNA. If retroviruses are present (as they commonly are in cells), the RNA can be reverse transcribed into DNA which can insert into the host chromosomes. This is what HIV, a retrovirus, does to human hosts (see AIDS, EVOLUTION OF). In this situation, it might not surprise scientists that *Rafflesia* might get new genes, not by evolving them itself but by acquiring them from its host.

New research indicates that when the mitochondrial *matR* gene, the 18S ribosomal gene, or the *PHYC* gene is used in a phylogenetic analysis (see CLADISTICS), *Rafflesia* is found to be closely related to other plants in the order Malphigiales. However, if the mitochondrial *nad1B-C* gene is used, *Rafflesia* is found to be closely related to grapevines. The anomalous result of the cladistic analysis using *nad1B-C* is best explained by saying that the *nad1B-C* gene, perhaps in the form of an RNA transcript, transferred from the *Tetrastigma* grapevine host to the *Rafflesia* parasite through the phloem. The host gene can then spread from the first parasite to many other parasites in the population, through normal sexual reproduction. Other research suggests that certain ferns obtained some of their mitochondrial DNA from flowering plants parasitic upon them.

Since *Rafflesia* lives only in *Tetrastigma,* the host genes will never go further than those two species. If a similar phenomenon occurs in a multi-host plant parasite like *Cuscuta,* might the parasite facilitate the transfer, however rarely, from one host species to another? Might parasites that act as disease vectors transfer genes from one species of animal to another? This is unlikely, because if a mosquito injected nucleic acid from another mammal species into a person's blood, the nucleic acid would probably be destroyed by the host's immune system.

There are some anomalies that have always puzzled geneticists. For example, there are some animals that produce cellulose. Cellulose is a major component of plant cell walls (and of cotton) and is assumed to be found only in plant and bacterial cells. But it is also the major component of the outer layer (the tunic) of tunicates, which are among the closest invertebrate relatives of the vertebrates (see INVERTEBRATES, EVOLUTION OF). An ancestor of these animals may have obtained the gene or genes for making cellulose by horizontal gene transfer from a bacterium, perhaps from cyanobacteria like those that are common in the food and environment of tunicates. Some scientists speculate that the few bacteria that produce cholesterol got the gene for it from eukaryotes. Scientists are only beginning to understand the many ways in which nature produces the genetic diversity upon which natural selection acts.

Further Reading

Arnold, Michael L. *Evolution Through Genetic Exchange.* New York: Oxford University Press, 2006.

Aminetzach, Yael T., et al. "Pesticide resistance via transposition-mediated adaptive gene truncation in Drosophila." *Science* 309 (2005): 764–767.

Davis, Charles C., and Kenneth J. Wurdack. "Host-to-parasite gene transfer in flowering plants: Phylogenetic evidence from Malphigiales." *Science* 305 (2004): 676–677.

———, William R. Anderson, and Kenneth J. Wurdack. "Gene transfer from a parasitic flowering plant to a fern." *Proceedings of the Royal Society of London B* 272 (2005): 2,237–2,242.

D'Costa, Vanessa M., et al. "Sampling the antibiotic resistome." *Science* 311 (2006): 374–377.

Doolittle, W. Ford. "Uprooting the tree of life." *Scientific American*, February 2000, 90–95.

Kazazian, Haig H., Jr. "Mobile elements: Drivers of genome evolution." *Science* 303 (2004): 1,626–1,632.

horses, evolution of Horses have been very important in human history, and the geological record of their fossils provides a beautiful illustration of evolutionary patterns.

Horses (including zebras), donkeys, and asses are *equines*. Equines, along with tapirs and rhinoceroses, are *perissodactyls* (odd-toed ungulates or grazing animals). The order Perissodactyla, like all the other extant mammalian orders, originated early in the TERTIARY PERIOD.

Major transitions that occurred during the evolution of horses included:

- *Increase in size.* The earliest identifiable ancestors of horses (*Hyracotherium*, in the Eocene epoch) were the size of a small dog.
- *Switch from browsing to grazing.* This transition is indicated by the fact that the earliest horse ancestors had teeth appropriate for browsing a wide variety of plant materials, while modern horses have teeth appropriate for grinding grasses. Grass is difficult to chew, as its cell walls contain a lot of silicon dioxide. A whole series of intermediate tooth forms are found in the horse lineage. Horses switched to grazing about the same time that ruminant mammal lineages (such as cows) did, which was also the time when extensive grasslands appeared upon the earth during a period of warm, dry climate.
- *Fusion of bones and loss of toes.* Modern horses run around on one middle toe (the hoof). Many of their ancestors had three toes. Even today, occasionally a horse is born in which the ancient genetic basis for having three toes is expressed (see DNA [EVIDENCE FOR EVOLUTION]).
- *Changes in leg joints.* In modern horses, the limbs can lock into position, allowing the horses to sleep standing up.

Primitive horses existed before North America separated very far from Eurasia. After the separation advanced, Eurasian horses became extinct. All subsequent horse evolution occurred in North America. *Mesohippus* and *Miohippus,* at 200 pounds (100 kg), were about 20 times bigger than the primitive *Hyracotherium* but smaller than modern horses. Although these horses had three toes on both front and back legs, the middle toe was the biggest one on each foot. Their leg bones were partly fused. These were all browsing animals.

The first grazing horses, such as *Merychippus,* still had three toes but the middle toes bore the weight of the body, the other toes being small and elevated. At a half ton (500 kg), these horses were the size of small modern horses. As conditions became even more arid, horses such as *Dinohippus* evolved. This genus had feet with one toe, although there were a few exceptions. *Equus* evolved into fully modern form less than five million years ago. Members of this genus spread from America throughout Eurasia, then became extinct in America near the end of the most recent ice age (see PLEISTOCENE EXTINCTION). There were no post-Pleistocene horses in America until the Spaniards reintroduced them in the 17th century. All wild horses in America are from escaped European animals.

The Ashfall Beds deposit in Nebraska contains a concentrated sample of intact mammal skeletons buried in volcanic ash 12 million years ago (see FOSSILS AND FOSSILIZATION). Among these mammals are horses. Within the population of *Pliohippus* horses, which all lived at the same place and time, some individuals had three toes, some had one. Variation existed within this population from which NATURAL SELECTION derived modern horse characteristics (see POPULATION GENETICS).

As the above description indicates, there is a wealth of intermediate animals between modern horses and the primitive *Hyracotherium* ancestor. As a matter of fact, there are so many that it is difficult to tell which evolutionary lineage led to modern horses. The evolutionary history of the horse is filled with numerous simultaneous lineages and extinctions.

Further Reading

Florida Museum of Natural History. "Pony express: Florida fossil horse newsletter." Available online. URL: http://www.flmnh.ufl.edu/ponyexpress/pe_newsletters.htm. Accessed July 20, 2005.

MacFadden, Bruce J. "Fossil horses—Evidence for evolution." *Science* 307 (2005): 1,728–1,730.

Hox genes *See* DEVELOPMENTAL EVOLUTION.

Hutton, James (1726–1797) Scottish *Geologist* James Hutton laid the foundation for UNIFORMITARIANISM in geology, a theory more fully developed by later geologists (see LYELL, CHARLES). Hutton was born June 3, 1726. Leaving his native Edinburgh, Hutton studied medicine but had difficulty obtaining employment. He decided to make a living by farming on inherited property. As he traveled through Europe to learn farming techniques, he studied the landscapes and began thinking about the processes of their formation. After he had established his farm in working order, he leased it and returned to Edinburgh to devote himself to scholarly pursuits, including geology.

Several observations made Hutton question CATASTROPHISM, the prevailing view of Earth history that attributed geological formations to a series of worldwide catastrophes. Hutton noticed that Hadrian's Wall had hardly eroded at all during the millennium and a half since its construction. This led Hutton to realize that geological processes such as erosion and the uplift of mountains were very slow. He also noticed that some features of the Earth's surface would have required a very long time to produce. In particular, Siccar Point, on the Atlantic Ocean near Edinburgh, was an UNCOMFORMITY in which an older, lower set of sediments had been deposited and turned to rock, and then a younger set of sediments had been deposited upon them and also turned to rock. This could not have occurred during a single biblical flood. Physicist Sir Isaac Newton and others had revealed a cosmos that appeared timeless and orderly; Hutton came to believe that the Earth was an ever-cycling machine that was just as timeless and orderly as the stars. Old continents erode, sediments are transformed into rock by heat and pressure of the deposits over them, new continents arise, the boundary between land and sea shifts, but the Earth continues. Hutton believed that the world had a beginning, a supernatural creation, and that it would have an end, but that neither the beginning nor the end was within the framework of natural cause and effect. Between the beginning and the end, the world operated by natural law. The Earth, he said, had "... no vestige of a beginning,—no prospect of an end."

In 1785 Hutton communicated his views to the Royal Society of Edinburgh in a paper entitled "Theory of the Earth, or an Investigation of the Laws Observable in the Composition, Dissolution and Restoration of Land upon the Globe." This became the basis of his 1795 book, *Theory of the Earth*. Many people found Hutton's attention to detail unreadable, and five years after Hutton died, John Playfair wrote a summary of Hutton's book, *Illustrations of the Huttonian Theory of the Earth,* in which form Hutton's views gained widespread attention.

Hutton's contributions were not limited to geology. He wrote a paper about rain which laid the framework for the current understanding of relative humidity and condensation. He also wrote about the nature of matter, fluids, cohesion, light, heat, and electricity. He wrote "An Investigation into the Principles of Knowledge" and "Elements of Agriculture." Hutton died March 26, 1797.

Further Reading

Baxter, Stephen. *Ages in Chaos: James Hutton and the True Age of the World.* New York: Forge, 2004.

Repcheck, Jack. *The Man Who Found Time: James Hutton and the Discovery of Earth's Antiquity.* New York: Perseus, 2003.

Huxley, Julian S. (1887–1975) British *Zoologist* Julian Sorell Huxley was born June 22, 1887. As the grandson of Thomas Huxley (see HUXLEY, THOMAS HENRY) and the son of writer Leonard Huxley, Julian Huxley grew up seeing scientists in action. During Sir Julian Huxley's early career, Darwinian evolution was fully accepted by scientists, but NATURAL SELECTION was not. MUTATION was considered to be an alternate method of the origin of species, rather than as raw material for the process of natural selection. In the 1930s, scientists began to integrate MENDELIAN GENETICS with natural selection. Several scientists explored the mathematical and theoretical basis of genetics and evolution (see FISHER, R. A.; HALDANE, J. B. S.; WRIGHT, SEWALL). Several other biologists, along with Julian Huxley (see DOBZHANSKY, THEODOSIUS; MAYR, ERNST; SIMPSON, GEORGE GAYLORD; STEBBINS, G. LEDYARD), gathered the scientific evidence that allowed the MODERN SYNTHESIS of genetics and evolution. Huxley's book, *Evolution: The Modern Synthesis,* was perhaps the central biological contribution to this synthesis.

Huxley also devoted great attention and effort to science education and to world issues in which science was involved. He was the first director-general of the United Nations Educational, Scientific and Cultural Organization (UNESCO) and one of the founders of the World Wildlife Fund. In particular he worked against racism; largely due to his efforts, the UNESCO statement on race reported that race was a cultural, not a scientific, concept, and that any attempts to find scientific evidence of the superiority of one race over another were invalid. His *Heredity, East and West* was one of the first exposés of LYSENKOISM. He collaborated with science fiction writer H. G. Wells to write *The Science of Life,* one of the first biology books for a popular audience. He also defended and wrote about the philosophy of humanism. One of his famous quotes about humanism and religion (in *Religion Without Revelation*) was that "... God is beginning to resemble not a ruler, but the last fading smile of a cosmic Cheshire Cat." Huxley made numerous television and radio appearances. He died February 14, 1975.

Further Reading

Huxley, Julian. *Religion Without Revelation.* New York: Harper, 1927. Repr. New York: New American Library, 1961.

———. *Evolution: The Modern Synthesis.* London: Allen and Unwin, 1942.

Wells, Herbert George, Julian S. Huxley, and George Philip Wells. *The Science of Life.* New York: Literary Guild, 1929.

Huxley, Thomas Henry (1825–1895) British *Evolutionary scientist* Thomas Henry Huxley was one of the first and most passionate defenders of Darwinian evolution, both in the scientific community and with the general public (see DARWIN, CHARLES; *ORIGIN OF SPECIES* [book]). Huxley thus earned the enduring description as "Darwin's bulldog." However, Huxley did not hesitate to disagree with Darwin on several points; in particular, he disagreed with Darwin about gradual transformation by NATURAL SELECTION as the principal mechanism of evolutionary change. As a zoologist and paleontologist, Huxley made many of his own contributions to evolutionary science. He also applied his scientific viewpoint to religion and ethics.

Born May 4, 1825, into a large and not affluent family, Thomas Huxley had limited formal education. However, he read voraciously in many scholarly fields. He won a scholarship to study medicine at a hospital in London. At age 21, he became the assistant surgeon on HMS *Rattlesnake,* during a voyage to map the seas around Australia. Huxley collected and studied marine invertebrates (see INVERTEBRATES, EVOLUTION OF). He became an expert in natural history the same way that Charles Darwin had on HMS *Beagle* and Joseph Hooker had on HMS *Erebus.* Huxley sent specimens and notes back to England, and when he returned in 1850 he found that these specimens and notes had made him a celebrity among English scientists, just as had happened with Darwin. He became acquainted with the leading natural scientists of his day (see LYELL, CHARLES; HOOKER, JOSEPH DALTON). Huxley supported himself on a Navy stipend and by writing popular scientific articles. In 1854 he was hired by the School of Mines as a lecturer, a position he retained until retirement.

Although he initially opposed evolution, especially the version proposed in the *Vestiges of Creation* (see CHAMBERS, ROBERT), Huxley became convinced of evolution by reading the *Origin of Species.* His reaction to it was something like, "How stupid of me not to have thought of that." Accustomed to lecturing and to popular writing, Huxley was ready to champion the theory that Charles Darwin was, due to a quiet nature and illness, reluctant to defend in public.

Huxley's most famous public moment was at the 1860 meeting of the British Association at Oxford University, where he debated Bishop Samuel Wiberforce, known as "Soapy Sam" because of the slippery debate maneuvers for which he was famous. Wilberforce knew little about science but had been coached by a famous paleontologist (see OWEN, RICHARD). Wilberforce could not resist a jab at Huxley, asking him whether he traced his descent from the apes on his mother's side or his father's side. Huxley muttered, "The Lord hath delivered him into my hands," referring to the harm Wilberforce had done to his own position by this statement. Huxley said, according to his own account, "If then the question is put to me would I rather have a miserable ape for a grandfather or a man highly endowed by nature and possessed of great means of influence and yet who employs these faculties and that influence for the mere purpose of introducing ridicule into a grave scientific discussion, I unhesitatingly affirm my preference for the ape." The debate was just one part of a series of speeches, and its importance was dramatically exaggerated in 20th-century accounts. Wilberforce has been portrayed as a vicious attacker of evolutionary scientists, while in reality Darwin and Wilberforce were on reasonably good terms. Wilberforce referred to Darwin as "a capital fellow." Furthermore, the main defense of Darwin's theory at this meeting was made by Joseph Hooker, not Huxley. However, the audience saw this meeting as much more than a scientific debate about zoology. The captain with whom Charles Darwin had sailed around the world walked around, holding up a Bible, saying, "The Book! The Book!" (see FITZROY, ROBERT).

Although he was a bulldog, Huxley was ready to admit mistakes and correct his viewpoints. He visited the United States to present lectures on the evolution of horses (see HORSES, EVOLUTION OF). Fossil horses from Europe seemed to indicate that the modern horse genus, *Equus,* had evolved there. American paleontologist Othniel C. Marsh, however, showed Huxley evidence that horses had evolved in North America, and some of them had migrated to Europe. Huxley completely and immediately rewrote his speeches. In another instance, Huxley became convinced that the ocean floor was covered with a formless ooze that might have been an amoebalike primitive organism, which he called *Bathybius haeckelii,* named after another famous paleontologist (see HAECKEL, ERNST). It turned out that the ooze was simply a chemical product of the reaction of mud with alcohol (used to preserve the specimens). Huxley also admitted this error.

Huxley took Darwin's theory further than Darwin had dared to propose in writing. Darwin wrote in the *Origin of Species* that "Much light will be thrown on the origin of man and his history." In 1963 Huxley published *Man's Place in Nature,* which explicitly presented the fossil evidence for the evolutionary place of humankind among the apes. Because little was known about human prehistory, aside from recently discovered skull fragments of NEANDERTALS, most of Huxley's evidence was to show the many anatomical similarities between humans and other primates. This brought him into conflict with Richard Owen, who insisted that the human brain had a structure, the hippocampus minor, that was not present in the brains of other primates. As more evidence became available about gorillas and chimpanzees, both recently discovered by Europeans, it became obvious that Owen was wrong.

Especially in his later years, Huxley gave much attention to ethics. He had rejected organized religion, yet was unwilling to conclude that God could not exist. The term *agnostic,* referring to someone who believes the existence of God cannot be known, was Huxley's invention. In *Evolution and Ethics,* Huxley claimed an evolutionary origin for human behavior but also noted that large brains give humans the ability to behave differently than other animals (see EVOLUTIONARY ETHICS). In saying that good ethics consists of resisting rather than embracing the tendencies of evolution, he placed his philosophy firmly against those of the social Darwinists (see SPENCER, HERBERT) and the eugenicists (see EUGENICS).

Huxley's passion for science and scholarship established a tradition in his family. One of his grandchildren (see HUXLEY, JULIAN) was one of the scientists who established the modern synthesis of evolution; another grandson, Andrew Huxley, won the Nobel Prize for his study of nerve impulses and muscle contraction; another grandson, Aldous Huxley, was a writer, most famous for *Brave New World.* Thomas Henry Huxley died June 29, 1895.

Examples of Interspecific Crosses in Animals and Plants (within Genera)

Parent	Parent	Hybrid
Jackass (*Equus asinus*)	Mare (*E. caballus*)	Mule
Jenny (*Equus asinus*)	Stallion (*E. caballus*)	Hinny
Male tiger (*Panthera tigris*)	Lioness (*P. leo*)	Tigon
Tigress (*Panthera tigris*)	Male lion (*P. leo*)	Liger
Leopard (*Panthera pardus*)	Lion (*P. leo*)	Leapon
Bluegill *(Lepomis macrochirus)*	Green sunfish *(L. cyanellis)*	Hybrid sunfish
Larkspur (*Delphinium hesperium*)	Larkspur (*D. recurvatum*)	*D. gypsophilium*
Penstemon (*Penstemon centrantifolius*)	Penstemon (*P. grinnellii*)	*P. spectabilis*
Japanese larch (*Larix kaempferi*)	European larch (*L. decidua*)	Dunkeld larch
Magnolia (*Magnolia denudata*)	Magnolia (*M. liliiflora*)	*M. soulangeana*
American sycamore (*Platanus occidentalis*)	European sycamore (*P. orientalis*)	London plane
Elm (*Ulmus glabra*)	Elm (*U. minor*)	*U. × hollandica*
Wingnut *(Pterocarya fraxinifolia)*	Wingnut *(P. stenoptera)*	*P. × rehderiana*
Linden *(Tilia cordata)*	Linden *(T. platyphyllos)*	*T. × europaea*
White poplar (*Populus alba*)	Aspen (*P. tremula*)	Gray poplar
Arbutus (*Arbutus unedo*)	Arbutus (*A. andrachne*)	*A. × andrachnoides*
Horse chestnut (*Aesculus hippocastanum*)	Horse chestnut (*A. pavia*)	*A. × carnea*

Further Reading

Gould, Stephen Jay. "Bathybius and Eozoon." Chap. 23 in *The Panda's Thumb: More Reflections in Natural History.* New York: Norton, 1980.

Huxley, Thomas Henry. *Man's Place in Nature.* London: Williams and Norgate, 1863. Reprinted with Introduction by Stephen Jay Gould. New York: Random House, 2001.

———. *Evolution and Ethics and Other Essays.* New York: Appleton, 1898.

Irvine, William. *Apes, Angels, and Victorians: Darwin, Huxley, and Evolution.* New York: McGraw-Hill, 1955.

hybridization Hybridization is the crossbreeding between different species of organisms. The term *hybrid* has also been used for crosses between varieties within a species (as in *hybrid corn* within the species *Zea mays*). The offspring with many heterozygous traits (see MENDELIAN GENETICS) produced by intraspecific crosses may produce *hybrid vigor.* However, evolutionary biologists generally restrict the term *hybrid* to crosses between species.

According to the biological species concept (see SPECIATION), distinct species do not normally produce fertile offspring. This is most obviously the case with crosses between donkeys and horses, resulting in mules, which are very strong but are sterile. Crosses between lions and tigers (such as ligers and tigons) have also occurred in zoos (see table above). It is possible that the elusive "marozi" of Africa is a naturally occurring hybrid between leopards and lions. Many other hybridizations are possible between feline species. While even fertile hybrids may be produced between different feline species, they are rare in the wild, which suggests that either the species nearly always avoid one another when mating and/or the hybrid offspring have reduced health or fertility (see ISOLATING MECHANISMS). Hybrids have also been reported between species of fishes. The bluegill *(Lepomis macrochirus)* can cross with the green sunfish *(L. cyanellis)*. Intergeneric hybrids, such as between pumas and leopards, and between dromedaries and llamas, have also been reported (see table on page 203). When a scientific name is assigned to a hybrid species, an × often precedes the specific name.

Hybrids between plant species in the same genus occur more often than hybrids between animal species. Hybridization in plants is not usually successful, for at least two reasons:

- The two species may pollinate at different times of the year. Among the oaks of North America, hybridization can occur among species of white oaks, and among species of red and black oaks (genus *Quercus*). However, all of the white oaks have acorns that mature during the same year, while most of the red and black oaks have acorns that mature the following year, after pollination. Because of these differences in reproductive pattern, hybrids between the two groups of oaks cannot occur. Hybridization between species of pine (the bishop pine *Pinus muricata* and the Monterey pine *P. radiata*), and between species of alders (the seaside alder *Alnus maritima* and the hazel alder *A. serrulata*) cannot occur, as pollination occurs at different times of the year.
- The two species may specialize on different habitats. Hybrids can and do occur within the white oak and within the black-red oak groups. Hybridization usually produces solitary unusual trees or scattered clusters with characteristics intermediate between the species. Occasionally there will be a widespread zone of hybridization between oak species. The oak species remain distinct probably because the hybrids are not adapted to either of the habitats of the parents. For example, even though post oak *(Q. stellata)* and bur oak *(Q. macrocarpa)* are both white oaks, hybrids between them would probably not grow well under natural

Examples of Intergeneric Crosses Reported from Animals and Plants

Parent	Parent	Hybrid
Leopard (*Panthera pardus*)	Puma (*Felis concolor*)	
Dromedary *(Camelus dromedarius)*	Guanaco *(Lama guanicoe)*	
Platygyra coral	Leptoria coral	
Wild wheat (*Triticum boeoticum*)	Goatgrass (*Aegilops speltoides*)	Durum wheat
Tetraploid wheat (*Triticum dicoccum*)	Goatgrass (*Aegilops squarrosa*)	Bread wheat
Bread wheat (*Triticum aestivum*)	Rye (*Secale cereale*)	*Triticale*
Radish (*Raphanus*)	Mustard (*Brassica*)	× *Raphanobrassica*
Cypress (*Cupressus*)	Cypress (*Chamaecyparis*)	× *Cupressocyparis*
Heuchera sanguinea	*Tiarella cordifolia*	× *Heucherella*
Purshia tridentata	*Cowania stansburiana*	*P. glandulosa*
Mandarin (*Citrus reticulata*)	Kumquat (*Fortunella*)	Calamondin
Cattleya intermedia orchid	*Laelia lobata* orchid	× *Laeliocattleya*
Laelia milleri orchid	*Brassavola nodosa* orchid	× *Brassolaelia*
Cattleya orchid	*Broughtonia* orchid	× *Cattleytonia*

conditions, since post oaks specialize on relatively dry, and bur oaks on relatively moist, habitats. The California black oak *(Q. kelloggii),* a tall, deciduous tree, can cross with the interior live oak *(Q. wislizenii* var. *frutescens),* an evergreen shrub, to produce the oracle oak *(Q. × morehus),* which is a small, partly deciduous tree with leaf shape and other characteristics intermediate between those of its parents (see figure below). All three species occur in the mountains of southern California, but the hybrid is rare. While most hybrids become extinct, recurrent origin maintains small populations of them.

If the hybridization event produces a plant with an odd number of chromosomes, the resulting plant is sterile. There are two ways in which this sterile hybrid can produce seeds.

- First, mutations may allow the plant to go ahead and produce seeds even without sexual reproduction. This process of *apomixis* produces seeds that are exact genetic copies of the sterile parent. The result may be a new species that never has sexual reproduction. Examples include the common dandelion *(Taraxacum oficinale)* and a South African sorrel *(Oxalis pes-caprae).*
- Second, if the reproductive cells undergo a spontaneous chromosome doubling, the resulting tetraploid plant (with chromosomes in groups of four) is no longer sterile: Each chromosome can match up with a copy of itself during MEIOSIS. The fertile hybrid plant can now produce seeds, and perhaps a small population of tetraploids. The hybrid tetraploid plants, though now fertile, have a different number of chromosomes than does either parent species. Therefore it cannot crossbreed with either of its parent species; it can only cross-pollinate with other members of the tetraploid population. The tetraploid population, reproductively isolated from the parental species, now functions as a new species.

Oak hybridization: (A) *Quercus kelloggii,* the California black oak; (B) *Quercus wislizenii,* the interior live oak; (C) the naturally occurring hybrid *(Courtesy of Wayne P. Armstrong)*

Examples of spontaneous chromosome doubling that has restored the fertility of the reproductive cells of the sterile plant are pickleweeds, tobaccos, mustards, and sunflowers:

- Cross-pollination between *Spartina maritima* (native to Europe) and *S. alterniflora* (native to California), two species of saltmarsh pickleweed, occurred when *S. alterniflora* was introduced into England. The hybrid, *S. × townsendii,* was sterile. Chromosome doubling in the sterile hybrid resulted in *S. anglia,* which now grows vigorously in English salt marshes.
- The common tobacco *Nicotiana tabacum* (with 48 chromosomes) can cross with another tobacco species, *N. glutinosa.* However, *N. glutinosa* has only 24 chromosomes. A sterile hybrid, with 36 chromosomes, results. Spontaneous chromosome doubling can occur in this sterile hybrid, resulting in seeds with 72 chromosomes. The result is the hybrid species *N. digluta.*
- Hybridization among three mustard species in the wild has produced three new species (see figure on page 204).

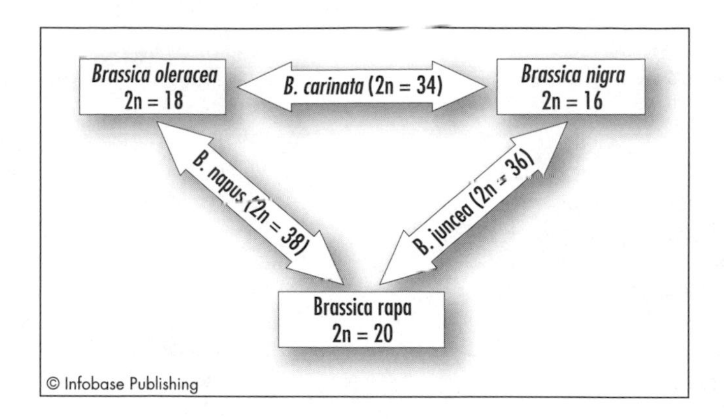

Three species of mustard *(Brassica carinata, B. juncea, B. napus)* have arisen as hybrids between *B. oleracea, B. nigra,* and *B. rapa.* Boxes represent original species; arrows represent hybrid species.

- The sunflowers *Helianthus deserticola* and *H. anomalus* have been shown to have hybrid origins. Botanist Loren Rieseberg has experimentally replicated the hybrid origin of *H. anomalus* in the greenhouse. By crossing *H. annuus* with *H. petiolaris,* Rieseberg produced three independent hybrid lines. Only four generations were required to select vigorous hybrids. The genetic markers (see QUANTITATIVE TRAIT LOCI) of the greenhouse hybrids closely resemble those in the wild populations of *H. anomalus.*

In plants, intergeneric hybrids (while still rare) are more common than in animals. Russian botanist Georgi Karpechenko crossed the radish *Raphanus sativus* with the mustard *Brassica oleracea,* producing a *Raphanobrassica* hybrid, in 1927. Unfortunately, the hybrid had the leaves of radish and the roots of cabbage. Wild *Raphanobrassica* hybrids have been reported from California. *Triticale,* a hybrid of wheat and rye, is reported to have the high yield of wheat and the hardiness of rye but is rarely grown on a commercial scale. Hybrids between plant genera are known from orchids and cypresses as well (see table on page 203). Among orchids, hybrids are possible among three or even more genera, although this almost always occurs under artificial conditions. In natural habitats, pollinator specificity usually prevents extensive crossings among orchids. For example, *Brassolaeliocattleya* is a cross among the genera *Brassavola, Laelia,* and *Cattleya.* When a scientific name is assigned to an intergeneric hybrid, an × may precede the generic name.

Natural hybrids tend to be rare, but hybrid plants produced for horticultural purposes can be vigorous. The reduced reproductive capacity of the horticultural hybrid is not a drawback because artificial propagation is available.

Perhaps the plant hybridizations that are most famous in the history of the world are the crosses that produced modern wheat. The wild wheat *Triticum boeoticum* had chromosomes in pairs (diploid, 14 chromosomes) and was bred by ancient farmers into the cultivated einkorn wheat *Triticum monococcum.* However, *T. boeoticum* accidentally cross-pollinated with the wild goatgrass *Aegilops speltoides.* The chromosomes from the two parents were incompatible, but chromosome doubling produced a fertile hybrid with chromosomes in groups of four (tetraploid, with 28 chromosomes), the wild *T. dicoccoides.* This was bred by ancient farmers into the emmer wheat *T. dicoccum* and into durum *(T. durum)* and other wheats. *T. dicoccum* also accidentally cross-pollinated with a wild goatgrass, this time *A. squarrosa.* Chromosome doubling turned a sterile hybrid into a species of wheat with chromosomes in groups of six (hexaploid, 42 chromosomes), which are today's major wheat species *T. spelta* (spelt) and *T. aestivum* (bread wheat) (see table on page 203). One of the distinguishing features of modern bread wheat is the gluten protein, which makes the flour sticky, allowing it to hold in bubbles of carbon dioxide during leavening. The gene for gluten apparently came not from the *Triticum* ancestor but from the *Aegilops squarrosa* weed with which emmer wheat accidentally hybridized.

Hybridization, aside from producing new plant species, can also facilitate the transfer of genes from one species of plant to another. Consider two plant species that hybridize and produce a new hybrid species. The two original plant species seldom hybridize with one another, but may be able to crossbreed more often with the intermediate hybrid species. Genes from the first plant species can enter into the hybrid population by cross-pollination and can then enter into the population of the second plant species, also by cross-pollination. The hybrid species has thus acted as a bridge over which genes have crossed, or introgressed, from one species into another. This concept of introgressive hybridization or *introgression* was first suggested by Edgar Anderson, an early 20th-century expert on the genetic history of crop species.

The relative rarity of hybridization between species suggests that a new species, once it evolves, has a coordinated team of genes that would be disrupted by mixing with a different set of genes. This may explain why evolutionary changes may occur so rapidly when a species first forms, followed by a period of relative stability (see PUNCTUATED EQUILIBRIA).

Further Reading

Gompert, Zachariah, et al. "Homoploid hybrid speciation in an extreme habitat." *Science* 314 (2006): 1,923–1,925.

Schwarzbach, Andrea, Lisa A. Donovan, and Loren H. Rieseberg. "Transgressive character expression in a hybrid sunflower species." *American Journal of Botany* 88 (2001): 270–277.

ice ages The term *ice ages* usually refers to the recurring periods of glaciation during the Pleistocene and Holocene epochs of the QUATERNARY PERIOD. During these times, glaciers advanced from the north across North America and Eurasia. The Quaternary period has been one of the coldest in all of Earth history, but glaciations also occurred many times in the past. For example, glaciations occurred in Gondwana (see CONTINENTAL DRIFT) 355 million to 280 million years ago and in Antarctica beginning 33 million years ago. Perhaps the most extensive ice ages occurred during the PRECAMBRIAN TIME (see SNOWBALL EARTH). This entry focuses upon the recurring Northern Hemisphere ice ages of the Quaternary period.

An *interglacial* period followed each ice age, in which the glaciers melted and retreated. Starting about 950,000 years ago, each ice age plus interglacial period lasted about 100,000 years. During the Pleistocene epoch, about 20 ice ages occurred, with the Holocene epoch consisting of the time since the peak of the most recent glaciation. In many cases, a more recent glaciation has obliterated the geological evidences of an earlier one. Several of the more extensive glaciations can be directly studied from geological deposits. The recent glaciations have been named for the location of their maximum extent, the same glaciations receiving different names in Europe and in North America (see table).

North American and European Glaciations

Approximate duration (thousand years ago)	North American	European
75–10	Wisconsinan	Würm
265–125	Illinoian	Riss
435–300	Kansan	Mindel
650–500	Nebraskan	Gunz

Major credit for the discovery of the Ice Ages goes to Louis Agassiz (see AGASSIZ, LOUIS) but Agassiz had predecessors. Jean de Charpentier, for example, observed that boulders that were out of place had to be moved by some massive force that did not involve the movement of the crust itself; he believed such a force could only have been caused by glaciers. The botanist Karl Schimper invented the concept of *Eiszeit,* which translates to Ice Age. Most scientists, such as Alexander von Humboldt, Roderick Murchison, and even Charles Lyell (see LYELL, CHARLES), the father of UNIFORMITARIANISM, opposed the idea when it was first proposed. Before this theory could be credible, a believable cause needed to be proposed and evidence for it presented. This was the same problem that evolutionary theory had, at about the same time. Despite the mystery surrounding what may have caused the ice ages, Agassiz assembled the evidence that at least one ice age had occurred. His evidence, including everything from boulders out of place to the rubble and bones found in caves, turned the tide of opinion. Interestingly, Agassiz believed in CATASTROPHISM, that Earth history consisted of a series of catastrophes. Earlier catastrophists had maintained that the Flood of Noah was the most recent of these catastrophes; Agassiz changed his mind, and those of other catastrophists, by concluding that this "flood" had been ice, not liquid water.

During the 1860s, scientific journals received papers about physics from a certain James Croll of Anderson's University in Glasgow. One of these papers suggested that variations in the orbit of the Earth caused ice ages to come and go. The calculations were oversimplified, but provided a major insight. It turns out Croll was not a professor, but a janitor who spent a lot of time in the library.

The Serbian mathematician Milutin Milanković undertook the necessary detailed calculations in the early 1900s to connect the ice ages to variations in Earth's orbit. He performed these calculations, which would take even a modern computer at least a few minutes or hours, by hand, over the course of several years, even while he was on vacation. The calculations were difficult because he took into account not only changes in Earth's orbit but also processes such as changes in tilt (the angle between the ecliptic, or plane of Earth's revolution around the Sun, and the angle of the axis of Earth's revolution; currently about 23 degrees) and precession (how the Earth spins like a top, with true north pointing to different parts of the sky at different times; true north currently points to Polaris, the North Star) (see figure). During World War I, Milanković was under house arrest. This allowed him to devote full time to the calculations; writer Bill Bryson calls him "the happiest prisoner of war in history." The dates that emerged from Milanković's calculations, for when ice ages were most likely to have occurred, did not correspond to what was then believed to be the times of glaciation. Therefore by the time he died he did not have the satisfaction of seeing his theory vindicated. Now that scientists have radiometric dates of several of the glaciations, they see that Milanković was largely correct. Modern evidence corroborates Milanković's belief that recent glaciations occurred about once every 100,000 years. Oxygen isotope ratios in the shells of microscopic fossils reflect the amount of water in the oceans. Evaporation removes the lighter isotope of water and the water is stored in glaciers; the remaining ocean water therefore is enriched in the heavier isotope (see ISOTOPES). Peak levels of oxygen isotope ratios occur about once every 100,000 years (see figure at right).

Variations in Earth's angle and orbit affect the ice ages because they shift the balance of solar energy going into and out of the planet. Short, cool summers resulting from less solar energy—because of a combination of distance from the Sun and angle of the sunlight—reduce snowmelt. The increased area covered with snow reflects more light into outer space, increasing the albedo (reflected light) of the planet. This further tilts the balance of climate toward an ice age. The accumulation of ice and snow in the northern latitudes eventually pushes the ice southward in great rivers of snow (glaciers). If the summers in the latitudes further south are cool enough that the ice does not melt as much as it advances, then the result is an advancing glacier. The rate of advance has been variable but has seldom exceeded three feet (1 m) a year. The most recent maximum glacial extent ended about 14,000 years ago. The Arctic Ocean was relatively ice-free during the glaciations and was the source of evaporated water for much of the snow that caused them. Extensive glaciations did not occur in the Southern Hemisphere because it contains mostly ocean, which does not change temperature as readily as land. Glaciers covered and still cover Antarctica, but there is virtually no land surface between Antarctica and the tip of South America. Therefore Antarctica produced, and produces, separate icebergs rather than advancing glaciers.

The Milanković cycle explains why cyclical glaciations occur today but does not explain why they did not occur prior to the Pleistocene. Continental drift may have been primarily responsible for beginning the Pleistocene-Holocene cycle of ice ages. The land that is now North and South

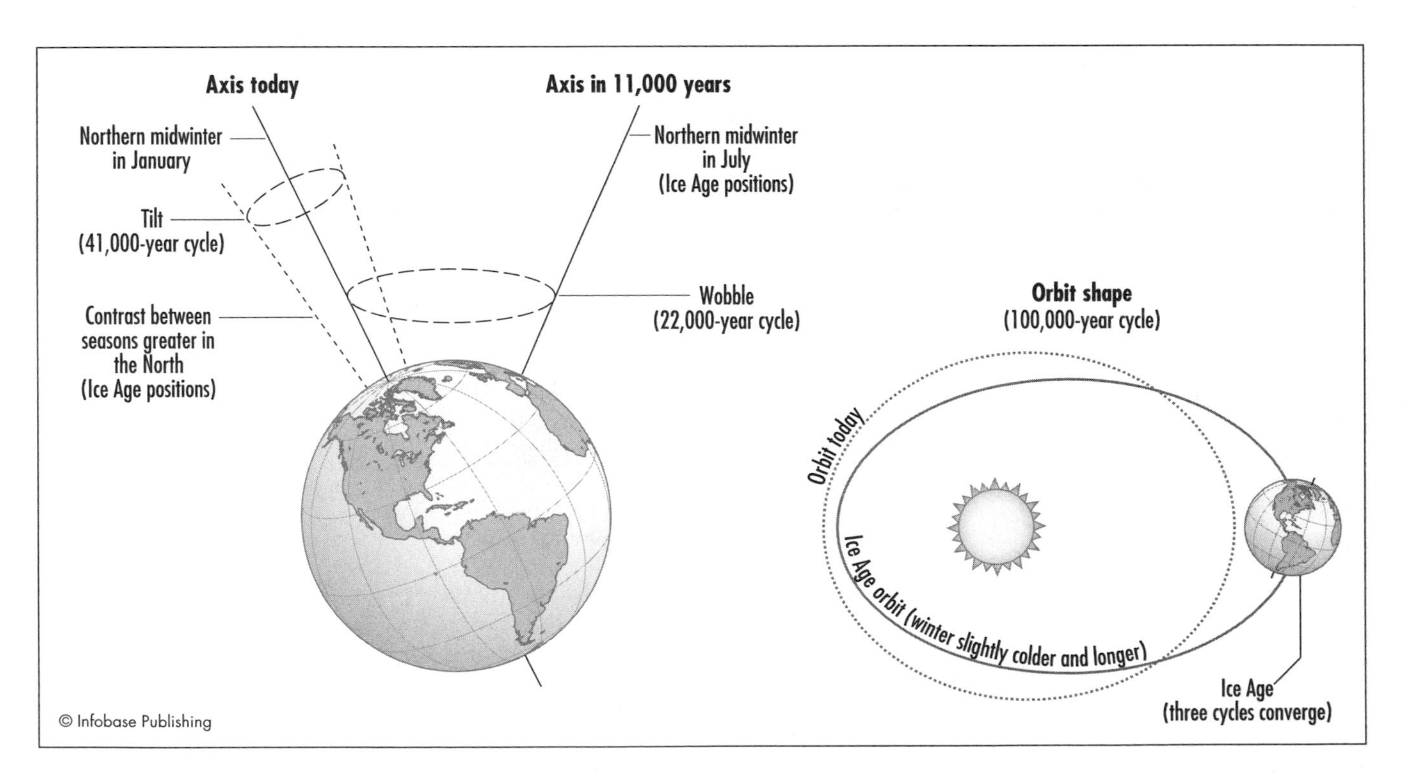

Ice ages result from the convergence of three cycles: changes in the tilt of the axis (currently about 23.5 degrees from vertical), the wobble of the axis, and the shape of the orbit. This convergence is called the Milanković cycle. ***(Modified from Bonnicksen)***

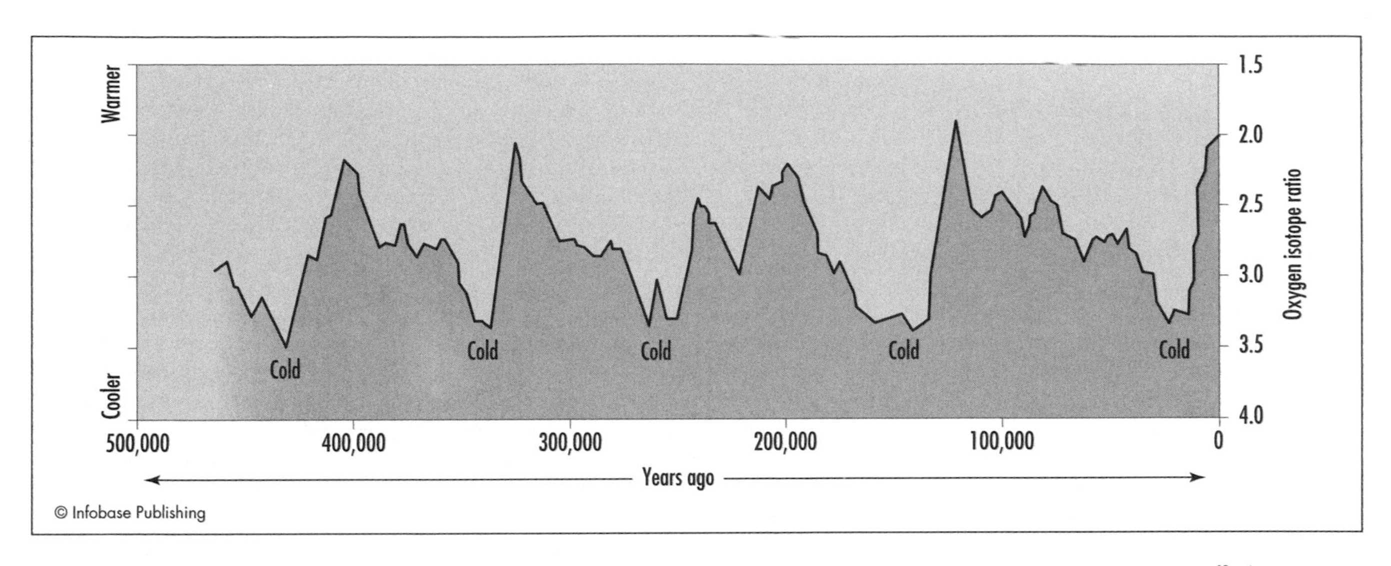

Oxygen isotope patterns indicate a clear pattern of ice ages about every 100,000 years during the Quaternary period. High percentages of ^{18}O (high oxygen isotope ratio) indicate cold, dry periods; lower percentages indicate warm, moist periods. ***(Based on an illustration in Harold Levin's* The Earth Through Time *© 2003; reprinted with permission of John Wiley & Sons, Inc.)***

America had not been connected since the breakup of Pangaea during the MESOZOIC ERA. A gap of ocean between them allowed a vast worldwide circulation of ocean water to flow into the Atlantic Ocean, forcing warm water into the Arctic Ocean. Water holds more heat than air, and this movement of heat from tropics to the arctic caused temperatures to be sufficiently warm that trees grew near the North Pole, even though they were in darkness half of the year. However, the Isthmus of Panama gradually arose, first as islands that slowed the flow of water, then as a land bridge that stopped it. Today, the Gulf Stream carries warm Caribbean water to the North Atlantic, causing the climate of Europe to be much warmer than the land at the same latitudes in North America (compare Labrador and England, for example). But the Gulf Stream is not big enough to warm up the Arctic Ocean. The planetary movements that produce the Milanković cycle occurred long before the Pleistocene, but until then they had not produced cyclical glaciations. In general, scientists know that continental movements are important in determining global climatic patterns but are not always sure how. Antarctica, for example, used to be covered with forests also. After it moved to be directly over the South Pole, it developed a cold climate and glaciers began to build up. Only in recent decades have they begun to melt. There was a 20-million-year lag time, as yet unexplained, between Antarctica covering the South Pole and the buildup of the glaciers.

As the glaciers advanced, they pushed piles of rubble ahead of them. These piles, after the glaciers began to retreat, formed moraines. Some moraines have resulted in major geographical landmarks, such as Long Island and Cape Cod. The type of rock that formed from glacial rubble (or till) is called tillite. Glaciers also pushed large rocks ahead of them, which produced scrapes in the underlying rock; these scratches are a major source of information about where the glaciers moved and when and were an important clue in the discovery of the ice ages themselves. Glaciers sometimes carried loads of till as they broke off to form icebergs. Anomalous piles of rocks, of many sizes and far out of place from their parent material, are called *dropstones* and occur at surprising places on the ocean floor. Glaciers tore large chunks of rock from some mountains, while going around others. The isolated mountains that rose above the sheet of ice were called *nunataks*. Sometimes whole mountain systems, such as the Black Hills of South Dakota, remained above the glaciers. The temperatures were sufficiently cold and dry that very few organisms survived on nunataks.

Glaciations during the Quaternary ice ages influenced the climate of North America very severely but also affected the tropical regions of the Earth. Although the tropics were a bit cooler, they were still suitable for the survival and growth of tropical grasses and forests. However, since so much of the water was removed from meteorological circulation, the global climate was drier. In the dry tropical regions, deserts and grasslands were extensive, while tropical forests consisted of isolated groves.

Once the summers began getting longer and warmer, melting exceeded advance, and the glaciers retreated. The most recent glacial retreat began about 12,000 years ago. Retreat was generally faster than advance, sometimes as much as 30 feet (10 m) a year. Therefore a person would readily notice the retreat of glaciers during a lifetime, and there were modern humans in both Eurasia and in North America to observe this. Many scientists warn that the GREENHOUSE EFFECT could produce similarly rapid temperature changes for which humans will be unprepared. The return of water to the meteorological cycle resulted in a wetter climate in tropical regions, in some regions even wetter than today. What is now the Sahara Desert, about 9,000 years ago, had extensive grasslands and herds of wild animals, which human tribes observed, carving images of them on rocks.

The vast amounts of water from melting glaciers had to go somewhere. Many rivers were much larger during the time of glacial melt than they are today. For example, the bluffs along the Missouri River, formed from fine soil called loess blown from land newly exposed by the retreating glaciers, are today over two miles (over 3 km) from the current banks of the river, implying that the river was about four miles (seven km) wider as the glaciers were melting upstream. Vast lakes formed from the meltwater, of which only remnants remain, such as the Great Lakes. Although these lakes are among the largest lakes in the world and contain about a fifth of the planet's freshwater, they are much smaller than Lake Agassiz, of which they are a remnant. In some cases, ice blocked the drainage of the lakes, and when the ice gave way, massive floods occurred downstream. This occurred with Lake Agassiz and produced a flood of water down the St. Lawrence River. So much freshwater drained during this flood that it altered the climate of the entire region for a few hundred years. Since freshwater is lighter than saltwater, this cold water remained at the surface of the north Atlantic and caused a temporary reversal of the warming trend that had been occurring. A similar flood, in western North America, produced huge ripples and other deposits known as the Scablands.

Within about 5,000 years (by 7,000 years before the present), the warming trend had reached a maximum, known as the *hypsithermal.* At this time, grasslands covered much of the land that is today forested in North America. This included a new kind of grassland, the tallgrass prairie, consisting of large deep-rooted grasses that grow only during warm summers. As conditions began to change once again, leading the Northern Hemisphere toward the next Ice Age, cool moist conditions returned to the north and the forests advanced southward. The tallgrass prairie, however, persisted in regions with a climate capable of supporting trees, because the accumulation of dead grass encouraged fires. The fires kill tree seedlings but merely create fertilizer for the grasses and wildflowers that resprout from underground stems and roots. The tallgrass prairie persisted until modern times, when American settlers destroyed them to create farms and cities. Today, barely 2 percent of the original tallgrass prairie remains.

The ice ages continue. During the last glaciation, 30 percent of North America was covered with ice. But 10 percent of it still is, and another 14 percent has permafrost, which is a permanent layer of ice beneath the soil. If past conditions occur again, another Ice Age is on the way. However, the greenhouse effect is causing the opposite trend to occur; glaciers and sea ice in North America, as well as Antarctica, are becoming smaller and thinner. It would be very nice for human civilization if the greenhouse effect precisely balanced the return of the ice ages; however, global temperature is rising (largely due to human production of carbon dioxide) 10 times faster than it did even during the period of rapid glacial retreat. The greenhouse effect is causing much of the permafrost to melt, resulting in soil subsidence and the buckling of highways.

The glaciations have occurred rapidly enough that very little evolution has occurred as a result of them. For the most part, species of plants and animals have dispersed to new locations, rather than evolving into new species, as a result of any single ice age (see BIOGEOGRAPHY). Sometimes, plant species have gotten trapped in limited locations as they have moved northward. The alpine tundras, at the tops of ranges such as the Rocky, Cascade, and Olympic Mountains and the Sierra Nevada, are an example. Some individual species have gotten trapped in regions to which they may not be well adapted, resulting in a relictual distribution pattern (see DISJUNCT SPECIES). One example is the Monterey cypress *(Cupressus macrocarpa)* of California. During the most recent Ice Age it survived in Mexico. As this species dispersed northward during glacial retreat, it migrated westward and ended up confined to a small area near the Monterey Peninsula of California. Had it turned eastward rather than westward, it might have found a much more extensive region in which to live.

In some cases, plant species dispersed northward after the last Ice Age more quickly than can be easily explained by the movements of their seeds. In Europe, the oak *Quercus robur* can spread rapidly, as jays carry their seeds, but *Q. petraea* seeds spread much more slowly. Recent research suggests that *Q. petraea* spread northward not by its seeds but by its pollen. The two oak species can cross-pollinate (see HYBRIDIZATION). *Q. robur* spread quickly northward as birds carried its seeds, and the wind carried pollen of *Q. petraea* northward. This pollen fertilized *Q. robur* individuals once they had matured. The resulting hybrid seeds grew into male-sterile trees, which produced only female flowers. When the hybrid trees matured, they, too, were pollinated by *Q. petraea* pollen, and some of the offspring were completely fertile *Q. petraea* trees. In this way, *Q. petraea* dispersed northward at the rapid pace of their windborne pollen, not the slow pace of their acorns.

Many extinctions occurred at the end of the last glaciation (see PLEISTOCENE EXTINCTIONS), far more than at the ends of previous ice ages. In many cases, trees were left without the animals that had previously dispersed their seeds (see essay, "What Are the 'Ghosts of Evolution'?").

In some cases, the overall climatic changes of the Pleistocene may have affected evolution, even if the individual glaciations did not. One example has direct relevance to humans. On at least two occasions starting about five million years ago, the spread of drier conditions across Africa caused forests to retreat and savannas to spread. These cycles of aridity may have been important stimuli in the evolution of HOMININS. Although BIPEDALISM evolved earlier (see AUSTRALOPITHECINES), a drying trend may have been the event that provoked a change in the lifestyle of some hominins, who began to make tools, eat more meat, and evolve larger brains (see *HOMO HABILIS*). A later drying trend may have stimulated the evolution of *HOMO ERGASTER.* Evolutionary shifts in other African mammals, such as antelope, occurred at about the same times. Some scientists, such as paleontologist Steven Stanley, consider humans to be "children of the Ice Age."

In Europe, during the Ice Age prior to the most recent, NEANDERTALS were the only humans. *HOMO SAPIENS* migrated into Europe and displaced the Neandertals during the interglacial period. The Neandertals were extinct by the time the most recent glaciation reached its maximum. Some anthropologists consider not just the genus *Homo* but the modern species *H. sapiens* to be children of the Ice Age, because the glacial advances forced previously scattered tribes to come in contact in southern Europe. This is the period of the great CRO-MAGNON cave paintings of southern Europe.

Further Reading

Bonnicksen, Thomas M. "The Great Cold." Chap. 1 in *America's Ancient Forests: From the Ice Age to the Age of Discovery.* New York: John Wiley, 2000.

Broecker, Wallace S. "Was the Younger Dryas triggered by a flood?" *Science* 312 (2006): 1,146–1,147.

Gould, Stephen Jay. "The freezing of Noah." Chap. 7 in *The Flamingo's Smile: Reflections in Natural History.* New York: Norton, 1985.

———. "The great Scablands debate." Chap. 19 in *The Panda's Thumb: More Reflections in Natural History.* New York: Norton, 1980.

Macdougall, Doug. *Frozen Earth: The Once and Future Story of Ice Ages.* Berkeley: University of California Press, 2004.

Petit, Rémy J., et al. "Hybridization as a mechanism of invasion in oaks." *New Phytologist* 161 (2003): 151–164.

Pielou, E. C. *After the Ice Age: The Return of Life to Glaciated North America.* Chicago: University of Chicago Press, 1992.

Raymo, Maureen E., L. E. Lisiecki, and Kerim H. Nisancioglu. "Plio-pleistocene ice volume, Antarctic climate, and the global $\delta^{18}O$ record." *Science* 313 (2006): 492-495.

Scher, Howie D., and Ellen E. Martin. "Timing and climatic consequences of the opening of Drake Passage." *Science* 312 (2006): 428–430.

Stanley, Steven M. *Children of the Ice Age: How a Global Catastrophe Allowed Humans to Evolve.* New York: Harmony Books, 1996.

inbreeding depression *See* MENDELIAN GENETICS.

inclusive fitness *See* ALTRUISM.

Inherit the Wind *See* SCOPES TRIAL.

instinct *See* BEHAVIOR, EVOLUTION OF.

intelligence, evolution of Intelligence is highly overrated. Most species do not have very much of it. Eighty percent of animal species are arthropods (see INVERTEBRATES, EVOLUTION OF), which do not have high levels of intelligence and show no evolutionary trend toward increasing it. High levels of intelligence evolved in the few species that do have it as a result of special circumstances that are still not understood.

Intelligence has proven impossible to define in a manner that all scientists can accept. If intelligence requires only a coordinated response to complex environmental factors, then plants have intelligence, as explained in the article by biologist Anthony Trewavas. Most scientists restrict the concept of intelligence to animals, which have nervous systems. Human intelligence includes consciousness, self-awareness, and the ability to form abstract concepts. Evolutionary scientists recognize a whole range of levels of animal intelligence.

There are evolutionary reasons why intelligence has evolved only in animals. Microbes and protists (see ARCHAEBACTERIA; BACTERIA, EVOLUTION OF; EUKARYOTES, EVOLUTION OF) could not have evolved intelligence, since they are primarily single-celled. Plants and fungi could not evolve intelligence, which requires a nervous system for the acquisition, processing, and storage of information.

In some groups of animals, such as the arthropods, a large body size and therefore a brain as large as that of a human is unlikely to evolve. But in many other animal lineages, intelligence could have evolved if it had provided a significant fitness advantage (see NATURAL SELECTION). The fact that high levels of intelligence have evolved so seldom during the history of animal life on Earth suggests that the circumstances that favor intelligence are rare. Therefore, as evolutionary biologist Stephen Jay Gould (see GOULD, STEPHEN JAY) pointed out, if complex life were to evolve all over again, it is unlikely that a human level of intelligence would evolve. Furthermore, science fiction notwithstanding, scientists should not expect to find highly intelligent life on every planet that is capable of supporting complex life. Gould strongly rejected the idea that structurally simpler organisms are evolving toward higher intelligence (see PROGRESS, CONCEPT OF).

High levels of intelligence require big brains, and big brains are very expensive. The human brain consumes a tremendous share of the food and oxygen that is available in the blood. Humans have an *encephalization quotient,* which compares human brain size with that of an average mammal of the same size, of about 6. The encephalization quotient of humans compared to other primates is about 3 (see ALLOMETRY). Big brains, and high intelligence, are simply not worth the cost for most animal species.

Biologists use the encephalization quotient rather than absolute brain size to compare animal species because brain size is correlated with body size. Big animals require bigger brains because their larger bodies have more sensory information to process (particularly the internal sensory information that informs the brain what is happening inside the body) and more muscles to control. In fact, beyond a certain size limit, centralized nervous control of a body may be biologically impossible. The largest dinosaurs appear to have had a somewhat large brain in their heads, that processed information to and from the front end, and a smaller swelling of the spinal cord near their tails, which processed information to and from the rear end. Nerve transmissions are rapid (particularly in nerve cells, or gray matter, that have fatty sheaths, or white matter) but apparently not rapid enough for the head end of the dinosaur to take full control of the entire body.

Intelligence is not perfectly correlated with brain size. Within the size range of the modern human brain, from about 80 to over 120 cubic inches (1,200 to 2,000 cc), there appears to be no correlation between brain size and intelligence. Men have larger brains than women, on the average, but are also larger than women. Even when differences of body size are taken into account, men still have about six cubic inches (100 cc) more brain volume than women. Extensive research, however, has failed to show any gender differences in intelligence. When modern humans first evolved over 100,000 years ago (see *HOMO SAPIENS*), their brains were as large as those of humans today, but these individuals displayed few of the behavior patterns (aside from tool use) that characterize modern human intelligence, at least not in a form that was preserved in the fossil record. By 30,000 years ago, modern humans produced abundant art and other physical evidences of intelligence. NEANDERTALS *(Homo neanderthalensis)* had brains as large as those of *Homo sapiens,* but even when they lived alongside modern humans in Europe 50,000 to 30,000 years ago they produced few evidences (again, aside from tool use) of modern human intelligence.

Evolutionary scientists disagree on two aspects of the evolution of human intelligence: the structure of human intelligence, and how human intelligence evolved.

Structure of Human Intelligence

All evolutionary scientists accept that human intelligence evolved along with the gigantic human brain. However, did intelligence, in general, evolve, or did specific components of intelligence evolve? If general intelligence evolved, but not its specific components, then the human brain must be extremely flexible during its development. If the specific components of intelligence evolved as well, then all humans regardless of culture and of how they are raised will display certain universal characteristics.

Certain human universals of intelligence and behavior are beyond question. One is language. The human brain seems to be "hardwired" for language acquisition, and specifically for language that consists of words and grammar. The occasional human child raised apart from human culture (such as "wolf children") develops rudimentary language components. All healthy humans have the Wernicke's area and Broca's area on the left side of the brain that function in language recognition and production, respectively. Anthropologists have examined hominin skulls closely for evidence of little bumps that might indicate these areas in the brains of *HOMO ERGASTER* but have not reached a firm conclusion as to whether they had them (see LANGUAGE, EVOLUTION OF).

Beyond this, the evolution of specific thought and behavior patterns has proved controversial. Sociobiologists (see SOCIOBIOLOGY) claim that specific thought patterns such as the fear of snakes and of strangers, and specific behavior patterns such as violence and polygamy, have evolved, and that specific neural networks for them exist in the brains of all humans. These neural networks are proclivities, rather than causes of action; therefore humans can learn to overcome innate tendencies that evolution has caused their brains to possess. Other scientists object strongly and assert that the human brain can be trained during childhood to think and act in an almost infinite variety of ways. Sociobiologists say that war is innate to the human brain; other scientists assert that humans can learn to be perfectly peaceful. The evidence is unclear: There are more universal behavior patterns (violence is one of them) than one would expect from a flexible brain, but even the one ability that all scientists admit is universal—language—is controlled by the brain in a decentralized way. The Wernicke's and Broca's areas are essential for language acquisition and production, but not sufficient. The neural networks that sociobiologists expect to find have not been clearly identified by brain studies. Positron emission tomography (PET scans) that reveal the parts of the brain that are active during certain mental functions have not shown a perfectly clear and consistent correspondence of specific behaviors with specific brain regions. Brain damage early in life can sometimes be overcome during development, as a different part of the brain takes over the function that the damaged portion would have governed. There are more universal behavior patterns than non-sociobiologists can explain, yet the brain is more flexible than sociobiologists might have expected.

Intelligence, though not its components, appears to have a genetic component. Psychologist Thomas J. Bouchard has measured IQ (intelligence quotient) of numerous people, most notably from many identical (monozygotic) and fraternal (dizygotic) twins. The correlation coefficients presented in the table are calculations of how much statistical variability can be explained by the comparison. Separate tests on the same person yield a close but not perfect correlation. The IQ scores of identical twins are more closely correlated than those of fraternal twins, which are more closely correlated than those of biological siblings; unrelated people living apart show no correlation of IQ with one another, on the average.

Twins usually share similar environments as well as the same genes. Bouchard's results also demonstrate the effects of environment as well as of genes on intelligence. IQ scores of identical twins raised in separate homes have a lower correlation than those of identical twins raised in the same home; the same pattern is true for fraternal twins and biological siblings. However, IQ scores of adopted children living in the same household show no closer correlation than do unrelated people living apart (see table).

Even twins raised in separate homes may share an environmental component to intelligence. Conditions of nutrition and stress that twins experience inside the womb may influ-

IQ Correlations[1] between Individuals

Same person tested twice	87%
Identical twins raised in same home	86%
Identical twins raised in separate homes	76%
Fraternal twins raised in same home	55%
Biological siblings	47%
Parents and children living together	40%
Parents and children living apart	31%
Adopted children living together	0%
Unrelated people living apart	0%

[1]Perfect correlation is 100%.

ence their intelligence. There is also a correlation between IQ and body left-right asymmetry. Left-right asymmetry is considered an indicator of stress and poor nutrition during fetal development. Those who reject a genetic component to intelligence (within the normal human range) interpret these results to indicate that intrauterine stresses cause lower IQ.

Human intelligence evolved, but did its components? Did neural networks for art, phobias, religion, and violence evolve? Or did evolution produce a brain that was simply big enough to develop these abilities and potentials? Evolutionary science can, at this point, proceed no further than the anatomy and physiology of brain science.

How Human Intelligence Evolved

One of the strongest factors that have caused some lineages of animals to evolve higher intelligence than others is their way of obtaining food.

- Among herbivores, intelligence is correlated with selectivity of food. A great deal of plant tissue is toxic. Many herbivores get around this problem by physiological means, such as complex digestive systems (see COEVOLUTION). Others get around the problem by means of intelligence. Howler monkeys eat leaves; they have long digestive systems that allow time for bacterial fermentation to break down many of the toxins. In contrast, spider monkeys selectively eat fruits and nuts, and have shorter digestive systems. Spider monkeys have larger brains than howler monkeys. Their relatively greater intelligence allows spider monkeys to find more food energy, and the energy requirement of their large brains requires them to do so. Among squirrels, the size of the hippocampus (the portion of the brain used to remember locations) is proportional to the amount of dispersion they use in food storage. Gray squirrels store their nuts more widely, and have a larger hippocampus, than red squirrels. The hippocampus swells in size in squirrels in the autumn, when food storage begins.
- Predators are usually more intelligent than herbivores, because they must seek and chase prey. Scavengers may also have higher intelligence than herbivores.

Brain size in human ancestors began to increase after the evolution of upright walking (see BIPEDALISM), and simultaneous with the evolution of tool use. AUSTRALOPITHECINES walked upright, but had brains no larger than those of chimpanzees, and made no stone tools (although, like modern chimps, they may have used sticks and rocks as tools). The first species of the human genus (see *HOMO HABILIS*) had larger brains and made more complex tools than the australopithecines. Both intelligence and tools may have allowed early humans to be effective scavengers and only much later to be effective hunters.

Once brain size had begun to increase, natural selection reinforced the increase. Since brains are expensive, early humans with larger brains needed more calories. There was therefore an evolutionary advantage to increased intelligence that would allow humans to select higher quality plant foods, to more intelligently locate animals killed by larger predators, and to hunt. This created a positive feedback situation in which larger brains created a demand for even larger brains.

A continued increase in brain size was inconvenient in hominins because of upright posture. Upright walking requires a narrow pelvis, with the legs directly beneath. This arrangement puts a constraint on the size of the birth canal. The head is the largest part of the baby that must pass through this canal. Therefore upright walking constrains the size of the head at birth. Unless natural or other selection provided a considerable advantage to a large brain, this would have placed an upper limit on human brain size. What actually happened, however, is that human babies began to be born at an earlier stage of development. If human embryonic development continued until the same stage that it does in other primates, human babies would be born after 14 months of gestation, rather than nine (see LIFE HISTORY, EVOLUTION OF). Human babies are still as helpless as embryos when they emerge, which would endanger their survival. The continued evolution of human intelligence allowed mothers to care for even these helpless infants. It has been suggested that the helplessness of the infant, and the preoccupation of the mother with its care, created an advantage for fathers to provide care and food for the mother and infant, which fathers in other primate species seldom do. According to this explanation, the evolutionary increase in brain size encouraged the evolution of family units and of high intelligence. The early birth of human babies, and high intelligence, are two features of NEOTENY, which is the preservation of juvenile characteristics into reproductive adulthood, and which characterizes the modern human species more than any other.

Once high intelligence had begun to evolve in humans, it would provide numerous advantages to individuals who possessed a greater degree of it than other individuals:

- Intelligent individuals were able to exploit their environment for resources more efficiently, as explained above.
- Intelligent individuals had a superior ability to understand and manipulate social relationships and kinship ties, as is observed in many primate societies. Called *Machiavellian intelligence* after the Renaissance Italian political philosopher Niccolò Machiavelli, such abilities would allow their possessor to become a dominant male and sire a large proportion of the offspring in the group. Dominant males in primate groups are often those who are best able to form alliances, rather than simply the strongest or the best fighters.
- Intelligent individuals could gather and analyze information about other human groups, the more effectively to outwit and outfight them.

Human intelligence went far beyond the practical aspects of hunting, gathering, fighting, and child care. Defense from outside threats, efficiency at obtaining resources, and success at gaining dominance within societies as explanations for the evolution of human intelligence are very male-biased, as it is primarily males that fight. What about female intelligence? Furthermore, as evolutionary biologist Geoffrey Miller has pointed out, Machiavellian intelligence is insufficient to explain the vast range of human intellectual capabilities. For example:

- The human ability for language far exceeds what is necessary for efficient communication during foraging and

hunting, for defense, and for the dynamics of dominance in groups; it enters the realm of art. Practically every language in the world, in its primitive state, is immensely and even wastefully complex.
- The human ability to create and appreciate art, music, and storytelling far exceeds what can be explained by natural selection for obtaining resources, is far more complex and costly than would be necessary to benefit one's family (through inclusive fitness), and is more complex than would result in a payback through reciprocal altruism (see ALTRUISM).

Miller claims instead that human intelligence evolved primarily by SEXUAL SELECTION. In sexual selection, the characteristics that prevail are those that induce one sex to choose certain individuals of the other sex. Usually, females choose among males on the basis of visible characteristics; sexual selection therefore favors those characteristics that the females like the most. This is why sexually selected characteristics reach their highest development in males in most animal species. These characteristics may have nothing at all to do with survival and may be entirely arbitrary. Because of this, sexual selection has taken a different direction in every species and has resulted in a prodigious variety of ornamentations in male animals, such as colorful feathers in male birds, as well as a prodigious variety of behaviors, such as courtship dances. Females use these visible characteristics to choose males because the genes of the males, which would be the most important thing for them to know about, are invisible. These characteristics therefore function as *fitness indicators,* that is, they provide a visible indication of a male's invisible genetic quality. In order to work, these indicators must reliably reflect the genetic quality of the male. For this reason, fitness indicators are nearly always very costly and usually wasteful. Only healthy male birds, with good metabolism and free from disease, can afford to produce costly and colorful feathers and dance and sing for hours and days on end.

It is generally the males that compete for access to females under conditions in which males can mate with many females. However, there is also some competition among females to choose the males with the highest fitness. Therefore sexual selection can cause the evolution of sexual characteristics in both males and females. While in most animal species sexual selection proceeds by female choice, in humans it proceeds mostly by mutual choice. Sexual selection has been responsible for a great number of characteristics in both men and women.

Miller attributes to sexual selection the major role in the evolution of all aspects of human intelligence. The mind, he says, is not a computer to solve problems, nor is it a Swiss army knife full of tools to help survival in different circumstances; it is a sexually selected entertainment system. Virtually every aspect of intelligence can serve as a fitness indicator. A person's abilities to create beautiful art, produce fine literature, tell clever jokes, accomplish feats of sport, communicate with the gods, or explain natural phenomena are indicators of intelligence and coordination and show that his or her genes are good—almost half of the genes are involved in the development of the brain—and that he or she is healthy. As mentioned above, in most species sexual selection favors male fitness indicators and female choice. But the ability of males to perform the acts of creativity and intelligence, and the ability of females to understand and judge them, are so closely intertwined that sexual selection has produced males and females of equal intelligence and with equal abilities to create art. Throughout history and in most cultures, it is usually the males who have "shown off" through creativity and sports, but females must have equal abilities in these areas or else the strategy would not work. Since courtship activity in many animal species represents a balance between male deception and female ability to see through the deception, natural selection favored females smart enough to recognize when males are faking. Although men are, on the average, stronger than women, due to sexual selection, the difference is less than that observed between males and females in many primate species (such as gorillas), perhaps because of the overwhelming importance that intelligence—the equal intelligence of men and women—has played in human evolution. The primacy of artistic expression over practical work in the evolution of human intelligence may be illustrated by what is considered to be the most practical product of early human intelligence: tools. Most hand axes were used as tools (see TECHNOLOGY), but some have been found that were much larger, more symmetrical, and more detailed in their production than was necessary—and which were never used. They may have been art.

The evolution of morality and kindness would have been more than a fitness indicator for sexual selection. A female who observed that a male was friendly and reliable—not just to his family, and not just to those who were likely to return the favor—would be able to conclude that he was very resourceful, particularly if the man, at great expense to himself, did great deeds of kindness and charity to the community as a whole. The female would conclude not just that he was resourceful but also that he was likely to be kind to her as well. This would have been important especially if the woman had a child from a previous relationship, as is frequently the case in tribal cultures. A stepchild has no genetic relatedness to the new father; but the mother, in choosing a new mate, would want a man who would be kind even to a child who was not his. Morality and kindness, as components of intelligence, may have evolved by means of indirect reciprocity (see ALTRUISM). In this way, Miller concludes, such good moral qualities as kindness, reliability, and generosity could have evolved from sexual selection. As Miller points out, not only was Ebenezer Scrooge (the main character in Charles Dickens's *A Christmas Carol*) mean to his nephew and a perfect example of the lack of reciprocal altruism to his fellow citizens, but also it was no wonder that he was single.

One alternative explanation is that males and females chose one another on the basis of youth (therefore length of remaining reproductive life), thereby favoring the evolution of neoteny. Larger brain size, therefore intelligence, may have been a side effect of the evolution of neoteny, according to this view, thereby making intelligence an exaptation rather than an ADAPTATION.

This does not mean that all of the manifestations of intelligence are motivated by sex. However, sexual selection can favor the evolution of intelligence-related traits even when they are motivated by things other than sex, such as a true love of one's fellow humans, or a sublime sense of aesthetics.

One might expect that, after many generations of sexual selection, all variability in fitness indicators would have been eliminated. The inferior traits would have died out, leaving only males and females with superior characteristics that the other sex finds appealing. But this point is never reached. If this point were approached, the fitness indicators would no longer be reliable. The whole point of a fitness indicator is that it reveals which individuals are superior in their health and potential fitness. If variability in one fitness indicator is eliminated, more variation in that indicator may evolve, or sexual selection may seize upon some other trait as a fitness indicator. Sexual selection never finishes its job. This is exactly what human intelligence looks like—a set of traits that has evolved to absurd lengths and keeps diversifying into new forms.

What remains unexplained is how intelligence began to evolve in the first place. The evolution of a little intelligence would create a demand for more, particularly through sexual selection. But how did it get started? Miller explains that sexual selection frequently produces characteristics that have no survival value. Intelligence could therefore have gotten started quite by chance, after which it evolved explosively by sexual selection. While natural selection may be unable to favor a partly evolved trait, sexual selection would be able to do so.

Although humans are uniquely high in intelligence, they are not unique among animals in the evolution of increased intelligence over time. Paleontologist Simon Conway Morris points out that intelligence, although less than that of humans, has evolved in cetaceans such as dolphins, and even in cephalopod mollusks such as the octopus. An octopus has a level of intelligence that, while below that of a human, takes human investigators by surprise. For example, an octopus can learn how to perform a complex task by watching another octopus perform it. The dolphin brain is structurally different from the human brain, as dolphins rely on sound much more than on sight, but it is approximately as large as the human brain and has even more of the convolutions that provide cortical surface areas for the integration of neurons involved in thought. In fact, dolphin brains may be superior to human brains in some aspects. A dolphin will allow half of its brain to sleep at a time, so that it can continue swimming while sleeping. This partial independence of the two brain hemispheres, which is one of the contributors to human intelligence, is more developed in dolphins than in humans. The discovery of high levels of intelligence in nonhuman species raises questions about ANIMAL RIGHTS.

Moreover, the evolution of intelligence in humans has occurred at least twice and possibly more often. Although Neandertals did not have art or religion, they did have brains as large as those of modern humans, and they evolved independently from *HOMO HEIDELBERGENSIS* ancestors. There is even some evidence that *Homo erectus* in Asia evolved slightly larger brains than their African *Homo ergaster* ancestors before becoming extinct (see *HOMO ERECTUS*). According to Conway Morris, the independent evolution of intelligence in mollusks, cetaceans, and two (possibly three) human lineages is an example of the pattern of CONVERGENCE that he says pervades evolutionary history.

Human intelligence is a many-faceted and extremely complex thing. Its current operation is not fully understood, therefore its evolution cannot be fully understood. Undoubtedly, many processes were at work in the evolution of intelligence. For example, one must not underestimate the central role that religion played in human evolution, including the evolution of intelligence (see RELIGION, EVOLUTION OF). Because intelligence is equally developed in all races, it seems certain that human intellectual abilities were fully evolved before *Homo sapiens* began to disperse from Africa more than 50,000 years ago. All of subsequent prehistory, and all of human history, has consisted of the cultural development of the human mind, using the biological abilities that had evolved prior to that time.

Further Reading

Bouchard, Thomas J., Jr., et al. "Sources of human psychological differences: The Minnesota study of twins reared apart." *Science* 250 (1990): 223–229.

Conway Morris, Simon. *Life's Solution: Inevitable Humans in a Lonely Universe.* Cambridge: Cambridge University Press, 2003.

Evans, Patrick D., et al. "*Microcephalin,* a gene regulating brain size, continues to evolve adaptively in humans." *Science* 309 (2005): 1,717–1,720.

Marcus, Gary. *The Birth of the Mind: How a Tiny Number of Genes Creates the Complexities of Human Thought.* New York: Basic Books, 2004.

Mekel-Bobrov, Nitzan, et al. "Ongoing adaptive evolution of *ASPM,* a brain size determinant in Homo sapiens." *Science* 309 (2005): 1,720–1,722.

Miller, Geoffrey. *The Mating Mind: How Sexual Choice Shaped the Evolution of Human Nature.* New York: Doubleday, 2000.

Miller, Greg. "The thick and thin of brainpower: Developmental timing linked to IQ." *Science* 311 (2006): 1,851.

Ridley, Matt. "Intelligence." Chap. 6 in *Genome: The Autobiography of a Species in 23 Chapters.* New York: HarperCollins, 1999.

Rumbaugh, Duane M., and David A. Washburn. *Intelligence of Apes and Other Rational Beings.* New Haven, Conn.: Yale University Press, 2003.

Trewavas, Anthony. "Aspects of plant intelligence." *Annals of Botany* 92 (2003): 1–20. Available online. URL: http://aob.oupjournals.org/cgi/content/full/92/1/1. Accessed September 7, 2005.

Wynne, Clive D. L. *Do Animals Think?* Princeton, N.J.: Princeton University Press, 2004.

intelligent design Intelligent design (ID) theory is minimalist creationism. It is a theory that claims there is irreducible complexity of biological systems and structures. According to this view, biological systems consist of many interacting components, none of which can work unless all the other components are present and operational. It is a form of CREATIONISM because, the defenders of intelligent design theory admit, only an intelligent Creator could have brought irreducibly complex systems into existence; irreducibly complex systems cannot be produced by a series of evolutionary steps.

However, ID theorists like to distance themselves from old-fashioned creationism.

The most popular statement of this position has been *Darwin's Black Box,* by Michael J. Behe, a biochemist at Lehigh University. Most of the following discussion is based upon Behe's book.

According to Behe, gone from ID are all the arguments that creationists usually present about gaps in the fossil record, young age of the Earth, and the Flood of Noah. Although most adherents of ID accept many or all of the tenets of creationism (for example, Jay W. Richards, a senior fellow at the Discovery Institute in Seattle, Wash., still uses the "gaps in the fossil record" argument), the debate sidesteps these issues. Few ID theorists accept the historical facts of human evolution; but their publications generally do not address it. Instead, it is enough for them to point out that human (and other) biochemistry is too complex to have arisen by evolution. Evolutionary scientists point out that the fossil record really does provide evidence that evolution occurred; Behe indicates that paleontology does not matter. Evolutionary scientists, starting with Darwin (see appendix, "Darwin's 'One Long Argument': A Summary of *Origin of Species*") and continuing through modern evolutionary scientists (see DAWKINS, RICHARD), point out the existence of intermediate stages in the evolution of the eye from the simple eyespot of a protozoan; Behe indicates that does not matter either. In his discussion of the bombardier beetle, Behe admits that other beetles have similar, and simpler, systems of defense; not surprisingly, Behe indicates that this does not matter either. All that matters, in his argument, is irreducible complexity on the biochemical and cellular level. Behe's examples of irreducibly complex systems include vision in a retinal cell; the explosive defense mechanism of the bombardier beetle; cilia and flagella (the whiplike mechanisms by which many single-celled organisms propel themselves); blood clot formation; and the mammalian immune response.

According to ID arguments, in order for evolution to produce biochemical complexity, uncountable billions of cells would have to die over the course of billions of years. As a matter of fact, that is exactly what happened during PRECAMBRIAN TIME. Almost 80 percent of evolutionary history occurred when life on Earth was primarily microbial. It looks from the Precambrian fossil record as if nothing much was happening, but this is because the significant events were taking place on a molecular scale that no fossil could reveal.

ID has two components, and there are major problems with both. First, there is the recognition of "irreducible complexity." The recognition of irreducible complexity involves a fallacy that Richard Dawkins has called "the argument from personal incredulity." Ever since the design argument in the early 19th century (see NATURAL THEOLOGY), creationists have argued that "I do not see how this could have evolved, therefore it could not have evolved." ID theory has simply continued this argument, at a biochemical and cellular level. This is no more convincing than when atheists claim, "I do not see how there could be a God, therefore God does not exist."

In fact, Behe's argument is what could be called "the argument from personal omniscience." It takes three forms:

- Behe makes it sound as if he assumes the omniscience of famous biochemists. Behe points out that Stanley Miller (see MILLER, STANLEY) has been doing origin-of-life simulations since 1953, the *Journal of Molecular Evolution* has been in print since 1971, and the most prominent biochemists in the world after all these years still have not figured out how complex systems might have evolved by gradual steps. Surely, if Stanley Miller cannot figure it out, then it could not have happened—which assumes the omniscience of Stanley Miller.
- Although Behe does not really consider himself omniscient, he did claim that ID theory (which everyone associates with him) is as important a contribution to science as anything contributed by Newton or Einstein.
- Behe makes it sound as if his readers are also omniscient. He describes a complex biochemical system, then he asks readers if they can imagine how it could have evolved in gradual steps. If they cannot, then it could not have happened that way.

Behe and other ID theorists also recognize irreducible complexity on the basis of biochemical and cellular systems as they now exist. They claim that simpler biochemical systems would not work, and would be of no use to the organisms even if they did:

- Behe cites the flagella of some eukaryotic cells as an example of irreducible complexity. However, there are some eukaryotic cells whose flagella lack one or more components yet still function. The sperm of eels, for example, lacks the central axis found in most eukaryotic flagella.
- Behe simply asserts that a system simpler than the irreducibly complex ones would be of no use to an organism. He uses the human immune system as an example. However, the phytoalexin response, by which plants fend off infection by fungi, is much simpler, and quite useful to the plant. Behe even cites an article by David Baltimore regarding the simpler, but still very effective, immune response of sharks, then disregards it.
- Behe overlooks the possibility that intermediate structures may have arisen as exaptations, having been useful for some other purpose than that which they now serve (see ADAPTATION).

Second, ID has difficulty with the process by which complexity has been generated. ID theorists usually insist that the only two alternatives for explaining what they call irreducible complexity is either gradual evolution or intelligent design, both from scratch. Problems with this approach include:

- *Substitutability of components of systems.* One creationist book from before the ID era, entitled *Evolution: Possible or Impossible?,* made arguments such as this: living cells use 20 different kinds of amino acids; the insulin protein has 51 amino acid residues, therefore the chances against the random origin of insulin are 20^{51}. The chance of one in 20^{51} essentially represents impossibility. Therefore, an insulin molecule could not have evolved in a single step. However, no evolutionary scientist believes that insulin arose in a single step. In addition, not all 51 amino acids have to be exactly the right ones. Bovine insulin, for example, differs from human insulin by two amino acids (about a four percent difference) yet for most diabetics bovine insulin works as well as human insulin. Substitutability of different proteins has been confirmed by genetic engineering, in which the protein used by one species can substitute for the corresponding (and structurally different) protein in another species (e.g., human genes in transgenic mice). Another example is found in the evolution of the eye. Crystallins are transparent, light-refracting proteins found in the lens of an eye. Animal species employ a wide variety of different protein types as crystallins. In all cases, the proteins were "co-opted" from different sources. In vertebrates, some crystallins evolved from heat shock and other stress proteins, while in insects, they evolved from proteins that are part of the external skeleton (see table). ID theory makes no allowance for the possible substitutability of different components of a supposedly "irreducibly complex" system.

- *Duplication and horizontal transfer of components of systems, or of entire systems.* New genes can arise from the duplication and modification of old genes (see DNA [RAW MATERIAL OF EVOLUTION]). Behe mentions this, then ignores it. Most ID theorists consider photosynthesis to be irreducibly complex. Photosynthesis of cyanobacteria, green algae, and plants has two phases. Some bacteria, however, have a simpler version of photosynthesis that closely resembles one of these systems (see PHOTOSYNTHESIS, EVOLUTION OF). It is certainly possible that cyanobacteria (the ancestors of chloroplasts; see SYMBIOGENESIS) obtained this subcomponent by HORIZONTAL GENE TRANSFER from these other bacteria. Behe admits that prokaryotic cells are simpler, and the role of symbiogenesis in the origin of eukaryotic cells, but refuses to admit this as an explanation for the supposed irreducible complexity of the eukaryotic cell. When Behe describes the complex cascade of reactions involved in the formation of a blood clot, he refers to the proteins involved without mentioning that these proteins are variant forms of serine proteases, enzymes that have functions other than blood clotting. The blood clotting genes arose as duplications of serine protease genes, followed by mutations.
- *A creator.* Behe and other ID theorists insist that you need not specify who or what created the irreducible complexity. Therefore, ID theorists insist that it is not creationism. As a result, ID theory posits a very large brick wall with a sign on it, "No scientific inquiry past this point." This is, ultimately, what makes ID nonscientific (see SCIENTIFIC METHOD). Of what kind of Creator might Behe be trying to persuade his readers? He refers to the cascade of events in blood clotting to be a Rube Goldberg apparatus, which makes it sound as if the irreducible complexities of the biological world are the product of a silly God. The Creator posited by ID also created parasites and allowed extinctions. Behe includes suffering and death as an aspect of the design of life, but he is no more successful than anyone else at explaining why a good God would have done this (see essay, "Can an Evolutionary Scientist Be Religious?").

Origins of Visual Crystallin Proteins in Different Animals (from True and Carroll)

Crystallin	Found in	Derived from
α	vertebrates	small heat shock proteins
β	vertebrates	similar to bacterial stress proteins
γ	vertebrates other than birds	similar to bacterial stress proteins
ρ	frogs	NADPH-dependent reductase
δ	turtles, lizards, crocodiles	arginosuccinate lyase
τ	turtles, lizards	α enolase
π	lizards	glyceraldehyde phosphate dehydrogenase
ε	crocodiles, birds	lactate dehydrogenase
μ	kangaroos	similar to bacterial deornithine aminase
η	humans	aldehyde dehydrogenase
ζ	guinea pigs, camels	alcohol dehydrogenase
λ	rabbits	hydroxyacyl-CoA deyhydrogenase
L	squids	aldehyde dehydrogenase
S	octopuses	glutathione S-transferase
Ω	octopuses	aldehyde dehydrogenase
0	octopuses	similar to yeast TSFI
droso-crystallin	fruit flies	insect cuticle protein
J1	jellyfishes	similar to chaperonin heat shock protein

Evolutionary scientists have been very vocal in attacking ID theory in general and Behe's book in particular. Among the reasons are:

- Behe implies that the evidence usually cited in favor of evolution does not matter (see above). It may very well be that paleontological and comparative evidence does not alter his particular argument. But by ignoring the evidence that evolution has, in fact, occurred, Behe perpetuates the misconception, prevalent in the general public, that such evidence does not exist.
- Behe does use one other creationist argument that has been repeatedly discredited. He suggests that, when evolutionary scientists disagree, they must all be wrong. He cites the controversy over symbiogenesis (see MARGULIS, LYNN), and the controversy over PUNCTUATED EQUILIBRIA (see ELDREDGE, NILES) to prove that all sides of these arguments must be wrong. He suggests, incorrectly, that Margulis and Eldredge have rejected natural selection. He compared the arguments of Margulis and the evolutionary biologist Thomas Cavalier-Smith, and said, "Each has pointed out the difficulties in each other's model, and each is correct." His conclusion is that "the natives are restless," that is, evolutionary scientists are becoming dissatisfied

with evolution, and the reason must be that in their heart of hearts they know that ID and irreducible complexity are true. If all evolutionary scientists agreed on everything, creationists could rightly suspect conspiracy; but because they do not, creationists infer that evolutionary scientists must be clueless.

- Behe uses a certain measure of ridicule, albeit muted, against anyone who would not agree with him. He reprints a Calvin and Hobbes cartoon about the little boy and his pet tiger who take imaginary journeys in a cardboard box. Behe says that when evolutionists cite examples of less complex systems it is just a "hop in the box with Calvin and Hobbes." Bruce Alberts (past president of the National Academy of Sciences) referred to the gradual evolution of series of enzyme reactions, such as A→B→C→D. Behe ridiculed him, saying that Alberts must have thought that it would be easy for C to evolve from D, since, after all, C and D are right next to each other in the alphabet. Behe not only makes evolutionary scientists look silly, but heartless. He tells the story of a child dying of a disease that results from a mutation in the membrane transport system. He leaves the reader to infer that only a heartless person would dream that such mutations could ever produce something good. He even accuses scientists of dishonesty, saying that Stanley Miller "jiggled the apparatus" until he got the results he wanted.
- Behe uses numerous folksy examples of irreducibly complex systems, from a mousetrap to roadkill to a bicycle to a man trying to swim with one arm. He implies that if one accepts evolutionary science, one must also believe that a chocolate cake could make and bake itself. Even though he criticizes evolutionary scientist Russell Doolittle for employing a "hail of metaphorical references," Behe uses 23 extended metaphors, some as long as four and a half pages, which add up to 36 pages of metaphorical material, fully 15.6 percent of his book. After the reader encounters so many of these analogies, he or she begins to accept his otherwise invalid argument from sheer repetition.

Intelligent design has replaced old-fashioned creationism as the weapon of choice of antievolutionists, many of whom are not familiar with the differences between ID and creationism. In the past, creationist activists called for the teaching of creationism alongside evolution in science classes, but now, creationist school board members, and even the president of the United States (on August 2, 2005), claim that ID should be taught alongside evolution in science classes. (President Bush's science adviser, John H. Marburger III, downplayed the president's comments and claimed that Mr. Bush only intended ID to be discussed as part of the "social context" of science; Marburger said that "evolution is the cornerstone of modern biology.")

In the minds of many people, ID is associated with the belief that there is a God who is the source of complexity and purpose in the universe and is the alternative to atheism and a purposeless universe. This is not, however, the way the ID theorists see it. They stop short of insisting on a belief in God. Although almost all of the staff of the Discovery Institute are Christians, and most are political conservatives, there is at least one fellow of the Institute who is reported to be neither. Perhaps more importantly, they are openly hostile to theistic evolutionists, who accept evolutionary science as part of God's universe. Discovery Institute fellow William Dembski, in a rebuttal to an article by physicist (and Christian) Howard Van Till, said that "theistic evolution remains intelligent design's most implacable foe."

Intelligent design theory represents a nonscientific assertion about the absence of evolution in the natural world. Many scientists believe that while God could have created "irreducible complexity," God did not do so, and that there is no way to study ID scientifically.

Further Reading

Behe, Michael J. *Darwin's Black Box: The Biochemical Challenge to Evolution.* New York: Free Press, 1998.

Bumiller, Elisabeth. "Bush remarks roil debate on teaching of evolution." *New York Times,* 3 August, 2005. Available online. URL: http://www.nytimes.com/2005/08/03/politics/03bush.html?pagewanted=print. Accessed 3 August, 2005.

Chang, Kenneth. "In explaining life's complexity, Darwinists and doubters clash." *New York Times,* 22 August, 2005. Available online. URL: http://www.nytimes.com/2005/08/22/national/22design.html. Accessed August 27, 2005.

Coppedge, James F. *Evolution: Possible or Impossible?* Grand Rapids, Mich.: Zondervan, 1973.

Dawkins, Richard. *The Blind Watchmaker: Why the Evidence of Evolution Reveals a Universe without Design.* New York: Norton, 1996.

Dembski, William A. *No Free Lunch: Why Specified Complexity Cannot be Purchased without Intelligence.* Lanham, Md.: Rowman and Littlefield, 2001.

Forrest, Barbara, and Paul R. Gross. *Creationism's Trojan Horse: The Wedge of Intelligent Design.* New York: Oxford University Press, 2004.

Krugman, Paul. "Design for confusion." *New York Times,* 5 August, 2005. Available online. URL: http://www.nytimes.com/2005/08/05/opinion/05krugman.html?ex=1123905600&en=f1c7eede740e674f&ei=5070&emc=eta1. Accessed August 5, 2005.

Miller, Kenneth R. "God the mechanic." Chapter 5 in *Finding Darwin's God: A Scientist's Search for Common Ground between God and Evolution.* New York: HarperCollins, 1999.

National Center for Science Education. *NCSE Reports*, September–December 2003 issue.

Perakh, Mark. *Unintelligent Design.* New York: Prometheus, 2003.

Richards, Jay W. "What intelligent design is—and isn't: The more scientifically sophisticated we get, the stronger the argument for intelligent design." Available online. URL: http://www.beliefnet.com/story/166/story_16659_1.html. Accessed August 28, 2005.

Shanks, Niall. *God, the Devil, and Darwin: A Critique of Intelligent Design Theory.* New York: Oxford University Press, 2004.

Thorson, Walter R. "Naturalism and design in biology: Is intelligent dialogue possible?" *Perspectives on Science and Christian Faith* 56 (2004): 26–36.

True, J. R., and Sean B. Carroll. "Gene co-option in physiological and morphological evolution." *Annual Review of Cell and Developmental Biology* 18 (2002): 53–80.

Van Till, Howard J. "*E. coli* at the No Free Lunchroom: Bacterial flagella and Dembski's case for Intelligent Design." Available online. URL: http://www.aaas.org/spp/dser/evolution/perspectives/vantillecoli.pdf. Accessed August 28, 2005.

West, John G. "Intelligent design and creationism just aren't the same." Available online. URL: http://www.discovery.org/scripts/viewDB/index.php?program=CRSC&command=view&id=1329. Accessed April 14, 2005.

Wilgoren, Jodi. "Politicized scholars put evolution on the defensive." *New York Times,* 21 August, 2005. Available online. URL: http://www.nytimes.com/2005/08/21/national/21evolve.html. Accessed August 27, 2005.

invasive species Invasive organisms enter a habitat in which they did not evolve and experience a population explosion (see POPULATION). While it is a natural process, species invasion has been accelerated by human activity. The invasive species often has a severe negative impact upon the community of species that it enters. Species invasion has affected the evolution of life on Earth. One well-known example is the invasion of placental mammals from North into South America following the formation of the Panama land bridge (see MAMMALS, EVOLUTION OF).

When individuals of a species enter into a community from outside, their most likely fate is death. The climatic and microclimatic conditions in the new habitat will probably be unsuitable for the introduced species. Even if the new species survives, its populations may grow slowly, and it may be eliminated by competition with the better-adapted native species.

In some cases, however, an invasive species (also called alien, or introduced, or exotic species) may find the new habitat to be ideal, not only for survival but as an area in which to proliferate wildly. The native range of the introduced species may have been very small, while its new range is very extensive. The Monterey pine *(Pinus radiata)* was native to three small forest patches in California, but upon its introduction to New Zealand it spread over large areas.

Human activity has had many significant worldwide impacts, including destruction of habitats, soil erosion, pollution, and depletion of resources. Many of these impacts are reversible: forests grow back, soil builds back up, and pollution degrades, after the departure of humans. However, once a new species has been introduced, and if it becomes a problematic invasive species, its presence is permanent and may resist even the most focused eradication efforts. Human activities that cause the rapid spread of introduced species include the following:

- Most of the major American agricultural and urban weeds, including the ubiquitous dandelion *Taraxacum oficinale,* came from Europe, probably because their seeds mixed with those of crop and garden species. Weeds thrive in disturbed areas, and humans have not only created many thousands of square kilometers of disturbed areas (such as farms and cities) but carry out a regular traffic among these disturbed areas. Many more weed species have made the journey from Eurasia to America than vice versa because agriculture has existed longer and been more extensive in Eurasia. Weeds evolved in areas of human disturbance from ancestors that specialized in natural disturbances (see AGRICULTURE, ORIGIN OF).
- Plants and animals may be brought to a new location as livestock, pets, or ornamentals. Ring-necked pheasants *(Phasianus colchicus)* were brought from China as game birds. Purple loosestrife *(Lythrum salicaria)* was an ornamental plant, brought to America from Europe. Some of these species have peculiarly interesting stories. Starlings *(Sturnus vulgaris)* were introduced by a man who wanted America to have all of the bird types mentioned in Shakespeare. One of the most persistent of American weeds, velvetleaf *(Abutilon theophrasti),* was brought to America because its beautiful immature fruits could be used to make decorations on slabs of butter, hence its other common names "stampweed" and "butterprint." In each case, organisms escaped and experienced rapid population growth. Many feral livestock animals now roam free, such as goats and pigs in Hawaii, and donkeys in Death Valley.
- Aquatic organisms live in bilgewater in ships that cross the ocean and escape when bilgewater is dumped. Eurasian watermilfoil *(Myriophyllum spicatum)* and the zebra mussel *(Dreissena polymorpha),* both of which are proliferating in many American waterways, may have been introduced in this way. The Asian tiger mosquito *(Aedes albopictus),* a vector of viral diseases, came to North America in pools of water that had accumulated in tires being shipped from Asia to the United States.
- New species are sometimes deliberately released to alter the natural environment. Australian melaleuca trees *(Melaleuca quinquinervia)* were released into the Everglades in order to dry them up for business and residential development.

In nearly all cases, the new species was introduced, while the predators and parasites that would normally have held its populations in check were not. Native species of the community may accidentally help the invader to spread; for example, muskrats help spread the underground stems of loosestrife.

The introduced species have many profound influences on the community they invade. They may displace native species or even drive them to extinction:

- Fire ants *(Solenopsis invicta)* are native to South America, but were introduced into the United States about 1930. They have spread throughout the southern states and have recently entered California as well. Now called Red Imported Fire Ants, they cause painful stings, kill some ground-dwelling wild animals, and damage some equipment and some agricultural crops.
- Ferocious African honeybees have displaced European honeybees (both varieties of *Apis mellifera*) from the tropical zones of the Americas, reaching as far north as southern Texas. African "killer bees" have had a longer history of exploitation by humans and other predators, and their evolutionary response has been a tendency to attack potentially dangerous animals. African bee queens were allowed to escape from a breeding experiment in Brazil in the 1960s.

Unable to tolerate cold winters, the African bees will probably not extend beyond the southernmost tier of states.

- The Nile perch *(Lates niloticus)* was introduced into Lake Victoria in Africa, where it has driven many of the native species of fish nearly to extinction.
- Mainland birds and other introduced animals have driven the native birds of Hawaii away from the coastlands, and many into extinction. The few remaining native Hawaiian birds are largely restricted to the highlands.
- European grasses, originally spread in the droppings of horses that escaped from the Spaniards, have displaced the native California grasses.
- Populations of melaleuca and loosestrife (see above) outgrow those of native plants because they produce millions of seeds.
- Introduced African grasses inhibit the germination of native Hawaiian tree seedlings.

Invading species may create many disturbances. Feral pigs in Hawaii create disturbances in the rain forest, in which the native plant species cannot grow well. Plant species on islands have few if any defenses against large browsing or grazing animals, as large animals seldom disperse to islands (see BIOGEOGRAPHY). However, introduced plants, such as the strawberry guava, thrive in the disturbances created by the feral pigs. When grasses are first introduced into a community that previously had none, they can create a fire hazard. Fires deplete the populations of species that are not fire-resistant and further encourage the spread of the grasses, which can grow back after the fires.

Invading plant species can inhibit the growth of native species. The leaves of the wax myrtle *(Myrica cerifera)*, which has invaded Florida wetlands, do not decompose well and have literally clogged up the aquatic food chain. The deep litter of casuarina trees *(Casuarina equisetifolia)* introduced into Hawaii from southeast Asia inhibits the growth of practically everything under their canopy. Kudzu vine *(Puereria montana)*, introduced from Asia, drapes over and kills many trees in the southeastern United States.

Exotic predators have disrupted whole food chains. For instance, the brown snake *(Boiga irregularis)* has eaten many of the birds on the island of Guam.

Invading species can have severe economic impacts on agriculture and even on public health, as well as on nature preserves. Much money and effort are spent in attempts to control their spread, often by means that are as dangerous as they are ineffective, such as pesticides or mass hunting campaigns. International cargo and luggage are carefully inspected to assure that there are no exotic plant or animal stowaways.

Scientists have investigated biological controls as a method of bringing escaped exotic species under control. Loosestrife is not a pest in Europe, where its populations are controlled by beetles. Scientists are now attempting to introduce those beetles to North America. Other scientists are trying to introduce parasitic flies, whose larvae develop inside the heads of Red Imported Fire Ants and cause the ants' heads to fall off, as a biological control agent. Biological controls, unlike pesticides, can be permanent.

Eventually, other species (either native or themselves introduced) will evolve ways of eating the invaders, bringing their populations into check (see COEVOLUTION). The time required for this process, however, is likely to make it meaningless to human concerns.

Further Reading

Agricultural Research Service, United States Department of Agriculture. "Biological control sends scurrying scourge scrambling." Available online. URL: http://www.ars.usda.gov/is/AR/archive/may03/biol0503.htm. Accessed September 22, 2005.

Burdick, Alan. "The truth about invasive species." *Discover,* May 2005, 34–41.

Cox, William. *Alien Species and Evolution: The Evolutionary Ecology of Exotic Plants, Animals, Microbes, and Interacting Native Species.* Washington, D.C.: Island Press, 2004.

National Agricultural Library. "Invasivespecies.gov: A gateway to federal and state invasive species activities and programs." Available online. URL: http://www.invasivespecies.gov. Accessed April 18, 2005.

Sax, Dov F., John J. Stachowicz, and Steven D. Gaines, eds. *Species Invasions: Insights into Ecology, Evolution, and Biogeography.* Sunderland, Mass.: Sinauer Associates, 2005.

Todd, Kim. *Tinkering with Eden: A Natural History of Exotic Species in America.* New York: Norton, 2002.

United States Department of Agriculture. "Animal and Plant Health Inspection Service: Safeguarding American agriculture." Available online. URL: http://www.aphis.usda.gov. Accessed April 18, 2005.

Vogt, J. T., Stanley A. Rice, and Steven A. Armstrong. "Seed preferences of the Red Imported Fire Ant (Hymenoptera: Formicidae) in Oklahoma." *Journal of Entomological Science* 38 (2003): 696–698.

invertebrates, evolution of Invertebrates are animals without internal skeletons and backbones. Invertebrates do not constitute a formal taxonomic group (see CLADISTICS) but are a category that contains all animals, except for the vertebrate lineage within the chordate phylum (see LINNAEAN SYSTEM). It is a distinction that human observers, as vertebrates, can hardly avoid making within their view of the world. As with other groups of organisms, the evolutionary history of invertebrates can be reconstructed by comparisons of DNA (see DNA [EVIDENCE FOR EVOLUTION]), comparisons of the structures of living forms, and from fossils.

Because of the internal skeletal system, some very large vertebrates have been able to evolve (see DINOSAURS). The external skeleton, and the various manners by which invertebrates breathe, have placed a size constraint on invertebrates (see ALLOMETRY). Therefore, with the exception of some large oceanic mollusks (giant clams and giant squid), and the *Anomalocaris* of the CAMBRIAN PERIOD (see BURGESS SHALE), all invertebrates have been and are relatively small.

All members of the animal kingdom are believed to be descended from an ancestral colonial protist. Because the most primitive animals are the sponges (see below), the most likely protistan ancestors are today represented by the choanoflagellates, which are protists that form clusters and grab

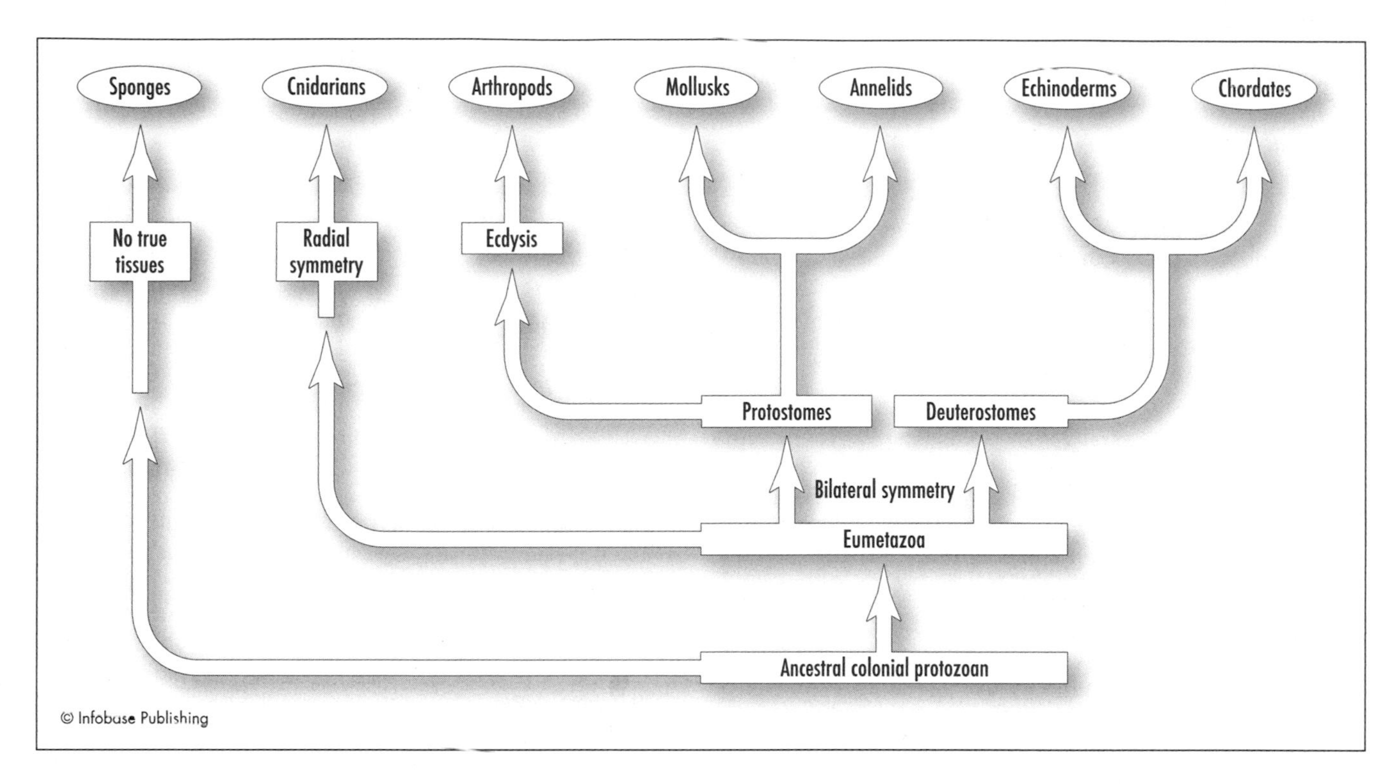

The major invertebrate groups can be distinguished on the basis of tissues, symmetry, and development. Many groups have been omitted from this diagram. Sponges have no true tissues; eumetazoans ("true animals") have tissues. Cnidarians have radial symmetry; the others have bilateral symmetry. The protostomes that have ecdysis (molting) evolved separately from those that do not. All of the vertebrates are within the chordate phylum.

food particles with whiplike structures. In the evolution of animals, the following significant divergences occurred (see figure). All of these divergences, and probably all animal phyla, had evolved by the end of the Cambrian period.

Sponges and placozoans vs. true animals. Sponges (Phylum Porifera) diverged from the Eumetazoa (true animals). The sponges are among the animals most like colonial protists. They have no true tissues. They resemble a colony of single cells, each of which is mostly on its own to obtain food. Each of the *choanocyte* cells has flagella that propel the water and grab food particles from it. Choanocyte cells closely resemble the cells of colonial choanoflagellates. The sponge cells usually rest on a framework of *spicules,* which is the tough matrix that remains when the cells die. One example of such a matrix is the familiar kitchen sponge. Sponges have some whole-organism characteristics that identify them as animals rather than clusters of protists. The entire sponge grows from one reproductive cell. Also, the choanocytes whip their flagella in unison, creating an orderly flow of water in through pores and out through a chimneylike structure at the top. Finally, the sponge produces amoebocyte cells that patrol the inner passages and eliminate invaders. The structural complexity of the sponge is a good example of EMERGENCE, as it results from the simple interactions of its component cells. The *placozoans* (Phylum Placozoa) are even smaller and simpler clusters of flagellated cells.

Radial vs. bilateral animals. The true animals diverged into the radially symmetrical cnidarians and ctenophores and the bilaterally symmetrical animals. In bilateral symmetry, there is a top and bottom, left and right, front and rear; in radial symmetry, there is at most only top and bottom.

- *Radially symmetrical animals.* The *cnidarians* (Phylum Cnidaria; formerly called coelenterates) include jellyfish, sea anemones, and corals. They have two tissue layers, an outer and an inner, with a noncellular material between them (the "jelly" of jellyfish). The sac formed by the inner layer serves as a digestive cavity. They have a network of nerves but no brains. They capture food by means of tentacles with stinging cells. They have tissues but no complex organs. Cnidarians alternate between a form that attaches to surfaces (the *polyp*) and the swimming sexual jellyfish stage (the *medusa*). What cnidarians lack in individual complexity they may make up for by forming complex colonial clusters. Most corals, and some jellyfish, obtain much of their food from single-celled algae that live inside their tissues (see COEVOLUTION). The *ctenophores,* or comb-jellies, share a common ancestor with cnidarians but, aside from a superficial similarity to jellyfish, are different in structure.
- *Bilaterally symmetrical animals.* All bilateral animals share a set of Hox genes that specify the front-to-back order of body parts (see DEVELOPMENTAL EVOLUTION) and are described below.

Protostomes v. deuterostomes. The bilaterally symmetrical animals diverged into two lineages that differ primarily in the manner in which their embryos develop. In *protostomes,* the early embryo cell divisions occur in a spiral fashion, whereas in *deuterostomes* they occur in a radial fashion. In protostomes, the developmental fate of each cell is determined from the beginning. In deuterostomes, the fate of the cells is not determined until about the eight-cell stage. Therefore, in deuterostomes, one fertilized egg cell can become up to eight identical offspring. Deuterostomes, unlike protostomes, can have identical twins, triplets, etc.

Protostomes: Ecdysozoans v. others. On the basis of DNA evidence, the protostomes are divided into two groups. The *ecdysozoans* secrete an external skeleton. As they grow, they split open the skeleton and squirm out of it, secreting a new one. This process, called *ecdysis,* characterizes arthropods and nematodes.

- *Arthropods* (Phylum Arthropoda) have hard external skeletons and jointed appendages. Appendages are legs or other leglike structures (such as the antennae of insects). Arthropods were among the earliest phyla of animals to have a burst of evolutionary diversification, in the Cambrian period, including the TRILOBITES, which evolved into numerous species, until becoming extinct at the end of the Permian period (see PERMIAN EXTINCTION). Today, the major arthropod groups are the crustaceans (including crabs, lobsters, shrimps, and terrestrial pillbugs); the centipedes and millipedes; the chelicerates (including spiders, scorpions, mites, and ticks); and the insects. In terms of sheer weight and number, as well as species diversity, arthropods rule the animal world (see BIODIVERSITY).
- *Nematodes* (Phylum Nematoda) are the roundworms. They are incredibly abundant inhabitants of the soil, especially near plant roots and in wetlands. Experts say that if everything in the world except nematodes vanished, one would still be able to see the Earth as a shimmering sphere of nematodes.

Among the other protostomes that are not ecdysozoans are:

- *Flatworms* (Phylum Platyhelminthes) include aquatic and parasitic forms. The parasitic forms may have extremely complex life cycles.
- *Mollusks and brachiopods. Mollusks* (Phylum Mollusca) have a mantle and guts over a single foot. *Bivalve* mollusks (e.g., clams) secrete two hard shells hinged together. They filter the water in which they live to obtain food particles. *Cephalopod* mollusks have a head (cephalo-) over a foot (-pod), with a beaked mouth in the middle of the foot, as in octopus and squid. *Gastropod* ("stomach-foot") mollusks include snails and slugs. Despite their superficial similarity to bivalves, the *brachiopod* "lampshells" are a distinct phylum, because unlike bivalves they have a specialized feeding apparatus called a *lophophore.* The present-day abundance of bivalves and relative scarcity of brachiopods may be a leftover effect of the Permian extinction, which nearly destroyed the brachiopods.
- *Annelids* (Phylum Annelida) are the segmented worms, such as many of the marine worms, the earthworms, and the leeches. Because annelids and insects both have segments, they were formerly thought to share a relatively recent common ancestor, but DNA analysis has forced a reevaluation of this idea.

Deuterostomes. The two phyla of deuterostomes are the echinoderms and the chordates.

- Adult *echinoderms* (Phylum Echinodermata, which means "spiny skin") have spiny or knobby outer layers and move along the rocks of the seashore by means of water-powered tube-feet. They include starfish and sea urchins. The larvae of echinoderms are bilaterally symmetrical, even though the adults look superficially radial.
- The *chordates* have a rod, or notochord, along the back, which strengthens the body and the main nerve cord. Invertebrate chordates include the lancelet, which looks superficially like a headless fish. The vertebrates evolved from an ancestor that resembled a lancelet (see FISHES, EVOLUTION OF).

Almost all of the animal phyla are represented in the Cambrian period in at least a rudimentary form. It is not clear how many of them were present before the CAMBRIAN EXPLOSION. Some of the EDIACARAN ORGANISMS from the period just before the Cambrian resemble jellyfish, while others resemble colonial cnidarians. All identifications of the Ediacarans, however, are controversial. Fossils of animal embryos are known from the Ediacaran period. DNA studies suggest that some of the earliest evolutionary divergences of animals occurred from 700 million to 1 billion years ago, although the fossil evidence for these divergences is scarce.

Most invertebrate phyla are completely aquatic. However, like the vertebrate lineage within the chordate phylum, several arthropod lineages have diversified into all the habitats of dry land.

Further Reading

Bottjer, David J. "The early evolution of animals." *Scientific American,* August 2005, 42–47.

Knoll, Andrew H., and Sean B. Carroll. "Early animal evolution: Emerging views from comparative biology and geology. *Science* 284 (1999): 2,129–2,137.

Museum of Paleontology, University of California, Berkeley. "Introduction to the Placozoa: The simplest of all known animals." Available online. URL: http://www.ucmp.berkeley.edu/phyla/placozoa/placozoa.html. Accessed April 18, 2005.

island biogeography *See* BIOGEOGRAPHY.

isolating mechanisms Isolating mechanisms are biological processes that cause *reproductive isolation,* in which populations cannot successfully interbreed if they are in contact. If two populations could interbreed if they were in contact but do not do so because they live in different locations, the populations are geographically isolated. Geographic isolation is not an isolating mechanism. Isolating mechanisms keep populations from interbreeding even if they live in the same place. Evolutionary biologists consider two populations that would readily interbreed and produce fertile offspring, if they were in contact, to be members of the same species. An isolating mechanism therefore represents the first step in turning separate populations into separate species (see HYBRIDIZATION; SPECIATION).

Geographic isolation can occur as CONTINENTAL DRIFT and the formation of new mountain ranges separate populations. Geographic isolation can also occur from local geographical features. In a large population that covers a wide area, all of the individuals may be potentially able to interbreed, but do not; by interbreeding with other individuals in the same locality, they form subpopulations or *demes* that are imperfectly isolated from one another. The effectiveness of geographic isolation between populations or between demes depends on how much dispersal occurs between them. The human species, after its origin about 100,000 years ago, spread all over the world and evolved geographic and racial differences. Travel was slow and difficult, and populations remained largely separate. However, this geographic isolation did not last long enough to produce different human species; all human races are able to interbreed, and frequently do. Today, geographic isolation of human populations has largely broken down, and some experts predict that distinctions between human races will disappear a few centuries from now.

As long as populations continue to exchange genes, for example through migration of individuals between them, they cannot begin the process of speciation. Two geographically isolated populations will almost inevitably evolve into different species. This may occur because environmental conditions (either climatic or biological) are different in the two populations, thus NATURAL SELECTION favors different characteristics in each population. But even if the environmental conditions are essentially the same for both populations, they will still diverge, because different mutations and gene combinations will occur by chance in each of them. It is highly unlikely that exactly the same history of adaptive events (genetic variation and natural selection) would happen in both isolated populations. If the two populations come back in contact, they do not interbreed, because isolating mechanisms have evolved separately in each.

In many cases, natural selection favors isolating mechanisms in the absence of geographical isolation. These isolating mechanisms prevent the populations from wasting their reproductive resources on crosses that would produce fewer or inferior offspring. Because natural selection occurs within populations, it is important to recognize that natural selection does not favor processes that drive populations apart. Instead it favors processes that maximize reproductive success within the resulting populations; reproductive isolation is a result, rather than a cause, of this process.

Prezygotic isolating mechanisms prevent interbreeding from occurring in the first place—that is, the zygote (fertilized egg) does not form. *Postzygotic* isolating mechanisms operate after the formation of the zygote.

Prezygotic isolating mechanisms include the following examples:

Differences in pollination mechanisms. If a mutation occurs in a population of flowering plants that changes the characteristics of the flowers, the new kind of flower may not be recognized by the same pollinator that visits the old kind of flower.

This is apparently what happened in California populations of two closely related species of monkeyflowers (genus *Mimulus*). *M. lewisii* has light purple flower that attract bees, while *M. cardinalis* has yellowish-red flowers that attract hummingbirds. Phylogenetic analysis (see CLADISTICS) indicates that the common ancestor of the two species was pollinated by bees. Apparently, mutations occurred that increased the production of red pigments (anthocyanins) and yellow pigments (carotenoids) and increased the amount of nectar production in the lineage that became *M. cardinalis*. The *M. cardinalis* flowers, though side by side with *M. lewisii* flowers, were reproductively isolated from them. This occurred because bees cannot see red, ignored *M. cardinalis,* and preferred *M. lewisii,* while hummingbirds can see red and preferred *M. cardinalis*.

Subsequent to the reproductive isolation, further evolutionary changes occurred in the two monkeyflower species. *M. cardinalis* evolved a long, tubular flower shape that matched the long break and tongue of the hummingbird, and *M. lewisii* retained the short, wide tubular flower shape, with a landing platform on the front, that the bees preferred. The changes in flower shape reinforced the initial reproductive isolation that had occurred because of a change in flower color and nectar production. While the change in flower color and nectar production was sufficient to separate one species into two, natural selection subsequently favored changes in flower shape that enhanced the success of each species with respect to its own pollinator. What was once one species of monkeyflower has evolved into two, side by side.

Although nobody was there to observe the original formation of the two monkeyflower species, evolutionary biologists Douglas Schemske and H. D. Bradshaw have experimentally re-created the sequence of events. They crossed the two monkeyflower species and produced a whole range of hybrids with intermediate shapes, nectar volumes, and colors. They also identified DNA markers that were associated with differences in pigment production, nectar production, and flower shape (see QUANTITATIVE TRAIT LOCI). They found that, among these hybrids, the DNA sequences that were associated with enhanced production of the two pigments and of nectar, not the DNA sequences associated with flower shape, were most effective in determining which pollinator visited each flower. Pigment and nectar production had started the reproductive isolation, and flower shape had reinforced it.

Another example of a change in pollination mechanism involves *Schiedea salicaria,* a small shrub that is native to Hawaii, investigated by evolutionary biologists Ann Sakai and Steve Weller. A few relatively minor genetic mutations appear to have caused some individuals in the population to rely more on pollination by wind, and less on pollination by insects. These two groups can cross-pollinate but, because they rely largely on different methods of pollination, they do not frequently do so. The individuals that have greater wind pollination characteristics tend to live in drier locations, where insect pollinators are less common. Though not completely isolated from one another, these two groups of plants have apparently started on the road to separation, perhaps eventually into two different species.

Natural selection often reinforces the dependence on different methods of pollination. For example, flowers that have characteristics intermediate in their appeal to hummingbirds and to bumblebees will probably not be pollinated very well by either hummingbirds or bumblebees.

Differences in timing of reproduction. Each species has its own reproductive season. Mutations that cause some individuals in a population to have a different reproductive season will prevent those individuals from interbreeding with others. Early in the evolution of alder trees (genus *Alnus*), populations diverged into some that pollinated in the autumn (subgenus *Clethropsis*) and those that pollinated in the spring (subgenus *Alnus*). These subgenera can no longer interbreed. In North America *Rhagoletis pomonella* flies whose maggots ate wild hawthorn berries, some populations evolved a preference for apples. An immediate reproductive isolation occurred because apple fruits mature three to four weeks earlier than those of hawthorns, with the result that the flies that hatch from apples are ready for reproduction weeks earlier than those that hatch from hawthorns.

Differences in animal mating behavior. Animal populations may be reproductively isolated because of differences in mating rituals. Evolutionary biologists Peter and Rosemary Grant have extensively studied the beak sizes and mating preferences of several species of DARWIN'S FINCHES on some uninhabited islands in the GALÁPAGOS ISLANDS. They have found that hybrids between finch species can occur, but occur rarely. Young male birds learn their mating songs from their fathers, and females use male mating songs to guide their choice of mates. Occasionally, a young male bird overhears and learns the song of a different species of finch nearby. When the young male bird grows up and sings, he attracts females of the wrong species.

Hawaii has hundreds of species of fruit flies (genus *Drosophila*), all of them unique to the Hawaiian Islands, and all of them having recently evolved there (the islands are from five million to one-half million years old). There may be scores of species living on a single island within the archipelago and close enough to one another to readily interbreed. They do not interbreed, however, because they have evolved different mating dances. The females of one species will not recognize the males of another species as being potential mates. Laboratory investigations of fruit flies (some species of which are the most intensively studied animals in the history of biology) indicate that small genetic mutations can cause major changes in mating behavior. Therefore the reproductive isolation between two groups of flies may have occurred quickly after a relatively minor genetic change.

Subsequent to the reproductive isolation caused by differences in mating dances in the Hawaiian flies, the populations evolved yet other differences, making them into recognizably different species. The species now differ in the anatomy of mating and reproductive anatomy, and other characteristics that make them look different to one another (for example, one species has its eyes on long stalks). These anatomical differences between species may or may not have anything to do with adaptation to their environments.

Specialization on different food resources. Specialization of animals on different food resources will result in reproductive isolation only if individuals crossbreed primarily with other individuals that have the same specialization. The most famous examples of this are of insect populations within the same species that specialize on new food plants. They not only eat the plants but live and mate upon them, isolated from the insects on a different kind of food plant. *Rhagoletis pomonella* flies (see above), native to North America, lay their eggs on the fruits of hawthorns, a wild bush that produces fruits that resemble small crabapples. When Europeans and white Americans began planting apples in eastern North America, some populations of this fly species began to lay their eggs in apples. The populations of flies that lived in apple trees could potentially interbreed with the populations that lived on hawthorns but seldom did so, even if wild hawthorns and cultivated apples were side by side. After a couple of centuries of isolation, the apple flies and the hawthorn flies will no longer interbreed even when given the opportunity to do so in the laboratory. This process was observed by naturalist Benjamin Walsh, an American acquaintance of Charles Darwin, in the 19th century, and was more fully investigated in the 20th century.

Another example of reproductive isolation occurring on newly introduced plant species involves the soapberry bug. Soapberry bugs have snouts about 0.36 inch (9 mm) in length that penetrate the fruit and consume contents from the seeds of balloon vines, native to the southern United States. However, in southern Florida, golden rain trees have been imported from Asia and used as ornamentals in the vicinity of balloon vines. Golden rain tree fruits are smaller, thus a smaller snout is needed to penetrate it and get to the seed. Some populations of soapberry bugs have specialized on golden rain trees and seldom interbreed with the populations that still live on balloon vines. The populations of bugs living on the golden rain trees have begun to evolve shorter beaks, about 0.28 inch (7 mm) in length. This process occurred recently, as golden rain trees were not extensively planted in Florida until the 1950s. In another example, fruit flies that resulted from the hybridization of a species of fly that lives and reproduces on blueberries and a species that lives and reproduces on snowberries have become reproductively isolated from both parental species by living on Japanese honeysuckle, which has been in North America for less than three centuries (see INVASIVE SPECIES).

Sometimes the isolation of populations on different food resources can result from just a small number of genetic changes. Two species of flies in the Seychelles archipelago of the Indian Ocean are *Drosophila simulans* and *Drosophila sechellia.* The first species avoids the poisonous morinda fruit, while the second species is attracted to it, for laying their eggs. Researchers found that the species differed by only a small number of genes—just enough to produce tolerance of and attraction to morinda fruits in *D. sechellia.*

Examples are not, however, limited to insects that live on and eat plants. Marine stickleback fishes invaded freshwater lakes of British Columbia after the most recent of the ICE AGES. Some of them had a tendency, apparently genetically based, to eat small plankton near the surface of the lakes; others preferred eating material that had settled to the bottoms of the lakes. Males and females mated preferentially in the same locations where they ate, which set them on the road to becoming two separate populations. Now, the surface- and the bottom-dwelling populations have begun to evolve unique adaptations to their new modes of life: The mouths of the bottom dwellers are larger and turned downward; those of the surface dwellers are smaller and not downturned.

In each of these cases, natural selection has probably favored the specialization on different food resources. If the populations interbred, the resulting offspring might be inefficient at exploiting either of the parental food resources.

Specialization on different symbionts. SYMBIOGENESIS is the origin of new species as a result of the symbiotic COEVOLUTION of two or more species. For example, a population of bacteria may take up residence in a new kind of host cell, for example an amoeba, and reproduce inside that host cell. Whenever the amoeba reproduces, the bacteria are passed on to the new amoebae. The amoeba represents a new environment for this population of bacteria. This population is isolated from the bacteria from which it originally came and adapts to a new set of conditions. The amoebae, which have evolved a tolerance for the bacteria, may also be isolated from other amoebae if the other amoebae cannot tolerate the bacteria. By depending on each other, the amoebae and the bacteria are isolated from other amoebae and bacteria. The result can be a new species of amoeba and a new species of bacteria. In time, the bacteria may degenerate to such an extent that they are no longer recognizable as a distinct species. The first stages of this process were actually observed to occur in a laboratory setting.

Differential success at fertilization. Even if individuals from two populations succeed in mating, their reproduction may still not be successful, if the male reproductive cells (sperm of animals, or pollen nuclei of plants) are inferior at fertilizing the female reproductive cells (eggs of animals, or ovules of plants). The sperm cells from one population may swim faster and suffer fewer deaths on their way to the egg than the sperm cells from another population; this will tend to isolate the two populations. The female reproductive tract of many vertebrates represents a hostile environment through which only the hardiest sperm can travel—that is, the sperm that are best adapted to those particular conditions. In flowering plants, pollen grains land on the stigma of the flower, germinate, and grow a pollen tube down into the ovary of the flower. Some pollen grains are better at doing this than others. Frequently, the stigma of a flower will not even permit the wrong kind of pollen—either pollen that is genetically identical to it, or pollen that comes from the wrong species—to germinate in the first place.

Postzygotic isolating mechanisms usually result because the offspring of matings between species (hybrids) or between populations have inferior reproduction. For example, hybrids between oak species can occur but are relatively rare; there are even reports of hybrids between the great cat species of Africa, although this is more common in zoos than in the wild. The rarity of such hybrids—they do not establish their own populations that persist and spread—suggests that they have inferior growth and reproduction.

On some of the uninhabited Galápagos Islands, species of Darwin's finches have postzygotic isolation, as well as the prezygotic isolation described above. The frequent shifts between rainy weather and drought keep the large-beaked finches and the small-beaked finches from successfully interbreeding. During times of rain, small seeds are abundant and small-beaked finches prosper; during droughts, large seeds are abundant and large-beaked finches prosper. Natural selection seldom favors beaks of intermediate size in these populations. However, on the Galápagos Islands that are inhabited, where human activity causes a continuous availability of seeds, the finches with intermediate beaks are successful. Natural selection seems to be favoring isolating mechanisms on the uninhabited islands, but not favoring them on the islands inhabited by humans.

Among plants, hybrids between species frequently occur but may result in offspring with incompatible chromosomes. If the chromosomes within the cells of the offspring are unable to match up with one another during meiosis, the offspring are sterile (see MENDELIAN GENETICS). However, among plants, sterile individuals may have options that are not available to animals.

- Many plants reproduce asexually, a process called *apomixis.* The embryo produced inside the seed is an exact genetic copy of the parent. The apomictic plants may constitute a separate species; they are reproductively isolated from both of the species that produced them, because they do not have sexual reproduction at all.
- Plants frequently undergo spontaneous doubling in the number of chromosomes. When the number of chromosomes doubles in an animal egg cell, the egg will probably die. Most animals tolerate only the most minor of chromosome changes; even then, as in Down syndrome or trisomy 21 in humans, a single extra chromosome results in sterility and retardation. However, a plant egg cell with doubled chromosomes may survive. A sexually sterile plant, if it undergoes chromosome doubling, may produce offspring that are fertile because the doubled chromosomes now form pairs. This allows meiosis to occur. The population of plants with doubled chromosomes, however, may not be able to interbreed with either of the ancestral populations. The hybrid is then instantly isolated from both parents and has characteristics different from both; it becomes an "instant species." This has been observed in the goatsbeard *Tragopogon* and in the sunflower *Helianthus.* This form of speciation is common in the plant kingdom. The study of chromosomes suggests that almost half of all plant species, including 95 percent of fern species, originated by chromosome doubling following hybridization.

The inferiority of offspring produced by interbreeding between two populations (a postzygotic isolating mechanism) may promote the evolution of prezygotic isolating mechanisms. Conversely, the benefit conferred by mating with conspecific individuals may promote the evolution of isolating mechanisms. Anatomical differences that prevent interbreeding may be favored by natural selection because they encourage individuals to mate only with other individuals that are similar to them. Traits that encourage mating among individuals that have very similar characteristics are called *specific mate recognition systems.* Specific mate recognition systems may allow individuals to avoid wasting their reproductive efforts on crossbreeding with an incompatible population.

Natural selection causes evolutionary changes within populations, as Charles Darwin (see DARWIN, CHARLES) first explained (see *ORIGIN OF SPECIES* [book]). However, Darwin did not adequately explain how one species could diversify into more than one. In order for one species to evolve into more than one, isolation is essential; and if isolation occurs, speciation may be inevitable. Evolutionary scientists since the time of Darwin have sought, and found, numerous examples of reproductive isolation, often resulting from isolating mechanisms.

Further Reading

Eldredge, Niles. "Of genes and species: Modern evolutionary theory." Chap. 5 in *The Pattern of Evolution.* New York: Freeman, 1999.

Malausa, Thibaut, et al. "Assortative mating in sympatric host races of the European corn borer." *Science* 308 (2005): 258–260.

Schilthuzen, Menno. *Frogs, Flies, and Dandelions: The Making of Species.* New York: Oxford University Press, 2001.

Stanton, Maureen L., and Candace Galen. "Life on the edge: Adaptation vs. environmentally mediated gene flow in the snow buttercup, *Ranunculus adoneus.*" *American Naturalist* 150 (1997): 143–178.

Weis, Arthur E., and Tanya M. Kossler. "Genetic variation in flowering time induces phenological assortative mating: Quantitative genetics methods applied to *Brassica rapa.*" *American Journal of Botany* 91 (2004): 825–836.

isotopes Isotopes are atoms of the same element that have different numbers of neutrons in the nucleus. The smallest (lightest) isotope contains about the same number of neutrons as protons. The number of protons defines the element. For example, all carbon atoms have six protons. Carbon-12 (^{12}C) has six protons and six neutrons, while ^{13}C has seven neutrons, and ^{14}C has eight neutrons.

Isotopes (see table) are useful in evolutionary studies for two reasons:

- Many isotopes are radioactive—that is, the extra neutrons destabilize the nucleus, which ejects particles and changes into another kind of atom at a constant rate. This makes radioactive isotopes useful for determining the ages of some rocks (see RADIOMETRIC DATING). ^{14}C is radioactive and is the basis of radiocarbon dating.
- Nonradioactive isotopes can be useful as indicators of environmental conditions or biological activity in ancient deposits, fossils, or remnants of organisms. ^{13}C is an example of a nonradioactive isotope.

Mass spectroscopy separates isotopes of one kind of atom from one another and measures them separately, allowing an *isotope ratio* to be calculated. The isotope ratio can be compared to standards to allow interpretation.

Carbon isotopes. Carbon dioxide (CO_2) in the air contains both $^{12}CO_2$ and a very small amount of $^{13}CO_2$. CO_2 reacts with calcium in seawater to produce calcium carbonate, which is limestone. Limestone can therefore be inorganic in origin, although massive deposits of calcium carbonate can be produced by microorganisms. The ratio of ^{13}C to ^{12}C in the air is defined as zero. When photosynthetic organisms remove CO_2 from the air to make it into sugar, the enzyme rubisco prefers ^{12}C. All of the organic compounds of the photosynthetic organisms, and the entire food chain of animals and decomposers, come from this sugar. Thus organic material has more ^{12}C, and less ^{13}C, than the air. Inorganic limestone, in contrast, tends to have a little more ^{13}C than does the CO_2 in air, because the heavier isotope sinks. Zero or positive numbers for the ratio (called $\delta^{13}C$, in parts per thousand) are inorganic, while negative numbers are organic. In the fossil record, inorganic limestone has a $\delta^{13}C$ of about zero, while for organic matter it is -25 parts per thousand. The Isua formation in Greenland (sedimentary rocks that are 3.8 billion years old, almost as soon as the Earth was cool enough for oceans to form) contains no fossil cells but does contain graphite with a significantly negative $\delta^{13}C$ ratio, suggesting it was organic in origin. This is the earliest evidence of life, and indicates that life originated very rapidly after the oceans formed (see ORIGIN OF LIFE). Frequently, a decrease in the $\delta^{13}C$ ratio precedes a period of glaciation, perhaps because photosynthesis removes a great deal of CO_2 from the air and reduces the GREENHOUSE EFFECT. Abrupt decreases in $\delta^{13}C$ are not, however, always associated with glaciations. A moderate $\delta^{13}C$ ratio may result from the photosynthesis of C_4 plants that live in warm regions (such as some grasses), because the enzyme that removes CO_2 from the air in these plants is not rubisco. However, C_4 plants have probably existed only for part of the CENOZOIC ERA (see PHOTOSYNTHESIS, EVOLUTION OF).

The higher $\delta^{13}C$ of C_4 plants can also serve as an indicator of an organism's diet. Laser studies of layers of enamel in the teeth of fossils of robust AUSTRALOPITHECINES indicates that they seasonally altered their diet between C_3 plants such as fruits and berries and C_4 plants such as grains. The most abundant agricultural C_4 plant in the world today is maize. Many processed foods contain high-fructose corn syrup, making maize (indirectly) the single greatest source of

Isotopes Used as Climate Indicators and Bioindicators

Element, common isotope	Less common isotope	If enriched, an indicator of:
Carbon, ^{12}C	^{13}C	^{12}C: photosynthesis ^{13}C: C_4 photosynthesis ^{13}C: inorganic source
Oxygen, ^{16}O	^{18}O	^{18}O: ocean evaporation, ice ages
Hydrogen, ^{1}H	^{2}H	^{2}H: ocean evaporation, ice ages
Nitrogen, ^{14}N	^{15}N	^{15}N: carnivorous diet
Sulfur, ^{32}S	^{34}S	^{32}S: bacterial metabolism
Strontium, ^{87}Sr	^{86}Sr	^{87}Sr: continental erosion

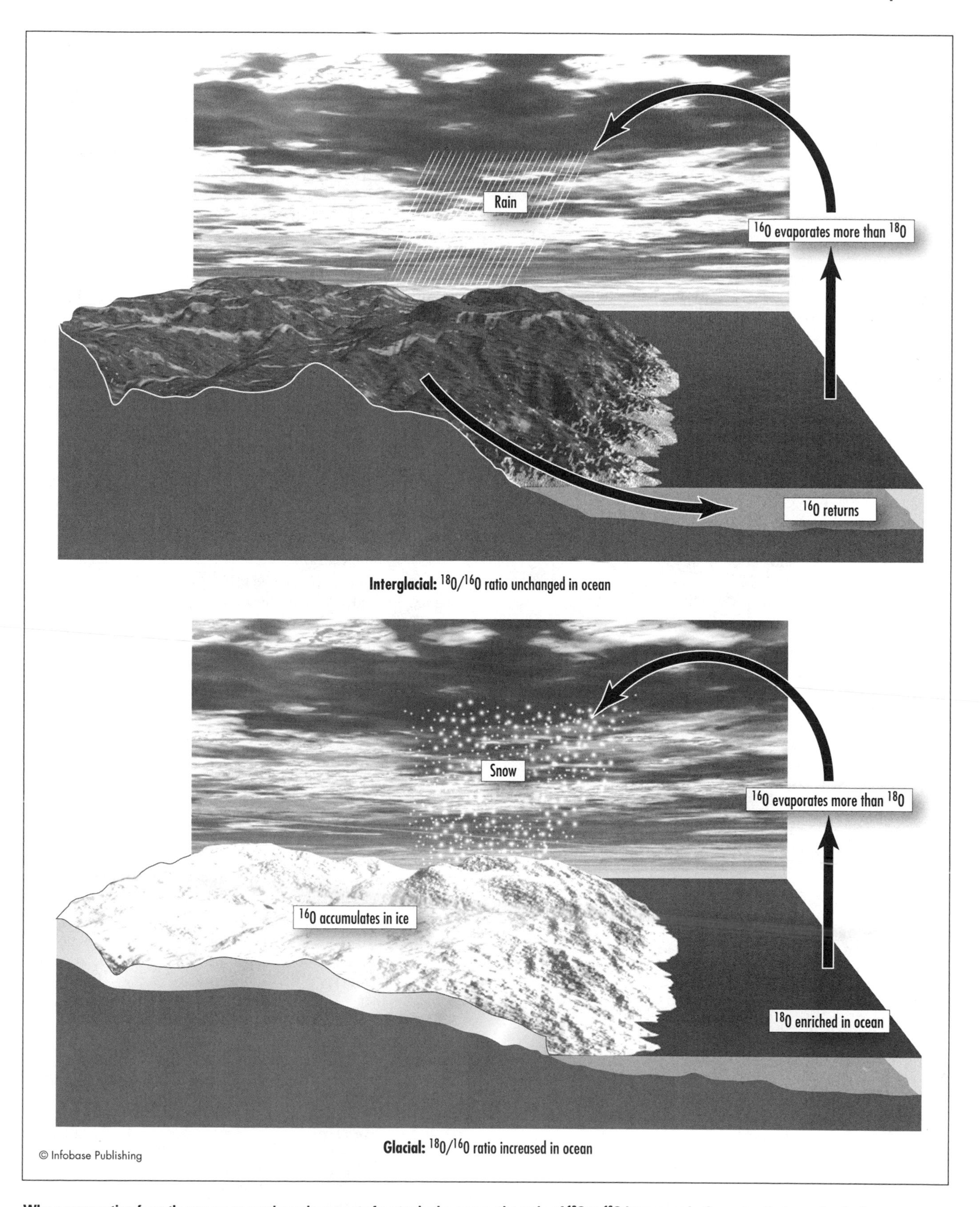

When evaporation from the ocean exceeds replacement of water in the ocean, the ratio of ^{18}O to ^{16}O increases in the water that remains in the ocean. This can occur when ice buildup occurs on continents.

food in developed countries. This has caused human bodies to have a measurably higher $\delta^{13}C$ than those of wild animals, prompting one researcher to call modern humans "corn chips on legs."

Since fossil fuel combustion, burning of forests, and landfills emit methane (CH_4) from biomass that contains a low $\delta^{13}C$, the $\delta^{13}C$ of methane in the atmosphere can serve as an indicator of these processes. Methane from ice cores can therefore indicate the extent of forest fires in the geological past.

Oxygen and hydrogen isotopes. Most oxygen atoms in seawater are ^{16}O, but some are ^{18}O. Evaporation of water (H_2O) from the ocean surface preferentially removes $H_2{}^{16}O$, as $H_2{}^{18}O$ is heavier and sinks. The evaporated water returns to the ocean as rain and from rivers. During times of glaciation, much of the $H_2{}^{16}O$ water is trapped in glaciers, and the weather is drier. Sediments that have high ^{18}O:^{16}O ratios were therefore deposited during times of glaciation and aridity (see figure on page 225). A similar process causes water that contains deuterium (^2H_2O) to accumulate in ice layers during times of glaciation and aridity.

Nitrogen isotopes. Bones of carnivorous animals tend to have more of the heavy isotope of nitrogen (^{15}N) relative to ^{14}N than do bones of herbivorous animals, since the heavier isotope accumulates with each level of the food chain. The nitrogen isotope ratio provides an indication of an animal's diet.

Sulfur isotopes. Gypsum is calcium sulfate ($CaSO_4$), usually of inorganic origin. Therefore it usually has a ratio of light sulfur (^{32}S) to heavy sulfur (^{34}S) that is the same as the seawater in which it is formed. However, sulfate-reducing bacteria produce iron pyrite, which precipitates out of the water. The bacterial enzymes that do this prefer ^{32}S to ^{34}S. Thus a lower than normal amount of ^{34}S in a deposit may suggest biological activity.

Strontium isotopes. Strontium atoms in seawater replace some of the calcium atoms in limestone. The ratio of heavy Sr (^{87}Sr) to light Sr (^{86}Sr) is always close to 0.71. However, the continents have relatively more of the heavy isotope. Thus, when sedimentary deposits show an increase in heavy Sr, this is taken as an indication of significant erosion from the continents. An increase in the strontium ratio occurred after the melting of the ice of SNOWBALL EARTH. It has also been increasing in the sediments around Eurasia for the past 20 million years as a result of the uplift of the Himalayas.

Other uses of isotopes. Geologists have used hydrogen isotope measurements from geological deposits to determine the time of mountain uplift. They have also used the buildup of a beryllium isotope on boulder surfaces to determine the timing of glacial retreat. Measurement of helium isotopes in dust allows a determination of how much of the dust has come from outer space. Isotope ratios can also be used as indicators of the geographical origin of deposits. Different oceanic areas have different neodymium ($^{143}Nd/^{144}Nd$) ratios. The study of neodymium ratios in deposits from the late QUATERNARY PERIOD has allowed the reconstruction of ocean circulation patterns at that time.

Isotopes are not the only molecular markers used in the fossil record. Other molecular markers include lipids, which are different in bacteria, archaea, cyanobacteria (which have 2-methylhopanes), and eukaryotes (in which steranes are derived from membrane cholesterol).

Different isotope ratios may measure the same phenomenon. Oxygen, hydrogen, and strontium isotope ratios are indicators of the climate that prevailed at the time the deposit was formed. The processes that influence oxygen, hydrogen, and strontium ratios are largely probably independent of one another. Therefore, in the great majority of cases in which oxygen, hydrogen, and strontium ratio estimates agree with one another, they provide independent, therefore believable, estimates of ancient climate.

Further Reading

Condon, Daniel, et al. "U-Pb ages from the neoproterozoic Doushantuo Formation, China." *Science* 308 (2005): 95–98. Summary by Kaufman, A. J., in *Science* 308 (2005): 59–60.

Eglinton, G., and R. Pancost. "Immortal molecules." *Geoscientist* 14 (2004): 4–16.

Lata, J. C., et al. "Short-term diet changes revealed using stable carbon isotopes in horse tailhair." *Functional Ecology* 18 (2004): 616–624.

Miller, Kenneth G., et al. "The Phanerozoic record of global sea-level change." *Science* 310 (2005): 1,293–1,298.

Mulch, Andreas, Stephan A. Graham, and C. Page Chamberlain. "Hydrogen isotopes in Eocene river gravels and paleoelevation of the Sierra Nevada." *Science* 313 (2006): 87–89.

Piotrowski, Alexander M., et al. "Temporal relationships of carbon cycling and ocean circulation at glacial boundaries." *Science* 307 (2005): 1,933–1,938.

Schaefer, Georg M., et al. "Near-synchronous interhemispheric termination of the last glacial maximum in mid-latitudes." *Science* 312 (2006): 1510–1513.

Schaefer, Hinrich, et al. "Ice record of $\delta^{13}C$ for atmospheric CH_4 across the Younger Dryas-Preboreal transition." *Science* 313 (2006): 1109–1112.

Siegenthaler, Urs, et al. "Stable carbon cycle-climate relationship during the late Pleistocene." *Science* 310 (2005): 1,313–1,317.

J

Java man *See* HOMO ERECTUS.

Johanson, Donald (1943–) American *Anthropologist* Donald Johanson is an anthropologist who has made some of the most important discoveries of the fossils of possible human ancestors. He is most famous for the 1974 discovery of a 3.2-million-year-old fossil of one of the AUSTRALOPITHECINES, an *Australopithecus afarensis* individual known to the world as "Lucy." This nearly complete skeleton left little to the imagination and demonstrated that before the HOMININ lineage began to use tools, and before their brains began to expand, they were walking upright (see BIPEDALISM).

Johanson was born June 28, 1943. As a teenager, Johanson read *Man's Place in Nature* (see HUXLEY, THOMAS HENRY), which inspired him to devote his life to the research of human evolution. After receiving his doctorate at the University of Chicago, Johanson undertook field research in Ethiopia (where he found Lucy) and Tanzania. He has continued to discover important hominin fossils, including numerous *A. afarensis* fossils from a single site (called the "First Family") and some specimens that may represent *HOMO HABILIS*.

Johanson has not only made discoveries and written scientific papers but has had a major impact on educating the public about human evolution. He founded the Institute of Human Origins, which is currently affiliated with Northern Arizona University in Flagstaff, in 1981. Besides writing several popular books, Johanson hosted and narrated a widely watched public television series about human evolution. For Johanson, human evolution is not merely interesting. He believes it is vital for the survival of humans and of all other species that people understand their evolutionary origins. Humans have deep biological roots in a world that they share with millions of other species, and therefore they must not treat the Earth like conquerors.

Further Reading

Johanson, Donald, and Maitland A. Edey. *Lucy: the Beginnings of Humankind*. New York: Simon and Schuster, 1981.

———, and James Shreeve. *Lucy's Child: The Discovery of a Human Ancestor*. New York: Early Man Publishing, Inc., 1989.

Jurassic period The Jurassic period (210 million to 140 million years ago) was the middle period of the MESOZOIC ERA (see GEOLOGICAL TIME SCALE). It followed the TRIASSIC PERIOD. The Mesozoic era is also known as the Age of Dinosaurs, because DINOSAURS were the largest land animals during that time. Dinosaurs became extinct during the CRETACEOUS EXTINCTION, which is one of the two most important MASS EXTINCTIONS in the history of the Earth.

Climate. In the mild climate of the Jurassic period, swamps and shallow seas covered widespread areas.

Continents. The continents formed by the breakup of Pangaea continued to separate (see CONTINENTAL DRIFT).

Marine life. All modern groups of marine organisms existed during the Cretaceous, except aquatic mammals. Large aquatic reptiles lived in the oceans.

Life on land. Gymnosperms dominated the terrestrial forests (see GYMNOSPERMS, EVOLUTION OF). Dinosaurs, some very large, flourished. The earliest known birds lived during the Jurassic period (see *ARCHAEOPTERYX;* BIRDS, EVOLUTION OF). Mammals were small and represented by few species compared to the dinosaurs (see MAMMALS, EVOLUTION OF).

Further Reading

Museum of Paleontology, University of California, Berkeley. "Jurassic period: Life." Available online. URL: http://www.ucmp.berkeley.edu/mesozoic/jurassic/jurassiclife.html. Accessed April 18, 2005.

Kelvin, Lord (1824–1907) British *Physicist* William Thomson, later Lord Kelvin, was the physicist who laid the foundation for the science of thermodynamics, which is the study of how energy moves and changes forms. Lord Kelvin influenced evolutionary science by developing the theoretical framework of thermodynamics, which allows scientists to understand how life works and how it could evolve; and by miscalculating the age of the Earth.

Born June 26, 1824, the son of a mathematics professor, Thomson entered Glasgow University at the age of 10. He graduated from Cambridge University, winning top prizes for mathematics and rowing. He also started a musical society at Cambridge. At age 22, he became a professor at Glasgow University, where he remained for 53 years. He wrote more than 600 scientific papers and had more than 60 patents. His research provided the basis for the development of refrigeration; the centigrade temperature scale (which, when based upon absolute zero, is measured in degrees Kelvin, or degrees K); telegram boosting devices; depth sounders; and an improved compass. He contributed greatly to the modern understanding of electromagnetism and wave theory of light. Perhaps his major accomplishment was to lay the foundation for the science of THERMODYNAMICS. The First Law of Thermodynamics stems largely from Thomson's work, which provided the basis for chemist J. Willard Gibbs's development of the concept of entropy, upon which the Second Law of Thermodynamics is based. Despite his immense scientific output over a lifetime, he did not receive peerage until 1892 when he was 68 years old.

The 1859 publication of Darwin's book (see DARWIN, CHARLES; *ORIGIN OF SPECIES* [book]) caused a tremendous stir among scientists. Thomson realized that thermodynamics might have something valuable to contribute to the evolution debate. In 1862 he published an article which said the Earth was 98 million years old, a calculation he based upon the rate at which the Earth cooled from an initially molten state. Earlier scientists had tried a similar approach to calculating the age of the Earth; Buffon (see BUFFON, GEORGES) had measured the rate at which heated metal balls cooled off and extrapolated this to an estimate of the age of the Earth: 75,000 years. Thomson, however, got the calculations right, based on modern thermodynamics. The problem was that his initial assumptions were wrong: that no new heat had been added to the volume of the Earth since its beginning. Thomson knew of no process (nor did anyone else) that could produce new heat in the Earth or keep the Sun burning for more than a few million years. This paper is still one of the most famous in the history of science, not only due to the spectacular error of one of the greatest minds of the 19th or 20th centuries but also because the entire paper was only a paragraph in length (plus a brief appendix of calculations). According to Kelvin, not only was the Earth less than a hundred million years old, but the Earth's surface would have been too hot for organisms to live upon it for most of that time.

The only process that any scientist at the time knew that could empower the Sun was combustion, as if the Sun were burning like a gigantic lump of coal. Thomson calculated that, given the size of the Sun, it could burn at most twenty million years. Therefore, even though the Earth might have been cooling off for almost a hundred million years, sunlight could not have been supporting biological life, and therefore evolution, for more than one-fifth of that period of time.

Twenty million years, or even a hundred million years, seemed a totally inadequate amount of time for Darwinian evolution to produce the entirety of species diversity on the Earth. Scientists scrambled for ways to accommodate this new information. Some of Darwin's associates (see HUXLEY, THOMAS HENRY) abandoned gradualism and embraced saltation, the production of major evolutionary innovation in single steps, to make evolution fit into the allotted time.

Thomson (by now Lord Kelvin) lived to see a discovery that would invalidate his assumptions. He was in the audience when physicist Ernest Rutherford spoke about the discovery of radioactivity by chemists Marie and Pierre Curie. Radioactive elements such as uranium (as determined by chemist Henri Becquerel) or radium (the Curies' discovery) emitted energy, which could keep the interior of the planet molten for billions of years. Thomson did not live to see the discovery of thermonuclear fusion as the explanation of the Sun's energy source. Lord Kelvin died December 17, 1907.

Further Reading

Gould, Stephen Jay. "False premise, good science." Chap. 8 in *The Flamingo's Smile.* New York: Norton, 1985.

Lindley, David. *Degrees Kelvin: A Tale of Genius, Invention, and Tragedy.* New York: Joseph Henry Press, 2004.

Thomson, William. "The 'Doctrine of Uniformity' in geology briefly refuted." *Proceedings of the Royal Society of Edinburgh* 5 (1866): 512–513. Available online. URL: http://zapatopi.net/kelvin/papers/the_doctrine_of_uniformity_in_geology_briefly_refuted.html. Accessed April 26, 2005.

Zapato, Lyle. "Kelvin is Lord!" Available online. URL: http://zapatopi.net/lordkelvin.html. Accessed April 26, 2005.

Kenyanthropus *See* AUSTRALOPITHECINES.

killer ape theory *See* AUSTRALOPITHECINES.

kin selection *See* ALTRUISM.

K/T event *See* CRETACEOUS EXTINCTION.

Laetoli footprints *See* AUSTRALOPITHECINES.

Lamarckism This theory is named for the French scientist Jean-Baptiste-Pierre-Antoine de Monet, Chevalier de Lamarck (1744–1829). Lamarck's theory is most remembered for the inheritance of acquired characteristics: That is, things that happen to an organism during its lifetime can be passed on to future generations. The example most often cited is that giraffes got their long necks because each generation stretched their necks a little further trying to reach for food in tall trees. This theory, now universally understood as false, is usually blamed on Lamarck. However, the inheritance of acquired characters was only part of his theory, and it was a view that was held widely by scientists before, during, and even long after his lifetime, including a few in the 20th century, long after the rediscovery of Gregor Mendel's work (see MENDELIAN GENETICS; LYSENKOISM).

Born in an aristocratic but impoverished family, Lamarck served in the army, and only when he had to quit for health reasons did he study medicine and biology. He was 44 when he was appointed to a position at the French Royal Gardens and 50 before he became a professor at the Natural History Museum in Paris. Although his reputation was overshadowed by that of paleontologist Georges Cuvier (see CUVIER, GEORGES), he published significant work in the classification of plants and insects; in fact, he invented the terms *invertebrate* and *biology*. He died in 1829, blind and poor.

His theory, as put forward in his *Philosophie Zoologique* of 1809, included more than the inheritance of acquired characteristics. He believed that microscopic life-forms were continually generated from soils and progressed to ever higher and more complex forms. In this way, organisms filled every niche of structure and function. One of these lineages had culminated in the human species. He thus added an evolutionary, or dynamic, dimension to the old concept of The Great Chain of Being (see SCALA NATURAE).

Although this theory was disproved by the work of Gregor Mendel (see MENDEL, GREGOR), Mendel's work remained unknown until the 20th century. Many scientists held beliefs about inheritance very similar to those of Lamarck. Even Darwin (see DARWIN, CHARLES) invented a theory of inheritance, called *pangenesis,* which was essentially the same as that of Lamarck.

Further Reading

Burkhardt, Richard W. *The Spirit of the System: Lamarck and Evolutionary Biology.* Cambridge, Mass.: Harvard University Press, 1995.

language, evolution of Language is a type of communication, involving words and grammar, that appears to be unique to humans. Evolutionary theories deal with (1) the evolution of language, that is, the human capacity for it; and (2) the evolution of languages, that is, the patterns of languages found throughout the human populations of the world.

The Evolution of Language Ability

No one has fully explained the evolutionary origin of human language ability. Many animal species have verbal and visual types of communication that are not true language:

- Sometimes even complex animal communications can occur without conscious thought. This appears to be the case with even the most complex vocalizations of birds. Many birds begin to sing when the pineal gland detects the lengthening days of spring and produces less melatonin. Bird vocalization is an example of what scientists call a fixed action pattern (see BEHAVIOR, EVOLUTION OF). The birds sing when the environment stimulates them to do so. It does not occur to the birds that their response to the stimulus may make no sense, for example a mockingbird singing at night in the winter because artificial lights have produced a mistaken idea of day length. Birdsong is not considered language, since it involves no conscious thought.

- In some cases, animal communications involve conscious concepts that may correspond to what we would call words. Vervet monkeys, for example, have different kinds of calls for aerial predators (such as raptors) and terrestrial predators (such as snakes), which evoke different kinds of responses (hiding down in the vegetation from the former, running away from the latter) in the hearers.

Human language represents an unprecedented level of complexity. Apes can make between 300 and 400 signs, at maximum, while a human child (when his or her brain is about the same size of those of nonhuman apes) knows at least 6,000 words. Human language always involves not just concrete concepts about immediate objects and events but abstract concepts as well (for example, the past or the future). True language also has a framework of grammar, which puts the concepts in relation to one another and into motion. All human languages, and no other forms of animal communication, have abstract words and grammar. The production of language apparently requires a large brain, particularly a large prefrontal cortex, which is unique to the genus *Homo*. An absolutely large brain, not just a relatively large one, is necessary for intelligence and language (see ALLOMETRY). The large brains of cetaceans allow complex communication. The expanded portions of the cetacean brain are different from those that have expanded during human evolution (see CONVERGENCE), and it is unclear whether their communication, however complex, contains the abstract elements of true language.

Language requires not only a large prefrontal cortex, to keep track of and relate all the complex concepts, but also specific brain structures:

- The *Wernicke's area* is essential for the understanding of language; damage to this area makes a person unable to understand language but does not inhibit the ability of the person to speak words. The victim may speak long strings of words that make no sense together. The part of the monkey brain that is homologous to the human Wernicke's area allows monkeys to recognize the calls of other monkeys.
- The *Broca's area* is essential for the production of language; damage to this area makes a person unable to speak, even though the victim may understand language perfectly. The part of the monkey brain that is homologous to the Broca's area allows monkeys to control facial muscles.

Although many fossilized crania of hominins have been found, and cranial capacity (brain volume) can be estimated for many, few of them have the detailed preservation necessary for the detection of Wernicke's or Broca's areas. Therefore anthropologists cannot know whether the earliest species such as *HOMO HABILIS* and *H. rudolfensis,* or intermediate species such as *HOMO ERGASTER, HOMO ERECTUS,* or *HOMO HEIDELBERGENSIS,* had language. There is even some doubt as to whether NEANDERTALS, despite the fact that their brains were as large as those of modern humans, had true language.

Earlier hominin species had the ability to produce some of the sounds involved in language. They apparently did not have the full range of language capacity that *Homo sapiens* possesses:

- The production of a range of vowel sounds requires a long larynx, or throat. AUSTRALOPITHECINES did not have a long larynx. Even Neandertals had an upper larynx, as preserved in some skulls, more closely resembling that of the earlier hominin species than that of modern humans. Therefore Neandertals, despite their strength, may have had squeaky voices with a very limited range of vowel sounds. However, more than just a long larynx is necessary for the production of a range of vowel sounds. The hyoid bone (the only bone in the human body that is not connected to another) is the site of attachment of several larynx muscles. Neandertals had hyoid bones indistinguishable from those of modern humans.
- The production of a range of consonants is the product of tongue and lips, not the larynx. Some scientists have pointed out that it is the consonants, rather than the vowels, that define language. Written Hebrew, for example, included only the consonants, until small marks were added later beneath the consonants to denote vowel sounds. In response to the claim that the inability to form the full range of vowels would have made Neandertal speech inadequate, a famous letter in a scientific journal said, "Et seems emprebeble thet ther speech wes enedeqwete bekes ef the lek ef ... vewels ..." thus indicating that consonants are essential, vowels merely helpful, in language. Anthropologists know very little about the complexity of consonant sounds that earlier hominin species may have produced. Chimpanzees appear unable to produce the same range of consonant-type sounds that humans can. Although scientists know nothing about the tongue and lips of *H. ergaster* or even of Neandertals, they do know something about the nerves that controlled their movements, because nerves pass through canals in the bones. *H. ergaster* apparently had fewer nerves controlling the muscles of the face and throat than do modern humans, but Neandertals had large openings for nerves that controlled the movement of the tongue.

The long larynx in adult humans is what allows adult humans to choke. When a person swallows, the epiglottis closes off the trachea; and food can slip into the trachea, blocking it. Other vertebrates, and human babies, with short tracheas, do not have this problem. Babies can swallow and breathe at the same time. It has been suggested that language ability was the principal factor in the evolution of the long trachea. If so, the evolution of language ability came at great cost: the risk of choking on food.

Language ability appears to be associated with specific genes:

- Williams syndrome, in which people have impaired mental ability but produce a constant stream of vivid words with rich vocabulary, is associated with a single mutation.
- SLI (Specific Language Impairment), in which the people have no instinct for grammar, is also associated with a specific mutation.

- Mutation of the FOXP2 gene impairs language ability. Studies of the FOXP2 gene suggest that this gene originated no more than 200,000 years ago. This suggests that earlier hominins, as well as Neandertals (which were in an evolutionary lineage distinct from that of modern humans) did not have full language ability.

All humans appear to be descended from ancestors for whom language ability was genetically encoded. Linguist Noam Chomsky referred to the "deep structure" of language in the human brain. Three lines of evidence suggest this conclusion:

- Children learn their first language, and often more than one, with astonishing ease. It is much more difficult for adults to learn a new language. In one notorious instance, a girl was kept imprisoned in a closet in a Los Angeles apartment until she was nearly an adult. Once she was discovered and rescued, therapy has allowed her to learn and use words, but she did not develop full grammatical ability.
- People with SLI (see above) have to think about each word in each sentence, rather than developing a habit for language construction. Anthropologist Myrna Gopick says that it is as if these people have no native language—they must learn their first language by rules just as normal adults learn a second language.
- When people with mutually unintelligible languages are in contact, they are often able to invent a *pidgin* language (the word *pidgin* is pidgin for "business") in which words easily pronounceable in both languages are strung together with minimal grammar. The language of the stronger trade partner, or colonizer, is usually dominant; therefore pidgin English is mostly English. Pidgin is never anybody's first language. Pidgin can then evolve into a full language, with grammar, at which time it becomes a *creole*, which (like Jamaican) can be a culture's first language. After a generation of contact, the pidgin language develops into a true creole, with grammatical rules. Anthropologist Derek Bickerton, studying cultures in Hawaii, found that it was children who invented the creole grammatical forms while playing together, which suggests that the invention of languages is an innate ability.

Primatologist Marc Hauser says that the first human languages evolved from primate calls. Anthropologist Philip Lieberman disagrees: He claims that language ability first took the form of gestures instead of verbal speech. He points out that apes are better at gesture communications than are other animals. Furthermore, a mental association between language ability and gesture ability is suggested by the fact that Parkinson's disease, which severely limits gestures, also limits the ability of the people to construct regular verbs—even though they can recall irregular verbs from memory. Finally, humans seem to learn sign languages, which include complete sets of grammatical rules, as readily as verbal languages. At several places independently around the world, groups of deaf children have invented their own sign languages without instruction.

Language ability is such a complex adaptation that there must have been very strong selection for it, once it had developed in rudimentary form (see GENE-CULTURE COEVOLUTION). Once humans had evolved the ability to use language, any individuals with less language ability would have been at a distinct disadvantage.

The origin of language seems to be inextricably tied with the origin of intelligence (see INTELLIGENCE, EVOLUTION OF). The Wernicke's and Broca's areas aside, intelligence is necessary for humans to have something to say. Some evolutionary scientists, such as Robin Dunbar, have noted a positive correlation between the size of the neocortex (the most recently evolved part of the brain) and the size of social groups. Small-brained primates get to know each other by social grooming, in which one will pick the parasites out of the fur of another. As social groups become larger, greater intelligence is needed to remember who is who. This can be important because an individual must remember who is a friend and who is an enemy, and to keep track of which individuals are not trustworthy reciprocators of altruism (see ALTRUISM). Complex vocal communications, and eventually language, substituted for grooming as a way of communication in large groups, and also as a medium for sharing gossip about different individuals—their characteristics, their allegiances.

Early 20th-century speculations about the origin of language included the following theories, which were given popular names, as recorded by linguists such as Mario Pei and Charlton Laird. According to these theories, languages began:

- as an imitation of animal communication: the *bow-wow* theory
- as emotional exclamations of fear, pain, or lust: the *pooh-pooh* theory
- as signals to help coordinate teamwork: the *yo-heave-ho* theory
- as song and dance: the *sing-song* theory
- or as onomotopoeia: the *ding-dong* theory

Most of these theories failed to take into account the incredible complexity of primitive languages, which is surprising since this pattern was well documented by the early 20th century. Not just some but all primitive languages that have been studied have astounding complexity. Within historical times, languages have simplified as they have been used more and more for commerce. This pattern of simplification through time from primitive complexity has been used by some religious writers (see CREATIONISM), including this author in his earlier life as a creationist, as evidence for the Tower of Babel theory. This is the Genesis story that God created the complexity of languages, then scattered people all over the world, from a Mesopotamian origin. The problem with this proposal, aside from its lack of supporting evidence, is that no truly primitive languages exist, therefore the current linguistic trend of simplification cannot lead scientists to understand the origin of language. What this trend does indicate is that the forces that selected for the origin of language ability were not the same as the ones operating today. Extrapolation into the historical past is valid; extrapolation into the evolutionary past is not. Modern English is much simpler than its immediate precursors such as Old English and Anglo-Saxon, and all the Romance languages (Spanish, French,

Italian, Portuguese, Romanian) are simpler than Latin. As anyone who has studied languages knows, English has much less complexity than other languages in verb conjugations for person, plurality, and tense; noun declensions; demonstratives (of which English has only "this" and "that"); articles (of which English has only "a/an" and "the"); and numeral classifiers (such as three *sheets* of paper or three *head* of cattle). Any explanation of the origin of language must take into account the complexity of languages that appears to surpass any practical necessity.

SEXUAL SELECTION may explain the origin of complex language ability, just as for the evolution of intelligence. Females may have selected mates that were not merely the strongest, or the most physically appealing, but also the most intelligent and altruistic; according to evolutionary scientist Geoffrey Miller, sexual selection was vital to the evolution of intelligence. Language would undoubtedly have been the major medium through which men communicated their intelligence to prospective mates. Sexual selection would also favor females who could understand the talk and reciprocate. If this is the case, then the earliest language would not have been practical but poetic and possibly musical.

Language may have begun not merely as love songs of individual suitors but as epic poems shared by an entire culture. Even today, the power of language in storytelling to bring tribes together (for example, around a campfire) and provide them a basis for unity is obvious. Tribal identity, and tribal distinction from other tribes, was undoubtedly crucial in the origin of bodily decorations, cave paintings, and perhaps all other forms of art and religion (see RELIGION, EVOLUTION OF). The revered shaman may have been both priest and storyteller. Words seem to have a magical power, a concept that is still maintained in modern religions as well as the folklore of enchantments, curses, spells, and blessings. If the words are connected to music—the almost magical power of harmony, and counterpoint such as rounds—they become even more powerful. One piece of evidence suggests the social, artistic, and magical origin of language comes from ancient Egypt. The Egyptians developed a primitive alphabet, which was used in business. Meanwhile, the priests continued to use hieroglyphics and thereby kept to themselves the power of religious mystery.

Language may have also arisen partly as play. Children are remarkably creative, and much human evolution can be attributed to the retention of juvenile characteristics into adulthood (see NEOTENY). Children can figure out ways of communicating with other children who speak a different language.

Because language contributes to tribal identity, each small band of humans probably developed their own language as they spread around the world from their African origin a hundred thousand years ago. When explorers penetrated the New Guinea highlands in recent centuries, they found tribes largely isolated from the rest of the world. Each tribe had a very different language from all the others. According to Jared Diamond, an evolutionary scientist who has traveled extensively in New Guinea, a large share of the world's languages (about a thousand languages) are spoken in New Guinea. He further suggests that this was the primitive Paleolithic condition of the whole planet. More recently, some languages spread and displaced others into extinction; the languages that spread became the progenitors of modern language families (see below).

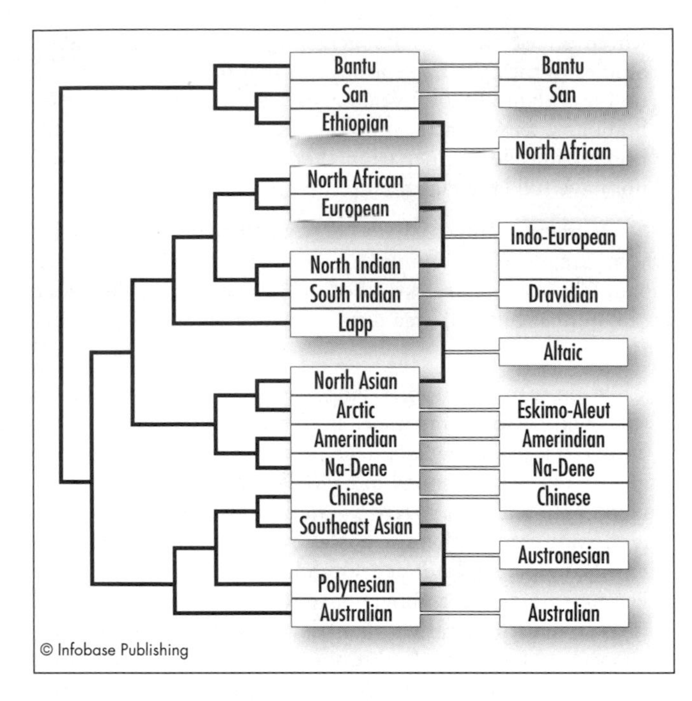

The work of Luigi Luca Cavalli-Sforza shows that there is an approximate similarity between the genetic evolution of human lineages and the languages spoken by these lineages. The differences indicate that some genetic groups have adopted languages from other genetic groups; for example, North Indians speak an Indo-European language but are genetically similar to South Indians. The recently developed diversity of Amerindian language families contradicts their relative genetic uniformity.

The Evolution of Language Diversity

Not only do humans have an innate ability to learn and invent languages, but the diversity of languages has resulted from cultural evolution that parallels recent human genetic evolution.

The earliest languages may have been click languages such as those of the San (Bushmen) of southern Africa and the Hadzane of Tanzania. These languages use not only vowels and consonants but also clicking sounds. The Hadzane live about 1,200 miles (2,000 km) from the San, and their language is unrelated to San except for the use of clicks. The split between Hadzane and the San languages may have occurred as long as 50,000 years ago, perhaps the oldest divergence in the evolution of extant languages. Anthropologists suggest that the primordial human language used clicks because click sounds, unlike voice sounds, can be used during hunting without scaring the prey away as a verbal language would. Some speakers of these languages sometimes just use clicks, with no verbal components, as if the verbal component is a later addition.

Humans spread all over the world from their African homeland. Once a tribe had a language that was their identity, they were unlikely to change it. As tribes migrated, and displaced or conquered other tribes, they took their languages with them. Languages have changed tremendously, and people have evolved slightly, since that time. Most of the genetic patterns that have resulted from recent human evolution consist of minor genetic variation (see MARKERS). Luigi Luca Cavalli-Sforza, a geneticist at Stanford University, has found a rough correlation between human genetic markers and languages (see figure). The major language groups correspond roughly to the major genetic groups.

Sometimes the correspondence between language and genetics is quite close. Both genetic markers and languages spread along with agriculture: Indo-European languages in Europe, Sino-Tibetan languages in China, and Niger-Congo languages in Africa. The linguistic association of Apaches and Navahos (Na-Dene) to the Athapascans of the Pacific Northwest is mirrored in the similarity of their mitochondrial DNA (see DNA [EVIDENCE FOR EVOLUTION]). The linguistic association between the Romany and Hindi languages cleared up the historical enigma of where the Gypsies came from, over a century before DNA tests could demonstrate the same thing. The Gypsies are apparently a tribe that left India and migrated into Europe about 2,000 years ago.

Sometimes the correspondences between languages and genes in human history are not as close or not necessarily reliable. Some of the language families that Cavalli-Sforza used in his analysis, such as Australian and (Native) American, are not linguistically related groups but geographical clumps. The American languages are of recent origin (probably the past 13,000 years) but constitute many linguistic families, while African languages are the oldest but constitute far fewer linguistic families.

Linguistic diversity has developed through the same two processes, vicariance and dispersal, that have stimulated the development of geographical patterns of species (see BIOGEOGRAPHY).

Vicariance occurs when an ancestral language spreads and is then divided into separate groups, and this has been the dominant process in the development of languages. Most of the languages spoken from Ireland to India have evolved from the language (called Indo-European) apparently spoken by people who spread out from northwestern Asia about 5,000 years ago. In each region it diverged into a different language family. The surviving Indo-European language families are:

- Indo-Iranian (today including Hindi)
- Greek
- Italic (today including Spanish, French, Italian, Portuguese, and Romanian)
- Celtic
- Germanic (today including English, German, Dutch, and the Scandinavian languages)
- Armenian
- Balto-Slavic (today including Russian, Polish, Czech, Serbo-Croatian, Bulgarian, and the Baltic languages)
- Albanian

Latin was ancestral to all surviving Romance languages, and during the Roman Empire it remained a coherent language. Once the empire fell, however, communication between regions was minimal, and Latin evolved into languages such as Spanish, Portuguese, French, Italian, and Romanian.

Dispersal is the long-distance spread of a people and language to a new location disjunct from its origin, as in the Gypsy example above. The Hungarian language emerged from the invasion of the Huns into Europe from northeast Asia. Although the Hungarian people are not Oriental, and the Hungarian language is not closely allied with North Asian languages, it is strikingly different from the Indo-European languages around it.

In some cases, a small number of conquerors spread and imposed their language on genetically different peoples. Alternatively, people may adopt new languages when they migrate, by choice or by force. Examples include:

- Finnish people are genetically Scandinavian but speak a language often classified as Altaic. Altaic-speaking men imposed their language on a Scandinavian population. A marker on the Y chromosome of Finnish people shows their association with North Asian speakers of other Altaic languages. The Lapps of Finland provide a separate example of an Altaic language replacing the original language.
- Some Southeast Asians are genetically related to the Chinese but speak Austronesian languages.
- North Indians are genetically similar to South Indians but speak the Indo-European language of ancient conquerors, while the South Indians still speak Dravidian languages.
- Ethiopians, genetically similar to other Africans, have adopted a North African language similar to those of peoples north of the Sahara Desert.
- African Americans have adopted the English language.
- Native Americans of Latin America have adopted Spanish and Portuguese languages.

Some cultural groups have genetically and linguistically complex origins. An excellent example is the Japanese. Even though the Japanese are primarily of Chinese genetic origin, the Japanese language is often classified as Altaic because of its grammar. The Japanese language is quite unlike the monosyllabic Chinese languages, but the Chinese immigrants added many of their words to the Altaic language that already had words for the same things. As a result, the Japanese language frequently has two different words, one of Altaic origin and the other of Chinese origin, for the same thing. For example, the Altaic *harakiri* and the Chinese *seppuku* denote the same kind of ritual suicide. The Chinese *(on)* words are used in more formal situations than the Altaic *(kun)* words, as when the Chinese *dai-* denotes "big" in formal names while the Altaic *okii* is used as the adjective. Even Japanese writing reveals the hybrid origin of the language. Written Japanese uses thousands of pictographic characters (kanji) derived from, and sometimes identical to, Chinese pictographs. Each kanji has a meaning and has both on and kun pronunciations. Written Japanese also uses syllabaries such as hiragana to connect and modify the kanji.

Some of the original linguistic diversity persisted into historic times. Etruscan was apparently a language unrelated to any other, although the Roman eradication of Etruscan culture was so thorough that little evidence remains. A small enclave of Basque culture and language near the border of France and Spain is unrelated to any other known language group.

Another parallel between biological species and languages is extinction. Most of the original languages have become extinct, just as have most species. The universal use of English, and widespread use of other commercial languages such as Indonesian, has caused many of the younger generation to treat their ancestral languages more as curiosities than as identities. Deliberate cultural efforts are underway to keep many Native American languages alive; for example, the Tsalagi (Cherokee) tribal newspaper is mostly in English but often has a column or two in the Cherokee language and script. Similar efforts keep Welsh and Irish alive. There are very few (or perhaps no) people who speak only Cherokee, Welsh, or Irish. Such languages are the cultural equivalent of endangered species in nature preserves. Manx Gaelic became extinct within the last century.

Further Reading

Anderson, Stephen R. "A telling difference." *Natural History,* November 2004, 38–43.

Bellwood, Peter, and Colin Renfrew, eds. *Examining the Farming/Language Dispersal Hypothesis.* New York: Cambridge University Press, 2002.

Cavalli-Sforza, Luigi Luca, et al. "Reconstruction of human evolution: Bringing together genetic, archaeological, and linguistic data." *Proceedings of the National Academy of Sciences USA* 85 (1988): 6,002–6,006.

Christiansen, Marten H., and Simon Kirby, eds. *Language Evolution.* New York: Oxford University Press, 2003.

Diamond, Jared M. "Bridges to human language." Chap. 8 in *The Third Chimpanzee: The Evolution and Future of the Human Animal.* New York: HarperPerennial, 1992.

Dunbar, Robin. *Grooming, Gossip, and the Evolution of Language.* Cambridge, Mass.: Harvard University Press, 1996.

Dunn, Michael, et al. "Structural phylogenetics and the reconstruction of ancient language history." *Science* 309 (2005): 2,072–2,075.

Laird, Charlton. *The Miracle of Language.* New York: Fawcett, 1973.

McCrone, John. *The Ape That Spoke: Language and the Evolution of the Human Mind.* New York: William Morrow, 1991.

Miller, Geoffrey F. *The Mating Mind: How Sexual Choice Shaped the Evolution of Human Nature.* New York: Doubleday, 2000.

Olson, Steve. *Mapping Human History: Discovering the Past through Our Genes.* New York: Houghton Mifflin, 2002.

Pei, Mario. *The Story of Language.* Philadelphia: Lippincott, 1949.

Lascaux caves *See* CRO-MAGNON.

lateral gene transfer *See* HORIZONTAL GENE TRANSFER.

Leakey, Louis S. B. (1903–1972) Kenyan *Anthropologist* Louis Seymour Bazett Leakey (see photo) had a tremendous impact on the scientific and popular understanding of human evolution. Born August 7, 1903, the son of English missionaries, Leakey grew up in Kenya. During his childhood he learned Kikuyu as fluently as English and was later initiated into the Kikuyu tribe. When he was 13, he discovered some stone tools and decided that he wanted to study the ancient people who had produced them. He went to Cambridge University, where he earned a degree in anthropology and archaeology in 1926. He conducted several expeditions in East Africa, which led to his Ph.D. in 1930. In 1936 he married Mary Nicol, a scientific illustrator, who would later make significant discoveries of her own (see LEAKEY, MARY).

When Louis Leakey was unable to verify the location from which some of his fossils had been found, some anthropologists dismissed his claims, and his chance for an academic career in England were diminished. He and Mary returned to Kenya, where he later became a curator at what is now the Kenya National Museum. The position paid little but

Louis Leakey examines a fossil fragment discovered by his son Richard at their base camp in Omo Valley, Ethiopia, in the late 1960s. *(Courtesy of Kenneth Garrett, National Geographic Society)*

allowed Louis and Mary Leakey to continue their research. Throughout the 1950s they excavated fossils from Olduvai Gorge in Tanganyika (now Tanzania). Mary Leakey discovered the first significant Olduvai HOMININ fossil in 1959. Louis Leakey named it *Zinjanthropus boisei* and claimed it as a human ancestor. Because of its huge teeth, it has been called Nutcracker Man. It is now classified in the genus *Paranthropus,* along with other robust AUSTRALOPITHECINES, which were a side branch of human evolution. Intriguingly, "Zinj" was found in the same deposit as primitive stone tools, which led Louis Leakey to believe that Zinj had made the tools. The National Geographic Society in the United States funded and publicized the findings, which allowed Louis and Mary Leakey to greatly expand their work. They found more hominin fossils, including some that much more closely resembled modern humans. Louis Leakey, together with collaborators, classified these latter fossils into a new species, *HOMO HABILIS* ("handy man"), so named because it, not Zinj, was the species that was most likely to have made the tools. The tools represented the most primitive type of stone implements; this type is called Oldowan, named after Olduvai Gorge (see TECHNOLOGY). From the original naming of *Zinjanthropus* to the discovery of *H. habilis,* Louis Leakey wanted to demonstrate that the human evolutionary lineage was separate from the australopithecines such as those discovered in Southern Africa (see DART, RAYMOND).

Mary Leakey continued work in Olduvai Gorge, while Louis Leakey traveled extensively. He was responsible for starting primatologist Jane Goodall's long study of chimpanzee behavior (see GOODALL, JANE), as well as a similar study of gorillas by primatologist Dian Fossey. He considered long-term, detailed studies of modern primates essential for understanding the behavior of human evolutionary ancestors. He was very popular in America where he gave speeches and raised money for his research. Accustomed to making bold projections that eventually yielded good results, Louis Leakey also took bold steps that failed, in particular his search for human ancestors in the Calico Hills of California, where humans did not exist until about 11,000 years ago.

Louis Leakey's son (see LEAKEY, RICHARD) had begun his own ambitious fieldwork to study human origins, leading to personal and professional tension with his father. Richard Leakey discovered the fossil skull ER 1470, which appeared to indicate that the human lineage was separate from that of the australopithecines, and thus to support Louis Leakey's views. This led to reconciliation, shortly before Louis Leakey's death on August 1, 1972.

Further Reading

Foley, Jim. "Biographies: Louis Leakey." Available online. URL: http://www.talkorigins.org/faqs/homs/lleakey.html. Accessed April 28, 2005.

Leakey, Louis S. B. *By the Evidence: Memoirs, 1932–1951.* New York: Harcourt Brace Jovanovich, 1974.

Morell, V. *Ancestral Passions: The Leakey Family and the Quest for Humankind's Beginnings.* New York: Simon and Schuster, 1995.

Willis, Delta. *The Leakey Family: Leaders in the Search for Human Origins.* New York: Facts On File, 1992.

Leakey, Mary (1913–1996) Kenyan *Anthropologist* Mary Douglas Leakey was the wife of anthropologist Louis S. B. Leakey (see LEAKEY, LOUIS S. B.) and made numerous important contributions to an understanding of human evolution.

Mary Douglas Nicol, born February 6, 1913, traveled through Europe as a child. Her visits to prehistoric sites such as CRO-MAGNON caves stimulated an interest in archaeology. This, together with her artistic talent, allowed her to work as an illustrator of archaeological artifacts in England. Louis Leakey asked her to illustrate one of his books. They married in 1937, and she spent the rest of her life excavating fossil HOMININS and other primates in Africa (see PRIMATES). In 1948 she discovered the fossil of *Proconsul africanus,* an ancestor of modern apes.

It was Mary Leakey who in 1959 found the specimen that Louis named *Zinjanthropus boisei* (now classified as a robust australopithecine in the genus *Paranthropus*). She reconstructed the fossil from hundreds of fragments. This was the discovery that earned Louis and Mary Leakey their fame and funding from the National Geographic Society. It was Mary Leakey who in 1961 found the remains of a large-brained hominin that lived at the same time as *Zinjanthropus,* which Louis Leakey described as *HOMO HABILIS.* This species is still considered the oldest member of the human genus, and the first to make stone tools.

Mary Leakey continued her work after Louis Leakey died in 1972. In 1976 she made one of the most important discoveries in the study of human evolution. About 30 miles (50 km) south of Olduvai Gorge at Laetoli, she discovered footprints preserved in what had been volcanic ash (see AUSTRALOPITHECINES). The footprints, made about three million years ago, appeared to have been made by *Australopithecus afarensis.* Mary Leakey died December 9, 1996.

Further Reading

Leakey, Mary. *Olduvai Gorge: My Search for Early Man.* New York: Collins, 1979.

———. *Disclosing the Past: An Autobiography.* New York: Doubleday, 1984.

Willis, Delta. *The Leakey Family: Leaders in the Search for Human Origins.* New York: Facts On File, 1992.

Leakey, Richard (1944–) Kenyan *Anthropologist* Born December 19, 1944, Richard Erskine Leakey is an anthropologist who has made tremendous contributions to the scientific understanding of human evolution. He was the second of the three sons of famous anthropologists in Kenya (see LEAKEY, LOUIS S. B.; LEAKEY, MARY). When young, Richard Leakey did not show intellectual promise, and he dropped out of high school. He was more interested in trapping wild animals and supplying skeletons to research institutions and museums. Then he started a safari business and taught himself how to fly small aircraft. These experiences were later to prove valuable in ways he did not anticipate. In 1964 Richard Leakey led an expedition to a fossil site he had seen from the air. He discovered that he enjoyed looking for fossils, and that without formal training in anthropology he received no credit for fossil discoveries. In 1965 Richard Leakey went to

Richard Leakey, in the fossil vault of the Kenya National Museum, displays four of the important fossil discoveries made by his team: from left to right, a robust australopithecine *(Paranthropus aethipicus)*, *Homo habilis*, Skull 1470 (probably *H. rudolfensis*), and *H. ergaster*. *(Courtesy of Kenneth Garrett, National Geographic Society)*

England to study for a degree. He soon returned to Kenya to lead safaris and scientific expeditions, and to work at the National Museum of Kenya. He never finished his degree.

Persistence and experience paid off despite the lack of formal academic training. Richard Leakey convinced the National Geographic Society, who had sponsored his parents' research, to sponsor his own excavations near Lake Rudolf (now Lake Turkana) in Kenya, and in 1968 he was appointed the director of the National Museum of Kenya. He found many HOMININ fossils at the Lake Rudolf site, including KNM-ER 1470, a relatively large-brained *HOMO HABILIS* that may represent a different species *(H. rudolfensis)*. Because this hominin had such a large brain at such an ancient date, it has raised the possibility that the AUSTRALOPITHECINES were not on the main line of human evolution, a position long held by Louis Leakey (see photo).

Richard Leakey recovered from a kidney disease that nearly killed him and returned to fieldwork and operated Kenya's museum system in 1980. In 1984 Kamoya Kimeu who worked with Leakey's team discovered one of the most important fossils in the study of human evolution: the Nariokotome Skeleton (called the Turkana Boy), a nearly complete specimen of *HOMO ERGASTER*. Leakey's team also discovered fossils of new species of robust australopithecines.

Years of fieldwork had convinced Richard Leakey that something had to be done to protect Africa's wildlife from extinction. He switched his focus from anthropology to conservation. As director of the Kenya Wildlife Service from 1989 to 1994, he undertook ambitious measures against elephant and rhinoceros poaching. He staged a mass burning of elephant tusks that had been confiscated from poachers, which brought the problem to world attention and helped to deflate the market for ivory. Political opposition forced his resignation, after a plane crash. Since 1994 Richard Leakey has been involved in an opposition political party and was elected to the Kenyan Parliament in 1997. After a career in anthropology, he had careers in conservation and then in politics.

In 1970 Richard married Meave Epps, a primate zoologist working in Kenya. Meave Leakey was soon to make her own numerous and important contributions to the study of human evolution. Since 1989 Meave Leakey has been the Leakey family leader in anthropological research. She and her team have discovered two hominin species: *Australopithecus anamensis* in 1995, and *Kenyanthropus platyops* in 2001. *K. platyops* has some characteristics that appear very modern despite its ancient age. Thus for the third time, a member of the Leakey family has discovered evidence that the australopithecines may not be on the main line of human evolution. Meave Leakey has been head of the Division of Paleontology at the Kenya National Museum since 1982 and has directed the museum's Turkana field projects since 1989.

Richard and Meave Leakey's daughter Louise completed a Ph.D. in paleontology in 2001 and may continue the Leakey family tradition of important contributions to evolutionary science.

Further Reading

Leakey, Richard E. *One Life: An Autobiography.* Salem, N.H.: Salem House, 1984.

Morell, V. *Ancestral Passions: The Leakey Family and the Quest for Humankind's Beginnings.* New York: Simon and Schuster, 1995.

Willis, Delta. *The Leakey Family: Leaders in the Search for Human Origins.* New York: Facts On File, 1992.

Lewontin, Richard (1929–) American *Geneticist* Born March 29, 1929, Richard Charles Lewontin has contributed greatly to an understanding of the connection between genetics and the evolutionary process. He pioneered the application of molecular techniques to the study of POPULATION GENETICS and evolutionary change. He has also been one of the most outspoken critics of what he considers the misapplication of genetic and evolutionary concepts to the complexities of human individuality and society.

After receiving a doctorate from Columbia University, Lewontin held faculty positions at North Carolina State University, the University of Rochester, and the University of Chicago. In 1966 Lewontin and geneticist J. L. Hubby published two papers that introduced the use of gel electrophoresis into the study of population genetics (see BIOINFORMATICS). He joined the faculty of Harvard University in 1973.

One of Lewontin's major contributions has been to criticize the overapplication of NATURAL SELECTION in evolutionary concepts. While natural selection has produced many evolutionary adaptations, there are many biological characteristics that are not the product of selection. Lewontin has made three contributions to this line of thought. First, his studies of population genetics revealed a great deal of DNA and protein variation that was not connected to survival or reproduction in wild populations. Second, together with a Harvard colleague (see GOULD, STEPHEN JAY), he pointed out that many characteristics were not themselves adaptations but were the indirect consequences of natural selection acting upon some other characteristic. This concept was formalized as exaptation by Gould and paleontologist Elisabeth Vrba (see ADAPTATION). Third, Lewontin has also been a tireless critic of SOCIOBIOLOGY. Lewontin believes that while the human brain and its mental capacities evolved, its particular characteristics did not, therefore it is useless to look for a genetic basis or an evolutionary reason for such characteristics as religion, fear of strangers, or intelligence. Lewontin sees belief in a genetic basis for differences in intelligence as the first step toward injustice. He was elected to, and resigned from, the National Academy of Sciences during the 1970s.

Further Reading

Gould, Stephen Jay, and Richard Lewontin. "The spandrels of San Marco and the Panglossian paradigm: A critique of the adaptationist programme." *Proceedings of the Royal Society of London* B 205 (1979): 581–598.

Hubby, John L., and Richard Lewontin. "A molecular approach to the study of genic heterozygosity in natural populations. I. The number of alleles at different loci in *Drosophila pseudoobscura.*" *Genetics* 54 (1966): 577–594.

Levins, Richard, and Richard Lewontin. *The Dialectical Biologist.* Cambridge, Mass.: Harvard University Press, 1987.

Lewontin, Richard. *The Genetic Basis of Evolutionary Change.* New York: Columbia University Press, 1974.

———. *It Ain't Necessarily So: The Dream of the Human Genome and Other Illusions.* New York: New York Review Books, 2000.

———. *The Triple Helix: Gene, Organism, and Environment.* Cambridge Mass.: Harvard University Press, 2000.

———, Steven Rose, and Leon J. Kamin. *Not in Our Genes: Biology, Ideology, and Human Nature.* New York: Pantheon, 1985.

———, and John L. Hubby. "A molecular approach to the study of genic heterozygoisty in natural populations. II. Amount of variation and degree of heterozygosity in natural populations of *Drosophila pseudoobscura.*" *Genetics* 54 (1966): 595–609.

life, origin of *See* ORIGIN OF LIFE.

life history, evolution of Life history refers to the patterns of growth, reproduction, and death in organisms. Evolutionary scientists study the life histories of organisms, attempting to explain them in terms of natural selection, relative to particular environments and ways of persisting in these environments. Certain sets of characteristics, called *strategies,* function successfully together; however, seldom does an environment or habitat have a single best life history strategy. Below are examples, rather than a complete overview.

Physiology and Growth

Organisms have limits. In particular, they have limited resources and must *allocate* those resources among functions. Frequently, to do more of one thing means to do less of another. Organisms accordingly *invest* resources into those functions that will allow them to obtain more resources and to reproduce more successfully.

Consider this simplified example of plant growth. The plant uses its food resources to produce leaves; this is an investment, because the leaves then allow the plant to make more food. However, it must also produce stems, to supply the leaves with water, and to hold the leaves up higher than the leaves of other plants; and roots, which obtain water and minerals for the leaves. If the plant allocates most of the food that the leaves make into the production of new leaves, neglecting its stems (for example, by making them flimsy) and producing few new roots, it may be able to grow very fast but will probably be unable to survive the winter. This is part of the life history strategy of a weed; it grows fast because it invests all its resources in growth, rather than maintenance. If the plant allocates more of the food made by the leaves into the production of wood, and perennial roots, it will grow more slowly but will be able to maintain a system of trunk, stems, and roots for many years. This is part of the life history strategy of a tree. An animal example would be to compare a butterfly and a turtle. The butterfly has wings that cannot repair themselves from damage, but this does not matter, as the butterfly dies very soon. The turtle, however, has a body built to last.

Consider the interactions of organisms with other species. If the plant produces lots of defensive chemicals in its leaves, it cannot grow as fast, since it has expended many of its resources on defense rather than growth. Often, the plants that live longer produce leaves that are more toxic, while short-lived plants have less toxic leaves. If insects attack, the plant with the defenses will not suffer as great a loss of leaf material; however, the faster-growing plant may rapidly grow back after the insects attack it. The plants and animals have evolved in response to one another (see COEVOLUTION).

Reproduction

Investment and allocation patterns can allow organisms to grow large and strong, but if they do not reproduce, NATURAL SELECTION eliminates their adaptations. The timing, amount, and pattern of reproduction are important parts of life history; in fact, they make the growth and physiology worthwhile from an evolutionary viewpoint. Reproductive investments are therefore more important than survival. For example, when spiderlings hatch, they may eat their own mother.

Consider this simplified example of plant reproduction. When the plant allocates its resources to reproduction, it can either produce lots of small seeds or a few large ones. Fast-growing plants frequently produce many small seeds; and as they store nothing back, they die, after a large burst of reproduction. This is part of the life history strategy of a weed. Plants that grow more slowly often produce fewer, larger seeds. Sometimes they save up resources for years, then reproduce in a large burst (*semelparous* species), and sometimes they reproduce a little bit every year or every few years (*iteroparous* species). The production of many small seeds is part of the weed life history, while the production of fewer, larger seeds is part of the tree life history. Animal examples of the two extremes include some invertebrates that produce thousands of tiny eggs, and humans, which produce very few but very expensive offspring.

Between the "weed" and "tree" extremes, and their animal counterparts, there is a whole continuum of life histories. There are no mammals that produce thousands of offspring, but mice produce many offspring and have brief lives, in contrast with elephants, which live for a century and produce fewer offspring. There are no birds that produce thousands of eggs, but mourning doves can produce several broods in a season and live only a couple of years, in contrast to the few, large eggs of the long-lived ostrich. There are no trees that live only a year, but some trees have relatively short lives and produce many small seeds; for example, the tree-of-heaven *(Ailanthus altissima)* grows fast, produces many small seeds, and dies after a few decades, in contrast to an oak tree that can live for centuries, producing relatively few large acorns throughout that time. One example of the trade-off between size and number of offspring is found among fishes. Fishes that are raised in hatcheries have a higher survival rate than fishes that hatch in the wild. As a result, small eggs are more likely to result in surviving offspring in a hatchery than in the wild. The recent practice of supplementing wild fish populations with individuals grown in fish hatcheries has therefore resulted in a reduction in the average egg size in wild fish species.

The weed life history strategy has been labeled r, meaning that the populations of these species spend most of their time in the rapid growth phase; natural selection favors characteristics that help them grow and reproduce rapidly. The tree life history strategy has been labeled K, meaning that the populations of these species spend most of their time in the phase of high population density; natural selection favors characteristics that help them compete with other individuals. The variables r and K, first used by ecologist Eric Pianka, come from the theoretical equation for population growth. Ecologist Philip Grime proposed a system in which plants (and other organisms) have three life history strategies: C for competitive (similar to K), R for ruderal (similar to r), and S for stress-tolerant, referring to organisms that grow slowly where resources are very limited or climatic conditions are extreme.

No single life history strategy is always better than another. The weed strategy and the tree strategy both work, but in different ways. Both minnows and sturgeons, and both mice and elephants, successfully make their living in the world. Although no one life history is superior over others, certain environments may favor one strategy over another.

- The tundra has few weeds. The growing season is so brief and chilly that a plant could not grow rapidly enough to reproduce successfully after only one year. If the plant must survive the winter, in order to have two or more growing seasons, it must have adaptations such as strong stems or underground storage organs, which force it to have slower growth.
- Deserts may have many plant species with a weedy life history. The brief wet springtime of a desert is warm enough that many plants can, in fact, complete their growth and reproduction within a few weeks. These plants may be extremely small. This is not the only successful strategy for a desert plant. Cactuses store water, and desert bushes such as mesquite and creosote grow very deep roots that allow them to survive the dry season so that they can grow for many years.
- In areas such as a forest disturbed by fires or storms, many weeds grow. Their small seeds germinate into small seedlings, but there is little competition from other plants (most of which were killed by the fire or storm). The seedlings grow rapidly and reproduce soon; they complete their lives before the forest grows back. Once the forest has grown back, however, weedy species are at a disadvantage,

since their tiny seeds could not grow well in the shade of the trees; now the *K* or *C* plants have the advantage.

Just as the human economy has evolved a great diversity of different ways of making a living, so the "economy of nature" has produced communities of plants and animals that make their living by means of many different life history strategies.

Charles Darwin, following the lead of economist Thomas Malthus, noted that all organisms have the capacity to reproduce far in excess of the space and resources available (see MALTHUS, THOMAS; *ORIGIN OF SPECIES* [book]). Darwin calculated that even a pair of elephants, if they bred as much as possible and all the offspring survived, could produce 19 million individuals in 750 years. The elephant represents the *K* end of the life history spectrum. Species at the *r* end of the spectrum have almost ridiculously large potential reproductive outputs, because the vast majority of their offspring die. If resources were unlimited and all offspring survived, a pair of houseflies could produce a population of 200 quintillion in five months, and a pair of starfish could produce a population in 16 years as great as the number of electrons in the visible universe. A single bacterium, dividing every few minutes, could cover the Earth two meters deep in bacterial slime in less than two days.

Organisms may increase or decrease their reproductive investments, depending on the immediate circumstances. During stressful years with inadequate resources, plants may produce fewer seeds, and animals smaller litters, than during years of abundance. Adjustments can be made even after the next generation has been initiated. Under conditions of inadequate resources or pollination, plants may abort seeds or fruits. Spontaneous abortions are common in animals. In golden hamsters, a mother may even eat some of her own offspring if resources become inadequate.

The life history of *Homo sapiens* and its immediate evolutionary ancestors is a particularly striking example of the *K* strategy. One of the major features of the evolution of the HOMININ lineage has been the prodigious increase in brain size; both relatively and absolutely, the modern human brain is outstanding in the animal kingdom. The first hominins to experience the evolutionary explosion in brain size were the early members of the genus *Homo* (see *HOMO HABILIS*). Once large brains and their associated intelligence had evolved, they provided an advantage that justified their tremendous cost. Much of this brain growth occurs during fetal development and continues well past childhood (see NEOTENY).

A human fetus, because of its large head, is born only with great difficulty. The passage of the head through the birth canal is the most prolonged, painful, and dangerous part of the birth process in humans. The birth canal cannot evolve to be much larger than it is, for otherwise a woman would not be able to walk. Therefore natural selection favored an increase in brain size, but also put constraints upon it. The evolutionary solution to this dilemma was for the fetus to be born early, while the head was still small. In chimpanzees, the ratio of the gestation period (in weeks) to the total life span (in years) is about 0.8. If the human gestation period were the same length of time, relative to the life span, as it is in chimps primates, the human gestation period would be about 15 months and birth would be impossible (see table). By being born after nine months' gestation, the baby's head is small enough to emerge, but the baby is still, in many ways, a fetus, having undergone only a little over half of the prenatal development that is normal for primates. Anthropologists estimate that this aspect of human life history (premature birth) began with the hominin species *HOMO ERGASTER*.

This is why babies are born in such a helpless condition. As a result, they require continual protection and care. The level of reproductive investment of a human mother is among the highest in the animal kingdom, for not only is the baby large when born (up to one-tenth the mother's weight), but it requires nursing and continual attention. The consequences do not stop there. Human mothers can take care of their babies by themselves, but the baby is much more likely to survive and be healthy if taken care of by both parents, and even more so with the help of grandparents (see below). A human male would be able to produce more offspring by having multiple mates but might produce more surviving and healthy offspring by staying home and helping out with just one mate. This may be the evolutionary basis for the human family unit. The cascade of evolutionary causation

Life Histories of Primates

Characteristic	Lemur	Macaque	Gibbon	Chimp	Human
Ratio of gestation (weeks) relative to life span (years)	1	1	1	0.85	0.54
Infancy*	6%	8%	10%	10%	10%
Subadult*	11%	12%	17%	16%	14%
Adult*	83%	80%	73%	74%	76%
Years to end of reproductive life	18	24	30	40	50
Post-reproductive years	0	0	0	0	20
Ratio of post-reproductive years to life span	0	0	0	0	0.3

*Percentage of years to end of reproductive life

is: selection for big brains → selection for premature births → selection for intense parental care of offspring → selection for parental cooperation in child care. Human societies have a full range of social structures, from monogamy to harems (one male with multiple females). The family unit has not always proven to be the arrangement that results in the greatest fitness in humans (see REPRODUCTIVE SYSTEMS).

Frequently, the allocation of resources that is best, from the viewpoint of the parents, is not the best from the viewpoint of the offspring. This concept was called *parent-offspring conflict* by evolutionary biologist Robert Trivers. It is a conflict of interest rather than open violence. Consider the example of reproductive rates. It may be in the interests of the parents to produce numerous offspring, as a hedge against environmental unpredictability (see above). But the existing offspring would benefit most from having all of the parents' attention and resources go just to them. This conflict of interest can play itself out in different ways:

- *Hormonal conflict.* The fetus produces an insulin-like growth factor that encourages the allocation of more resources to the fetus, even at the expense of the mother's health. The mother, however, has a gene that inactivates the fetal gene for this growth factor (see EVOLUTIONARY MEDICINE).
- *Emotional and physical conflict among siblings.* A child often displays jealousy when a new baby sibling arrives. In many bird species, the bigger chicks push smaller chicks out of the nest, or even peck the smaller ones to death in full view of the parents.
- *Deception.* The offspring benefit from being deceptive about their needs, for example by signaling a greater degree of hunger than they are actually experiencing.
- *Marsupial life history.* The babies of marsupial mammals are born at a very early stage of development and must crawl from the birth canal to a pouch, where they attach to nipples. Most of their fetal development occurs in the pouch, not the uterus. In placental mammals, most of the development occurs in the uterus, and when the baby is born, it is usually nearly ready for survival (see MAMMALS, EVOLUTION OF). The marsupial life history may provide an advantage for the mother under conditions of variable resource availability. In times of scarcity, when raising offspring would be a significant burden and possibly dangerous to the mother, the mother marsupial can simply discard the offspring. Except under special circumstances, however, placental mammals cannot abort their own offspring. From the viewpoint of the offspring, the marsupial life history is never an advantage; a womb is always safer than a pouch.

In all such examples of parent-offspring conflict, natural selection favors a life history that balances the needs of both.

Most animals are either males or females. In many fishes, however, an individual animal can change from one gender to another. This makes the life histories of these fishes very unusual. SEXUAL SELECTION favors large males; therefore a small, young male would be at a disadvantage in competition with the few, large, dominant males. In these fishes, there are many small and medium-sized females. Some females, when they become large, change into males. These individuals are maximizing their reproductive output by being female when they are young and small, and male when they are older and larger.

Death

Evolution has failed to produce immortal organisms. Every organism runs the risk of accidents leading to injury or death. Furthermore, mutations accumulate in the chromosomes of organisms. Radiation in the environment, and even the very oxygen in the air, induces mutations. The longer an organism lives, the greater risk it runs of death. It is therefore inescapable that evolution has been unable to produce immortal organisms.

It is quite clear, however, that cells and organisms ought to be able to live much longer than they actually do:

- Animal populations contain genetic variability that would allow an increased life span. Researchers have found the *indy* ("I'm not dead yet") gene in the fruit fly *Drosophila melanogaster* and a similar gene in the roundworm *Caenorhabditis elegans.* This gene extends life span by reducing metabolic rate. Fruit flies have been experimentally bred that have life spans twice as long as those of wild fruit flies.
- Germ (germline) cells in animals, and most plant cells, are potentially immortal. Early in embryonic development, animal cells differentiate into *somatic* cells, which become nonreproductive tissues and organs, and which eventually die, and *germ* cells, which are passed on to the next generation in the form of eggs, sperm, or spores. (These are not to be confused with pathogens such as bacteria, which are popularly called germs.) Among the somatic cells, *stem cells* may also be potentially immortal. White blood cells, the basis of the immune system, have a much greater capacity to reproduce than most other cells of vertebrates. Cancerous white blood cells (as in leukemia) may have no physiological constraints on life span. A leukemia patient, Henrietta Lacks, died in 1951, but some of her "HeLa" white blood cells are still alive in medical laboratories all over the world.

The inescapable conclusion is that natural selection has actually favored organisms that die. The reason seems to be that reproduction itself poses a survival and reproductive cost on the parent. In order to reproduce more, organisms must take metabolic, behavioral, and fitness risks.

- *Metabolic risks.* A higher metabolic rate may allow greater reproductive output. Animal species with faster metabolism do not live as long as species with slower metabolism (see ALLOMETRY). Experimental breeding of fruit flies has shown that the longer-lived strains have lower metabolism and lower reproductive output. Food restriction can cause an individual, whether a fly or a human, to live longer. An exception is that most bats have rapid metabolism and long life spans.
- *Behavioral risks.* An animal that takes risks may be injured but may reproduce more than a cautious animal.

- *Fitness risks.* By repressing the binding of insulin, the hormone Klotho acts as an anti- aging hormone in mice. Underexpression of the gene for this hormone accelerates aging in, and overexpression extends the life span of, mice. However, the longer-lived mice have lower fertility.

The expense of producing a long-lived body may so greatly impair reproduction that natural selection will favor shorter-lived organisms, if this allows them a greater total reproductive output. For animals, the continual renewal of components that are damaged, such as the beak of a woodpecker, would be enormously expensive—and that expense would come at the cost of reproductive fitness. Since very old individuals would be rare even in a population of biologically immortal animals (most would already have died from accidents), natural selection has not favored the evolution of adaptations that prolong the lives of very old individuals (see essay, "Why Do Humans Die?"). The reptiles with some of the longest life spans are turtles, and the mammals with some of the longest life spans (aside from humans) are porcupines, both of which are relatively sluggish and well protected from injury. Therefore natural selection will not eliminate mutations that bring harm only to older individuals. One reason is that, in most populations, few individuals live to be old. Another reason is that even those individuals that do live to be old will have very little reproductive life left ahead of them. Scientists such as biologist Peter Medawar have pointed out that *pleiotropic* genes, which have more than one effect, may be responsible for aging *(senescence)*. Such genes may enhance fitness in younger individuals even if they also reduce the survival of older individuals.

Those few organisms that do live for a very long time (all of them trees) either possess very expensive adaptations, or else grow slowly. An example of the former is the giant sequoia *(Sequoiadendron giganteum)*, many of which are more than 2,000 years old, and which produce very thick fire-retardant bark. An example of the latter is the bristlecone pine *(Pinus longeava)* in California. Many of these pines are more than 4,500 years old but have grown less than 30 feet (10 m) tall during that time, because of the harshly cold and dry environment in which they live. Many of the world's oldest and largest trees, such as the ancient lime (linden) trees of England and Germany, are not actually the original trees but are the outgrowth of branches or roots that the original tree produced, from which the original trunks have been lost.

Virginia opossums *(Didelphus virginiana)* on the mainland of the Southeastern United States have evolved in the presence of predators, but on Sapelo Island there are no predators. The island opossums have evolved delayed old-age senescence (as measured by the breaking time of their collagen fibers). They also reproduce relatively more in their second year of life, and less in their first year, compared to mainland opossums. This pattern suggests that under more dangerous conditions, natural selection favors earlier reproduction and favors (or at least does not disfavor) more rapid aging.

Humans are unusual not only in the enormous amount of care they give to offspring but also in the enormously prolonged post-reproductive life span. In most animals, once reproduction is finished, the individuals die. A comparison among different primates shows that chimps and humans have slightly longer infancy and juvenile periods, relative to the life span up to the end of reproduction, than do other primates. But the most striking difference is the human post-reproductive period, which adds another one-third to the life span (see table on page 241). In humans, the post-reproductive life span can be almost as long as the juvenile and reproductive life spans combined. The human with the longest life span, verified by a birth certificate, was Jeanne Calment, who died in Arles, France, at the age of 122 in 1997.

The explanation usually given for this is that the older individuals possess a wealth of knowledge that can prove immensely valuable to individual families and to the tribe as a whole. This body of knowledge can be so large that it takes a lifetime to teach. A bird can learn its repertory of songs and how to forage in a single season, but human knowledge fills a lifetime longer than the reproductive life span. Since the invention of writing, this knowledge can be passed from one generation to another impersonally, but through most of human evolutionary history old people, not books or databases, were the repository of tribal knowledge, especially about the uses of the many wild species. Science writer Natalie Angier calls grandmothers the "Alexandrian libraries for preliterate tribes." This cultural knowledge would directly benefit the wise old person's descendants.

Evolutionary scientists have had a particularly difficult time explaining the origin of menopause. Many evolutionary scientists consider menopause to be a nonadaptive side effect of the evolution of human life history. Menopause evolved accidentally, and since very few old women survived in prehistoric societies, there was no selection against menopause. In contrast, some evolutionary biologists, such as Kristin Hawkes, say that menopause is adaptive. Dubbed the "grandmother hypothesis," this theory says that a woman may have greater inclusive fitness (see ALTRUISM) by devoting herself to the nurture of her grandchildren than by continuing to have her own children, once she is old. This may explain why older women seldom reproduce, but how could natural selection have favored the sudden enforced cessation of reproduction? The reason may be that continued reproduction, even continued reproductive ability, in older women may present physiological risks:

- *Breast cancer.* There is a correlation between the number of menstrual cycles and the risk of breast cancer, since the hormones that the woman's body produces after ovulation (estrogen and progesterone) stimulate breast cell division. Continuation of menstrual cycles into old age could therefore represent a risk. Some evolutionary scientists conclude from this that menopause reduced breast cancer incidence in older women. Other evolutionary scientists point out that women have far fewer menstrual cycles in primitive societies than in civilized societies, because they were usually pregnant or lactating. Breast cancer incidence is much lower in societies with high reproductive rates. In

Why Do Humans Die?

The traditional religious and modern creationist explanation (see CREATIONISM) about why humans are mortal is that God cursed the human race in response to Adam's sin. In the biblical book of Genesis, Adam was the first man, and God created Eve from one of his ribs. Adam ate the fruit that God had forbidden him to eat (many traditionalists blame Eve for giving him the fruit), and he was transformed from an immortal into a mortal. The entire human race is descended from Adam and has inherited his mortality. Aside from the many difficulties of this attempted explanation, it is not even an explanation. It leaves the question unanswered: What actually happened to introduce the physiological processes of aging and death into the human body?

The evolutionary explanation is that mortality is an inescapable part of being an organism. In addition, natural selection has actually favored the evolution of mortality. That is, a limited life span confers greater fitness than would an unlimited life span, if the latter were even possible.

Aging is the result of senescence. Senescence is not death but is a programmed, gradual breakdown of biological processes, which leads to death. The sequence of events in senescence is in the cells, in animal bodies, and the product of natural selection.

Aging and Death Are in the Cells

1. Aging of cells. As cells, and lineages of cells, become older, the chromosomes begin to lose the DNA caps at their ends (known as *telomeres*). A telomere is a nucleotide sequence repeated about 2,000 times. The sequence is TTAGGG in most animals and fungi; plant telomeres have an extra T; and protist ciliates have TTTTGGGG or TTGGGG telomeres. Each time a cell duplicates, its chromosomes duplicate, and each time this happens, portions of the telomere caps are not replicated. This occurs because the polymerase enzyme, which copies the chromosome, cannot start right at the end. An 80-year-old person has telomeres about five-eighths as long as those of a newborn. An enzyme called telomerase restores telomeres, but this enzyme is not active in most cells.

Most cells are capable of only a limited number of cell divisions, whether in an animal body or in a test tube tissue culture. Skin cells, for example, begin losing their smoothness as they go through more divisions, which explains why old people have wrinkled skin, and why the skin of people who spend a lot of time in the sun is wrinkled: Ultraviolet light damages skin cells, which undergo divisions to repair the damage, more than is the case with a person who does not spend much time in the sun.

There appears to be a correlation between telomere loss and aging. Evidence includes:

- Genetically engineered skin cells that produce telomerase live longer in culture.
- Genetically engineered mice that could not produce telomerase aged rapidly.
- Werner's syndrome is a type of progeria (rapid aging) in which children undergo rapid senescence and die during their second decade. One component of this syndrome is rapid telomere loss.
- The storm petrel lives 30 years, which is one of the longest life spans for a bird. Unlike those of almost all other animals, the telomeres of the storm petrel do not shorten with age.

The correlation between telomere loss and aging does not mean that telomere loss causes aging. Telomere loss could be an effect rather than a cause of aging. Furthermore, the loss of telomere nucleotides is not the only cellular change that accompanies aging. Aging is also correlated with oxidative stress. Oxygen gas (O_2) is very reactive. When they come in contact with water, oxygen molecules can produce highly dangerous molecules such as peroxide ions and superoxide radicals. Cells have enzymes that protect them from these dangerous molecules. Superoxide dismutase changes superoxide radicals into peroxide, and catalase changes peroxide into water. Antioxidants (such as some of the vitamins) also inactivate oxygen radicals. Flies that have a more active form of superoxide dismutase, and worms that produce high levels of cellular antioxidants, have longer life spans. Without such protective enzymes and antioxidants, a cell would quickly die. Oxygen gas quickly kills all anaerobic cells, which do not have protective enzymes. But even these enzymes and antioxidants cannot protect a cell perfectly.

There are some cells in an animal body that remain eternally young. Among these are:

Stem cells. While most *embryonic stem cells* differentiate into tissues and organs, some cells remain undifferentiated and genetically young. In children and adults, these *adult stem cells* retain the ability to divide and differentiate at a later time, as part of the process of healing and regeneration in damaged tissue. Most adult stem cells have only a limited ability to differentiate. For example, satellite cells in muscle tissue can differentiate only into various kinds of muscle cells. Nervous tissue, even in the brain, contains some stem cells that can differentiate into new nerve cells. Bone marrow stem cells produce new red and white blood cells. If scientists understood the biochemical differences between stem cells and other cells of the body, they might be able to transform mortal cells into immortal ones; at this time, however, there is no promise of such a breakthrough. Such cells, though genetically immortal, would eventually die for other reasons, explained below. As stem cells make up only a very small fraction of the cells of the body, the body undergoes senescence and carries the stem cells down into the abyss of death with it.

Germ cells. The cells involved in the production of eggs and sperm also remain eternally young, not losing their telomeres, in the body of an animal. These germ cells specialize very early in embryonic development. Germ cells represent an immortal cell line throughout all animal generations. The germ cells of any individual, however, are housed within a mortal body, making their eventual death as certain as that of stem cells.

Cancer cells. One of the problems with trying to make cells immortal is that there is a fine distinction between immortality and cancer on the cellular level. Cancer cells are cells that have lost the ability to stop reproducing, and the ability to differentiate into different kinds of cells with specific functions. This is why cancer forms tumors of blob-like cells that keep spreading. Cancer cells may be immortal but are within a mortal body, the death of which they hasten.

It may be impossible for an organism to consist only of immortal cells. First, cells appear to lose their immortality by differentiating into specialized functions such as those of skin, muscle, and nerve cells—the very specialization that stem, germ, and cancer cells do not have. Second, processes that promote cell immortality can easily slip into a destructive mode. Telomerase may stimulate cancer. The more active form of superoxide dismutase causes ALS (amyotrophic lateral sclerosis, or Lou Gehrig's disease) in humans when it functions in nerve cells. As the years roll by, telomeres shorten, oxidative damage occurs, and the molecular armamentarium of the cells appears unable to stop these processes from eventually killing an individual.

2. Inevitability of cancer. Among the many genes in a eukaryotic cell are *tumor-suppressor genes.* These genes become active when a cell experiences a potentially carcinogenic mutation. These TSGs cause the mutated cell to commit biochemical suicide, thereby preventing the cell from causing cancer. Benzopyrene in cigarette smoke mutates the p53 TSG in lung cells, making these cells vulnerable to cancer. This is the principal reason that smoking causes lung cancer. Mutations in the p53 gene are also responsible for cancers in response to ultraviolet light and to certain fungal toxins. There are other TSGs, which protect cells from other kinds of cancer.

Because humans are diploid organisms (see MENDELIAN GENETICS), each cell has two copies of each TSG. If a mutation occurs in one of these copies in a cell, the other can serve as a backup in that cell. Eventually both will mutate, as the body is bombarded by radiation and exposed to carcinogenic chemicals (including many that are natural components of food or the environment). Cancer therefore appears to be inevitable after a long enough period of time. People who inherit mutated TSGs from their parents have an increased risk of cancer. If a person inherits one copy of a mutated TSG, cancer can result quickly in a cell because there is no functional backup TSG. This is what happens in women who inherit one mutated copy of the BRCA1 gene, a TSG that protects them from breast cancer. These women are at greatly increased risk of breast cancer. As the years roll by, cells lose their tumor-suppressor genes and cancer results.

3. Inevitability of mutations. Many other mutations accumulate in chromosomes, besides those in the tumor-suppressor genes. In time, the accumulation of mutations would ruin the genes. Mutations cannot be prevented.

One of the major causes of mutations is the very air that animals breathe! Among the results of oxidative stress, described above, is DNA mutation. Eventually, oxygen gas would create so many mutations that the genes would stop working. At the same time, oxygen is essential for most cells to release energy from food and put it to use. Oxygen gas is essential to all higher life-forms (anaerobic life-forms are all simple and slow), yet it would eventually kill all of these life-forms.

Another major cause of mutations is radiation, especially ultraviolet radiation from the sun. The ozone layer shields the world from much, but not all, ultraviolet radiation. Sunlight is necessary for life, but this same sunlight would eventually kill all cells and cell lineages by creating mutations and ruining their genes. As the years roll by, mutations accumulate.

Therefore, in a world of oxygen and ultraviolet radiation, it appears that no individual organism could live forever. Natural selection does get rid of many mutations that would otherwise build up through generations of exposure to oxygen and radiation. Mutations do accumulate in germline cells—and they are the source of the genetic variation (see DNA [RAW MATERIAL OF EVOLUTION]; MUTATIONS; POPULATION GENETICS) upon which NATURAL SELECTION acts.

Aging and Death Are in the Body

From ages 30 to 75, the average male human loses 44 percent of his brain weight, 64 percent of his taste buds, and 44 percent of his lung capacity. The older man's heart output is 30 percent lower, and his nerves 10 percent slower. His brain receives 20 percent less blood. During exercise, the older man absorbs 60 percent less oxygen into his blood. Just as cells accumulate molecular damage, so the organs accumulate damage due to wear and tear. Repair mechanisms exist but are imperfect.

Aging Is the Product of Natural Selection

The longer an organism lives, the more likely it is to meet its demise from factors unrelated to its own health. In the wild, it would be practically inconceivable for an animal to live a thousand years without experiencing an accident or disaster or cancer. Some plants live a long time but experience much damage during that time. Giant sequoia trees, many of them more than two millennia old, almost all show signs of fire damage. Even in an imaginary population of physically immortal animals, there would be very few extremely old ones. (The immortals of mythology avoided this problem by not being physical.) Natural selection would favor the ability of their bodies to repair themselves for at most a few decades. Any genes that especially promoted the repair of cells in very old individuals would confer little benefit, since there would be so few of these individuals. Besides, repair mechanisms have costs and risks associated with them. Examples include:

- The cells that have the greatest ability to regenerate are the very cells that have the greatest risk of cancer—which is, after all, simply cell division that has gone out of control. Examples include skin and intestinal cells.
- The same physiological processes that protect younger individuals can accumulate and cause problems in older individuals. Urate molecules in the blood act as antioxidants, but they also accumulate in joints, causing gout.

Early in the 20th century, evolutionary biologist J. B. S. Haldane (see HALDANE, J. B. S.) pointed out that natural selection cannot eliminate mutations that are deleterious to individuals who have passed their reproductive life span. Biologists Peter Medawar and George Williams expanded this concept. *Pleiotropic* genes have more than one effect. If the effect of a gene on young individuals is good but the effect on older individuals is bad, natural selection will actually favor the gene, despite its deleterious effects on older individuals. Natural selection has favored a long, but not vast, period of post-reproductive life in humans (see LIFE HISTORY, EVOLUTION OF).

Natural selection favors characteristics that promote successful reproduction, even at the expense of individual survival. This is why evolutionary fitness is defined as successful reproduction. Organisms that devote the most effort and resources to reproduction when they are young, even if they "wear themselves out" in doing so, can produce more offspring than can organisms that reproduce fewer offspring over a longer period of time. This has been confirmed

(continues)

Why Do Humans Die? *(continued)*

by experimental comparisons of closely related breeds of animals that differ in their reproductive patterns. As evolutionary biologist Richard Dawkins (see DAWKINS, RICHARD) points out,

> *... everybody is descended from an unbroken line of ancestors all of whom were at some time in their lives young but many of whom were never old. So we inherit whatever it takes to be young, but not necessarily whatever it takes to be old.*

The first two reasons listed above (cellular and organ damage) make it appear that aging is physically unavoidable. Even if a genetically immortal organism could exist, it is unlikely that such an organism would be lucky enough to avoid accidents, infections, and other threats forever. But it is the third reason—the action of natural selection—that explains why death is inevitable. Natural selection has favored the evolution of senescence, the gradual journey toward death, in all organisms.

Religious fundamentalists believe that Adam and Eve were physically immortal until Adam "fell" into sin. In order for this to be possible, the entire world of physics and chemistry would have been different before Adam's fall; oxygen and ultraviolet light, for example, would not have caused mutations. The Bible does not say that God changed all the laws of physics and chemistry at the moment Adam fell. Furthermore, in Genesis, God told Adam, "*In the day* that you eat [of the tree], you will die." The fact that Adam did not physically die that very day indicates that the "death" referred to in Genesis was not physical senescence and disease but was a spiritual concept. The pre-Adamic immortality insisted upon by religious fundamentalists is not only impossible but is also not clearly derived from the very scripture upon which they base their beliefs.

Further Reading

Flores, Ignacio, et al. "Effects of telomerase and telomere length on epidermal stem cell behavior." *Science* 309 (2005): 1,253–1,256.

Herbig, Utz, et al. "Cellular senescence in aging primates." *Science* 311 (2006): 1,257.

Kirkwood, Thomas B. L., and S. N. Austad. "Why do we age?" *Nature* 408 (2000): 233–238.

Medawar, Peter. *An Unsolved Problem in Biology.* London: H. K. Lewis, 1952.

Rose, Michael R. *The Long Tomorrow: How Advances in Evolutionary Biology Can Help Us Postpone Aging.* New York: Oxford University Press, 2005.

Scaffidi, Paola, and Tom Misteli. "Lamin A-dependent nuclear defects in human aging." *Science* 312 (2006): 1,059–1,063.

Schriner, Samuel E., et al. "Extension of murine life span by overexpression of catalase targeted to mitochondria." *Science* 308 (2005): 1,909–1,911. Summary by Miller, Richard A., *Science* 308 (2005): 1,875–1,876.

Westendorp, R., and Thomas B. L. Kirkwood. "Human longevity at the cost of reproductive success." *Nature* 396 (1998): 743–746.

Williams, George C. "Pleiotropy, natural selection and the evolution of senescence." *Evolution* 11 (1957): 398–411.

early human societies, even old women would not have had enough menstrual cycles to create a risk of breast cancer.

- *Difficulty during childbirth.* An older woman would be at greater risk of difficulty during childbirth and would be less likely to live long enough to nurture her own child to maturity.

Older men experience reduced sperm production, but no discrete point in their lives at which sexual reproduction ceases. This may reflect the fact that sperm production is less risky than childbirth.

Natural selection has favored, under some conditions, characteristics (including a brief life span) that goes with the *r* life history strategy. But natural selection has also favored limited life spans for species with the *K* life history strategy. The short lives of weeds and mayflies, therefore, just like the long but finite life of a human, are not evolutionary shortcomings but are products of natural selection.

Further Reading

Charlesworth, Brian. "Fisher, Medawar, Hamilton and the evolution of aging." *Genetics* 156 (2000): 927–931. Available online. URL: http://www.genetics.org/cgi/content/full/156/3/927. Accessed April 20, 2005.

Clutton-Brock, T. H. *The Evolution of Parental Care.* Princeton, N.J.: Princeton University Press, 1991.

Conti, Bruno, et al. "Transgenic mice with a reduced core body temperature have an increased life span." *Science* 314 (2006): 825–828. Summarized by Saper, Clifford B., "Life, the universe, and body temperature." *Science* 314 (2006): 773–774.

De Magalhães, João Pedro. "Why are we allowed to age?" Available online. URL: http://senescence.info/evolution.html. Accessed April 20, 2005.

Elgar, Mark A. "Evolutionary compromise between a few large and many small eggs: Comparative evidence in teleost fish." *Oikos* 59 (1990): 283–287.

Ellison, P. T. *On Fertile Ground: A Natural History of Human Reproduction.* Cambridge, Mass.: Harvard University Press, 2003.

Hawkes, Kristin. "Grandmothering, menopause, and the evolution of human life histories." *Proceedings of the National Academy of Sciences USA* 95 (1998): 1–4.

Hrdy, Sarah Blaffer. *Mother Nature: Maternal Instincts and How They Shape the Human Species.* New York: Ballantine, 2000.

Kirkwood, Thomas B. L. and S. N. Austad. "Why do we age?" *Nature* 408 (2000): 233–238.

Kurosu, Hiroshi, et al. "Suppression of aging in mice by the hormone Klotho." *Science* 309 (2005): 1,829–1,833.

Medawar, Peter. *An Unsolved Problem in Biology.* London: H. K. Lewis, 1952.

Pinkston, Julie M., et al. "Mutations that increase the life span of *C. elegans* inhibit tumor growth." *Science* 313 (2006): 971–975.

Trivers, Robert L. "Parent-offspring conflict." *American Zoologist* 14 (1974): 249–264.

Williams, George C. "Pleiotropy, natural selection and the evolution of senescence." *Evolution* 11 (1957): 398–411.

———. *Sex and Evolution.* Princeton, N.J.: Princeton University Press, 1975.

Linnaean system The Linnaean system is used by biologists to classify species of organisms into nested categories. The system was invented by the Swedish botanist Karl Linné (see LINNAEUS, CAROLUS) in the 18th century. Prior to Linnaeus, the classification of organisms was in disarray.

First, each species was usually identified by long Latin descriptions. For example, the common ground cherry was called *Physalis amno ramosissime ramis angulosis glabris foliis dentoserratis,* which means, in part, that it had angled stems and the leaves were hairless and had toothed margins. Linnaeus shortened these long names to two-word names (in this case, *Physalis angulata,* a name it still uses). Thus Linnaeus established the practice of *binomial nomenclature* ("naming system that uses two names").

Second, the same species could be given different descriptive names by different botanists. Linnaeus standardized the names. Was *Rosa sylvestris alba cum rubore, folio glabro* the same as *Rosa sylvestris inodora seu canina?* Linnaeus standardized it to *Rosa canina.* Mistakes still happen, in which the same species is accidentally given two different names, but it is now much less common.

These innovations were accepted by the biological community with a sense of relief, but this was not the most important innovation of the Linnaean system. Before Linnaeus, there was no recognizable basis for classification. Some zoologists classified animals on the basis of terrestrial vs. aquatic, others on the basis of big vs. small. French paleontologist Georges Buffon (see BUFFON, GEORGES) organized animals according to their usefulness in the human economy.

Linnaeus brought order into this chaos. In classifying plants, Linnaeus recognized that reproductive characteristics were less variable than the characteristics of leaves and stems. He therefore classified plants on the basis of the numbers and condition of the stamens (male parts) and pistils (female parts) of their flowers. Consistent with the male bias of science and society, Linnaeus's major axis of classification was the number of stamens: 1, 2, 3, 4, 5, 6, 7, 8, 9, 10, 12, and many individual stamens; the grouping of the stamens (tetradidymous, dididymous, monadelphous, diadelphous, polyadelphous); the fusion of stamens to one another or to pistils; plants with separate male and female flowers on the same tree (monoecious) or different trees (dioecious) or a mixture of arrangements. His last class, the cryptogams (*crypto-* means hidden), consisted of plants that had no obvious flowers (see SEEDLESS PLANTS, EVOLUTION OF). Within the male axis, he divided plants on the basis of their female parts. He applied his system in a rigid fashion, resulting in classifications that are surprising to modern botanists. For example, he placed oaks, beeches, hazelnuts, and cattails into the same category; modern botanists recognize cattails as very different from the other three (see ANGIOSPERMS, EVOLUTION OF).

The central feature of the Linnaean system is that it is a nested hierarchy. The specific epithet (the second of the two words in the name) referred to the species, and the genus name (the first word) was given to a set of closely related species. As the system was modified by later scientists, closely related genera were clustered into families; closely related families into orders; closely related orders into classes; closely related classes into phyla; phyla into the major kingdoms, of which there were originally only two: plants and animals. Linnaeus inspired the scientific world because he brought, or revealed, order within the diversity of organisms on the Earth which was just then being thoroughly explored.

The Linnaean system seemed to correspond closely to the concept of the Great Chain of Being (see SCALA NATURAE). All things, from rocks to plants to animals to humans to angels to God, were believed to be connected like a tapestry of fibers. Scientists were not surprised to find intermediate entities—for example, a sponge as intermediate between animal and rock, or beings halfway between humans and apes—because it was all part of the created tapestry. Linnaeus had no thought of evolutionary origins of organisms; this is revealed by the fact that he applied his system to rocks as well as to organisms. Linnaeus tried to make the system as natural as possible by using characteristics that were a realistic reflection of the Chain of Being.

While Linnaeus gave no thought to evolution, the fact that the natural world could be so readily classified into nested categories struck Charles Darwin as being a major piece of evidence favoring evolution (see *ORIGIN OF SPECIES* [book]). Despite naturalists' belief in the Scala Naturae, there actually were not many intermediates between modern entities. Sponges are not really half-rock, half-animal. Dogs, wolves, coyotes, and foxes form a group of similar canines; cats, lions, tigers, leopards, and cheetahs form a group of similar felines; and there are no cat-dogs. Organisms, as Linnaeus classified them, display a pattern like branches of an evolutionary tree.

Today, biologists attempt to classify organisms using the Linnaean system but try to make the system as natural as possible by defining genera, families, orders, classes, and phyla on the basis of evolutionary divergence (see CLADISTICS). The species themselves are recognized when possible on the basis of reproductive isolation (see SPECIATION). When a group of species share many features with one another, scientists conclude that they evolved from a common ancestor more recently than species that share few features.

The Latin (or Greek) generic and specific names may describe the characteristics that make the genus or species different from other species or genera; they may describe visible features of the organism, or the location in which it lives; they may be the ancient Greek or Latin names for the organisms; or they may be a tribute to a scientist who studied the organisms. For example, dogs are *Canis familiaris*—genus *Canis,* species *familiaris. Canis* is a Latin word for dog. Other members of the genus *Canis* include *Canis lupus,* the wolf (*lupus* is Latin for wolf), and *Canis latrans,* the coyote (*latrans* comes from the Latin for barking).

This system of groupings also explains the higher levels of the Linnaean system (see figure on page 248):

- Genera that evolved from a common ancestor most recently, and are therefore most similar to one another, are

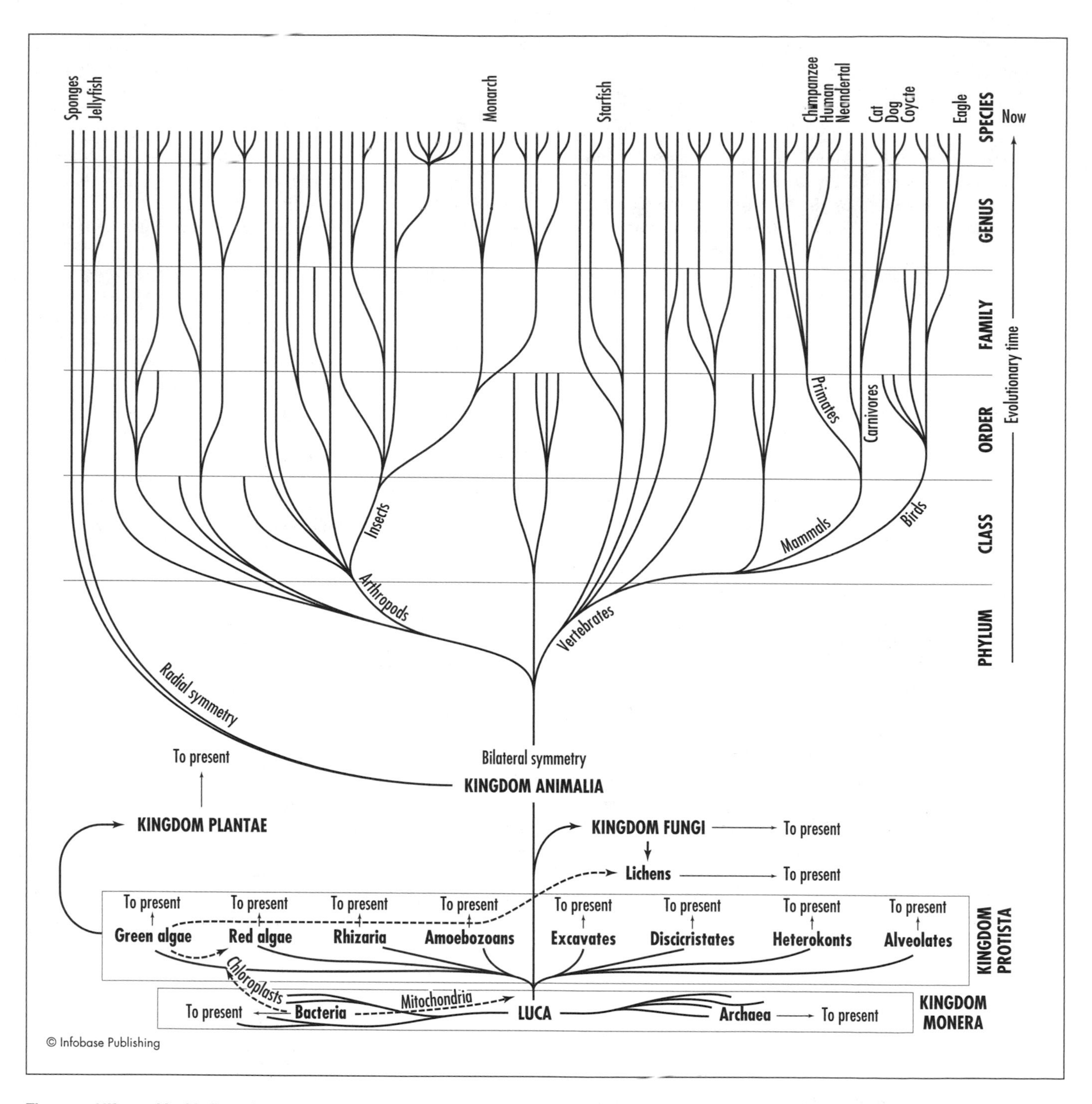

The tree of life used in this figure brings the Linnaean system together with modern evolutionary science, with a focus on animals. The last universal common ancestor (LUCA) evolved into bacteria, archaebacteria, and eukaryotes. Most biologists classify the bacteria and archaebacteria into Kingdom Monera. The eukaryotes diversified into many lineages, most of which are classified into Kingdom Protista. One eukaryotic lineage evolved into the modern plant kingdom, and another into the modern animal and fungus kingdoms. Dashed lines represent symbiogenesis.

grouped together into the same *family.* The canid family, Canidae, contains all the doglike animals, such as foxes, along with the true dogs.

- Families that evolved most recently from a common ancestor, and are therefore most similar to one another, are members of the same *order.* Canids, along with cats, bears, raccoons, weasels, badgers, skunks, and otters, are members of the Order Carnivora, the carnivores. The families in the carnivore order share common features such as the kinds of teeth they possess.

- Orders that evolved most recently from a common ancestor, and are therefore most similar to one another, are members of the same *class.* Carnivores, along with all other mammals, are members of the Class Mammalia. The orders within the mammal class share common features such as the production of hair and milk.
- Classes of animals or protists that evolved most recently from a common ancestor, and are therefore most similar to one another, are members of the same *phylum* (plural *phyla*). In plants, fungi, and bacteria, this category has traditionally been called a *division.* Mammals, along with all other vertebrates (which possess a backbone), and those other animals that have a flexible rod instead of a backbone, are members of the Phylum Chordata.
- Phyla or classes of organisms that evolved most recently from a common ancestor, and are therefore most similar to one another, are members of the same *kingdom.* Most biologists recognize five kingdoms of organisms.

In recent years, biologists have clustered kingdoms into three *domains.* Two of the domains, often placed into the Kingdom Monera, consist of single-celled organisms commonly called bacteria:

- The *archaebacteria* (Domain Archaea) live in environments such as deep ocean volcanic vents and salt slush that, while they seem extremely inhospitable to humans, are similar to the conditions found on the early Earth (see ARCHAEBACTERIA).
- The *eubacteria* are mostly the bacteria that live in soils and water, and on or in larger organisms, on the Earth's surface. Some of them are green and make food energy from sunlight by photosynthesis (see BACTERIA, EVOLUTION OF).

The other three kingdoms consist of organisms that either are, or are constructed from, a more complex form of cell called *eukaryotic* and make up the Domain Eucarya (see EUKARYOTES, EVOLUTION OF):

- The *protists* are either single-celled or, if multicellular, are not organized into fully integrated bodies.
- The *fungi* secrete digestive enzymes, then absorb the digested food molecules individually from the environment.
- *Plants* make food from small molecules in the environment, using energy from the Sun.
- *Animals* consume food as particles or chunks rather than as individual molecules.

The protists do not all have a common ancestor that differs from the common ancestor of fungi, plants, and animals; therefore it is not a *monophyletic* group. Many biologists no longer recognize Kingdom Protista as a formal category.

Early in the 20th century, some biologists applied the Linnaean system at least roughly to communities, particularly of the plants that dominated the landscape. Therefore, to ecologists such as Frederick Clements, a forest of pines (genus *Pinus*) was quite literally a *Pinetum,* and an oak forest (genus *Quercus*) was a *Quercetum.* By the middle of the 20th century, however, biologists realized that each species of organism had its own geographical range and set of habitat preferences. A pine forest was not a coherent entity but rather a collection of interacting species that happened to be in the same place.

Further Reading

Kastner, Joseph. "An Adam and his apostle." Chap. 2 in *A Species of Eternity.* New York: Dutton, 1977.

Linnaeus, Carolus (1707–1778) Swedish *Botanist* Carolus Linnaeus was the Swedish botanist who invented modern taxonomic classification using Latin names (see LINNAEAN SYSTEM); he changed his own name, Karl Linné, into Latin.

Linné was born May 23, 1707. Like many other important intellectuals in world history, Linnaeus showed little promise as a student. His father apprenticed him to a cobbler, which inspired Karl to go back to school and improve his grades. He studied medicine, but his passion, starting in his twenties, was to catalog all the known species of organisms.

Linnaeus was passionate about two other things as well. First, he was passionate about his own greatness. He wanted *Princeps Botanicorum* (prince of botanists) on his tombstone and considered his method of classification to be "the greatest achievement in the realm of science." Second, he was passionate about sex. He used a sexual system, based upon anthers and pistils, to classify plants. But he went further. His terms for the anatomy of mollusks included vulva, labia, pubes, anus, hymen, even though the structures he described had no relation to the corresponding parts of human anatomy. Among the names he gave plants were *Clitoria, Fornicata,* and *Vulva.* His descriptions of plant reproduction were very florid. He described the parts of the flower, correctly, as sexual organs but went further: "The flowers' leaves serve as a bridal bed ... the bridegroom with his bride might there celebrate their nuptials ... then it is time for the bridegroom to embrace his beloved bride and surrender himself to her." Predictably, some scholars disliked Linnaeus's sexual classification system. A professor in St. Petersburg, Johann Siegesbeck, wrote that God would not have created the natural world based on this "loathsome harlotry." Linnaeus's response was to name a small, ugly weed *Siegesbeckia* (now *Sigesbeckia*) in his honor. In his personal life, Linnaeus was very conservative, despite his florid writings.

The order that Linnaeus brought into biology by this system, however, inspired nearly all scientists. In particular, a number of young men were his disciples and explored all over the world, at great personal risk, to find new species of plants to classify with the Linnaean system. Peter Kalm explored Canada during the height of the tensions between France and England; Frederik Hasselquist explored Egypt; Peter Osbeck went to China; Pehr Löfling went to Spain; and Lars Montin went to Lapland, all spending their own fortunes to collect plants and send them back to Linnaeus. Though not a disciple of Linnaeus, botanist John Bartram went into the wilderness of the American southeast looking for plants and was one of the first white visitors to the Cherokee tribe.

Another reason that Linnaeus's system caused a flurry of botanical exploration is that it allowed naturalists to name plants after themselves or other naturalists whom they admired. The genus name of a small herbaceous plant of the north woods, *Linnaea borealis,* honors Linnaeus, and the species name of *Abutilon theophrasti,* the velvetleaf weed, honors the Greek philosopher Theophrastus. Both amateur naturalists and professional scientists yearned for the immortality that would result from having a plant named for them.

Even though he did not even think about evolution, Linnaeus included in his writings some intermediate forms based on hearsay. He was willing to believe them because he accepted the concept of the Great Chain of Being (see SCALA NATURAE). For example, he wrote about strange species of humans such as *Homo ferus,* a feral human on all fours and unable to talk, and *Homo caudatus,* man with a tail. Linné died on January 10, 1778.

Further Reading

Gould, Stephen Jay. "Ordering nature by budding and full-breasted sexuality." Chap. 33 in *Dinosaur in a Haystack: Reflections in Natural History.* New York: Harmony, 1995.

Schiebinger, Londa. "The private lives of plants: Sexual politics in Carl Linnaeus and Erasmus Darwin." Pages 121–143 in M. Benjamin, ed., *Science and Sensibility: Gender and Scientific Enquiry 1780–1945.* New York; Oxford University Press, 1991.

———. "The loves of the plants." *Scientific American,* February 1996, 110–115.

living fossils Living fossils is the popular term given to modern multicellular organisms that closely resemble forms that have been known from the fossil record for at least several million years (see figures and table). During their time on Earth, these organisms have experienced stasis (see PUNCTUATED EQUILIBRIA) and stabilizing selection (see NATURAL SELECTION). While in most cases evolution has favored change and diversification, in these cases it has favored stability. For more information about these examples, see GYMNOSPERMS, EVOLUTION OF; SEEDLESS PLANTS, EVOLUTION OF; INVERTEBRATES, EVOLUTION OF; FISHES, EVOLUTION OF.

"Living Fossil" Plants and Animals

Organism	Taxonomic category	Relatively unchanged since
Plants:		
Dawn redwood (Metasequoia)	gymnosperm seed plant	Pliocene Epoch
Ginkgo (Ginkgo)	gymnosperm seed plant	Jurassic period
Monkey puzzle (Auracaria)	gymnosperm seed plant	Triassic period
Cycad (Cycas)	gymnosperm seed plant	Permian period
Horsetail (Equisetum)	pteridophyte plant	Permian period
Animals:		
Horseshoe crab (Limulus)	arthropod	Triassic period
Fairy shrimp (Triops)	arthropod	Silurian period
Lampshell (Lingula)	brachiopod	Silurian period
Coelacanth (Latimeria)	lobe-finned "fish"	Cretaceous period

Lucy *See* AUSTRALOPITHECINES.

Lyell, Charles (1797–1875) Scottish *Geologist* Sir Charles Lyell changed the science of geology into its modern form by leading scientists away from CATASTROPHISM and toward UNIFORMITARIANISM, a theory that had been proposed earlier (see HUTTON, JAMES). Although uniformitarianism is no longer accepted in the form that Lyell proposed it, it was Lyell who led geologists to understand that the history of the Earth has occurred by the operation of natural laws over a long period of time. Lyell's geological theories laid the groundwork for the development of evolutionary science (see DARWIN, CHARLES).

Born November 14, 1797, Charles Lyell had a keen interest in natural history as he grew up. His father was a botanist, and young Lyell collected and studied insects. Lyell became interested in geology when he entered Oxford University and attended lectures by Rev. William Buckland. Buckland stimulated Lyell's imagination, not the least by his unusual behavior. Buckland performed his fieldwork wearing academic gowns, kept many wild animals at his house, and ate any animal he could find. Guests might be treated to baked guinea pig, battered mice, roasted hedgehog, or boiled sea slug. Although Lyell was to publish works that overturned Buckland's theories, Lyell learned from Buckland to be fearless in choosing direct observation and experience over tradition. Lyell trained for a law career but, after going on a geological expedition through Scotland with Buckland, gave up his legal career for science.

After a geological expedition in Europe, Lyell began writing the *Principles of Geology: An Attempt to Explain the Former Changes of the Earth's Surface by Reference to Causes now in Operation.* The first two volumes were published in 1830 and 1832. These two volumes began a tide of success, going into a second edition before the third volume was published in 1833. The *Principles* went through 11 editions, each updated and revised with much new information, during Lyell's lifetime, and he was working on the 12th edition when he died. Lyell's *Elements of Geology,* focusing upon FOSSILS AND FOSSILIZATION, first appeared in 1838 and went through six editions during Lyell's lifetime. Lyell traveled extensively, gathering new observations; his visits to the United States, where he studied the rate of erosion of Niagara Falls and the rate of sedimentation in the Mississippi Delta,

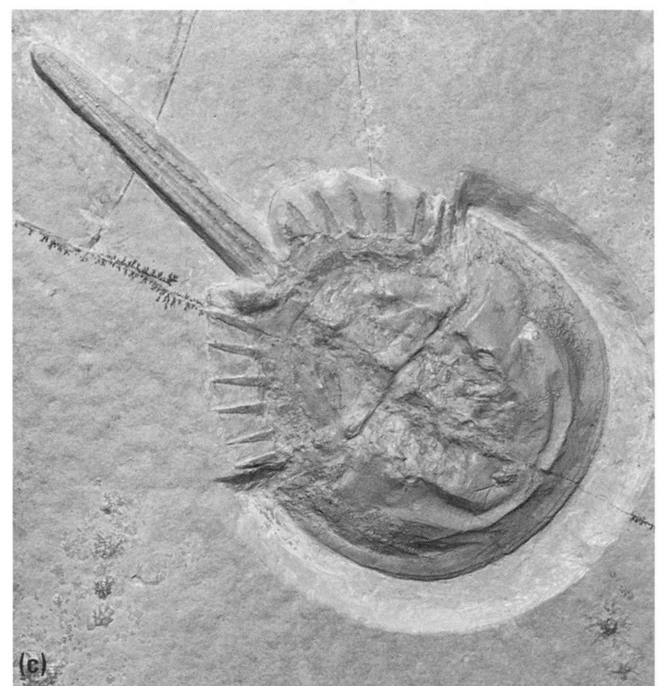

Some modern species closely resemble their ancient evolutionary ancestors. (a) A fossilized leaf of *Ginkgo adiantoides* closely resembles a modern *Ginkgo biloba* leaf; a fossilized horseshoe crab (b) closely resembles a modern horseshoe crab (c); the modern dawn redwood (d) was known only from fossils until it was discovered in China during the 20th century. ***(Photographs courtesy Marli Miller/Visuals Unlimited (a); Volker Steger/Photo Researchers, Inc. (b); Rich Reid, National Geographic Society (c); Stanley A. Rice (d))***

were particularly valuable. A large measure of Lyell's success came from his training as a lawyer, which allowed him to make clear and convincing arguments.

Lyell's work, and Lyell himself, made Charles Darwin's scientific career possible. When the first volume of the *Principles* appeared in 1830, Darwin read it eagerly. Darwin received the second volume while on his voyage aboard HMS *Beagle*. Lyell's book opened Darwin's mind to the creative power of natural forces acting over long periods of time. Lyell was also largely responsible for promoting Darwin's early scientific career, while Darwin was thinking but not publishing about evolution. When Darwin's book came out in 1859 (see ORIGIN OF SPECIES [book]), Lyell did not at first accept evolution. Lyell's uniformitarianism required an endless cycle of geological formation and erosion, but also of the recurrence of species. Lyell's view of life was cyclical, whereas Darwin's was linear. By the time Lyell's *The Geological Evidence of the Antiquity of Man* appeared in 1863, however, he had begun to agree with Darwin, because of continuing discoveries of very old stone tools and human skeletal fragments. In this and many other ways, Lyell, though having virtually built a discipline of science with his own hands, kept his mind open to the new insights of younger scientists. Lyell died February 22, 1875.

Further Reading

Gould, Stephen Jay. "In the midst of life ..." Chap. 12 in *The Panda's Thumb: More Reflections in Natural History.* New York: Norton, 1980.

Lysenkoism Lysenkoism is the doctrine of agriculturalist Trofim Lysenko, who dominated Soviet biological science during the Stalinist era. The early years of the Soviet state were plagued with social upheaval and, in the early 1920s, poor harvests. Wheat was usually planted in the autumn, when it produced leaves; after overwintering, the wheat plants would reproduce in the spring for an early summer harvest. The Russian winters frequently killed the overwintering wheat. Lysenko, a plant breeder in Azerbaijan, demonstrated that if the wheat seeds were stratified (kept in cool moist conditions for a few weeks), they could be sown in the spring and would reproduce in time for an autumn harvest. Stratification is now known among plant physiologists as one of the standard ways of influencing the germination and developmental characteristics of seeds. If Lysenko had stopped here, he might today be revered as the man who helped to save Soviet agriculture, but he went further. He claimed that this stratification process actually changed the seeds in a way that could be inherited. Stratify the first generation, he said, and all the subsequent generations would have the new, convenient trait. His genetic theory was essentially the same as that of French biologist Jean Baptiste de Lamarck (see LAMARCKISM), which had been discredited years earlier by most scientists.

What happened next is an illustration of a government imposing its philosophy on science and on its technological application. Lysenko's Lamarckism happened to resonate well with Soviet philosophy, which claimed that individuals and whole societies could be changed if forced to change, and the change would be permanent. If humans, why not all species? Moreover, Lysenko adopted just part of the Mendelian view (see MENDELIAN GENETICS) as European and American scientists understood it at that time: that heritable changes could occur by big, sudden leaps (mutations). This also pleased the Soviet authorities, still proud of their Bolshevik Revolution that appeared to them to have, in the single year of 1917, propelled Russia from the Middle Ages into the modern world. Lysenko's doctrine was proclaimed to be truth; evidence to the contrary was suppressed, and experimental results were forced to fit into a Lysenkoist interpretation.

The principal Russian scientist to disagree with Lysenko was geneticist Nikolai Vavilov. (Due to secrecy imposed by Soviet authorities, scientists outside Russia knew scarcely anything about what was happening there.) Vavilov had done extensive research regarding genetic variation in crop species (local varieties and wild relatives) and provided great insights into the processes of domestication that had produced these crops. The thing that emerged most clearly from his research was that in order to breed crops, and in order to save them from disease, it is necessary to save the genes. The researcher must travel extensively, gather seeds or other plant reproductive parts (such as potato tubers), and keep them alive. One cannot just grab some seeds, like Lysenko, and force them to change into what one needs them to be. Vavilov spent time in prison for his beliefs. He died during the Nazi siege of Leningrad (now once again St. Petersburg). Some of his fellow geneticists starved within reach of bags of potatoes, which they were saving for the future of agriculture. Vavilov was one of the small number of scientific martyrs.

Lysenko's claims proved utterly disastrous for Soviet agriculture. In the winter of 1928–1929, several million acres of wheat, that had been prepared and planted by the Lysenko method, died, contributing to widespread famine. As he became older, Lysenko made even more absurd claims. He claimed to have created a hornbeam tree that bore hazelnuts; to have developed a wheat plant that bore rye; and to have seen cuckoos hatching from warbler eggs.

After Stalin's death in 1952, Soviet leadership had to rethink many things about domestic and external policy. While they never openly repented for their Lysenkoist errors, the Soviet political and intellectual leadership moved away from Lysenkoism and adopted the same kind of genetics that was proving successful in the West—particularly with the breakthrough the next year by Watson and Crick in demonstrating the structure, and genetic efficacy, of DNA (see DNA [RAW MATERIAL OF EVOLUTION]). Lysenkoism was just the most recent major example of many cases (see SPENCER, HERBERT) in which Lamarckism was embraced because it seemed fair—if one works hard, one ought to be able to change—unlike the New Synthesis of Darwinism, in which natural selection could act only upon the genetic variability that had been provided by chance.

Further Reading

Gould, Stephen Jay. "A Hearing for Vavilov." Chap. 10 in *Hen's Teeth and Horse's Toes.* New York: Norton, 1983.

macroevolution Macroevolution refers to the evolution of major new characteristics that make organisms recognizable as a new species, genus, family, or higher taxon (see SPECIATION). Divergence of an evolutionary lineage into two or more lineages has also been called *cladogenesis* ("origin of branches"). In contrast, *microevolution* refers to small changes within an evolutionary lineage (also called *anagenesis*). Microevolution usually occurs by NATURAL SELECTION but can also occur as a result of other processes such as genetic drift.

Controversy has long surrounded the concept of macroevolution. To many evolutionary scientists, the distinction between microevolution and macroevolution is artificial; macroevolution is simply what happens after microevolution has occurred long enough. Ever since the MODERN SYNTHESIS, in which Darwinian natural selection was integrated with MENDELIAN GENETICS, most biologists have assumed that gradual evolutionary change (microevolution), given enough time, can produce major evolutionary differences. Other scientists claim that macroevolutionary patterns cannot be completely explained by microevolution. Scientists who studied the fossil record expected to find evidence, in all lineages, of gradual change over time. As paleontologist Niles Eldredge explains, he and evolutionary biologist Stephen Jay Gould, in their independent lines of research (Gould with snails, Eldredge with trilobites) expected to find a record of gradual change. Instead they found that species remained essentially unchanged for millions of years (stasis or equilibrium), then became extinct. Rapid evolutionary change occurred when a new species originated, as a punctuation to the history of life. They called this pattern PUNCTUATED EQUILIBRIA (see ELDREDGE, NILES; GOULD, STEPHEN JAY). Microevolution, in contrast, has seldom been controversial; even creationists conspicuously accept it (see CREATIONISM).

Evolutionary biologist Richard Dawkins pointed out that there is no reason to expect natural selection to always proceed at the same rate (see DAWKINS, RICHARD). His most memorable example of this was the story of the Israelites crossing the desert of Sinai during the Exodus. According to the biblical story, it took the Israelites 40 years to travel a couple of hundred miles. This does not mean that they traveled a few miles every year, or a few feet per day. Instead, long periods of encampment alternated with a few long expeditions. Similarly, microevolution proceeds sometimes rapidly, sometimes slowly; sometimes by directional selection, sometimes by stabilizing selection. In this way gradualism could produce a punctuated equilibrium pattern.

Not quite, Eldredge and Gould responded. Punctuated equilibria were not merely random patterns of fast and slow evolution. The punctuation of rapid evolution occurred in most cases at the time a species originated, followed in each case by a long period of stasis. After a population separates into two or more populations, evolution would then proceed more rapidly than at other times. This would occur, they pointed out, when a population is marginalized (for example, a small population is driven into a new and unfamiliar habitat) or when the entire species experiences a major habitat change. This is why, said Eldredge and Gould, speciation often occurred in conjunction with major pulses of species turnover. Species do not respond to minor environmental changes by evolution; instead they migrate to new locations. In contrast, major crises of environmental change (e.g., a relatively sudden cooling or drying of the environment) would cause the extinction of some species, and quick evolutionary change in others. Cooling and drying in East Africa about two and a half million years ago resulted in the evolution of many new species of mammals, including the first members of the human genus, *Homo*.

The origin of complex adaptations has presented a challenge to evolutionary theory. How could gradual microevolution have produced the macroevolution of a complex adaptation such as the vertebrate eye? For natural selection to work, each of the gradual steps needs to provide a significant benefit, or else natural selection would actually eliminate it. This is still one of the favorite arguments of creationists (see

INTELLIGENT DESIGN). Further investigation has shown that either the intermediate forms could have been functional, or that the complex adaptation may have arisen as an exaptation (see ADAPTATION):

- *Functional intermediate forms.* In one famous scientific paper, a computer model simulated the origin of the eye. Starting from a simple optically sensitive surface, the computer program generated random mutations and selected the ones that improved visual acuity at each step. The simulation produced the model of an eye not strikingly different from the vertebrate eye.
- *Exaptation.* Complex flight feathers could not have evolved by gradual steps. But the first feathers might have functioned to hold in body heat, a function for which simple feathers are adequate. If these feathers were brightly colored in one sex, a gradual increase in their complexity might have resulted from SEXUAL SELECTION. As the feathers became even more complex, they might have allowed a male bird to jump and glide in sexual courtship, even if it could not fly. From these intermediate stages of complexity, fully complex flight feathers could have gradually evolved. This might explain why the earliest bird fossils (see *ARCHAEOPTERYX*) have complex feathers without having other adaptations for flight.

Some macroevolutionary patterns may have resulted from processes that would not be apparent on a microevolutionary scale. Niles Eldredge and Stephen Jay Gould have proposed such a mechanism, called *species sorting.* Some species have a tendency to produce more new species than do others, perhaps because natural selection favors mating or cross-pollination with near neighbors. Many extinctions, particularly mass extinctions, occur by chance rather than as a result of natural selection (see EXTINCTION; MASS EXTINCTIONS). The more "speciose" species, those that left more descendant species, would survive mass extinctions more often than the less speciose species. Natural selection has not in this case favored speciation, but enhanced speciation is the result.

Thus macroevolution may not be simply microevolution acting over a long period of time but may result from punctuated equilibria and species sorting.

Further Reading

Dawkins, Richard. *The Blind Watchmaker: Why the Evidence of Evolution Reveals a Universe Without Design.* New York: Norton, 1986.

Eldredge, Niles. *Reinventing Darwin: The Great Debate at the High Table of Evolutionary Theory.* New York: John Wiley, 1995.

Nilsson, Dan-Eric, and Suzanne Pilger. "A pessimistic estimate of the time required for an eye to evolve." *Proceedings of the Royal Society of London* B 256 (1994): 53–58.

Malthus, Thomas (1766–1834) British *Economist* Thomas Robert Malthus, economist, philosopher, and clergyman, practically invented the modern study of POPULATION, on which all of evolutionary biology depends. Charles Darwin (see DARWIN, CHARLES) read Malthus's *Essay on the Principle of Population* which was published in 1798. Malthus's interest was only in humans, but it was from Malthus's essay that Darwin got an insight that was essential to NATURAL SELECTION.

Born February 13, 1766, Thomas Malthus studied at Cambridge. He was ordained an Anglican minister, and was also Professor of History and Political Economy at the East India College in Hertfordshire. Malthus's main argument was that human populations grow in an exponential fashion (for example by doubling), while resources increase in a linear fashion (at a constant rate) if at all. Therefore, Malthus concluded, human populations will always exceed resources such as food and shelter. This was not good news to social reformers. Malthus's doctrine indicated that all the efforts of social reformers to relieve the misery of the poor are doomed to failure, unless population growth is limited. Malthus was notorious for his opposition to the Poor Laws, which provided assistance to poor families, and his support of the Corn Laws, which caused food to be more expensive. He reasoned that welfare to the poor was only a temporary solution to the problem of privation. Malthus's doctrine also directly contradicted the religious concept of divine providence, which maintains that God created a world that could sustain humans. Instead, Malthus claimed (in the first edition of his *Essay,* an argument dropped from later editions) that suffering and privation were part of a divine plan to continually improve the world. In a world with unlimited resources, everyone would succeed. But in a Malthusian world of population growth amid limited resources, those individuals that had the opportunities, abilities, and motivation would be the ones to succeed, while the others would perish.

Darwin realized that Malthus's doctrine also applied to animals and plants. In a crowded world, only those individuals with superior abilities would survive and reproduce. Because traits are inherited, the next generation would resemble those superior individuals. This is natural selection. But the world remains crowded, and only the superior among the superior individuals would survive and reproduce. Therefore, natural selection does not produce adaptations that are simply good enough, but selects for continual improvement. With this line of reasoning, Darwin realized that natural selection could cause even the simplest organisms of the past to eventually develop into the complex organisms on the Earth today. Another evolutionary scientist (see WALLACE, ALFRED RUSSEL) also reached the same conclusion, independently of Darwin, after reading Malthus's essay.

Malthus's argument has proven true in nonhuman species but incorrect in human populations. Recent experience has shown that providing assistance to nations with high levels of poverty is associated with a decrease, not an increase, in population growth rate; and that economic success is not associated with genetic differences.

Malthus died December 23, 1834.

Further Reading

Literary Encyclopedia. "Malthus, Thomas." Available online. URL: http://www.litencyc.com/php/speople.php?rec=true&UID=2902. Accessed April 30, 2005.

mammals, evolution of The Class Mammalia consists of vertebrates that have hair and produce milk. Linnaeus (see LINNAEUS, CAROLUS) classified mammals into one group, and even invented the name, based on the Latin *mammae,* referring to breasts. Mammals also have four-chambered hearts and teeth clearly differentiated into incisors, canines, and molars. There are or have been in the past about 5,000 genera of mammals. The surviving 1,135 genera contain about 4,700 species. This is far less than the three-quarter million species of beetles.

Mammals evolved from *synapsids,* one of the lineages of reptiles. One synapsid reptile of the late PALEOZOIC ERA was *Dimetrodon,* which had a huge sail of bony extensions along its back. Although *Dimetrodon* itself is unlikely to have been the ancestor of mammals, it may have shared a characteristic that may have been widespread among synapsids: warm blood. Modern mammals generate body heat internally (are *endothermic*) and usually maintain a constant body temperature (are *homeothermic*). *Dimetrodon* may have used its sail to warm its blood in the morning (by absorbing sunshine) and cool it off in midday (by radiating heat through the skin surface). This may have helped *Dimetrodon* to be, while not homeothermic, at least partly endothermic.

Synapsids included the pelycosaurs and the *therapsids* (from the Greek for "nurse"). Compared to earlier reptiles, therapsids had fewer skull bones, teeth that were more differentiated, and multiple sets of replacement teeth rather than just juvenile "milk teeth" replaced by adult teeth. One group of therapsids was the *cynodonts* (from the Greek for "dog teeth"), with teeth even more differentiated (doglike) than those of earlier therapsids. Small pits in the facial bones of cynodonts may have been openings for extensive blood vessels, and this has led some scientists to conclude that cynodonts were fully warm-blooded. Cynodonts such as *Morganucodon* and *Sinocodon,* which lived during the JURASSIC PERIOD, represent nearly perfect "missing links" (actually not missing) between earlier synapsid reptiles and modern mammals. *Morganucodon* is considered one of the earliest mammals, even though it is intermediate between synapsid reptiles and modern mammals. The very fact that so many "missing links" have been discovered means that the line between reptiles and mammals is somewhat arbitrary. Like modern mammals, *Morganucodon* had juvenile milk teeth replaced by adult teeth; *Sinocodon* retained the ancestral reptilian pattern of multiple replacement sets of teeth.

The intermediate status of the cynodonts is illustrated in three ways:

- *Lower jaw.* In early synapsid reptiles, as in modern reptiles, the lower jaw consists of several bones, one of which is the dentary bone; but in mammals the lower jaw consists entirely of the dentary bone. In cynodonts, the dentary bone was intermediate in size (see figure at right).
- *Ear bones.* In early synapsid reptiles, as in modern reptiles, the inner ear has a single bone, the stapes, which conducts vibrations to the cochlea. The mammalian inner ear consists of three bones: the incus, which corresponds to the reptilian quadrate bone; the malleus, which corresponds to

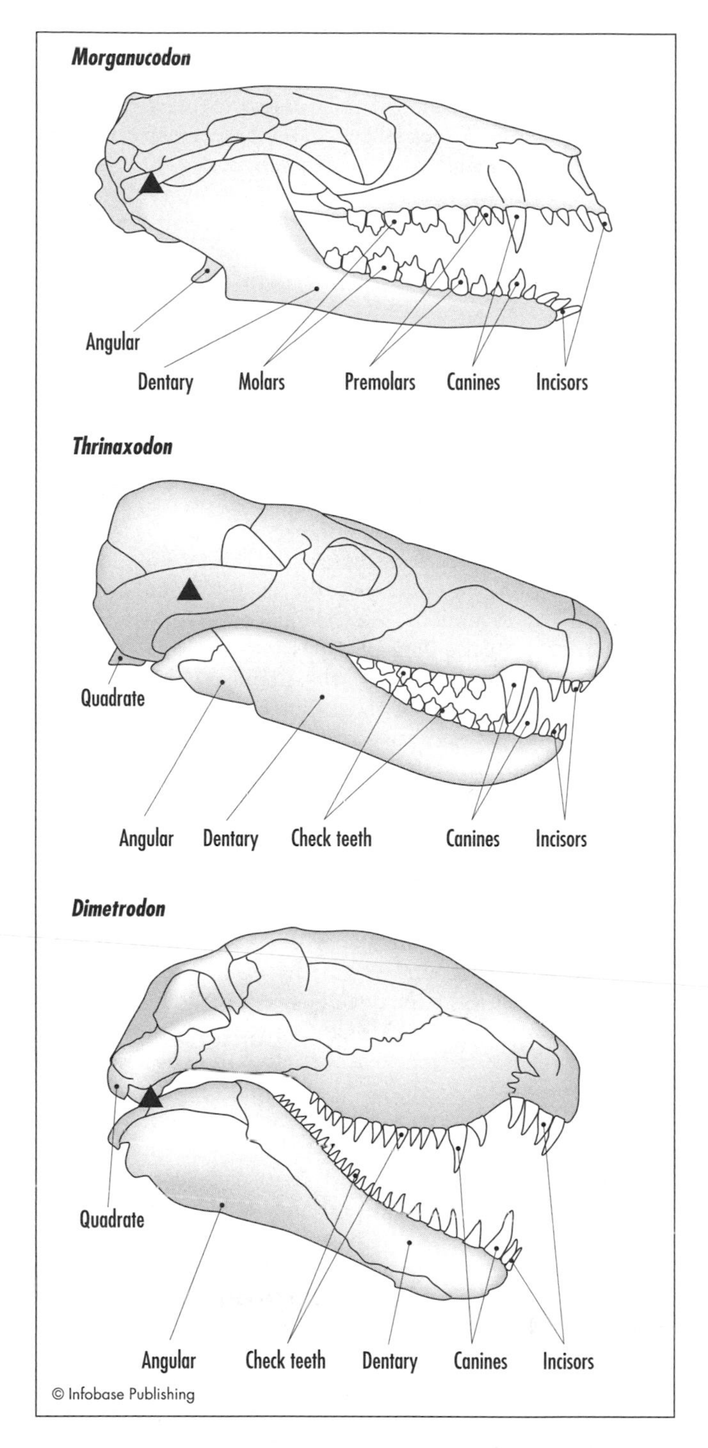

Intermediate forms in the evolution of mammals have been found. The mammal-like reptile *Dimetrodon* had cheek teeth but no molars; its jaw contained large dentary and angular bones, and the jaw joint was between the articular bone of the jaw and the quadrate bone of the skull. It had only one bone, the stapes, inside the ear. In the early mammal *Morganucodon,* the cheek teeth had become premolars and molars; its jaw consisted mostly of the dentary bone, the angular bone being small; the dentary bone connected directly with the skull. The quadrate and articular bones had become the incus and malleus bones inside the ear of *Morganucodon. Thrinaxodon* was intermediate in all of these respects between *Dimetrodon* and *Morganucodon.* The triangle indicates the location of the jaw joint. *(Adapted from Benton)*

the reptilian articular bone; and the stapes. In cynodonts, these bones are intermediate between the upper jaw bones of reptiles and the inner ear bones of mammals. In fact, juvenile opossums (genus *Didelphis*) have inner ear bones that closely resemble those of *Morganucodon*.

- *Jaw joint.* In early synapsid reptiles, as in modern reptiles, the quadrate bone articulates with the articular bone of the skull, but in mammals it is the dentary bone that articulates with the skull. This means that during the evolution of mammals, the location of the jaw joint had to change from quadrate-articular to dentary-skull. In cynodonts, both jaw joints were functional.

Because the term *reptile* usually excludes mammals and birds, "reptile" is not a coherent taxonomic group (see CLADISTICS; REPTILES, EVOLUTION OF). However, mammals are a coherent taxonomic group, representing all of, and only, the modern descendants of the cynodonts.

By the Jurassic period, there were three major lineages of mammals: the monotremes, the multituberculates, and the therians. The modern descendants of the Jurassic monotremes are the duck-billed platypus and the spiny anteater (echidna) of Australia and nearby islands. They retain the primitive feature shared by most (but not all) reptiles of laying eggs. The mothers produce milk but do so from glands rather than from breasts. The multituberculate lineage became extinct during the Oligocene epoch of the TERTIARY PERIOD. Other modern mammals evolved from the Jurassic therian lineage.

By the CRETACEOUS PERIOD, the therian lineage had already split into its two modern branches: the *marsupials,* which bear their young at a very early developmental stage, and the young nurse inside of a pouch; and the *placentals,* in which the fetus develops inside the uterus of the mother and is nourished through a placenta. The placenta, a structure vascularized by both the mother's and the fetus's blood vessels, allows the mother's blood to come close to but not in contact with the fetus's blood. This prevents the mother's body from launching an immune attack against the fetus. Therefore the fetus can stay inside the mother's body until highly developed. When born, the offspring has a relatively high chance of survival. In marsupials, on the other hand, the fetus must be born before the mother's body launches the immune attack. The fetus must crawl from the birth canal to a nipple, where it attaches and nurses until it is able to survive on its own. It is protected inside a pouch of skin.

Most of the evolutionary history of mammals occurred during the MESOZOIC ERA, alongside the dinosaurs. The dinosaurs were, on the average, larger and ruled the day, while most mammals were smaller and were probably most active at night. While dinosaurs probably often ate mammals, occasionally a large mammal would eat a dinosaur. In 2005 scientists published an account of a three-foot- (1-m-) long *Kepenomamus robustus* mammal that had eaten a baby dinosaur 130 million years ago.

The CRETACEOUS EXTINCTION caused the extinction of the dinosaurs. When the dust settled, the surviving mammals encountered a world in which dinosaur competitors had been cleared away. Within 10 million years, during the Paleocene epoch of the Tertiary period, the modern orders of mammals, and many that are now extinct, had evolved. Because this was a time of rapid evolution (which can be considered a punctuation; see PUNCTUATED EQUILIBRIA), it is difficult to reconstruct those early steps, whether from fossils or from DNA (see DNA [EVIDENCE FOR EVOLUTION]). Although all modern orders of mammals existed in the early Tertiary, the dominant animals were in families that no longer exist, and the representatives of modern orders did not look like their modern counterparts. That is, modern orders existed, but modern families and genera did not. In contrast, most of the genera of early Tertiary trees would have been recognizable to an experienced naturalist.

Surviving and recently extinct marsupials included or include many forms that closely resemble placental mammals. The marsupial counterparts of bats, mice, dogs, and cats represent a striking case of CONVERGENCE. Kangaroos can be considered the marsupial equivalent of deer, a convergence of lifestyle but not of appearance as much as in the other examples.

Modern placental mammals include:

- Edentates such as sloths, anteaters, and armadillos
- Insectivores such as shrews and hedgehogs
- Bats, including fruit bats and pollinator bats with excellent night vision, and other bats that rely primarily on echolocation to capture flying insects
- Primates, with excellent vision (see PRIMATES)
- Carnivores, including dogs, cats, weasels, raccoons, bears, seals, and walruses
- The elephants and sea cows
- Odd-toed ungulates such as rhinoceroses and horses (see HORSES, EVOLUTION OF)
- Even-toed ungulates such as pigs, hippopotamuses, camels, deer, antelopes, and cows; the cetaceans such as whales and dolphins are close relatives of the hippopotamus (see CLADISTICS; WHALES, EVOLUTION OF)
- Rabbits
- Rodents, which comprise half of modern mammal species, not only mice and rats but also porcupines, beavers, and squirrels. The continuously growing incisors allow gnawing, which, along with their high reproductive rate, has made rodents very successful.

Competition between marsupials and placentals. Wherever marsupials and placentals have been in contact, the placentals seem to eventually prevail:

- In the Mesozoic era, marsupials were widespread in the world. After the Oligocene, however, most marsupials were found in Australia and South America. Placental mammals may have outcompeted them in other places.
- After the Panama land bridge formed, connecting North and South America for the first time since the breakup of Pangaea during the Mesozoic era, many marsupial mammals migrated south from North America and outcompeted marsupials in South America. Few South American marsu-

pials successfully moved northward, the opossum being the lone North American marsupial success story.

- When modern humans arrived in Australia thousands of years ago, they brought dogs, which became the wild dingos; Europeans brought many placental mammals such as dogs, cats, and rodents a couple of hundred years ago. Competition with placentals has driven most Australian marsupials (except kangaroos) near or into extinction.

This should not, however, be taken as evidence of inherent inferiority of the marsupial reproductive system. Some observers think that marsupials lost in competition because they had not been, in southern continents, subjected during their evolution to the climatic extremes that had occurred in the northern continents. However, the marsupial reproductive system seems well suited to variable and occasionally harsh conditions. In placentals, short of spontaneous abortion, the mother has to carry the fetuses to full term, even under stressful conditions such as famine. The infants may die soon after birth, but not until after the placental mother has made a tremendous prenatal investment in them. In marsupials, however, the fetus is born at a very early stage, and when food is scarce this is the stage at which they can be abandoned. Marsupials, in addition, have a lower metabolic rate during rest, which may help them conserve scarce energy. Because marsupials appear superior under the stressful conditions that all lineages have experienced at some time or other, the reasons why the placentals have largely prevailed over the marsupials are not clear.

Further Reading

Benton, Michael. "Four Feet on the Ground." Chap. 4 in Gould, Stephen Jay, ed., *The Book of Life: An Illustrated History of the Evolution of Life on Earth.* New York: Norton, 1993.

Gould, Stephen Jay. "Sticking up for marsupials." Chap. 28 in *The Panda's Thumb.* New York: Norton, 1980.

Hopson, J. A. "The mammal-like reptiles: A study of transitional fossils." *American Biology Teacher* 49 (1987): 16–26.

Jehle, Martin. "Marsupials: A southern success story." In *Paleocene Mammals of the World.* Available online. URL: http://www.paleocene-mammals.de/marsupials.htm. Accessed September 28, 2005.

Kielan-Jaworowska, Zofia, Richard L. Cifelli, and Zhe-Xi Luo. *Mammals from the Age of Dinosaurs: Origins, Evolution, and Structure.* New York: Cambridge University Press, 2004.

Rich, Thomas H., et al. "Independent origins of middle ear bones in monotremes and therians." *Science* 307 (2005): 910–914.

Margulis, Lynn (1938–) American *Evolutionary biologist* Lynn Margulis (see photo) is one of the few scientists alive today who has changed some of the basic assumptions of a whole field of study. She persistently defended the endosymbiotic theory of the origin of mitochondria and chloroplasts that had been proposed decades earlier by Russian botanists and by American biologist Ivan Wallin. She gathered evidence from many studies that finally established this theory. Mitochondria and chloroplasts are the evolutionary descendants of bacteria. She is currently testing the theory that eukaryotic cilia, flagella, and microtubules are also the evolutionary descendants of bacteria (see SYMBIOGENESIS). She has presented evidence that symbiogenesis is an important, perhaps the most important, source of evolutionary novelty and diversification. Symbiosis (for example, with microorganisms living inside the roots of plants) made plant life on land possible, and plants then created the habitats in which animals lived. Margulis says, "Symbiogenesis was the moon that pulled the tide of life from its oceanic depths to dry land and up into the air." She also explains that, because all life-forms interact so closely, they produce a global living system that regulates some physical conditions (such as temperature and atmospheric gas composition) of the entire planet (see GAIA HYPOTHESIS). To Margulis, symbiosis is pervasive throughout the living world, from the smallest cellular components to the global ecosystem. She says, "Symbiosis is not a marginal or rare phenomenon. It is natural and common. We abide in a symbiotic world."

Lynn Margulis is most famous for her research into symbiogenesis. *(Courtesy of University of Massachusetts at Amherst)*

Margulis has always been a creative thinker, willing to take bold steps to pursue new theories. As Lynn Alexander, she was an early entrant, at age 14, into the University of Chicago. She pursued a general liberal arts degree that emphasized independent study and the reading of original publications rather than textbooks. While at Chicago she met another precocious and creative student of science. Astronomer Carl Sagan was already a graduate student at age 19 (see SAGAN, CARL). Both she and Sagan, who later married, were to become two of the most creative thinkers and prolific writers in modern science, and both were to contribute in different ways to an understanding of planetary atmospheres as well as of evolution.

Lynn and Carl Sagan moved to Wisconsin. She studied genetics at the University of Wisconsin, Madison, then continued her studies at the University of California at Berkeley, where she completed her doctorate in 1965. She became especially interested in cytoplasmic inheritance, in which genetic traits were passed from one generation to another in the cytoplasm rather than by the nucleus. She found out that early 20th-century geneticists Hugo DeVries (see DEVRIES, HUGO) and Carl Correns knew that not all of a plant cell's genes were in the nucleus and had discovered that chloroplasts also had genes. But they and most geneticists since them had divided genetic inheritance into "nuclear vs. unclear," dismissing cytoplasmic inheritance as unimportant. After 15 rejections and losses, Lynn Sagan's paper (see Further Reading) was published in 1967. Her persistence and her refusal to accept a simplistic doctrine eventually led to a revolution in biology.

Lynn Margulis joined the biology faculty of Boston University in 1966. The departmental chair, biologist George Fulton, described Margulis as "the only instructor who was paid half time and worked time and a half." Margulis left Boston University and in 1988 joined the faculty, first in biology then in geosciences, at the University of Massachusetts at Amherst, where she is currently a Distinguished Professor. She believes strongly that students in her courses should go outdoors and directly observe organisms in their contexts of symbiotic interaction. Margulis is a member of the National Academy of Sciences and became president of Sigma Xi, the Scientific Research Society, for a year beginning in 2005. She has written numerous books and articles with her son, the writer Dorion Sagan.

Further Reading

Dolan, Michael F., and Lynn Margulis. *Early Life,* 2nd ed. Sudbury, Mass.: Jones and Bartlett, in press.

Margulis, Lynn. *Symbiosis in Cell Evolution: Microbial Communities in the Archaean and Proterozoic Eons,* 2nd ed. New York: W. H. Freeman, 1993.

———. *Symbiotic Planet: A New Look at Evolution.* New York: Basic Books, 1998.

———. "Mixing it up: How I became a scientist." *Natural History,* September 2004, 80.

———, and Dorion Sagan. *Mystery Dance: On the Evolution of Sexuality.* New York: Touchstone, 1992.

———, and ———. *Slanted Truths: Essays on Gaia, Symbiosis, and Evolution.* New York: Copernicus, 1997.

———, and ———. "The beast with five genomes." *Natural History,* June 2001, 38–41.

———, and Karlene V. Schwartz. *Five Kingdoms: An Illustrated Guide to the Phyla of Life on Earth,* 3rd ed. New York: W. H. Freeman, 1998.

Sagan, Lynn. "On the Origin of Mitosing Cells." *Journal of Theoretical Biology* 14 (1967): 225–274.

Sciencewriters. "Sciencewriters: About the Authors." Available online. URL: http://www.xsnrg.com/sciencewriters/authors.htm. Accessed April 12, 2005.

markers Genetic markers are segments of DNA that can be used to trace patterns of inheritance. Genetic markers may or may not have any genetic function. In higher organisms, well over 90 percent of the DNA conveys no genetic information. NONCODING DNA, while not useful as a source of genetic information in cells, can be used by scientists to trace genetic ancestry (see DNA [EVIDENCE FOR EVOLUTION]).

Chromosomal DNA is reshuffled each generation, as MEIOSIS separates pairs of chromosomes, and fertilization brings them back together in new combinations—one chromosome from the mother, and one from the father, in each pair (see MENDELIAN GENETICS). Chromosomal DNA, therefore, is inherited by complex routes through both the mother and the father. However, DNA within mitochondria is inherited only from the mother, and DNA of the Y chromosome in humans is inherited only from the father.

Agriculture entered Europe from the Middle East about 7,500 years ago (see AGRICULTURE, EVOLUTION OF). Historians have debated whether this occurred because of a progressive invasion of Europe by farmers from the Middle East, or because agricultural techniques diffused from one population to another from the Middle East. Analysis of genetic markers from skeletons of the earliest farmers showed little similarity of their mitochondrial DNA with that of modern Europeans. Modern European males have Y chromosome markers that resemble those now common in the Middle East. These facts suggest that immigrant Middle Eastern males mated with resident females in Europe about 7,500 years ago, bringing agriculture with them.

Three famous examples of the use of DNA markers are the identification of the remains of the last czar of Russia, Aaron's Y chromosome, and Thomas Jefferson's descendants.

The last czar. Part of the noncoding DNA consists of *tandem repeats,* which are short, meaningless sections of DNA repeated over and over, and occurring right next to one another on the same chromosome. There are no surviving samples of DNA (e.g., from hair) from the last Russian royal family—Czar Nicholas II, the Czarina Alexandra, four daughters, and the son Alexis, who were executed during the Bolshevik Revolution in 1917 and whose remains were buried in an undisclosed location. The Czarina Alexandra was the granddaughter of Queen Victoria. Scientists studied DNA samples from modern descendants of Victoria and found a section of tandem repeated noncoding DNA (a marker) that most of them shared. When an unmarked grave was found in Russia that was believed to be that of the last royal family, researchers took DNA samples from the bones and found this tandem repeated DNA in some of them, thus confirming that these bones are very likely to be those of the last czar and his family.

Aaron's Y chromosome. According to Jewish tradition, high priests are all male and are all descendants of the first high priest, Aaron, brother of Moses. These males are known as kohanim, and today their last names have such variants as Cohen, Cohn, and Kahn. Researchers obtained DNA samples (from cheek cells) from modern Jewish men who identified themselves as kohanim, in Israel, North America, and England. About half of these shared a DNA sequence (a marker) in their Y chromosomes that had to come from a common ancestor about a hundred generations back—which was approximately the time that Aaron is believed to have lived. This Cohen Modal Haplotype may very well have come from Aaron's Y chromosome. (A *haplotype* is a DNA sequence that is not broken up by sexual recombination; in many cases, haplotypes are genes, but in this case it is nearly the entire Y chromosome. *Modal* refers to the most frequent set of sequences found in this region of the chromosome in these men.)

The Lemba tribe in southern Africa, though phenotypically black, claimed a tradition of Jewish ancestry. In 1997 researchers studied Y chromosomes from men in this tribe. Two-thirds of the Y chromosomes in this sample were of Middle Eastern, rather than Bantu African, origin. And most of the Y chromosomes contained the Cohen Modal Haplotype, suggesting that they were not merely of Jewish descent but of Jewish priestly descent. In contrast, the mitochondrial DNA in this tribe was not of Jewish origin, strengthening the suggestion that this tribe is descended from Jewish men, not Jewish women.

Jefferson's descendants. Many historians believe that American president Thomas Jefferson produced offspring through his slave and friend Sally Hemings. Researchers found a DNA marker that was shared between known white descendants of Thomas Jefferson and some black Americans who also had a family tradition of Jeffersonian descent. While most historians accept this as proof that Thomas Jefferson had illegitimate offspring, some point out that the DNA marker could just as easily have come from Thomas Jefferson's brother, who was also present at the places and times when the offspring would have been conceived. This example illustrates both the power and the limitations of DNA markers in genetic research.

Markers are also used in cases of paternity identification and criminal investigation. Markers, not usually being subject to evolutionary forces, often change rapidly over time. A major application of markers to evolutionary science is to short-term studies of POPULATION GENETICS. Tracing the parentage of individual animals and plants in populations helps to reveal the patterns of mating and of reproductive success in these evolving populations.

Further Reading

Ayres, Debra R., and Donald R. Strong. "Origin and genetic diversity of *Spartina anglica* (Poaceae) using nuclear DNA markers." *American Journal of Botany* 88 (2001): 1,863–1,867.

Drayna, Dennis. "Founder mutations." *Scientific American,* October 2005, 78–85.

Haak, Wolfgang, et al. "Ancient DNA from the first European farmers in 7500-year-old Neolithic sites." *Science* 310 (2005): 1,016–1,021. Summary by Michael Balter, *Science* 310 (2005): 964–965.

Olson, Steve. *Mapping Human History: Discovering the Past through Our Genes.* New York: Houghton Mifflin, 2002.

Shreeve, James. "The greatest journey." *National Geographic,* March 2006, 60–73.

Mars, life on Life-forms resembling the bacteria found on Earth may have once existed on Mars. When the evidence for this was first announced, some people considered it the most important discovery in the history of science. The evidence came from a Martian meteorite that was found in Antarctica. Had the evidence been confirmed, it would have indicated that the evolution of life from simple molecules was not an isolated event on the Earth. If life originated on two planets in the same solar system, it is quite likely to have occurred many times on many other planets throughout the universe.

Mars is the planet most similar to Earth, in temperature and chemical composition, in the solar system. Jupiter and Saturn consist largely of cold liquids, while Venus is extremely hot. Some moons of Jupiter and Saturn are more likely to have conditions suitable for the origin of life, although they too are very cold. Scientists have long believed that, if life ever existed outside of the Earth, it would have been on Mars or on one of these moons.

In recent years, astronomers have presented evidence for several planets orbiting other stars in the Milky Way galaxy. The first photograph of such a planet was published in 2005. Because these planets are usually too far away to be directly observed, the evidence for their existence is usually the gravitational force that they exert upon the stars around which they revolve. From the movements of these stars, the mass and orbit of the unseen planets can be calculated. Most of these planets appear to be large and gaseous and very close to their stars ("hot Jupiters"). However, this does not mean that Earth-like planets are rare; small planets like Earth are much less likely to be detected by such methods. The presence of planets may also be inferred by periodic slight decreases in star luminosity, perhaps produced when the planet passes between the star and the human observer.

The U.S. National Aeronautics and Space Administration (NASA) sent two spacecraft that landed on Mars in 1976. These spacecraft sent back photographs and data, none of which indicated life on Mars. A burst of carbon dioxide production from soil was thought at first to indicate microbial life on Mars, but scientists later decided that the carbon dioxide could have been produced by an inorganic reaction. Other spacecraft sent to Mars in 1997 and 2004 have sent back data that reinforce the conclusion that Mars is lifeless at the present time.

But was there life on Mars in the past? Mars was a very different planet in the past than it is at present. When first formed, it was warm and had liquid water. The evidence for this comes from the following sources:

- Satellites orbiting Mars have photographed geological patterns that appear to have been produced by water erosion (see photo on page 260).

- The spacecraft *Opportunity* and the rovers that it delivered to the Martian surface sent back evidence in 2004 that the chemistry and structure of some Martian rocks probably resulted from percolation by water in the distant past.
- Satellite photographs of Martian gullies, published in 2006, showed that over a four-year time span, a flow of liquid water had filled a gully with ice and mineral deposits.

What happened to the water on Mars? Being a smaller planet than Earth, Mars cooled down more quickly from its initially molten condition. With less gravitation, it was unable to hold onto a significant atmosphere. The Martian atmosphere is about one percent as dense as Earth's. Mars apparently no longer has liquid magma circulating in its interior; therefore, it has no magnetic field that can deflect solar particles. The solar particles have scoured away much of the atmosphere, including the water vapor.

In 1996 scientists from NASA and several universities announced direct evidence that there may once have been life on Mars. They obtained the evidence from one of more than 50 meteorites that apparently came from Mars: a rock called ALH84001, so named because it was specimen number 1 found in 1984 in the Allen Hills of Antarctica. Meteorites from Mars are the next best thing to actually bringing rocks back from Mars.

How could scientists conclude that the meteorite came from Mars?

- Scientists had to confirm that the rock really was a meteorite. It was found on top of an ice field in Antarctica, far from rock outcrops. Scientists therefore concluded that the rocks at this location were meteorites. At the location where the rocks were found, the ice slowly rises and the water molecules enter the atmosphere, leaving the meteorites behind. Therefore, over thousands of years, meteorites have accumulated at this location in relatively large numbers.
- Scientists had to confirm that the meteorite really came from Mars. Most meteorites are small asteroids (see ASTEROIDS AND COMETS). Some meteorites are volcanic in origin, which means they must have come from a planet. ALH84001 is of volcanic origin. Could it have come from a volcano on the Earth? An analysis of gas bubbles trapped within the rocks closely matched the atmosphere of Mars. The rocks were therefore identified as meteorites from Mars.

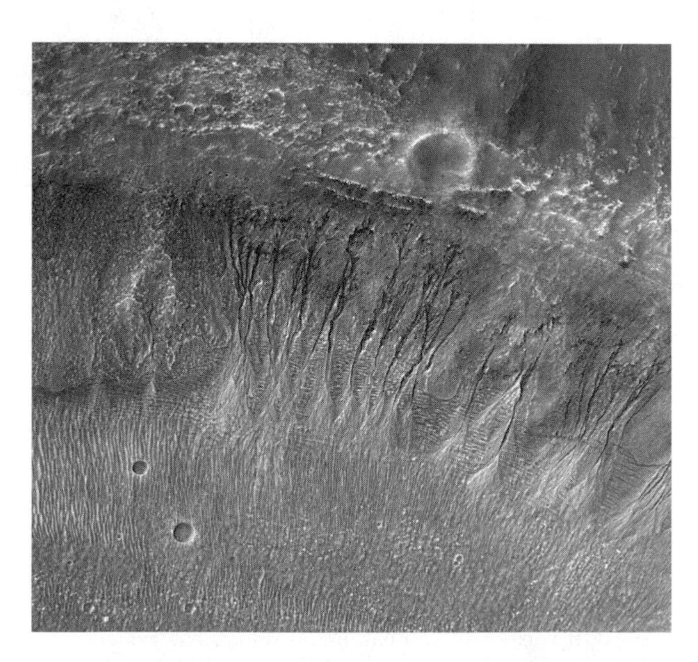

This photograph taken in 2004 by the *Mars Global Surveyor* shows gullies coming from the sides of an impact crater. These gullies were most likely formed by melting water, either from groundwater or from snow. The scale bar is about 980 feet (300 m). *(Courtesy of NASA)*

How could volcanic rocks from Mars have reached the Earth? The most likely explanation is that a large meteoric impact on Mars about 15 million years ago blasted some volcanic rocks up into space. Some of these volcanic rocks crossed the orbit of the Earth and became meteorites. The ALH84001 meteorite fell to the Earth about 13,000 years ago.

What evidence of life can be found in these meteorites? NASA scientists presented two types of evidence: chemical and fossil.

Chemical evidence. ALH84001 contains organic compounds known as polycyclic aromatic hydrocarbons (PAHs), similar to some of the compounds found in petroleum. Although terrestrial PAHs are usually produced by biological activity (as in petroleum), they are also found in interstellar space (see ORIGIN OF LIFE).

Is there any direct evidence that the PAHs in the Martian meteorite were formed by biological activity? The PAHs were found closely associated with carbonates in the meteorite. Carbonates (for example, in limestone and the shells of invertebrates) are often produced by biological processes—but can also be produced inorganically as dissolved CO_2 reacts with calcium in water. The carbonates have been dated to be 3.6 billion years old, younger than the 4.5-billion-year-old rock itself. The carbonates, and PAHs, may have entered the rock after it formed, perhaps from prolonged soaking in water that contained microbes.

The scientists considered the possibility that the PAHs could be contaminants from petroleum fumes in the Earth's polluted atmosphere. But, they concluded, if this were the case, there would be more PAHs on the outside than the inside of the meteorite. Since the outside of the meteorite had less PAH than did the inside, the scientists concluded that the PAHs could not have come from a relatively brief exposure to atmospheric PAHs on earth.

Fossil evidence. Scanning electron micrographs of the meteorite revealed structures that look like microorganisms (see photo on page 261). Although it is possible that they are crystal-like structures formed under lifeless conditions, they resembled Earth's microorganisms so much that some observers could not resist believing them to be the fossils of once living cells.

Since the initial announcement of the possible Martian microbes, enthusiasm for the discovery has diminished. The

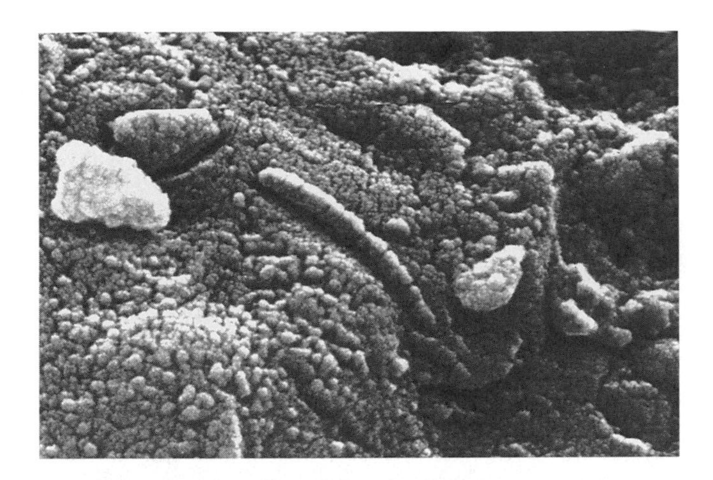

A scanning electron micrograph from meteorite ALH84001 shows structures that researchers initially thought were bacteria-like life-forms, although they are much smaller than bacteria on Earth. ***(Courtesy of NASA)***

fossilized structures are a hundred times smaller than the smallest known bacteria on Earth, being more similar in size to viruses. Viruses cannot live outside of cells. Is it possible that such small organisms could have lived in a shallow Martian pond billions of years ago? Every piece of evidence for life in the Martian meteorite can be explained in other ways. Most scientists remain unconvinced that the organic chemicals were produced by Martian structures. It is often said that extraordinary claims (such as life on Mars) require extraordinary evidence. This evidence is good, but not good enough for most scientists.

Further Reading

Jet Propulsion Laboratory, National Aeronautics and Space Administration, Pasadena, Calif. "Mars Meteorites." Available online. URL: http://www2.jpl.nasa.gov/snc/index.html. Accessed July 10, 2006.

Kerr, Richard A. "New signs of ancient life in another Martian meteorite?" *Science* 311 (2006): 1,858–1,859.

Malin, Michael C. et al. "Present-day impact cratering rate and contemporary gully activity on Mars." *Science* 314 (2006): 1,573–1,577.

McKay, D. S., et al. "Search for past life on Mars: Possible relic biogenic activity in Martian meteorite ALH84001." *Science* 273 (1996): 924–930. Available online. URL: http://www-curator.jsc.nasa.gov/curator/antmet/marsmets/SearchForLife/SearchForLife.htm. Accessed April 25, 2005.

Morton, Oliver. "Mars: Planet Ice." *National Geographic,* January 2004, 2–31.

mass extinctions Mass extinction occurs when many species become extinct at roughly the same time. EXTINCTION may result if populations do not have the genes that allow them to adapt to environmental changes, that is, from "bad genes"; but extinction may also occur suddenly as a result of a worldwide disaster, for which no populations possess genetic resistance, that is, from "bad luck," to use paleontologist David Raup's terminology. Global mass extinctions are usually considered to have resulted from "bad luck." Therefore true mass extinction events may result from rare catastrophic events rather than an acceleration of normal extinction patterns.

Most of the geological periods (see GEOLOGICAL TIME SCALE) ended with global changes in climate and the origin of recognizably different sets of species. Five of these were mass extinctions (see BIODIVERSITY) (see table). Two of the mass extinction events brought the first two eras of the Phanerozoic Eon: the PERMIAN EXTINCTION ended the PALEOZOIC ERA, and the CRETACEOUS EXTINCTION ended the MESOZOIC ERA. A mass extinction, especially of the EDIACARAN ORGANISMS, may have occurred at the end of the PRECAMBRIAN TIME. A new analysis in 2005 suggests that the Ordovician, Permian, and Cretaceous mass extinctions occurred against a background of rapid evolution and therefore were caused by global catastrophes, whereas the Devonian and Triassic mass extinctions occurred against a background of ongoing extinction and may have been caused by an amplification of climatic trends that were already occurring.

Many scientists describe the extinctions that are now occurring throughout the world as a result of human activities, such as the destruction of natural habitats, as the "sixth mass extinction." Extinctions are in fact now occurring at a rate that would justify this description. Before the appearance of humans, the global extinction rate was about one per million species per year; now the global extinction rate is one per thousand per year.

The most famous example of a mass extinction event is the Cretaceous extinction event 65 million years ago. At that time, many species of organisms, most famously the dinosaurs, became extinct. Exceptionally high levels of the element iridium in the geological deposits 65 million years ago suggest that the Earth was hit by an asteroid (see ASTEROIDS AND COMETS); asteroids usually contain higher levels of this element than do the rocks of the Earth's crust. The greatly eroded remnants of this impact can still be seen in the Chicxulub crater of Yucatán.

Even the Cretaceous extinction event, however, had more than one cause. By the time the asteroid hit the Earth, populations of many dinosaur and other species had already been in decline for millions of years. Moreover, a massive volcanic eruption occurred about the same time in what is now India,

Major Extinctions in the History of Life

Geological era: End of	Millions of years ago	Percent of families dying	Percent of genera dying	Percent of species dying
Ordovician	439	26	60	85
Devonian	367	22	57	83
Permian	251	51	82	95
Triassic	208	22	53	80
Cretaceous	65	16	47	76

producing a 200,000 square mile (over 500,000 square km) lava flow called the Deccan Traps. Several interacting causes, the relative importances of which are still debated, brought about the Cretaceous extinction.

The Cretaceous extinction was not the biggest extinction event in Earth history. The mass extinction that brought the Permian period, and the whole Paleozoic era, to an end 250 million years ago was, in the words of paleontologist Douglas Erwin, the "mother of all extinctions." At this time, at least half of all families of organisms became extinct. Because some families contain many species, it has been estimated that up to 95 percent of all species became extinct at that time! Because this event was further back in time, it is more difficult to study: There has been more time for evidence to be lost, and it is more difficult to calculate precise dates for the events that occurred at that time. Despite this, evidence has been found of an asteroid impact at the end of the Permian period. Nevertheless, it is difficult for scientists to determine to what extent these events associated with the Permian extinction occurred simultaneously.

Perhaps the biggest unsolved mystery regarding mass extinction events is what appears to be their recurring pattern. Paleontologists David Raup and J. J. Sepkoski have calculated a 26-million-year cycle of recurring mass extinctions. The pattern is not perfect: Mass extinctions have not occurred every 26 million years, nor have they occurred at precisely this interval. However, their results are statistically significant. Scientists have been unable to explain normal geological events that might cause such a recurring synchronous pattern. Earth scientists Marc Davis, Piet Hut, and Robert Muller have suggested that there is a small companion star to the Sun, which sweeps a mass of comets and asteroids along with it. The orbit of this star, they speculate, brings it close enough to the earth every 26 million years to bombard the Earth with comets and asteroids. Even those that suggest the existence of this "Nemesis star" admit that it has not been seen or otherwise detected.

Solar flares, which would flood the Earth with intense radiation, have been suggested as a possible cause of mass extinctions. Scientists know little about the timing of these solar flares through the history of the solar system, and these flares would leave no evidence of having struck the Earth, other than the mass extinction itself. Because they cannot think of a way to investigate this possibility, most scientists dismiss solar flares as a cause of extinction events.

Following each mass extinction event, the diversity of species has not only recuperated but increased. The best fossil record that is available is that of marine invertebrates (see FOSSILS AND FOSSILIZATION). This record indicates a steady increase in worldwide species diversity, especially after mass extinction events (see BIODIVERSITY). One example of evolution following a mass extinction is the adaptive radiation of mammal species after the extinction of the dinosaurs (see ADAPTIVE RADIATION). Many scientists fear that species diversity may not recover from the mass extinction that is now occurring as a result of human activity, because unlike past mass extinctions, the cause of extinction (human activity) is continuing unabated.

Some extinction events, although not considered mass extinctions, have still had an important effect on the evolutionary history of life. The Hemphillian extinction event about five million years ago produced relatively few extinctions, but among them were many grazing animals. A diversity of horse species was reduced to just one.

Terrestrial plants may not respond to mass extinction events as rapidly or as markedly as marine and terrestrial animals. Most plant extinctions have been caused by long-term climatic changes. While many plants became extinct during the five mass extinction events, their response was delayed until several million years after each of the extinction events.

Life has recuperated from mass extinctions largely because asteroid collisions and other planetary disasters have been relatively rare for the past one or two billion years. According to earth scientists Peter Ward and David Brownlee, humans should, so to speak, thank their lucky star for this: The Sun is a relatively stable star, compared to most in the universe, and the planet Jupiter (which can be considered a star that never ignited) has swept part of the solar system free of most asteroids except those in the asteroid belt. Craters on the moon appear to be mostly three billion to four billion years old, indicating that during this time period comets and asteroids were very common in the solar system. Even though bacterial life appeared on Earth soon after its formation, frequent bombardment of the Earth during that period may have delayed the appearance of complex cells until about a billion and a half years ago (see ORIGIN OF LIFE). The evolution of complex life-forms would have been impossible on a planet subjected to frequent mass extinctions.

Further Reading

Courtillot, Vincent. *Evolutionary Catastrophes: The Science of Mass Extinction.* Trans. Joe McClinton. New York: Cambridge University Press, 2002.

Davis, M., P. Hut, and R. Muller. "Extinction of species by periodic comet showers." *Nature* 308 (1984): 715–717.

Hallam, Anthony. *Catastrophes and Lesser Calamities: The Causes of Mass Extinctions.* New York: Oxford University Press, 2004.

———, and P. B. Wignall. *Mass Extinctions and Their Aftermath.* New York: Oxford University Press, 1997.

Raup, David M. *Extinction: Bad Genes or Bad Luck?* New York: Norton, 1991.

Villier, Loïc, and Dieter Korn. "Morphological disparity of ammonoids and the mark of Permian mass extinctions." *Science* 306 (2004): 264–266.

Ward, Peter, and David Brownlee. *Rare Earth.* New York: Copernicus Books, 2000.

———, and Alexis Rockman. *Future Evolution.* New York: Freeman, 2001.

Maynard Smith, John (1920–2004) British *Evolutionary biologist* John Maynard Smith was one of the leading evolutionary theorists of the 20th century. He contributed greatly to an understanding of how the process of evolution works. Maynard Smith's contributions to evolutionary science included:

- Application of mathematical principles of game theory to natural selection
- Theories about the major transitions in evolution, e.g., from replicators to chromosomes to single-celled organisms to multicellular organisms, and the evolution of sex

John Maynard Smith was born January 6, 1920. When he was a student at Eton College, he read some of the writings of an alumnus (see HALDANE, J. B. S.) that provided a mathematical basis to Darwinian evolution. While the modern theory of evolution was taking shape (see MODERN SYNTHESIS), Maynard Smith was an engineering student at Cambridge University. After completing his degree in 1941, he worked in aircraft design.

Maynard Smith could not stay away from evolution. He entered University College, London, to study genetics with Haldane. After completing his doctorate, he continued teaching at University College from 1952 until 1965. He published a popular book about evolution in 1958. Besides sharing with Haldane an interest in the genetic basis of evolution, Maynard Smith also shared with Haldane a devotion to communism. Both, however, left the party (Haldane in 1950, Maynard Smith in 1956) in response to news about Soviet brutalities and especially LYSENKOISM, a philosophy that was directly antagonistic to the MENDELIAN GENETICS upon which evolution is based. Maynard Smith was the dean of the University of Sussex from 1965 to 1985, and he shared the 1999 Crafoord Prize (awarded by the Royal Swedish Academy of Sciences in areas not covered by the Nobel Prize) with evolutionary scientists Ernst Mayr and George C. Williams (see MAYR, ERNST). John Maynard Smith died April 19, 2004.

Further Reading

Lewontin, Richard. "In memory of John Maynard Smith (1920–2004)." *Science* 304 (2004): 979.

Maynard Smith, John. *Did Darwin Get It Right? Essays on Games, Sex and Evolution.* London: Chapman and Hall, 1988.

———. *The Evolution of Sex.* Cambridge, U.K.: Cambridge University Press, 1978.

———. *Evolution and the Theory of Games.* Cambridge: Cambridge University Press, 1982.

———. *The Theory of Evolution.* London: Penguin Books, 1958.

———, and Eörs Szathmáry. *The Major Transitions in Evolution.* New York: Oxford University Press, 1997.

———, and ———. *The Origins of Life: From the Birth of Life to the Origin of Language.* Oxford: Oxford University Press.

Mayr, Ernst (1904–2005) American *Ornithologist, Evolutionary biologist* Ernst Mayr (see photo) was the last survivor of the group of scientists who brought Darwinian NATURAL SELECTION and MENDELIAN GENETICS together in the MODERN SYNTHESIS, which is the modern understanding of how evolution works (see DOBZHANSKY, THEODOSIUS; SIMPSON, GEORGE GAYLORD; STEBBINS, G. LEDYARD). Even to his 100th birthday, Mayr continued to write books that synthesized the understanding of evolution and the meaning of biology.

Ernst Mayr played a major role in studying the evolutionary process and the history of evolutionary science. His career spanned most of the 20th century. ***(Courtesy of Harvard University)***

Like many other important scientists (see DARWIN, CHARLES; WILSON, EDWARD O.), Mayr gained his first insights by traveling extensively in the natural world. Born July 5, 1904, Mayr was educated in Germany and began work for his Ph.D. at the Natural History Museum in Berlin in 1925. The Zoological Museum of Berlin, the American Museum of Natural History in New York, and Lord Rothschild's Museum in London sent Mayr to New Guinea in 1928 to complete a thorough ornithological survey. Travel is still difficult in New Guinea, but Mayr did it at a time when intertribal warfare continued to make it dangerous for any outsider to visit. He survived tropical diseases, a descent down a waterfall, an overturned canoe, and managed to climb to the summits of all five major mountains.

Mayr's observations of the geographical variation among populations within bird species were the basis for two of his major contributions to biology (besides the Modern Synthesis). First, he clarified the *biological species concept* which defined species as containing all the organisms that could potentially interbreed under natural conditions (see SPECIATION). Second, he explained geographical isolation of peripherally isolated populations, one of the major mechanisms of speciation. The mathematical forebears of the Modern Synthesis (see HALDANE, J. B. S.; FISHER, R. A.) had explained the theoretical basis for evolutionary change, but not for speciation. This was mainly Mayr's contribution.

Mayr got a job in 1930 at the American Museum of Natural History, where he organized their collection of bird specimens. While there, in 1942, he wrote *Systematics and the Origin of Species,* the taxonomic counterpart to *Genetics and the Origin of Species* by geneticist Theodosius Dobzhansky. He moved to Harvard University's Museum of Comparative Zoology in 1953 as the Alexander Agassiz Professor of

Zoology. He became director in 1961. He retired as director in 1970 and from the museum in 1975. This gave him the opportunity to become a prolific writer, particularly on the historical development of biological science.

Throughout his career, Mayr emphasized the allopatric mode of speciation that results from geographical isolation, which he considered the predominant mode of speciation in birds and mammals. However, he recognized that many other types of speciation are possible, especially in microorganisms. Rather than feel threatened by the emergence of sympatric speciation, speciation by hybridization and polyploidy, HORIZONTAL GENE TRANSFER, and SYMBIOGENESIS, he embraced them: At age 100, he wrote, "There are whole new worlds to be discovered with, perhaps, new modes of speciation..." He died February 3, 2005.

Further Reading

Margulis, Lynn. "Ernst Mayr, Biologist extraordinaire." *American Scientist* 93 (2005): 200–201.

Mayr, Ernst. "80 years of watching the evolution scenery." *Science* 305 (2004): 46–47.

———. *The Growth of Biological Thought: Diversity, Evolution, and Inheritance.* Cambridge, Mass.: Harvard University Press, 1982.

———. *One Long Argument: Charles Darwin and the Genesis of Modern Evolutionary Thought.* Cambridge, Mass.: Harvard University Press, 1991.

———. *This Is Biology: The Science of the Living World.* Cambridge, Mass.: Harvard University Press, 1997.

———. *What Evolution Is.* New York: Basic Books, 2001.

———. *What Makes Biology Unique?: Considerations on the Autonomy of a Scientific Discipline.* New York: Cambridge University Press, 2004.

meiosis Eukaryotic cells (see EUKARYOTES, EVOLUTION OF) have nuclei that contain chromosomes. Chromosomes are the structures that contain DNA (see DNA [RAW MATERIAL OF EVOLUTION]). When a cell reproduces itself, the chromosomes of its nucleus replicate, then the rest of the cell divides. There are two types of cell and nuclear division. In *mitosis,* two cells result, which have the same number of chromosomes as the original cell. Mitosis occurs in most of the tissues and organs of organisms as they grow and as they replace old cells. *Meiosis* is a type of nuclear and cell division that occurs during sexual reproduction. During meiosis, the nuclei reduce their chromosome number by half when they divide. Meiosis produces the sex cells: the *spores* of fungi and plants, and the *eggs* and *sperm* of animals. Eggs and sperm are collectively called *gametes.* When an egg and a sperm fuse together in the process of *fertilization,* they form a *zygote.* Fertilization restores the original chromosome number. In every generation in which sexual reproduction occurs, meiosis alternates with fertilization.

Chromosomes of almost all eukaryotic cells occur in pairs. Each organism inherits one member of each pair from its female parent, the other from its male parent. What is it about the chromosomes that make them "a pair"? They function as a pair because each chromosome carries genes that code for the same traits as the other chromosome in the pair. Consider a chromosome in a plant cell that carries the gene for red flowers. The chromosome with which it is paired also has the gene for flower color, at the same location on the chromosome—but not necessarily the same form of the gene. The other chromosome may have the same form of the gene, for red flowers; or it may have another form of the gene, coding for white flowers. Different forms of the same gene are called *alleles.* The two chromosomes of this pair are *homologous* to one another (they are *homologs*). Chromosomes in pairs are homologous to one another just as people are homologous to one another. People have eyes, noses, mouths, chins, arms, etc., at the same relative locations on their bodies, but the forms of these organs can be very different. Chromosomes that form pairs with one another are homologous, usually not identical, just as people are homologous, usually not identical.

Because the chromosomes of most eukaryotic cells occur in pairs, the cells are called *diploid* (*di-* denotes two), sometimes abbreviated 2N. Spores and gametes have chromosomes that are unpaired; they are called *haploid,* sometimes abbreviated 1N or simply N. Diploid cells of humans normally have 46 chromosomes, in 23 pairs; human haploid cells have 23 chromosomes, with no pairs. Meiosis produces haploid cells from diploid cells; fertilization unites haploid cells back into diploid cells.

Meiosis consists of two divisions (referred to as Meiosis I and Meiosis II). In the first division, one cell becomes two; in the second, two cells may become four. The stages of the cell cycle during each division have the same names in meiosis and in mitosis (interphase, prophase, metaphase, anaphase, telophase).

Interphase. During the *interphase* that precedes meiosis, each chromosome replicates and forms two *chromatids.* However, chromosome replication is not quite completed. Each chromosome contains short stretches of DNA that have not yet been duplicated.

First division. The first division of meiosis separates homologs from one another:

- The homologs recognize and line up next to each other as the first phase, *prophase I,* begins. This happens because the short unduplicated segments of DNA in one chromosome recognize corresponding segments of its homolog. There are special proteins whose only function is to promote the recognition between homologs so that they can come together during meiosis. Because this stage of meiosis requires pairs of chromosomes, cells with unpaired sets of chromosomes cannot undergo meiosis.
- Once the homologs have paired with one another, chromatids of the two homologs may cross one another at one or more locations, break, then rejoin. This process is called *crossing over.* At this time special proteins that function only during meiosis clip the DNA of each homolog at the same location and fuse the DNA strands back together, having switched the strands of the two homologs. As a result of crossing over, some alleles that had been on one homolog now occur on the other homolog. That is, some

of the alleles that the organism received from its male parent switch places with alleles that the organism received from its female parent. Now the two chromatids of each chromosome are no longer identical to one another. Crossing over therefore contributes to genetic recombination, the production of new combinations of alleles,

- In *metaphase I,* the chromosomes line up in the middle of the cell. Each chromosome is paired with its homolog, and each homolog is attached to a separate protein strand.
- During *anaphase I,* the protein fibers pull the homologs toward opposite poles, a process completed in *telophase I.*

Therefore, during the first meiotic division, the homologous chromosomes are separated; this does not occur during mitosis.

For each pair of homologs, chance determines which homolog goes to each of the two poles. As a result, some of the chromosomes that originally came from the individual's mother, and some of the chromosomes that originally came from the individual's father, will be pulled toward the same pole. This results in the *independent assortment* of traits from the two parents to their offspring (see MENDELIAN GENETICS). In this way, as well as by crossing over, meiosis contributes to the production of genetic diversity.

Second division. During metaphase II, the chromosomes again line up in the middle of the cell and protein strands attach to the chromatids. During anaphase II, the chromatids are pulled apart. Therefore during the second meiotic division, chromatids are separated, just as during mitosis. The end result of meiosis is usually the production of four haploid nuclei, usually in four cells. Each of the nuclei contains one chromosome from each homologous pair.

This is the process of meiosis as it occurs in the production of sperm or the small spores of plants. However, in the production of eggs or larger spores, each phase of meiosis can produce cells of unequal size. The smaller cells are called *polar bodies* and die. This allows the resulting egg, or the larger spore, to be nearly as large as the original cell at the beginning of meiosis. In some cases, some homologs are better able to secure the attachment of the protein fibers, ensuring that they, rather than their corresponding homologs, are the ones that end up in the egg and not in a polar body (see SELFISH GENETIC ELEMENTS).

In some instances, both homologs are accidentally pulled toward the same pole. This process, called *nondisjunction,* can result in a haploid cell with one or more extra chromosomes. Because such a cell usually cannot function properly in humans, the cell usually dies. However, in human women, nondisjunction of chromosome number 21 sometimes produces an egg cell with 24, rather than 23, chromosomes. This egg, once fertilized, can develop into a person with trisomy 21 or *Down syndrome.* Since chromosome 21 is one of the smallest chromosomes, a person whose cells have an extra chromosome 21 has mental and physical handicaps, but this chromosomal arrangement is not lethal; in fact, the person can often, with help, lead a normal life. Nondisjunction of other chromosomes is usually fatal to humans and other animals. Almost a third of all fertilizations end in the death

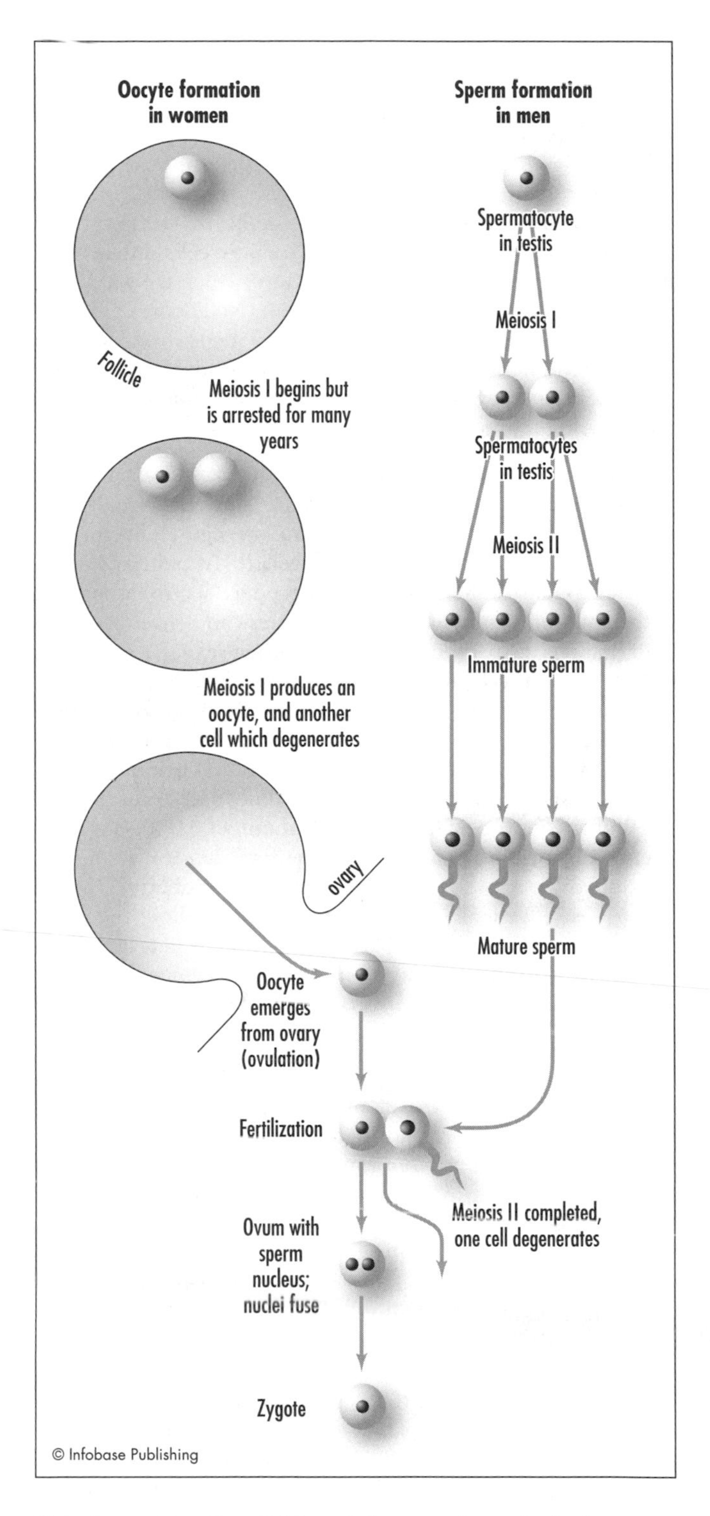

Meiosis consists of two divisions. In human males, meiosis of a primary spermatocyte produces four sperm. In human females, the first meiotic division of a primary oocyte produces a secondary oocyte and a polar body, which eventually degenerates. The secondary oocyte is released during ovulation and may be fertilized by a sperm. Only after fertilization does the second division of meiosis produce an ovum and another polar body, which degenerates. The fertilized ovum is now a zygote.

of the fertilized egg or young embryo, often due to chromosomal abnormalities such as nondisjunction. In plants, complete nondisjunction sometimes occurs, resulting in a diploid egg cell that survives.

In animals, the cells that will ultimately become gametes arise from *germ cells* that assume their specialized function very early in the development of the embryo. The nonsexual cells are *somatic* cells. Some of the somatic cells differentiate into *gonads* or reproductive organs, into which the germ cells migrate during embryonic development. Male gonads (*testes*) contain male germ cells, and female gonads (*ovaries*) contain female germ cells. In most animals, male and female germ cells occur in separate male and female individuals. The germ cells are all diploid at this stage.

- Male germ cells undergo mitosis, producing more male germ cells. Then many of them become *spermatocytes* which undergo meiosis and produce sperm (see figure). Sperm cells are very small, and swim with a flagellum. At puberty, human testes contain millions of cells that produce sperm throughout the adult life of the male. Human testes can deliver hundreds of millions of sperm cells at a time during sexual activity.
- Female germ cells undergo mitosis, producing more female germ cells, during the fetal development of the human female. These cells, inside the ovaries, begin the first division of meiosis, becoming *oocytes*. Oocytes are much larger than sperm and are not motile. Almost all of the oocytes a human female will ever have, about 800,000, are present at birth. The oocytes begin the first division of meiosis prior to birth but do not complete it until puberty. Oocytes are released, usually one each month, throughout the woman's reproductive life. When the oocyte is released from the ovary, the second division of meiosis begins, but is not completed until the oocyte is fertilized by a sperm cell.

Normally, when a sperm fertilizes an egg, the two haploid gametes become one diploid zygote. However, in plants, if a normal, haploid sperm nucleus fertilizes a diploid egg nucleus produced by nondisjunction, the result is a zygote that develops into a *triploid* (3N) organism. Such a plant will have chromosomes in groups of three rather than in pairs. The same thing would happen if an unusual diploid sperm nucleus fertilized a normal haploid egg nucleus. If a diploid sperm nucleus fertilizes a diploid egg nucleus, the zygote grows into a *tetraploid* (4N) plant, with chromosomes in groups of four. Nondisjunction in plants can produce gametes that will result in zygotes with chromosomes in groups of five (5N, or *pentaploid*), six (6N, or *hexaploid*), or even higher numbers. These zygotes usually develop into perfectly healthy plants; indeed, plants with doubled chromosome numbers can be especially vigorous. Organisms with chromosomes in groups greater than two are called *polyploids*. In contrast to plants, polyploid animals are very rare, because a polyploid animal zygote usually fails to develop. However, in some animals such as some amphibians, polyploidy has occurred, resulting in very large chromosome numbers.

Because chromosomes must form pairs during meiosis, polyploids that have an odd number of chromosomes (3N, 5N, etc.) cannot complete meiosis; they are sterile. Dandelions (*Taraxacum oficinale*), despite their abundant flower production, cannot carry out sexual reproduction, because they are triploid. They produce triploid egg cells, then triploid seeds, that develop without fertilization. As everyone who mows lawns would conclude, the triploid condition of the dandelion does not reduce its vigor. Likewise, cultivated bananas are triploid and must be propagated by cuttings rather than by seed. Polyploid plants that have an even number of chromosomes (4N, 6N, etc.) can often carry out normal meiosis, because no homologs remain unpaired during meiosis.

Occasionally, a sperm and an egg come together whose chromosomes are so different that they cannot function as homologous pairs. The sperm or egg may come from a mutant individual within a species, or may come from two different species. Although the chromosomes cannot function as pairs, they are frequently able to carry out normal gene expression. *Hybrid* organisms, whose chromosomes do not match up precisely, may develop into fully healthy organisms (see HYBRIDIZATION). However, because in many cases their chromosomes do not form homologous pairs, the process of meiosis cannot be completed. Therefore many hybrid animals (as the mule, which is a cross between a horse and a donkey) are sterile. In plants, chromosome doubling can allow cells of a sterile hybrid 2N to produce a 4N cell that can produce fertile 4N plants. Since these 4N plants cannot cross-breed with the 2N plants that produced them, these 4N plants function as a new species. Speciation by polyploidy can occur within a single generation (see SPECIATION).

Meiosis is the essential process that allows almost all individuals in all species to produce genetically variable offspring, which is essential to the continued evolution of each lineage (see SEX, EVOLUTION OF).

Mendel, Gregor (1822–1884) Austrian *Monk, Geneticist* Raising peas in a garden was not the main responsibility of Gregor Mendel, a monk in the monastery at Brünn, now Brno in the Czech Republic. But it was Mendel's close observations of and experiments with these peas that led him to discover some basic patterns of inheritance of physical characteristics that have become the foundation of the modern sciences of genetics and evolution (see figure). Mendel's work was not recognized by leading scientists of his day. In particular, at a time when even the leading scientists believed in blending inheritance, Mendel realized that traits were passed on in a particulate fashion, which was the clue that was needed to connect natural selection with genetics. He is one of the few people in history whose name has become an adjective (see MENDELIAN GENETICS).

Johann Mendel was born July 27, 1822, into a peasant family in Heinzendorf (now in Austria) in Silesia (most of which is now part of Poland). He learned gardening and grafting from his father, which was to prove valuable to him, and to the future of science. He was doing well in school when his father was permanently injured by a falling tree. His father, however, believed in his abilities and sold the farm so he could pay for his son to finish school and go to Olmütz University. Johann Mendel entered the priesthood (where he took on the name Gregor) to continue his education. He tried being a parish priest but was not successful; and he failed the exam for being a science teacher. He went back for more education, this time studying physics and mathematics at the University of Vienna. He was ordained and entered the monastery at Brünn. He had been recruited to the monastery upon the recommendation not of clerics but of a professor of physics at Vienna, because the monastery, with greenhouses, a big herbarium, and a library with 20,000 volumes, was looking not just for a monk but for someone who would conduct agricultural research. Several monks at Brünn were involved in research with crop plants and livestock animals for the benefit of the local peasant economy. The monastery supplied Mendel with two full-time assistants to help with his research.

Mendel began his experiments with pea plants in 1856, which grew into a study involving many thousands of plants, on which he kept careful records of cross-pollination. In one case he obtained 14,949 dominant to 5,010 recessive plants, a ratio of 2.98 to 1, which is very close to the theoretical 3:1 ratio that Mendel expected. In fact it was closer than scientists (or especially genetics students in college biology laboratories) could ever expect to get. Some historians have suggested that Mendel's results were influenced by what he was expecting to see.

Mendel presented his results at a meeting of the Natural History Society of Brünn, and the published results were distributed to more than a hundred other scholarly societies in Europe, including the Royal Society and the Linnaean Society in England. His work went largely unnoticed, except for a brief review by the German botanist Wilhelm Focke. One leading botanist, Karl Wilhelm von Nägeli, believed so strongly in blending inheritance that he concluded that Mendel must be wrong and ignored his work. Nägeli's 1884 book about evolution and inheritance patterns makes no mention of Mendel. Nägeli even suggested to Mendel that he study hawkweeds, which Mendel did. Mendel did not find his 3:1 ratio in hawkweeds. Today scientists know why: The plants reproduce without sexual recombination. Nägeli's main claim to fame today, therefore, is in ignoring, then misleading, the "Father of Genetics."

Gregor Mendel, the monk who discovered the laws of heredity, holds a plant. ***(Courtesy of James King-Holmes, Science Photo Library)***

Mendel studied many other things, such as sunspots, bees, and mice. In 1868 Mendel assumed a leadership role in his monastery, which became embroiled in legal problems, and this left him no time for research.

Mendel was not the first investigator to suspect that inheritance proceeded in a particulate, rather than blending, fashion. Pierre Maupertuis, director of the French Academy of Sciences during the 18th century, had a surgeon friend, Jacob Ruhe, who had six fingers, and whose family had several other six-fingered members (polydactyly). Maupertuis said that traits such as polydactyly must be passed from one generation as a discrete unit rather than by blending. Besides the fact that his ideas were ahead of their time, Maupertuis's reputation was skewered by the acerbic pen of the writer Voltaire.

Furthermore, Gregor Mendel was not the first investigator to study the inheritance patterns of characteristics in peas. According to geneticist Richard Lewontin (see LEWONTIN, RICHARD), earlier scientists such as Alexander Seton and John Goss in 1822 and Thomas Knight in 1823 observed segregation of green and pale seeds in second generations of pea crosses, the very species and trait Mendel studied. Louis Vilmorin in 1856 counted results from individual crosses, even reported the same 3:1 ratios that Mendel saw. But these investigators did not draw the conclusion of particulate inheritance that Mendel did. Charles Darwin observed what we now understand to be Mendelian patterns in his study of snapdragons, and he speculated about particulate inheritance ("crossed forms go back ... to ancestral forms") but did not take it any further.

Mendel died January 6, 1884, without ever knowing that his work would prove important as the foundation for the science of genetics. In addition, his work provided the key for understanding how natural selection could work (see NATURAL SELECTION; MODERN SYNTHESIS). Charles Darwin studied the review that Focke had written of Mendel's work but apparently missed its significance. Mendel had a German translation of Darwin's *Origin of Species* but also apparently failed to see the connection. If Darwin had realized what Mendel's work had meant, he might have been able to answer the challenges of the engineer and scientist Fleeming Jenkin and other skeptics and would not have wasted his time inventing the now discredited theory of pangenesis.

Three botanists (Hugo de Vries, Carl Correns, and Erich von Tschermak von Seysenegg; see DEVRIES, HUGO) rediscovered Mendel's work about 1900. About the same time, Walter Sutton, an American geneticist, noticed that chromosomes behaved like Mendelian elements during sexual cell division. Nearly 20 years after Mendel's death, the world discovered him.

Further Reading

Henig, Robin Marantz. *The Monk in the Garden: The Lost and Found Genius of Gregor Mendel, the Father of Genetics*. New York: Mariner, 2001.

Mendelian genetics

Mendelian genetics is the study of the inheritance pattern of traits from one generation to the next, as first explained by Gregor Mendel (see MENDEL, GREGOR). Mendelian genetics is the foundation of modern genetics and, as such, one of the foundations of evolutionary science.

Inheritance Is Not Lamarckian

Even in prehistoric times, people understood that offspring resembled their parents. Not only did plants and animals reproduce "after their own kind," to use the phrase from the biblical book of Genesis, but the offspring resembled the parents more closely than they resembled other members of the same species. Although ancient people did not have a modern concept of species, they recognized a similar concept; anthropological studies have shown that tribal words for plants and animals closely align with modern species definitions. Before recorded history, ancient people had applied their rudimentary understanding of inheritance patterns well enough to produce all of the most important crop plant and livestock animal species, from wheat and rice to cattle and sheep, upon which the world food supply still depends (see AGRICULTURE, EVOLUTION OF).

Previous to Mendel's work, scientific and popular opinions about inheritance patterns had not been systematically studied and were subject to hearsay and confusion. Most importantly, many ancient people believed that environmental conditions could induce changes in organisms (this is correct) that can be passed on to future generations (this is incorrect). This is called *the inheritance of acquired characteristics*. The biblical character Jacob placed striped tree boughs near the water troughs where goats mated, and the goats produced striped offspring. The writer of that Genesis account, and all contemporary readers, assumed that striped offspring resulted because the goats looked at the striped boughs. (Despite the dependence of CREATIONISM upon biblical literalism, creationists have not embraced the inheritance of acquired characters as their model for genetics.) In reality the striped pattern of the goats probably resulted from crossbreeding between light and dark goats.

Belief in the inheritance of acquired characteristics persisted well into the age of modern science. This belief is often named after the French biologist Jean-Baptiste de Lamarck, although his beliefs were the same as those of most of his contemporaries about inheritance patterns (see LAMARCKISM). Mendel's work demolished the inheritance of acquired characteristics. One particularly tragic example of belief in the wrong theory of inheritance was that of Trofim Lysenko, who espoused Lamarckism in the Soviet Union decades after it had been discredited by all competent scientists (see LYSENKOISM).

Inheritance Is Particulate, Not Blending

Most people, from ancient times up into 19th-century science, also believed in *blending inheritance*. First, according to this view, the offspring were intermediate between the parents for each trait. Second, the conditions of the two parents could not be retrieved. It is called blending inheritance because it is like blending two colors of paint.

Mendel blazed a new trail away from these beliefs and toward an understanding that inheritance is *particulate*, that is, the traits are passed on as units rather than blending. First, Mendel was able to control the crossbreeding of his experimental organisms. The garden peas that he used in his experiments had flowers that did not open fully; therefore he could transfer pollen from one flower to another with a little brush, without having the pollen blown randomly about by the wind. He could prevent the flowers from pollinating themselves by removing the immature male parts from the flowers which he intended to pollinate. Second, he was very systematic and organized. He defined the traits that he was studying—flower color, seed color, seed surface texture—and he labeled the flowers and seeds so that he could be certain which seeds came from which flower, and which plant of the next generation had come from which seed of the previous generation. Rather than making generalized observations, he counted the numbers of plants that displayed the characteristics he was studying.

Mendel began by identifying true breeds. For example, some pea plants produced purple flowers generation after generation; others consistently produced white flowers. Once he had identified these true-breeding plants, he crossbred them. Blending inheritance theory would lead to the expectation that the resulting flower color should be halfway between purple and white. Instead, all of the resulting seeds grew into plants that had purple flowers. Conventional wisdom would have interpreted this to mean that the white color was recessive and had been lost in the dominant purple color like a drop of white paint in a can of purple. But Mendel carried the experiment on into another generation, crossbreeding the purple-flowered plants that had resulted from mixed parentage. The white flower trait reappeared in the third generation. The white trait had not been lost, or blended into the purple trait, but had retained its individuality. Traits, Mendel discovered, are particulate. Mendel was not the first investigator to notice this phenomenon, but he used it as the basis of a theory of inheritance rather than as just an interesting observation.

The white trait had reappeared in only one-fourth of the third-generation offspring. This is where Mendel's mathematical acumen came into play. Mendel proposed that each individual plant had two copies of each characteristic. Today geneticists would say that chromosomes, and therefore genes, come in pairs (see DNA [RAW MATERIAL OF EVOLUTION]), and that these individuals are therefore *diploid*. In the true-breeding plants, both copies were the same; today geneticists say that they were *homozygous*. One of the original parents was homozygous for purple flowers, the other homozygous for white flowers. White and purple are two *alleles* of the same gene. The true-breeding plants could be denoted *AA* (pure-breeding purple) and *aa* (pure-breeding white). The hybrid plants had one copy from each parent; today geneticists say that they were *heterozygous* (denoted *Aa*). The purple trait was dominant and hid the white trait in the flowers of heterozygous plants.

When plants produce pollen and ovules, the alleles are separated. This also occurs when animals produce eggs and sperm. Each sex cell carries just one allele and is therefore *haploid*. This separation is caused by the process of MEIOSIS. Normal duplication of the cell nucleus, called *mitosis,* results in two nuclei that are identical to the original nucleus. Meiosis, in contrast, produces nuclei that have only half as many chromosomes as the original nucleus. Some pollen and ovules produced by Mendel's heterozygous peas carried one copy of *A*, others carried one copy of *a*. Sex cells fuse together into a *zygote,* a process called *fertilization*. When Mendel crossed two heterozygous plants, they received one allele from each parent. There were four possible outcomes (see figure).

- two purple-flower alleles, one from each parent
- two white-flower alleles, one from each parent
- a white allele from the first parent and a purple allele from the second parent
- a purple allele from the first and a white allele from the second parent

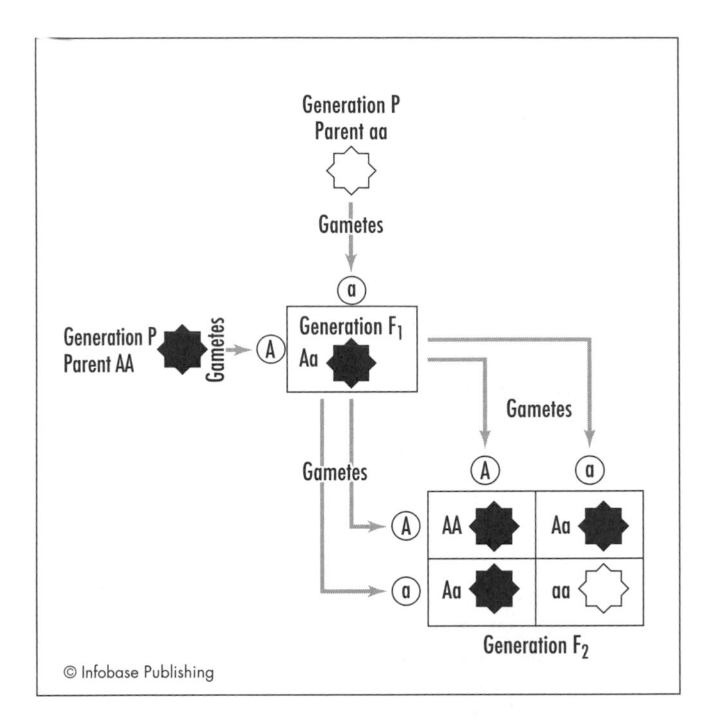

This diagram shows what happened with Mendel's experiment with the colors of pea flowers. Pure-breeding purple-flowered peas (parental P generation) had only allele A; pure-breeding white-flowered peas had only allele a. The hybrids of the F_1 generation were all Aa. When the hybrids were crossed, they produced gametes, which came together in three possible combinations: AA (purple flowers), Aa (purple flowers), and aa (white flowers).

Only one of these four outcomes would produce plants with white flowers. This would explain why only one-fourth of the plants had white flowers: the famous *Mendelian ratio* of 3:1 dominant trait to recessive trait in the third generation. Mendel also found a 3:1 ratio when he paid attention to seed color (yellow was dominant over green) or seed texture (smooth was dominant over wrinkled). Mendel did not know what was causing this to happen, but he had cracked the code of inheritance patterns.

Mendel's results drew little attention. One reason is that Mendel hardly had time for a complete research program. Once he was promoted to a leadership position, Mendel had no time at all for scientific studies. Another reason is that the Mendelian ratio did not show up in every trait. When other researchers tried similar experiments, they did not observe the expected 3:1 ratio. Mendel himself worked on hawkweeds after finishing his garden pea studies; this time, he failed to find interpretable results. He died thinking that he had found an interesting, but not very important, pattern of inheritance.

A 3:1 Mendelian ratio is not found in the inheritance patterns of all traits for several reasons. Among them are:

- Some traits do in fact show what looks like a blending inheritance pattern. In some plant species, purebred red flowers crossed with purebred white flowers produce seeds that grow into pink-flowered plants. The red trait is *codominant* with the white trait. The red trait dominates over the white trait; white is simply the absence of red. The red color genes are present in only half their previous number in the heterozygotes; because there is not enough red pigment to produce a pure redness, a pink color results. This is not blending inheritance, because the red and white alleles are still separate. In later generations, red and white flowers can emerge again among the offspring, whereas pink paint will never be either red or white again. This will not be observed unless the investigator continues the experiment past the first generation.
- Most traits are influenced by more than one gene. Each of the genes may have a 3:1 ratio, but the joint result of all of the genes will not. Most of these *polygenic* traits show a whole range of outcomes. Human height, for example, shows a whole range of values, from a few short people to a lot of medium-sized people to a few tall people. In humans, there are relatively few traits that have a simple Mendelian inheritance pattern. These include blue vs. brown eyes; attached vs. unattached earlobes; presence vs. absence of a widow's-peak hairline; presence vs. absence of the hitchhiker's thumb; the ability to taste the bitter chemical phenylthiocarbamide; and a considerable list of metabolic and other genetic diseases. Most human traits, including other traits related to eye color, ear structure, hair patterns, and thumbs, not to mention traits like intelligence that are so complex that they cannot even be defined, result from more than one gene.
- Almost all traits are influenced by the environment, especially the experiences of the embryo during its development (see ADAPTATION). The characteristics of organisms result from both genes and environment. The pattern of gene expression is determined by the environment at many different levels. This occurs in addition to the direct effects of

the environment on the organism. The range of measurable organism characteristics under a range of different conditions is called the *norm of reaction.* Height in humans is influenced not only by several genes but also by nutrition during fetal development and childhood.

Most traits that show a Mendelian pattern are simple examples of MUTATION: For example, blue eyes result from a mutation that inactivates the gene that creates the brown pigment in the iris of the eyes.

Ratios other than 3:1 can also result for *sex-linked* traits. Men have one X and one Y chromosome, while women have two X chromosomes. The genes on the Y chromosome do not match up with those on the X chromosome. The X chromosome of the man, therefore, may express all of its genes, regardless of whether they are dominant or recessive, because there is no second allele to hide them. Many recessive alleles on the X chromosome, like most other recessive alleles in humans, are associated with genetic deficiencies. Some are serious, such as hemophilia (the inability to stop bleeding), while others are less serious, such as the inability to visually distinguish red and green. These genetic disorders are much more common in males, because whenever the defective allele is present there will never be a dominant one to mask its effects. Women, however, would need to receive two recessive alleles, one from each parent, in order to have the disorder. For hemophilia, this was unlikely in the past, as the father would have had to be a hemophiliac, and hemophiliacs seldom survived to sexual maturity. Injections of blood proteins now allow hemophiliacs to survive and reproduce. Many other metabolic mutations on the X chromosome remain deadly despite the advances of modern medicine, and it is mostly men that they affect.

What advantage might an organism obtain by being diploid? The diploid condition of a cell is like a truck with double tires. If one tire blows out, the other will still support the truck. Similarly, if one allele is defective, the other can still produce the protein and, at least partially, compensate for it.

Once geneticists George Beadle and Edward Tatum had discovered that there was a one-to-one correspondence between genes and proteins (or protein components), biologists could understand that Mendelian inheritance patterns worked perfectly well on the DNA level. Many traits resulted from the complex interactions of proteins, but the proteins often express themselves by a Mendelian 3:1 ratio. This has been confirmed by numerous studies of proteins, most of them associated with metabolic deficiencies. Mendel was vindicated, on the molecular level. Recently, geneticists have begun to realize that a single gene can result in several to many different proteins, as the expression of a single gene is altered in different ways. This new science of *epigenetics* reveals an even greater diversity of raw material for evolution.

Mendelian genetics made modern evolutionary theory possible. Among Darwin's many critics was the engineer Henry Charles Fleeming Jenkin, who wrote that NATURAL SELECTION could never cause a rare, favorable trait to become common. All good traits, when they first start, must be rare. But a rare good trait will disappear into a mediocre population like a drop of white paint in a bucket of red (actually Jenkin used a different, and racist, analogy). (Jenkin was apparently referring to rare, large mutations, rather than minor genetic variation.) Having no Mendelian alternative to the blending theory, Darwin was at a loss for an answer. It was largely in response to this challenge that he contrived his pangenesis theory, which was basically Lamarckism. Because genes are particulate, natural selection can cause a rare and good allele to become common in a population, and do so very quickly.

Mendel's work, forgotten during his lifetime, was rediscovered independently by three scientists in 1900: Erich von Tschermak von Seysenegg, Hugo DeVries, and Carl Correns (see DEVRIES, HUGO). Strangely enough, when Mendelian inheritance patterns were first rediscovered, they were used as a criticism rather than a support of natural selection. This is because these researchers, and most of their contemporaries, focused only on big mutations that would cause major changes in plants and animals, even the sudden origin of new species, rather than small mutations that are the fuel for gradual changes in plant and animal species. The unification of Mendelian genetics and Darwinian natural selection, called the MODERN SYNTHESIS, had to wait until the 1930s and the work of Theodosius Dobzhansky working with fruit flies, Ernst Mayr working with birds, George Gaylord Simpson working with fossils, and G. Ledyard Stebbins working with plants (see DOBZHANSKY, THEODOSIUS; MAYR, ERNST; SIMPSON, GEORGE GAYLORD; STEBBINS, G. LEDYARD).

Linkage and Epistasis Affect Inheritance

Mendel's calculations assumed that each trait had *independent assortment.* This means that the alleles of one trait could go to any of the offspring, unaffected by the alleles of any other trait. Consider the example of flower color and seed color in Mendel's peas. If these traits assort independently, plants with purple flowers could produce either green or yellow peas, and plants with white flowers could produce either green or yellow peas.

Studies of inheritance patterns, shortly after the rediscovery of Mendelian genetics in the early 20th century, revealed that many traits did not assort independently. The traits were linked, thus forming *linkage groups.* If pea plants with white flowers always had green seeds and plants with purple flowers always had yellow seeds, this would indicate a linkage between flower and seed color. (This is a hypothetical example, using real traits.) Further studies indicated that the number of linkage groups was similar to the number of chromosomes. Therefore, traits that are linked were those traits that are encoded on the same chromosome. Chromosomes are, roughly speaking, linkage groups. In the hypothetical pea plant example, the purple-flower allele would be on the same chromosome as the yellow-seed allele, and the white-flower allele on the same chromosome as the green-seed allele.

Linkage groups are not permanent. During meiosis, pairs of chromosomes are not only separated from one another, but chromosomes can exchange segments, a process called crossing over. Whenever a portion of one chromosome breaks off and changes places with the corresponding part of the other chromosome, new linkage groups are formed. The purple-flower allele would no longer be on the same chromosome as the yellow-seed allele but become linked to the green-seed allele, if crossing over occurred in the portion of the chromosome that is between their respective locations. The breakup of linked traits is called *recombination.*

Several factors affect linkage between two traits:

- *Distance.* Crossing over will disrupt the linkage of two traits more often if they are farther apart from one another on the chromosome. Two traits that are very close together may remain linked for a very long time, while two traits that are far apart will be separated in a few generations. This general pattern has allowed geneticists to construct linkage maps, which specify the relative locations of genes on each chromosome.
- *Time.* Traits distant from one another on a chromosome are separated after a few generations, while it takes many generations to separate traits that are closely linked. If genes for two traits are relatively close together on a chromosome, the origin of the trait may be recent, whereas if the traits are not linked, their origin may have occurred further into the past. The same reasoning can be applied to a trait and noncoding DNA which is close to it (see MARKERS). Linkage disequilibrium can therefore be used as a measure of the evolutionary age of the trait.
- *Recombination hotspots.* Some parts of a chromosome (hotspots) experience crossing over much more frequently than others.

Linkage can obscure the study of natural selection. Consider again the linkage between flower color and seed color in peas. Suppose that natural selection acted upon flower color, causing white flowers to become more common, and purple flowers less common, in the pea plant population. The green-seed color trait would be dragged along with the white flower color by linkage. An investigator would therefore also discover that green seeds are becoming more common, and yellow seeds less common, in the population. The investigator might assume, erroneously, that natural selection favors green seeds.

Early 20th-century mathematical studies of Mendelian genetics, such as those of evolutionary biologist R. A. Fisher (see FISHER, R. A.), assumed that different genes had an additive effect: Each contributed separately and equally to the characteristics of an organism. Geneticist Sewall Wright (see WRIGHT, SEWALL) introduced the concept of *epistasis.* Epistatic genes affect one another's expression. If a scientist knows the separate effects of two epistatic genes, that scientist cannot predict the combined effect of those genes. Numerous examples of epistasis have been found. One recent study example found that mutations that had drastically bad effects on one species of fruit fly, *Drosophila melanogaster,* were the very same mutations that were found in normal, healthy fruit flies of another species, *D. pseudoobscura.* The researchers explained that epistatic genes allowed *D. pseudoobscura* to compensate for the negative effect of the mutation. They also found that about 10 percent of the mutations that are lethal in *D. melanogaster* are present in wild flies and mosquitoes. Similarly, about 10 percent of mutations that are lethal in humans are present in some other mammal wild type. Epistasis has been found in different strains of HIV (see AIDS, EVOLUTION OF).

Mendelian Genetics Affects Humans

Now that the Mendelian basis of inheritance is known, it is common for people to think of genes as determining the fates of individuals. Modern biologists consider this view, called *genetic determinism,* to be incorrect, especially for characteristics such as intelligence which are strongly influenced by the environment and actions of the individual. Even some genetic disorders can often be treated by environmental therapy and are therefore not genetically determined. Children with PKU (phenylketonuria) may die if they eat food that contains significant amounts of the amino acid phenylalanine; they cannot metabolize this amino acid, and it instead becomes a toxic side product that poisons the brain. The solution is simple: The children should not eat food that has phenylalanine. Hemophilia results from the inability to make a certain blood clotting protein. Hemophiliacs can now inject the clotting protein into their blood, entirely compensating for their missing protein. Now hemophiliac men not only survive but can even have children, with the result that the number of hemophiliac women is now increasing. Some other mutations, such as fragile X syndrome (in which the X chromosome breaks) or the allele that causes cystic fibrosis, are much harder to overcome. The mucus buildup that accompanies cystic fibrosis causes respiratory infections. This is treated by laboriously removing and attempting to reduce the mucus buildup and continually battling the infections. The Human Genome Project has, as one of its main objectives, the identification of all human genes, therefore of all genetic disorders, so that the defective or missing protein can be identified and perhaps replaced.

Mendelian genetics has therefore allowed scientists to understand the process of evolution and has in many cases given them the tools to bring genetic disorders and defects under control, thus benefiting the lives of many individuals.

Further Reading

Bonhoeffer, Sebastian, et al. "Evidence for positive epistasis in HIV-1." *Science* 306 (2004): 1,547–1,550. Summarized by Michalakis, Yannis, and Dennis Roze, *Science* 306 (2004): 1,492–1,493.

Gould, Stephen Jay. "Fleeming Jenkin revisited." Chap. 23 in *Bully for Brontosaurus.* New York: Norton, 1991.

Kulathinal, Rob J., Brian R. Bettencourt, and Daniel L. Hartl. "Compensated deleterious mutations in insect genomes." *Science* 306 (2004): 1,553–1,554.

Mesolithic *See* TECHNOLOGY.

Mesozoic era The Mesozoic era (the era of "middle life") is the second era of the Phanerozoic Eon, or period of visible multicellular life, which followed the PRECAMBRIAN TIME in Earth history (see GEOLOGICAL TIME SCALE). The Mesozoic era began with the mass extinction event that occurred at the end of the PERMIAN PERIOD of the PALEOZOIC ERA (see MASS EXTINCTIONS; PERMIAN EXTINCTION). The Mesozoic era consists of three geological periods, the TRIASSIC PERIOD, the JURASSIC PERIOD, and the CRETACEOUS PERIOD.

The single great continent of Pangaea broke apart during the Mesozoic era, to form two groups of continents, one of which would later become North America and Eurasia, the

Stanley Miller did one of the first experiments that simulated the origin of the organic molecules from which all life-forms are constructed. ***(Courtesy of Bettmann/Corbis)***

other of which would later become South America, Africa, Australia, and Antarctica (see CONTINENTAL DRIFT).

Extinction altered the previous balance of dominance among lineages of organisms. The tree-sized club mosses, horsetails, and ferns that had dominated the forests of the middle Paleozoic era declined in importance and were extinct by the beginning of the Mesozoic era (see SEEDLESS PLANTS, EVOLUTION OF). Forests, mostly of conifers (see GYMNOSPERMS, EVOLUTION OF), covered most of the land surface of the Earth. One of the major lineages of arthropods, the TRILOBITES, became extinct. Before the Mesozoic, both mollusks and brachiopods were common, but the mollusks have been more common ever since (see INVERTEBRATES, EVOLUTION OF). Mammals and birds evolved from reptilian ancestors (see BIRDS, EVOLUTION OF; MAMMALS, EVOLUTION OF). The mammal-like reptiles and their mammalian descendants were less common in the Mesozoic than the mammal-like reptiles had previously been (see REPTILES, EVOLUTION OF), and the DINOSAURS evolved into a great diversity of species, many of them large. There were also many large flying reptiles and large marine reptiles.

The Mesozoic era came to an end when an asteroid collided with the Earth and created cold, dark conditions in which many species, especially large dinosaurs, could not survive (see CRETACEOUS EXTINCTION).

microevolution *See* MACROEVOLUTION.

Miller, Stanley (1930–) American *Biochemist* Stanley L. Miller (see photo), born March 7, 1930, is a biochemist who played the greatest role in starting the experimental study of the ORIGIN OF LIFE. He was a graduate student at the University of Chicago in 1953; his graduate adviser, biochemist Harold C. Urey, discouraged him from undertaking an experiment that would delay the completion of his doctoral degree. But Miller performed the experiment, which was to make him famous and transform the study of the origin of life. The results took the scientific community by surprise and almost did not get published.

In the 1930s, Russian biochemist Aleksandr I. Oparin and British biologist J. B. S. Haldane (see HALDANE, J. B. S.) had proposed that life originated from chemical reactions in the ocean before there was any oxygen gas in the atmosphere. In the 1950s, most scientists thought the primordial atmosphere of the Earth was like that found today on Jupiter, rich in ammonia, methane, and hydrogen. But few thought that the reactions that produced the first biological chemicals could be replicated experimentally. Most chemists thought that such a simulation would produce a mixture of so many different kinds of chemicals that it could not be analyzed, but Stanley Miller confined water, ammonia, methane, and hydrogen in glassware and introduced an electric spark. After a week, the inside of the glassware was coated with a layer of material as impervious to analysis as anyone might have predicted, but the water was a different story. Rather than a random mix of chemicals, the water contained high yields of amino acids and other organic acids, many of them the same as are found in living cells today. This was the first, and very exciting, evidence that life may have begun from chemical reactions.

Since this original experiment, Miller and many other biochemists have worked on numerous difficulties that still need to be explained, such as the origin of ribose, a sugar necessary for the nucleic acid RNA, and the manner in which large molecules can be polymerized from smaller molecules (see ORIGIN OF LIFE). Miller's initial success opened the possibility that the mystery of life's origin would be quickly solved, but it has turned out to be one of the most difficult ongoing areas of evolutionary research.

Stanley Miller joined the faculty of the University of California at San Diego in 1958. A member of the National Academy of Sciences, Miller continues his research into the origin of life at that institution.

Further Reading

Henahan, Sean. "From primordial soup to the prebiotic beach: An interview with exobiology pioneer, Dr. Stanley L. Miller, University of California, San Diego." Available online. URL: http://www.accessexcellence.org/WN/NM/miller.html. Accessed April 30, 2005.

mimicry Mimicry occurs when NATURAL SELECTION favors characteristics that make organisms (the *mimics*) resemble other organisms or objects (the *models*). By resembling models, the mimics are often able to avoid being eaten by predators, or are able to obtain pollination services. Most populations have a range of characteristics that affect their appearance. If some members of the population even slightly resemble a model, they may benefit from this resemblance; each generation, the incipient mimics are selected to more and more closely resemble the model. After a long period of evolution, the resemblance can sometimes be very striking but need not be so. Many predators and pollinators do not have very good eyesight, and an incipient mimicry may be perfectly adequate to begin the process. Male animals, in particular, may rush to mate with artificial objects, supplied by scientists, that (to human eyes) hardly resemble the female of the species at all. Mimicry occurs when a mimic evolves in response to predators or pollinators. Natural selection also favors the ability of the predator or pollinator to distinguish mimic from model. Therefore mimicry is an example of COEVOLUTION.

Camouflage mimicry. Camouflage is widespread among animals. Many mammals, for example, have dappled coats that help them to blend in with vegetation and shade, and many lizards are colored like rocks. Many insects have striking resemblance to sticks or leaves (see figure). Most animals are counter-shaded, which means that their backs are darker than their bellies. Mimicry is less common but not unknown among plants. Living-stones, which are succulent plants of the genus *Lithops* in the deserts of Southern Africa, resemble green rocks and thus avoid being eaten. Young leaves of many plant species, while still tender and not yet toxic, are often reddish or brownish in color, which may make herbivores mistake them for dead leaves.

Pollination mimicry. There are numerous examples of flowers that resemble, to certain wasps and flies, either their food or the female of their species. Male wasps pollinate some species of orchids by attempting to mate with the flowers, and female flies lay eggs in flowers that look and smell like rotting meat.

Mullerian mimicry. Named after 19th-century German zoologist Fritz Muller, this form of mimicry occurs when poisonous prey species evolve to resemble one another. The entire set of species is both mimic and model. They converge upon a set of characteristics such as warning coloration that allows the predator to recognize them as poisonous. Black alternating with white, red, yellow, and/or orange are widespread warning colorations among animals.

The caterpillar of the large maple spanworm moth (*Prochoerodes transversata*) resembles a stick and thus avoids predators. *(Courtesy of Milton Tierney/Visuals Unlimited)*

Batesian mimicry. Named after a 19th-century British naturalist who first observed this phenomenon in the Amazon rain forest (see BATES, HENRY WALTER), this form of mimicry occurs when nonpoisonous prey (mimics) evolve a resemblance to poisonous animals (models). The mimic therefore benefits from the protection afforded by the poisonous models, without themselves having to expend resources to make themselves poisonous. Examples include stingless flies that resemble bees and wasps, and the viceroy butterfly (*Limenitis archippus*) that resembles the poisonous monarch butterfly (*Danaus plexippus*). If the mimic becomes too common, relative to the model, the system becomes unstable, because when the predators occasionally eat the mimic it will not suffer the consequences that originally selected for the ability to recognize the model.

Mimicry provides numerous excellent examples of natural selection. Batesian mimicry formed an important piece of evidence that Charles Darwin cited in his original presentation of evolutionary theory (see DARWIN, CHARLES; *ORIGIN OF SPECIES* [book]).

Further Reading

Kaiser, Roman. "Plants and fungi use scents to mimic each other." *Science* 311 (2006): 806–807.

Ruxton, Graeme D., Thomas N. Sherratt, and Michael P. Speed. *Avoiding Attack: The Evolutionary Ecology of Crypsis, Warning Signals and Mimicry.* New York: Oxford University Press, 2005.

missing links "Missing links" refers to fossil evidence of evolutionary transitions that has not been found. The term makes little sense today, as practically all "missing links" have been found. This is particularly amazing when one considers the low probability that any individual, or even any species, might be preserved (see FOSSILS AND FOSSILIZATION).

Fossils have been found that confirm most of the transitional states between major (and minor) groups of organisms. For example:

- The transition of reptiles to birds required the evolution of feathers, the loss of tail vertebrae, the loss of teeth, and the development of the furcula or breastbone, among other things. A missing link might have feathers but have reptilian skeletal features such as tail vertebrae and teeth. Numerous fossils of feathered dinosaurs have been found, as well as primitive birds that had teeth (see BIRDS, EVOLUTION OF). The most famous of these fossils is *ARCHAEOPTERYX*.
- The transition of reptiles to mammals required numerous changes, some of which can be preserved in fossils. One of these is the transition of some reptilian jawbones into mammalian ear bones, and the relocation of the jaw joint. A missing link might have intermediate jawbones and two jaw joints, one reptilian and one mammalian. A series of such fossils, the therapsids or mammal-like reptiles, has been found that displays these features, including some with two jaw joints (see MAMMALS, EVOLUTION OF).
- The transition of terrestrial mammals to whales required, among other things, the loss of hind legs, the transition of front limbs from legs into paddles, and the migration of the nostrils to the top of the head. A missing link might have small limbs and a skull intermediate between that of terrestrial mammals and modern whales. A series of such fossils has been found (see WHALES, EVOLUTION OF).

It is not always certain that the transitional fossil or fossils are actually the direct ancestors of the modern forms. *Archaeopteryx* is a good example. There were other birds that had more modern characteristics and that lived soon after *Archaeopteryx*. Therefore *Archaeopteryx* itself may not have been an ancestor of modern birds. It was probably a cousin, not an ancestor. *Archaeopteryx* and the true ancestor of modern birds descended from a common ancestor, and *Archaeopteryx* had evolved less than the ancestor of modern birds at the time that *Archaeopteryx* lived. *Archaeopteryx* is very clear confirmation of a transitional state between Mesozoic dinosaurs and modern birds. Feathered dinosaurs lived at the very time that one would expect them to have lived, if birds evolved from dinosaurs, but scientists are not sure which feathered dinosaur might have been the true ancestor.

In some cases, the evolution of a complex adaptation might appear improbable or even unbelievable. This is one of the major claims of INTELLIGENT DESIGN. In such cases, the discovery of a fossil "missing link," or the living descendant of a "missing link," can demonstrate that the complex adaptation did, in fact, evolve, even if an observer could not have imagined how. One example of many is the complex set of adaptations displayed by desert cacti such as the saguaro of the American southwest. Such cacti can survive and flourish in desert areas because they carry out photosynthesis and store water in thick green stems rather than leaves. In order to do this, cacti underwent several evolutionary modifications. Their leaves were reduced to small and temporary structures that did not carry out photosynthesis; stomata, the pores normally found on leaves, developed on stems instead; bark development was suppressed on the stems, allowing them to be soft and green rather than hard and brown; instead of many small branches, they developed large, thick branches that stored water; and they developed a modified form of photosynthesis known as CAM (see PHOTOSYNTHESIS, EVOLUTION OF). So complex are these adaptations that some observers are tempted to attribute them to special creation. But, in fact, there are several species of plants of the genus *Pereskia* which are almost perfect intermediates between desert cacti and their ancestors. They live in dry mountainous areas of Central and South America. *Pereskia* plants have leaves, even though they are succulent leaves; some have stem stomata, some do not; some have delayed bark formation, some do not. All of them are bushes or trees, rather than having succulent branches. Even their flowers are intermediate between those of desert cacti and their ancestors. Many have succulent leaves, and some of them can use CAM under drought conditions. They closely resemble the "missing links" that must have existed around 30 million years ago.

Throughout this encyclopedia, transitional forms are discussed. This next section focuses on "missing links" that seem

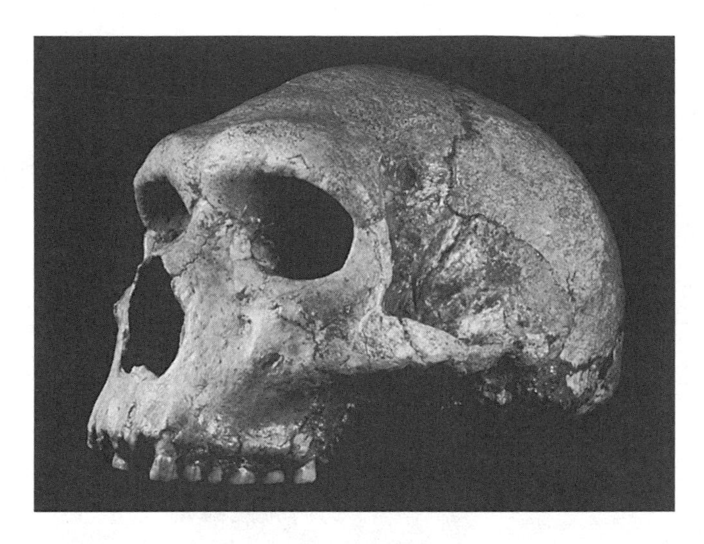

A skull found at Petralona in Greece is 200,000 years old and has characteristics intermediate between *Homo heidelbergensis* and Neandertals. *(Courtesy of Jeffrey H. Schwartz)*

to draw the most attention from people who are opposed to evolutionary science: "missing links" in human evolution.

The major species of humans that lived between 2 million and 0.1 million years ago represent a series of "missing links" between AUSTRALOPITHECINES and modern humans:

- *HOMO HABILIS* represents a "missing link" between australopithecines and *Homo ergaster.* The KNM-ER 1470 skull has a larger brain capacity than most *H. habilis* and many classify it as a different species, *H. rudolfensis.*
- *HOMO ERGASTER* represents a "missing link" between *H. habilis* and *H. heidelbergensis.* Very little is left to the imagination regarding this species, for the Nariokotome skeleton is nearly complete. This individual lived 1.6 million years ago and had a brain capacity of about 57 cubic inches (910 cc). *H. ergaster* lived at a time and in a place that would make it transitional between the earliest species of the genus *Homo* and later species such as Neandertals and modern humans. Was the ancestor of all modern humans someone in the tribe of which the Nariokotome boy was a member? Or someone in another tribe nearby, or a few hundred kilometers away? Scientists cannot be certain, nor can they ever be certain that a new hominin discovery will not show *H. ergaster* to be a side-branch rather than an ancestor. But what they can be certain of is this. There were part-human, part-ape animals running around exactly in the place (Africa) and exactly at the time (a million and a half years ago) that one would expect them to be, if modern humans really did evolve from apelike ancestors. ("Ape" is here used to refer to non-hominin apes.) One look at the Nariokotome skeleton and it becomes extremely difficult to believe that humans arose miraculously without ancestors.
- *HOMO HEIDELBERGENSIS* represents a "missing link" between *H. ergaster* and the two large-brained human species. Numerous specimens of this species have been found.

There are even no-longer-missing links that connect *H. ergaster* with *H. heidelbergensis,* and *H. heidelbergensis* with the two human species that evolved from it: NEANDERTALS and *HOMO SAPIENS:*

- Specimens found at Grand Dolina in Spain and Ceprano in Italy are intermediate, in size and characteristics, between *H. ergaster* and *H. heidelbergensis* (see table).
- Specimens found in several locations in Europe are intermediate, in size and characteristics, between *H. heidelbergensis* and *H. neanderthalensis* (see photo at left and table on page 277). Some anthropologists refer to them as pre-Neandertals.
- Specimens found in Africa are intermediate, in size and characteristics, between *H. heidelbergensis* and *H. sapiens* (see photo below and table).

In fact, there is a continuum of brain sizes among the hominins intermediate between *H. ergaster* and later hominin species (see figure on page 276). The figure shows the maximum length and breadth of crania from hominins usually classified as *H. ergaster* (1,2); hominins usually classified as *H. heidelbergensis* or intermediate between it and Neandertals (3,4,11,12,14); Asian *HOMO ERECTUS* (5,6,7,8,13); and

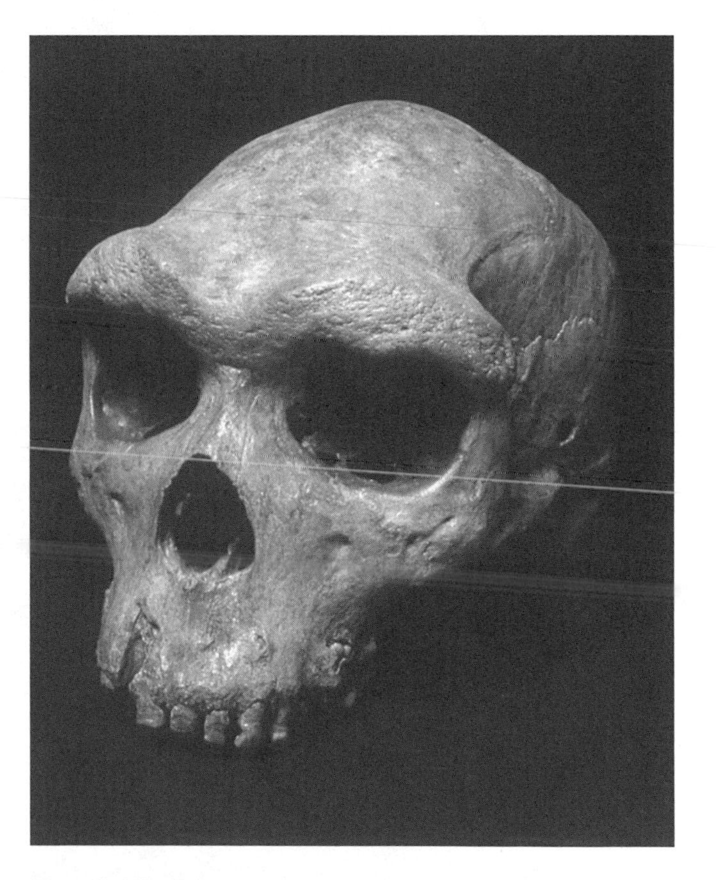

Also called Rhodesian man, the Kabwe skull was discovered in the Broken Hill mine near Kabwe in Zambia. It has characteristics intermediate between *Homo heidelbergensis* and modern humans. *(Courtesy of John Reader/Science Photo Library)*

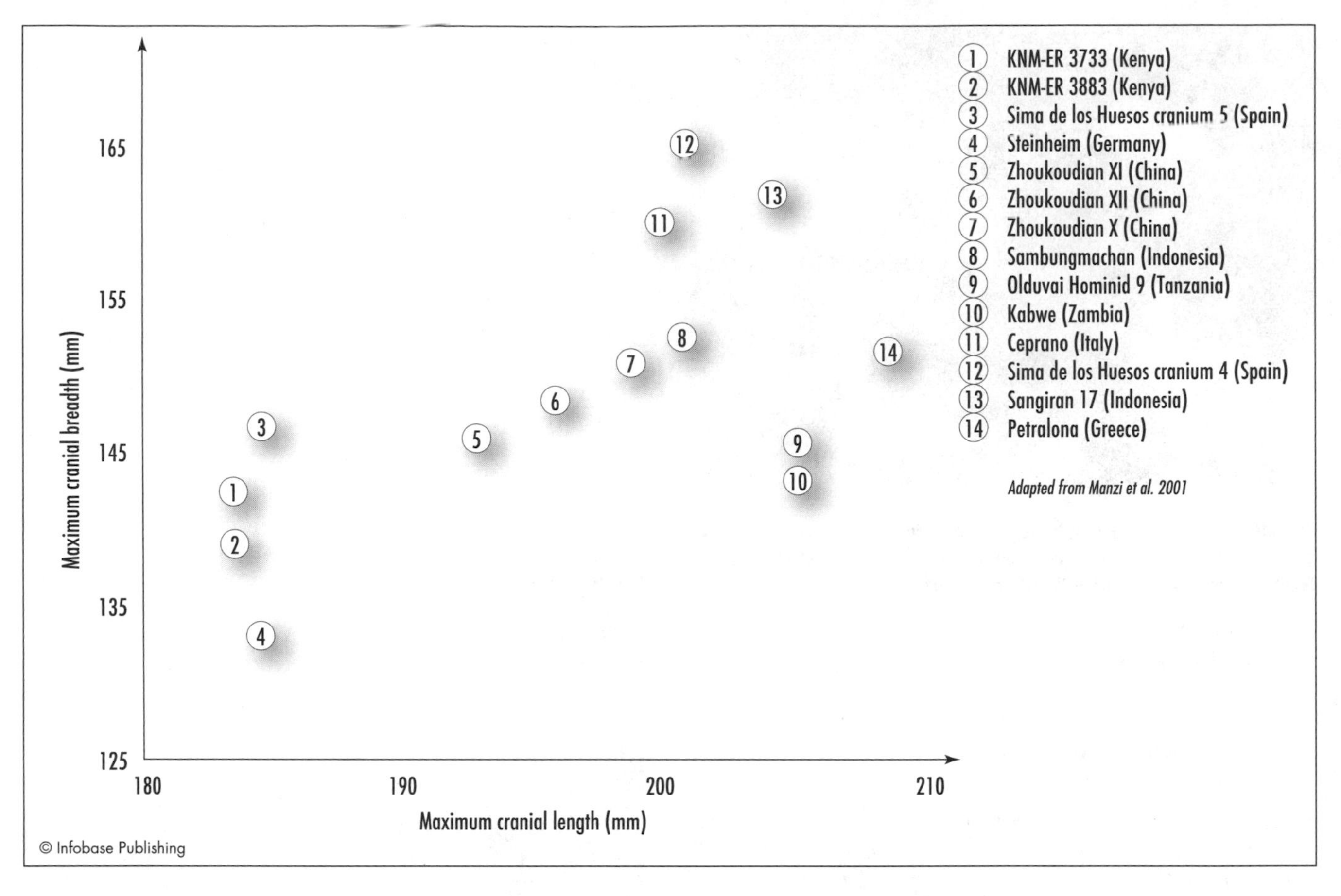

Hominin skulls have been found that display the entire range of both length and breadth of crania. These skulls do not represent a single lineage of hominins; the *Homo erectus* skulls (5,6,7,8,13) are a separate lineage from the *H. heidelbergensis* leading to Neandertals (3,4,11,12,14) and from the *H. heidelbergensis* leading to modern humans (10).

the immediate ancestors of modern humans (9,10). This figure also illustrates what appears to be an example of CONVERGENT EVOLUTION with independent increases in brain size in Africa, Europe, and Asia. There is an approximate time order also, with the smaller skulls usually being older. These crania represent intermediate states but do not represent a single evolutionary lineage.

In conclusion, it is evident that the links between modern humans and the common ancestor of humans and chimpanzees are no longer missing (australopithecines; *H. habilis; H. ergaster; H. heidelbergensis*), and that the links between the major hominin species are no longer missing (between *H. ergaster* and *H. heidelbergensis;* between *H. heidelbergensis* and Neandertals; between *H. heidelbergensis* and modern humans). The "missing links" have been found

Further Reading

Brass, M. "The antiquity of man: Exploring human evolution and the dawn of civilisation." Available online. URL: http://www.antiquityofman.com/archaeology_links_homininarticles.html. Accessed April 29, 2005.

Edwards, Erika J., Reto Nyffeler, and Michael J. Donoghue. "Basal cactus phylogeny: Implications of *Pereskia* (Cactaceae) paraphyly for the transition to the cactus life form." *American Journal of Botany* 92 (2005): 1,177–1,188.

Foley, Jim. "Fossil hominids: The evidence for human evolution." Available online. URL: http://www.talkorigins.org/faqs/homs/. Accessed April 29, 2005.

Kreger, C. David. "A look at modern human origins." Available online. URL: http://www.modernhumanorigins.com/. Accessed April 29, 2005.

Manzi, G., F. Mallegni, and A. Ascenzi. "A cranium for the earliest Europeans: Phylogenetic position of the hominid from Ceprano, Italy." *Proceedings of the National Academy of Sciences USA* 98 (2001): 10,011–10,016.

Museum of Natural History, Smithsonian Institution, Washington, D.C. "Human origins program site map/contents page." Available online. URL: http://www.mnh.si.edu/anthro/humanorigins/sitemap.htm. Accessed April 29, 2005.

Tattersall, Ian, and Jeffrey Schwartz. *Extinct Humans.* New York: Westview Press, 2001.

mitochondrial DNA *See* SYMBIOGENESIS.

modern synthesis The modern synthesis is the reconciliation of Darwinian NATURAL SELECTION with MENDELIAN

Fossil Hominin Skulls Intermediate between Major Hominin Species

Specimen	Age (thousand years ago)	Cranial capacity (cubic cm) (estimate)
Transitional to *Homo heidelbergensis:*		
Gran Dolina, Spain	780	1,000
Ceprano, Italy	800	1,000
Between *Homo heidelbergensis* and Neandertals:		
Arago, France	450	1,150
Sima de los Huesos, Spain	300	1,150
Petralona, Greece	200	1,220
Between *Homo heidelbergensis* and modern humans:		
Bodo, Ethiopia	600	1,100
Ndutu, Tanzania	400	1,100
Sale, Morocco	400	1,000
Florisbad, South Africa	250	1,280
Kabwe, Zambia	175	1,280
Laetoli, Tanzania	150	1,350

GENETICS that characterizes the modern understanding of evolution. The phrase comes from the title of evolutionary biologist Julian Huxley's 1942 book *Evolution: The Modern Synthesis* (see HUXLEY, JULIAN S.).

When Charles Darwin proposed natural selection as the mechanism of evolutionary change (see *ORIGIN OF SPECIES* [book]), neither he nor any other scientist understood the mechanism of heredity. Scientists generally held a vague notion called *blending inheritance* in which offspring displayed a blend of traits from the parents. This blend, however, was considered inseparable: As engineer Fleeming Jenkin pointed out, to Darwin's dismay, if a new trait appeared, natural selection would never have a chance to favor it, because it would get blended in with the other traits and vanish. This argument would not be true, however, if hereditary traits were particulate in nature; that is, if they were genes. Genes could be hidden but could never be blended in; rather than being like drops of paint, genes are like beads, which can reappear. If their distinct identity is maintained, then natural selection can favor them. The particulate nature of genes was being investigated by an Austrian monk (see MENDEL, GREGOR). Mendel published his work shortly after Darwin published the *Origin;* but since most scientists who studied inheritance patterns clung to the blending theory, Mendel's work was ignored and Darwin apparently never heard of it.

It was not at all obvious to the scientists who rediscovered Mendel's work that Mendelian genetics was compatible with natural selection. Hugo DeVries (see DEVRIES, HUGO) was a botanist who studied inheritance patterns in primroses, and William Bateson a zoologist who studied population variation in molluscs. During background research, DeVries (as well as geneticists Carl Correns and Erich von Tschermak von Seysenegg) discovered the obscure publications of Gregor Mendel about 1900. To DeVries and Bateson, however, Mendel's work seemed to indicate that heritable differences produced large discontinuities: big differences between one breed of primrose and another, and big differences between mollusks that inhabited different regions of the Aral Sea, adapted to different salinity levels. New genes, called mutations, had big effects that were immediately visible, they believed. This belief was inconsistent with the slow, gradual changes that Darwin said would be produced by natural selection acting upon a population. Meanwhile, scientists who believed that evolutionary changes were gradual rejected Mendelian principles. This was because most traits really did look like blending inheritance rather than Mendelian discontinuities. Mendel crossed plants that produced green peas with those that produced yellow peas; most offspring produced yellow peas, a few produced green peas, nothing in between. But with most traits, offspring were intermediate between the parents.

To resolve this situation, it was first necessary to demonstrate that there was someplace that genes could be located. Microscopists had discovered chromosomes in the nuclei of cells. Geneticist Thomas Hunt Morgan studied genetic traits of fruit flies (which just happened to have some cells with giant chromosomes) and was able to associate the inheritance of these traits with specific chromosomes. Geneticist Herman J. Muller bombarded cells with X-rays and induced mutations, which were heritable. Next, it was necessary to demonstrate (mathematically) that many apparently continuous (seemingly blended) traits could result from the concurrent action of many genes. Ronald A. Fisher (see FISHER, R. A.) in England did this, in a complex paper that was at first misunderstood and rejected but, upon publication, was seen to bridge the gulf between distinct genes and continuous traits. He published his ideas in *The Genetical Theory of Natural Selection.* Further work in England (see HALDANE, J. B. S.) and America (see WRIGHT, SEWALL) expanded on Fisher's ideas. Theoretical concepts were united with laboratory and field data by zoologists (see DOBZHANSKY, THEODOSIUS; MAYR, ERNST), paleontologists (see SIMPSON, GEORGE GAYLORD), and botanists (see STEBBINS, G. LEDYARD).

Assumptions that accompanied the modern synthesis were:

- Evolution occurs when natural selection acts upon small genetic variations within populations; therefore evolutionary changes (*microevolution*) are small.
- Large evolutionary changes (*macroevolution*) can be explained by microevolution occurring over long periods of time.

These assumptions, together called *gradualism,* appear to be a direct consequence of the modern synthesis. Most scientists further believed that gradualism would reflect not just the process of evolution but also the pattern of evolution over long periods of time. Disagreement over this point emerged with the proposal of PUNCTUATED EQUILIBRIA in 1972 (see ELDREDGE, NILES; GOULD, STEPHEN JAY). While these scientists did not reject gradualism as an evolutionary mechanism, they did reject it as a macroevolutionary pattern. They claimed that major evolutionary changes occurred when rapid microevolution (directional selection) occurred (punctuations), followed by long periods of stasis (stabilizing selection). Evolutionary scientists are increasingly accepting punctuated equilibria.

Another assumption that accompanied the modern synthesis in the minds of most scientists was that the variation upon which natural selection acted was supplied entirely by mutations in existing genomes. At the same time that the modern synthesis was forming among Western scientists, some Russian geneticists such as Boris Kozo-Polyansky were proposing that a few major evolutionary innovations had occurred by what is now called SYMBIOGENESIS. Lynn Margulis (see MARGULIS, LYNN), building upon the work of Russian scientists such as Kozo-Polyansky and upon the work of the American biologist Ivan Wallin, demonstrated that mitochondria and chloroplasts were the evolutionary descendants of endosymbiotic bacteria. While symbiogenesis in no way contradicts the mechanism of natural selection, it does present new possibilities for the origin of new genetic variation upon which natural selection acts. Margulis continues to point out that symbiogenesis may be much more common than evolutionary scientists have generally appreciated.

The modern synthesis not only brought Mendelian genetics together with Darwinian evolution but has also revolutionized conservation biology. Rare species often suffer from a lack of adequate genetic diversity in their populations; this affects both their genetic characteristics (harmful mutations show up in the organisms) and their ability to keep evolving in response to environmental changes and other species (particularly parasites) (see EXTINCTION; RED QUEEN HYPOTHESIS). Modern ecologists, as well as modern evolutionary scientists, can thank the pioneers of the modern synthesis, because they now know that to save a species, one cannot merely take two of every kind onto an Ark but must save whole populations.

Further Reading

Dobzhansky, Theodosius. *Genetics and the Origin of Species.* New York: Columbia University Press, 1937.

Fisher, R. A. *The Genetical Theory of Natural Selection.* Oxford: Oxford University Press, 1930.

Haldane, J. B. S. *The Causes of Evolution.* London: Longmans, Green, 1932.

Huxley, Julian S. *Evolution: The Modern Synthesis.* London: Allen and Unwin, 1942.

Mayr, Ernst. *Systematics and the Origin of Species.* New York: Columbia University Press, 1942.

———, and William B. Provine, eds. *The Evolutionary Synthesis: Perspectives on the Unification of Biology.* Cambridge, Mass.: Harvard University Press, 1998.

Simpson, George Gaylord. *Tempo and Mode in Evolution.* New York: Columbia University Press, 1944.

Stebbins, G. Ledyard. *Variation and Evolution in Plants.* New York: Columbia University Press, 1950.

Wright, Sewall. "Evolution in Mendelian populations." *Genetics* 16 (1931): 97–159.

molecular clock A molecular clock technique uses changes in biological molecules as a measure of the passage of time. Two species that differ only slightly in their molecular makeup diverged from a common ancestor more recently than two species that differ greatly in their molecular makeup. Molecules such as DNA (see DNA [EVIDENCE FOR EVOLUTION]) can thus be used to reconstruct evolutionary history (see CLADISTICS). If the assumption is made and confirmed that the molecules change at a constant rate over time, the degree of divergence between two species can also be used as a molecular clock to indicate how many years ago the divergence occurred. The technique was first proposed by chemists Emile Zuckerkandl and Linus Pauling in 1962.

The rate at which molecules change over evolutionary time can be influenced by the population size and the generation time. In larger populations, in which there is less genetic drift (see FOUNDER EFFECT; POPULATION GENETICS), the molecules may change more slowly over time. If the molecules change each generation, the changes would occur more rapidly in species with short generation times. As Japanese geneticist Tomoko Ōta pointed out, species with large populations tended to have short generation times, and species with small populations tended to have long generation times. It is therefore possible that the effects of population size and generation time on the rate of molecular evolution effectively cancel one another out.

Among the difficulties encountered by the molecular clock hypothesis are:

- The rate of change may not be constant over time (the clock speeds up or slows down).
- The rate of change is different for different kinds of molecules (some clocks are faster or slower than others). For example, among proteins, fibrinopeptides evolve faster than globins, which evolve faster than cytochrome c, which evolves faster than histones. Histones are components of chromosomes and are constrained from evolving rapidly because their exact structure is important in the processes of cell division such as mitosis and MEIOSIS.
- The rate of change can be influenced by NATURAL SELECTION. Molecular clock studies use neutral variations that experience genetic drift rather than natural selection. For example, the DNA sequences used in molecular clock studies often come from NONCODING DNA.

The clock must be calibrated. To do this, investigators must correlate the time at which the divergence of the variant forms of the molecule began with a date in the fossil record provided by RADIOMETRIC DATING. One problem with this approach is that the divergence of the molecules begins earlier than the visible differences among organisms in the fossil record. A molecular clock can be calibrated only by a minimum age (that is, the time of divergence must be older than the date determined from the fossil record). Once a molecular clock is calibrated, it can be used for comparisons among species for which the fossil record is inadequate.

Linkage disequilibrium can also be used as a molecular clock (see MENDELIAN GENETICS). Two DNA sequences (such as MARKERS) that are linked on the same chromosome will eventually be separated by crossing over; meanwhile they are in linkage disequilibrium. Over time, the markers approach greater equilibrium as more crossing over events separate them. How far away they are from equilibrium can be used to estimate how long it has been since one of the markers originated. For example, there are two markers that are very close to the CCR5 allele that confers resistance to HIV (see AIDS, EVOLUTION OF). Geneticists estimate that the observed degree of disequilibrium between the markers and the allele would have developed in about 28 generations, or just under 700 years. The evolutionary importance of this result is that the CCR5 mutation may have arisen as a defense against bacteria during the Black Death, which occurred about 700 years ago.

Molecular clocks are a widely accepted method of evolutionary study but are not accepted without thorough calibration.

Further Reading

Ho, Simon Y. W., et al. "Accuracy of rate estimation using relaxed-clock models with a critical focus on the early metazoan radiation." *Molecular Biology and Evolution* 22 (2005): 1,355–1,363.

Langley, C. H., and Walter Fitch. "An estimation of the constancy of the molecular rate of evolution." *Journal of Molecular Evolution* 3 (1974): 161–177.

Zuckerkandl, Emile, and Linus Pauling. "Molecular disease, evolution, and genic heterogeneity." In M. Kash and B. Pullman, eds., *Horizons in Biochemistry.* New York: Academic Press, 1962.

mutations Mutations are alterations in DNA. They create new genetic variation that allows evolutionary innovation (see DNA [RAW MATERIAL OF EVOLUTION]).

Most mutations are neutral—that is, they have no effect on the evolutionary process:

- Most mutations occur in cells that will die when the organism dies. For example, a mutation in a muscle cell of an animal will be lost when that cell dies. Some of these *somatic* mutations may induce cancer, if the mutation causes the cell to lose control of cell division. Some somatic mutations in plants may persist if the part of the plant that contains the mutation is used for vegetative propagation. This is how the mutation for seedless oranges has persisted: The original mutation produced a branch with seedless oranges, and pieces of the branch were grafted onto other orange trees, from which further grafts were propagated, until there are now many thousands of orange trees that produce fruit without seeds. But usually the only mutations that will be passed into future generations are the *germ line* mutations, which occur in eggs or sperm, or in the cells that produce eggs and sperm. Such a mutation may end up in a fertilized egg, from which an organism develops, and will be found in every cell, including the germ line cells, of the offspring.
- Most germ line mutations are also neutral in their effect. This is because most of the DNA in eukaryotic cells does not encode genetic information. Mutations in the NONCODING DNA usually do not matter, since the information in this DNA is not used to construct proteins. Mutations may accumulate in noncoding DNA, acting as a measure of evolutionary divergence (see DNA [EVIDENCE FOR EVOLUTION]).
- Even within genes, many mutations are neutral. DNA encodes genetic information in codons. There are 64 possible codons but only 20 kinds of amino acids which these codons can specify. That is, in DNA language, there are 64 different words to specify only 20 different meanings. If a mutation occurs that changes one codon to another, without changing the amino acid that it specifies, the resulting protein will not be changed. This often occurs when a mutation changes the third base in the codon. Chloroplast DNA extracted from 20-million-year-old leaves of *Taxodium* and *Magnolia* and compared to modern chloroplast DNA indicate that most of the mutations of *Magnolia,* and all of the mutations in *Taxodium,* that occurred during the past 20 million years were in the third bases of codons (see FOSSILS AND FOSSILIZATION). Mutations in the third base of a codon are not always neutral. If a mutation changes a codon from one that matches a common transfer RNA to one that matches an uncommon one, the efficiency of translation may be reduced.
- Many mutations are almost neutral. If a mutation in a gene causes a different amino acid to be placed in a certain position in a protein, the protein will be changed—but perhaps not significantly. If one amino acid substitutes for another amino acid that is chemically similar (for example, if leucine substitutes for isoleucine), the protein shape may be almost identical to what it had previously been. Even a major amino acid change may have no effect on a protein if it occurs someplace out of the way—in a position that is on the outside of the protein and away from the active site, which is the location on the protein where the chemical reaction occurs. Many proteins exist in a great variety of forms known as *isozymes.* For example, cells of all organisms contain the protein cytochrome c. In all organisms, cytochrome c does the same job, but its structure is different in many different species. It therefore exists in hundreds of different forms, due to mutations. Often an isozyme from even a distantly related species can function well if inserted in place of an organism's normal enzyme. For example, many human genes work well when inserted into other animals such as mice.

Some mutations, however, can cause major changes in the protein specified by a gene:

- If a major amino acid substitution occurs inside of a protein, it may alter the entire shape of the protein. If a major amino acid substitution occurs near the active site of the protein, the function of the protein may be altered. This is what happened with the hemoglobin mutation that causes sickle-cell anemia. A mutation in the DNA caused a valine to substitute for a glutamic acid in the resulting protein, in position 6 out of 146 amino acids. This change was enough to cause the hemoglobin to crystallize under acidic conditions. This mutation can cause severe medical problems but also confers resistance to malaria (see BALANCED POLYMORPHISM).
- In most cases, such a mutation will disrupt the function of the protein, but in some cases a mutation may improve the

function of the protein. A mutation that causes the protein to function poorly at normal temperatures but better at hot temperatures may be detrimental, except in organisms whose environments are getting hotter or that are migrating into hotter environments.

- Some mutations involve more than a simple amino acid substitution. In some cases, when a base is lost from or added to the DNA, the entire frame of reference is changed. For each codon "downstream" from the mutation, the base that used to be number one now becomes number two or number three. This *frameshift mutation* would cause almost every amino acid downstream from the mutation to be different, which could result in an entirely different protein. Some mutations occur when an entire chunk of DNA is eliminated, duplicated, or moved to a new location. If the chunk of DNA in a new location is ignored, it simply becomes one more of the many chunks of noncoding DNA with which the chromosomes are already crowded. But the translocated chunk may cause a mutation in its new location.

While mutations may occur uniformly over time, the mutation rate per generation is higher in species that have longer generation times. Fruit flies *(Drosophila melanogaster)* have 0.14 mutations per genome per generation, while humans have 1.6 mutations per genome per generation. A fruit fly sperm is the product of only about 25 cell divisions, while a sperm from a 30-year-old man is the product of more than 400 cell divisions.

Evolutionary scientists often assume that mutations occur randomly. This, however, may not be entirely true. Cells seem to be, in some cases, able to control the frequency with which mutations occur, allowing an increased rate of mutations during times of environmental instability. For example, during times of stress, the mutation rate in bacteria has been found to increase. Increased mutation rate during times of stress has been observed in yeast and in maize as well as bacteria. Bacteria have an *S.O.S. response* by which they survive sudden increases in mutation. In addition to this, during stress, bacteria tend to release and absorb small circular chunks of DNA known as plasmids, some of which have genes that allow the bacteria to handle the stress better. Antibiotic resistance genes, for example, are often found on plasmids; and bacteria swap plasmids more rapidly when antibiotics are present than at normal times (see RESISTANCE, EVOLUTION OF). However, modern evolutionary theory insists that mutations occur randomly with respect to the needs of the organism. For example, an animal may need, or greatly benefit from, a mutation that would cause an enhanced ability to run away from predators. Such a mutation, however, is no more or less likely to occur than is any other mutation.

Mutations in developmental control genes can have large effects on the structure of the resulting organism (see DEVELOPMENTAL EVOLUTION). Most such mutations are lethal, but if the organism with such a mutation survives and breeds, it might be the founder of a new species, genus, or even higher group.

Evolutionary change tends to be conservative, largely because most big mutations are bad. Evolutionary modification occurs within existing patterns and structures, rather than inventing entirely new innovations. Adaptations tend to be, according to evolutionary biologist Stephen Jay Gould, "jury-rigged contrivances," rather than designs from a master mind. Because most big mutations are bad, there are some things evolution simply cannot do and has not done.

Further Reading

Michel, Bénédicte. "After 30 years of study, the bacterial SOS response still surprises us." *PLoS Biology* 3 (2005): e255. Available online. URL: http://biology.plosjournals.org/perlserv/?request=get-document&doi=10.1371/journal.pbio.0030255. Accessed July 10, 2006.

mutualism *See* COEVOLUTION.

Nariokotome skeleton *See* *HOMO ERGASTER*.

natural selection Natural selection is the process by which evolution produces ADAPTATION. Other processes (such as genetic drift) cause evolution to occur, but in a random direction. Only natural selection guides the evolutionary process in a direction that produces adaptation to environmental and social conditions. Natural selection is the process that was first clearly explained by Charles Darwin (see DARWIN, CHARLES; *ORIGIN OF SPECIES* [book]), although some earlier writers such as W. C. Wells, Edward Blyth, and Patrick Mathew had presented fragments of the idea. Natural selection was independently discovered by Wallace (see WALLACE, ALFRED RUSSEL).

Darwin not only demonstrated that evolution had occurred but also explained how it occurred. Evolution became believable for the first time. Darwin's contemporaries largely accepted his demonstration that evolution had occurred, but the scientific community did not embrace natural selection for another 80 years (see MODERN SYNTHESIS).

SEXUAL SELECTION, which was also first elucidated by Charles Darwin, operates in a manner similar to natural selection. Many evolutionary scientists consider it a subset of natural selection. However, sexual selection usually produces arbitrary and sometimes outlandish characteristics that are not usually considered adaptation.

Both Darwin and Wallace alluded to ARTIFICIAL SELECTION, the process by which humans breed crops and livestock in desired directions, in crafting the phrase *natural selection*. In both cases, some individuals are selected to propagate, and others are not selected to propagate, only in natural selection it is nature rather than humans that is doing the selecting. Both artificial and natural selection can produce genetic changes within populations and can cause divergence of populations into different varieties. Either could eventually cause varieties to be different enough that they would constitute different species. Artificial selection, however, has not gone on long enough for this to happen.

Natural selection occurs within populations. Individual organisms cannot evolve because they cannot undergo whole-body genetic changes (see MENDELIAN GENETICS). If MUTATIONS occur in *somatic* cells, which make up most of the body of an animal, the cells that contain the mutations will die when the organism dies. Only those mutations that occur in *germ line* cells, either eggs or sperm or the cells that produce them, can be passed on to the next generation. Therefore the genetic effects of evolution can only be observed in the next generation of the population.

Natural selection occurs because individuals within populations differ from one another in how successfully they reproduce. Evolutionary theories before Darwin (see LAMARCKISM; CHAMBERS, ROBERT) were incorrectly based upon the transformation of an entire population or species. Because some individuals in the population reproduce more successfully than others, over time the population begins to resemble the successful individuals. Ernst Mayr (see MAYR, ERNST) said that by formulating the theory of natural selection, Darwin introduced population thinking into evolution, to replace typological thinking (see EVOLUTION)—that is, evolution occurs in populations of diverse individuals, rather than being the transformation of one type into another.

Process of Natural Selection

Ernst Mayr summarized the process of natural selection as five facts and three inferences that can be derived from those facts. All of these facts are part of the science of ECOLOGY, the study of the relationship between organisms and their living and nonliving environments. Therefore by elucidating the process of natural selection, Darwin was not only making evolution believable for the first time but also inventing the science of ecology while he was at it.

Fact 1. Populations can grow exponentially. Because reproductive groups of individuals can multiply, populations grow in curves, not in lines (see POPULATION). In a population of bacteria, one bacterium can produce two, but one million bacteria can produce two million bacteria during cell division. This allows populations to grow explosively. This will occur, of course, only in an unlimited environment. As Darwin wrote:

> There is no exception to the rule that every organic being naturally increases at so high a rate, that, if not destroyed, the earth would be covered by the progeny of a single pair. Even slow-breeding man has doubled in twenty-five years, and at this rate, in less than a thousand years, there would literally not be standing-room for his progeny. Linnaeus has calculated that if an annual plant produced only two seeds—and there is no plant so unproductive as this—and their seedlings next year produced two, and so on, then in twenty years there should be a million plants.

Darwin did some calculations for the *Origin of Species* (incorrectly, as it turned out) on the reproductive rate and population growth of elephants in an unlimited environment:

> The elephant is reckoned the slowest breeder of all known animals, and I have taken some pains to estimate its probable minimum rate of natural increase; it will be safest to assume that it begins breeding when thirty years old, and goes on breeding till ninety years old, bringing forth six young in the interval, and surviving till one hundred years old; if this be so, after a period of from 740 to 750 years there would be nearly nineteen million elephants alive, descended from the first pair.

Fact 2. Populations sometimes grow, and sometimes become smaller, but tend to remain stable. Therefore something must be preventing population growth, either by repressing reproduction or by causing death. Darwin wrote:

> Lighten any check, mitigate the destruction ever so little, and the number of the species will almost instantaneously increase to any amount.

Darwin insisted on evidence, and on experimentation where possible. He dug and cleared a small plot of ground at his estate and laboriously marked each seedling that germinated. The total was 357. Of these, 295 seedlings died, mainly because slugs and insects ate them. He also counted the number of bird nests on his property in the spring of 1854 and again in the spring of 1855 and estimated that during the intervening winter nearly four-fifths of the birds had died.

Fact 3. Resources are limited. Ultimately, if reproduction continued unabated, "the world could not contain" the offspring, as Darwin wrote. However, resources are limited by much more than simply room in the world. Organisms need resources and ecosystem services that provide food, water, shelter, and oxygen, and process their wastes. Each kind of organism is adapted to live in only a limited range of conditions; therefore, for a polar bear, it does not much matter how big the world is, but only how big the arctic is, and for species that are limited to a single island or mountain or swamp, the world is pretty small. The disparity between limited resources and exponential population growth in human societies had been famously pointed out by Malthus (see MALTHUS, THOMAS), and Darwin simply applied it to nature: "It is the doctrine of Malthus applied with manifold force to the whole animal and vegetable kingdoms ..."

These three facts lead to *Inference 1,* which is the struggle for existence. In nearly all natural populations, there are not enough resources for all of the individuals to have all of the resources they need. Today scientists say that *competition* occurs in nearly all natural populations nearly all of the time. Sometimes competition is strong enough that scientists can actually observe the struggle. In a jungle, vines grow over trees, and trees crowd together, as if they are fighting for access to the light. In a desert, the bushes are evenly spaced across the landscape, because there is not enough water for any bush to grow close to another bush. Predators fight over a kill, and birds chase one another out of the choice territories. But even when it cannot be seen, competition can be inferred. Competition is the unavoidable result of exponential population growth and limited resources. As Darwin wrote:

> Nothing is easier than to admit in words the truth of the universal struggle for life, or more difficult—at least I have found it so—than constantly to bear this conclusion in mind ... We behold the face of nature bright with gladness, we often see superabundance of food; we do not see or we forget, that the birds that are idly singing round us live on insects or seeds, and are thus constantly destroying life; or we forget how largely these songsters, or their eggs, or their nestlings, are destroyed by birds or beasts of prey; we do not always bear in mind, that, though food may now be superabundant, it is not so at all seasons of each recurring year.

Darwin's theory of evolution by means of natural selection severely shook the complacency of European science, which had comfortably assumed the divine creation and long-term stability of species. His invention of ecology had at least as great an affect on scientific and popular thought. Many people thought of the natural world as God's vast and orderly garden, full of life. Darwin showed the natural world to be violent and full of death. To his readers, Darwin had shaken their view of the world—and they had not even gotten past Chapter 3 (The Struggle for Existence) of the *Origin!* The poet Alfred, Lord Tennyson, reflected the sadness not only of the death of individuals but the extinction of species in his poem *In Memoriam,* which is the origin of the famous phrase "red in tooth and claw." (Tennyson published the poem before Darwin wrote his book.) The fact of extinction may have forced scholars to think about the fact that disease and death—often the result of the struggle for existence—was much more painful than it needed to be in a perfectly ordered world. Darwin experienced this firsthand through the death of his mother when he was young. Even without evolution, religious scholars had, and still have, to struggle with the "problem of natural evil" (see essay, "Can an Evolutionary Scientist Be Religious?").

Fact 4. Each individual in a population is unique. Each population contains a tremendous amount of diversity. Humans can easily observe this diversity in their own populations, but it is true of the populations of all other species as well, even if humans cannot recognize it. Darwin learned this not from other scientists but from animal breeders, particularly pigeon fanciers. All the different breeds of pigeon had been selected from ancestral stocks of rock pigeons *(Columba livia).* These breeds included some with outlandish feathers, anatomy (for instance, the ability to inflate their crops massively), or behavior (such as the ability to tumble during flight). Since all of these breeds had come from rock doves, then all of their characteristics had to be present, but hidden, within the rock dove populations. Breeding brought out these traits, distilling them as one might distill brandy from wine. The first two chapters of the *Origin* described how artificial selection and the observation of natural populations showed the tremendous amount of diversity that they contained.

Fact 5. Much of the variability in populations is heritable. Darwin knew nothing about genetics. He was apparently unaware of the writings (see MENDEL, GREGOR) that explained how traits could be passed from one generation to another. But every plant and animal breeder knew that this was the case. There was confusion about the mechanism. Many scientists held varieties of Lamarckism, or the inheritance of acquired characteristics. Many also believed in "blending inheritance" in which parental traits blended permanently in their offspring, like blending different colors of paint. Mendel, in contrast, was one of the researchers who demonstrated that traits were inherited in a particulate rather than a blending fashion; that is, a trait could disappear in one generation and reappear in another, like mixing different colors of marbles rather than of paint. Darwin himself ended up accepting a variation of Lamarckism, which he called *pangenesis,* a theory that today most evolutionary scientists consider to be his only major blunder.

The first inference, plus facts 4 and 5, give rise to *Inference 2,* which is natural selection itself. In order to reproduce, an individual needs to obtain resources beyond just what is needed for survival. Because of the struggle for existence (inference 1), not all individuals in the population can reproduce, or even survive. If all individuals in the population were alike, then the ones that prevailed would be due either to chance, or to historical contingency, which might just be another form of chance. But since individuals are unique, then some of the individuals have what it takes to survive and reproduce more than others. For example, in a moist environment, some of the plants may have larger leaves and shallower roots than others. The plants with larger leaves could grow faster because they would have greater photosynthesis, and they would not waste their resources producing deep roots that are not needed in a moist environment. But if the trait cannot be passed on to the next generation, the superior survival and reproduction of the plants with larger leaves would not matter, since their offspring would not have leaves any larger than those of other plants.

Natural selection is the differential reproduction of organisms in a population for reasons not due to chance. That is all that it is. If some genetically distinct individuals consistently reproduce more than others, natural selection occurs. Scientists may be able to determine what caused the superiority of some individuals over others (see DARWIN'S FINCHES) or they may not (see PEPPERED MOTHS). Either way, natural selection has occurred. As Darwin described it,

> Owing to [the] struggle [for existence], variations, however slight and from whatever cause proceeding, if they be in any degree profitable to the individuals of a species, in their infinitely complex relations to other organic beings and to their physical conditions of life, will tend to the preservation of such individuals, and will generally be inherited by the offspring ... I have called this principle, by which each slight variation, if useful, is preserved, by the term Natural Selection...

Inference 2 leads directly to *Inference 3,* which is that if natural selection continues long enough, evolutionary change results in the population.

The traits that confer superiority in natural selection, therefore in evolution, are not the same everywhere. There are countless pathways to success, in the natural world as in the human economy. The above example described successful plants in moist environments as having large leaves and shallow roots. In an arid environment successful plants would have small leaves (which waste less water than large leaves) and deep roots (that may be able to tap into new sources of water). In most cases, evolution can go in more than one direction, depending on the circumstances. To use one example close to home, the major pattern of evolution in the human lineage has been the increase in brain size (see AUSTRALOPITHECINES; *HOMO HABILIS; HOMO ERGASTER; HOMO SAPIENS*) but in at least one case, human brain size decreased during evolution (see FLORES ISLAND PEOPLE).

Types of Natural Selection

Evolutionary scientists generally classify natural selection into three types (see figure on page 284).

1. In *stabilizing selection,* the average individuals reproduce better than the extreme individuals. For example, in an environment with medium humidity, very large-leaved and very small-leaved individuals in a plant population would grow more slowly and reproduce less. Stabilizing selection reinforces the mean or average trait value by eliminating outlying variation. Stabilizing selection may be common in populations once a particularly successful combination of genes has been assembled. According to some evolutionary scientists,

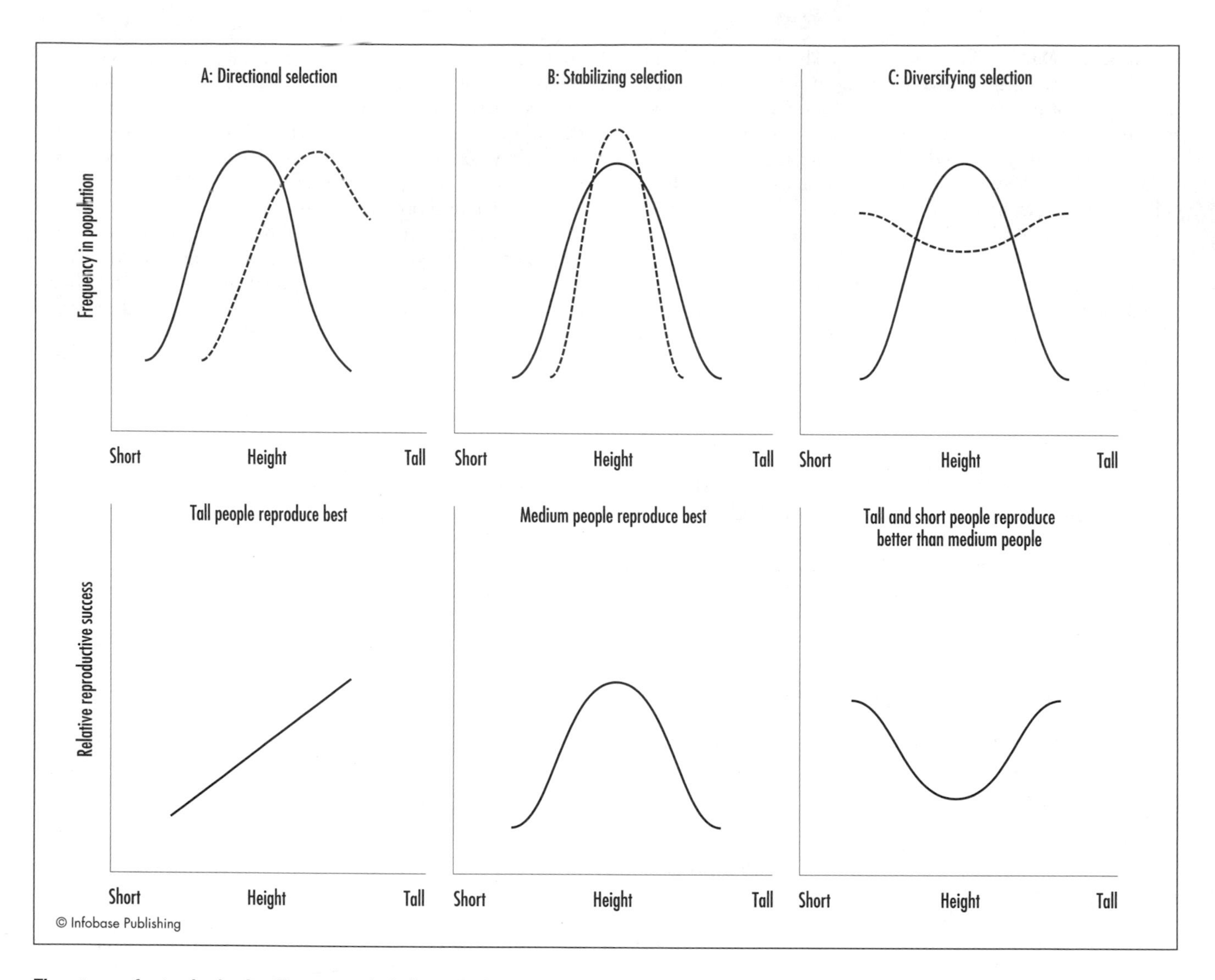

Three types of natural selection. The top graphs indicate the frequency of people of different height categories in a population, with a few short people, many medium-sized people, and a few tall people. The solid line is before, the dashed line after, natural selection has occurred. If tall people reproduce more than medium or short people, directional selection causes the average height in the population to increase. If medium people reproduce best, stabilizing selection causes the average height of the population to remain unchanged. If medium people reproduce worst, diversifying selection causes the population to become dimorphic—with short people, tall people, and relatively few medium-sized people.

stabilizing selection has been the experience of nearly every species during most of its existence, resulting in long periods of evolutionary stasis (see PUNCTUATED EQUILIBRIA).

2. In *directional selection,* individuals at one of the extremes reproduce better than either the average individuals or the individuals at the other extreme. If the environment of the population of plants described above became wetter, the larger-leaved individuals would grow and reproduce better than the individuals with either medium or small leaves. A drier environment would favor the individuals with small leaves over all the others. Either of these would be directional selection.

3. In *diversifying selection,* the average individuals are at a disadvantage and the extreme individuals reproduce the most. During intense competition, the average individuals may compete most strongly. Individuals that are very small or very large might escape direct competition. As a result, the population may begin evolving in two directions at once. This process cannot continue as long as the individuals can all interbreed. If a reproductive barrier arises between the small and large individuals, they may begin to diversify into two species (see REPRODUCTIVE ISOLATION; SPECIATION). Darwin neatly explained how a population could evolve, perhaps eventually becoming a new species. Natural selection does not itself explain how one species can evolve into two or more species. Reproductive isolation must also occur.

Frequency-dependent selection is a special instance in which an individual has greater reproductive success not because of better adaptation to environmental conditions but simply from being rare. The most commonly cited example of frequency-dependent selection is that of sex ratios (see SEX, EVOLUTION OF). If males are rare in a population, the average male will leave more offspring than the average female. Conversely, rare females will leave more offspring, on the average, than common males. Whichever sex is rare enjoys greater reproductive success. Frequency-dependent selection is considered to be the main process by which most animal populations maintain an equal mix of males and females.

Limitations of Natural Selection

Natural selection is not all-powerful. Below are examples of limitations upon the power of natural selection:

- *Physical constraints.* Natural selection cannot cause the evolution of characteristics that violate the laws of physics and chemistry. For example, natural selection cannot produce a large animal with skinny legs (see ALLOMETRY). Also, the tallest trees in the world are about 350 feet (about 70 m) tall. Trees much taller than this are not possible, because there is a maximum height to which wood can conduct water, and the tallest redwoods may be very close to this maximum height. Large animals with wheels cannot evolve, for the wheel could not receive nutrition from the rest of the body through the axle.
- *Phyletic constraints.* Once an evolutionary lineage has adopted a certain adaptation, this adaptation itself puts constraints on evolutionary change. It is not possible for an animal to have an exoskeleton and to be large. Once arthropods evolved exoskeletons (see INVERTEBRATES, EVOLUTION OF), large size could no longer evolve in their lineage.
- *Absence of genetic variation.* Natural selection can only choose from among the genetic characteristics that are available in the population. These characteristics arise by chance mutations.
- *Linkage of traits.* Natural selection may be unable to select traits separately, if they are linked either on the same chromosome (as linkage groups; Mendelian genetics) or in development. In flowers, petal size and stamen size share developmental events, and natural selection cannot cause one to become bigger or smaller without the other.
- *Plasticity and dispersal.* Organisms may respond to changes in environmental conditions by plasticity or acclimation, or by dispersal to a new location. These are ways in which individual organisms can avoid becoming the victims of natural selection.
- *Post-reproductive individuals.* Natural selection cannot eliminate harmful mutations that have their effect when all or most of the organism's reproduction has been completed (see LIFE HISTORY, EVOLUTION OF). This is sometimes called the "kingpin principle," based on a story about automobile manufacturer Henry Ford. He had his engineers study old Ford cars and find which part of the old cars was in the best condition. They reported that the kingpins were just like new. Ford told his factory managers to stop wasting their time making such good kingpins. There was no need for kingpins to last forever in a car that did not. Similarly, in organisms, there is no advantage for genes that promote the health and vigor of post-reproductive individuals.
- *The future.* Natural selection cannot anticipate the future and prepare for it.
- *Group selection.* Natural selection cannot favor characteristics that benefit the group but reduce the fitness of the individual that possesses them. Some adaptations result in harm for the individual (for example, worker bees that die after stinging a victim, and a mother spider that dies when her spiderlings eat her) but these are compensated by an increase in inclusive fitness (see ALTRUISM). When pathogens infect a host, natural selection may favor those selfish lineages of pathogens that reproduce most rapidly, even if the host dies as a result. Natural selection may work against such selfish pathogens if the death of the host results in the death of the entire pathogen population (see GROUP SELECTION) or if the illness of the host causes other hosts to stay away from the sick individual, thus preventing the dispersal of any pathogens of that population to new hosts (see EVOLUTIONARY MEDICINE). But natural selection cannot result in individuals sacrificing their own fitness within a population.

Natural Selection Observed in Nature

Below are just a few examples of natural selection that have been observed in nature.

Stabilizing selection. Evolutionary biologists Arthur Weis and Warren Abrahamson studied insects that lay their eggs inside of plant stems. The grub releases a compound that causes the stem to produce a round swelling called a gall. The grub can develop, well fed, inside this gall. There are dangers from predators, however. Gall wasps can recognize galls and pierce their ovipositors into the gall, laying their eggs on the grub, and the wasp larvae eat the grub. Birds can also recognize galls and tear them open, to eat the protein-rich grub. Gall wasps prefer smaller galls; wasp parasitism therefore selects for larger galls. Birds prefer larger galls; bird predation therefore selects for smaller galls. The result is what looks like stabilizing selection for medium-sized galls but is actually a compromise of two processes of directional selection.

Directional selection. Soapberry bugs (*Jadera haematoloma*) pierce fruits of certain plants with their long beaks. They are native to the keys of Florida, where they obtained their food from native balloon vines. After 1925, landscapers introduced the golden rain tree, an Asian relative of the balloon vine but with smaller fruits, into central Florida. Soapberry bugs began to use the new trees as a food source. In central Florida, natural selection favored bugs with shorter beaks. Over the next few decades, the central Florida bug populations evolved shorter beaks, while the bugs in the keys retained their long beaks. Evolutionary scientists Scott Carroll and Christin Boyd documented this change by examining bug collections that had been made before and after the introduction of golden rain trees to central Florida.

Diversifying selection. In populations of African black-bellied seedcrackers (a type of bird), individuals have either large or small beaks. These two beak sizes correspond to two different kinds of seeds (large v. small) that the birds eat. Juvenile birds have greater variation in beak size than adults. Evolutionary biologist Thomas Bates Smith found that the juveniles that survived were those that specialized most narrowly upon the available seed sizes. Among the large-beaked individuals, those with beaks too large or too small died, and among the small-beaked individuals, those with beaks too large or too small also died. Diversifying selection thus reinforced the large and the small beak sizes in this bird population.

Natural Selection Experiments

Occasionally, natural selection can be verified in natural populations by manipulative experiments, as in these examples:

Anole lizards. Anoles are lizards that live in the Caribbean. Some very small Caribbean islands have no natural populations of anoles. Evolutionary biologist Jonathan Losos introduced anole lizards onto some of these islands. The islands differed in the relative amount of plant cover. Within 10 to 14 years, recognizably different forms of anole had apparently evolved on different islands.

Guppies. Guppies live in freshwater pools on the island of Trinidad. In some pools, there are relatively large predators that eat guppies of all sizes; in other pools, there are only small predators, which eat only small guppies. The guppies that live with large predators have a more rapid life cycle, growing faster and reaching reproductive maturity sooner, than those that live with small predators. Evolutionary biologist David Reznick found some pools that had small predators, no large predators, and also no guppies. This was the perfect opportunity for an experiment. He predicted that if fast-growing guppies that had previously experienced large predators were exposed to this new environment, they would evolve into slower-growing forms. He performed the experiment and obtained the expected results.

Importance of Natural Selection to All of Society

Evolution in general, and natural selection in particular, has been a subject of little interest to many students (and professors) in the health sciences and agriculture. Recent developments, however, have shown that it was disastrous for medical and agricultural professionals to ignore the process of natural selection (see RESISTANCE, EVOLUTION OF). In particular, natural selection has led directly to the evolution of antibiotic-resistant bacteria. Natural selection in pathogenic bacteria is fast-paced, outrunning the ability of the medical industry to produce new antibiotics; it costs a lot of money; and it kills thousands of people. The evolution of pesticide resistance in mosquitoes, flies, ticks, lice, and other disease vectors presents yet another public health menace. Quite literally, what a person does not know about evolution can kill him or her. Natural selection also occurs everywhere and all the time in agricultural fields, as insects evolve resistance to pesticides and weeds evolve resistance to herbicides. The response of many farmers is to spray more, with less success, which results in both the poisoning of the environment and the loss of agricultural production. Medical and agricultural professionals, however, have found some degree of success by using the process of natural selection in their favor rather than letting it work against them. By using antibiotics and pesticides sparingly rather than abundantly, and by using mixtures of several types rather than a single type, they have been able to slow down the evolution of resistance. As ecologist Rachel Carson indicated, Darwin could not have imagined such a vindication of his theory.

Further Reading

Balanyá, Joan, et al. "Global genetic change tracks global climate warming in *Drosophila pseudoobscura*." *Science* 313 (2006): 1773–1775.

Bates Smith, Thomas. "Disruptive selection and the genetic basis of bill size polymorphism in the African finch *Pyrenestes*." *Nature* 363 (1993): 618–620.

Carroll, Scott P., and Christin Boyd. "Host race radiation in the soapberry bug: Natural history with the history." *Evolution* 46 (1992): 1,052–1,069.

Jones, Kristina N., and Jennifer S. Reithel. "Pollinator-mediated selection on a flower color polymorphism in experimental populations of *Antirrhinum* (Scrophulariaceae)." *American Journal of Botany* 88 (2001): 447–454.

Losos, Jonathan B. et al. "Rapid temporal reversal in predator-driven natural selection." *Science* 314 (2006): 1111.

Mayr, Ernst. *What Evolution Is.* New York: Basic Books, 2001.

Reznick, David N., et al. "Evaluation of the rate of evolution of natural populations of guppies (*Poecilia reticulata*)." *Science* 275 (1997): 1,934–1,937.

Sabeti, P.C., et al. "Positive selection in the human lineage." *Science* 312 (2006): 1614–1620.

Turner, Paul E. "Cheating viruses and game theory." *American Scientist* 93 (2005): 428–435.

Weiner, Jonathan. "Evolution in action." *Natural History,* November 2005, 47–51.

Weis, Arthur E., and Warren G. Abrahamson. "Evolution of host plant manipulation by gall makers: Ecological and genetic factors in the *Solidago-Eurosta* system." *American Naturalist* 127 (1986): 681–695.

natural theology Natural theology is the use of information from the natural world to support a belief in the past and present action of a supernatural Creator. The central argument of natural theology is the argument from design. Its most famous presentation was by the British theologian William Paley in 1802. Despite all the evidence presented in this encyclopedia that evolution has occurred, and all the evidence that reveals how evolution occurs, a large proportion of the American public still adheres to the design argument in essentially its pre-Darwinian form as presented by Paley (see CREATIONISM; INTELLIGENT DESIGN).

William Paley (1743–1805) obtained a degree from Cambridge in 1763 and was ordained as an Anglican priest in 1767. He worked his way up through the Anglican hierarchy. He was well known for his writings in theology and his opposition to the slave trade. He used commonsense arguments to defend traditional Christianity, which made all of his writings, though neither original nor philosophically defensible, very popular. Paley published *Reasons for Contentment: Addressed to the Laboring Part of the British Public* in 1793. In 1794 he wrote *A View of the Evidences of Christianity*. His most famous book was *Natural Theology: or, Evidences of the Existence and Attributes of the Deity, Collected from the Appearances of Nature,* which he published in 1802. Although the very rationalist philosophers at whom the book was aimed did not accept his arguments, it remained a popular textbook at Cambridge for decades. Charles Darwin (see DARWIN, CHARLES), while a theology student at Cambridge, loved the book and claimed to have practically memorized it.

Paley's argument is presented here in his words, as they remain the clearest statement of the argument from design. Paley began his argument:

> In crossing a heath, suppose I pitched my foot against a stone and were asked how the stone came to be there, I might possibly answer that for anything I knew to the contrary it had lain there forever; nor would it, perhaps, be very easy to show the absurdity of this answer. But suppose I had found a watch upon the ground, and it should be inquired how the watch happened to be in that place, I should hardly think of the answer which I had before given, that for anything I knew the watch might have always been there. Yet why should not this answer serve for the watch as well as for the stone? Why is it not as admissible in the second case as in the first? For this reason, and for no other, namely, that when we come to inspect the watch, we perceive—what we could not discover in the stone—that its several parts are framed and put together for a purpose, e.g., that they are so formed and adjusted as to produce motion, and that motion so regulated as to point out the hour of the day; that if the different parts had been differently shaped from what they are, of a different size from what they are, or placed after any other manner or in any other order than that in which they are placed, either no motion at all would have been carried on in the machine, or none which would have answered the use that is now served by it.

Paley then considered several possible objections to the argument:

> Nor would it, I apprehend, weaken the conclusion, that we had never seen a watch made—that we had never known an artist capable of making one—that we were altogether incapable of executing such a piece of workmanship ourselves, or of understanding in what manner it was performed ... Neither ... would it invalidate our conclusion, that the watch sometimes went wrong or that it seldom went exactly right. It is not necessary that a machine be perfect in order to show with what design it was made ... Nor ... would it bring any uncertainty into the argument, if there were a few parts of the watch, concerning which we could not discover or had not yet discovered in what manner they conduced to the general effect...

Paley proceeded, through the rest of the book, to show that the human body appears designed for the purpose of sustaining life, just as the watch appears designed for telling time. He says that it is just as much "a perversion of language" to attribute the origin of the human body to mere natural law as it would be to attribute the watch to mere metallic law. Perhaps Paley had in mind the evolutionary speculations that some naturalists had published (see DARWIN, ERASMUS) which claimed that evolution just happened, but provided no explanation as to how it could have happened. In effect, Paley argued, design requires a Designer.

One of Paley's arguments is not used today by defenders of intelligent design. Paley said that his argument was not diminished if some component is not essential for the operation of the watch. Intelligent design theorists, in contrast, insist that biological systems have "irreducible complexity" and will not work if even one of the components is missing.

Though many scientists and science educators today dismiss Paley's arguments, these arguments were very reasonable for the state of knowledge that existed in 1802. But that was soon to change. Little did young Charles Darwin suspect, as he read Paley, that he would be the one to propose a process that could produce design without an intelligent designer (see NATURAL SELECTION). The effectiveness of natural selection in producing the appearance of design is the inspiration for the book title *The Blind Watchmaker* (see DAWKINS, RICHARD).

Further Reading

Paley, William. *Natural Theology.* Available online. URL: http://www.hti.umich.edu/cgi/p/pd-modeng/pd-modeng-idx?type=HTML&rgn=TEI.2 &byte=53049319. Accessed May 3, 2005.

Neandertals Neandertals were a human species that lived in Europe and adjacent regions of Asia between about 120 thousand and 30 thousand years ago. The original spelling was *Neanderthals* (still a preferred spelling in Britain) but the updated spelling reflects the German pronunciation and new German spelling. Neandertals were apparently not ancestral to modern humans (see *HOMO SAPIENS*), and represent a separate line of evolution from an African species of HOMININ, probably *HOMO HEIDELBERGENSIS.* Most anthropologists today accord them their own species status as *Homo neanderthalensis.* Because of their large brains, they are usually called humans despite a lack of direct evolutionary connection to the modern human species. Neandertals have gotten a bad reputation far beyond what the evidence requires. They were probably the origin of the ferocious "cave man" image in popular culture, and "Neanderthal" is a favorite epithet of some politicians to describe their putatively stupid opponents.

The earliest discovery of Neandertal remains were skeletal fragments of a child found at Engis Cave in Belgium about 1830. Much more famous was the 1848 discovery of a Neandertal cranium at Forbes Quarry in Gibraltar. Neither of these discoveries was originally recognized as representing a different species of human. Only much later were the Gibraltar specimens, which were lying in a museum, recognized as also being Neandertal remains.

Then in 1856, some obviously human limb bones, pelvis, and portion of a cranium were found by workmen excavating the Feldhofer Grotto in a limestone cliff in Germany. The cliff overlooked the Neander Valley (Neander-thal in German) through which the Düssel River flows toward its confluence with the Rhine at Düsseldorf. The valley was named after Joachim Neander, a 17th-century teacher and theologian who became famously controversial when he refused to take Holy Communion. The Neander Valley was to become famously controversial once again. The workmen brought the bones they had unearthed to a local schoolmaster and amateur naturalist, Johann Fuhlrott. Fuhlrott contacted Hermann Schaaffhausen, an anatomist at the University of Bonn. The cranium was human but had features, particularly a prominent eyebrow ridge of bone, not found in modern people. Fuhlrott and Schaaffhausen jointly presented the news of the discovery to scientists in Bonn in 1857, suggesting (two years before Charles Darwin published his famous book) that a "barbarous and savage race" had once occupied the cave (see DARWIN, CHARLES; *ORIGIN OF SPECIES* [book]). When Huxley wrote *Man's Place in Nature* in 1863, he included a sketch of the Neander Valley specimen as evidence of the animal origins of the human species (see HUXLEY, THOMAS HENRY). William King, student of Charles Lyell and a professor of geology at Queen's College in Belfast, Ireland, proposed the scientific name *Homo neanderthalensis* for this race of people, recognizing them as a separate species.

Some scholars who maintained the separate creation of modern humans dismissed the skeletal fragments as being merely deformed modern individuals. (To this day, both creationist books and some evolution museums present reconstructed Neandertals in modern dress and grooming, with the caption, "Have you seen this man?"). August Mayer, a pathologist at the University of Bonn, claimed as late as 1864 (after the publication of both Darwin's and Huxley's books) that the bones were from a Cossack soldier who was wounded in the Napoleonic war of 1814 and had climbed into the cave to die. Huxley retorted that it was remarkable that a mortally wounded soldier would crawl 60 feet up a cliff, remove all clothing and belongings, seal the cave, and bury himself under two feet of soil. Another anthropologist noted a poorly healed arm fracture and speculated that the Neandertal had grown his brow ridge because of frowning in pain. The anatomist Rudolf Virchow claimed that the Neandertal specimens were merely modern people.

But it was not long before nearly everyone recognized that the Neandertals were a very ancient race of people. Their physical features (described below) contrasted noticeably not only with those of modern humans but also with the CRO-MAGNON "cave men" that also lived in Europe. In scientific theory and in the popular mind, Neandertals were seen as the crude ancestors of the refined and noble Cro-Magnon. According to most scholars through the middle of the 20th century, Neandertals evolved into modern Europeans. The current scientific explanation is quite different and is presented below.

Where Did Neandertals Come From?

Between 300,000 and 200,000 years ago, the hominins that lived in Europe (usually classified as *Homo heidelbergensis*, named after the Heidelberg Man fossil; see MISSING LINKS) had begun to evolve characteristics that were similar to Neandertals; in particular, larger brains. By about 120,000 years ago, throughout Europe and adjacent regions of Asia, the set of characteristics that scientists associate with Neandertals had evolved. Neandertals, therefore, probably evolved in Europe and migrated to adjacent areas.

What Did Neandertals Look Like?

Scientists have recovered bones of more than 275 individuals, and stone artifacts from many sites occupied by Neandertals:

They had brains as large as those of modern humans. The shape of the Neandertal brain, however, was different. The frontal lobe was smaller (their foreheads sloped more than those of modern humans) but the occipital lobe (in back) was larger. Some people cite the smaller frontal lobe as evidence of lesser intelligence, and others cite the larger occipital lobe as evidence that Neandertals had enhanced sensory perception (e.g., smell), but this is speculative.

Their skulls had distinctive features. In particular, they had supraorbital processes, otherwise known as brow ridges. Earlier hominins also had brow ridges, but the double Neandertal brow ridge is unique. It is generally assumed that these thick ridges protected the eyes from injury, to which Neandertals were especially prone (see below). Other distinctive features included the lack of a prominent chin, and very large teeth. Wear marks on the teeth suggest that they actually used their teeth as vise clamps by placing objects of stone or wood or bone in their mouths and working on them with stone tools.

They were relatively short. Neandertals were less than six feet (2 m) in height—well within the range of modern humans, but shorter than the average human. Their limbs were shorter and stockier. The explanation usually given for this is that they evolved in a cold climate. Modern humans such as the Inuit who are physically adapted to cold climates also have short limbs. This feature reduces the amount of surface area that is exposed to cold conditions and therefore reduces the loss of body heat. The Neandertals became extinct during the most recent of the ICE AGES, but probably not because it was an ice age. They had survived a previous ice age in Europe. They had survived equally well in the relatively warm interglacial period following the previous ice age.

They were very strong. Not only were their limbs short, but the muscle attachment sites (recognizable on fossil bones) were quite extensive. In fact, their muscles were apparently so strong that continual use of muscles caused the bones to bend.

Aside from these things, scientists know little about what the Neandertals looked like. Reconstructions usually show them as looking like modern humans except for the skeletal features listed above. It is assumed they wore animal skins (they certainly had tools suitable for cutting and scraping the skins) because of the cold climate, but any animal skins they may have worn have decomposed. Some writers suggest that they had a thick coat of hair and therefore did not need clothes. The consistent lack of clearly verifiable hybrids between *Homo sapiens* and *Homo neanderthalensis,* both in Europe and in the Middle East, despite 20,000–40,000 years of coexistence, suggests that *Homo sapiens* did not recognize Neandertals as potential mates.

How Did Neandertals Live?

Scientists have little evidence for reconstructing Neandertal life; but sometimes a lack of evidence can reveal important conclusions:

Neandertals were primarily hunters and scavengers. Some anthropologists calculate that Neandertals would have needed many more calories than a typical modern human to survive—5,000 or more calories per day, compared to about 2,000 for modern humans. A diet very high in meat would probably have been necessary. Analyses of coprolites (see FOSSILS AND FOSSILIZATION) also suggest a diet high in meat. There is no evidence that Neandertals had bows and arrows. Many of their exquisite stone tools were undoubtedly used as spear points. Anthropologists have found actual spears (without stone points) used by the *Homo heidelbergensis* ancestors of Neandertals over 300,000 years ago. Neandertal skeletal remains have numerous injuries; indeed, no Neandertal individual remains have been found without significant injury. The pattern of injuries in their remains has been compared with those experienced by modern rodeo riders. From this, scientists conclude that the Neandertals attacked and killed large animals (whose bones have been found in Neandertal caves) by rushing at them with spears. In one cave, the Moula-Guercy Cave in France, evidence has been found of Neandertal cannibalism, but cannibalism was uncommon.

Neandertals made stone tools. Their tools represented the Mousterian culture, the third major era of tool-making after the Oldowan tools used by *HOMO HABILIS* and related species, and the Acheulean tools used by *HOMO ERGASTER* and *Homo heidelbergensis* (see TECHNOLOGY). Mousterian tools were well made (worked on two surfaces and very sharp) and suitable for diverse purposes. Neandertals made these tools primarily out of local rocks, in contrast to the Cro-Magnon who would transport their favorite rock materials with them for many miles. Throughout much of what is now France, Neandertals made more advanced stone tools (called Châtelperronian tools) that resembled some of those made by the Cro-Magnon people who lived in the same region at the same time. It is not clear whether the Neandertals invented the Châtelperronian tools themselves or copied them from the Cro-Magnon or stole them. The Cro-Magnon toolkit was considerably more diverse, however, and included tools of wood and bone. Among the bone tools that *Homo sapiens* possessed but Neandertals did not were needles for sewing clothes.

Neandertals used fire. Hearths have been discovered from many Neandertal sites, although the evidence is thin that they cooked their meat. The fire may have been principally for warmth, and to keep predators away.

Neandertals did not have what anthropologists recognize as advanced social culture. The contrast, in this sense, with *Homo sapiens* (whether in the Middle East, western Asia, or Europe) could not be greater. Most striking was the artwork that *Homo sapiens* produced and which, as far as is known, the Neandertals totally lacked. The Cro-Magnon produced extensive cave paintings, produced figurines of stone, bone, and clay, and drilled holes in bones and shells to make necklaces. In particular, the Cro-Magnon (as well as *Homo sapiens* in other areas) buried their dead with artifacts, sometimes thousands of them. This is considered by most anthropologists as evidence that the lives and cultures of *Homo sapiens* were (and indeed still are) pervaded by religion, while the Neandertals were, as anthropologist Daniel Lewis Williams calls them, congenital atheists (see RELIGION, EVOLUTION OF). However, other scholars point out that entire cultures (for example, of Native Americans) produce religious artifacts entirely of perishable materials. What would scholars know of 18th-century Dakota or Anishinabe religion if all they had were their stone tools? The very few possible remnants of Neandertal art and religion are controversial. For example, an animal bone with two holes found from a Neandertal deposit may have been drilled by Neandertals for decoration, or may have been bitten by a lion. The "old man of Shanidar" whose remains were found in a cave in Iraq was reputedly buried with flowers as part of a Neandertal religious ceremony, but some investigators claim that the flower pollen found in the grave was brought in by burrowing animals. The claim that Neandertals created altars around cave bear skulls has been even more controversial. However, archaeologist João Zilhão points out that the Châtelperronian Neandertal sites contained animal teeth drilled for use on necklaces, while Cro-Magnon sites used primarily shell or bone for beads, which suggests that the Neandertals independently invented the practice of making necklaces. It is unclear whether Neandertals had language (see LANGUAGE, EVOLUTION OF).

Neandertals had social cooperation. The above evidence suggests that the level of cooperation within Neandertal societies may not have exceeded that of their immediate hominin ancestors. They lived in small groups in caves and had to cooperate in their dangerous style of hunting. Some investigators claim Neandertals did not practice advance planning. The caves they occupied were mostly in valleys, unlike the mountaintop caves inhabited by Cro-Magnon. The resulting image is of Cro-Magnon who would plan their hunting trips by watching for distant migrating herds of animals, and of Neandertals that would just go out and look for food if they got hungry. Here again, the "old man of Shanidar" may reveal something. He had been severely injured, but survived for many years. This could not have happened unless other members of his tribe had taken care of him. This level of ALTRUISM has not been found in any animal species except *Homo neanderthalensis, Homo sapiens,* and possibly *Homo ergaster.*

Neandertals lived short lives. The Neandertal who was oldest at death, the "old man of Shanidar," was about 40 years old when he died. All *Homo sapiens* societies, in the prehistoric past as now, included some individuals older than this. Because *Homo sapiens* societies, especially after the beginning of agriculture, had high levels of infant mortality, there may have been no difference between the average life spans of Neandertals and modern humans until recent centuries. However, many anthropologists point out that old people in human societies are important repositories of cultural information, a repository that Neandertals would not have had (see LIFE HISTORY, EVOLUTION OF).

What Happened to the Neandertals?

Neandertals died away from most of Europe beginning about 40,000 years ago. The most recent known Neandertal skeleton, 27,000 years old, was found in Zafarraya Cave in Spain. The disappearance of the Neandertals paralleled the movement of *Homo sapiens* across Europe during that time. The conclusion seems inescapable that modern humans drove the Neandertals into extinction. This would not have required warfare and slaughter; indeed, there is no evidence that the Cro-Magnon killed any Neandertals at all. Few anthropologists would want to place a wager that a Cro-Magnon, even with bows and arrows (which Neandertals did not have), could win a fight against a Neandertal. But most anthropologists conclude that the superior social cohesion of modern humans, reinforced by art and religion, allowed *Homo sapiens* societies to displace Neandertal societies from the environments that had the most resources. Neandertals were marginalized into areas with little food, where their reproduction declined and where they eventually became extinct.

For several decades, most anthropologists have concluded that Neandertals did not evolve into any modern group of humans, nor are any of their genes surviving in our populations. A true test of this hypothesis, however, would require a sample of Neandertal DNA. DNA represents a repository of ancient evolutionary events (see DNA [EVIDENCE FOR EVOLUTION]). *Homo sapiens* has one of the lowest levels of DNA variability of any species, which most scientists take as evidence that all humans have evolved from a recent African origin, perhaps only 100,000 to 200,000 years ago, and that even since that time the entire human species has gone through a genetic bottleneck (see POPULATION GENETICS) in which there were few survivors. If Neandertals and *Homo sapiens* evolved separately from hominins such as *Homo heidelbergensis,* then Neandertal DNA should be very different from the DNA of any modern human group. If Europeans evolved from Neandertals, while other humans evolved separately from earlier hominins, then Neandertal DNA would be within the range of modern human DNA variability. The chances of finding Neandertal DNA seemed pretty slim. However, geneticist Svante Pääbo and others have found enough fragments of mitochondrial DNA from three different samples of Neandertal bone, including one from the original bones found in the Neander Valley, that they could compare it with the corresponding mitochondrial DNA in modern humans. Differing from modern human mitochondrial DNA by 27 out of 379 of the nucleotides, Neandertal DNA was completely outside the range of modern humans.

Some anthropologists have suggested that the Neandertals disappeared by interbreeding with modern humans. Most anthropologists have not been convinced that the 24,000-year-old skeleton of a child found in Portugal was a hybrid between Neandertals and modern humans. The supposed Neandertal characteristics were not very clear, and there is a gap of thousands of years between the age of this specimen and the age of the last known pure Neandertal. Neandertals and *Homo sapiens* lived in adjacent caves in the Middle East (for example, both Neandertal and human caves have been found near Mt. Carmel and Bethlehem in Israel) for about 40,000 or 50,000 years. There is no evidence that modern humans and Neandertals interbred during this long period. The 50,000-year coexistence of Neandertals and moderns in the Middle East has puzzled some scholars, but others have pointed out that during this time even the modern humans had only the same Mousterian tools upon which Neandertals also relied.

Therefore the Neandertals live on at most only in legend but probably not even there, for most popular conceptions of them postdate the discovery of their bones in the 19th century. The Neandertals could be said to represent a parallel experiment in humanization. They evolved big brains and social life independently of modern humans but did not have art and religion. They represent the closest thing scientists may ever find to what an ape, with advanced intelligence but without art and religion, would be like.

Further Reading

Finlayson, Clive. *Neanderthals and Modern Humans: An Ecological and Evolutionary Perspective.* New York: Columbia University Press, 2004.

Gore, Rick. "Neandertals." *National Geographic,* January 1996, 2–35.

Klein, Richard G., and Blake Edgar. *The Dawn of Human Culture: A Bold New Theory on What Sparked the "Big Bang" of Human Consciousness.* New York: John Wiley, 2002.

Krings, M., et al. "Neandertal DNA sequences and the origin of modern humans." *Cell* 90 (1997): 19–30.

———, et al. "DNA sequence of the mitochondrial hypervariable region II from the Neandertal type specimen." *Proceedings of the National Academy of Sciences USA* 96 (1999): 5,581–5,585.

Mellars, Paul. *The Neanderthal Legacy.* Princeton, N.J.: Princeton University Press, 1995.

Ovchinnikov, I. V., et al. "Molecular analysis of Neanderthal DNA from the northern Caucasus." *Nature* 404 (2000): 490–493.

Richards, M. P., et al. "Neandertal diet at Vindija and Neandertal predation: The evidence from stable isotopes." *Proceedings of the National Academy of Sciences USA* 97 (2000): 7,663–7,666.

Shreeve, James. *The Neandertal Enigma: Solving the Mystery of Modern Human Origins.* New York: HarperCollins, 1995.

Stringer, Christopher, and Clive Gamble. *In Search of the Neanderthals.* London: Thames and Hudson, 1993.

Trinkaus, Eric. "Neandertal mortality patterns." *Journal of Archaeological Science* 22 (1995): 121–142.

Neolithic *See* TECHNOLOGY.

neoteny Neoteny is the retention of juvenile characteristics in adult animals. Neoteny has been an important part of human evolution.

As juvenile animals grow, they experience both physical and reproductive maturation. They usually go together: Individuals develop adult physical characteristics and reproductive maturity at the same time. But this does not always occur. In *heterochrony* (Greek for "different times"), physical and reproductive maturation occur at different times. Most examples of heterochrony are of animals that become reproductively mature while still retaining juvenile characteristics, a process called *paedomorphosis* (Greek for "juvenile form").

Paedomorphosis may occur in either of two ways. *Progenesis* occurs when physical juveniles become reproductively mature. Reproductive maturity causes physical maturation to stop; therefore these animals retain juvenile characteristics until they die. Examples include several kinds of salamanders. The juvenile phase of the tiger salamander (genus *Proteus*) is called the axolotl, which has external gills that allow underwater existence. Under normal (moist) environmental conditions, the juvenile phase develops into the adult phase. Under stressful (dry) conditions, an axolotl can become sexually mature. This premature reproduction produces fewer offspring than normal adult reproduction, but it has the advantage of allowing reproduction to occur even when the ponds in which the salamanders live dry up early in the season. This flexibility of development allows the individuals of this species to reproduce well during moist years and reproduce at least a little in dry years. The mud puppy (genus *Necturus*) always becomes reproductively mature in the juvenile phase.

In the late 19th century, zoologist Walter Garstang used progenesis (although the term had not yet been invented) to explain the origin of the vertebrates. The invertebrates known as tunicates (see INVERTEBRATES, EVOLUTION OF) have an adult phase that attaches to surfaces such as rocks. It siphons water in through one orifice and out through another and filters food from the water. It certainly does not resemble a vertebrate. The juvenile phase of the tunicate looks like a small headless fish. It has a mouth and gill slits in front, and along its back, just where human backbones and spinal columns are found, it has a cartilage rod (a *notochord*) with the major nerve running along it. Garstang proposed that the first vertebrates evolved from animals similar to tunicates that became reproductively active in the juvenile phase, after which the former adult phase never recurred.

The baby chimpanzee (left) has a face with features more closely resembling those of humans than does the mother chimpanzee (right). The mother chimpanzee is using a stick as a tool to collect insects. ***(Courtesy of Jim Moore/Anthro-Photo)***

In contrast to progenesis, *neoteny* occurs when reproductive maturity takes place at the normal time but physical development slows down. The effect is the same: reproductively mature individuals that resemble juveniles.

The principal example that is used to illustrate neoteny is the human species. Even before the acceptance of evolutionary theory, Georges Buffon (see BUFFON, GEORGES) and Etienne Geoffroy St. Hilaire recognized the remarkable similarity between the skulls of humans and of juvenile apes, but not of adult apes (see photo on page 291). Their work was developed in an evolutionary context by the Dutch anatomist Louis Bolk, whose major work was published in 1926. Bolk's arguments were later expanded by zoologist Sir Gavin de Beer. Human characteristics contrast not just with those of adult apes but with those of most adult vertebrates. Here are just a few of the similarities between adult humans and juvenile nonhuman vertebrates:

- The head is relatively large. Everyone who has seen baby guppies, puppies, kittens, calves, and birds knows that juvenile vertebrates have relatively large heads. (This is true even of some invertebrates such as grasshoppers.) Related to this, the brain is relatively large in adult humans and in juvenile nonhuman vertebrates. Brain development slows very soon after birth in nonhuman apes but continues for years in humans. The continuation of embryonic brain growth into human adulthood gives humans the largest encephalization quotient of any animal (see ALLOMETRY). Not surprisingly, the bones of the human skull retain the juvenile ability to expand during this time of brain growth.
- The face is relatively vertical. During adult development in nonhuman apes, the facial angle extends outward, resulting in a forehead that slopes back from the nose and mouth. Related to this, the foramen magnum (the hole in the skull through which the spinal cord connects with the brain) is underneath the skull in embryonic vertebrates and in adult humans but is in the back of the skull in adult nonhuman vertebrates. It is essential that the foramen magnum be at the bottom of the skull for upright posture (see BIPEDALISM).
- The skull has no pronounced brow ridges. All adult nonhuman apes, but no newborn apes, have ridges of bone above the eyes (also called supraorbital processes). The lack of brow ridges distinguishes adult humans from every other adult modern ape and from all other known adult hominins in the fossil record.
- The body has relatively little hair. Juvenile apes are born with little hair aside from that on the head. Adult humans have very little hair other than on the head, prompting zoologist Desmond Morris to call the human species the Naked Ape.

The human brain grows so rapidly during fetal development that a human baby must be born at a stage of development that is much earlier than that of other apes (see LIFE HISTORY, EVOLUTION OF). Human babies are, effectively, helpless embryos when born.

The juvenile characteristics of adult humans also include behavior. Most young animals are playful, as they explore their environments and practice at behaviors that they will need when they are adults. Humans remain playful all of their lives, as evidenced by activities as diverse as musical creativity, scientific research, and sports. This neotenous playfulness has resulted from but also enhanced the evolution of the brain (see INTELLIGENCE, EVOLUTION OF) and has made the evolution of human culture possible (see RELIGION, EVOLUTION OF; LANGUAGE, EVOLUTION OF).

The most important features that characterize modern humans—large brains, upright posture, and creativity—have resulted from neoteny. Neoteny serves as a unifying principle for the suite of changes that have occurred in human evolution, resulting perhaps from a few simple changes in genes that control development (see DEVELOPMENTAL EVOLUTION). Because all of these differences result from neoteny, it is not necessary to find an adaptive explanation for every human feature. One famous example is the chin. Humans are unique among apes in having chins. Some anthropologists have speculated that the chin evolved for some adaptive function, perhaps as a threat display. Neoteny explains that the human chin did not evolve in response to natural selection. In effect, the human chin is not an entity; it is the product of differences in growth rates of bones that occurred during neoteny. There is no need to assign any adaptive function at all to the chin.

In modern humans, females display more neotenous features (in head size, facial angle, and lack of body hair) than males. Traditionally, males (including male scientists) have considered themselves to be more advanced than females. But if evolutionary advancement is tied to neoteny, then the case could be made that women are more advanced than men.

What would happen if human physical development continued past reproductive maturity, rather than stopping? This question can only be addressed through fiction. Evolutionary biologist Julian Huxley (see HUXLEY, JULIAN S.) injected hormones into a neotenous amphibian. This caused the amphibian to develop into an adult form that had never been observed in nature. His brother, novelist Aldous Huxley, had a grasp on evolution far beyond most other writers of fiction and put his brother's discovery to use. In his short novel *After Many a Summer,* Aldous Huxley tells the story of a rich couple who discovered the secret of immortality, but they could not stop the continued development of their bodies. They were discovered in a cave; at nearly 200 years of age, they looked like chimpanzees.

Further Reading

Gould, Stephen Jay. "The Child as Man's Real Father." Chap. 7 in *Ever Since Darwin: Reflections in Natural History.* New York: Norton, 1977.

new synthesis *See* MODERN SYNTHESIS.

noncoding DNA All of the enzymes that work in the human body, or in any other species, are encoded in genes made of DNA (see DNA [RAW MATERIAL OF EVOLUTION]). Genes make up less than 10 percent of the DNA in human chromosomes, and the same is true of most other eukaryotic species. The remaining DNA, over 90 percent, is referred to as noncoding DNA. Noncoding DNA is popularly called "junk DNA," but geneticists have found that some of the noncoding DNA performs important functions in cells. Unlike genes, the useful portion of the noncoding DNA has functions that are not as dependent upon the precise nucleotide sequence. Over time, MUTATIONS accumulate in the noncoding DNA much more than in the genetic DNA. NATURAL SELECTION does not usually eliminate mutations from the noncoding DNA. This is why the noncoding DNA is important in reconstructing the evolutionary relationships among species (see DNA [EVIDENCE FOR EVOLUTION]).

Noncoding DNA can occur between genes or within genes. Within genes, the noncoding stretches of DNA are called *introns* and the stretches of DNA that specify the sequence of amino acids are called *exons*. During gene expression, the intron RNA is removed. Introns take up a great deal of room inside of some genes. For example, the section of DNA that contains the gene for the human protein dystrophin (a protein that is an essential component in muscle cells) has more than two million nucleotides. Within this gene are more than 50 introns. When the introns are removed, only about 11,000 of the original two million nucleotides remain.

There are three broad categories of noncoding DNA:

Pseudogenes. An old gene or copy of a gene that has lost its PROMOTER is no longer expressed but is still present in the DNA. These leftovers from the past are copied and recopied but not used. They are like computer files that have been deleted but are still present on the computer disc. Occasionally, these old genes can be reactivated. It is estimated that humans have 6,000 pseudogenes. For example, humans have three hemoglobin pseudogenes. Some pseudogenes are *processed pseudogenes* that have been reverse-transcribed from RNA and inserted back into the chromosome. Geneticists draw this conclusion because processed pseudogenes have no introns.

Repeated segments. If, during normal DNA replication, two strands do not line up properly, short segments of one of the strands may get copied over and over, resulting in repeated segments. Some of these repeated segments are next to each other on a chromosome *(tandem repeats)*, while others may be scattered among several chromosomes. *Variable number tandem repeats* (or minisatellites) and *short segment repeats* (or microsatellites) generally contain no useful information. They are like a page filled with one word, repeated over and over. Each individual within a population may have a unique number of these repeats.

Transposable elements. Some segments of DNA encode or used to encode proteins that allow these segments to move around within and among the chromosomes. These *transposable elements* (also called *transposons* or *jumping genes*) can insert into new locations, where they may disrupt the expression of essential genes. About one in 700 human mutations is caused by a transposon inserting in a location where it interferes with gene function.

Transposable elements are placed in two classes. Class I transposable elements move around via RNA intermediates. To accomplish this, they require reverse transcription (the information in RNA must be transcribed back into DNA), which requires the *reverse transcriptase* enzyme. Reverse transcriptase is the hallmark of retroviruses such as HIV (see AIDS, EVOLUTION OF). These transposons are called *retrotransposons.* When retrotransposons move around, they leave a copy of themselves behind in the original location. Most retrotransposons have lost their ability to transpose and just remain in place, generation after generation. Almost 15 percent of human DNA consists of (mostly inactivated) retrotransposons. They include the following:

- *Long interspersed nuclear elements* (LINEs) may contain the gene for reverse transcriptase and may be able to replicate themselves. Often called HERVs (human endogenous retroviruses), they may be remnants of retroviruses that infected eukaryotic cells long ago. Human DNA contains about a hundred copies of the gene for making the enzyme reverse transcriptase. Human chromosome 22, even though it is one of the smallest human chromosomes, has more than 14,000 LINEs.
- *Long terminal repeats* (LTRs) are also left over from retroviruses and are found on some retrotransposons. Sometimes an LTR is accompanied by the entire genome of a retrovirus that lost the ability to make capsules, and sometimes it is just the terminal repeat sequences that remain after the virus itself has departed. When the entire genome of the yeast was sequenced, it was discovered to have 52 complete virus genomes and 264 LTRs that viruses left behind. Maize has 10 families of LTRs, each in 10,000 to 30,000 copies in the genome.
- *Retrosequences* do not encode their own reverse transcriptase. They include short interspersed nuclear elements (SINEs) of mammals, which were produced by reverse transcription but do not include a copy of the reverse transcriptase gene. Alu, a human SINE sequence, closely resembles certain genes involved in protein transport across membranes. A haploid complement of human DNA contains about 1,090,000 Alu sequences. Alu sequences make up about 10 percent of human DNA.

Class II transposable elements have a DNA intermediate. This type of transposon is more commonly found in bacteria, but there are some in eukaryotic cells as well.

How can a eukaryotic cell limit the potentially explosive spread of transposable elements? First, the fact that they are so common indicates that cells have not been entirely successful at containing their spread. Second, crossing over (see MEIOSIS) breaks up transposable elements, which then stop transposing. Third, cells can inactivate transposons (for example through methylation) just as they inactivate genes.

Noncoding DNA, even if much of it originated from nucleic acids that went out of control, has some important functions in the cell. The first is alternative splicing. As the gene is expressed, different numbers of introns may be removed. This results in more than one kind of protein being produced from a single gene. The second is gene regulation.

Some regulatory molecules, such as small interfering RNA molecules (RNAi), come from noncoding DNA; and some regulatory molecules bind to noncoding DNA sequences. Third, some noncoding DNA is important in the functions of the chromosomes themselves: for example, centromeres, to which the protein strands attach during cell division, and telomeres, the caps at the ends of the chromosomes that may be important in protecting them from age-related damage.

Noncoding DNA has been very important in evolution. First, many instances of gene duplication (making an extra copy of a gene) have resulted from transposable elements taking genes with them when they transpose. Second, the vertebrate immune system produces up to a million different kinds of antibodies even though the nucleus does not have a million antibody genes. Antibodies result from the cutting and splicing of a smaller number of genes. This process resembles what happens during transposition and, suggests geneticist Alka Agrawal, may have evolved from it.

Transposable elements not only move from one place to another within the nucleus of a eukaryotic cell but can sometimes move from one cell to another, in a different individual of the same species, or even in a different species (see HORIZONTAL GENE TRANSFER). For example, old world monkeys (see PRIMATES) have a virus gene that they share only with six species of cats and which must have resulted from horizontal gene transfer.

Genes whose PROMOTERS cease to function become pseudogenes, which are also an important component of noncoding DNA. Pseudogenes are no longer used but persist as evidence of evolutionary ancestry. Mammals have many genes that allow the detection and discrimination of scents; mice have and use 1036 of these genes. Primates, in contrast, rely more heavily on sight than on scent. Humans, for example, use only 347 scent genes. Human chromosomes still retain many of the other scent genes, but they are pseudogenes.

The genome (the stored genetic information) of a eukaryotic cell has been compared to a library, or a computer database. But when considering the vast amount of noncoding DNA in the nucleus, some geneticists consider the genome to more closely resemble an attic, with useful items buried amid junk.

Further Reading

Agrawal, Alka, Q. M. Eastman, and D. G. Schatz. "Transposition mediated by RAG1 and RAG2 and its implications for the evolution of the immune system." *Nature* 394 (1998): 744–751.

Alberts, Bruce, et al. *Molecular Biology of the Cell,* 4th ed. New York: Garland Science, 2002.

Lynch, Michael, Britt Koskella, and Sarah Schaack. "Mutation pressure and the evolution of organelle genomic architecture." *Science* 311 (2006): 1,727–1,730.

Ochman, Howard, and Liliana M. Davalos. "The nature and dynamics of bacterial genomes." *Science* 311 (2006): 1,730–1,733.

Ordovician period The Ordovician period (510 million to 440 million years ago) was the second period of the PALEOZOIC ERA (see GEOLOGICAL TIME SCALE). It followed the CAMBRIAN PERIOD and preceded the SILURIAN PERIOD.

Climate. Oceans were warm around equatorial continents, but cool around the mass of southern continents. There was, however, little if any multicellular life on the continents to be affected by terrestrial climate.

Continents. Oceans separated the continents of Laurentia, Baltica, Siberia (these three continents today dominating the Northern Hemisphere) and Gondwana (today, the southern continents).

Marine life. Almost all organisms lived in the oceans and freshwaters at this time. The diversity of marine animals more than tripled over that of the Cambrian period. This was partly due to an increasing specialization in the ways that the animals obtained food. It was also partly due to the formation of the first reefs, though the corals that formed them were of classes that are now extinct, and many reefs were formed by sponges. When the reefs formed, they created many new habitats for other swimming and crawling and sessile animals. Most animal species were invertebrates (see INVERTEBRATES, EVOLUTION OF). The most abundant group of arthropods were the TRILOBITES; modern forms of arthropods such as crabs were rare. The only vertebrates were fishes (see FISHES, EVOLUTION OF), jawless descendants of the early fishes of the Cambrian period. *Astrapsis* had no jaws or fins; only its tail protruded outside a bony armor that covered its body. This four-inch- (10-cm-) long fish presumably used its simple tubelike mouth to filter food from the ocean floor. Seaweeds underwent evolutionary diversification.

Life on land. Terrestrial life-forms have not been clearly confirmed from the Ordovician period. Spores characteristic of primitive land plants have been found from the later part of this period.

Extinctions. One of the great MASS EXTINCTIONS occurred at the end of the Ordovician period. It is estimated that 26 percent of families, representing 83 percent of species, became extinct by the end of the Ordovician. While the continent that later became North America and Europe was equatorial, and surrounded by tropical reefs, the large Gondwanan continent (which today exists as South America, Africa, Australia, Antarctica, and fragments of other landmasses) was at the South Pole. Not only did Gondwanan weather become colder but ocean levels dropped as glaciers formed around the pole, and this devastated tropical reefs. The return of warm weather, and the melting of glaciers, toward the end of the Ordovician may also have caused stagnation of deep-sea waters.

Further Reading

Museum of Paleontology, University of California, Berkeley. "The Ordovician." Available online. URL: http://www.ucmp.berkeley.edu/ordovician/ordovician.html. Accessed May 1, 2005.

origin of life One of the greatest challenges to evolutionary science has been to explain how life began. To approach this question, one must consider what life is, and when, where, and how it began.

What

What is life? This question has come up in the discussion of whether the chemicals and structures in ALH84001, the meteorite from Mars (see MARS, LIFE ON), did or did not constitute evidence that there was life on Mars billions of years ago. Unfortunately, this question has proven very difficult to answer. Is artificial intelligence a form of life? If robots can construct copies of themselves, does that mean they are alive? Is the Earth alive?

Ancient writings, such as the Bible, make numerous references to life. It is on this basis that creationists (see CREATIONISM) have launched numerous attacks on the scientific hypotheses of the origin of life. However, these ancient documents did not mean the same thing as biologists mean when they refer to life today. The Book of Genesis, for example, said nothing about bacteria; and the Genesis term *breath of life* referred to animal life. Even today, people use the word many different ways. In the debate about whether cloned human embryos are alive, the disagreement is not over biological life (which they certainly have) but human life. This confusion is even greater when investigating the origin of life on the early Earth.

The following are some of the characteristics that most scientists have in mind when they refer to life (see BIOLOGY):

A. Life consists of complex carbon-based molecules. All life on Earth is based upon the element carbon (C). Is it possible that life in a different part of the universe might be based upon a different element—or perhaps on pure energy, rather than matter? The second possibility cannot now be investigated. However, the element silicon (Si) has some chemical properties similar to carbon. Might life somewhere in the universe be based upon silicon instead of carbon? This is unlikely to occur, because:

- Silicon is much heavier than carbon, therefore an active, moving organism made out of complex molecules with silicon would need to live in a world with much less gravity. Such a planet would probably also have insufficient gravity to retain an atmosphere.
- Carbon is able to cycle through our planet, from CO_2 in the air, through photosynthesis, through the food chain, through decomposition, and back into the air, because CO_2 is a gas. SiO_2, however, is a mineral (quartz); a food chain probably could not be based upon it.
- Silicon-based molecules would have limited complexity. Although some molecules involved in biological processes are simple (such as nitrous oxide that functions in the human body), most biologically active molecules are complex.

Perhaps, therefore, carbon-based life is the only possibility. Fortunately, there is a lot of carbon in the universe.

B. Life operates in a water medium. A solid medium would be too slow, and a gaseous medium too chaotic, for life processes. Water appears to be the only liquid abundant enough in the universe and suitable as a medium for biological processes, although some scientists speculate that liquid methane (as on one of Saturn's moons) may be a suitable medium.

C. Life-forms obtain energy and matter from the environment and process it into new forms, releasing waste products.

D. Life-forms have the ability to organize themselves, rather than depending on externally imposed structure.

E. Life-forms have the ability to respond to environmental conditions and information.

F. Life-forms have a genetic system, which stores information for structures and functions:

- The genetic information controls the growth and maintenance of the organism.
- The genetic information is copied when reproduction occurs.
- Mutations in the genetic information allow evolution to occur.

G. Life-forms are enclosed within membranes that keep their processes distinct and separate from the environment. Today, all life processes are encapsulated within cells. Cell membranes prevent biological systems from being disrupted and dispersed into the environment around them.

One of the simplest definitions of life, yet still complex, was offered by biochemist Leslie Orgel. He refers to life-forms as Citroens (making humorous reference to the French automobile Citroën): Complex Information Transforming Reproducing Objects That Evolve by Natural Selection. Within a century, therefore, the scientific community has changed from considering natural selection as a hypothesis to making natural selection a part of life's very definition.

When

In his famous 1954 article, biochemist George Wald said that time "is the hero of the plot." Given enough time (billions of years between the formation of the Earth and the first complex life-forms), the impossible becomes possible, the possible becomes likely, and the likely becomes inevitable. Scientists now know that this is not what happened. Life could not have originated slowly, over the course of billions of years; it had to originate quickly:

- Cells were probably in existence by 3.5 billion years ago, which was not very long after the oceans themselves formed. Evolutionary scientist J. William Schopf has found evidence, albeit controversial, of cells in rocks of the 3.5-billion-year-old Apex chert of Australia. From these, and other rocks of similar age from around the world, Schopf claims to recognize at least 11 different kinds of cells, which resemble modern cyanobacteria (see BACTERIA, EVOLUTION OF).
- The oldest sedimentary rocks in the world—3.8 billion years old—come from the Isua formation of Greenland. These rocks contain no fossils but do contain carbon compounds with an isotope ratio that suggests that it is of biological origin (see ISOTOPES).
- Until 3.9 billion years ago, the solar system was filled with errant ASTEROIDS AND COMETS that crashed into the planets. On Earth, evidence of this bombardment has been largely erased by erosion; however, on the Moon, the craters have been preserved. Most of the craters were produced more than 3.9 billion years ago, although a few craters such as Tycho were produced more recently. Collisions with asteroids may have vaporized any oceans, and any life they might have contained. The collision between the Earth and another planet, 4.4 billion years ago, which ejected part of the Earth's crust (which became the Moon), would certainly have done so. The water now in the oceans was delivered later, perhaps by comets, which consist mostly of ice.

It appears that the origin of life is bracketed between 3.9 billion years ago, before which life would have been exterminated, and 3.8 billion years ago, by which time life was already in existence.

Are Humans Alone in the Universe?

When considering the evolution of intelligent life in the entire universe, scientists are confronted by two stupefying vastnesses: the improbability of the evolution of complex life, and the prodigious expanses of the universe. In a universe with possibly a hundred billion galaxies, each with billions of stars, even the least probable event may be expected to occur occasionally, even frequently.

Humans will probably never know whether or not the universe contains other life-forms with a human level of intelligence. Even if there were thousands of other civilizations in the universe, could they contact humans, or humans contact them? Light and other forms of photonic transmission require millions of years to travel among galaxies. Human observers would learn about these other civilizations only if they had evolved to the extent that they could have sent messages millions of years ago; and by now, they may no longer exist. Another problem is the ability to recognize a generalized transmission from another civilization. This is the idea behind SETI—the Search for Extraterrestrial Intelligence—and the novel *Contact* (see SAGAN, CARL): one has to look everywhere for anything that might be a signal, perhaps an irregularity in what had been considered the lifeless throbbing of a pulsar.

Because of the enormous resources that deep space communication and travel would require, the fact that humans have neither been visited nor contacted from outer space means little. Science fiction speculates about the possibility of travel through wormholes or through other dimensions, but any travel through a wormhole or a black hole would probably result in the destructive scrambling of whatever goes through it.

The equation first formulated by astronomer Frank Drake allows a rough estimate of how many planets with advanced civilizations might exist in the Milky Way galaxy. The equation consists of a series of probabilities: the fraction of stars that have planetary systems, the number of planets in a system that have ecological conditions suitable for life, the probability that life will evolve, the probability that intelligence will evolve, and the probability that advanced technology will develop. The problem with these calculations lies with assigning values to the probabilities. Cosmologists disagree greatly on even the order of magnitude of some of them.

There is little doubt that simple life could be widespread in the universe. Bacteria and archaebacteria on Earth survive in a tremendous variety of circumstances that humans would consider destructive to life (see ARCHAEBACTERIA; BACTERIA, EVOLUTION OF). Any conceivable bacterial life might require a solid planet with water. But there might be billions of such planets in the universe, and many of them might have life that resembles bacteria. Bacterial life-forms may have evolved even on Mars (see MARS, LIFE ON). However, the conditions that would allow the evolution of complex life-forms, beyond the bacterial stage, might be vanishingly rare. In order for complex life to evolve on a planet, the planet must have relative stability for a long period of time. The emergence of complex life on Earth was preceded by nearly three billion years of microbial evolution. How likely is the existence of other planets which, like the Earth, have had billions of years of relatively stable conditions? It depends on the answers to questions such as the following:

How typical is the galaxy in which humans live? It is true that there are a lot of galaxies. The first deep field photographs from the Hubble telescope, in 1995, showed 1,500 galaxies from an area of deep space just one-thirtieth the area of the full moon. However, the Milky Way galaxy seems to be an unusually calm place to live:

- Gamma ray bursts from colliding neutron stars would destroy any life for many hundreds of light-years around, perhaps life in an entire galaxy. These bursts are common enough in the universe that Earth should be hit by one every 200 million years, but Earth has apparently not experienced any such radiation during the entire three and a half billion years of life.
- The Milky Way galaxy is a disc galaxy with orderly revolutions of stars; but many galaxies are elliptical galaxies, which have less stable star orbits, and whose stars experience many collisions.

The Milky Way galaxy has had relatively few stellar collisions and their attendant bursts of radiation. This has contributed to a long period of stable conditions that have allowed complex and intelligent life to evolve on Earth.

How typical is the solar system? The Sun and its planets appear to be very unusual in the Milky Way galaxy, for reasons such as these:

- *The habitable zone.* The solar system is in the habitable zone of the Milky Way galaxy. If the Sun were nearer to the center of the galaxy, all the stars would be very close together, and the Sun would be dangerously close to many neutron stars and black holes. If the Sun were near the edge of the galaxy, it may have produced few or no atoms larger than helium.
- *The size of the Sun.* Many stars are giants and burn out as quickly as 10 million years rather than the 10 billion that the Sun will persist. Ten million years would not be enough time for complex life to evolve. Many stars are hotter than the Sun and would produce much ultraviolet radiation, which would destroy life on its planets. Many stars are small enough that they just convert hydrogen to helium, perhaps up to bismuth. This is mostly what is happening in the Sun now. However, the supernova that produced the solar system generated many heavier elements. Because of this, according to geologists Peter Ward and Donald Brownlee, the Sun has a 25 percent greater amount of heavy elements than a typical star its size.
- *The stability of the Sun.* Many stars fluctuate in energy output; in contrast, the Sun has been mild and stable, not producing the bursts of energy that could easily wipe out advanced life-forms. Binary star systems are very common (two-thirds of sun-sized stars are in binary or multiple groups), but complex life could not evolve on planets associated with binary systems because climatic conditions would be very unstable.
- *The rarity of asteroids.* Early in the history of the solar system, the Earth, Moon, and Mars were bombarded by asteroids. This bombardment stopped about four billion years ago (see ASTEROIDS AND COMETS). The first evidence of sediments on Earth comes from this time, when the Earth cooled and the oceans formed. Part of the reason this bombardment stopped is that

(continues)

Are Humans Alone in the Universe? *(continued)*

Jupiter has so much gravity that it has stabilized the asteroid belt and licked up many comets and asteroids from the solar system. Jupiter is just near enough to clear away asteroids from the Earth's path, but far enough away not to bother the Earth. Jupiter disturbs just a few asteroids into the path of the Earth, punctuating the history of life (see MASS EXTINCTIONS) but not often enough to destroy life.

- *The stability of planetary orbits.* As of January 2007, 199 planets had been detected that revolve around other stars. The same Doppler effect that causes the red shift of expanding universe (see UNIVERSE, ORIGIN OF) also shows that these stars are being tugged by the gravity of planets going around them. The sizes and distances of these planets from their suns can be calculated; if the residual variation shows a pattern, it can be assumed that the star has two or more planets. In 2005 a photograph of a large gaseous planet around a distant star was published. It may be impossible to detect planets as small as the Earth around other stars, either by the Doppler effect or photographically. The planets so far detected are frequently "hot Jupiters" (as big and gaseous as Jupiter, close to the star, perhaps representing a failed binary system). Such huge planets would destabilize the orbits of smaller planets, causing them to crash into the large planet, into the star, or be thrown into interstellar space. A solar system of hot Jupiters would be very unstable, and any earthlike planets would certainly experience wild swings of climate which would prevent the evolution of complex life. The eccentricity (departure from a circular orbit) of the Earth is only 0.0167, which is just enough to produce climatic effects. Some of the planets detected around other stars have an eccentricity as high as 0.93. Such planets, even if otherwise suitable, would have wild swings of climate that would prevent the evolution of complex life.
- *The availability of comets.* Comets may have played an important role in bringing water to the primordial Earth (see ORIGIN OF LIFE). How can scientists know whether other solar systems even have comets? The Oort cloud extends halfway to the next star (two light-years into outer space). Some of those comets may be as likely to go to Proxima Centauri as to the Sun. Based on chemical analysis, all the comets known in our solar system seem to be from the Oort cloud and Kuiper Belt. Apparently there are no comets from Proxima Centauri coming into the solar system, suggesting that Proxima Centauri has nothing that corresponds to the Oort cloud. The star Beta Pictoris appears to have comets falling into it, based on bursts of different colors of light. But how typical is this of solar systems?
- *The Moon.* The Moon stabilizes the 23-degree tilt of Earth. This stable tilt is what causes the regular alternation of seasons. Billions of years ago, the Moon was closer and the Earth spun faster, resulting in extreme winds. The Moon has been moving away from the Earth, and the Earth's rotation has been slowing down, so that winds are not now deadly. The Moon has allowed the Earth to have stable conditions for long enough that complex life has evolved. A bigger or smaller moon, or multiple moons, would be unlikely to produce this effect.

Therefore solar systems like the one in which humans live, suitable for the evolution of complex and intelligent life, might be very rare in this or any other galaxy.

How typical is Earth? The Earth may be a very unusual planet in the Milky Way or any other galaxy.

Where

There are three possibilities for where life originated:

A. *Life evolved on Earth only.* According to the Rare Earth hypothesis (see essay, "Are Humans Alone in the Universe?"), the conditions necessary for complex life are so uncommon that the Earth may in fact be the only planet on which complex life has evolved. The authors of this hypothesis assert that, although complex and intelligent life might be unique to the Earth, bacterial life might be common in the universe.
B. *Life evolved someplace else, and was then transported to the Earth.* This hypothesis is called *panspermia* ("seeds everywhere"). Swedish chemist Svante Arrhenius brought up this idea early in the 20th century. Biochemists Francis Crick and Leslie Orgel have written about the extraterrestrial origin of the organic molecules that produced life on Earth, even though life itself evolved here. Astronomers Fred Hoyle and Chandra Wickramasinghe have quite seriously suggested that the first cells came to the Earth from outer space, from a "life cloud". The Martian meteorite ALH84001 contains organic molecules, and structures that may be bacterial fossils. Despite these suggestions, it is unlikely that Martian bacteria would have survived being ejected from Mars, the journey through outer space, and falling through the atmosphere. It is therefore unlikely that meteorites brought the molecules of life to the Earth.
C. *Life evolved on Earth, but also in other places.* This idea is popular among scientists because the universe is, in fact, full of organic molecules. On the Earth today, all organic molecules have a biological origin: Even petrochemicals are the products of plants that died millions of years ago. But organic molecules can be produced during the same processes that form stars and solar systems. The most common elements in universe are hydrogen, helium, oxygen, carbon, nitrogen, and neon; all but the last of these predominate in organic molecules:

- Spectral analyses of starlight through nebulae reveal the existence of at least 62 kinds of organic molecules in those nebulae. Organic chemicals (for example, naphthalene) are common in nebulae, which has led evolutionary biolo-

- *Chemical composition of the Earth.* Most known planets from other solar systems are gaseous. But if the solid planets were like many asteroids, which have a great deal of carbon and water, then the Earth would have either a massive GREENHOUSE EFFECT from carbon dioxide, or would have been completely covered with deep oceans, not having the shallow waters that appear necessary for the evolution of life.
- *Size of the Earth.* If the Earth had been a little smaller, it would by now have completely cooled and lost its magnetic field. This would have allowed cosmic radiation to rip away the atmosphere and water and life. This appears to be what happened on Mars. On a larger planet, gravitation would be so strong that complex life might not be possible. On a larger planet, all geological forms might collapse underneath an ocean; not only would there be no terrestrial life but also no erosion of nutrients into the ocean. As a result, the entire ocean would be nutrient-poor, just as the middle of the oceans on Earth is today.
- *Habitable zone of the solar system.* Within the solar system, Earth is in just the right place, the only planet in the habitable zone. If the Earth were one percent further away from the sun, it would experience a runaway ice age; if it were five percent closer, it would experience a runaway greenhouse effect, sometime during its history. If the Earth were closer to a smaller star, the star's gravity would hold the Earth in an orbit in which one side would always face the sun—just as the moon always faces the Earth. Under such conditions, one side of the Earth would burn up, the other would freeze, and extremely strong winds would result.

As some observers have said, humans should "thank their lucky star," but also their extremely lucky solar system and Moon. Without the concatenation of all of these unlikely events, the Earth would have had an extremely unstable history, losing its oceans, or fluctuating in temperature so greatly that only bacterial life could have survived. On a cosmic scale, even the nearly complete freezing of the Earth that occurred most recently about 700 million years ago is a mild occurrence (see SNOWBALL EARTH).

According to Peter Ward and Donald Brownlee, when you consider all of these factors, it is possible that the Earth is the only planet that has been stable enough for complex life to evolve—even in the entire universe. At the very least, they claim, the universe is not like *Star Trek,* full of humanoids with whom humans can make contact. Some people have used these very same data to claim that the Earth has been prepared for our arrival by a higher intelligence (see ANTHROPIC PRINCIPLE). However, such a principle is unnecessary. While it is unlikely for all of these lucky things to have happened right here in this part of this galaxy, it could very well be that humans exist and think about such things in this place and not somewhere else simply because this place is where the luck happened to occur. At the same time, it might also mean that the rest of the universe, even if chock-full of bacteria, is a very lonely place for creatures with higher intelligence.

Further Reading

Basalla, George. *Civilized Life in the Universe: Scientists on Intelligent Extraterrestrials.* Oxford, U.K.: Oxford University Press, 2005.

Conway Morris, Simon. *Life's Solution: Inevitable Humans in a Lonely Universe.* Cambridge University Press, 2003.

Jackson, Randal. "PlanetQuest: The search for another Earth." Jet Propulsion Laboratory, California Institute of Technology. Available online. URL: http://planetquest.jpl.nasa.gov/index.cfm. Accessed April 24, 2006.

Sagan, Carl. *Cosmos.* New York: Random House, 1980.

Ward, Peter D., and Donald Brownlee. *Rare Earth: Why Complex Life is Uncommon in the Universe.* New York: Copernicus, 2000.

gist Simon Conway Morris to comment that the universe "smells faintly of mothballs." There are at least 27 kinds of organic molecules, some quite complex, in the tails of Halley's and Hale-Bopp comets.

- Organic molecules are common in the carbonaceous chondrite asteroids left over from the initial formation of our solar system. Even in the 1830s it was known that carbonaceous meteorites contained organic molecules. The Murchison meteorite, a carbonaceous meteorite that fell in Australia in 1969, contained at least 74 kinds of molecules, of which eight are amino acids found today in living cells, as well as fatty acids, glycerol, and purine and pyrimidine bases (found in nucleic acids such as DNA). This was also true of the Tagish Lake meteorite, which fell in Canada in 2000. Careful analysis discounted the possibility that these molecules were terrestrial contaminants.

The nebular and comet-tail chemicals are very sparse—only a few molecules per cubic meter. How could they be concentrated and delivered to the Earth? An immense amount of comet dust rains on Earth: about 40,000 tons per year. However, it is unlikely that a significant amount of organic material would have survived on meteorites. The presence of organic molecules in outer space does not explain where terrestrial organic molecules came from. It demonstrates that the universe has produced immense amounts of the very kinds of organic molecules from which life is made—and this could have happened on the early Earth as easily as anyplace else in the universe. Because organic molecules are so common in the universe, the presence of PAH (polycyclic aromatic hydrocarbons) in the famous Mars meteorite is not itself evidence of life.

The study of life outside of the Earth is called astrobiology ("star-life"), formerly called exobiology ("outside life"). Astronomer Jonathan Lunine notes that since the discovery of ALH84001, astronomers and the National Aeronautics and Space Administration (NASA) have had a renewed interest in astrobiology and even in the possibility of panspermia.

How

Assuming that life evolved from organic molecules that formed on the Earth, how could this have occurred? The question is not new. Charles Darwin wrote a letter to Joseph Hooker, dated February 1, 1871, in which he made his famous reference to life originating in a "warm little pond":

> It is often said that all the conditions for the first production of a living organism are now present, which would ever have been present. But if (& oh what a big if) we could conceive in some warm little pond with all sorts of ammonia & phosphoric salts,—light, heat, electricity &c present, that a protein compound was chemically formed, ready to undergo still more complex changes, at the present day such matter would be instantly devoured, or absorbed, which would not have been the case before living creatures were formed.

Scientists disagree about how easy this process would have been. Biochemist Christian de Duve said "The universe was pregnant with life ... we belong to a Universe of which life is a necessary component, not a freak manifestation." Others, however, point out the many difficulties of producing the first life-forms: first, in producing the molecules themselves, then in assembling the complex organic systems in which these molecules function, then in the formation of the first cells.

A. Producing the molecules. Biological molecules can form from inorganic molecules. Jöns Berzelius, one of the most famous chemists of the early 19th century, said in 1827 that organic molecules could not be made from inorganic sources. The very next year, his friend and former student Friedrich Wöhler synthesized urea, known at that time only from kidneys, simply by heating ammonium cyanate. Louis Pasteur showed that life comes only from life today and under normal conditions (the law of biogenesis), but many 19th- and early 20th-century scientists speculated about the abiotic origin of life *(abiogenesis),* especially these questions:

1. *What was the energy source?* To make simple inorganic molecules into complex organic molecules, energy is necessary. The early Earth had numerous sources of energy for the synthesis of organic molecules: lightning, ultraviolet light, cosmic radiation, and heat from asteroid bombardments and volcanoes, to name a few.
2. *What was the atmosphere like?* During the mid-20th century, the atmosphere of the early Earth was assumed to be *reducing,* consisting of molecules such as methane, ammonia, water vapor, and hydrogen. This was a reasonable assumption, as these molecules are common in the atmospheres of the large gas planets such as Jupiter. By the late 20th century most scientists believed that the early Earth atmosphere was *neutral,* consisting largely of carbon dioxide. Recent evidence suggests that the atmosphere might have contained a substantial amount of methane after all. Based on studies of siderite minerals from the earliest sedimentary rocks, some scientists have suggested that the atmosphere could not have contained enough CO_2 to have produced the GREENHOUSE EFFECT that was known to have prevailed on the Earth at that time, and that methane, a potent greenhouse gas, might have been largely responsible for the warmth of the atmosphere and Earth. The only fact on which there is universal agreement is that the atmosphere of the early Earth contained virtually no oxygen.
3. *Where did life originate on the Earth?* Many scientists assume this must have occurred in shallow seas. One problem with this hypothesis is that the same ultraviolet radiation that would provide energy to organic syntheses would also have destroyed the molecules. Some scientists, such as geologist John Corliss, have asserted that life originated in deep ocean vents, where water meets lava. This has the disadvantage that the heat itself could have destroyed the organic molecules as easily as it created them. Some bacteria that live under those conditions today have special adaptations that prevent the heat from disrupting their molecules. The deep sea vents have abundant ferrous ions, which would have produced a reducing environment and encouraged the synthesis of organic molecules.
4. *What is the evidence for the origin of organic molecules on the early Earth?* Many scientists, starting with Aleksandr Ivanovich Oparin in Russia and J. B. S. Haldane in England (see HALDANE, J. B. S.), have suggested that organic molecules came into existence under the conditions of the primordial Earth. During the second half of the 20th century, numerous simulations of early Earth conditions have been conducted in laboratories. The earliest, and still most famous, of these simulations was conducted by biochemist Stanley Miller in 1953 (see MILLER, STANLEY). He put a mixture of reducing gases (including methane and ammonia) into glassware that circulated the gases through water and past an electric spark. Organic molecules accumulated in the flask. Rather than being an incomprehensible mixture of molecules, the product was dominated by a few molecules: formic acid, glycine, glycolic acid, alanine, lactic acid, acetic acid, and propionic acid, in descending order. In much smaller quantities were urea, aspartic acid, and glutamic acid. Four of these are amino acids. In 1960, chemist John Oró conducted a similar experiment that produced adenine. All of these molecules are used by organisms today. Amino acids are the building blocks of proteins, which are largely responsible for the complexity of life processes. The popular press got the impression that Miller had made life in the laboratory, a claim he never made. The world was astounded that the building blocks of life could be so easily produced by simple chemical reactions. Many other simulations have been performed since Miller's original experiment. Miller's experiment was not the first synthesis of its kind. In 1913, Walther Löb produced amino acids from wet formaldehyde plus an electric discharge. Löb was simply studying the chemistry, rather than investigating the origin of life.

   Scientists are a long way from figuring out how life might have arisen by spontaneous chemical reactions on an inorganic Earth.
5. *Producing all of the necessary molecules.* Although many organic molecules can be formed in laboratory simulations, many of the molecules that are essential to life today have not been formed in these simulations. Ribose, a component of RNA (see below), would have been especially difficult to form. In addition to these molecules, the simulation experiments have also produced a tar-like "goo" that covers the inside surfaces of the flasks. As biochemists Stanley Miller and Antonio Lazcano say, "The primordial

broth may have been a chemical wonderland, but it could not have included all the compounds and molecular structures found in the simplest of present-day microbes."

A particular problem has been in designing a "one-pot reaction" in which the scientist puts inorganic chemicals in and gets living cells out. In order to obtain the different molecules necessary for life, very different conditions of acidity and temperature are needed. For example, freezing conditions are necessary to preserve adenine and guanine, while for cytosine and uracil warm evaporative conditions are needed. Biologist Robert Shapiro has been particularly eager to point out how implausible these prebiotic syntheses would have been, requiring an unbelievable configuration of heat and cold, salinity and purity, acidity and neutrality—a series of unlikely events occurring in just the right order to produce even a few of the chemical building blocks of life. The chemical simulations are, according to Simon Conway Morris, "...highly artificial, if not contrived." The outcome of the syntheses depends strongly on the starting conditions. The reactions would not work at all in a neutral atmosphere, rather than a reducing atmosphere (see above). Stanley Miller used a continuous spark plug; but when brief, stronger blasts of electricity (which simulate lightning) were used, the results were disappointing. When the order of the glassware components changed, the results were different. Some experiments would not have worked without doubly purified water, available in chemical laboratories but not on the primitive Earth. So far, there has been little promise from one-pot reactions in explaining the origin of all of the chemicals of life, converging in one place at one time.

6. *Origin of "handedness" in biological molecules.* Organic molecules can have left- v. right-handed mirror images, exactly the way human left and right hands are identical but opposite to one another. Organic molecules produced by nonliving processes are almost always *chiral,* an equal mixture of the two forms. However, proteins in organisms use left-handed amino acids and right-handed nucleic acids. A protein made of all left-handed, or all right-handed, amino acids is more stable than a protein made of a mixture. But, how did the preponderance of left-handed amino acids ever get started in the first place? There was great excitement when it was discovered that meteorites could have an excess of left-handed amino acids—but this excess was only 7–10 percent. It has been suggested that ultraviolet light polarized the amino acids before the meteorites reached the Earth. It is still unclear whether this could have happened on the primordial Earth. It has been demonstrated that divalent cations (such as the calcium ion Ca^{++}) can produce an uneven mixture of the two forms of organic molecules.
7. *Synthesis of large molecules from small precursors.* The production of small organic molecules may have been easy on the primordial Earth, just as in nebulae and comets, but how could small molecules have assembled into large ones, such as the proteins and nucleic acids needed by modern cells? Large molecules tend to dissociate into smaller ones in water. There are two processes that may have allowed the formation of large molecules on the primordial Earth:

- *Activated precursors.* Regular amino acids do not polymerize into proteins in water very well, but amino acids in the amide form do. As biochemist James Ferris explains, carboxyanhydrides can form into protein-like molecules in water. When this happens, he points out, the resulting protein-like molecules form only the correct type of peptide bonds that characterize biological proteins; and, moreover, the resulting molecules, up to 10 amino acids in length, contain only left-handed components.
- *Adsorption on mineral surfaces.* Small molecules bumping into one another in watery swirls would be unlikely to form the complex molecules characteristic of life. In many chemical reactions, solid surfaces allow the orderly catalysis of reactions. The catalytic converter of an automobile uses surfaces of the metal palladium to catalyze the reaction that eliminates carbon monoxide from auto exhaust. J. D. Bernal, a British biochemist, first proposed the possibility that the synthesis of large molecules from small ones may have occurred on clay mineral surfaces. Clays certainly provide an enormous amount of surface area for such reactions to take place. Leslie Orgel calls this possibility "life on the rocks" and has demonstrated that RNA molecules of length up to 40 bases can be produced on a mineral surface. RNA molecules of this size would be long enough to get the *RNA world* (see below) started.

B. Assembly of complex chemical systems. Modern life, even of the simplest bacteria, is too complex to have arisen directly. In particular, the elegant genetic system of DNA (see DNA [RAW MATERIAL OF EVOLUTION]) must represent an advanced system not found in the first life-forms. In modern cells, DNA stores genetic information, which is transcribed into RNA, which directs the formation of proteins. As microbiologist Carl Woese proposed in 1967 (see WOESE, CARL R.), scientists realize that a simpler genetic system of life must have preceded that of DNA in the modern cell. Once modern DNA cells came into existence, they would have outcompeted the more primitive form which, therefore, no longer exists. Suggestions (not all mutually exclusive) include:

- Some scientists, such as Stuart Kaufmann, Gunter Wächtershäuser, and Christian de Duve, have proposed that metabolic systems (such as glycolysis, which releases energy from sugar so that organisms can use the energy to operate) preceded genetic systems. This suggestion has met with skepticism among most scientists because, at some point, a genetic system would have to take over the control of the cell.
- The first genetic system may have been formed of clay mineral crystals rather than of organic molecules. In this scenario, proposed by British scientist Graham Cairns-Smith, organic molecules such as RNA helped the mineral crystals in their replication. Later, the minerals served as scaffolding for the reactions of the organic molecules, which later took place without the scaffolding.
- The first genetic molecule may have been something like TNA (threonucleic acid) or PNA (peptide nucleic acid). The formation of such molecules could have occurred more readily on the early Earth (in particular, TNA contains no

ribose); it works in a fashion similar to DNA and RNA; and it can bind with DNA and RNA, which means that it could have transferred its genetic information to RNA during the evolution of a new RNA-based life-form. In 2000, Israeli scientists proposed that the first genetic system may have consisted of lipid-like molecules that can not only form cell-like structures that grow, but also pass information into the new cells.

- It is widely accepted, following the lead of biochemist Manfred Eigen, that an RNA-based genetic system would have preceded a DNA-based system, for several reasons. First, modern cells make DNA bases out of RNA bases. Second, all modern cells use modified ribonucleotides as the basis for some of their essential metabolic chemistry, such as ATP and NADH, and some modern enzymes need small RNA molecules to help them carry out their reactions. Third, biochemists Thomas Cech and Sidney Altman, who won the Nobel Prize for Chemistry in 1989, showed that RNA can act as an enzyme (they are called *ribozymes*). That is, RNA can be both genotype and phenotype. Some organisms, such as the protist *Tetrahymena thermophila,* use ribozymes. Many scientists consider the ribosome, which is built of both RNA and protein, and in which it is the RNA that has catalytic activity, to be a remnant of the time when RNA was the blueprint of life, and in the laboratory, RNA can catalyze its own reproduction. Strings of RNA containing guanine result when RNA molecules containing cytosine are used as a template. Leslie Orgel calls this the "molecular biologist's dream." More than 20 different ribozymes have been produced from random RNA mixtures followed by a selection experiment, including a ribozyme that synthesizes a nucleotide from a base, and another ribozyme that makes more RNA.

Chemist Walter Gilbert proposed the *RNA world* scenario in 1986. In this scenario, the primordial seas were filled with RNA molecules that replicated themselves and therefore constituted a primitive form of life. Later, according to evolutionary biologists John Maynard Smith (see MAYNARD SMITH, JOHN) and Eörs Szathmáry, these RNA molecules were assisted by amino acids, which resulted in the origin of the genetic code. Still later, complex DNA replaced simple RNA as a more stable form of genetic information—but living cells never got rid of the RNA completely. This process would occur more efficiently if the RNA molecules consisted of bases that were all left- or all right-handed, which would explain the origin of handedness in the genetic molecules. In RNA selection experiments, smaller RNA molecules replicate faster than larger ones, which makes it difficult to explain how complex molecules could have arisen in an RNA world. The last universal common ancestor (LUCA) of all cells that are alive today stored genetic information in DNA, but the original life-form may have used RNA or an even earlier chemical basis.

C. Formation of the first cells. The final step in the origin of life would be to explain how the life reactions could have been isolated into cells—a step necessary to keep the waves of the ocean from separating and diluting them. Most modern cell membranes are made from phospholipids, which are molecules that can bridge the gap between fatty and watery molecules. Lipid-like molecules are today found in sea foam. Some scientists suggest that this sea foam, in shallow primordial ponds, may have formed the first cell membranes. Other scientists have formed *micelles,* which are clusters of molecules that carry out chemical reactions and replicate themselves, in the laboratory, and they propose that the first cells may have resembled these micelles.

At some point, the origin of cells must be explained. Life may have been in operation for a long time before it was compartmentalized into cells, according to Carl Woese, who has described a *life state* that preceded *life-forms.*

In conclusion, scientific research has illuminated many possibilities for the origin of life, in particular answers to the questions of when and where; but for now scientists will have to be satisfied with not knowing a definite answer to the questions of how.

Further Reading

Busemann, Henner, et al. "Interstellar chemicals recorded in organic matter from primitive meteorites." *Science* 312 (2006): 727–730.

Chen, Irene A. "The emergence of cells during the origin of life." *Science* 314 (2006): 1558–1559.

Davies, Paul. *The Fifth Miracle: The Search for the Origin and Meaning of Life.* New York: Simon and Schuster, 1999.

Dick, Steven J., and James E. Strick. *The Living Universe: NASA and the Development of Astrobiology.* New Brunswick, N.J.: Rutgers University Press, 2004.

Hazen, Robert M. *Genesis: The Scientific Quest for Life's Origin.* Washington, D.C.: Joseph Henry Press, 2005.

Kasting, James F. "When methane made climate." *Scientific American,* July 2004, 78–85.

Lazcano, Antonio. "The origins of life." *Natural History,* February 2006, 36–41.

Lunine, Jonathan. *Astrobiology: A Multi-disciplinary Approach.* Upper Saddle River, N.J.: Addison-Wesley, 2004.

Marty, Bernard. "The primordial porridge." *Science* 312 (2006): 706–707.

Schopf, J. William. "Microfossils of the early Archaean Apex chert: New evidence of the antiquity of life." *Science* 260 (1993): 620–646.

———, ed. *Life's Origin: The Beginnings of Biological Evolution.* Berkeley: University of California, 2002.

Shapiro, Robert. *Origins: A Skeptic's Guide to the Creation of Life On Earth.* New York: Bantam, 1986.

Tian, Feng, et al. "A hydrogen-rich early Earth atmosphere." *Science* 308 (2005): 1,014–1,017.

Wald, George. "The origin of life." *Scientific American,* August 1954, 44–53.

Warmflash, David, and Benjamin Weiss. "Did life come from another world?" *Scientific American,* November 2005, 64–71.

***Origin of Species* (book)** *On the Origin of Species by Means of Natural Selection* was Charles Darwin's famous 1859 book that is considered the founding document of modern evolutionary science. It is considered one of the most important books ever written. Evolutionary scientist Ashley

Montagu said, "Next to the Bible, no work has been quite as influential, in virtually every aspect of human thought, as *The Origin of Species.*"

Shortly after returning from his voyage around the world, marrying, and settling down in England, Darwin formulated his theory of NATURAL SELECTION as the mechanism by which evolution had occurred. His acceptance of UNIFORMITARIANISM (see LYELL, CHARLES), the numerous observations of BIOGEOGRAPHY and fossils that he had made during his voyage, and the principles of population biology (see POPULATION) that he had read (see MALTHUS, THOMAS), all converged in his mind upon natural selection. However, he was reluctant to present his ideas in public. Earlier presentations of evolution (see LAMARCKISM) had claimed that evolution had occurred but had not presented a mechanism for it. This was the principal reason that evolution did not have scientific credibility. Darwin became even more convinced of this when he saw the chilly reception and outright hostility occasioned by the 1844 book *Vestiges of Creation* (see CHAMBERS, ROBERT), which presented evolution without explaining how it worked. As theologian William Paley had explained, saying that complex design simply happened by natural law was not an explanation (see NATURAL THEOLOGY). The creationism that prevailed in the early 19th century at least had the advantage of explaining how organisms had been designed—God did it—while evolution, as presented in the *Vestiges,* simply said it just happened. Darwin had figured out a mechanism, but he wanted to assemble all the evidence for every part of his theory before presenting it in public. He conducted research for many years, filled several notebooks with information, and began long manuscripts, all intended to eventually form his big book of evolution. Darwin told some close associates, such as Sir Charles Lyell and Sir Joseph Hooker (see HOOKER, JOSEPH DALTON), about his theory.

Darwin did not get the chance to write his big book. In 1858 he received a letter from a young British naturalist who was working in Southeast Asia (see WALLACE, ALFRED RUSSEL), which described the theory of natural selection almost exactly as Darwin had conceived it decades earlier. Unlike Darwin, Wallace was ready to publish. Had Wallace published his article in a scientific journal rather than sending it to Darwin, scientists might be referring to evolution as Wallace's theory. Darwin now knew that he could not delay in presenting his theory, which Wallace had independently proposed. Lyell and Hooker were able to vouch that Darwin had thought of natural selection before receiving Wallace's letter. The paper in which natural selection was presented to the scientific world contained an essay Darwin had written the previous decade and Wallace's letter and was read to the Linnaean Society on July 1, 1858. Neither Darwin, who was ill, nor Wallace, who was also ill and still in Southeast Asia, were present. Why the paper aroused little curiosity or discussion remains unexplained.

Darwin wrote his book in a hurry. He referred to it as the "briefest abstract" of his ideas, to be followed someday by his big book that would provide all the information. His abstract was more than 400 pages in length. Because Darwin was trying to be brief and clear, *Origin of Species* remains one of the masterpieces of scientific writing, much more readable and much more widely read than the portions of his big book that he did finish (such as *Variation of Plants and Animals under Domestication*). The first printing of *Origin of Species,* 1,250 copies printed by publisher John Murray, sold out on the first day, November 24, 1859.

Darwin undertook a monumental task in writing this book. First, he demonstrated that heritable variation exists in plant and animal populations. Next, he presented the Malthusian argument for the struggle for existence, as applied to plants and animals, not just to humans. Chapter 3 of the *Origin,* "Struggle for Existence," therefore became the founding document of the science of ECOLOGY. Next, Darwin brought these together in an explanation of natural selection. Having presented his theory of how evolution works, Darwin then presented the evidence that evolution had occurred throughout the history of the Earth, from the order of fossils in the fossil record, to the biogeography of modern organisms, to the vestigial evidences of evolution to be found in rudimentary organs and in embryos. In his concluding chapter, he refrained from making anything more than the gentlest reference to human evolution ("Much light will be thrown on the origin of man and his history"). His final statement has become one of the most famous in biology:

> There is grandeur in this view of life ... whilst this planet has gone cycling on according to the fixed law of gravity, from so simple a beginning endless forms most beautiful and most wonderful have been, and are being evolved.

The word *evolution* had often been used to describe the playing out of a pre-ordained history of the world (see EVOLUTION). Since Darwin did not mean to imply that the direction of natural history was preordained, he avoided the word *evolution,* preferring instead the phrase "descent with modification." The last word is the only time in the book that Darwin used a version of the word *evolution.*

Darwin issued six editions of *Origin of Species,* the last one in January 1872. Each time, he incorporated new information and recent discoveries. He also added a great deal of material to answer critics. He devoted an entire chapter to answering criticisms, many raised by zoologist St. George Jackson Mivart, whose 1871 book *The Genesis of Species* may have been one of the most influential challenges to *Origin of Species,* although it is today largely forgotten.

Even though *Origin of Species* is an abstract of Darwin's thought, most of even highly educated modern people have not actually read it. The author of this encyclopedia, like most evolution educators, strongly encourages every person to read the complete *Origin of Species.* In the event that one does not have time to do so, the author has included a summary of *Origin of Species* as an appendix to this encyclopedia.

Further Reading

Darwin, Charles. *On the Origin of Species by Means of Natural Selection,* 1st ed. London: John Murray, 1859. Available online. URL: http://pages.britishlibrary.net/charles.darwin/texts/origin1859/origin_fm.html. Accessed May 3, 2005.

———. *On the Origin of Species by Means of Natural Selection,* 6th ed. Reprinted with introduction by Julian S. Huxley. New York: Times Mirror, 1958.

Desmond, Adrian, and James Moore. *Darwin: The Life of a Tormented Evolutionist.* New York: Warner, 1992. First published, London: Michael Joseph, 1991.

Harper, John L. "A Darwinian approach to plant ecology." *Journal of Ecology* 55 (1967): 247–270.

Orrorin *See* AUSTRALOPITHECINES.

Owen, Richard (1804–1892) British *Paleontologist* Richard Owen was a prominent British vertebrate paleontologist of the 19th century. Born July 20, 1804, and first trained as a physician, Owen became a world authority on the anatomy of living and fossil vertebrates.

The Royal College of Surgeons asked Owen to organize and expand their vast collection of specimens. In this capacity Owen had first rights to any animal that died at the London Zoo. Owen's wife came home once and found a recently deceased rhinoceros in the front hall. When *ARCHAEOPTERYX* was discovered, the specimen was sent to Owen. Owen was the first to recognize the existence of DINOSAURS as a separate group of prehistoric animals, coining the term in 1842. Despite his reputation as an expert, he did publish a few errors, such as his 1848 hypothesis of the manner in which vertebrae are modified to form skulls of different species.

Despite his fame (he became Sir Richard Owen), much of his legacy is questionable. Owen was the only person whom Charles Darwin was known to hate. Owen also had a tendency to exaggerate his personal accomplishments. He listed himself as professor of comparative anatomy and physiology at the Government School of Mines, in *Churchill's Medical Dictionary,* when in reality this post was held by someone else (see HUXLEY, THOMAS HENRY). Owen had several disputes with other scientists (for example, Hugh Falconer) who claimed that Owen took credit for their discoveries. Amateur paleontologist Gideon Mantell found the first dinosaur tooth that alerted scientists that these reptiles were different from any other known orders; Mantell discovered and named several fossil species. When Mantell submitted manuscripts that described these new fossil species, Owen had them rejected—then published descriptions of these species himself, renaming them. Mantell by this time had suffered a crippling accident which left his spine deformed. From pain and rejection, Mantell committed suicide. His spine was removed and sent to the Royal College of Surgeons, where it was preserved and placed under the care of none other than Sir Richard Owen. Owen even anonymously wrote an obituary for Mantell, discrediting him as mediocre.

Owen is perhaps best remembered for his role in two famous debates. In an 1860 meeting at Oxford University that included an attack on Darwinian theory by Bishop Samuel Wilberforce, it was an open secret that Wilberforce had been coached by Richard Owen. The other debate was with Huxley over whether or not the brains of apes and humans differed in kind (as he insisted) or merely in degree (as Huxley insisted). Owen claimed that human brains had a hippocampus minor, and ape brains did not. Huxley turned out to be right.

Eventually, as Owen's questionable acts became known, he was voted off of some of his academic leadership positions. The second half of his career was spent in building up the British Natural History Museum, which is clearly the greatest ongoing public benefit of his work.

Before his death on December 18, 1892, Owen had one final influence. He opposed the erection of a statue of Charles Darwin in the Natural History Museum. He failed, but today his statue is at the base of the stairway to the main hall of the Museum, while the statues of Darwin and Huxley are in small upstairs galleries.

Paleolithic *See* TECHNOLOGY.

paleomagnetism Paleomagnetism refers to the ancient records of the Earth's magnetic field (*paleo-* means ancient). The magnetic field of the Earth is believed to result from the circulation of liquid metal in the inner Earth. Volcanic rocks on the Earth's surface contain iron minerals which, like compass needles, orient themselves in line with the magnetic field of the Earth while the lava is cooling. The *remnant magnetism* of the rock can be measured. If the Earth's magnetic field does not change, and the rock does not move, the iron minerals in the rock will be oriented parallel to the magnetic field. However, both of these things change:

- *Reversal of magnetic fields.* The polarity of the Earth's magnetic field has changed frequently during Earth history. During the past 35 million years, the polarity of the Earth's magnetic field has reversed (as indicated by the iron minerals in rocks) more than a hundred times. When this occurs, the north and south magnetic poles switch positions. The periods between magnetic reversals have averaged, during the Cenozoic era, about a half million years apart; however, they can be as long as several million years or as short as a few thousand years apart. The magnetic field has not reversed for almost a million years. The Earth's magnetic field has been diminishing in intensity during the past century, leading some scientists to predict that another magnetic reversal is about to occur, within a few more millennia. The cause of the reversals is unknown.
- *Movement of the continents* (see CONTINENTAL DRIFT; PLATE TECTONICS). The Earth's magnetic field has not only a north-south component but also a vertical component. It is nearly horizontal at the equator and vertical at the poles. Many volcanic rocks on continents in polar regions have horizontal magnetic fields, which indicates that the continents were near the equator when the rocks were formed and have drifted to their present locations.

Within any one location, the pattern of reversals produces a barcode-like signature of magnetic patterns. The ages of different formations can be determined by lining up their pattern of reversals with one another. Paleomagnetism, along with RADIOMETRIC DATING, allows geologists to provide very accurate age determinations for volcanic deposits and, therefore, the fossil-bearing deposits that are between them.

A strong magnetic field protects the Earth from cosmic radiation. While the reversal is occurring, the strength of the magnetic field is temporarily reduced. This allows higher levels of cosmic radiation to reach the Earth's surface. There are, however, no patterns of increased extinction in connection with reversals of the magnetic field during the history of the Earth. While a reversal of the magnetic field would be unlikely to cause disaster to the natural world, it would be likely to disrupt many electronically based economic activities of modern humans.

Paleozoic era The Paleozoic era (the era of "ancient life") is the first era of the Phanerozoic Eon, or period of visible multicellular life, which followed the PRECAMBRIAN TIME in Earth history (see GEOLOGICAL TIME SCALE). The Paleozoic era began with a massive diversification of species (see CAMBRIAN EXPLOSION) and ended with the Earth's greatest extinction event (see PERMIAN EXTINCTION; MASS EXTINCTIONS). The Paleozoic era consists of six geological periods (see CAMBRIAN PERIOD; ORDOVICIAN PERIOD; SILURIAN PERIOD; DEVONIAN PERIOD; CARBONIFEROUS PERIOD; PERMIAN PERIOD).

During the middle of the Paleozoic, a great deal of continental landmass was centered in the Southern Hemisphere. By the end of the Paleozoic, all the continents coalesced into a single continent, Pangaea. Climatic conditions were cold and dry in the middle and along the southern edge of Pangaea during the Permian period.

During the first two periods, almost all life-forms were aquatic. Probably every animal phylum (including vertebrates in the Chordata) existed by the end of the Cambrian period, as well as seaweeds. Diversity of marine species was even greater during the Ordovician. A global mass extinction occurred at the end of the Ordovician. Evolutionary diversification continued in the oceans during the Silurian period. The first small land plants proliferated in wetlands during the Silurian. By the Devonian period, forests that consisted mostly of relatives of modern club mosses and horsetails (see SEEDLESS PLANTS, EVOLUTION OF) covered the extensive wetlands, and fishes diversified in the fresh waters and oceans (see FISHES, EVOLUTION OF). Amphibians (see AMPHIBIANS, EVOLUTION OF) evolved and terrestrial arthropods such as insects (see INVERTEBRATES, EVOLUTION OF) diversified during the Devonian. During the Carboniferous and Permian periods, forests dominated by the first seed plants (see GYMNOSPERMS, EVOLUTION OF) spread beyond the wetlands, and reptiles (see REPTILES, EVOLUTION OF) diversified. There were no birds or mammals.

The Paleozoic era came to an end when massive volcanic eruptions were followed by an asteroid that collided with the Earth, creating temperature extremes and acid rain that led to worldwide extinctions.

Paley, William *See* NATURAL THEOLOGY.

panspermia *See* ORIGIN OF LIFE.

Paranthropus *See* AUSTRALOPITHECINES.

parasitism *See* COEVOLUTION.

Peking man *See* *HOMO ERECTUS.*

peppered moths A species of moth, *Biston betularia,* lives in Europe and was for many years cited as the best example of NATURAL SELECTION in action. The adult moths have two color forms: the light (forma *typica*) form, which has colored flecks against an almost white background ("peppered") on its wings and body, and the dark form (forma *carbonaria*), which is nearly black (see photos at right). The dark color is controlled by a single dominant allele (see MENDELIAN GENETICS). Moths with one or two copies of the allele are dark; they are light only in the absence of the allele (they have a recessive allele that does not stimulate production of dark color).

Insect collectors began to notice the appearance of dark forms of *Biston betularia* in England in the middle of the 19th century. A specimen was found in 1848 near Manchester, in 1860 in Cheshire, in 1861 in Yorkshire, spreading to Westmorland by 1870, to Staffordshire by 1878, and to London by 1897. Whenever the dark form appeared, it increased in abundance until it became more common than the light form. Between 1848 and 1898, dark moths increased from less than 1 percent to more than 99 percent of the moth population near Manchester. Dark *Biston betularia* showed up in continental Europe as well: 1867 in the Netherlands, 1884 in Hanover, and 1888 in Thuringia. Scientists began to notice a clear pattern: The places where the dark moths appeared and became common were industrialized areas of England

Peppered moths come in light and dark forms. The dark form blends in with dark tree bark (such as bark darkened by industrial soot) better than the light form (a). The light form blends in with light tree bark (or bark covered by lichens) better than the dark form (b). During periods of industrial pollution, the dark forms largely replaced the light forms. This has been attributed to camouflage protecting them from bird predation. ***(Courtesy of Michael Willmer Forbes Tweedie/Photo Researchers, Inc.)***

and continental Europe, but they did not usually appear in rural areas of Cornwall, Scotland, and Wales. The conclusion seemed inescapable that the spread of the dark moths was associated with the dark clouds of industrial smoke from the coal-fired factories. A similar trend was occurring in other insect species, and in the industrialized areas of North America as well. This widespread trend for dark insects to become more common in polluted regions has been called *industrial melanism* (*melanic* means black or dark). Among the changes that occurred as a result of industrial pollution were: First, the light-colored lichens that grew on tree bark began to die, exposing the dark bark underneath; and, second, the smoke itself sometimes smudged the bark, particularly noticeable on light-colored bark such as on the birch tree.

Starting in the 1950s, air pollution laws were enacted in Europe and in North America, particularly in response to increasing evidence of the diseases caused by pollution and from "killer fogs" that reduced visibility to nearly zero and caused human deaths. The air became cleaner, and at the same time the dark moths began to decline in abundance and the light moths began to return. In northwest England, where 90 percent of the *Biston betularia* adults were dark in 1959, only 30 percent were dark by 1989. Similar declines occurred wherever dark insects were common. The increase, then the decrease, in frequency of the dark allele in the insect populations was so clear-cut that natural selection must have been the cause, according to most scientists then as now. The increase in the frequency of the dark allele between 1848 and 1898 was so rapid that, according to calculations by a geneticist and mathematician (see HALDANE, J. B. S.), the dark form must have had a 50 percent advantage over the light form during that time—one of the largest selection coefficients ever known.

One of the pioneers of POPULATION GENETICS was Oxford biologist E. B. Ford. His main contention, which he defended his entire career, was that the genetic variation found in populations was maintained by natural selection—that is, the various allele forms of a gene in a population were beneficial, not random. Toward the end of his career, his view was strongly challenged by measurements of protein variability in populations and by the neutral model of genetic diversity. Ford believed that the spread of industrial melanism was a clear example of natural selection in action, and that it should be studied. He found, and employed at Oxford, a medical doctor who was a moth enthusiast to do the necessary work: Henry Bernard Davis Kettlewell.

The explanation tested by Kettlewell was that the dark moths were well camouflaged against the tree trunks from which lichens and been killed and which were smudged by smoke; under these conditions, the dark moths had an advantage, and became more common than the light moths. The light moths were well camouflaged against the light-colored lichens and against unsmudged birch bark. The camouflage would protect the moths primarily against predation by birds. The birds were the agent of natural selection in the moth populations. When J. W. Tutt first presented this explanation in 1896, natural selection was not a popular concept among biologists; but by the 1950s, following the MODERN SYNTHESIS, natural selection had become a popular explanation of evolutionary change.

The study of *Biston betularia* that Ford and Kettlewell planned promised to be immensely valuable to evolutionary science. Sewall Wright's studies of genetic drift (see FOUNDER EFFECT; WRIGHT, SEWALL) suggested that natural selection was not always, and was perhaps seldom, acting in wild populations. Previous studies of natural selection, for example, of the banding patterns in the snail *Cepea nemoralis* by evolutionary biologists Arthur Cain and Philip Sheppard, had shown that genetic changes could occur in wild populations. But they had not been able to demonstrate any adaptive value to the snail coloration patterns. Ford and Kettlewell hoped to go far beyond Cain and Sheppard, and to elucidate a complete story of natural selection in the wild, including an explanation of why it happened.

In order to study the evolution of industrial melanism in *Biston betularia,* it would be necessary to obtain a lot of moths and do a lot of fieldwork. Kettlewell was just the person to do it. He loved the outdoors. Above all, he was intensely familiar with moths, which he pursued for long hours in the field under all conditions. His huge hands were extremely deft at manipulating caterpillars and moths quickly without hurting them.

Kettlewell began his study in the summer of 1953, in Beacon Wood, near Birmingham, a polluted region of England. He had raised a large number of peppered moths from caterpillars. He marked each individual moth before releasing it. He released equal numbers of light and dark moths, during the daytime. At night, he turned on a lamp that attracted many insects. Among the insects that came to the lamp were many peppered moths; and some of them were the ones he had previously marked. At first, the number of moths that returned was very low. After a few nights, however, his moth recapture numbers increased, and the results were gratifying: He recaptured three times as many dark moths as light moths. This appeared to demonstrate that the light moths had died far more often than the dark moths in this polluted woodland. In 1955 Kettlewell performed an almost identical experiment, but this time in a woodland in Dorset, which was far from sources of pollution. The results were again gratifying: This time he recaptured twice as many light moths as dark ones, suggesting that the dark moths had died far more often than the light moths in this relatively pristine woodland. Raising the moths, and conducting the experiment, took a tremendous amount of work, and Kettlewell, getting little sleep during this time, damaged his health from overwork.

By conducting two experiments rather than stopping with one, Kettlewell avoided some problems in interpreting his results. There were other possible explanations for the results of the first experiment. Was it possible that one of the forms of the moth simply flew away more than the other? Alternatively, was it possible that one of the forms of the moth was more strongly attracted to the light than the other? Kettlewell's opposite results from the same technique in the second experiment discounted these possibilities.

Some data did not fit Kettlewell's hypothesis. Dark moths were becoming more common in the relatively clean forests of East Anglia, which did not fit the air pollution explanation. Was this a problem? Scientists generally conduct statistical tests upon samples, rather than expecting every datum to conform to the hypothesis (see SCIENTIFIC METHOD). A few exceptions did not invalidate the overall hypothesis.

There remained some skepticism, however, as to what was causing the differential survival of the light and dark moths. Ford and Kettlewell claimed that it was predation by birds. Light moths camouflaged themselves against lichens and light bark, and dark moths camouflaged themselves against dark bark. One problem that was raised to Kettlewell's results was, how was a moth to know what its own color was? How would a light moth know that it was light and seek light-colored bark on which to hide? Kettlewell's explanation implied that natural selection was acting not only upon the color of the moths but upon the behavior of the moths. This was not a major problem, as the whole point was that moths settled on tree trunks regardless of their color, and suffered the consequences of it.

All doubt as to the effectiveness of bird predation on the moths seemed to be put to rest when Niko Tinbergen brought his movie camera. Tinbergen was a pioneer not only in the study of animal behavior (he later shared a Nobel Prize with Konrad Lorenz and Karl von Frisch; see BEHAVIOR, EVOLUTION OF) but also in making nature documentaries. When Kettlewell set his moths out on the trunks of trees, Tinbergen was waiting in a blind with his camera. Birds came and ate the moths. Tinbergen's film footage of bird predation on *Biston betularia* was shown widely to scientific audiences, and this appeared to prove the hypothesis of Ford and Kettlewell: Everyone could see natural selection in action, on film! The peppered moth story was almost universally accepted and became standard fare of every biology textbook. The photographs seemed to prove that a light moth was very conspicuous against a dark trunk but was nearly invisible against lichens or light bark. Sewall Wright, who did not frequently recognize natural selection at work in natural populations, wrote in 1978 that the peppered moth story was ". ... the clearest case in which a conspicuous evolutionary process has been actually observed."

It was many years before doubts began to surface regarding the Kettlewell experiment. Among the scientists who noticed problems was biologist Theodore Sargent at the University of Massachusetts at Amherst. They knew that the peppered moths did not in fact rest on tree trunks during the day; they would instead rest underneath branches, remaining hidden during the daytime. Kettlewell had experimentally placed the moths, sometimes in unnaturally high concentrations, on the trunks. And he had to do so in the daytime, since he ran the light at night for collecting the moths. If Kettlewell had released the moths at night they would have gone straight to the light, proving nothing. Sargent pointed out that birds can learn, and teach one another, new foraging strategies. When some birds discovered the moths in Kettlewell's experiments, they began eating them, and other birds imitated them. What Kettlewell had done, according to Sargent, was to create a bird feeder. Kettlewell had unwittingly trained the birds to pick moths off of tree trunks where they were not normally found. It was hardly surprising that the birds would eat the more visible moths more than the less visible ones, but this certainly did not prove that predation by birds was the reason that dark moths had replaced light moths in the woodlands throughout England and continental Europe. A similar phenomenon is well known to biologists who research small mammals in fields and forests. These scientists use live traps baited with food to capture small mammals, which they measure, mark, and release. After a couple of times, the mammals begin to learn that a trap is a good and safe place to get food. They become what field biologists call *trap-happy*. Because these individuals seek out the traps, trapping no longer provides a random sample of the small mammals in the field. Field biologists take special precautions against making their research animals trap-happy. Were Kettlewell's birds doing something similar to trap-happy mammals?

Scientists began to notice other problems with Kettlewell's work:

- The principal predator upon the moths may be bats, catching the moths by sonar at night, rather than birds during the day. If so, the color of the moths would not matter.
- There is a problem with claiming that dark moths have no place to hide in an unpolluted woodland. Dark moths can always find a dark space to hide from birds.
- The return of the light moths after air pollution was brought under control was very rapid. The dark moths had almost completely replaced the light ones by the early 20th century; but now, the light moths are increasing in abundance so fast that the dark forms may approach extinction by the early 21st century. The lichens, against the surfaces of which the light moths supposedly obtain camouflage, are no longer being killed by air pollution as much, but it takes decades for lichens to grow back. The death of lichens may have helped explain the decline of the light moths; regrowth of lichens perhaps cannot explain the return of the light moths.

Such doubts about the peppered moth story accumulated. A 1987 article in *New Scientist* announced the "exploding" of the "myth of the melanic moths." Nobody doubted that dark moths had become more common in polluted decades and light moths more common after air pollution was brought under control, or that this was an example of natural selection; but many scientists began to dismiss bird predation as the cause of this change. Perhaps the dark moths were better able to resist the pollution; if this is the case, the dark color was simply a corollary to the evolution of resistance to chemicals, not very different from the numerous examples of the evolution of pesticide resistance in moths and other insects (see RESISTANCE, EVOLUTION OF). If this is the case, natural selection may be acting primarily at the caterpillar stage rather than the adult stage. The rapidity of the change between dark and light forms in the moth populations also suggests that it was not caused by direct action upon the color genes. Theodore Sargent found that he could get otherwise identical moths to develop light and dark forms depending on the maturity of the food materials that they ate. Perhaps the change from dark to light, and from light to dark, in the moth populations was an induction of genes that were already present (plasticity) rather than an actual change in the genetic makeup of their populations (see ADAPTATION). If the color change in the moth populations resulted from induction, not only was it not caused by bird predation, but it was not even natural selection.

The peppered moth research conducted by H. B. D. Kettlewell was a monumental amount of work. It is no wonder that nobody has tried to do the entire project again, this time correcting the flaws that have been uncovered. Evolutionary biologist Michael Majerus continues research into this system. There are many variables that simply cannot be controlled in these natural populations, particularly the fact that once the moths are released they could fly anywhere. In order to study natural selection, it may be necessary to study a population of organisms in which the entire population can be continuously monitored. This is precisely what evolutionary biologists Peter and Rosemary Grant have done with DARWIN'S FINCHES. Their studies show that the body and beak size in finch populations change as the availability of large vs. small seeds changes. They can study every bird, and they can explain the relationship between beak size and seed size. Their studies, rather than those upon the peppered moths, are now considered to be perhaps the best example of natural selection in action in the wild.

Further Reading

Grant, Bruce S. "Fine tuning the peppered moth paradigm." *Evolution* 53 (1999): 980–984.

Hooper, Judith. *Of Moths and Men: The Untold Story of Science and the Peppered Moth*. New York: Norton, 2002.

Majerus, Michael. *Melanism: Evolution in Action*. New York: Oxford University Press, 1998.

Permian extinction The Permian extinction was the biggest mass extinction event in the history of life (see MASS EXTINCTIONS). It has been known since the work of geologist John Phillips in the 1840s that the end of the Permian period was marked by a major turnover in marine animal species; only recently, however, have the magnitude, and the causes, of this turnover become apparent. According to dates published in 2004, the Permian extinction occurred 252.6 million years ago and marked the end of the PERMIAN PERIOD and the PALEOZOIC ERA. Paleontologist Douglas Erwin called it the "mother of all extinctions." Approximately 90 percent of species died during and immediately after this event (in the early TRIASSIC PERIOD). In contrast, the extinction event that occurred at the end of the Cretaceous, and which exterminated the dinosaurs and many other species (see CRETACEOUS EXTINCTION) killed only about 50 percent of the species on Earth. Life on Earth recovered from the Permian extinction event, but it took nearly 100 million years to regain its former diversity. Though the major groups of organisms that had dominated the Earth during the Permian (such as seed plants, mollusks, fishes, amphibians, and reptiles) survived, most of the species did not; the Triassic period began with a vastly impoverished set of species, and the later Triassic had an almost entirely new set of species within these groups.

Geological deposits that span the boundary between the Permian and Triassic periods are found in China, the Karoo Basin of South Africa, and in Russia. The interpretation of the information from these deposits needed to await the refinement of RADIOMETRIC DATING methods. There was worldwide evidence of a massive turnover in species makeup, but only by precise radiometric dating could scientists be sure that this turnover had happened at the same time all over the world, at or immediately after the Permo-Triassic boundary. It also awaited geopolitical developments that allowed scientific teams from Russia and America, and China and America, to work together on detailed studies.

Consider the geological evidence of what happened at the end of the Permian:

- One geological indicator is the dark-colored deposits at the Permo-Triassic boundary. The dark color is an indicator of the lack of oxygen gas (O_2) at the time and place the sediments were deposited: The iron is dark green instead of bright red, and undecomposed organic matter darkens it further to black. Such deposits occur today at the bottoms of many ponds and swamps. Some of the deposits at the Permo-Triassic boundary contain pyrites, an iron-sulfur mineral that forms only in the absence of oxygen. The worldwide black slime layer suggests worldwide anoxia, a radical event in Earth history. Today, oxygen levels in the atmosphere remain very steady at about 21 percent. Some estimates put the atmospheric oxygen content as high as 35 percent during the CARBONIFEROUS PERIOD. Toward the end of the Permian period, however, the oxygen content may have dropped as low as 15 percent. According to geologists Raymond Huey and Peter Ward, this would have restricted vertebrates to very limited habitats at low elevation where air pressure was highest. Almost half of the land area of the earth would have been, they claim, uninhabitable by large animals. The black "death bed" layer at the end of the Permian represents between 10,000 and 60,000 years of time.
- Another geological indicator is the deposit, during the very early Triassic, of coarse sediments that were washed down violent rivers; this sometimes includes very large boulders that were moved several hundred kilometers from their point of origin. The structure of river channels also suggests that rapid erosion was taking place, more rapid than practically anyplace on the Earth now, and it was occurring everywhere. This would indicate a worldwide reduction of forest cover.
- Another geological indicator is the worldwide shift in oxygen and carbon isotope ratios. The ratio of oxygen isotopes acts as a permanent record of temperature. The seemingly worldwide decrease in oxygen isotope ratio (see ISOTOPES) suggests a massive global warming, averaging nine degrees F (16°C). There was also a global shift in carbon isotope ratios. Plant photosynthesis prefers one carbon isotope over another as it removes carbon dioxide from the air to make sugar; thus organic matter has a different carbon isotope ratio from inorganic carbon-containing molecules. The worldwide decrease in carbon isotope ratios at the Permian boundary suggests a worldwide decrease in plant growth.

There is also fossil biological evidence:

- During the Permian, there had been fossils of plant parts from the ferns and conifers that were the dominant plants at that time. Fossil pollen was most abundant, since pollen is produced in large quantities and spreads far from the source plant and has a coat that resists decay, but at the Permo-Triassic boundary, fungus spores are very abundant. This may mean that the world was rotting. Something had killed most of the plants, and mold was growing all over them.

- During the early Triassic recovery period, plant pollen reappeared, but it was mostly pollen from small plants such as club mosses that could grow in recently ravaged landscapes. Usually, in every geological period (until recently), coal was forming from vast swamps of plants somewhere in the world. The sediments of the first 20 million years of the Triassic, however, stood out for the absence of coal deposits. The "spike" of fungus spores, and the "coal gap," are biotic indicators of a severe devastation of forest growth.
- The Permo-Triassic boundary is also characterized by a "chert gap." Chert is a rock formed from the ooze of billions of microscopic shelled animals accumulating on the ocean floor. The ooze from which chert forms is found somewhere in the deposits of almost every geological age—except at the Permo-Triassic boundary.
- Stromatolites reappeared in the fossil record immediately after the Permo-Triassic boundary. Stromatolites are layers of cyanobacteria and were common in PRECAMBRIAN TIME but became scarce after the evolution of aquatic grazing animals. Today they are found only in shallow seas where the water is too salty for most of the animals that would otherwise eat them. The reappearance of stromatolites at the end of the Permian suggests that many aquatic grazing animals had died.

All these things appear to have happened right at the Permo-Triassic boundary. Geologists also studied the fossil species that lived in the late Permian and early Triassic, to reconstruct how the living world had, in fact, changed at that boundary.

The Permo-Triassic deposits in China were formed under shallow marine (continental shelf) conditions. They contain a record of what happened to marine organisms. Most species of marine organisms that are preserved as fossils are microscopic plankton with calcium or other hard shells. Their shells preserve well, and the organisms themselves were already living in or near the sediments (see FOSSILS AND FOSSILIZATION). When scientists reconstructed the durations of more than 300 marine animal species from the Chinese rocks, they found that 94 percent of them vanished from the record soon before, at, or soon after the boundary. Furthermore, the deposits right before the boundary show evidence of extensive burrowing by a large number of marine worms; no such evidence occurs at or immediately after the boundary. Therefore toward the end of the Permian and at the beginning of the Triassic, there was a collapse in marine life.

Scientists who studied the Chinese deposits have pointed out that many species may appear to have vanished before the end of the Permian, simply because by chance their fossils were not preserved during the last few thousand years of their existence; and that many species may appear to have vanished after the end of the Permian, simply because by chance their fossils were washed upward from the shallow ocean floor and got mixed in with later, Triassic deposits. These arguments would suggest that the Permian extinction was a relatively sudden event, rather than spread out over a long period.

Regardless of how sudden the event was, it had a drastic effect on marine animal life. A Permian seafloor community full of coral reefs, plankton, and many invertebrate animals including the last of the fading dynasty of TRILOBITES was replaced by an early Triassic seafloor community consisting of very few species. For example, whole early Triassic rocks are formed from small shells of just a few species of mollusks that lived right after the disaster. One of the types of animals that survived was the *Lingula* brachiopod shell. Its survival tells a story. It has survived 500 million years and survives today, because it burrows deep in the mud and can survive long periods with severely restricted food and oxygen. Among the swimming animals, many species of bony fishes and sharks also died, to be replaced by other, smaller fish species. Both among the mollusks and the fishes, the survivors were smaller than those that had dominated the Permian seas. In short, according to paleontologist Michael Benton, the survivors appeared to be those which, partly because they were small, may have needed less oxygen. Paleontologist Douglas Erwin disputes the correlation between ability to tolerate anoxia and success at surviving the Permian extinction.

Meanwhile, what was happening on land? The Karoo Basin deposits in South Africa formed from shallow lake sediments that washed northward from a mountain range, now worn away, in what is now Antarctica. The vertebrate fossils in the late Permian Karoo deposits include two amphibian and 72 reptile species. These diverse reptiles included some that ate plants, some that ate insects, some that ate one another; some large, some small; a complex community of interacting reptile species. However, the early Triassic deposits have only 28 terrestrial vertebrate species. Since most of them are not the same species that existed in the late Permian, this represents a drastic turnover of species composition. Rather than a diverse community of interacting species, many of the vertebrate skeletons from the early Triassic are of one kind of reptile: *Lystrosaurus,* whose generalist eat-anything niche has made some scientists call it the pig of the ancient reptile world. The Karoo animal world soon after the Permian extinction was crawling with *Lystrosaurus* and had a few other, rare, reptiles. This pattern is seen in deposits not only from South Africa but also from South America, Antarctica, India, China, and Russia. A complex Permian community of interacting species that specialized on their means of living was replaced by a worldwide Triassic landscape dominated by one kind of animal that did a little bit of everything. Overall, 27 of 29 reptile families and six of nine amphibian families became extinct. There was a similar turnover in plant species: 140 plant species, in 11 families, from the Permian were replaced by fewer than 50 species, in only three families, in the Triassic.

Therefore, at the end of the Permian, something killed most of the plants; partly as a result, the atmosphere nearly ran out of oxygen, and the landscape eroded severely. What could have caused such a severe event? Several causes have been suggested:

Loss of continental shelf area. One early suggestion, from the 1950s, was that smaller continents collided together to form the world supercontinent of Pangaea (see CONTINENTAL DRIFT). Since many species of organisms known from the fossil record lived in shallow water, the collision of several continents into one would have drastically reduced the amount of coastline environment in which they lived. A massive continent, furthermore, would have had more severe terrestrial climates. Oceans ameliorate temperature extremes, which is why the summers are hotter and winters are colder in Minneapolis than in Seattle. The interior of Pangaea, then, would have had extreme climatic conditions that would have caused at least some extinctions. Finally, species that had evolved separately might have been thrown together into a cosmopolitan mix, in which a few "winners" would outcompete "losers" that had formerly been dominant in their own separate lands. While it is beyond dispute that Pangaea formed during Permian times, many scientists have concluded that the resulting changes in terrestrial climate and marine habitat would not have been great enough to cause the Permian extinction. It certainly would not account for the worldwide loss of oxygen and massive death of vegetation.

Asteroid impact. Could the extinction have been caused by an asteroid, as was the case with the Cretaceous Extinction? Scientists are much less likely to discover evidence of such an impact in deposits 250 million years old than in deposits 65 million years old. However unlikely it might seem, evidence of a large asteroid impact was recently found: the Bedout Crater off the coast of Australia. The date of this crater corresponds closely to the time of the Permian extinction.

Volcanic eruptions. A massive set of volcanic eruptions known as the Siberian Traps occurred at the Permo-Triassic boundary. ("Traps" comes from the Swedish for staircase or steps, referring to the successive layers of lava flow.) These eruptions continued for a few million years. They cover an area of Siberia larger than the entire European Community. The gases ejected from these eruptions could have caused worldwide devastation. Sulfur dioxide (SO_2) reacts with water to produce sulfuric acid, a component today of acid rain. Severe acid rain may have killed much of the vegetation on land and photosynthetic organisms in the oceans, causing the collapse of food chains in both. Other large volcanic eruptions during Earth history were not associated with mass extinctions, but the Siberian Trap eruptions may have had a worldwide impact because they ejected more sulfur than other volcanic eruptions. The lava from these eruptions is rich in sulfur minerals, suggesting that these eruptions would have ejected even more sulfur dioxide than most volcanic eruptions observed today. Volcanoes also eject large amounts of carbon dioxide (CO_2), a greenhouse gas that leads to global warming (see GREENHOUSE EFFECT). Isotope ratios suggest a nine degree F (16°C) increase in global temperature. The resulting death of plants could have caused the worldwide plunge in oxygen levels.

Release of methane. Paleontologist Michael Benton suggests, though without direct evidence, that the moderate global warming caused by the volcanoes caused yet another set of catastrophes to occur: gigantic global burps. Deep underneath continental shelf sediments, especially in polar regions, there are today (and may have been during the Permian) very large deposits of methane hydrate, which is an ice-like combination of water and natural gas. Pressure and cold temperature keep the methane in a solid state, but if the ocean waters became warm, the methane might evaporate explosively, bubbling quickly through the ocean and into the atmosphere. There is evidence that such "burps" have occurred in the geologically recent past: an eruption 55 million years ago apparently caused a global warming of about 3–4°F (5–7°C) over a 10,000 year period. Methane, though short-lived in the atmosphere, is a strong greenhouse gas. By enhancing the global warming that was already going on, these methane eruptions could have started a positive feedback loop in which more global warming caused the release of even more methane. Methane reacts with oxygen, and this would have worsened the already dire problem of anoxia. When methane reacts with oxygen, it produces carbon dioxide, which is also a greenhouse gas.

The Permian extinction had some permanent effects on the history of life. During the Permian, brachiopod shells were relatively common, and mollusk shells were relatively rare; while a few million years after the extinction event, bivalve mollusk shells evolved into an explosive radiation of new species, while the brachiopods remained, and remain, rare. This occurred not because, under normal seafloor conditions, mollusks were superior to brachiopods, but because mollusks either got lucky, or else most of them were better adapted to the post-disaster world of the earliest Triassic.

The implications of the Permian extinction for the future of the Earth are both good and bad. Many biologists believe that human activity is causing a sixth mass extinction. Humans are causing species to become extinct at a rate as great as what occurred in previous mass extinctions (see BIODIVERSITY). The bad news is that continued destruction of vegetation can, in fact, lead to massive soil erosion and, if this destruction continues yet further, can even contribute to a depletion of oxygen from the atmosphere. At the end of the Permian, fossil evidence of plants (e.g., pollen) nearly vanished, and so did oxygen. There is no need to speculate that life on Earth is strongly dependent on the work of plants: The "experiment" has in fact been done. The good news is that the Earth can recover from an almost total extinction event. No matter what humans do to the Earth, and probably no matter what happens to it from other causes, except for the final explosion of the sun several billion years from now, the Earth will recover. But this good news is also bad. It would take, from the viewpoint of human history, forever for the Earth to recover. The human economy is dependent upon the continued stability and smooth operation of natural systems. For humans, even a slight change in global temperature would spell agricultural and economic disaster. Because civilization depends upon the exploitation of natural resources, even the slightest interruption of these resources would send humankind, very few of whom know how to survive in the wild, into a desperate tailspin. Humans would probably not have survived the Permian extinction.

Further Reading

Benton, Michael J. *When Life Nearly Died: The Greatest Mass Extinction of All Time.* London: Thames and Hudson, 2003.

Department of Geological Sciences, University of California, Santa Barbara. "Evidence of meteor impact found off Australian coast." Available online. URL: http://beckeraustralia.crustal.ucsb.edu/. Accessed March 22, 2005.

Erwin, Douglas H. *Extinction: How Life on Earth Nearly Ended 250 Million Years Ago.* Princeton, N.J.: Princeton University Press, 2006.

Grice, Kliti, et al. "Photic zone euxinia during the Permian-Triassic superanoxic event." *Science* 307 (2005): 706–709.

Huey, Raymond B., and Peter D. Ward. "Hypoxia, global warming, and terrestrial late Permian extinctions." *Science* 308 (2005): 398–401.

Ward, Peter D., et al. "Abrupt and gradual extinction among Late Permian land vertebrates in the Karoo Basin, South Africa." *Science* 307 (2005): 709–714.

Permian period The Permian period (290 million to 250 million years ago) was the sixth and final period of the PALEOZOIC ERA (see GEOLOGICAL TIME SCALE). It followed the CARBONIFEROUS PERIOD. At the beginning of the Paleozoic era, almost all life was aquatic; by the end of the Paleozoic era, forests filled with vertebrates and insects covered large areas of dry land. The Permian period had the greatest diversity of species of the entire Paleozoic era. It ended with the PERMIAN EXTINCTION, the most important of the MASS EXTINCTIONS in the history of the Earth.

Climate. The climate of the Permian was cooler and drier than that of earlier periods of the Paleozoic, especially in the center of the massive continent of Pangaea. Today, deserts with extreme temperatures are found in the middle of the Eurasian continent and were probably found in the middle of Pangaea as well. Glaciation occurred in the southern portions of Pangaea. Then global warming, caused initially by volcanic eruptions, occurred near the end of the Permian period, which contributed to massive extinctions.

Continents. All of the continents collided to form one massive continent called Pangaea.

Marine life. Marine invertebrate life had reached its greatest diversity to that time, with many species in all of the phyla that are known from the fossil record. Some invertebrates that are now rare, such as stalked echinoderms and brachiopods, were common in the Permian. Reefs, with an abundance of species, were common. The exception was the trilobites, which had gradually declined from being the dominant marine invertebrates from the Cambrian period onward. Diversity was also high among the vertebrates. Many species of fishes had evolved in both marine and freshwater environments, including bony fishes, sharks, lungfishes, and coelacanths (see FISHES, EVOLUTION OF). There were no aquatic reptiles or aquatic mammals.

Life on land. Forests covered much of the land surface:

- *Plants.* The tree-sized horsetails and club mosses that had characterized the "coal swamps" of the Carboniferous were close to extinction (see SEEDLESS PLANTS, EVOLUTION OF), replaced by tree-sized ferns, seed-ferns (a type of plant now extinct), and conifers (see GYMNOSPERMS, EVOLUTION OF). There were no flowering plants.
- *Animals.* The drier conditions encouraged the diversification and spread of reptiles. There were two major groups of reptiles: the synapsids, and the diapsids. There were no dinosaurs, but some of the reptiles of the Permian were rather large. None of these reptiles were of recognizably modern groups; however, the diapsids were the ancestors of most of the modern animals commonly called reptiles, as well as extinct reptiles such as archosaurs and dinosaurs (see REPTILES, EVOLUTION OF). The diapsids were also the ancestors of birds. The synapsids are today extinct except for one line of descendants: mammals. Some Permian synapsids grew as large as hippos. The most famous Permian synapsid was *Dimetrodon,* whose ribs formed a huge "sail" along its back. Some synapsids had many mammal-like characteristics; for example, unlike diapsids and modern reptiles, they had several kinds of specialized teeth. They may also have had whiskers: their skulls have small nerve openings where these whiskers would have been. If they had whiskers, they had hair, and therefore were probably warm-blooded. The dissipation of body heat may have been the main function of the sail along the back of *Dimetrodon.* In other words, although these synapsids were not mammals, they closely resembled them. A nearly complete set of synapsid transitional forms has been found between reptiles and mammals (see MAMMALS, EVOLUTION OF).

Extinctions. The Permian extinction brought many evolutionary lineages to an end and severely reduced the abundance of many others. Scientists will never know what might have happened if evolution could have continued uninterrupted. Would the synapsids have developed into the dominant vertebrates, ushering in the age of mammals 250 million rather than 65 million years ago? Would this have prevented the evolution of the dinosaurs? These questions will never be answered.

Further Reading

Kazlev, M. Alan. "Permian Period." Available online. URL: http://www.palaeos.com/Paleozoic/Permian/Permian.htm. Accessed May 2, 2005.

Museum of Paleontology, University of California, Berkeley. "Permian Period." Available online. URL: http://www.ucmp.berkeley.edu/permian/permian.html. Accessed May 2, 2005.

phenotypic plasticity *See* ADAPTATION.

photosynthesis, evolution of Photosynthesis is the process by which some bacteria, some protists, and most plants make food using carbon dioxide (CO_2) and the energy from sunlight.

Autotrophic cells and organisms make their own food from carbon dioxide, some of which they use themselves (see RESPIRATION, EVOLUTION OF) and some of which gets eaten by *heterotrophic* cells or organisms that cannot make their own food. The two main categories of autotrophic cells and organisms are the *chemotrophic* cells and the *photosynthetic* cells and organisms, which differ from one another in the way that they obtain the energy that they need for making small carbon dioxide molecules into large organic molecules.

- Chemotrophic bacteria obtain energy from inorganic chemical reactions. Some inorganic reactions such as the oxidation of ferrous into ferric iron (rusting) release energy, and there are bacteria that can capture and use this energy. Chemotrophic bacteria use many different inorganic sources of energy, mostly minerals. The importance of bacteria in the production of mineral deposits has only recently been recognized. They do not require sunlight.
- Photosynthetic bacteria and eukaryotic cells obtain energy from light, usually sunlight. They use pigments such as *chlorophyll* to absorb the light energy.

Photosynthesis has two stages: the stage in which light energy is captured, and the stage in which carbon dioxide is fixed into carbohydrate.

The light capture phase. The light capture phase uses light energy to create an electrical current which travels through proteins that have been modified with metal atoms. These reactions therefore require not only a source of light but a source from which to obtain electrons.

In the most primitive forms of photosynthesis, found in some bacteria (see BACTERIA, EVOLUTION OF), the source of the electrons can be hydrogen sulfide (H_2S), which becomes elemental sulfur (S) after the electrons are removed; or hydrogen gas (H_2), or organic molecules such as succinate or malate. The bacteria that use hydrogen or organic molecules can live as heterotrophs in darkness or as autotrophs when light is available. All of these bacteria live in ponds in the absence of oxygen. Because this means that they cannot be near the surfaces of the ponds, they must be able to absorb light very efficiently. Their pigment, bacteriochlorophyll, can absorb infrared radiation that is past the red end of the visible light spectrum. The light absorption system consists of a single system of reactions.

In the more advanced form of photosynthesis, used by cyanobacteria, the source of the electrons is water (H_2O), which becomes elemental oxygen gas (O_2) after the electrons are removed. Light energy is absorbed by chlorophyll molecules.

The light absorption system of the more advanced form of photosynthesis consists of two cycles of reactions that work in series. One of these cycles resembles the single cycle of some of the primitive bacteria; the other cycle resembles the single cycle of other primitive bacteria. The two-cycle advanced system may have resulted from the genetic merger of two different kinds of bacteria (see HORIZONTAL GENE TRANSFER).

Some ARCHAEBACTERIA (such as *Halobacterium salinum,* which lives in very salty water in the absence of oxygen) have a pigment system that absorbs light and uses it to pump ions across membranes. These bacteria are heterotrophic, but they use light as an energy source during times when food molecules are scarce. The pigment molecule is bacteriorhodopsin, similar to one of the visual pigments found in the vertebrate eye. Since the light energy is not used to make carbohydrates, it is not considered a form of photosynthesis.

The bacteria with the primitive light absorption system do not produce oxygen, and today grow in habitats in which the water contains no oxygen (*anaerobic* conditions). This is consistent with their origin in the Archaean era (see PRECAMBRIAN TIME) when the atmosphere of the Earth did not yet contain oxygen. The bacteria with the more advanced light absorption system produce oxygen. Most evolutionary scientists have concluded that it was the oxygen produced by cyanobacteria that filled the atmosphere with oxygen gas during the Proterozoic era and created the *aerobic* conditions that now prevail upon the Earth. Photosynthesis continues to put oxygen gas into the atmosphere. Earth's atmosphere is unique among known planets in having an oxygen atmosphere, and no inorganic process is known that could create an oxygen atmosphere on a planet. Such an atmosphere is a clear indicator of photosynthesis and therefore of life (see GAIA HYPOTHESIS).

The carbon dioxide fixation phase. The carbon dioxide fixation stage involves a cycle of reactions called the Calvin cycle, which produces carbohydrates such as sugar. All advanced photosynthetic cells (and chemosynthetic cells as well) use this or a similar set of reactions. The enzyme that fixes the carbon dioxide, called *rubisco,* may be the most abundant enzyme in the world.

Photosynthesis occurs in the chloroplasts of many protists (see EUKARYOTES, EVOLUTION OF) and almost all plants. The set of light- and carbon-fixing reactions in chloroplasts is nearly identical to the reactions in cyanobacteria. This is not coincidence. Chloroplasts are the evolutionary descendants of cyanobacteria that moved into and formed a mutualistic association with primitive eukaryotic cells (see SYMBIOGENESIS). Therefore it can be said that nearly all of the photosynthesis in the world is conducted by cyanobacteria—either free-living cyanobacteria in oceans, ponds, and rivers, or cyanobacteria that have evolved into chloroplasts inside of the cells of eukaryotic algae and plants.

Most plants have only the Calvin cycle as the carbon dioxide fixation stage of photosynthesis. They are called C_3 plants because the carbon dioxide is first made into a three-carbon molecule. However, some plants (just over 10 percent of plant species) have different forms of carbon fixation.

- C_4 plants have two carbon fixation cycles that both occur during the daytime. The first cycle fixes carbon dioxide into an acid (which has four carbon atoms, hence the name). The second cycle is the Calvin cycle. Therefore it appears that a new carbon fixation cycle has been added onto the Calvin cycle that was used by the ancestors of the C_4 plants. C_4 photosynthesis has evolved as many as 31 times in separate lineages of plants and therefore is found in slightly different forms (see CONVERGENCE). All of the enzymes involved in the first cycle (which produces the acid) are used in other parts of the cell for other purposes. The enzyme that captures the carbon dioxide, *PEP carboxylase,* produces organic acids in all the cells of the plant. Therefore the evolution of C_4 photosynthesis did not require the origin of new chemical reactions but borrowed them from other parts of the cell.

- CAM plants (which have *Crassulacean Acid Metabolism*) also have the same two carbon fixation cycles, but the first one (acid production) occurs at night, and the second one (the Calvin cycle) occurs in the daytime. Like C_4 photosynthesis, CAM appears to have evolved more than once, for it is found in some species in several plant families (such as the cactus, jade plant, and lily families) but not in their common ancestors.

Both C_4 and CAM pathways are adaptations to hot, dry conditions:

- Rubisco, the enzyme that fixes carbon dioxide in the Calvin cycle, is inhibited by oxygen. However, PEP carboxylase, the enzyme that fixes carbon dioxide in the first cycle of C_4 photosynthesis, is not inhibited by oxygen. On a hot dry day, leaves often close their pores (stomata) to prevent excessive water loss. This causes a buildup of oxygen inside the leaves. This oxygen buildup causes the photosynthesis of C_3 plants to slow down but does not inhibit the photosynthesis of C_4 plants. Therefore, in hot dry climates, C_4 plants can continue making food under conditions that would cause C_3 plants to stop making food. Because the carbon fixation in C_4 plants has two cycles rather than just one, it is more expensive; but in hot dry climates the extra expense is worth the cost. However, in cool moist climates, the expense is not worth the cost. Therefore C_4 plants are relatively more common in hot dry climates (usually grasslands) and C_3 plants in cool, moist climates. C_4 photosynthesis has also evolved in some aquatic plants, because carbon dioxide is often scarce in water. The PEP carboxylase helps them obtain more carbon dioxide than would rubisco acting alone.
- CAM plants open their pores and absorb carbon dioxide from the air at night, when conditions are relatively cool and moist. They make it into acid and store the acid until dawn. Then they close their pores and use the acid as a source of carbon dioxide from which the Calvin cycle can produce sugar during the day. This has the advantage of allowing the CAM plants to open their pores at night rather than during the day, but it has a cost: They have to store the acid in their tissues. Most CAM plants are succulents (such as cactuses, jade plants, and agaves). As with C_4 photosynthesis, CAM is worth the cost in hot, dry climates (usually deserts) but not worth the cost in cool, moist climates.

The fact that most C_4 plants are grasses (which did not become abundant until the middle of the CENOZOIC ERA), and that C_4 has evolved several times, suggests that it is of recent origin. CAM appears to also be of recent origin.

Before oxygen gas was abundant in the atmosphere of the Earth, ultraviolet radiation bombarded the surface of the planet. This would probably have killed most life-forms at or near the surface of the ocean. Most scientists believe that the first life-forms evolved below the ocean surface, perhaps at deep sea vents (see ORIGIN OF LIFE). Oxygen gas produced by photosynthesis would have overcome the problem of ultraviolet radiation and allowed photosynthetic organisms to grow at or near the surface. The obvious problem is, what were the photosynthetic organisms doing when the only safe place for them to live was deep in the ocean? The great sensitivity of anoxic photosynthetic bacteria to light was mentioned above. In 2005 scientists announced the discovery that anoxic green sulfur bacteria could carry out photosynthesis over 200 feet (80 m) below the surface of the Black Sea, under conditions that appeared to human observers to be totally dark. Sensitive measurements indicated that the volcanic vents in this region produced light in the visible range, and that the bacteria could use this faint light for photosynthesis. These bacteria were also, unlike other green sulfur bacteria, not killed by oxygen gas. In two ways—their ability to use very faint light and their ability to tolerate oxygen—these bacteria may represent what the first photosynthetic cells were like.

Further Reading

Bohannon, John. "Microbe may push photosynthesis into deep water." *Science* 308 (2005): 1,855.

Nasa Astrobiology Institute. "Terrestrial Powerhouses." Available online. URL: http://www.nai.arc.nasa.gov/news_stories/news_detail.cfm?article=old/powerhouses.htm. Accessed May 1, 2005.

Whitmarsh, John, and Govindjee. "The Photosynthetic Process." Available online. URL: http://www.life.uiuc.edu/govindjee/paper/gov.html#32. Accessed May 1, 2005.

Piltdown man Piltdown man was a fraudulent fossil human that was planted in a gravel quarry in England in 1911. The perpetrator/s of the hoax, which was revealed in 1953, were never definitely identified, although the person who had greatest opportunity was Charles Dawson, a lawyer and amateur archaeologist who was involved in other fraudulent schemes. At that time, the only other fossil humans, aside from those that were clearly in the modern human species (see CRO-MAGNON), were the NEANDERTALS in mainland Europe and Java man (see *HOMO ERECTUS;* DUBOIS, EUGÈNE). Throughout the early period of the study of human evolution, the question persisted: Which came first, upright posture or a large brain? The Neandertal and Java man skeletons suggested that apelike features of the skull persisted even after erect posture had evolved in the human lineage. Piltdown man was, for four decades, accepted as evidence that brain expansion may have preceded the evolution of the rest of the skeleton. Also, this ancient expansion of brain capacity just happened to occur in what is now England, which at the beginning of the 20th century still considered itself the most intellectually advanced nation in the world.

In 1911 Dawson visited the Piltdown gravel quarry near his home and asked the quarrymen to be on the watch for fossils. When he inquired again, they gave him a fossilized bone they had found and which he identified as a fragment of parietal bone, from the upper side of a humanlike skull. Later, Dawson himself found another bone in the quarry, this time a piece of frontal bone from a humanlike skull. He contacted prominent British paleontologist Arthur Smith Woodward, who looked at the fossils himself in 1912. At the quarry, they were joined by the young French cleric and amateur archaeologist Teilhard de Chardin (see TEILHARD DE CHARDIN, PIERRE). As RADIOMETRIC DATING techniques had not been developed, it was unclear just how ancient the Piltdown gravel was. Teilhard de Chardin found an elephant tooth, which demonstrated the antiquity of the gravels, since elephants lived in what is now England many thousands of years earlier when the climate had been tropical. But it was Dawson who just happened to find yet another piece of humanlike skull. The three investigators kept finding more skull fragments, which happened to fit perfectly together. Then, with a single stroke of his geologist's hammer, Dawson sent a fossil jawbone flying into the air. All of the pieces of bone were stained with iron, like much of the gravel in which they were found. The similarity of staining convinced the investigators that all the pieces belonged to the same skull. This could have been proven if the part of the jawbone that articulates with the skull had been preserved, but it was missing.

An exciting announcement of the discovery emerged in the newspapers before Dawson and Smith Woodward presented the actual fragments for other scientists to observe, at the 1912 meeting of the British Geological Society in December. The premature public announcement may have influenced the scientists in attendance toward believing Dawson. The reconstructed skull astounded everyone. The brain was somewhat small but well within the modern human range. The jawbone was U-shaped like that of an ape, rather than V-shaped like a human jaw; and the two remaining teeth also appeared to be very apelike. If this were the case, the scientists concluded, then Piltdown man must have lived long before the Neandertals (as Piltdown man had a more primitive jaw). Both Neandertals and Java man must have been sidelines of evolution, lineages that never developed into modern humans and eventually became extinct (which turned out to be true, but was the only correct conclusion these scientists reached). Human evolution, it appeared to them, began with the brain and began in what is now England. They named Piltdown man *Eoanthropus dawsoni,* or Dawson's dawn-man.

Some doubts remained. German anthropologist Franz Weidenreich said the jawbone looked an awful lot like that of an orangutan. Smith Woodward said that if only they had been able to find a canine tooth, then they could be more certain that Piltdown man really was intermediate between humans and apes. Smith Woodward predicted that this tooth would be intermediate between the large canine tooth of an ape and the small canine tooth of a human. In the 1913 season in the quarry, Teilhard de Chardin found exactly this tooth, and it looked exactly as Smith Woodward had predicted. The tooth was stained just like the other fragments.

For four decades, Dawson's name was enshrined among the great investigators of human evolution. In 1953 investigators revealed that all of the bones were recent in age, and had been stained to look old; some had even been filed to have a convincing shape. The skull fragments were from a modern human, and the jawbone was, as Weidenreich had suspected, from an orangutan. Some historians have suggested that Dawson intended Piltdown as a joke, not a serious hoax, and that he planned to show the world how scientists can be as gullible as anyone else. Historians will never know what he intended to do, as he was killed in 1917 in World War I. Evolutionary biologist Stephen Jay Gould suggested that Teilhard de Chardin was also involved in the hoax; this, also, will remain unknown.

For most scientists and historians, the Piltdown episode remains an interesting story of how investigators need to be skeptical when new evidence appears to contradict previous evidence; as many scientists say, extraordinary claims require extraordinary evidence. Creationists keep the Piltdown story very much alive, implying that the entire fossil record of humans is a patchwork of unreliable fragments and, quite possibly, hoaxes (see CREATIONISM). An episode such as the Piltdown affair could not happen today. Microscopic, chemical, and radiometric methods available today would immediately reveal a modern Piltdown to be filed, stained, and modern.

Further Reading

Gould, Stephen Jay. "Teilhard and Piltdown." Section 4 in *Hen's Teeth and Horse's Toes.* New York: Norton, 1983.

Walsh, John Evangelist. *Unraveling Piltdown: The Science Fraud of the Century and Its Solution.* New York: Random House, 1996.

plate tectonics The movement of the crustal plates of the Earth's surface is called plate tectonics. The Earth's surface is fractured into eight major plates and several smaller ones (see figure on page 316), which move relative to one another. The plates appear to be propelled by flowing magma underneath the surface. The ocean floors are formed of heavier material, while the continents, formed of lighter rocks, sit on the surfaces of the heavier rocks. The lighter continental rock accumulated during Earth's early history, especially during the Archaean Eon about 3 billion years ago. Plate tectonics is now understood as the driving force of CONTINENTAL DRIFT. The theory of continental drift, now confirmed by the mechanism of plate tectonics, has had a revolutionary impact on explaining the evolutionary patterns of life on Earth.

Oceans can form or widen because new ocean floor crust forms along the fracture (a *mid-ocean ridge*) where crustal plates separate. Underwater volcanic eruptions exude new plate material along the mid-ocean ridge. As a result, the youngest crust material is right next to the mid-ocean ridge, and the oldest material is furthest from the ridge. Geologists can determine the actual ages of volcanic rocks by radioactivity (see RADIOMETRIC DATING) and the relative ages of rocks by determining the orientation of the magnetic fields of the rocks (see PALEOMAGNETISM). Both the ages, and the magnetic field patterns, of the rocks of the Atlantic Ocean floor form bands of increasing age away from the Mid-Atlantic Ridge. When geologists Drummond Matthews and Fred Vine

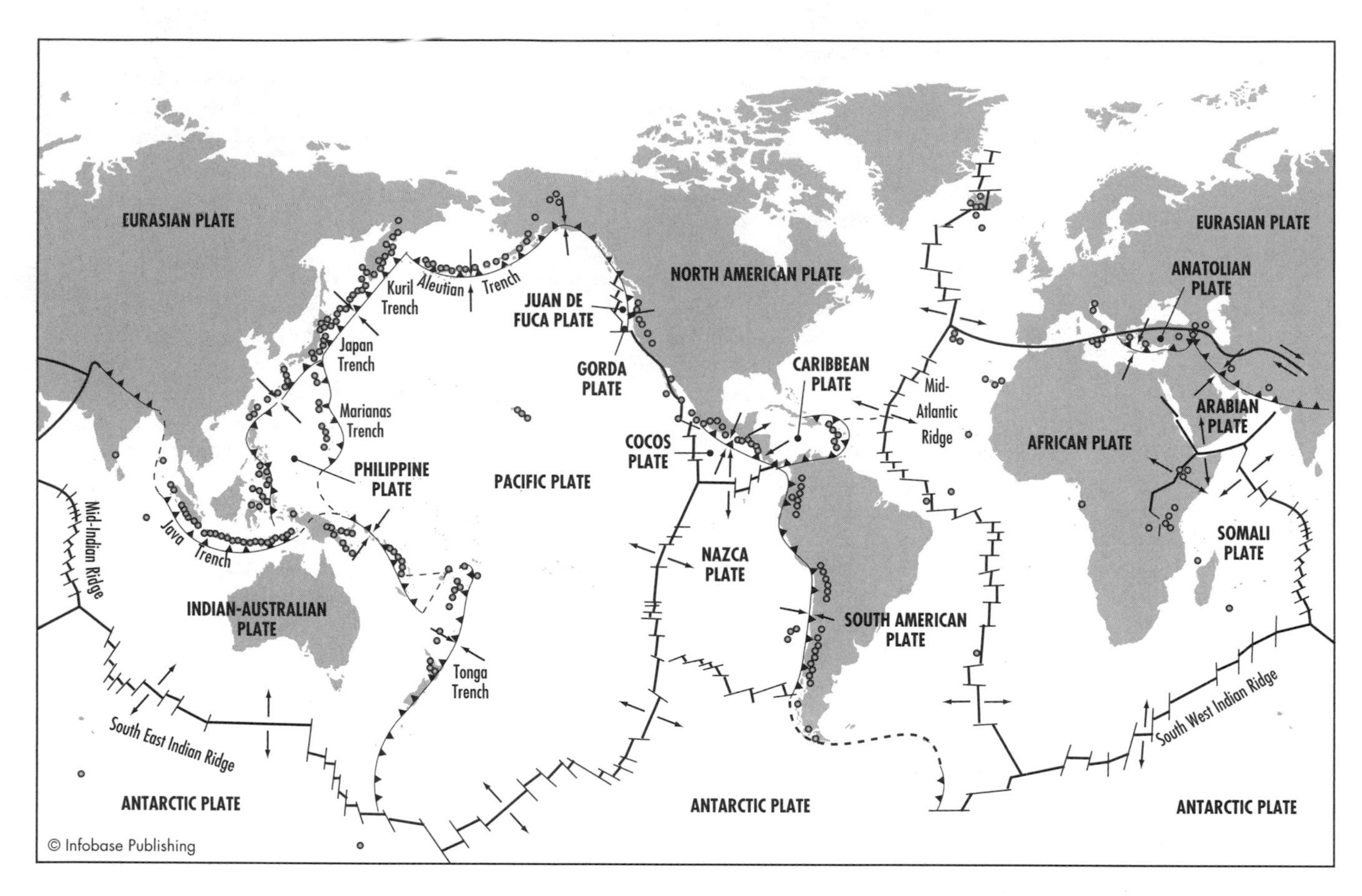

The Earth's surface consists of eight major and several smaller crustal plates. Arrows indicate direction of plate movement. Most volcanoes (dots) occur near plate interfaces. Most earthquakes also occur in the same locations as the volcanoes.

from Cambridge University presented this evidence in 1963, their paper was denied publication at first, but when the evidence was finally accepted, it became perhaps the single most important factor in convincing geologists that the theory of plate tectonics was correct.

The Eurasian Plate and the North American Plate have been separating for more than 120 million years, at about the same rate that fingernails grow. Eurasia and North America, parts of which remained connected until about 50 million years ago, are now separated by the North Atlantic Ocean. During this same time, South America and Africa have separated. The Mid-Atlantic Ridge runs along the ocean floor like a gigantic backbone. In Iceland, the Mid-Atlantic Ridge emerges.

When crustal plates collide, two things happen. First, one of the plates is thrust underneath the other plate, a process known as subduction. The subducted materials melt and recirculate in the magma. This happens primarily with the heavier materials that make up the ocean floors. Because ocean floor crust is recirculated by subduction, no portion of the ocean floor is much older than 200 million years. Second, the continents, composed of lighter material, usually move up rather than down, forming mountain ranges. Because continental rock is less frequently subducted than ocean floor material, many continental rocks are billions of years old. One of the subduction zones occurs where the Pacific Plate is thrust beneath the Indian-Australian Plate, forming the Marianas Trench, which at over seven miles (11 km) is the deepest part of the ocean. The plate that carries the subcontinent of India has collided with the Eurasian Plate, resulting in the relatively rapid and dramatic rise of the Himalayas.

Two plates can rub against one another. For example, the Pacific Plate is moving northwest and rubbing against the North American Plate, along the famous San Andreas fault line.

Geological events accompany tectonic movements, such as:

- All three kinds of plate movements can produce volcanic eruptions. The locations of volcanoes correspond closely to the fault lines where plates meet. Most famous is the *ring of fire* around the Pacific Ocean, with volcanoes in South America, Mexico, the western United States, Alaska, Japan, and southeast Asia. Volcanic gas consists largely of carbon dioxide and water vapor, from the ocean water and limestone that are subducted along with ocean floor rock.
- Earthquakes occur as plates move. When two plates rub against one another, as in the San Andreas fault, the movement is uneven; they can stick, then slip suddenly, resulting in major earthquakes. For this reason, frequent small

earthquakes are a welcome process that reduces the risk of major earthquakes.

Earthquakes and volcanic eruptions can also occur in the middle of crustal plates, usually at volcanic *hot spots* where magma rises with great strength. The Hawaiian Islands, in the middle of the Pacific Plate, formed in this manner. The hot spot is currently located underneath the southeastern side of the main island where eruptions are forming a new island, already named Loihi even though it remains thousands of feet below the ocean surface. It continues to contribute to eruptions of Mauna Kea and Mauna Loa, the large volcanoes on the main island. The Pacific Plate is moving northwest. The other islands used to be directly over the hot spot, but as the plate moved, the volcanoes moved with it and became extinct. Therefore, as one moves northwest along the chain of islands, one encounters successively older volcanoes. Other hot spots have caused eruptions of lava and hot water, as well as earthquakes, in the Rocky Mountains (the geysers of Yellowstone in Wyoming, and the extinct volcanoes of Capulin in New Mexico) and along the New Madrid fault in northwest Missouri.

One cannot assume that continental drift always occurs at the same rate. The rate of crustal movement may be influenced by the rotation of the Earth, which causes matter to migrate centrifugally toward the equator. However, continental drift will continue to occur. In about 10 million years Los Angeles will pass by San Francisco (or what remains of them), and in 60 million years Los Angeles will be falling down a trench just south of Alaska.

Further Reading

Oreskes, Naomi. *Plate Tectonics: An Insider's History of the Modern Theory of the Earth.* New York: Westview, 2003.

Pleistocene epoch *See* QUATERNARY PERIOD.

Pleistocene extinction During the latter part of, or immediately after, the Pleistocene epoch (see TERTIARY PERIOD), many species of large mammals became extinct at various locations around the world. Not without controversy, these extinctions have been attributed at least in part to hunting by humans.

At the end of the most recent ICE AGE, North America was the home of a much greater range of large mammals than is known from historical times, for example (see table):

- In addition to grizzly, black, and brown bears, there was also a short-faced bear.
- In addition to cougars, there were lions and cheetahs resembling African species, and a saber-toothed cat.
- In addition to the extant species of beavers, bison, and wolves, there was a larger species of each.
- There were large mammal species otherwise known only from the Old World, such as camels and horses.
- There were large mammal species different from their modern relatives, such as mammoths, mastodons, gomphotheres, and giant sloths.

The disappearance of these large mammals had a tremendous effect on the ecology of North America. In particular, numerous plant species produce fruits that can no longer be eaten in their entirety by any extant species of animal (see essay, "What Are the 'Ghosts of Evolution'?").

Some North American Mammals That Became Extinct at the End of the Pleistocene Epoch

Category	Common name	Scientific name
Bear	Short-faced bear	*Arctodus simus*
Beaver	Giant beaver	*Castoroides ohioensis*
Bovines	Shrub ox	*Eucerotherium collinum*
	Harlan's muskox	*Boötherium bombifrons*
Camels	Yesterday's camel	*Camelops hesternus*
	Large-headed llama	*Hemiauchenia macrocephala*
Cats	Saber-toothed cat	*Smilodon fatalis*
	American lion	*Panthera atrox*
	American cheetah	*Miracinonyx trumani*
Elephant-like	American mastodon	*Mammut americanum*
	Columbian mammoth	*Mammuthus columbi*
Ground sloths	Jefferson's ground sloth	*Megalonyx jeffersonii*
	Shasta ground sloth	*Nothrotheriops shastensis*
	Harlan's ground sloth	*Glossotherium harlani*
Horses	Mexican horse	*Equus conversidens*
	Western horse	*Equus occidentalis*
Pig	Flat-headed peccary	*Platygonus compressus*
Pronghorn	Diminutive pronghorn	*Capromeryx minor*
Wolf	Dire wolf	*Canis dirus*

There are several reasons that these extinctions are blamed on human hunting:

- The extinctions were overwhelmingly of large mammals that would have been prime targets of hunters. Small mammals suffered far lower extinction rates.
- The extinctions occurred mainly in Australia and North and South America, which had not been previously inhabited by humans (see table on page 318). The large mammals would not have evolved, or learned the fear of humans, which would have made them irresistibly easy targets. The humans who arrived in Australia and North and South America were fully modern and had advanced hunting capabilities. In Eurasia and Africa, in contrast, humans and large mammals had coevolved for thousands or millions of years, and the first humans in those regions had only primitive hunting capabilities.
- The timing of the extinctions corresponded closely to the first arrival of humans: about 50,000 years ago in Australia, and about 11,000 years ago in North and South America.
- Similar extinctions have occurred within historical times as a result of the first contact of humans with inexperienced large mammals and birds, such as the eradication of the moa birds by the first Polynesians in New Zealand, the

Percent of Mammal Genera Becoming Extinct upon Arrival of First Humans (from Barnosky)

Continent	Extinction
Africa	18%
Australia	88%
Eurasia	36%
North America	72%
South America	83%

eradication of elephant birds and giant lemurs by the first immigrants to Madagascar, and the eradication of passenger pigeons and near-eradication of bison by white Americans.

Difficulties have been noted, and some responses formulated, regarding this "Pleistocene overkill hypothesis":

- In North America, the arrival of humans was not the only thing that was happening 11,000 years ago. The most recent Ice Age was also ending, accompanied by rapid climatic changes. Defenders of the overkill hypothesis point out that the climate was becoming warmer, which should have encouraged the survival, not the extinction, of animals. The abundance and diversity of animals during the Ice Ages, however, suggests that a warmer climate was not necessarily an improvement for these animals. Any climatic change that would have altered the food base for these animals may have undermined their populations.
- Some large mammal species were already in decline by the time humans arrived, for example the large bison of North America.
- Evidence is lacking for massive kills of large mammals. Arrowheads have been found in mammoth bones, but only occasionally, and mass grave sites of large animals are very uncommon.
- Native Americans, on whom the Pleistocene overkill in the New World is blamed, do not have a mass slaughter approach to hunting. In fact, some Native American traditions are famous for spiritual preparation and gratitude before killing an animal, as well as careful use of the animal. However, the Native Americans that the Europeans encountered beginning just before the 16th century were perhaps very different from those that arrived in America at the end of the last Ice Age. In particular, Native Americans may have developed a conservation ethic in response to the collapse of the hunting stock caused by their careless ancestors.

Paleontologist Anthony Barnosky performed what may be the most thorough analysis of the Pleistocene extinctions to date. He confirmed that large mammals died in disproportionately large numbers, and at times that corresponded (within a thousand years) to human arrival. He concluded that neither climate change nor overhunting could by themselves have caused the extinctions. Together, however, they could have produced the Pleistocene extinctions. It would not have been necessary for either climate or hunters to have killed every last animal. Once a population becomes sufficiently small, it becomes vulnerable to extinction. Overhunted animal populations were vulnerable to extinction from climate changes, and climate changes may have reduced animal populations to the point that hunting could send them spiraling to oblivion.

As the Earth enters into what many scientists consider the sixth of the MASS EXTINCTIONS as a result of human activity, the lessons of the Pleistocene extinctions may be essential. Now, as then, it is not necessary to kill many animals in order to drive them to extinction. At the end of the Pleistocene, climate change reduced their habitats and made them vulnerable to hunting. Today, habitat destruction for agriculture and human habitation is the principal threat to the survival of most threatened species (see EXTINCTION).

Further Reading

Barnosky, Anthony D. "Assessing the causes of late Pleistocene extinctions on the continents." *Science* 306 (2004): 70–75.

Martin, Paul S. *Twilight of the Mammoths: Extinctions and the Rewilding of America.* Berkeley: University of California Press, 2005.

———, and R. G. Klein. *Quaternary Extinctions: A Prehistoric Revolution.* Tempe: University of Arizona Press, 1984.

Miller, Gifford H., et al. "Ecosystem collapse in Pleistocene Australia and a human role in megafaunal extinction." *Science* 309 (2005): 287–290.

Mosimann, J. E., and Paul S. Martin. "Simulating overkill by Paleoindians." *American Scientist* 63 (1975): 304–313.

Pielou, E. C. *After the Ice Age: The Return of Life to Glaciated North America.* Chicago: University of Chicago Press, 1992.

population All of the members of one species within a defined area, such as a pond, a forest, a nation, or the world, constitute a population. Within this area, the individuals of the population are potentially able to interbreed, which is generally considered to define species membership. This is the *biological species concept* (see SPECIATION).

Diversity in Populations

Being members of the same species is the only thing that individuals in a population can be sure to have in common. Populations, whether of *Streptococcus pyogenes* bacteria or *Homo sapiens* humans, contain great diversity:

- Individuals within a population differ in age and size.
- Individuals within a population differ in sexual characteristics. Most animals are either male or female; in some animal species, however, individuals have both male and female organs. Most plants have reproductive structures (such as flowers) that contain both male and female components; however, in some plant species, there are separate male and female reproductive structures on the same individual, and in some other species, there are separate male and female plants. In fungus and many protist species, there are many different mating types that cannot be easily designated as male or female.

- Individuals within a population differ genetically (see POPULATION GENETICS). Though in most cases all members of a population have the same genes on their chromosomes, individuals differ in the alleles for the genes that they possess (see MENDELIAN GENETICS). Everyone is aware of the astounding and obvious diversity within and between human populations; and yet humans have less genetic diversity than most other species. There is no species in which all the genetic diversity, in all its populations, could be contained within a single pair of individuals gathering into an ark. This genetic diversity is the raw material upon which NATURAL SELECTION and other evolutionary processes work. Evolution, therefore, occurs within populations, because populations contain genetic diversity.

Growth of Populations

Populations grow *exponentially*. In bacteria, one individual divides to become two, each of these two divides again, producing four. Each generation, the population size can double. After 10 generations, there would be 2^{10} (which is 1024) bacteria (the 10 is the exponent). The graph of population growth over time is therefore a curve, not a straight line. In species that reproduce sexually, the calculations are more complex but still follow an exponential pattern. The human mind, accustomed to think in linear terms, can be surprised by population growth. If a nation's human population *doubling time* is 20 years, then a simple linear calculation of the resources that it will need 20 years from now, based on previous years, will fall far short of what it will actually need for a population that is twice as large as it is now.

Some populations grow faster than others:

- Populations with a high *fertility rate* grow faster than those with a lower fertility rate. Fertility rate is the average number of offspring per female. Populations of rabbits, in which females typically have numerous offspring more than once a season, grow faster than human populations. In many poor countries, the fertility rate is much higher (in some cases, up to eight children per family) than in richer countries. A fertility rate of about two results in a population in which births and deaths balance one another—a population that remains the same size.
- Populations with a low *mortality rate* grow faster than those with a higher infant mortality rate. In many populations, most deaths occur among young individuals. In human populations, the death of individuals during their first year is called *infant mortality.* A plant may produce a thousand seeds but only one may grow; the 99.9 percent mortality rate offsets the high fertility rate. Although humans have a lower fertility rate than most other species, they also have a low mortality rate. Even the highest infant mortality rates among human populations are less than 15 percent.
- Populations with a low *age at first reproduction* grow faster than those with a higher age at first reproduction. Rabbit populations grow faster than human populations not only because rabbits have high fertility rates but also because rabbits begin to reproduce during their first year of life.

Exponential growth cannot continue indefinitely. Among the factors that cause population growth to slow down, or even stop, or cause populations to shrink, are:

Density-independent factors. These are factors that would occur with about the same intensity regardless of how large or dense the population is. Weather events (such as floods or droughts) are usually density-independent.

Density-dependent factors. These factors occur with greater intensity in populations that are larger or denser. These include:

- *Disease.* Parasites can spread more rapidly when organisms are crowded together with one another, with their wastes, and with their trash.
- *Competition.* At low density, there is enough room (and therefore enough resources) for all the individuals. However, at high density, resources (such as light, water, and soil space for plants, or food items for animals) can become scarce for the average individual. As a result of competition among individuals, some obtain more resources than others. For example, in a population of plants, there may be a few large ones, which produce most of the seeds, and many small ones that do not produce at all. In an animal population, a dominant male may be responsible for most of the reproduction. In all these cases, not all of the individuals will reproduce as much as they could. Competition can take either subtle or violent forms.

The *carrying capacity* of a population is the number of individuals for which there are adequate resources for survival and reproduction. In practice, it is usually impossible to calculate the carrying capacity of a population, because the individuals can adjust their resource use. At high density, plants adjust their growth and animals adjust their behavior in such a way that resources are used more efficiently. This is perhaps most evident with human beings. The American population uses about one-third of the world's resources. If all human populations used as much energy and materials as Americans, the human carrying capacity of the world would be about 800 million. The current world population, about six and a half billion, far exceeds this because most people in the world use far fewer resources, and use them more efficiently, than Americans. Scientists therefore do not know what the world carrying capacity for humans is, or whether the human population has exceeded it.

The fact that populations grow exponentially but resources do not, an insight first developed by Thomas Malthus (see MALTHUS, THOMAS), was the foundation upon which the theory of NATURAL SELECTION was built (see DARWIN, CHARLES; WALLACE, ALFRED RUSSEL). Malthus used the principles of population to explain human misery; Darwin and Wallace used it to explain the creativity of evolution.

Further Reading

Brown, Lester R., Gary Gardner, and Brian Halweil. *Beyond Malthus: Nineteen Dimensions of the Population Challenge.* New York: Norton, 1999.

Vandermeer, John H., and Deborah E. Goldberg. *Population Ecology: First Principles.* Princeton, N.J.: Princeton University Press, 2003.

population genetics Population genetics is the study of genetic variation in populations. Genes, encoded in DNA, are the only information that passes from one generation to the next in most species (see DNA [RAW MATERIAL OF EVOLUTION]), although many animal species, especially humans, transmit cultural information from one generation to the next. In eukaryotic cells (see EUKARYOTES, EVOLUTION OF), most of the genes are in the nucleus, although there are also many *cytoplasmic factors* (genes in chloroplasts and mitochondria; see SYMBIOGENESIS). Genes (the genotype) affect the characteristics of organisms (the phenotype); however, phenotype changes that occur during the life of an organism cannot be transferred back to the genotype. That is, acquired characteristics cannot be inherited (see LAMARCKISM). Evolution occurs in populations. Therefore, the study of genetic variation in a population is the study of the raw material available for evolution in that population. The only evolution that can occur is what the genetic variation permits.

A POPULATION is defined as a set of organisms, all of the same species, that can interbreed with one another because there are no reproductive barriers (see ISOLATING MECHANISMS) and because they all live in the same place. A local population in which interbreeding is not only possible but frequent is considered a *deme*. The genes of the organisms in a deme, or in a population, are therefore considered to be a single *gene pool*.

Populations Have Genetic Variability

In most species, an individual organism can have at most two alleles for a given *locus* (plural *loci*) or location on a chromosome: It has two copies of one allele if it is homozygous, and one copy of each of two alleles if it is heterozygous (see MENDELIAN GENETICS) for that locus. A population can have many alleles for each locus. Usually, individual eukaryotic organisms are heterozygous at only about 5–15 percent of their loci, but populations usually contain more than one allele at one-quarter to one-half of the loci. In many laboratory populations, one allele predominates (the *wild type*) over a few rare mutant alleles, which produce defects in the phenotype. In many wild populations, for many loci, there is more than one common allele (a *polymorphism*).

Populations can differ from one another in the types of alleles that they have for a locus. Even if they have the same alleles, they can differ in the relative frequencies of these alleles. For example, in the MN blood protein group, NN individuals (homozygous for allele N) are the most common genotype among Australian aborigines; MM individuals (homozygous for allele M) are the most common genotype among Navaho Native Americans; and the heterozygous MN individuals are the most common genotype among Caucasians in the United States (see table). The N allele strongly predominates in Australian aboriginal populations, and the M allele in Navaho populations, while the two alleles occur at approximately of equal frequency among Caucasians in the United States. The three populations differ genetically, even though all of them have the same two alleles for the locus.

Because evolution depends upon the variation in the gene pool, one would expect that evolution would proceed more rapidly in a population that has greater genetic variability. This is the idea behind the *fundamental theorem of natural selection* proposed by Fisher (see FISHER, R. A.). Fruit flies (genus *Drosophila*) have a short generation time and can be raised easily in the laboratory. Geneticists such as Thomas Hunt Morgan and Theodosius Dobzhansky (see DOBZHANSKY, THEODOSIUS) pioneered the use of fruit flies for genetic and evolutionary studies. Experiments with *Drosophila* indicate that populations that have greater genetic variability grow faster than populations with less genetic variability when that population is exposed to new environmental conditions (for example, a change in temperature), which is consistent with Fisher's theorem.

One difficulty that geneticists often encounter when studying populations is how to define an individual. The etymology of the word itself suggests that an individual is an entity that cannot be divided without killing it. With vertebrates, it is easy enough to recognize an individual. With plants and some invertebrates, however, it is not as easy, as in these examples:

- Plants and starfish can be cut up into pieces, and a new organism can grow from each piece. Are plants and starfish individuals?
- Many plants and invertebrates produce seeds and eggs asexually, resulting in numerous organisms that are genetically identical. Should these *clonal* organisms be considered individuals?
- Some plants spread by asexual means, for example by underground branches. In this way, an entire field of goldenrod plants *(ramets)* might be a single genetic individual *(genet)*. The connections among the ramets might be maintained, or might not, depending upon the species and circumstances. Should ramets be considered individuals?
- It may make more sense for a tree to be considered a population of branches rather than an individual. *Somatic*

Frequencies of Genotypes, and of Alleles, for the MN Blood Proteins in Three Populations of Humans

	Genotypes of Individuals			Alleles	
Population	MM	MN	NN	Allele M	Allele N
Australian aborigines	3.0%	29.6%	67.4%	17.8%	82.2%
Navaho Native Americans	84.5%	14.4%	1.1%	91.7%	8.3%
U.S. Caucasians	29.2%	49.6%	21.3%	53.9%	46.1%

mutations can occur in buds. A branch that grows from a bud that has a somatic mutation is genetically different from the other branches of the tree. Each branch of a tree may differ in its ability to resist herbivores (see COEVOLUTION). Insect herbivores, which have several generations in a single year, can evolve much faster than trees. It has been suggested that the generation of somatic mutations allows trees to keep up with the evolutionary pace of the insects that eat their leaves. If this branch produces seeds, the genetic differences can be passed on to the next generation. Does a tree function as a genetic and even as an evolutionary population?

Measuring Genetic Variability in Populations

It has often proved very difficult to measure variability for any given gene, or overall genetic variability, in populations. Several approaches have been used:

Phenotypic traits. One approach is to estimate variability of a measurable phenotypic trait (e.g., area of a leaf, or length of a bird's beak). This variability can be expressed as a variance, as a standard deviation, as a standard error, or as a coefficient of variation. This variability among individuals, however, is caused partly by genetic differences among the individuals (the *heritability*) and partly by differences among the individuals in the environmental conditions they have experienced. Variance, as originally defined by Fisher, is the average squared deviation from the average value; because of the square component, variance is symbolized as s^2. The heritability component of the variance is symbolized as h^2. Genetic and environmental variability is also symbolized by the letter V. Thus, total phenotypic variability (V_P) consists of genetic variability (V_G) and environmental variability (V_E). Genetic variability can itself be divided into two components: the variability that results from diversity of alleles (additive genetic variability, V_A) and the variability that results from one allele being dominant over another (dominance variability, V_D). Therefore $V_G = V_A + V_D$. Heritability can be expressed as either broad-sense or narrow-sense heritability:

- *Broad-sense heritability* is $V_G/V_P = V_G/(V_G+V_E)$, the proportion of phenotypic variation that is genetic.
- *Narrow-sense heritability* quantifies only the additive genetic variation and is therefore $V_A/V_P = V_A/(V_A+V_D+V_E)$. Narrow-sense heritability can also be quantified as the proportion of total variability in offspring that can be explained by the variability in the parents in a statistical analysis.

Heritability calculations assume that parental and offspring environments are uncorrelated; if they are correlated, environmental effect can be mistaken for genetic inheritance. Researchers can sometimes avoid this problem experimentally. They can, for example, switch eggs among bird nests, with the result that the offspring of each parent experiences each of the nest environments. This cannot be done in all experimental systems.

Heritability is a measure of genetic variation among members of a population; it is not the same as inheritance. If there is only one allele for a locus in the entire population, and that allele has a strong effect on a trait, then the trait is very strongly inherited but has a heritability of zero. Heritability is not a fixed number for a population, since it depends upon the amount of plasticity. Consider a genetic disease that is caused by a single allele and that is invariably fatal; the heritability would be 100 percent. But as soon as a medicine is discovered that cures the disease, the heritability drops to zero percent.

Proteins. Another approach is to measure the variability of proteins. Differences among proteins can be very important, as enzyme proteins control all the chemical reactions in an organism. Different forms of the same enzyme, determined by different alleles at the same locus, are called *allozymes.* An individual that is homozygous for one of the alleles of the gene that encodes this protein will produce just one allozyme; an individual heterozygous for this gene can produce two allozymes; the individuals in the population can collectively produce more than two allozymes. However, two allozymes, while they differ in structure, may not differ in function. That is, they may have exactly the same effect on the phenotype of the organism and on its ability to survive and reproduce. Therefore the study of allozymes may reveal greater variability than the study of phenotypic traits.

Proteins can be separated from one another by *electrophoresis.* A mixture of proteins is placed at one end of a gelatin-like sheet (usually composed of the complex sugar agarose). An electric current creates positive and negative poles on this sheet. Proteins, which have a slight negative charge, diffuse through the molecules of the gel toward the positive pole. Because they must diffuse through an obstacle course of agarose molecules, the smaller proteins move faster than the larger proteins. Therefore electrophoresis separates proteins on the basis of size as well as of charge. If a type of protein from an individual is placed on a gel, the protein will show up as a band, after the appropriate staining is used to make it visible. If the individual is homozygous for that allozyme, one band shows up; an individual heterozygous (producing two allozymes) will have two bands. Electrophoresis was first used to study the protein variability of populations by Richard Lewontin and John Hubby (see LEWONTIN, RICHARD) in the 1960s.

DNA. Another approach is to measure the variability of the DNA itself. Measuring the overall variability of the DNA itself would seem to be an ideal quantification of genetic variability. However, since most of the DNA in a eukaryotic cell is NONCODING DNA, a measure of overall DNA variability may be meaningless as an indicator of how much evolution may be possible in that population. Even measuring the variability in the DNA of a specified gene may not be very meaningful, as much of this variability may not produce differences in proteins. However, differences between populations, or differences between species, in their DNA (whether it codes for proteins or not) is routinely used as a measure of how much evolutionary divergence has occurred between them (see DNA [EVIDENCE FOR EVOLUTION]).

Measurements of genetic variability can be used to reconstruct population and species history. For example, biologists

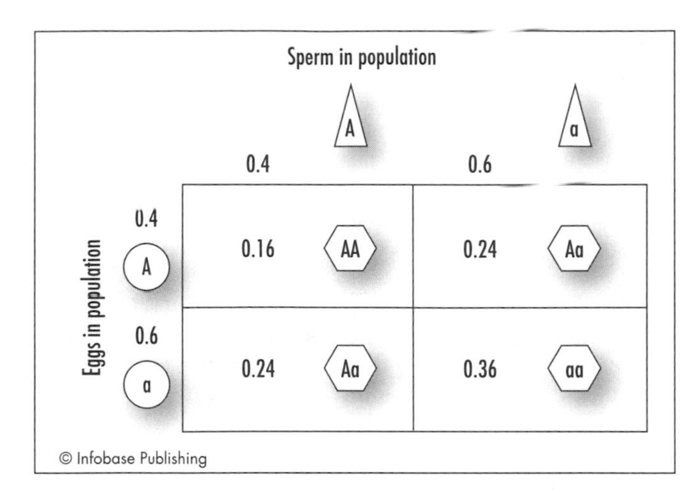

If allele A is present in 40 percent of the gametes (eggs and sperm) of a population, and allele a in 60 percent of the gametes, the resulting frequencies of the organisms will be 16 percent AA, 48 percent Aa, and 36 percent aa, under conditions of the Hardy-Weinberg equilibrium.

took blood samples from 20 birds representing six of the 13 species of DARWIN'S FINCHES on the GALÁPAGOS ISLANDS. They measured the number of alleles in these 20 birds for a single locus, for the MHC II B class 1 gene. In 20 birds, there can be at most 40 alleles. These birds had 21 alleles. Since the rate at which this gene mutates has been estimated, the researchers concluded that the common ancestor of all these birds lived at least 10 million years ago. But the islands are not that old. The original immigrant finches may have lived on one of the islands that has eroded away and migrated to their current locations. Furthermore, in order for the finches on the Galápagos Islands to have this much genetic variability, the original population would have had to contain at least 40 genetically distinct individual birds.

Genetic Changes in Populations

In order to determine whether or not natural selection is occurring in a population, biologists need to determine what the frequency of each allele (for a given locus) would be in the absence of natural selection, that is, if the population was at *equilibrium* rather than being influenced by natural selection.

Consider a simple example in a species of animal, such as a kind of fish, that has external fertilization and in which a certain locus has just two alleles, *A* and *a*. Forty percent of the sperm and eggs released into the water by the fish in this population carry allele *A*, while 60 percent of the sperm and eggs carry allele *a*. The frequency of allele *A* is $p = 0.4$, and the frequency of allele *a* is $q = 0.6$. Because there are only two alleles, $p + q = 1$. If the eggs and the sperm come together randomly, you would expect the following results (see figure).

1. Since 40% of the eggs, and 40% of the sperm, carry allele *A*, there is a $40\% \times 40\% = p^2 = 16\%$ chance that the zygote will have the homozygous *AA* genotype.
2. Since 60% of the eggs, and 60% of the sperm, carry allele *a*, there is a $60\% \times 60\% = q^2 = 36\%$ chance that the zygote will have the homozygous *aa* genotype.
3. There is a $40\% \times 60\% = pq = 24\%$ chance that a sperm carrying allele *a* will meet an egg carrying allele *A;* there is also a 24% chance that the opposite will happen, a sperm carrying allele *A* will meet an egg carrying allele *a*. There is therefore a $24\% + 24\% = 2pq = 48\%$ chance that the zygote will have the homozygous *Aa* genotype.
4. All of the zygotes will have one of the three genotypes *AA, Aa,* or *aa*. Therefore $p^2 + 2pq + q^2 = 16\% + 48\% + 36\% = 100\%$.

In a population that is at equilibrium, one would expect these results simply due to genetic recombination: 16% *AA*, 48% *Aa*, and 36% *aa*. This principle was first explained by G. H. Hardy, a mathematician, and W. Weinberg, a biologist, and is therefore called the *Hardy-Weinberg equilibrium*. If the population differs from this pattern, the departure from equilibrium could be caused by a number of possibilities:

1. The population may not be very large. Random departures from equilibrium can occur in small populations (such as genetic drift; see FOUNDER EFFECT).
2. The mixture of gametes may not be occurring randomly among individuals in the population. This could be caused by nonrandom mating patterns, in which individuals within the population prefer to mate with one genotype more than with another.
3. Mutations have occurred.
4. Migration (of one genotype more than of another) has occurred into or out of the population.
5. Natural selection has occurred.

In this last possibility, one of the alleles may be less frequent than its equilibrium value because natural selection has "selected against" individuals that carry this allele. The Hardy-Weinberg equilibrium is the background against which natural selection, defined as changes in allele frequencies in a population, might be detected.

If natural selection operates against an allele, the allele will become less common in the population. Strong selection against a dominant allele (an allele whose presence always shows up in the phenotype) may quickly eliminate that allele. (This rarely occurs, because an allele that produces an inferior phenotypic trait would probably never have become dominant in the first place.) Strong selection against a recessive allele (such as a lethal mutation), however, may never eliminate the allele from the population. This is because the recessive allele can hide in the heterozygotes; the heterozygotes are therefore *carriers* of this allele. A lethal recessive will decline in frequency according to this formula:

$$q_n = q_0/1+nq_0,$$

in which q is the frequency of the lethal recessive allele, q_0 is the initial frequency of the allele, and q_n is the frequency after n generations. Starting with a frequency of 25 percent, the allele will decline to a frequency of 8 percent after eight generations. Even after a hundred generations, the frequency of this recessive allele will still be just under 1 percent. When the allele is at very low frequency, it may be lost from the population by random events such as genetic drift.

Sometimes a recessive allele might be lethal in the homozygous form but might actually confer a survival or reproductive advantage in the heterozygous form. This can lead to a BALANCED POLYMORPHISM because the balance between advantage and disadvantage can cause more than one form of the allele (a polymorphism) to be maintained in the population. The most famous example is sickle-cell anemia, which is caused by a recessive mutation in one of the hemoglobin genes. The mutation causes anemia, because the blood cells that have this hemoglobin are sickle-shaped and are destroyed. However, the malaria parasite cannot reproduce in these blood cells. Therefore, in regions (usually central Africa) where malaria is common, people who are homozygous for the mutant die from anemia but people who are homozygous for normal hemoglobin often die of malaria. People who are heterozygous for the mutation strike a balance between anemia and malaria, avoiding malaria and having relatively mild anemia. Natural selection therefore favors this mutation, because there are many more heterozygous people than homozygous recessive people.

Another possible example of a balanced polymorphism involves cystic fibrosis. Individuals homozygous for this mutation usually die at a young age; the carriers of the mutation, however, are largely unaffected. Population geneticists have been puzzled as to why this mutation is so common in certain (often northern European) populations. They concluded that there must be a heterozygote advantage, and it probably is connected with resistance to certain diseases.

Why Do Populations Have So Much Variability?

When natural selection began to gain popularity as the mechanism of evolution (see MODERN SYNTHESIS), geneticists were attracted to the idea that natural selection maintained this genetic variability. Diversifying selection in a heterogeneous environment would be able to do so (see NATURAL SELECTION). They considered that there had to be a good evolutionary reason why there was so much genetic variation. The reason cannot be that populations need it so that they can evolve. Natural selection can produce only those characteristics that provide a benefit to individuals, not to whole populations (see GROUP SELECTION). The primacy of natural selection in producing and maintaining genetic variability was central to the contributions of Fisher and Haldane (see HALDANE, J. B. S.) to the Modern Synthesis. In contrast, Wright (see WRIGHT, SEWALL) emphasized random events such as genetic drift in the course of evolution.

With the advent of electrophoresis, it became apparent that there was a lot more genetic variability in populations than could be seen in the phenotype. Populations appeared to accumulate neutral variation—genetic variation that was neither beneficial nor harmful. Beginning in the late 1960s, the *neutral theory of evolution,* proposed by Japanese geneticist Motō Kimura, emphasized the primacy of neutral genetic variation in populations. This offered a challenge to the selectionist arguments of Fisher and Haldane and was more consistent with the genetic drift of Wright.

Inbreeding

Inbreeding occurs when close genetic relatives interbreed. Most individual organisms have a few lethal recessive mutations that originated in relatively recent ancestors. In the heterozygous state, these mutations have no effect on the phenotype. So long as the individual does not mate with another individual that carries the same mutation (that is, a close relative descended from the recent ancestor that had the mutation), the mutation will not appear in the homozygous form. When two carriers interbreed, there is a one-in-four chance that the offspring will receive a double dose of the mutation and die as a result.

Small populations often have reduced genetic variability. There is a greater risk that individuals in these small populations will crossbreed with another individual that carries the same lethal mutation. This is one of the principal problems with species that are close to EXTINCTION—even if they are rescued and given a safe habitat in which to live, these species (such as cheetahs or the Hawaiian goose) will have poor reproductive success. The genetic variability that they lost is impossible to replace in the short run and perhaps forever.

Inbreeding may occur in humans for cultural reasons. In isolated villages, populations are often founded by a small number of people, and thus with limited genetic variability. Most people mate with other people in their village, with the result that they often share lethal recessive mutations with their mates. One example comes from the Dominican Republic village of Salinas (population about 4,000), where about one out of every 90 girls turns into a boy when reaching puberty. This happens so frequently that the villagers have a name for it: "guevedoche," or "penis at twelve." A similar phenomenon occurs among the Sambia tribe of Papua New Guinea, where they call it "kwolu-aatmwol" in their native tongue or "turnim-men" in pidgin. Women have two X chromosomes, while men are XY (see SEX, EVOLUTION OF). Gender, however, turns out to be due to more than just sex chromosomes. The guevedoche and kwolu-aatmwol boys had Y chromosomes and produced testosterone like regular boys; however, they lacked an enzyme called 5-α-reductase. The gene encoding this enzyme was defective. This enzyme changes testosterone into a related molecule (DHT) that stimulates genital development in the fetus. Therefore, during gestation and childhood, these boys had genitals that appeared female. However, at puberty, testosterone itself stimulates and maintains masculinity; therefore, at puberty, the boys developed male genitalia. All of the guevedoche boys in the Dominican Republic can trace their ancestry to one woman, Altagracia Carrasco; she was probably the person who carried the mutation. Inbreeding can also occur in religious communities that remain apart from the rest of society. Some religious sects such as the Amish have elevated levels of occasionally lethal genetic diseases.

Summary

Natural populations have genetic variability. Even those species that produce offspring without sexual recombination have at least some genetic variability due to mutation. Genetic variability in populations can be measured on phenotypic traits, or on proteins. The Hardy-Weinberg equilibrium expresses the mixture of genes that will result in the absence of natural selection. Recessive lethal mutations can be eliminated only slowly from populations; lethal mutations may be selected if they provide a benefit to heterozygotes; and lethal mutations can show up more frequently in small populations, often of endangered species, than in large populations.

Further Reading

Avise, John C. *Population Genetics in the Wild.* Washington, D.C.: Smithsonian, 2002.

Ayala, Francisco J. *Population and Evolutionary Genetics: A Primer.* Menlo Park, Calif.: Benjamin-Cummings, 1982.

Freeman, Scott, and Jon C. Herron. "Mechanisms of evolutionary change." Part 2 of *Evolutionary Analysis,* 3rd ed. Upper Saddle River, N.J.: Pearson Prentice Hall, 2004.

LeRoi, Armand. *Mutants: On Genetic Variety and the Human Body.* New York: Viking, 2003.

Lewontin, Richard. *The Genetic Basis of Evolutionary Change.* New York: Columbia University Press, 1974.

Ruder, Kate. "Genomics in Amish country." Genome News Network. Available online. URL: http://www.genomenewsnetwork.org/articles/2004/07/23/sids.php. Accessed October 5, 2005.

Stebbins, G. Ledyard. *Processes of Organic Evolution.* Englewood Cliffs, N.J.: Prentice Hall, 1971.

Precambrian time Precambrian time refers to the history of the Earth before the CAMBRIAN PERIOD of the PALEOZOIC ERA (see GEOLOGICAL TIME SCALE). Precambrian time makes up about almost 90 percent of Earth history. Geologists have divided Precambrian time into:

- *Hadean Eon* (4.5 billion–3.8 billion years ago), when the Earth was forming from solar system dust (see UNIVERSE, ORIGIN OF). The Earth experienced massive bombardments (see ASTEROIDS AND COMETS). Oceans had not yet formed.
- *Archaean Eon* (3.8 billion–2.5 billion years ago), when the Earth was mostly covered by oceans, and the atmosphere contained almost no oxygen gas. Bacteria evolved early in the Archaean Eon (see ARCHAEBACTERIA; BACTERIA, EVOLUTION OF; ORIGIN OF LIFE). A tremendous amount of biochemical evolution occurred during this time, which was invisible in the fossils.
- *Proterozoic Eon* (2.5 billion–0.5 billion years ago), when continents began to form, and oxygen began to accumulate in the atmosphere (see PHOTOSYNTHESIS, EVOLUTION OF). Although oxygen buildup began in the atmosphere and shallow oceans at the beginning of the Proterozoic, the deep oceans may have remained anoxic until less than a billion years ago. In the middle of the Proterozoic Eon, the first complex cells formed (see EUKARYOTES, EVOLUTION OF).

Toward the end of the Proterozoic Eon, multicellular organisms existed: seaweeds, animal embryos, and EDIACARAN ORGANISMS. The Ediacaran period has been recently defined as lasting from about 600 million to about 540 million years ago.

Further Reading

Kennedy, Martin, et al. "Late Precambrian oxygenation: Inception of the clay mineral factory." *Science* 311 (2006): 1,446–1,449.

Kerr, Richard A. "A shot of oxygen to unleash the evolution of animals." *Science* 314 (2006): 1529.

primates Primates constitute the class of mammals (see MAMMALS, EVOLUTION OF) that includes lemurs, monkeys, and apes. Humans are apes, therefore primates. The term *primate,* used by scientists for several centuries, suggests that this group, which includes humans, is the most important among mammalian classes. While scientists no longer rank classes by their importance relative to humanity, the term *primate* continues in use.

Biologists classify primates into approximately 230 living species of prosimians, tarsiers, and anthropoids. Primates, in general, share these characteristics:

- They have the largest brains among mammals, after adjustment for body size (see ALLOMETRY).
- They often live in large, complex social groups. Large brains aid survival but are considered to be most important for social interactions (see INTELLIGENCE, EVOLUTION OF).
- They take a longer time to reach maturity than most other mammals, a characteristic related to the evolution of intelligence.

The first primates lived in trees, and most still do, except for a few groups such as baboons and humans that have (independently from one another) evolved the ability to live on the ground. This adaptation is complete enough that in humans, uniquely among primates, the lower limbs (legs) are relatively long. Some, like galagos, have remarkable leaping abilities: Though less than a foot and a half (50 cm) in length, galagos can leap over six feet (2 m). Leaping from branch to branch in trees has encouraged the evolution of important characteristics that primates possess more than most other mammals:

- *The ability to grasp.* This results primarily from an opposable thumb, rare in other mammals. Primates not only have the ability to grasp branches in trees, but the juvenile primates can efficiently cling to the fur of their mothers. Most primates also have opposable big toes, although primates that have evolved a ground-based existence have lost this characteristic. Nonhuman apes have opposable big toes; humans do not; the AUSTRALOPITHECINE ancestors of humans had big toes that were intermediate in their adaptation for grasping branches.
- *Excellent vision.* Nocturnal primates have big eyes. For example, each of the tarsier's eyes is larger than its brain. Diurnal primates have inherited big eyes from nocturnal ancestors. Visual acuity is important not only for night vision but for jumping around in trees. For prosimians, smell remains an important sense, while most anthropoids rely primarily on vision.

- *Depth perception.* Primates have both eyes on the front of the face, a characteristic most people notice immediately in all primates. This allows binocular vision and depth perception, which allows primates to grasp branches while leaping.
- *Color vision.* While not the only mammals with color vision, primates (primarily anthropoids) benefit from color vision because it allows some primates to recognize ripe fruits in colors such as red against a background of green leaves, in the absence of a well-developed sense of smell.

This entry follows the classification system, which is not exhaustive:

Order: Primates
- Suborder Prosimii (prosimians)
  - Infraorder Lemuriformes (lemurs)
  - Infraorder Lorisiformes (lorises)
- Suborder Tarsiiformes (tarsiers)
- Suborder Anthropoidea
  - Infraorder Platyrrhini (New World monkeys)
  - Infraorder Catarrhini (Old World apes and monkeys)
      - Family Cercopithecidae (Old World monkeys)
    - Superfamily Hominoidea (apes)
      - Family Hylobatidae (gibbons)
      - Family Hominidae
        - Subfamily Ponginae (orangutans)
        - Subfamily Homininae
          - Tribe Gorillini (gorillas)
          - Tribe Hominini (chimps and humans)

Prosimians are the more primitive group of primates and are generally smaller than anthropoids. Most prosimians are lemurs that live in Madagascar, but others live on the African continent and in southeast Asia. *Lemurs* range in size from the pygmy mouse lemur (weighing about an ounce, or 30 g) to the indri (weighing about 20 pounds, or 9 kg). Some lemurs are diurnal, others nocturnal. *Galagos* and *lorises* live on the African continent; lorises also live in India and southeast Asia. Many zoologists classify prosimians into the subclass Strepsorhini.

Tarsiers are small, nocturnal primates that live in southeast Asia. Though they look to many people very similar to the prosimians, they share some features with the anthropoids, such as skull characteristics, and big hands and feet. Many zoologists classify tarsiers together with anthropoids into the subclass Haplorhini.

Anthropoids (monkeys and apes) have the largest brains among primates and include two families of New World monkeys, found in the tropical rain forests of Central and South America; the Old World monkeys, the lesser apes, and the great apes, all of which originated in Africa or Asia. Most monkeys and apes have longer arms than do prosimians, which allows them not to merely jump around in trees but to swing from branches, a unique kind of locomotion called *brachiation.* Most anthropoids are diurnal. Most have highly developed social patterns, which may be the principal reason for their braininess. For example, when howler monkeys of the Amazon recognize a threat (for instance, in the form of a human intruder), they may form into a circle in the branches over the intruder's head, and it is time for the human to leave.

New World monkeys have evolved separately from other primates for at least 20 million years, largely as a result of the separation of the African and South American continents (see CONTINENTAL DRIFT). All live in trees. Their nostrils are flat and face sideways, which is why they are called *platyrrhines* (flat-nosed). The most noticeable distinction is that they have long *prehensile* tails that have muscles and even fingerprint skin that allows these monkeys to grasp branches with their tails. New World monkeys have smaller thumbs (spider monkeys have none at all) than Old World monkeys and apes. The marmosets and tamarins of South America differ from other New World monkeys in having only two, rather than three, molars and are classified separately.

The Old World monkeys and the apes are *catarrhines,* so named because their nostrils are narrow and face downward. *Old World monkeys* live primarily in trees, except baboons, but spend part of their time on the ground. They are much less likely to have long tails, and these tails are not prehensile. Their rump skin pads help them sit on the ground but are also useful in sexual signaling: colors indicate species, and estrus, which is the female readiness for mating. The *colobine* subfamily of Old World monkeys includes the langurs of Asia and the colobus monkeys of Africa. They have relatively longer legs, tails, and thumbs than other Old World monkeys. Although they eat a variety of foods, their large multichambered stomachs allow the digestion of leaf material. A diet of leaves, most of which are toxic, requires the assistance of microbes in mammals such as the colobus monkey, the cow, and the marsupial koala; even the marine iguanas of the Galápagos Islands, which live on seaweed, require microbial symbionts (see COEVOLUTION). The *cercopithecine* subfamily of Old World monkeys includes the macaques of Africa and Asia (some extend their range a little bit into Europe), and the baboons and mangabeys of Africa. Cercopithecines have less complex stomachs and eat less leaf material. The Japanese macaque monkeys live farther north than any other nonhuman primate. Their intelligence and cultural propagation of knowledge has been studied as a model for the evolution of intelligence (see GENE-CULTURE COEVOLUTION).

Apes have no tails and have larger brains than other anthropoids. Apes have at least a limited ability to walk on two legs (see BIPEDALISM). *Lesser apes* include the gibbons and siamangs of southeast Asia which have extremely long arms with which they brachiate. "Lesser" refers to the smaller size of these apes than the great apes. Lesser apes, unusual among animals, have a lifelong monogamous mating system (see REPRODUCTIVE SYSTEMS). *Great apes* (family Hominidae; the *hominids*) include the orangutans of southeast Asia, and the gorilla, chimpanzee, bonobo, and humans that originated in Africa. Because great apes do not have rump pads, they generally build nests in trees or on the ground for sleeping.

With males weighing as much as 160 pounds (about 80 kg), the orangutans of Borneo and Sumatra are the largest arboreal primates. Orangutans are the least social of apes. They forage separately for fruit. Males and females interact only to mate, and the only social bond among orangutans is between mother and infants. With males weighing as much as 350 pounds (about 180 kg), the gorillas are the largest primates. The three subspecies of gorillas (the mountain gorilla, and the eastern and western lowland gorillas) live in the rain forests of central Africa, in small groups that eat plant materials. In both orangutans and gorillas, males weigh about twice as much as females, which is characteristic of the harem. On at least two occasions, gorillas have been observed to use wooden tools.

The apes commonly called chimpanzees consist of two species: the true chimpanzees *(Pan troglodytes)* and the pygmy chimpanzee or bonobo *(Pan paniscus)*. True chimps range from dry woodlands to rain forests throughout central Africa; pygmy chimps live in a small area south of the Congo River. They eat a wide variety of foods, including leaves, fruits, seeds, insects, and occasional red meat. They spend part of their time in trees and part on the ground (bonobos less than true chimps). Chimps exhibit behaviors that scientists once thought were found only in humans, such as the use of tools (although they do not fashion stones into tools) and warfare. Most of these behaviors were revealed only when researchers spent long periods of time with chimp societies (see GOODALL, JANE). True chimps are very aggressive, in contrast to bonobos, who live in small bands and engage in frequent, human-like copulation. As primatologist Frans DeWaal says, "The chimpanzee resolves sexual issues with power; the bonobo resolves power issues with sex." Humans share a common ancestor with chimps and bonobos that lived only about five million years ago (see DNA [EVIDENCE FOR EVOLUTION]).

Primates similar to today's prosimians evolved shortly after the CRETACEOUS EXTINCTION that wiped out the DINOSAURS and other large reptiles. Some other highly specialized orders of mammals, such as bats and whales, also evolved in the Paleocene epoch of the CENOZOIC PERIOD (see MAMMALS, EVOLUTION OF). When dinosaurs were no longer the dominant land animals, mammalian evolution exploded in diversity. One of the earliest primates, *Purgatorius,* may have evolved before the end of the Cretaceous. *Purgatorius,* and the *plesiadapiform* primates that evolved during the Paleocene, lived in Europe and North America (which were still connected at that time). They resembled shrews. During the ensuing Eocene epoch, *adapid* and *onomyid* primates evolved. It is unclear which of these groups may have been the ancestors of monkeys and apes. Both groups of Eocene primates had larger brains, shorter faces, and a better ability to grasp than Paleocene primates, and they relied on sight more than smell. The first true monkeys evolved during the ensuing Oligocene epoch, in the Old World. Oligocene primates from what is now the Fayyum deposit in Egypt (such as *Proconsul*) may represent the first true apes.

Sometime after seven million years ago, perhaps associated with the spread of drier conditions across Africa, the human evolutionary lineage (now usually called the HOMININ lineage) separated from the chimpanzee lineages. The earliest distinctive characteristic of the hominins was upright posture and walking on two legs, although numerous distinguishing features (such as very large brains, fashioning stones into tools, use of fire, use of language) arose later in the hominin line. The earliest known hominins were *Orrorin* and *Sahelanthropus* from west Africa and *Australopithecus* primarily from east Africa. Many australopithecines (and related hominins such as *Kenyanthropus*) are known from the period between five million and two million years ago in east Africa. The embarrassing plenitude of fossil information reveals a complex bush of hominin lineages, rather than a clear line of human ancestry. About the only thing regarding which anthropologists are certain is that the *robust* australopithecines (genus *Paranthropus*) did not lead to modern humans. From one of the *gracile* australopithecines, either known or yet to be discovered, the genus *Homo* evolved about two million years ago in east Africa (see *HOMO HABILIS; HOMO ERGASTER*). Today, *HOMO SAPIENS* is the only surviving species of hominin.

The fact that humans are primates means that the physiology of the other primates most nearly resembles that of humans. It is for this reason that primates, especially monkeys and (even better) chimps, are favored for medical experimentation. The fact that monkeys and nonhuman apes have large brains and complex behavior means that they may also experience such experimentation as cruelty, more than other laboratory animals such as mice, rats, and dogs. For both reasons, medical experimentation upon primates has become particularly controversial (see ANIMAL RIGHTS).

Most nonhuman primates survive in remote areas away from human contact, and a large proportion of them are endangered. Most primates are endangered from habitat destruction, but they are also endangered by predation by humans as "bushmeat" and for traditional medicine. Only a few hundred mountain gorillas survive, in the Virunga Mountains of east Africa.

Further Reading

De Waal, Frans B. M. *My Family Album: Thirty Years of Primate Photography.* Berkeley: University of California Press, 2003.

Rowe, Noel. *The Pictorial Guide to the Living Primates.* East Hampton, N.Y.: Pogonias Press, 1996.

———. "Primate Conservation, Inc." Available online. URL: http://www.primate.org. Accessed May 3, 2005.

Seiffert, Erik R. "Basal anthropoids from Egypt and the antiquity of Africa's higher primate radiation." *Science* 310 (2005): 300–304.

Stewart, Caro-Beth, and Todd R. Disotell. "Primate evolution—in and out of Africa." *Current Biology* 8 (1998): R582–R588.

Walker, Alan, and Pat Shipman. *The Ape in the Tree: An Intellectual and Natural History of Proconsul.* Cambridge, Mass.: Harvard University Press, 2005.

progress, concept of Progress is a concept that evolution inevitably leads toward humans, or at least toward greater complexity. In the minds of most people, this concept is inseparable from evolution. If evolution has occurred, then it must have started with simple cells and progressed to complex cells and more complex multicellular organisms over the course of billions of years. Nobody can deny that early evolutionary history was dominated by simple organisms, while complex organisms came into existence only later. This is abundantly demonstrated by the fossil record (although this could not be the case if the fossils were the products of a single gigantic flood; see CREATIONISM). The controversy is over whether this pattern should be called "progress." Is progress simply something that happened to occur, or is it something that had to occur? That is, is progress a side effect, or is it a process? Progress has been recognized in both the complexity, and the diversity, of organisms.

Progress as a process within evolution. The concept that progress is part of the process of creation predates evolution. Catastrophists (see CATASTROPHISM) believed in a series of creations in which each new era had more complex and diverse organisms than the previous era. With the ascendancy of evolutionary science, the notion of progress, as a necessary and built-in process, was not abandoned. Some people (see SPENCER, HERBERT) assumed that evolutionary change had to be progressive. This notion has been perhaps best characterized in the 20th century in the writings of philosopher-scientists Henri Bergson, Pierre Teilhard de Chardin (see TEILHARD DE CHARDIN, PIERRE), and Pierre Lecomte du Noüy.

When evolutionary scientists began to understand how the process of evolution actually worked (see MODERN SYNTHESIS), they could find no mechanism that could generate inevitable progress. Henri Bergson wrote that there was a mysterious "life force" *(élan vital)* that caused evolutionary progress. In a famous retort, evolutionary biologist Julian Huxley (one of the architects of the Modern Synthesis; see HUXLEY, JULIAN S.) said that someone who knew nothing about engines might assume that a train running on its tracks was propelled by a mysterious *"élan locomotif."* Although the concept keeps coming back in some popular writings, evolutionary scientists have abandoned progress as a process, simply because nobody can think of how such a process could actually occur on the molecular level.

Progress as a result of evolution. Most observers will admit the overall pattern of progress that has resulted from evolution. However, they attribute this to NATURAL SELECTION acting upon random variation. It occurs because there are more ways for evolutionary diversification to produce greater complexity than there are ways for it to produce greater simplicity; that is, there are more possible complexities than there are simplicities. Paleontologist Stephen Jay Gould strongly denounced the concept of progress, claiming that life today, as it has been for the last three and a half billion years, is dominated by the bacterial mode (see BACTERIA, EVOLUTION OF). Bacteria constitute the overwhelming share of the mass, diversity, and history of life. Complexity has evolved in multicellular organisms, against this bacterial background. This has occurred by means of what he called the "drunkard's walk," in which random stumbling results in progress because the drunkard bounces off of one wall and keeps going but can no longer continue if he falls into the gutter. What scientists call progress has resulted not so much from an increase in the mean level of complexity, as from an increase in the variance of complexity, with one direction (the direction of simplicity) truncated. Most observers believe that Gould was making a distinction without a difference; what Gould described is, in fact, what most evolutionary scientists would call progress that has resulted from natural selection acting on random variation.

Some evolutionary scientists have claimed that progress is an inevitable product of evolution. A recent example of this is Simon Conway Morris, a British paleontologist who figured prominently in the interpretation of Cambrian fauna (see BURGESS SHALE). He claimed neither miraculous creation nor a mysterious *élan* within evolutionary history. He pointed out that progress has occurred because there are only a certain number of evolutionary pathways that will work. Because light provides a good source of information about the environment, it is no surprise that vision, even acute vision with complex eyes, will evolve over and over during evolutionary history. The evolution of complex eyes from simple eyespots is progress, but it has resulted, over and over again in separate lineages, because natural selection has favored it in each case. Another example he provides is the evolution of intelligence: It has evolved separately in hominins and cetaceans, because natural selection has favored it in both cases, and evolved separately in *Homo sapiens* and *Homo neanderthalensis,* for the same reason (see INTELLIGENCE, EVOLUTION OF; *HOMO SAPIENS;* NEANDERTALS). Progress is inevitable, because natural selection has caused a *convergence* of adaptation upon a relatively small number of workable solutions to problems of survival (see CONVERGENCE). As a result, wherever life may have a chance to evolve for a long time in the universe, Conway Morris claims, it is inevitable that something resembling a human being would come into existence.

Evolutionary scientists will continue to study the processes by which complex adaptations have evolved and converged, regardless of whether the result of such processes is eventually considered to be progress or not.

Further Reading

Bergson, Henri. *Creative Evolution.* Translated by Arthur Mitchell. New York: Henry Holt, 1911. Reprint, New York: Dover, 1998.

Conway Morris, Simon. *Life's Solution: Inevitable Humans in a Lonely Universe.* London: Cambridge University Press, 2003.

Dixon, Dougal. *After Man: A Zoology of the Future.* New York: St. Martins, 1981.

du Noüy, Pierre Lecomte. *Human Destiny.* New York: Longmans, Green and Co., 1947.

Gould, Stephen Jay. *Full House: The Spread of Excellence from Plato to Darwin.* New York: Harmony Books, 1996.

Ruse, Michael. *Monad to Man: The Concept of Progress in Evolutionary Biology.* Cambridge, Mass.: Harvard University Press, 1996.

Teilhard de Chardin, Pierre. *The Phenomenon of Man.* New York: Harper, 1959. Reprint, New York: Harper Perennial, 1976.

Ward, Peter, and Alexis Rockman. *Future Evolution: An Illuminated History of Life to Come.* New York: Henry Holt, 2001.

promoter A promoter is a position on a chromosome that indicates the beginning of a gene. A gene is the DNA that instructs the cell how to make one or more proteins or protein components (see DNA [RAW MATERIAL OF EVOLUTION]).

Each chromosome contains thousands of genes, each with its own promoter site or sites. The promoter is like a lock; the gene cannot be put to use unless the promoter is unlocked by a protein key. The *expression* of the gene, which results ultimately in the production of a protein, begins when an enzyme of *RNA polymerase* binds to a promoter.

Most of an organism's cells contain an entire set of the genetic information for the organism. For example, a skin cell contains the genes not just for skin but also for nerves, blood, muscles, and glands. In theory, a human being could be cloned from a single cell, so long as that cell contains a full nucleus (red blood cells do not have nuclei, and sperm and ova have only half of the chromosomes). But the skin cell uses only a small amount of the DNA that it contains. How does it select the genes that it uses, and ignore the others? There are different categories of promoters. Skin cells, for example, unlock and activate only those genes that have the skin cell promoters. The successive activation of genes, each controlled by their own promoters, is what orchestrates the process of development from a fertilized egg to an embryo to a fully formed organism (see DEVELOPMENTAL EVOLUTION).

Cells can also block the genes for which they have no current use. *Inhibitor* molecules can bind to a site very near the promoter, which will prevent the RNA polymerase from binding to the promoter and unlocking it. In some cases, the protein that is produced by the expression of the last gene in a set is itself part of the inhibitor that blocks the expression of those genes. In this way, RNA polymerase can bind to the promoter if the protein is scarce, but if the protein is abundant the RNA polymerase cannot bind. The cell therefore makes more protein when that protein is scarce and stops making it when that protein is abundant. This is an example of *end-product inhibition* that helps to maintain a constant supply of the protein in the cell.

Promoters are important not only in the development of individual organisms but in the evolution of populations and species. If a promoter is lost or damaged, or the cell loses the ability to recognize a certain promoter, the gene remains locked and the cell ignores it. The gene is still present, but it might as well not be. The cells of most species contain a great deal of DNA that is never used for guiding the production of proteins (see NONCODING DNA). Much of the noncoding DNA consists of old unexpressed genes *(pseudogenes)*. Many such ancient genes are piled up, like items in an attic, in the chromosomes of the cells of most complex species. In this way, a gene that is "lost" is like a file on a computer disc that is deleted: The file itself is not destroyed, but just the information that tells the computer where to find the file.

Occasionally, a promoter or its control proteins may mutate, causing the gene to be used in a different context. In this way, organisms have been able to "co-opt" genes for new uses. One examples is the crystallin proteins that make up the lenses of eyes. Different lineages of animals have reassigned different kinds of proteins, from metabolic enzymes to heat shock proteins to exoskeleton proteins, to a new use as crystallins. These proteins are now expressed in the lenses of eyes rather than being used for their original functions (see INTELLIGENT DESIGN). Another example is the evolution of C_4 and CAM photosynthesis (see PHOTOSYNTHESIS, EVOLUTION OF). The enzymes involved in these processes have been borrowed from other uses within plant cells. Because proteins can be co-opted, in part by changes in promoters, it is not necessary for evolution to "start from scratch" to produce the genes for a new function.

Occasionally, ancient genes can be reactivated or artificially stimulated. Sometimes, horses with extra toes are born. Modern horses have one toe (the hoof) while the evolutionary ancestors of modern horses had more toes. The genes involved in the production of extra toes, once active in the horse lineage but lost, are in those cases reactivated. Ancient hidden genes can also be artificially stimulated. In 1980 developmental biologists E. J. Kollar and C. Fisher grafted a portion of mouse embryo onto a chick embryo. The mouse cells stimulated the chick embryo to develop rudimentary teeth (see photo below). Millions of years ago, birds had teeth, but today they have only hardened lips (beaks). The little chick, with teeth, was the first bird in millions of years to have

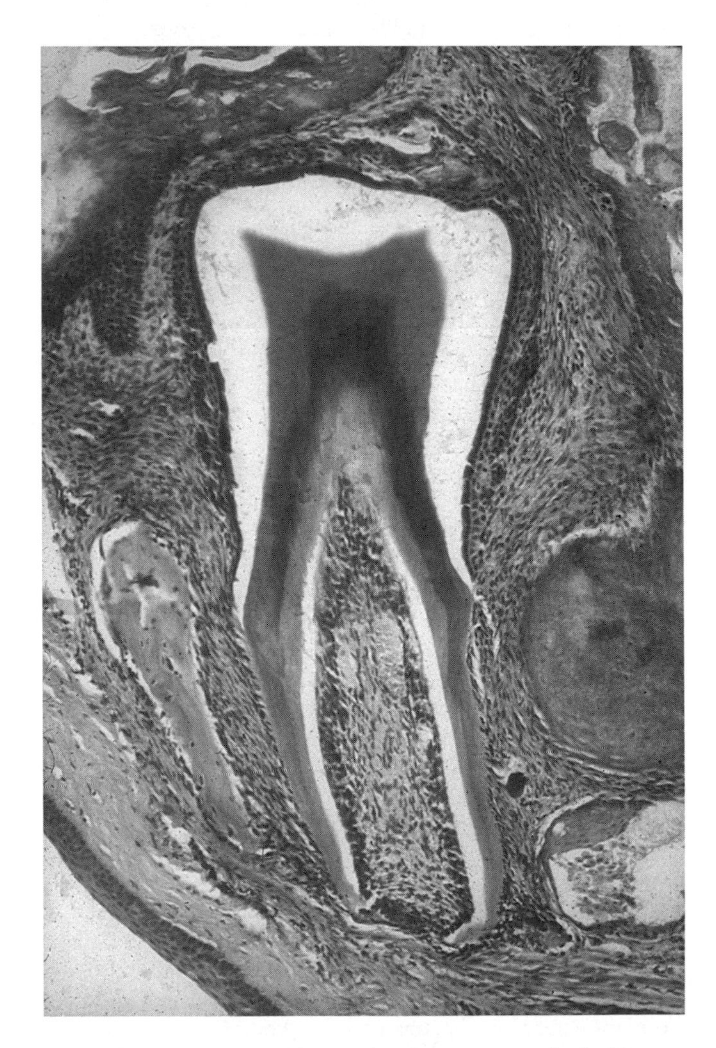

This tooth developed in a chicken embryo because genes that had been silent for millions of years were experimentally activated. ***(Courtesy of E. J. Kollar and C. Fisher)***

teeth. They were not mouse teeth, but bird teeth, produced by ancient bird tooth genes that were copied each generation but which lay unused in the bird chromosomes. Ancient genes, without their promoters, are therefore important as (1) evidence for evolutionary ancestry (see DNA [EVIDENCE FOR EVOLUTION]), and (2) repositories of genetic variability that can occasionally resurface and upon which natural selection can again act.

Further Reading

Coen, Enrico. *The Art of Genes.* New York: Oxford University Press, 1999.

Gould, Stephen Jay. "Hen's Teeth and Horse's Toes." Chap. 14 in *Hen's Teeth and Horse's Toes.* New York: Norton, 1983.

Kollar, E. J. and Fisher, C. "Tooth induction in chick epithelium: Expression of quiescent genes for enamel synthesis." *Science* 207 (1980): 993–995.

Rice, Stanley, and John McArthur. "Computer Analogies: Teaching Molecular Biology and Ecology." *Journal of College Science Teaching* 32 (2002): 176–181.

protists *See* EUKARYOTES, EVOLUTION OF.

punctuated equilibria Punctuated equilibria are patterns of evolution in which long-term equilibria of species are punctuated by relatively rapid speciation events. It is an alternative to *gradualism,* a pattern in which gradual directional change within species eventually leads to speciation. Punctuated equilibria were first proposed in 1972 by two American paleontologists (see ELDREDGE, NILES; GOULD, STEPHEN JAY), which they intended as a description of the pattern that they observed in the fossil record. Gradualism had been the pattern that evolutionary scientists expected to reconstruct from the fossil record, ever since the writings of Darwin (see DARWIN, CHARLES). Both gradualism and punctuated equilibria are descriptions of the patterns rather than explanations of the processes that may cause them. There has been controversy both about the pattern of evolution itself, and the processes that may cause the pattern. Much of the controversy has emerged from confusion between pattern and process.

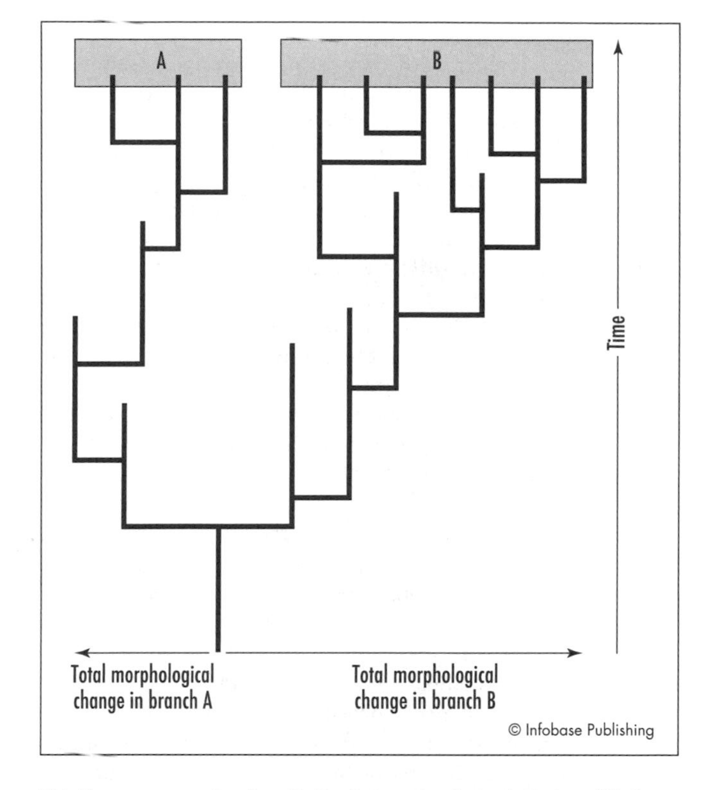

This figure represents a hypothetical example of punctuated equilibria in two branches from a common ancestor. Each vertical line represents the persistence of a species in time. Horizontal lines represent the evolutionary origin of a species, which is not instantaneous but occurs rapidly enough that, on this timescale, the slope is not visible. Lineage B, perhaps because of a different breeding system, speciates more often than lineage A. Two results of this could be that (1) lineage B has more species at the present time than lineage A (shaded boxes); and (2) lineage B has undergone more morphological change (horizontal axis) than lineage A, even though no morphological change at all occurred within any species during its time of existence. Species sorting can produce patterns of evolutionary change and speciation even if evolutionary change does not occur within any species.

Pattern

Prior to the proposal of punctuated equilibria, evolutionary scientists assumed that the characteristics of all species evolved gradually, tracking the gradual changes of the environments in which they lived. This would produce an evolutionary tree in which the branches gradually curved. This was in fact the pattern that they observed, on a very coarse scale, for example over the course of hundreds of millions of years, from widely spaced points on evolutionary trees that were provided by the fossil record. They further assumed that this pattern of gradualism also held on a fine scale, for example over the course of millions or tens of millions of years. They could not confirm gradualism on a fine scale from the fossil record, because, they claimed, so much of it was missing. This was the same argument, "the extreme imperfection of the fossil record," that Darwin had used (see *ORIGIN OF SPECIES* [book]). However, Gould and Eldredge pointed out that the fossil record indicated most evolutionary lineages remained virtually unchanged for long periods of time, and that when evolutionary transformation did occur, it occurred rather quickly. This produced an evolutionary tree with branches that were straight or bent at nearly right angles, rather than branches that gradually curved (see figure). They believed that paleontologists had finally found enough fossils to conclude that the pattern of evolutionary change was not gradual; paleontologists should, they said, believe what they see in the fossil record, rather than appealing to missing fossils of organisms that may not ever have existed.

For most groups of organisms, the fossils are insufficient to distinguish a gradualistic from a punctuated pattern of the origin of any given species. There are a few fossil assemblages, however, that are sufficiently complete that it may be possible to discern the pattern of change over time. Perhaps the most famous of these is the TRILOBITES. Trilobites were marine, abundant, had hard shells, evolved into many species, and persisted for most of the PALEOZOIC ERA. For all these reasons, their fossil record may be more complete than that of any other group of organisms. When the pattern of trilobite evolution was reconstructed, the result was one that both gradualists and punctuationalists could cite as evidence. A small amount of gradual change appeared to occur in all of the species; a significant amount occurred in some, as a gradualist would point out. But the rate of change seemed very slow, relative to the differences among the species, and in some cases the changes fluctuated back and forth, resulting in no net transformation of characteristics, as punctuationalists would point out. In the end, the gradualists have been unable to establish any unbroken series of gradual changes leading to the diversity of species that we see today in the world. The pattern appears, to many evolutionary scientists, to be of species that appear somewhat suddenly and persist relatively unchanged for millions of years before becoming extinct.

The punctuations in the fossil record have often been cited by critics of evolutionary science (see CREATIONISM) as evidence that the evolutionary transitions did not occur; they often refer to them as "gaps" in the fossil record. Most of them claim that such gaps could be bridged only by a Creator inventing new and complex designs (see INTELLIGENT DESIGN). However, the punctuations that appear to have produced new species are in reality quite small. Are the creationists indeed willing to propose millions of independent creations of each species, spread out over billions of years? Between older and more recent species, transitional forms have been found (see MISSING LINKS; WHALES, EVOLUTION OF). These "missing links" appear to have had a punctuated evolutionary origin.

Does the observation of evolutionary changes in populations today offer any insight into which pattern may be more prevalent? Studies of DARWIN'S FINCHES on the GALÁPAGOS ISLANDS show that characteristics such as body and beak size in birds change gradually, tracking environmental changes such as wet vs. dry years, and that these changes are of a magnitude as great as the differences between species. However, these changes exhibit no consistent directional change—the beaks evolve to be larger, then evolve back to being smaller—and the formation of new finch species has not been verified. On the other hand, studies of *Mimulus* flowers in western North America show that a small number of genetic changes can immediately result in the flower having a different pollinator, resulting in immediate reproductive isolation (see QUALITATIVE TRAIT LOCI; SPECIATION). Although the *Mimulus* species can still cross-pollinate, they are usually classified as different species. These observations seem very long-term to human observers, but over evolutionary time they must be considered very brief. Both of these examples can be interpreted as either gradualism or punctuated equilibrium.

Process

Although punctuated equilibrium theory denotes a pattern, not a process, the credibility of the pattern hinges largely on the discovery of a process that could produce it.

The punctuation proposed by Gould, Eldredge, and other evolutionary scientists may appear sudden on the scale of geological time but is not instantaneous. In the fossil record, a speciation event that occurred over the course of 10,000 years would be difficult or impossible to detect; it would be indistinguishable from an instantaneous origin. Some critics of punctuated equilibrium have equated it with the "hopeful monster" theory of geneticist Richard Goldschmidt, or the saltation model proposed by Hugo DeVries (see DEVRIES, HUGO) in which a major evolutionary transition occurs within a single generation. This is an incorrect comparison. The punctuation event, if observed on a human timescale, would certainly appear slow and gradual and require at least several thousand generations—but probably not several million.

The processes necessary for punctuated equilibrium would not, in fact, be any different from those already understood to occur in populations. Instead of slow directional selection over millions of years, the pattern would be (1) rapid directional selection followed by (2) a long period of stabilizing selection (see NATURAL SELECTION). Defenders of punctuated equilibrium insist that there are good reasons for believing that directional selection can be rapid, and that stabilizing selection should be the norm.

Why would directional selection appear rapid? Two reasons have been proposed.

1. *Speciation often occurs during events of relatively rapid and major environmental change.* One major example would be the rapid origin of many diverse types of invertebrate animals near the beginning of the CAMBRIAN PERIOD (see INVERTEBRATES, EVOLUTION OF; CAMBRIAN EXPLOSION), and the rapid origin of many diverse types of mammals, some of them very large, from relatively few, small mammalian ancestors, during the Paleocene epoch (see CENOZOIC ERA; TERTIARY PERIOD; MAMMALS, EVOLUTION OF). The rapid evolution of invertebrates in the Cambrian period may have been spurred by the origin of grazing animals and the end of SNOWBALL EARTH. With the extinction of the dinosaurs, newly evolved mammalian forms encountered very little competition (see DINOSAURS; CRETACEOUS EXTINCTION). Directional selection would occur rapidly when severe environmental changes selected the most extreme individuals in the evolving populations. Alternatively, speciation may occur not as a result of a whole species experiencing rapid environmental change but as a result of part of the species dispersing to, and being isolated in, a new environment: This "peripherally isolated population," though not the entire species, then experiences a rapid environmental change. Not all major periods of evolutionary innovation occur in conjunction with catastrophic environmental change. Flowering plants, and modern bony fishes, both evolved at times of relatively stable climate (see ANGIOSPERMS, EVOLUTION OF; FISHES, EVOLUTION OF).

An example closer to home is the evolution of human brain size. The brains of AUSTRALOPITHECINES and similar HOMININ species were only about 25 cubic inches (400 cc) in volume, similar to the brains of modern chimpanzees, and they did not make and use stone tools. The brains of early *Homo* species, such as some attributed to *HOMO HABILIS,* were larger, about 50 cubic inches (800 cc), and they made crude stone tools. The brains of *HOMO ERGASTER* and other *Homo* species were yet larger, more than 60 cubic inches (1,000 cc), and they made more advanced stone tools. The brains of NEANDERTALS and *HOMO SAPIENS* were and are the largest, more than 90 cubic inches (1,500 cc), and these species made the most advanced tools. This sounds like gradualism, until one considers that australopithecine brain size changed relatively little during the one million to two million years that each species persisted. The same was true of the early *Homo* species, the middle *Homo* species, and the most recent *Homo* species. Most strikingly, the earliest *Homo sapiens* had brains of fully modern size (and perhaps capability) 100,000 years ago. While some cases of brain size increase have occurred (e.g., the brain size of *HOMO ERECTUS* such as Peking man was greatest just before its extinction, and later Neandertals had larger brains than early Neandertals), these changes were much smaller than the differences in brain size between the species. Apparently, australopithecines had an equilibrium brain size, then early *Homo,* after its punctuated origin, had a larger equilibrium brain size, then later *Homo* species, after punctuated origins, had the largest equilibrium brain sizes.

Another possibility is that the appearance of a new species in the fossil record represents the migration of that species from a different location, rather than the origin of that species. Most environmental changes cause species to migrate rather to evolve. This is what happened after the most recent of the ICE AGES.

2. *Speciation often occurs in small populations.* Evolutionary change can occur more rapidly in small populations, if sufficient genetic variability is available. This is partly due to the reduced competition: In a large population, individuals with new characteristics would have to compete with many others, while in a small population, these novel individuals may survive and breed well, so long as they can handle the new environment. Another reason is that processes other than natural selection (such as genetic drift; see FOUNDER EFFECT) can occur in small populations but not in large ones. A small population, on the other hand, would not be likely to leave much or any fossil record. Just by chance, the first fossil formed (and if a paleontologist is lucky enough to find it) would already have been noticeably different from the ancestral population—different enough to be called a new species.

Why would stasis and stabilizing selection be the norm, once a species originates? Genes (see DNA [RAW MATERIAL OF EVOLUTION]) seldom function in isolation. Almost every gene works together with other genes to produce an organism. Natural selection, therefore, favors genes that work in teams. Most MUTATIONS produce genes that do not work even in isolation, and of the ones that do work, they are ineffective members of teams. Therefore, evolutionary scientists expect that over time, most small, isolated populations will become extinct. Once a good team of genes has evolved, it is stable (and experiences stabilizing selection) because most departures from it are inferior in function. This explains why even the gradual, directional changes seen in the fossil lineages of trilobites involve mostly changes in size or number of existing parts, rather than major changes in the parts themselves.

Another reason that stasis would be the norm for a species is that gradual evolutionary change may be occurring in individual populations, with many of the populations evolving in different directions from the others. As members of these populations crossbreed, their different evolutionary directionalities partially cancel one another, resulting in stasis for the species as a whole.

Yet another reason that stasis would be the norm is that species usually respond to relatively minor environmental changes not by evolving but by migrating to a new location. A large, widespread environmental change is necessary to induce evolutionary change, which may then occur rapidly, according to the punctuated equilibrium model.

Species sorting

Punctuated equilibria, on a fine scale, could produce what appears to be gradualism on a coarse scale through a process known as species sorting. It has been called species selection, but since natural selection operates on individuals and not on groups (see GROUP SELECTION), this term is not preferred. As Niles Eldredge explains it, think of a species as a real evolutionary entity: It is born (speciation), it lives, it dies (extinction); and before it dies, it may produce progeny (speciation). One species can be more fit than another if it produces more new species.

1. *Species sorting can produce the appearance of directional change.* Consider the example presented above regarding the evolution of human brain size. The directional increase in human brain size occurred because early *Homo* survived longer than the australopithecines, and later species of *Homo* survived longer than the early species. The overall increase in brain size resulted from rapid directional selection when each species originated but also from differential persistence of the species.

2. *Species sorting can affect the pattern of evolution.* Consider two lineages that are more or less equally well adapted to environmental conditions, and whose populations consist of individuals with more or less equal fitness. The only difference between the two lineages is that one of them is more "speciose" than the other—that is, one lineage tends to fragment into more new species than the other. This could occur because its offspring tend to disperse further, into new and isolated habitats, or because it is very good at evolving ISOLATING MECHANISMS. In the more speciose lineage, genes do not mix as well as in the less speciose lineage. This relatively simple difference in dispersal or gene mixture is enough to cause one lineage to produce more species than the other (see figure on page 329). If each species is equally likely to survive, perhaps by chance, into the next geological era, the lineage that produces more species is less likely to die out. This is how species sorting, quite apart from the quality of adaptation, can cause one lineage to persist longer than another. A possible real example is the ginkgo tree (see GYMNOSPERMS, EVOLUTION OF). During the MESOZOIC ERA, there were numerous ginkgo species, only one of which *(Ginkgo biloba)* survives today (see LIVING FOSSILS). Had the ginkgo lineage been represented by only a few species during the Mesozoic, the entire lineage might be extinct today.

Gradualism was very clearly the idea that Charles Darwin had in mind (see appendix, "Darwin's 'One Long Argument': A Summary of *Origin of Species*"). Darwin took a chance when he declared that if evolution did not occur in gradual steps, one could "rightly reject my whole theory." Huxley (see HUXLEY, THOMAS HENRY) told him, "You load yourself with unnecessary difficulty ..." by insisting on gradualism. According to the defenders of punctuated equilibria, Huxley was right.

The difference between gradualism and punctuated equilibrium is that the former posits gradual directional selection, and the latter posits periods of stabilizing selection punctuated by periods of rapid directional selection. Both models appeal to the same Darwinian processes of natural selection, plus some genetic drift. Though the proponents of each approach defend their views passionately, the impression that outside observers often get—that punctuationists are undermining Darwinian evolution, and that the theory is in crisis—is quite incorrect.

Further Reading

Eldredge, Niles. *Reinventing Darwin: The Great Debate at the High Table of Evolutionary Theory.* New York: John Wiley, 1995.

Gould, Stephen J. *The Structure of Evolutionary Theory.* Cambridge, Mass.: Harvard University Press, 2002.

———, and Niles Eldredge. "Punctuated equilibrium comes of age." *Nature* 366 (1993): 223–227.

Sheldon, P. R. "Parallel gradualistic evolution of Ordovician trilobites." *Nature* 330 (1987): 561–563.

quantitative trait loci Quantitative trait loci (QTLs) are locations on chromosomes, identified by markers, that have a statistical association with certain traits. In some cases, a recognizable trait is caused by a single gene at a single locus on a chromosome (see DNA [RAW MATERIAL OF EVOLUTION]; MENDELIAN GENETICS). The examples of such traits in humans are few and trivial, for example blue vs. brown eyes, and whether or not the thumb points backward (hitchhiker's thumb). Most traits are the product of two or more loci. It would require a tremendous amount of work to identify these loci. Many evolutionary biologists who might benefit from such knowledge do not have the facilities and expertise in molecular genetics to perform such studies. Instead, they identify MARKERS that have a statistical association with those loci.

Following the tradition of Gregor Mendel, researchers cross two genetically different organisms within a species, or individuals of two closely related species, producing an F_1 generation in which the traits mix together. Then they crossbreed the members of the F_1 to produce an F_2 generation in which the traits may separate back out. Using these F_2 individuals, the researchers seek statistical correlations between the traits and genetic markers. The markers that are most strongly correlated with the traits—and there is usually more than one such marker—are considered QTLs. QTLs indicate how many loci influence a trait, which of them are most important, and where they are found on the chromosomes.

In one experiment, evolutionary biologists H. D. Bradshaw and Douglas Schemske crossed two closely related species of *Mimulus* flowers (see HYBRIDIZATION). These two species grow together in the Sierra Nevada of California but do not hybridize because they have different pollinators (see ISOLATING MECHANISMS). Bees pollinate *M. lewisii,* while hummingbirds pollinate *M. cardinalis. M. cardinalis* flowers have more red and yellow pigment, produce more nectar, and are a different shape from *M. lewisii* flowers. Which floral characteristic—color, nectar, or shape—was initially responsible for the evolutionary divergence of these two species? In the F_2 generation after Bradshaw and Schemske hybridized the two species, there was a whole range of flower types that differed in color, nectar production, and shape. Bradshaw and Schemske also identified 66 QTL markers. They then planted the F_2 generation of flowers in the field and observed the responses of pollinators to them. They found that the flowers visited most by hummingbirds were those that had QTL markers associated with flower color and nectar production. The QTL markers associated with flower shape did not have a strong effect on hummingbird visitation. This demonstrates that flower color and nectar production (perhaps controlled by DNA associated with these QTL markers) were responsible for the initial evolutionary divergence of these two species; changes in flower shape later reinforced the divergence.

The use of QTLs allows researchers to identify the locations of DNA sequences that may be important in evolutionary changes such as SPECIATION. Once they known where to look, researchers can begin to identify the genes and control sequences that are responsible for the observed evolutionary events.

Further Reading

Bradshaw, H. D., Jr., et al. "Quantitative trait loci affecting differences in floral morphology between two species of monkeyflower *(Mimulus).*" *Genetics* 149 (1998): 367–382.

Schemske, Douglas, and H. D. Bradshaw, Jr. "Pollinator preference and the evolution of floral traits in monkeyflowers *(Mimulus).*" *Proceedings of the National Academy of Sciences USA* 96 (1999): 11,910–11,915.

Quaternary period The Quaternary period (beginning about two million years ago) is the current period of Earth history (in the CENOZOIC ERA; see GEOLOGICAL TIME SCALE). It follows the TERTIARY PERIOD. The Cenozoic era is also known as the Age of Mammals, because large mammals dominated many parts of the Earth after the extinction of the DINOSAURS following the CRETACEOUS EXTINCTION. The first epoch of the Quaternary period was the Pleistocene epoch, which lasted until about 10,000 years ago, at which time the Holocene (Recent) epoch began.

Climate. The Quaternary period marks the beginning of recurring ICE AGES, in which glaciations and interglacial periods alternate with one another; a glaciation occurs about once every 100,000 years. Around 20 ice ages occurred during the Pleistocene epoch. The Holocene epoch began with the interglacial period following the most recent glaciation.

Continents. The continents were very close to their current locations.

Marine life. Marine life was very similar to what is now found in the oceans.

Life on land. The ecological communities of the continents were very similar to those found today. One major difference was that all of the continents (except Antarctica) were as filled with large species of mammals as Africa is today. At the end of the Pleistocene epoch, many of these large mammal species became extinct, in some places as many as 80 percent of them. Researchers attribute this to a combination of climate change and overhunting by humans (see PLEISTOCENE EXTINCTION).

Many evolutionary scientists and geologists now divide the Cenozoic era into the Paleogene (Paleocene, Eocene, Oligocene) and the Neogene (Miocene, Pliocene, Pleistocene, Holocene) rather than the traditional Tertiary and Quaternary periods.

radiometric dating Radiometric dating uses the measurement of radioactive decay of atoms to determine the ages of volcanic rocks and some organic materials. *Radioactive* elements are those with unstable nuclei: large atoms, or ISOTOPES of smaller atoms. Atoms can decay through the loss of an alpha particle (consisting of two protons and two neutrons), loss of a beta particle (electron), or the capture of a beta particle (which turns a proton into a neutron).

Each of the radioactive elements decays at its own constant rate (see table). After a certain amount of time, half of the radioactive atoms will decay. This time is called the *half-life* of the element. Beginning with 100 percent of a sample of radioactive atoms, 50 percent will remain after one half life, 25 percent after two half-lives, 12.5 percent after three half-lives, and so on. Each of the radioactive isotopes is most useful for a different span of time. If it decays very slowly, it cannot be used to determine the age of geologically young material, as an immeasurably small amount of decay will have occurred. If it decays rapidly, it cannot be used to determine the age of geologically old material, as an unmeasurably small amount of the original material remains.

The ratio of product to original isotope indicates the time since the formation of the material that contains them. The assumption must be made that the element that is the product of decay was not originally present. This assumption is frequently cited by creationists as an invalidation of radiometric dating (see CREATIONISM), but geologists make this assumption only when it is valid. For example, potassium-argon dating may be performed on intact volcanic material because the initial amount of argon (a gas) can be assumed to be zero. The two uranium-lead dating techniques are not performed on entire samples of volcanic rock, in which some lead might initially have been present, but on zircon crystals. Zircon crystals contain zirconium; uranium atoms can substitute for zirconium, but lead atoms cannot. Therefore, when zircon crystals form, they contain uranium atoms and no lead atoms. Any lead atoms in a zircon crystal can be legitimately assumed to have been produced by radioactive decay of uranium. In fact, because zircon crystals contain both isotopes of uranium, radiometric dating can be performed twice on the same crystal to assure that the dates are consistent. Previously, geologists had to selectively cull zircon crystals that had appeared to be altered by weathering, since weathering

Radioactive Isotopes and Half-lives

Radioactive isotope	Half-life (years)	Effective dating range (years)	Product
Rubidium 87 (^{87}Rb)	48.6 billion	0.01–4.6 billion	Strontium 87 (^{87}Sr)
Thorium 232 (^{232}Th)	14.1 billion	0.01–4.6 billion	Lead 208 (^{208}Pb)
Uranium 238 (^{238}U)	4.5 billion	0.01–4.6 billion	Lead 206 (^{206}Pb)
Potassium 40 (^{40}K)	1.3 billion	0.0001–4.6 billion	Argon 40 (^{40}Ar)
Uranium 235 (^{235}U)	71.3 million	0.01–4.6 billion	Lead 207 (^{207}Pb)
Uranium 234 (^{234}U)	245,000	1,000–800,000	Thorium 230 (^{230}Th)
Carbon 14 (^{14}C)	5,730	100–100,000	Nitrogen 14 (^{14}N)

would cause some of the lead to diffuse out. In 2004 a new method was published in which acid-pressure treatment of zircon crystals is used to remove the weathered portions of zircon crystals. Radiometric dating of geological deposits can only be used on volcanic rocks. Therefore sedimentary deposits can only be dated by determining the ages of the volcanic rocks above and below them.

The uranium-lead method was developed by chemist Clair Patterson, starting in 1948. One of the major problems he encountered was environmental lead contamination. Patterson had to design a special clean laboratory and methods for obtaining specimens that were not contaminated. He proved from an examination of ice cores from glaciers that the lead contamination began about 1923, which was the year that tetraethyl lead began to be added to gasoline. In this way, Patterson not only developed a valuable scientific technique but also brought a major problem of environmental pollution to light.

The radiocarbon method can be used on organic material. Organisms incorporate ^{14}C along with other carbon isotopes (^{13}C and ^{12}C) into their bodies. This process stops when the organism dies. Because ^{14}C is only about 0.1 percent of the total carbon in the original specimen, this method cannot be easily used on specimens older than 50,000 years.

The radiocarbon method was invented by chemist Willard Libby, who received the Nobel Prize in 1960 for this discovery. It has proven immensely valuable but there are several problems that have to be taken into account:

- Due to an inaccuracy in the formula that was originally used, all raw radiocarbon dates are too young by three percent. This is why radiocarbon dates are reported in scientific literature as raw or as corrected.
- Another problem is that older organic specimens may be contaminated by carbon from decomposing litter, making the determined ages too young. Alternatively, ^{14}C can be leached out of a specimen. Only a slight amount of leaching of ^{14}C in or out of a specimen can lead to inaccurate age determination. A one percent increase in ^{14}C can make a 67,000-year-old bone appear to be 37,000 years old.
- ^{14}C is produced by cosmic radiation as a neutron enters the nucleus of a nitrogen atom. The amount of ^{14}C in the atmosphere is influenced by the Earth's magnetic field, which varies and reverses (see PALEOMAGNETISM). This is why radiocarbon dates must be calibrated. The standard against which this calibration is performed, called IntCal98, uses tree rings, which are reliable to within 30 years, to calibrate specimens up to 11,800 years before the present, then uses data from marine sediments to calibrate older specimens. Fluctuations in the ratio of ^{12}C to ^{14}C make some periods of history difficult to calibrate, such as the period 750 to 400 BPE, and the Younger Dryas period following the most recent of the ICE AGES, 13,000 to 11,900 years ago.
- Even a diet high in fish can make radiocarbon dates of bones too old.

Uranium 234 (^{234}U) decays into thorium 230 (^{230}Th) relatively quickly and is therefore not very useful for radiometric dating of rocks. Corals, however, incorporate a small amount of uranium into their shells, and some of this is ^{234}U. The decay of ^{234}U into ^{230}Th can be used to determine the age of old coral specimens, in the range of a few thousand to about 300,000 years. In 2004 this method was used to determine the ages of Hawaiian archaeological deposits in which coral was used as a building material.

Particles released by radioactive decay can leave fission tracks in surrounding material. The older the material is, the more fission tracks will be found in it. Fission tracks have also been used as a method of radiometric dating.

Buried objects, such as bones too old for radiocarbon dating, can absorb radioactive material from the deposits that surround them. There are three techniques (thermoluminescence, optically stimulated luminescence, and electron spin resonance) by which the amount of absorbed radioactivity can be measured.

Further Reading

Bard, Edouard, Franke Rostek, and Guillamette Ménot. "A better radiocarbon clock." *Science* 303 (2004): 178–179.

Mundil, Roland, et al. "Age and timing of the Permian mass extinctions: U/Pb dating of closed-system zircons." *Science* 305 (2004): 1,760–1,763.

Robbins, Michael W. "How we know Earth's age." *Discover,* March 2006, 22–23.

recapitulation Biologists who studied the development of animal embryos in the 18th and 19th centuries noticed the similarities between the stages of embryonic development and the range of animal diversity. In this view, embryonic development recapitulates, or retells the story of, animal history. In 1816 the German anatomist Friedrich Tiedemann published evidence that the human brain, during embryonic development, passed through stages that resembled those of simpler vertebrates. He wrote, "We therefore cannot doubt that nature follows a uniform plan in the creation and development of the brain in both the human fetus and in the sequence of vertebrate animals."

Once evolution had been accepted (see DARWIN, CHARLES; *ORIGIN OF SPECIES* [book]; *DESCENT OF MAN* [book]), some scientists translated Tiedemann's sequence of divine pattern into the belief that embryonic development retraced the evolutionary history of vertebrates. The most famous of these evolutionary scientists (see HAECKEL, ERNST) claimed that humans go through fish, amphibian, and reptile stages during embryonic development. His famous slogan for this hypothesis was "Ontogeny recapitulates phylogeny." Haeckel published drawings of embryonic development in a fish, a salamander, a tortoise, a chick, a hog, a calf, a rabbit, and a human. The early embryos of each were virtually identical in structure, while the later embryos showed the distinctive features of each animal. Haeckel was so convinced of recapitulation that his drawings were idealized rather than realistic. In particular, early vertebrate embryos are not as similar to one another as Haeckel depicted. Haeckel's drawings were immediately recognized as inaccurate, and some scientists declared them fraudulent.

Recapitulation is no longer considered an essential evolutionary hypothesis. Evolutionary scientists do not claim that the human embryo actually goes through a fish stage. Recapitulation is a visible pattern that results from other processes:

- Because embryonic development proceeds from simple to complex, it is inevitable that young embryos of any vertebrate species will resemble one another more than will the older embryos. Zoologist Karl Ernst von Baer formulated this hypothesis, still called von Baer's law, in the 19th century.
- Because embryonic development occurs in a sheltered, wet environment, such as inside of an egg or a uterus, embryos of vertebrate species may resemble one another because of their adaptation to a similar environment.
- The early development of the spinal cord means that it will be larger than, and project from, the rest of the embryo, creating the appearance of a tail.
- Some genes have in fact been retained from earlier stages of evolution (see DEVELOPMENTAL EVOLUTION). The human embryo does not literally have gill slits during early development but expresses some of the same genes that begin the production of gill slits in fishes.

Although recapitulation is not an evolutionary process, it does reveal some interesting analogies between evolution and embryology. Animal evolution began with single-celled protists (see EUKARYOTES, EVOLUTION OF). The earliest animal-like multicellular protists may have been hollow balls of relatively unspecialized cells, resembling the blastula state of embryonic development. The earliest true animals may have resembled cnidarians (see INVERTEBRATES, EVOLUTION OF) with an external tissue layer and a tissue-lined gut with a single opening, which also closely resembles the blastula state of embryonic development. Past these stages, however, the comparison between evolution and development is unclear. The best examples of recapitulation may therefore have been at far earlier stages than those that Haeckel illustrated.

The emphasis that Haeckel placed upon recapitulation in his popularization of evolution may have had a major influence upon the theories of psychologist Sigmund Freud. Freud claimed that cultural evolution paralleled the developmental stages of children. Therefore, people in primitive societies were more childlike than people in more advanced societies. Freud considered modern human neuroses to be holdovers of earlier evolutionary stages. Belief in the inheritance of acquired characteristics (see LAMARCKISM) provided Freud with a mechanism by which cultural experiences could be imprinted on the genes that controlled development. Freud's disciple Sándor Ferenczi took the analogy even further. He interpreted the penis as a fish that was heading back into the primeval ocean of the womb, and the postcoital relaxation as a return to the tranquillity of the Precambrian ocean. Most scientists consider the analogies used by Freud and Ferenczi to be even less realistic than that used by Haeckel.

Further Reading

Gould, Stephen Jay. *Ontogeny and Phylogeny.* Cambridge, Mass.: Harvard University Press, 1977.

———. "Freud's evolutionary fantasy." Chap. 8 in *I Have Landed: The End of a Beginning in Natural History.* New York: Harmony Books, 2002.

———. "*Abscheulich!* (Atrocious)." Chap. 22 in *I Have Landed: The End of a Beginning in Natural History.* New York: Harmony Books, 2002.

———. "The great physiologist of Heidelberg." Chap. 27 in *I Have Landed: The End of a Beginning in Natural History.* New York: Harmony Books, 2002.

reciprocal altruism *See* ALTRUISM.

red queen hypothesis The red queen hypothesis was proposed by evolutionary biologist Leigh Van Valen in 1973 as an important component of evolutionary theory. Van Valen found, from a study of the fossil record, that EXTINCTION within evolutionary lineages (not counting MASS EXTINCTIONS) occurred at constant rates. From the viewpoint that evolution produces a continuous improvement in the ADAPTATION of a lineage, this seemed puzzling: Modern lineages were no better at surviving than ancient ones. Van Valen interpreted this to mean that the environment was always changing, with the result that evolutionary lineages could never finish the process of adapting to them. This is mainly true of the biological environment, the other species with which each species interacts (see COEVOLUTION).

In writer Lewis Carroll's *Alice in Wonderland,* Alice and the Red Queen must continually run as fast as they can just to stay where they are, because the world is moving past them just as rapidly. This metaphor reflects what is happening in the process of evolution. When species evolve in response to one another, evolutionary changes in one species may select for evolutionary changes in other species. This may result in two types of evolutionary change:

- *Directional change.* If a species of herbivore evolves the ability to eat certain plants despite their defenses, the plant species benefits from evolving more effective defenses. If predators evolve the ability to detect prey despite their camouflage, the prey benefits from evolving more effective camouflage. Over time, directional change occurs as the defenses of plants and prey accumulate in response to herbivores and predators. This explains why plants have such an astounding array of defensive chemicals.
- *Fluctuating change.* Pathogens have very short generation times and can evolve rapidly; when a pathogen evolves the ability to infect a host that had previously resisted it, the host species benefits from evolving new defenses against it. Some members of a population of a vertebrate host may evolve the ability, through their immune systems, to defend themselves from a pathogen; in response, mutants arise in the pathogen population that can elude the host defenses. Defenses can be gained and lost, resulting in fluctuating changes.

In the red queen evolutionary scenario, there is never "enough" evolution; the plants and the prey never arrive at the point when they have enough defense, and the host never evolves a permanent defense. The herbivores, predators, and pathogens never evolve permanent ways of taking advantage of their victims.

The red queen hypothesis applies particularly to the evolution of sex (see SEX, EVOLUTION OF). An asexual lineage may be perfectly able to evolve adaptations to climatic conditions. But sexual reproduction continually produces new combinations of, for example, defensive chemicals, allowing sexual organisms to more effectively "outrun" their predators or pathogens than asexual organisms.

Further Reading

Van Valen, Leigh. "A new evolutionary law." *Evolutionary Theory* 1 (1973): 1–30.

religion, evolution of *Homo sapiens* is a religious species. This is one of the few behavioral universals of the human species. Though there is individual variation in the strength of this trait, no society is without it; those that tried to eliminate it have failed. Religion is ineradicable. There is clearly no biological basis for particular religions, but it is likely that there is a biological basis for the capacity for religion.

NEANDERTALS apparently did not have religion. The contrast between the religious artifacts of *Homo sapiens* and their absence in *Homo neanderthalensis* in their regions of overlap could hardly be greater. While *Homo sapiens* had intricate burials (a burial of two children in Sungir, Russia, contained 10,000 shell beads, each of which took from one to three person hours to prepare), Neandertals apparently dug the shallowest possible graves to keep the body from stinking. The Shanidar burial of a Neandertal with flowers, even if it is confirmed, is a rare and isolated instance. Neandertal pendants associated with the Châtelperronian culture have been mostly dismissed as imitative of the Aurignacian culture of the CRO-MAGNON, and even if they were not, their connection with religion is unclear. Neandertal caves totally lack the wall art that abundantly and resplendently represents Cro-Magnon religious experience. Anthropologist David Lewis-Williams calls Neandertals "congenital atheists." Yet Neandertals had brains at least as large as those of modern humans. Other groups of *H. sapiens,* all over the world, had a similar abundance of religious practice. The evolution of religion in the modern human species may have involved a change in brain quality unconnected with a change in brain size, which has remained unaltered in *H. sapiens* for at least 100,000 years.

Specific areas and functions of the brain have been implicated in the mystical, religious experience. Neurologists Andrew Newberg and Eugene d'Aquili have been particularly active in researching the brain activities associated with religion. They point out, for example, that when humans enter an altered state of consciousness, the orientation association area of the brain (which in the left lobe is associated with the sensation of having a limited body, and in the right lobe with the sense of space that a person occupies) contribute to out-of-body experiences.

Religion is not pathological, but brain pathologies can help scientists to understand its neurological basis. Stimulation of the right temporal lobe by electrodes (or pathological stimulation by epilepsy) produces experiences closely paralleling the *near-death experience* of passing through a tunnel toward the light. Similar effects are also produced by the drug ketamine. The patterns of brain activity during religious ecstasy are similar, according to Newberg and d'Aquili, to those of other forms of ecstasy, for example sexual ecstasy. Stimulation of the amygdala can create a sensation of awe, which is also part of the complex of religious feelings.

Of course, religion does not consist only of such experiences. Religions also have content: belief in life after death, Supreme Being or beings, universal mythological story lines, etc. The human brain cannot not think; therefore, when religious experience first evolved, humans struggled to make sense of it (see INTELLIGENCE, EVOLUTION OF). A side effect of intelligence may have been the susceptibility to altered mental states, induced by natural compounds (such as psilocybin in some mushrooms), sensory deprivation (as in a cave), ritual rhythmicity, or particularly vivid dreams, which further contributed to the circumstances that favored the evolution of religion.

Religion would probably have been a local aberration in early human populations had it not provided some evolutionary advantage. In modern tribal societies, shamans who have (and can confirm) exceptional religious experiences have considerable social power, which can translate into greater resources and reproductive opportunities; no doubt this was also the case during the prehistory of *Homo sapiens.* David Lewis-Williams points out that this would be the same in the Lascaux cave as in modern charismatic Christianity: In a church as in a cave, those individuals who receive "showers of blessing" are revered by the others. This would be the within-population fitness advantage that is necessary for NATURAL SELECTION. Once the trait was established or at least common within a tribe, this tribe would have advantages over tribes that did not possess it—for example, social cohesion and identity that allowed them to prevail in conflicts. Individuals within the tribe would benefit from membership in a tribe in which religion was established by biology and/or culture (see GROUP SELECTION).

Although all *Homo sapiens* groups have religion, there was a striking development of religion when *Homo sapiens* encountered *Homo neanderthalensis* in Europe, and later when the most recent ICE AGE forced Northern European tribes southward where they encountered tribes that already lived in Southern Europe. Religion then functioned in tribal identity. There were geographical differences in types and styles of artwork—for example, cave painting vs. pendants vs. sculptures. Different caves specialize on different animals, reflecting differences in established traditions.

There have been many different attempted explanations of the Cro-Magnon cave paintings. Some anthropologists say that the paintings were sympathetic magic to promote successful hunting. However, as David Lewis-Williams points out, the set of animals in the paintings is not the same as the animals that the people ate (for example, they did not eat bears and lions), and only 15 percent of the paintings show animals with spears. They obviously performed a religious function, but what was it? Lewis-Williams points out that the paintings resemble what people at that time would have seen during hallucinations. Some of the geometric symbols resemble the images seen during migraines. The animal images are not in their natural habitats, and appear to be floating, with no particular orientation with respect to one another or the ground, and they sometimes lack hooves.

To the Cro-Magnon, as to many recent tribal peoples, the wall of a cave may have represented an interface between the outer world and the underworld. When the prehistoric people entered a cave, they were literally entering the underworld. In the darkness they would hallucinate from sensory deprivation, and possibly also from the high levels of carbon dioxide. Then when the lamps were lit, they would paint the images they had seen. Sometimes the artists would also paint their hands, leaving either positive or negative images of them, as a mark of direct contact with the underworld. Hand contact with the wall, and the process of spit-painting, were part of the overall religious experience. The visionary quality of the paintings is particularly evident in the deepest recesses of the caves, where quick sketches of many animals overlapped.

Next would come the communal aspect. The solitary visionary could then lead other people down into the cave, where his (or her) paintings would represent visual evidence that they had, indeed, made contact with the world of spirit animals. Unlike the modern view of the paintings with harsh light, the Cro-Magnon would have seen them flickeringly illuminated by small lamps. The shamans could have manipulated the visitors by suddenly surprising them with a previously hidden image. They also had flutes, and there is evidence that they struck the sides of stalactites—creating ritualistic sounds. As Lewis-Williams says, "The caves, if not the hills, were alive with the sound of music." Perhaps only the shamans went deep into the caves to paint or carve (some paintings are as much as a kilometer from the entrance), and perhaps only their associates followed them later; the shallower reaches of the cave served as the assembly rooms for the general population, and it is in such places that the large, vivid images are found. The shamans' claim that they had seen these visions would be credible to the general population, all of whom would at least have had (to them inexplicable) dreams.

With the advent of civilization, religion was usually dominated by priests and kings and used as a tool of oppression of the masses, as Marxist and other sociologists have pointed out. The power structure that provided an advantage of some people over others in a society could also allow one city-state to dominate another. This role of religion continues to this day. At the same time, there has been a parallel lineage of prophets who have visions and criticize social norms. In ancient Israel, for example, practically all the prophets were outcasts who lived in huts and caves and were sorely hated by the priests and kings.

The foregoing does not mean that there can be no such thing as revealed religion from a higher deity, in which many scientists believe (see essay, "Can an Evolutionary Scientist Be Religious?"). Rather, evolution explains the neurological basis that made human spirituality possible.

Religion will be with the human species as long as it exists. Not just a religious sense of reverence, but specific and even fundamentalist religious doctrines, seem to be here to stay. With a rush of Enlightenment optimism, Thomas Jefferson said, "There is no young man alive today who will not die a Unitarian." But it is the most fundamentalist forms of religion that are now spreading the most rapidly in the world. Today, as during the Paleolithic, people want the assurance that they have a degree of magical control over their health, wealth, and fate, and over other people; and they want to be told what to believe, rather than to face the dangers of the unknown.

Further Reading

Atran, Scott. *In Gods We Trust: The Evolutionary Landscape of Religion.* New York: Oxford University Press, 2002.

Bering, Jesse M. "The cognitive psychology of belief in the supernatural." *American Scientist* 94 (2006): 142–149.

Broom, Donald. *The Evolution of Morality and Religion.* New York: Cambridge University Press, 2004.

Dawkins, Richard. *The God Delusion.* New York: Houghton Mifflin, 2006.

Dennett, Daniel C. *Breaking the Spell: Religion as a Natural Phenomenon.* New York: Viking Penguin, 2006.

Hamer, Dean H. *The God Gene: How Faith Is Hardwired into Our Genes.* New York: Doubleday, 2004.

Lewis-Williams, David. *The Mind in the Cave: Consciousness and the Origins of Art.* London: Thames and Hudson, 2002.

——— and David Pearce. *Inside the Neolithic Mind.* London: Thames and Hudson, 2005.

Newberg, Andrew, Eugene d'Aquili, and Vince Rause. *Why God Won't Go Away: Brain Science and the Biology of Belief.* New York: Ballantine, 2001.

Paul, Gregory S. "Cross-national correlations of quantifiable societal health with popular religiosity and secularism in the prosperous democracies: A first look." *Journal of Religion and Society* 7 (2005): 1–17. Available online. URL: http://moses.creighton.edu/JRS/2005/2005-11.html. Accessed October 7, 2005.

Shermer, Michael. *How We Believe: The Search for God in an Age of Science.* New York: Freeman, 2000.

Sosis, Richard. "The adaptive value of religious ritual." *American Scientist* 92 (2004): 166–172.

reproductive systems Reproductive systems are adaptations of organisms that promote crossbreeding with other organisms in the population. Reproductive systems are part of the LIFE HISTORY of an organism, which is the pattern of growth and reproduction, from birth to death. Sexual reproduction allows new gene combinations to be produced (see SEX, EVOLUTION OF). Many adaptations have evolved that enhance reproductive success, differently for males and females (see SEXUAL SELECTION).

Across the animal kingdom, individuals typically seek mates that are genetically unrelated to them. Humans do this not only through social convention (the "taboo" against incest is one of the few nearly universal human morals) but also subconsciously. Experiments by Swiss evolutionary biologist Claus Wedekind, in which women were asked to rate the attractiveness of the scent of shirts worn by men (whose

Catkins of male flowers (a) and catkins of female flowers (b) are produced by separate male and female trees in *Populus deltoides*, the cottonwood. *(Photographs by Stanley A. Rice)*

identity was concealed), showed that the women chose men who had immune system proteins most different from theirs.

Evolution depends on the availability of genetic variation, therefore genetic recombination, in populations (see NATURAL SELECTION); thus the importance of reproductive systems can hardly be overemphasized. Since natural selection favors only those characteristics that benefit individuals, the genetic variation promoted by reproductive systems must confer advantages on individuals, not just on populations (see GROUP SELECTION). This entry considers seed plant and animal examples, but other kinds of organisms (e.g., fungi) also have adaptations that promote crossbreeding.

Plant Adaptations That Promote Crossbreeding

Seed plants produce male structures called *pollen,* which contain sperm nuclei, and female structures called *ovules,* which contain egg nuclei and which eventually become seeds. Plant reproductive systems help to assure that pollen travels from one plant to a different plant, where the sperm nucleus fertilizes the egg nucleus. In conifers, the reproductive structures are cones (see GYMNOSPERMS, EVOLUTION OF); in flowering plants, the reproductive structures are flowers (see ANGIOSPERMS, EVOLUTION OF).

Some plant species have separate male and female individuals. In many junipers, for example, male trees produce small brown cones that release pollen into the air, while female trees produce berrylike greenish cones that contain ovules. These greenish cones eventually produce seeds. In cottonwoods, male trees produce dangling catkins of male flowers that release pollen into the air, while female trees produce dangling catkins of female flowers that contain ovules (see photos above). These female catkins eventually produce capsules that release seeds that blow away in the wind. In these *dioecious* plants (dioecy), crossbreeding is inevitable, since there are no individuals that have both male and female parts.

Some plant species have separate male and female reproductive structures, but these structures are on the same individual. In pines, for example, the tips of lower branches produce small brown male cones that release pollen into the air, while the tips of upper branches produce small purplish female cones that contain ovules. After fertilization, the female cones grow larger and eventually become mature brown cones, from which the seeds are released. In birches, dangling catkins of male flowers release pollen into the air from branch tips, while further back on the branches spikes of female flowers contain ovules, from which seeds eventually fall. In these *monoecious* plants (monoecy), crossbreeding usually occurs, since the pollen is more likely to land on the female parts of a different plant than on the female parts of the same plant.

Most flowering plant species have male structures *(stamens)* and female structures *(pistils)* in the same flowers. In these *hermaphroditic* flowers, it would be very easy for the pollen from the stamens to come in contact with the pistils of the same flower, which would be a failure of crossbreeding. However, many flowers have adaptations that prevent this from happening:

- In many flowers, the stamens and the pistils are not active at the same time. In *protandrous* flowers (Greek for "male first"), the stamens release pollen, then wither away; only afterward does the pistil become receptive to pollen. In *protogynous* flowers (Greek for "female first"), the pistils receive pollen, then after the seeds are fertilized the stamens open and release pollen. In the first example, the flower acts as a father first, then as a mother; in the second example, the flower is a mother first, then a father. A flower cannot pollinate itself, because it does not release and receive pollen at the same time.
- In many flowers, the pollen cannot germinate unless it receives the correct chemical signal. If the pollen lands on the pistil of the same plant, it does not receive the correct signal, therefore it never germinates. These *self-incompatible* flowers cannot fertilize themselves. In some cases, flowers may be self-incompatible when they first open, but if they do not get pollinated, they become self-compatible. This allows them to produce some seeds rather than none, in the event that they do not cross-pollinate.
- In some plant species, the flowers contain both male and female parts but the flowers of different individuals have different structures. In some of the flowers the stamens are short and the pistils are long. In others, the pistils are short and the stamens are long. If a pollinating insect visits the first flower, it gets pollen on its head, not on its back. If the insect visits another flower on the same plant, the pollen on its head does not come in contact with the pistil. But if the insect visits another plant, with the other kind of flower, the pollen on its head rubs off on the short pistil; and the long stamens put pollen on the insect's back. If the insect then visits the first kind of flower again, pollen rubs off from its back onto the long pistil. The insects can visit whichever flowers they wish, but pollen is transferred successfully only between flowers with different structures. In this case, the plant cannot pollinate itself, even if the insect visits flower after flower on the same plant (see figure below).

Ancestral flowering plants appear to have been hermaphroditic. What caused dioecy to evolve in flowering plants? Dioecy promotes crossbreeding among individual plants in a population, but this does not prove that the enhancement of crossbreeding was the reason that dioecy evolved in the first place. In order to address this question, it is important to find a species of plants in which some individuals or populations are hermaphroditic and some dioecious. Botanist Summer

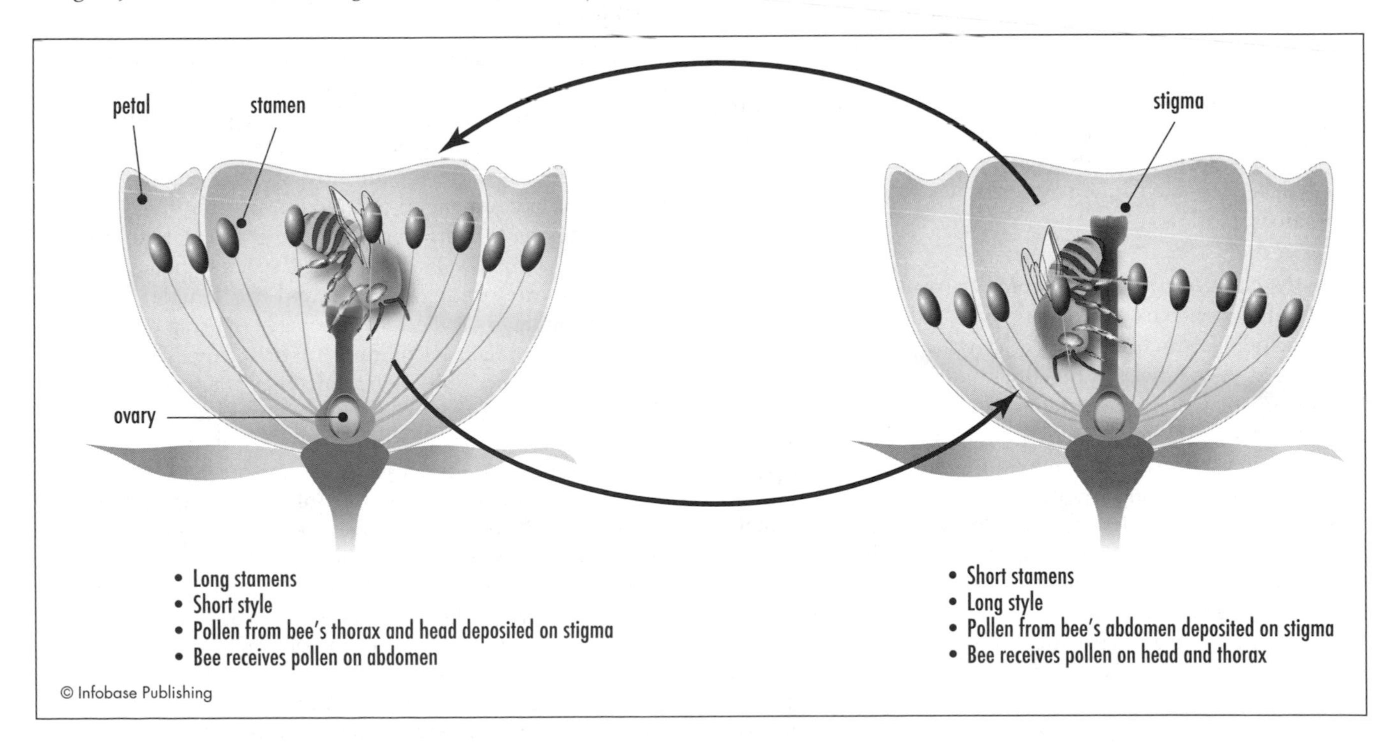

In many species of angiosperms, the flowers come in more than one form. In this example, the pollinator can carry pollen only from one form of flower to another form. This prevents self-pollination.

Scobell has investigated populations of the cactus *Echinocereus coccineus,* in which populations that live along the Rocky Mountain corridor are usually hermaphroditic, while populations that live in desert regions at lower elevation are usually dioecious. Her analysis indicates that the occurrence of dioecy is related to the abundance of hummingbirds. Where there are more hummingbirds (which pollinate the red flowers), the hummingbirds promote out-crossing and the cactus populations are usually hermaphroditic; where hummingbirds are scarce, the cactus populations appear to compensate for this with a dioecious reproductive system.

Plants have evolved numerous adaptations that facilitate pollination. Plants that rely on the wind for pollination produce huge amounts of pollen, sometimes enough to turn the sky yellow (as in ponderosa pine forests in early summer). Angiosperms such as cottonwoods and birches that rely on wind pollination have flowers with small or no petals and do not produce nectar. In contrast, plants that rely on animals to carry their pollen produce flowers with colorful petals and produce nectar that attracts and feeds the animals. The structure of the flower not only permits pollination by certain animals but excludes other animals. Long, tubular flowers, for example, can be pollinated by animals with long tongues; animals with short tongues cannot get down into the tube. The correct pollinators are rewarded; the wrong pollinators, which might carry the pollen to the wrong plant, are excluded. In most cases, the plant rewards the animal and the animal pollinates the plant. However, in some cases, pollination is not mutually advantageous. The animal may steal nectar without touching the stamens, therefore without pollinating the flower; or the plant may trick the animal into pollinating its flowers, without rewarding it, and maybe even killing it. To a certain extent, the evolutionary history of plants has been the increasingly successful adaptation to new climatic conditions. But most of the astounding diversity of plant species, therefore most of the story of plant evolution, has been due to the COEVOLUTION of plants and their pollinators.

Animal Adaptations That Promote Crossbreeding

Nearly all animals are either male or female. The few exceptions include earthworms, which have both male and female parts. Even they, however, breed with one another.

In some cases, as in certain mollusks and fishes, an animal can change from one sex to another. In mollusks that form clusters, an individual may adjust its sex according to the other individuals around it. In fishes as in many other vertebrates, dominant males often get to do all of the breeding. Smaller individuals tend to be female, then when they grow to be large, some of the females will transform into males. Most animals, however, remain one sex all of their lives.

Because most animals are one sex or the other, crossbreeding seems inevitable. But scientists seek evolutionary reasons why most animals are male or female rather than both. The general understanding is that the male function requires the successful delivery of sperm to as many females as possible, while the female function requires not just the receipt of sperm but the production of eggs and, in many cases, the care of the young. The male function may be more efficiently performed by animals that specialize upon behaviors that maximize their reproduction at the expense of other males. The female function may be more efficiently performed by animals that specialize upon behaviors that maximize their care of the offspring. This occurs because a female usually cannot increase reproductive output by having more mates, while a male can—because sperm are cheap. There would therefore be an advantage to an animal that specializes upon just one sexual function and does it well. Plants seldom specialize upon just one sex, because they are limited in the kinds of behavior that they can have. A male cottonwood tree can release pollen into the wind but cannot fight other male cottonwood trees, or perform mating dances that allow female cottonwood trees to choose among them.

Evolutionary biologists have traditionally believed that males benefit from maximizing the number of mates, but females do not, because while the female is pregnant she cannot produce a greater number of offspring by mating again. This is called the Bateman principle, after the geneticist A. J. Bateman who first elaborated it. As a result, there is a continual "battle of the sexes" in which the male wants more mates and the female wants to keep the male from pursuing them. This is expressed in the ditty variously attributed to George Bernard Shaw, William James, Ogden Nash, or Dorothy Parker:

Hoggamus, higgamus,
Men are polygamous,
Higgamus, hoggamus,
Women monogamous.

This pattern has not been supported by research in wild populations. What matters is not male v. female, but which sex invests more in each of the offspring. The number of offspring produced as a function of the number of mates is called a Bateman gradient. In most species, females provide most of the investment in production of offspring. In these cases, males have a steeper Bateman gradient than females (see figure on page 343). In some species, males provide most of the care of the offspring, and it is females that have a steeper Bateman gradient. Moreover, offspring quality may matter just as much as number of offspring. Females may be able to enhance the quality of their offspring by having additional mates, even if they cannot increase the number of offspring.

Males in many animal societies compete with one another for females and tend to be more violent than females, regardless of the type of reproductive system. Sociobiologists (see SOCIOBIOLOGY) attribute human aggression, from individual aggressions to full-scale wars, to a genetically based violent behavior pattern in humans, although many other scientists insist that these behaviors are learned and can be unlearned. Males usually fight other males. In some cases, males are violent toward females. More often, males treat females as resources, and the violence is toward using rather than greatly harming them. Even in humans, the primary historical pattern is for conquering armies to kill the men and rape the women.

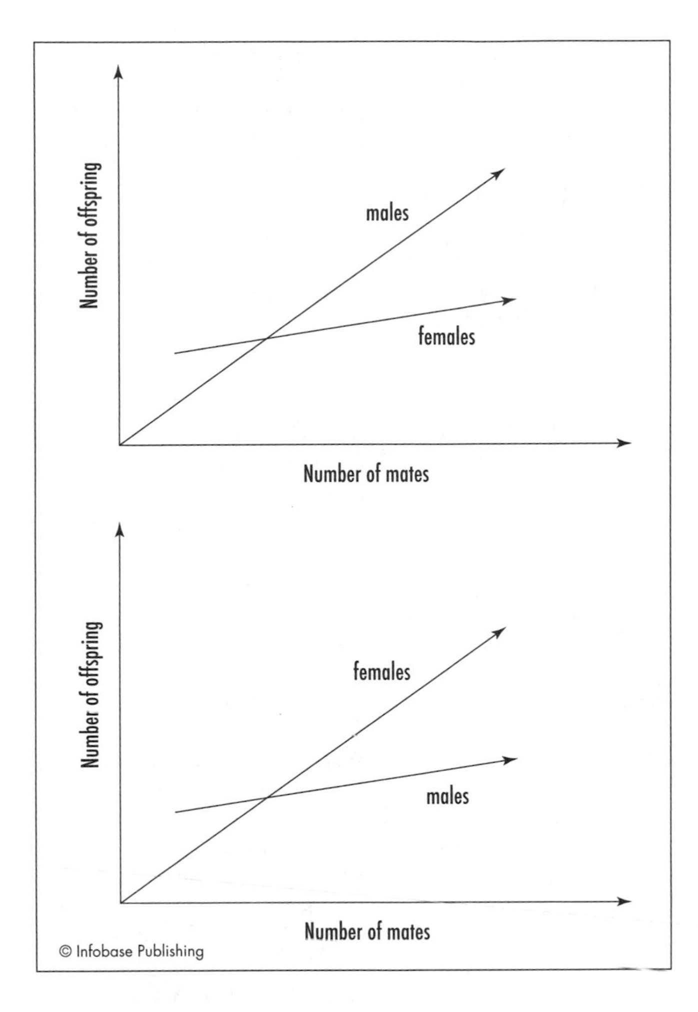

Two different hypothetical Bateman gradients. The graph on the top represents the situation in most animal species, in which males provide little care for the offspring, and can afford to produce many offspring. Males often have no mates or offspring, but when they do, increasing the number of mates increases reproductive success. Females nearly always have offspring, but increasing the number of mates does not help to increase the number of offspring very much. The situation is reversed in the graph on the bottom, which represents the situation in broad-nosed pipefish, in which the males care for the offspring.

Males are often violent toward offspring, though usually not toward their own. Male lions will often kill the cubs of a female he has just acquired. If the female expended time and resources nurturing the cubs, this would reduce her ability to nurture his offspring. One-third of all deaths of infant gorillas are caused by a new dominant male killing them. Even though infanticide is much less common among humans than gorillas or lions, men tend to show less care toward stepchildren than toward their own children. According to behavioral scientists Martin Daly and Margo Wilson, "the truth about Cinderella" is that there is an evolutionary reason why stepchildren receive less care than biological children.

Reproductive systems of most animals can be roughly classified into (1) *monogamy*, where one male and one female are mated for life or at least for a long time; (2) *polygamy*, which includes (2a) animal societies that have a dominant reproductive individual and (2b) *promiscuity*, where individuals mate with other individuals without forming even temporary pair bonds. Animal reproductive systems form a continuum among these categories:

Monogamy. Among the explanations that have been offered for monogamy are:

- *Expensive offspring*. Monogamy is advantageous only when neither the male nor the female could obtain greater reproductive success by finding more mates. This may happen when the offspring are expensive and require considerable care during upbringing, as in humans and some birds. Human offspring are notoriously expensive to produce, vulnerable at birth and during infancy, and dependent upon long years of training, particularly for social interactions. Social interactions may have been the primary stimulus for the evolution of language and of large brains (see INTELLIGENCE, EVOLUTION OF; LANGUAGE, EVOLUTION OF). It would be uncommon without organized society or extended family structures for a single mother to raise children by herself, and even then it is extremely difficult.
- *Certainty of paternity*. Within a monogamous couple, the male can be certain that he is the father. In many animal species, such as lions, the male will kill their mate's offspring if these offspring were sired by other males. This may be in the interest of the male, but not of the female. Monogamy might help to prevent infanticide.

Monogamy is an example of one of the behavior patterns that humans call "virtues" and that turn out to provide an evolutionary advantage. Lifelong and strict monogamy, however, does not yield to this simple evolutionary explanation. In

Monogamous Animal Species

Category	Scientific name	Common name
Invertebrates	*Lysiosquilla sulcata*	Mantis shrimp
	Reticulitermes flavipes	Termite
Birds	*Coragyps atratus*	Black vulture
	Corvus monedula	Jackdaw
	Pygoscelis antarctica	Chinstrap penguin
	Aceros undulatus	Wreathed hornbills
	Cygnus columbianus bewickii	Bewick's swan
Mammals	*Modoqua kirkii*	Kirk's dik-dik (antelope)
	Peromyscus californicus	California mouse
	Microtus ochrogaster	Prairie vole
	Phodopus campbelli	Djungarian hamster
	Hystrix indica	Indian crested porcupine

fact, there is no widely accepted general theory of monogamy, but rather a list, and not a very long one, of special instances of it, widely spaced across the animal kingdom. Humans are not on the list because virtually all human cultures are either polygamous or serially monogamous rather than strictly monogamous (see table on page 343).

Many birds and mammals were considered to be, as far as bird-watchers and animal watchers could tell, monogamous. DNA analyses of offspring, however, reveal that broods of bird and mammal species previously considered monogamous are sired by more than one father. One example is the pied flycatcher. A male flycatcher will often have a primary nest with a primary female, to which he provides most of his efforts such as helping with providing food to the offspring. Meanwhile, he may also have a secondary nest with a secondary female, to which he provides less assistance. The advantage to the male is not surprising; a polygamous male flycatcher produces eight chicks, while a monogamous male produces only five, on the average. The offspring of the monogamous males tend to be smaller. The direction of cause and effect is unclear in this case: Does monogamy produce small size, or does small size produce monogamy (that is, is it only the smaller males that consent to being monogamous)? The primary females average five surviving offspring, the secondary females only three. Why, then, would a female flycatcher consent to being a secondary female? One possible reason is that being the secondary female of a dominant male (which may have a superior territory and superior genes) is still better than being the primary female of a subordinate male (which may have an inferior territory and inferior genes). Another possible reason is that birds, like humans, often hide their infidelities: the secondary female may not know she is secondary; the secondary nest is usually a considerable distance away from the primary nest. Furthermore, this arrangement is not perfect for the dominant male. One-third of all copulations are between a female and an intruding male, usually when the dominant male is away at his secondary nest! One-fourth of all nestlings, even in the primary nest, are the offspring of an intruding male.

In some animal societies, a dominant and mostly monogamous breeding pair produces most of the offspring. Termites typically have a king and queen that produce all of the offspring in the colony. Among wolves, a dominant pair produces all the offspring, but subordinate males and females help with the care of the pups of the dominant pair. Subordinate females may even experience a pseudopregnancy. The subordinate members of the pack may benefit either through inclusive fitness (see ALTRUISM) or may someday become dominant.

Polygamy: societies with dominant males. In some animal species, dominant males keep a harem of females, with the subordinate males on the fringes of society. The dominant male benefits by having many offspring that he can be certain are his. The subordinate males may later become dominant males. The females benefit from the resources commanded by the dominant male. In these species, males tend to be much larger than females and have adaptations (such as big canine teeth) for fighting each other for dominant status. Strictly from the viewpoint of reproduction, there is no reason for males to be larger than the females that bear most of the cost of producing and caring for offspring. Males with harems usually do not have anatomical specializations for giving an advantage of the delivery of their sperm. The penes of male gorillas, for example, are small; however, a male gorilla can be reasonably certain that he is the sole mate of the females in his harem. In a promiscuous breeding system, such as that of chimpanzees, males are not much larger than females but have large penes, which may help deliver their sperm more efficiently into the female reproductive tract than those of other males who have just mated, or will soon mate, with that female.

Animal societies with dominant males usually also have males that sneak into the harem and inseminate some of the females while the dominant male is not looking. These *satellite males* do not produce as many offspring as the dominant male, but they may produce more than any of the subordinate males except those that are fortunate enough to someday become dominant.

Polygamy: societies with one or a few reproductive females. In some insect societies, as in honeybees *(Apis mellifera),* only the queen mates and lays eggs. The other females are sterile workers. This system has evolved only when the males are haploid and the females are diploid, which causes the workers to have a close genetic relatedness to one another. During a mating flight, up to several thousand males (drones) chase one virgin queen, who mates with one or several, and stores their sperm for a lifetime of fertilization. The drones then die. Even though the worker females do not reproduce, their genes are passed on to the next generation because they are closely related to the queen. Scientists estimate that this breeding system has evolved at least 17 separate times among insects.

In some vertebrate societies, such as naked mole rats *(Heterocephalus glaber)* and spotted hyenas, a dominant female suppresses reproduction by other females by hormones and by behavioral influences. In the spotted hyena *(Crocuta crocuta),* the females are dominant. The clitoris is long and penis-like. Baby hyenas must be born through this structure, which tears it; 10 percent of female hyenas die during their first childbirth, and half of the firstborn young are stillborn. Scientists have puzzled over the possible adaptive significance of this system. It may be that natural selection has favored high levels of androgens in the females, which promote aggressive behavior, rather than for the reproductive system itself. Animal societies represent the entire range between equitable reproduction among females to all of the reproduction being performed by one female.

Polygamy: promiscuity. Promiscuity allows a male to produce more offspring by mating with more females (offspring quantity). It may also allow a female to choose genetically superior males (offspring quality). Avoidance of infanticide has also been offered as an explanation for promiscuity as well as for monogamy. Just as a male in a monogamous pair, or with a well-guarded harem, can be sure the offspring is his, a male in a promiscuous species cannot be sure that an infant is not his. When a male in a promiscuous animal society takes on a new mate, therefore, he will not kill the offspring, because they just might be his.

Other reproductive and related activities. Evolutionary explanations are also available for reproductive systems that seem very strange to human observers. In sea horses and some other fishes and amphibians, males may carry the fertilized eggs in their mouths or in special pouches until they hatch. These examples of "fathers giving birth" are not simply weird stories of nature. In aquatic animal species with external fertilization (where females deposit eggs and the males deposit semen on the eggs), there is at that moment no question of paternity. However, if both of them leave, some other male could deposit semen, which might affect the outcome of paternity in at least some of the hatchlings. The male, by gathering the eggs and carrying them, can prevent other males from fertilizing them. Male brooding, far from being an example of males showing care to their offspring, may actually be an example of evolution in response to male–male competition.

Evolutionary hypotheses might also help to explain homosexual behavior. Though not a reproductive system, it is derived from reproductive systems that the animals already have. Homosexual behavior, both between males and between females, is widespread in the animal kingdom. Examples include bonobos, penguins, dolphins, macaques, baboons, and rhesus monkeys. In some cases, it is connected with male aggressive behavior, as in razorbill birds. Animals with homosexual proclivities can pass on their genes, if they are homosexual only part of the time; or, their relatives can pass on the genes if the genes are not expressed. Its widespread occurrence among animals suggests that homosexual behavior may have a genetic basis.

Concealed ovulation. Humans are one of 32 primate species with concealed ovulation. In many animal species, the female advertises ovulation, sometimes with conspicuous swellings, scents, and behaviors. The males, whether in a harem or a promiscuous breeding system, know which females to mate with and when, to maximize their chances of producing offspring. However, in humans, men do not know when women ovulate—nor do the women. Therefore men cannot choose to mate with women only during ovulation.

Two major explanations for the evolution of concealed ovulation have been offered. One, proposed by evolutionary biologists Richard Hamilton and Katharine Noonan, has been called the "daddy at home" theory. Concealed ovulation keeps the male home, because he cannot know when his mate is fertile and has to keep copulating with her over long periods of time in order to produce offspring. If he stays home he is more likely to help provide for, and raise, the offspring. Concealed ovulation might also discourage the man from finding other mates for reproductive purposes, because he cannot know which of them, at any given time, are fertile. This explanation posits an association of concealed ovulation with monogamy.

The other explanation, offered by evolutionary biologist Sarah Blaffer Hrdy, has been called the "many fathers" theory. She suggests that concealed ovulation makes it impossible for a male to know which offspring are really his, or, more to the point, which ones are not, therefore which ones to kill. The females never benefit from having their offspring killed, therefore concealed ovulation might be the female's way of keeping males from killing her offspring. Male gorillas, in a harem system, know which offspring are theirs, and they kill others; male vervet monkeys, in a promiscuous system, do not know which offspring are theirs and seldom kill them. This explanation suggests an association of concealed ovulation with promiscuity.

Nothing could be easier, it would seem, than to simply look at the data to see which explanation is correct. It turns out not to be quite so simple. Swedish evolutionary biologists Birgitta Sillén-Tullberg and Anders Møller gathered data on primate breeding systems and the occurrence of concealed ovulation. They found that nearly all monogamous primates (10 out of 11) had concealed ovulation. This would seem to confirm the daddy-at-home theory. However, the reverse is not true. Of the 32 primate species with concealed ovulation, 22 are not monogamous. According to Sillén-Tullberg and Møller, concealed ovulation may have evolved eight different times in different primate lineages, and it may not be possible to formulate a single explanation for it.

Sex Ratios

In most populations of animals, and of plants that have separate sexes, there is an equal number of males and females. The sex ratio (number of males per female) is about 1:1. This occurs because if one of the sexes is less abundant, each individual of that sex can have more offspring than an individual of the other sex, which creates a selective advantage for the less abundant sex. This is an example of frequency-dependent selection. This occurs only when offspring of the two sexes are about equally expensive for the parents to produce. In some special cases, the sex ratio departs from 1:1.

- Among honeybees, the queen is related to each of her sons and daughters by a relatedness of 0.5. The queen would therefore benefit from producing an equal number of sons and daughters. The workers, however, may be related to one another by a factor of 0.75, but to their brothers by only 0.5. The queen may lay equal numbers of male (haploid) and female (diploid) eggs, but the workers kill many of the males, resulting in a sex ratio considerably less than 1:1.
- In some insects, the larvae parasitize hosts by living inside of them. When they become mature, the only other members of their species that they may encounter may be those that emerged from the same host, which are likely to be their siblings. In such situations, it is advantageous for the parent who lays the eggs to produce only enough males to fertilize the females, resulting in a sex ratio considerably less than 1:1.

Reproductive systems, in general, encourage the production of genetically varied offspring. These systems can take on many and varied forms over evolutionary time and can be influenced by factors that are unrelated to reproduction. Reproductive systems therefore defy complete explanation or even adequate classification.

Further Reading

Asa, Cheryl S., and Carolina Valdespino. "Canid reproductive biology: An integration of proximate mechanisms and ultimate causes." *American Zoologist* 38 (1998): 251–259.

Blanckenhorn, W. U. "Does testis size track expected mating success in yellow dung flies?" *Functional Ecology* 18 (2004): 414–418.

Boyd, I. L. "State-dependent fertility in pinnipeds: Contrasting capital and income breeders." *Functional Ecology* 14 (2000): 623–630.

Daly, Martin, and Margo Wilson. *The Truth about Cinderella: A Darwinian View of Parental Love.* New Haven, Conn.: Yale University Press, 1999.

Diamond, Jared. *Why Is Sex Fun? The Evolution of Human Sexuality.* New York: Basic Books, 1997.

Einum, S., and I. A. Fleming. "Does within-population variation in egg size reduce intraspecific competition in Atlantic Salmon, *Salmo salar?*" *Functional Ecology* 18 (2004): 110–115.

Gjershaug, Jan Ove, Torbjörn Järvi, and Eivin Røskaft. "Marriage entrapment by 'solitary' mothers: A study on male deception by female pied flycatchers." *American Naturalist* 133 (1989): 273–276.

Hrdy, Sarah Blaffer. *The Woman That Never Evolved.* Cambridge, Mass.: Harvard University Press, 1981.

Jones, Adam G., DeEtte Walker, and John C. Avise. "Genetic evidence for extreme polyandry and extraordinary sex role reversal in a pipefish." *Proceedings of the Royal Society of London B* 268 (2001): 2,531–2,535.

Judson, Olivia. *Dr. Tatiana's Sex Advice to All Creation.* New York: Owl Books, 2002.

Kullberg, C., et al. "Impaired flight ability prior to egg-laying: A cost of being a capital breeder." *Functional Ecology* 19 (2005): 98–101.

Locher, R., and R. Baur. "Nutritional stress changes sex-specific reproductive allocation in the simultaneously hermaphroditic land snail *Arianta arbustorum.*" *Functional Ecology* 16 (2002): 623–632.

Marshall, Diane L., and Pamela K. Diggle. "Mechanisms of differential pollen donor performance in wild radish, *Raphanus sativus* (Brassicaceae)." *American Journal of Botany* 88 (2001): 242–257.

Pasonen, Hanna-Lenna, Pertti Pulkkinen, and Markku Käpylä. "Do pollen donors with fastest-growing pollen tubes sire the best offspring in an anemophilous tree, *Betula pendula* (Betulaceae)?" *American Journal of Botany* 88 (2001): 854–860.

Sherman, Paul W., et al. "The eusociality continuum." *Behavioral Ecology* 6 (1995): 102–108.

Sillén-Tulberg, Birgitta, and Anders Pape Møller. "The relationship between concealed ovulation and mating systems in anthropoid primates: a phylogenetic analysis." *American Naturalist* 141 (1993): 1–25.

Wedekind, Claus, et al. "MHC-dependent mate preferences in humans." *Proceedings of the Royal Society of London B* 260 (1995): 245–249.

reptiles, evolution of Reptiles are primarily terrestrial vertebrates, with amniotic eggs, but without feathers or hair. An *amniotic* egg has a shell and a series of fluid-filled membranes that feed and protect the developing embryo. Largely because of the amniotic egg, most reptiles are very well adapted to life on dry land. Reptiles are considered the dominant animals of the MESOZOIC ERA, sometimes called the Age of Reptiles, even though during the Mesozoic as throughout Earth history bacteria have been more diverse and abundant.

Reptiles, as the term is usually used, form a paraphyletic group, which includes some but not all of the lineages that diverged from a common ancestor (see CLADISTICS). Since the bird and mammal lineages evolved from reptile lineages, they would also be considered members of a monophyletic reptile group (see BIRDS, EVOLUTION OF; MAMMALS, EVOLUTION OF). Many evolutionary biologists now distinguish between the synapsids and the reptiles. The synapsid group is today represented only by mammals, while the reptile group is represented by birds and those animals commonly called reptiles. The reptiles, mammals, and birds (amniotes) represent a lineage within the four-legged vertebrates (see AMPHIBIANS, EVOLUTION OF). The major lineages of reptiles are:

- The *synapsids* included the pelycosaurs such as *Dimetrodon* of the PERMIAN PERIOD and the therapsids of the TRIASSIC PERIOD and JURASSIC PERIOD. The mammals, first known from the Jurassic period, evolved within this lineage. The name refers to the single opening of the dermal bones of their skulls.
- The *anapsids* are turtles and their relatives and ancestors, first known from the Jurassic period. They have shells (upper carapace and lower plastron) in which ribs are fused with dermal tissue. The name refers to the lack of an opening in the dermal bones of their skulls.
- The *diapsids* include all other reptiles. The birds evolved from within this lineage. The name refers to the two openings of the dermal bones of their skulls.

The diapsids diverged into the archosaurs and the lepidosaurs:

- The *archosaurs* diverged into the crocodiles and their relatives, first known from the Jurassic period; the pterosaurs, flying reptiles of the Jurassic period and CRETACEOUS PERIOD; and the DINOSAURS. Birds represent the surviving lineage of dinosaurs.
- The *lepidosaurs* diverged into the tuataras, of which only one lineage survives on islands off of New Zealand (see LIVING FOSSILS), and the lizards and snakes, first known from the Jurassic period.

Most reptiles lay eggs on land. Even marine turtles return to land to lay their eggs. As noted above, these eggs usually have hard shells and sufficient food reserves to allow the embryo to develop under dry conditions. In some snakes, the eggs hatch inside the mother's body, allowing the live birth of young.

resistance, evolution of Bacteria, insects, and weeds have evolved resistance to the chemical methods used to control them. The most important modern example of natural selection, both in terms of the clear and easily studied process, and of importance to human health and the economy, is the evolution of antibiotic resistance in bacteria, pesticide resistance in animals, and herbicide resistance in weeds. As biologist Rachel Carson wrote in 1962, "If Darwin were alive today the insect world would delight and astound him with its impressive verification of his theories of the survival of the fittest. Under the stress of intensive chemical spraying the weaker members of the insect populations are being weeded out. Now, in many areas and among many species only the strong and fit remain to defy our efforts to control them."

Antibiotics are chemicals (often produced by fungi) that kill bacteria. Pesticides kill animals, and herbicides kill weeds. Within populations of bacteria, animals, and weeds, individual organisms may be resistant to the chemicals that kill most of the other organisms.

- Within a population of bacteria, penicillin may be deadly to most of the individuals, but a few of them have a genetically based ability to survive exposure to penicillin. This resistance may result from an altered mechanism of bacterial cell wall formation.
- Within a population of insects, certain individuals may have a genetically based ability to resist pesticides that will kill most members of their species. This resistance may result from an increased production of defensive enzymes, the production of an improved defensive enzyme, or a behavioral tendency to detect and avoid the pesticide. For example, some cockroach populations have evolved an aversion to sugar—not a good thing for them, under normal conditions, but it does cause them to avoid sugar-baited pesticide traps.
- Within a population of weeds, certain individuals may have a genetically based ability to resist herbicides that would kill most members of their species. This resistance may result from a mutation that alters the membranes of the chloroplast, the structure within plant cells that carries out photosynthetic food production.

Under normal conditions (in the absence of antibiotics, pesticides, and herbicides), the mutant (resistant) individuals have inferior reproduction:

- Insects that expend much of their metabolic reserves producing resistance enzymes have less energy left over for reproduction than normal insects.
- Rats that are resistant to the rat poison warfarin suffer a high incidence of bleeding and require higher levels of vitamin K in their diets; therefore, under normal conditions, they bleed to death and die of vitamin deficiency more frequently than normal rats.
- Weeds with mutated chloroplasts have inferior photosynthetic food production compared to normal weeds.

Under these conditions, natural selection operates against the resistant individuals (see figure). However, when antibiotics, pesticides, or herbicides are common and persistent in the environment of the populations, natural selection operates in favor of these very same resistant individuals. The heavy and persistent use of chemicals, therefore, creates the perfect environment for the evolution of bacteria, animals, and weeds that resist the very chemicals that humans use to control their populations.

Individual bacteria, animals, and weeds generally cannot change their resistance. They either are resistant or are not. The individuals survive and reproduce, or they do not survive; therefore only the populations and species can evolve resistance (see NATURAL SELECTION).

Antibiotic resistance. Antibiotic resistance can be detected by measuring the zone of inhibition, in which bacteria will not grow near the antibiotic on a culture plate. The overuse of antibiotics has resulted from the tendency of doctors to prescribe antibiotics for mild bacterial infections, and for viral infections that cannot be controlled by antibiotics. This evolution can occur very rapidly. Consider these examples:

- *Intestinal infections.* A study of Swiss hospitals showed that between 1983 and 1990, when only 1.4 percent of the patients were receiving routine antibiotic administrations, there were no samples of the bacterium *E. coli* that could resist any of the five kinds of fluoroquinolone antibiotics. (*E. coli* is a bacterium that lives in human intestines and is usually harmless, but it can cause infections in people with impaired immune systems.) However, between 1991 and 1993, during which time 45 percent of the patients were receiving routine antibiotic administrations, 28 percent of the *E. coli* samples were resistant to all five of the antibiotics.
- *Sexually transmitted diseases (STDs).* From 1993 to 2001 Hawaii experienced a rapid increase in the incidence of gonorrhea bacteria *(Neisseria gonorrhoeae)* that resist fluoroquinolones. Gonorrhea had originally been treated with penicillin, but when the bacteria evolved resistance to penicillin, doctors substituted tetracycline, which also became

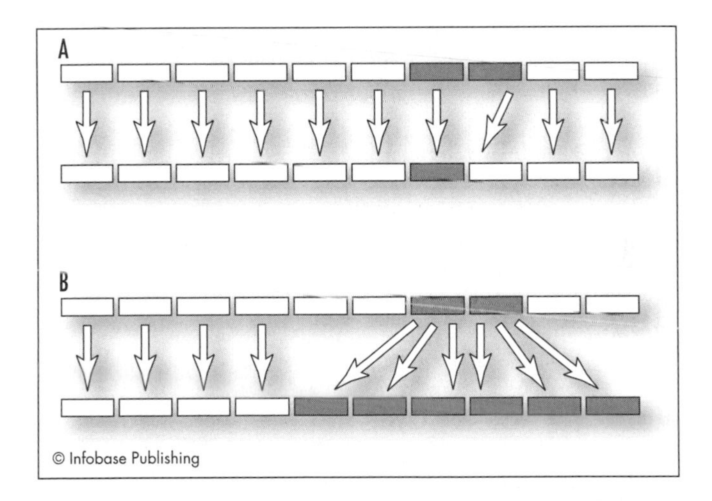

Natural selection of resistance. In A and B, the top line of boxes is one generation, the bottom line of boxes is the next generation, and arrows represent reproduction. (A) In the absence of chemicals, the resistant organisms (shaded boxes) reproduce less often than the nonresistant organisms (white boxes); natural selection operates against them. In this example, the proportion of resistant organisms in the population declines from 20 percent to 10 percent. (B) In the presence of chemicals, the resistant organisms reproduce more often than the nonresistant organisms; natural selection operates in their favor. By chance, a few organisms that are not resistant to the antibiotics will reproduce. In this example, the proportion of resistant organisms in the population increases from 20 percent to 60 percent.

ineffective and was replaced by fluoroquinolines, which are now becoming ineffective also.

- *Food poisoning.* Forty percent of the samples of the food poisoning bacteria of the genus *Campylobacter* in Finland could resist fluoroquinolones in 1996, but 60 percent were resistant in 1999. In America antibiotic-resistant *Salmonella* was present in only 5 percent of samples in 1997 but in 10 percent of samples in 2001.
- *Spinal infections.* In 1996, 10.4 percent of samples of the meningitis bacterium *Streptococcus pneumoniae* resisted penicillin, but by 2001, 51.5 percent of the samples were resistant to penicillin; resistance to macrolide antibiotics increased from 16.5 percent to 30.0 percent during that time.
- *General infections.* Over 90 percent of the strains of "staph" *(Staphylococcus aureus),* a common cause of infections, now resist penicillin and related antibiotics.

Another reason that antibiotic resistance can spread rapidly through bacterial populations is that bacteria can transfer pieces of DNA from one to another, and this transfer can occur even from one bacterial species to another (see BACTERIA, EVOLUTION OF; HORIZONTAL GENE TRANSFER). In eukaryotic species that are totally separate, resistance cannot evolve until the resistance mutations occur within their separate populations; but in bacteria, one resistant species can donate resistant genes to another species!

This is why dozens of species of bacteria have become resistant to one kind of antibiotic; many are resistant to more than one. Some strains of bacteria resist all antibiotics except vancomycin, an antibiotic rarely used because it is difficult to administer and because of its severe side effects. But even vancomycin resistance has occurred in bacteria. At first, it was only in the harmless intestinal bacterium *Enterococcus faecium.* When these bacteria transferred their resistance genes to more harmful bacteria, the result was what public health officials refer to as a *superbug.* This happened in 1996, when the first case of intermediate-level vancomycin resistance was reported in staph bacteria in Japan. As of 2002 the United States had eight confirmed cases. The year 2002 also saw the first case of staph bacteria that were completely resistant to vancomycin, rather than having intermediate resistance. Six cases have now been reported in the world.

When a bacterium resistant to penicillin (or a similar drug such as the widely used amoxicillin) infects a person, the physician can administer a different antibiotic, saving the life of the patient if the second antibiotic is administered in time, and if the bacterium does not resist it also. Because the trial of each new antibiotic on a patient wastes precious time, a *multidrug-resistant bacterial strain* threatens the patient's life.

Because hospitals represent a fertile breeding ground for the evolution of antibiotic resistance in bacteria, patients often become infected with these bacteria when they are in the hospital. As one observer noted, a hospital is a good place to go to get sick. About two million Americans experience *nosocomial* (hospital-acquired) infections each year, and more than half of these infections resist at least one antibiotic. Medical scientists Richard P. Wenzel and Michael B. Edmond estimate that between 17,500 and 70,000 people die from nosocomial infections each year in the United States.

Antibiotic resistance has also resulted from the routine addition of antibiotics to livestock feed. Some of the *Salmonella* bacteria, normally present in livestock, have evolved resistance to the antibiotics used in the feed. In some cases, these bacteria have spread to people and caused infections, as a result of contact or improper food handling. The *Campylobacter* mentioned previously is also a common contaminant in supermarket meats. The use of antibiotics in livestock feed has contributed to the evolution of resistance to some antibiotics that may be crucial for human health. The heavy use of the antibiotic Baytril in poultry feed encouraged the evolution of bacteria that could resist Cipro, since the two antibiotics are biochemically similar to one another. Cipro is considered one of the most important antibiotics that can be used in response to bioterrorist attacks, for example with anthrax. The U.S. federal government banned the use of Baytril in poultry feed in August 2005.

Antibiotic resistance can even evolve within a population of bacteria in a single host individual. This frequently happens during infections, such as tuberculosis, that require a long period of antibiotic treatment. This explains why it is often more difficult to treat a relapse than to treat the original infection in a patient. HIV can evolve resistance to the drugs that are used against it, even within the body of a single host (see AIDS, EVOLUTION OF).

Bacteriologist Stuart Levy has brought together a set of recommendations for the judicious use of antibiotics, for the protection of individuals, and to maintain the effectiveness of antibiotics:

For individuals:

- Wash fruits and vegetables before consuming them.
- Avoid raw eggs and undercooked meat.
- Use antibacterial soaps only when needed to protect patients with immune deficiency.
- Complete the full course of prescribed antibiotics.

For doctors:

- Wash hands thoroughly between patients.
- Do not prescribe antibiotics unnecessarily, e.g., for viral infections.
- Prescribe antibiotics that target the narrowest possible range of bacteria.
- Isolate patients with multiple-drug-resistant strains of infectious bacteria.

Pesticide and herbicide resistance. The extensive and heavy use of pesticides (see photo on page 349) has selected pesticide-resistant insects and rats; the extensive and heavy use of herbicides has selected herbicide-resistant weeds. The first DDT-resistant mosquitoes in Pakistan were detected in 1965, just five years after DDT use began in the region. The swift evolution of resistance occurs with each new pesticide. In the last half decade, more than 520 species of insects and mites, 273 weed species, 150 plant diseases, and 10 rodent species have developed genetic resistance to at least one pesticide or herbicide.

A U.S. soldier sprays DDT on a civilian shortly after World War II. Originally, DDT was effective against mosquitoes that spread malaria and lice that carry typhus. The World Health Organization claims that the use of DDT saved 25 million lives. However, shortly after the widespread use of DDT began, insect populations began to evolve resistance to it. ***(Courtesy of the Centers for Disease Control and Prevention, Public Health Image Library)***

Many insect species are now becoming resistant to all major classes of insecticides. An insect that is resistant to one pesticide is often resistant to other related, chemically similar pesticides. In addition to promoting the evolution of resistant strains of bacteria, animals, and weeds, the over-use of chemicals in medicine, insect control, and agriculture has resulted in the buildup of toxicity in the environment and in the food chain.

The evolution of pesticide resistance in insects and rat populations has implications both for public health and for agriculture. Pesticide sprays are heavily used to control many species of insects, such as the mosquitoes that spread diseases like malaria and yellow fever. Insect pests (such as weevils and caterpillars) cause millions of dollars of agricultural yield loss. When the sprays become less effective (because many, though not all, of the insect populations evolve resistance), public health officials may respond by increasing the application of sprays, which has two effects: greater environmental contamination, and more evolution of resistance.

The evolution of herbicide resistance in weeds also portends greater agricultural yield loss (due to competition between weeds and crops) and greater environmental contamination from stepped-up herbicide applications. In recent years, crops (such as maize, soybeans, and potatoes) that have been genetically engineered to resist herbicides have come on the market, and large acreages are now devoted to them. Farmers can now spray massive amounts of herbicides on their fields, to kill the weeds, without fear of killing the crops. This encourages the heavy use of herbicides—which is precisely what the companies that sell the seeds of these crops want, since the herbicides are manufactured by the same companies that sell the seeds. Unfortunately, this will result in even greater environmental contamination.

There is some concern that the genetic basis of herbicide resistance, which has been put into crops by genetic engineering, could spread to wild weed species by crossbreeding (see HYBRIDIZATION). While this has not happened extensively, it has occurred: Herbicide resistance has been transferred from crops to weeds within the crucifer (mustard) family by cross-pollination. Once the resistance genes are in the wild weed populations, natural selection will favor their spread whenever herbicides are used.

An evolutionary understanding of medicine, public health, and agriculture demands that antibiotics, pesticides, and herbicides be used sparingly in order to minimize environmental contamination and to prevent the evolution of resistance in the very species of organisms humans are trying to control. By restraining the use of chemical control agents, they will remain effective when they are really needed in an emergency.

The dangerous emergence of bacteria, pests, and weeds that resist the chemicals that we use to control them has resulted from our overuse of chemical control agents. The overuse of antibiotics, pesticides, and herbicides resulted not only from overlooking the laws of ecology but also the laws of evolution. As noted in the Introduction to this encyclopedia, "what you don't know about evolution can kill you."

Further Reading

Amábile-Cuevas, Carlos F. "New antibiotics and new resistance." *American Scientist* 91 (2003): 138–149.

Brookfield, J. F. Y. "The resistance movement." *Nature* 350 (1991): 107–108.

Carson, Rachel. *Silent Spring.* New York: Houghton Mifflin, 1962.

Department of Health and Human Services, Washington, D.C. "Centers for Disease Control and Prevention." Available online. URL: http://www.cdc.gov. Accessed May 4, 2005.

Enright, Mark C. "The evolutionary history of methicillin-resistant *Staphylococcus aureus* (MRSA). *Proceedings of the National Academy of Sciences* 99 (2002): 7,687–7,692.

Hegde, Subray S., et al. "A fluoroquinolone resistance protein from *Mycobacterium tuberculosis* that mimics DNA." *Science* 308 (2005): 1,480–1,483. Summarized by Ferber, Dan. "Protein that mimics DNA helps tuberculosis bacteria resist antibiotics." *Science* 308 (2005): 1,393.

Levy, Stuart B. "The challenge of antibiotic resistance." *Scientific American,* March 1998, 32–39.

Monnet, Dominique L., et al. "Antimicrobial drug use and methicillin-resistant *Staphylococcus aureus,* Aberdeen, 1996–2000." *Emerging Infectious Diseases* 10 (2004): 1,432–1,441. Available online. URL: http://www.cdc.gov/ncidod/eid/vol10no8/02-0694.htm. Accessed May 4, 2005.

Regoes, Roland R., and Sebastian Bonhoeffer. "Emergence of drug-resistant influenza virus: Population dynamical considerations." *Science* 312 (2006): 389–391.

Smith, P., et al. "A new aspect of warfarin resistance in wild rats: Benefits in the absence of poison." *Functional Ecology* 7 (1993): 190–194.

Stix, Gary. "An antibiotic resistance fighter." *Scientific American,* April 2006, 80–83.

Tiemersma, Edine W., et al. "Methicillin-resistant *Staphylococcus aureus* in Europe, 1999–2002. *Emerging Infectious Diseases* 10 (2004): 1,627–1,634. Available online. URL: http://www.cdc.gov/ncidod/EID/vol10no9/04-0069.htm. Accessed May 4, 2005.

WeedScience.org. "International Survey of Herbicide Resistant Weeds." Available online. URL: http://www.weedscience.org/in.asp. Accessed May 4, 2005.

Wenzel, Richard P., and Michael B. Edmond. "The impact of hospital-acquired bloodstream infections." *Emerging Infectious Diseases* 7 (2001): 174–177. Available online. URL: http://www.cdc.gov/ncidod/eid/vol7no2/wenzel.htm. Accessed October 7, 2005.

respiration, evolution of Respiration is the process by which cells transfer energy from food molecules to ATP. ATP (adenosine triphosphate) is almost universally used as the molecule that puts energy directly into enzyme reactions. And since almost all biological reactions are controlled by enzymes, ATP is called the "energy currency of the cell." Cellular respiration, which produces ATP, occurs in many bacteria (see BACTERIA, EVOLUTION OF), and in the cytoplasm and mitochondria of eukaryotic cells (see EUKARYOTES, EVOLUTION OF). Food molecules store energy over relatively long time periods, while ATP puts energy to immediate use. Therefore mitochondria are like power plants, which transfer the energy from long-term storage (coal, natural gas) into immediately available forms like electricity.

The earliest bacteria, during the Archaean eon (see PRECAMBRIAN TIME), were anaerobic: not only was there no oxygen gas in the environment, but oxygen gas would have been deadly to them. Their descendants are the anaerobic bacteria that today can only live in mud and in the intestines of animals. These bacteria, as well as a few anaerobic protists and invertebrates (see INVERTEBRATES, EVOLUTION OF), use a set of reactions called *glycolysis* to break down glucose sugar molecules and release some energy from them into ATP. In these organisms, glycolysis is followed by *fermentation.* In some cases, as with yeasts, fermentation produces ethyl alcohol (ethanol); in other cases, as with *Lactobacillus* bacteria that make milk into yogurt, fermentation produces lactic acid (lactate). In some cases, the cells of organisms that rely upon oxygen can revert temporarily to a dependence on glycolysis and fermentation, when oxygen is not available. Muscle cells, for example, can revert to fermentation when they work so fast that the blood cannot supply sufficient oxygen to them. Muscle pain results from the buildup of lactate, and the muscles develop an *oxygen debt.* The muscles stop working after a few minutes of emergency oxygen debt. Ethanol and lactate, however, still contain most of the energy that was in the original glucose.

During the Proterozoic era of the Precambrian, cyanobacteria began producing oxygen gas, which gradually accumulated to become abundant in the oceans and atmosphere (see PHOTOSYNTHESIS, EVOLUTION OF). This was a crisis for anaerobic bacteria, which could survive only in the places where they are found today. However, at this time, some bacteria evolved a set of reactions *(aerobic respiration)* that not only tolerated oxygen gas but actually made use of it. A cycle of chemical reactions (the Krebs cycle or citric acid cycle) breaks down the product of glycolysis into carbon dioxide, which diffuses into the environment as a waste product. Products from the Krebs cycle produce electric currents in membranes. From both the cycle and the currents, large amounts of ATP are produced. However, in the electric currents, the electricity has to have someplace to go. It goes into oxygen molecules, converting them (along with hydrogen ions) into water molecules.

The primitive anaerobic bacteria had only glycolysis and fermentation. The more advanced bacteria, and mitochondria, have both glycolysis and aerobic respiration. The more advanced reactions appear to have been added onto the more primitive reactions, perhaps by horizontal gene transfer or by an ancient genetic merger between two kinds of bacteria, one with glycolysis (like the anaerobic bacteria still extant) and one with aerobic respiration (now extinct). Respiration occurs in the mitochondria of many protists and almost all eukaryotic cells. The set of glycolysis reactions in the cytoplasm, and of aerobic respiration reactions in mitochondria, are nearly identical to the corresponding reactions in bacteria. This is not coincidence. Mitochondria are the evolutionary descendants of bacteria that moved into and formed a mutualistic association with primitive eukaryotic cells (see SYMBIOGENESIS).

Further Reading

Raymond, Jason, and Daniel Segrè. "The effect of oxygen on biochemical networks and the evolution of complex life." *Science* 311 (2006): 1,764–1,767. Summary by Falkowski, Paul G., "Tracing oxygen's imprint on Earth's metabolic evolution." *Science* 311 (2006): 1,724–1,725.

Sagan, Carl (1934–1996) American *Astronomer* Carl E. Sagan was famous as an astronomer who adopted an evolutionary perspective, and as a popularizer of science, particularly evolution. His books and media productions served as the major source of evolutionary science for many thousands of people.

Born November 9, 1934, Carl Sagan was an undergraduate and then graduate student at the University of Chicago, where he earned a doctorate in astronomy and astrophysics in 1960. He taught at Harvard until 1968, when he moved to Cornell University, where he quickly advanced to the rank of full professor in 1971.

Astronomy. Sagan's evolutionary contributions to astronomy included:

- *Planets other than Earth that might have life.* Sagan was among the first to propose that some moons such as Jupiter's Europa and Saturn's Titan might possess oceans and thus be suitable for the ORIGIN OF LIFE. These proposals were later confirmed by spacecraft. He considered that planetary systems, such as the chemistry and circulation of an atmosphere, would not only influence life but be influenced by life. This interactive view of evolution and environment (see GAIA HYPOTHESIS) was more fully developed by chemist James Lovelock, and by Sagan's first wife, who is one of the most significant modern evolutionary biologists (see MARGULIS, LYNN).
- *The search for extraterrestrial life.* Sagan believed that intelligent life might be common in the universe, a belief that has been recently challenged (see essay, "Are Humans Alone in the Universe?"). He designed a plaque that would be able to communicate information about humans to intelligent beings regardless of their language. This plaque was included with a Voyager spacecraft that eventually left the solar system. Sagan also was a major proponent of the use of radio telescopes to listen for nonrandom signals from outer space (SETI, or the Search for Extraterrestrial Intelligence).

Popularization of science. Sagan's contributions to the popularization of evolutionary science included *Cosmos,* a highly acclaimed 13-part PBS series aired in 1980, which presented not only extraterrestrial evolution but also the story of evolution on Earth. He wrote popular books about evolution, including *The Dragons of Eden* and *Broca's Brain.* His novel *Contact,* about humans making contact with an extraterrestrial intelligence, was a best seller and was adapted into a prizewinning movie in 1997. Sagan frequently appeared on *The Tonight Show* and became associated with the distinctively delivered phrase "billions and billions," referring to the vast number of galaxies in the universe, and which was eventually used as a book title. Sagan also considered many traditional religious beliefs to be dangerous to the intellectual development of individuals and nations. In contrast to frozen religious dogma, science corrects itself, as he explained in *The Demon-Haunted World,* in which he presented some of his own mistakes as an example of how scientists change their minds when new evidence shows them to be in error (see SCIENTIFIC METHOD).

Environmental issues. Sagan's intense interest in environmental issues stemmed from his study of astronomy. He believed that intelligent life may originate frequently in the universe, but the Fermi Paradox (based upon an offhand statement by physicist Enrico Fermi) indicated that if intelligence was common in the universe, humans should have already been contacted by these other intelligent creatures. To Sagan, this indicated a strong likelihood that advanced civilizations tend to destroy themselves before they have a chance to develop space travel or have time to send and receive communications over the vast universal distances. Among the environmental threats that Sagan thought might destroy humankind were the following:

- *Nuclear winter.* Sagan was one of the authors of the famous 1983 "TTAPS Paper," named after the initials of the five authors, which predicted that nuclear winter might follow a major nuclear exchange. Previous scholars who investigated the possible consequences of nuclear war focused on the destruction of human life and civilization. Sagan and coauthors claimed that the billows of black smoke would block sunshine and cause worldwide freezing, and a collapse of food production, in countries not bombed by nuclear weapons. Once a considerable portion of the Earth was covered by ice, the reflection of sunlight would prevent it from melting and maintain a permanent frozen state. The TTAPS paper was controversial, as was the paper that followed it in the journal of the American Association for the Advancement of Science, which was coauthored by more than 20 scientists and claimed that a nuclear winter might result in the extinction of all life on Earth. Subsequent research has shown that a major nuclear exchange was more likely to result in a "nuclear autumn" than a nuclear winter. Nuclear winter would be similar to what happened in the CRETACEOUS EXTINCTION, which was at the time a still-controversial theory, and in SNOWBALL EARTH, which had not yet been publicized and remains controversial today. Sagan's advocacy of nuclear winter theory was one of the main reasons that he spoke out and performed acts of civil disobedience against the development of outer-space nuclear weapons during the presidency of Ronald Reagan.
- *Global warming.* Sagan had studied the way carbon dioxide causes an intense GREENHOUSE EFFECT on Venus, and he was familiar with the evidence that an excessive greenhouse effect, caused by human activity, was beginning on Earth.

Carl Sagan contributed to the public understanding of science by bringing together insights from different fields of study, and by his zeal for the importance of science. Sagan died December 20, 1996.

Further Reading

Davidson, Keay. *Carl Sagan: A Life.* New York: John Wiley and Sons, 2000.

Sagan, Carl. *Billions & Billions: Thoughts on Life and Death at the Brink of the Millennium.* New York: Random House, 1997.

———. *Broca's Brain: Reflections on the Romance of Science.* New York: Random House, 1979.

———. *Contact.* New York: Simon and Schuster, 1985.

———. *Cosmos.* New York: Random House, 1980.

———. *The Demon-Haunted World: Science as a Candle in the Dark.* New York: Ballantine Books, 1997.

———. *The Dragons of Eden: Speculations on the Evolution of Human Intelligence.* New York: Bantam, 1977.

———. *Pale Blue Dot: A Vision of the Human Future in Space.* New York: Ballantine, 1997.

Turco, R. P., O. B. Toon, T. P. Ackerman, J. B. Pollack, and Carl Sagan. "Nuclear winter: Global consequences of multiple nuclear explosions." *Science* 222 (1983): 1,283–1,292.

Sahelanthropus *See* AUSTRALOPITHECINES.

scala naturae Also known as The Great Chain of Being, the scala naturae (scale of nature) was an ancient and medieval classification of all created objects that arranges them from lower to higher and makes connections among all of them. Inanimate objects (such as the four elements air, water, earth, and fire) are lowest on the scale, organisms without consciousness (such as plants) are a little higher, animals a little higher, humans yet higher, and angelic beings highest. The scala naturae was often interpreted to show connections among the created objects. From this viewpoint, sponges were between animals and rocks, and scholars expected that animals intermediate between humans and animals should exist.

Among the earliest sources of the scala naturae are the Greek philosopher Plato and the writings of the fourth century C.E. theologian and philosopher Augustine. Augustine's two major reasons for laying the foundation of the scala naturae were the principle of plenitude and a defense of theodicy. *Plenitude* refers to the completeness of creation: It contains all levels of complexity and all intermediates among types. The world, in order to operate, requires all these levels and intermediates. While scholars in the Augustinian tradition considered plants below animals and animals below people, they knew that animals could not exist without plants or people without a world of plants and animals. *Theodicy* is a subdiscipline of theology which investigates the reasons why a good and all-powerful God would allow suffering and death. Although the scala naturae never offered a complete explanation of human suffering and death, it did explain why animals and plants had to die: The world would not operate without having them as food.

Some observers have noted a similarity between the scala naturae and the ecological understanding of nature, with food chains and complex webs of interaction, which began late in the 19th century. If this similarity is real, it did not lead ancient or medieval scholars to understand actual ecological processes. Other observers have noted a similarity between the scala naturae and evolution, especially when they find out that the scala naturae included animal-human intermediates. This similarity is accidental. First, ancient and medieval scholars had no concept of the development of one form of organism from another, or of organisms from inanimate beginnings. Second, the evolutionary process has produced a TREE OF LIFE with branches that, except for SYMBIOGENESIS, separate from one another rather than forming a network as depicted in the scala naturae.

Further Reading

Lovejoy, A. B. *The Great Chain of Being.* Cambridge, Mass.: Harvard University Press, 1936.

scientific method Darwin's friend Thomas Henry Huxley (see HUXLEY, THOMAS HENRY) called the scientific method "organized common sense" or "simply common sense at its best," which is almost completely true. There is no established list of steps or rules in the scientific method, but certain features are universally recognized by scientists, some of which are presented here.

Science investigates only physical causation and occurrences. This does not mean that scientists believe only in physical processes. Many scientists are religious people (see essay, "Can an Evolutionary Scientist Be Religious?"). However, scientists do not introduce non-physical causation into scientific explanations or investigation (see figure at right). Scientists recognize that science is one way of knowing—a very powerful and successful way, but not the only way. As paleontologist Stephen Jay Gould frequently pointed out, science has been extremely successful at explaining how the world works and how it evolved but is not always very useful in helping to generate values and ethics. It is just not designed to do that. Perhaps it may have been said best by the astronomer Galileo: Science tells humans "how the heavens go, not how to go to Heaven." Scholars may use the scientific method to investigate the reliability of religious texts, or the mental basis of religious experience (see RELIGION, EVOLUTION OF), but usually do not presume to use science as a basis for their religious experiences. Therefore, most scientists dismiss the recent INTELLIGENT DESIGN challenges to evolutionary science because the Intelligent Designer, admitted by everyone to be a supreme supernatural being, cannot "itself" be investigated scientifically. Belief in a supreme spiritual being is, as it has always been, a legitimate human activity, but it is not science. In this way, as in many others listed below, the scientific method is useful only because of its limitations. If science tries to be everything, it will end up being nothing. If no physical explanation is possible, the scientist concludes that he or she does not have a scientific explanation—even if he or she privately believes a religious explanation.

Some scholars (see WILSON, EDWARD O.) have attempted to bring ethical and religious concepts into a unified field of scientific knowledge, an approach that Wilson calls *consilience.* At the present time, however, this concept has not gained universal acceptance.

Science investigates only repeatable occurrences. This is the principal reason that science is limited to physical processes and data. Miracles happen only once; or, if they happen more than once, they cannot be counted on to happen on any investigable schedule. Some creationists claim that creation science is scientific in the same sense that forensic science is scientific. The occurrences of murder are unique and unpredictable, but forensic science can investigate them. However, this reasoning is invalid. Forensic science is based upon facts and processes that are well understood from many repeated observations: regarding rates of bodily decomposition, symptoms of trauma, the physics of footprints, etc. No such database exists for the results of miracles.

Scientific research tests hypotheses. This is probably the main feature that distinguishes science from other approaches to knowledge. Scientific research poses statements that can be tested—that is, for which a clear answer can be obtained. It is the testing of specific hypotheses that distinguishes science from a mere "walk in the woods," in which everything is observed, or the accumulation of knowledge for its own sake. Scientific observations and measurements are directed toward testing hypotheses. The scientific method can take idle speculation and turn it into an investigation that yields a clear answer—an answer to a testable hypothesis. Scientists usually try to keep hypotheses as simple as possible, in order to test them easily and clearly.

THEN A MIRACLE OCCURS . . .
S. Harris
"I THINK YOU SHOULD BE MORE EXPLICIT HERE IN STEP TWO."

This cartoon illustrates what makes scientific inquiry different from other "ways of knowing." ***(Courtesy of Sidney Harris)***

The scientific method of hypothesis testing need not be restricted to subjects generally called scientific. For example, the scientific method can help a person to determine why the oil light comes on in an automobile (see figure on page 356):

- Hypothesis 1: The car needs more oil. To test the hypothesis: check the oil level. If this hypothesis is not confirmed, try another hypothesis.
- Hypothesis 2: The oil has no pressure (the pump is defective). To test this hypothesis: check the oil pump. If this hypothesis is not confirmed, try another.
- Hypothesis 3: The oil light is illuminating improperly. To test this hypothesis, check the oil light itself.

Yet other hypotheses can be suggested: a clogged oil pipe, for example. This example demonstrates that the scientific approach can be used in daily life for almost anything (except miracles).

Where do hypotheses come from? This is the part of the scientific process that is closest to being an art. Hypotheses are generated by the minds of scientists who are very, very familiar with the systems they are studying. In the previous example, the more one knows about cars, the more hypotheses one can generate about why the oil light came on. Years of study and direct contact with the natural world give scientists the knowledge that is necessary for generating new hypotheses: to know what is or might be possible, and to know what has been studied before. In order to become more familiar with the facts, and to know what has been done before, scientists need to keep up with a truly prodigious set of published books and articles. Hardly any scientist can do this just by reading in the library anymore. Scientists usually use computer databases to search for articles relevant to their field of study.

Aside from familiarity, hypotheses can come from almost anywhere. The scientist may be quite unaware of how the idea popped into his or her head. Sometimes the ideas come from a relaxed mind, even a dream. This occurs rarely but famously. The 19th-century chemist Friedrich Kekulé was puzzling over the chemical structure of benzene. He had a dream of a snake biting its tail, forming a ring structure. When he awoke, he realized that the benzene molecule might have a ring structure. The dream was a metaphor which then had to be tested. (It turned out that Kekulé was right.) Once the metaphor has suggested a hypothesis, its usefulness is spent and, like scaffolding, abandoned. Frequently, the hypothesis is suggested by a metaphor from another field—for example, hypotheses about the way plants grow may arise from a study of architecture.

Hypothesis testing uses both inductive and deductive logic. Scientists induce general principles from looking at a range of facts, then they deduce conclusions from these general principles. The deductions can then suggest further hypotheses, which can be inductively tested, and so on. Scientists do not, as did some ancient philosophers, begin from first principles and deduce their conclusions without reference to data from the real world.

Philosopher Karl Popper insisted that scientific hypothesis testing permitted only falsification, rather than affirmation. That is, scientists can only show a hypothesis to be false, but never to be true; therefore, they remain uncertain about all scientific assertions. If the hypothesis is "dogs have four legs," it only takes one three-legged dog to disprove the hypothesis, but the scientist will never know the hypothesis to be true unless he or she examines all dogs that have ever existed or will ever exist anywhere. Most scientists, however, do not accept this idea in a strictly literal fashion. Most scientists will accept the statement "dogs have four legs" as a *provisionally* true statement.

Scientists assemble hypotheses into theories. Theories are broad, sweeping statements about the overall patterns of physical processes. If hypotheses are the bricks, a theory is the house. Theories cannot be tested; they are too big to be tested. But once enough of the constituent bricks have been tested, scientists begin to accept the theory as well. Darwin's theory of natural selection cannot be tested as a whole, but such a large number of individual hypotheses have been tested—natural selection has proven successful in so many individual cases—that the theory is now accepted by virtually all scientists. Evolution itself can be considered a macrotheory, consisting of smaller theoretical components (see EVOLUTION).

Philosopher Thomas Kuhn considered major theoretical frameworks to be *paradigms.* Scientists are ready and willing to change their minds about little things, but they resist changes in paradigms. As 18th-century philosopher David Hume said, extraordinary claims require extraordinary evidence. Once enough hypotheses have been tested, and a critical point is reached, the community of scientists experiences a relatively rapid change of thinking which Kuhn called a "paradigm shift" and which constitutes a *scientific revolution.*

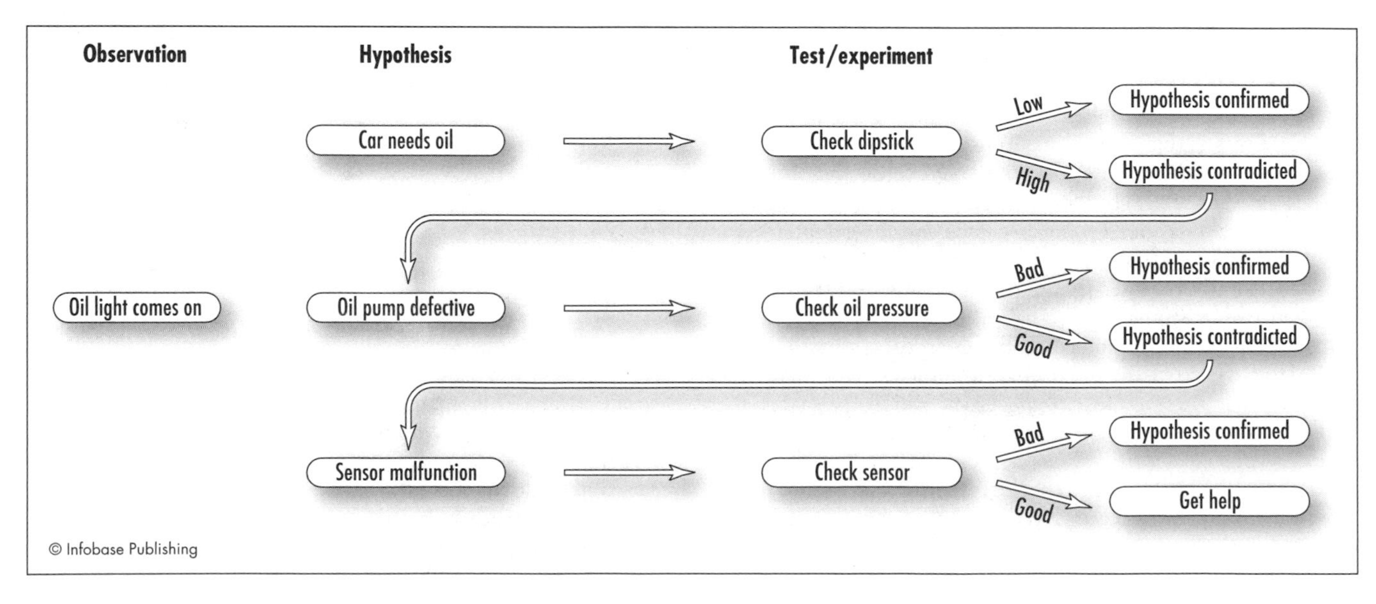

The scientific method can be applied to figuring out why the oil light comes on in an automobile dashboard. Three hypotheses are sequentially tested.

The testing of hypotheses accumulates gradually, but changes in theory can be rapid.

Scientists accept the simplest hypothesis that will explain the observations. When given a choice between a simple, straightforward explanation, and a complex one (especially one that requires numerous assumptions), scientists will choose the former. This is referred to as *Occam's Razor,* named after William of Occam (or Ockham), a medieval English philosopher and theologian.

Scientific research uses null hypotheses and statistical analysis to determine whether the results might have occurred by chance and will accept the results only if they are very unlikely to have occurred by chance. In everyday life, observers frequently notice patterns in events and objects. Millions of years of evolution have given the human brains the habit of looking for patterns. However, these patterns may be the product of imagination rather than a component of reality. Scientists are no different from any other people in having brains that can deceive them into believing false patterns, but they take special precautions to prevent this from happening—a set of precautions usually absent from nonscientific ways of knowing. For example, three good days on the stock exchange may look like a trend, and some investors would take it for one. A statistical analysis may show that such a three-day streak could readily occur by chance. The scientific method is, therefore, like a self-imposed yoke: It restricts scientists from wandering off in erroneous directions as cows or humans are wont to do; and in the process of restricting them, the scientific method allows scientists to do the useful work of pulling the cart of knowledge forward.

In order to test a hypothesis about a process, a scientist must specify what would happen if that process were *not* occurring. This *null hypothesis* is therefore the alternative to the hypothesis the scientist is investigating. When experiments are involved, the null hypothesis is usually investigated by a *control,* which is just like the experimental treatment in every way except for the factor being investigated. One of the earliest, and most famous, examples of a null hypothesis control was from Italian scholar Francesco Redi's 16th-century experiment that tested the hypothesis of biogenesis. Biogenesis asserts that life comes from preexisting life. If maggots appear in rotting meat, it must be because flies laid eggs there. The null hypothesis was that life need not come from preexisting life—that is, maggots can arise spontaneously from rotting meat, even in the absence of flies. Redi took two pieces of meat, put them in two jars, but covered one of the jars with a screen that excluded flies. Both pieces of meat rotted, but only the meat in the open jar produced maggots.

Almost anything can happen by chance, once in a while. In the case of the flies and the maggots, the results are pretty clear. But in many or most other scientific investigations, the results are far less clear. How can a scientist be reasonably sure that the results did not "just happen" by chance? The science of *statistics* allows the calculations of probability to be applied to hypothesis testing.

There are two kinds of error that a scientist can make regarding these probabilities. The first kind of error (called Type I error) occurs when the scientist concludes that the results were due to chance, when in reality the hypothesis was correct. The scientist failed to find something that was real. This is not considered a serious error, because later investigation may allow more chances to discover the truth. The second kind of error (of course, Type II error) occurs when the scientist concludes that the hypothesis was correct ("Eureka!") when in reality the results were due to chance. This is a more serious kind of error, because the scientist and his or her peers around the world might conduct further investigations and waste time and effort under the misguided notion that the hypothesis was correct. Therefore scientists try their best to avoid Type II error. Since they can never be totally sure, scientists universally accept a probability of 5 percent as the generally acceptable risk for Type II error. If the probability is less than 1 in 20 ($p < 0.05$) that the results could have occurred by chance, then scientists generally believe the results. The calculations of probability are quite complex, and most scientists let computers do these calculations.

Scientists take special precautions to avoid biased observations. Biased observations occur when the scientist expects certain results, even wants them to occur, and then tends to favor them when he or she sees them. To avoid this very human tendency to see what they want to see, scientists use objective measurements—temperature, weight, voltage, for example—rather than subjective assessments and often design their studies to be *blind.* That is, the scientist may not even know the sources of some of the specimens that he or she measures. One scientist may gather specimens and label them with simply a number. Another scientist receives the specimens, identified only by their label numbers, and makes measurements on them. This is called a blind experiment because the scientist who is making the measurements cannot be biased by the knowledge of where the specimens came from. This is particularly important in drug tests, because patients may report feeling better, and may actually improve, if they think they have received the drug (this is called the *placebo effect*). Even if the scientist does not tell the patient which pills are real and which are sugar pills, the scientist can subconsciously communicate the information, for example by the tone of voice. In such tests, therefore, a *double blind* procedure is routinely used, in which neither the investigator nor the subjects know which pill is which.

Scientific research frequently involves experimentation. When possible, scientists conduct experiments. In an experiment, the scientist imposes conditions upon the phenomena being studied, so that, to the greatest extent possible, only one factor is allowed to vary. In a laboratory, all conditions such as lighting, temperature, and humidity can be controlled. In the field, conditions may be quite variable, but if the experimental treatment and the control are side by side, the variability of all factors except the one being studied might be the same and therefore cancel out of the analysis.

Experiments are not always possible. Sometimes the arena of investigation is simply too big. How can one conduct an experiment with a whole mountain? Actually, some ecologists in the 1970s studied the effects of clear-cutting, strip cutting, and burning on the flow of nutrients in stream water on entire mountainsides. Such large experiments are, however, very rare. Sometimes experimentation is not ethical. Sometimes an experiment would be too disruptive to the system, which must be left undisturbed. Some whole fields of science are largely nonexperimental. Paleontology (the study of fossils) and astronomy are almost completely nonexperimental.

Scientists distinguish correlation from causation. Just because two variables appear to have a common pattern does not mean that one causes the other. For example, in 1993 a large public health study showed a correlation between smoking and diabetes. However, this does not mean that one caused the other. It could be that a third factor was causing both of them. Even if one is the cause and the other is the effect, it is not always possible to tell which is which. Does smoking cause diabetes? (This might occur from tissue damage due to chemicals in cigarette smoke.) Or does diabetes cause smoking? (This could occur because sick people seek solace in cigarettes.)

Sometimes the distinction between cause and effect can be crucial. Scientists who study human populations know that there is a correlation between a country's wealth and its population growth rate: Poor countries have rapid population growth. But which causes which? If one says that population growth causes poverty, then the solution is to restrict food and medical aid—if the people eat more they will just have more kids. But if one says that poverty causes population growth, then the solution is exactly the opposite—if the people eat more they will have fewer kids. One cannot distinguish these two possibilities from the correlation itself. More information is needed—for example, what happens in countries when food and medical aid is provided, and how is it different from countries in which no such aid is provided? Though this experiment has not been done, both of these situations have occurred and can be analyzed scientifically. In numerous instances, food and medical aid have been provided to countries which then experienced slower population growth. At last, it is possible to conclude not just that poverty and rapid population growth are correlated, but that poverty causes rapid population growth.

Causation, in the scientific sense, must be both necessary and sufficient. The cause is necessary when the effect will not occur unless the cause operates. Something that is necessary, however, may not be sufficient. The cause is sufficient if it, acting alone, can bring about the effect. Sometimes scientists study multiple causation, in which several factors may be necessary, but only the entire set of causes is sufficient.

The scientific approach differs from the reasoning used, for example, by politicians. Consider the hypothesis that a certain government policy has proven effective at reducing poverty. The government institutes the policy, and after a few years poverty rates decrease, and the politicians take the credit for it. But did the government policy actually cause poverty to decrease, or is this an accidental correlation? On the other hand, suppose that poverty stays the same or even increases, which might be accepted as evidence that the policy failed. But it could be that the policy is actually working, having caused poverty to increase less than it otherwise would. Either way, the government can take credit; and either way, critics can claim that the government failed—whichever spin they prefer. It depends on the null hypothesis: If the policy has no effect, what would have happened? In such cases, one cannot have a control: If the government chooses to fight poverty, it cannot do so in only certain places.

Scientific research is usually limited to measurable phenomena. Whenever possible, scientists select a type of measurement that correctly indicates what they want to know. Suppose that a scientist wants to know how much employee stress was created by a certain workplace situation. Rather than just watching the employees, or asking them if they felt stress (qualitative indicators), investigators measure something: They either have the employees indicate how much stress they feel on a numerical scale, or they measure blood pressure or skin cortisol levels as indicators of stress. Rather than just saying that the plants look healthy, a scientist will make measurements that realistically reflect health: the leaf area or seed production of the plant, for example. Scientists are careful to assure that the observation or measurement techniques do not themselves influence the results. For example, the presence of an observer might cause animals to behave differently than when the observer is absent.

Because of the necessity of valid measurements, scientists usually specialize within a narrow area. A scientist must be an expert in certain measurement techniques. In earlier centuries, a scientist could conduct valid research in more than one area. Robert Brown was a botanist, but by looking at pollen on a water drop under a microscope discovered Brownian motion, the movement today known to be caused by atoms; and he also discovered that plant cells have nuclei. Today few scientists could operate in these three different fields of study (botany, chemistry, cell biology).

Scientific research should yield conclusions that are applicable beyond the bounds of the investigation. If the sample upon which the scientist conducts his or her investigation is sufficiently broad, the results can be generalized to the world as a whole. But if the scientist investigates only organisms of one species, he or she cannot generalize to other species. If only one cultural group of people, or one age group, or one geographical area, is studied, then the results may not be valid for other people. Scientists work hard and spend a lot of money to make sure their studies have *external validity.*

Scientists share their results internationally through meetings and publications. Scientists are much closer to forming an international community than people engaged in commerce or politics. Even during the Cold War between the United States and the Soviet bloc of countries, much scientific research (none of it, of course, involving military secrets) was shared between the enemy groups. Much scientific research is conducted by international teams. Scientists announce their results at meetings by presenting papers or posters, and they publish their results in journals. The articles are reviewed anonymously to make sure that the results are properly obtained (usually an objective process) and that the results are sufficiently interesting to be published (usually a subjective process). Only articles that more than one reviewer approve are published in major scientific journals. This imperfect process is nevertheless one of the fairest ways of disseminating reliable information.

Scientists must confront ethical issues. Hypotheses are strictly physical and do not require ethical positions to be taken; however, scientists are human and work in a society where ethical issues cannot be avoided.

One example comes from the history of RADIOMETRIC DATING. When chemist Clair Patterson invented the uranium-lead method of radiometric dating, he found lead contamination everywhere in the environment. This contamination would cause inaccurate measurements, so he developed a clean laboratory and methods to avoid contamination. His initial purpose was not to be an environmentalist, but when his studies of ice cores proved that massive environmental lead contamination had begun in 1923, the very year that tetraethyl lead began to be added to gasoline, he publicized his findings. As a result, his funding was canceled by government agencies, even by the U.S. Public Health Service.

Scientists had, in fact, investigated tetraethyl lead prior to its approval for addition to gasoline. However, the investigations were invalid because they were prejudiced: The investigators set out to prove that lead was safe. In medical research, volunteers swallowed lead, then a doctor sampled their urine. When no lead was found in the urine, investigators concluded that lead was not toxic. This was invalid because the human body accumulates lead; this is the very reason it would not be excreted in the urine, and this is the very thing that makes it dangerous. The reason that such faulty research was accepted was that there was big money in it: The study was funded by manufacturers of lead additives. For 40 years, studies such as these gave lead additives a clean bill of health. This represents a clear ethical violation that is now admitted by most of the scientific community—and, at last, tetraethyl lead is no longer added to gasoline.

Many scientists insist that it is not enough merely to be ethical. Science, they believe, must serve the public interest. A premier example of such a scientist was George Washington Carver, who conducted monumental research into the cultivation and uses of peanuts and sweet potatoes, among other things. His outspoken purpose was to improve the lives of rural farmers. His research was not taken seriously by much of the scientific community at the time; because of widespread American prejudice against black intellectuals, Carver could find employment only at a black college, the Tuskegee Institute in Alabama. He considered that this was a worthy place in which to pursue his research in the public interest, and that the Tuskegee Institute was performing an essential function in the world of education. Today the value of his research is recognized by all scientists.

An opposite example occurred in Nazi Germany. Relatively fewer biologists supported the Nazi party than did the population as a whole. Nevertheless, half of the biologists at academic institutions did support the Nazi party. Not just in Nazi Germany but even in the United States and other free countries, the pseudoscience of EUGENICS was used to oppress minorities such as immigrants from southern Europe.

Science must be as free from political or economic influence as possible. Scientific investigation cannot be expected to yield trustworthy results if its conclusions are determined, or influenced, in advance by nonscientific motives. It is understandable that, for example, a pharmaceutical company would like to demonstrate the effectiveness of a drug in which it has invested. Companies are required to follow standard scientific procedures of honesty (see below) in conducting and reporting research about their potential products.

Scientific research must also be free of political pressure. It is the prerogative of governments to decide policy and law but not to dictate scientific facts. Governments have frequently exerted pressure on scientists to reach certain conclusions. One of the most extreme examples occurred in the Soviet Union in the middle of the 20th century (see LYSENKOISM). The United States has been a world leader in scientific research because its conclusions have not been constrained by government dictates until recently. The administration of President George W. Bush has suppressed and manipulated scientific research on the GREENHOUSE EFFECT, cancer, and sexually transmitted diseases, and it has even openly rejected the scientific basis of evolution. Two examples are:

- The Environmental Protection Agency Web site devoted to the greenhouse effect says that scientists are highly uncertain about whether the greenhouse effect will have any significant effect on natural or human systems (see Further Reading), while in reality no such uncertainty exists. A 2006 federal report finally admitted the reality of global warming, but the EPA "uncertainties" Web site remained active.
- The federal government (in 2002–2003) and the state governments of Mississippi, Kansas, Texas, and Louisiana promoted the suggestion that abortion causes breast cancer, which is contradicted by scientific research.

In some cases, the federal government has taken more direct action. In June 2005, Representative Joe Barton (R-Tex.), chair of the Energy and Commerce Committee, launched an investigation of three prominent scientists who had published research about the greenhouse effect. The scientists, the National Academy of Sciences, and the American Association for the Advancement of Science, as well as Representative Sherwood Boehlert, the Republican chair of the House Science Committee, have objected to Barton's actions; some representatives denounced it as harassment of the scientists.

Under previous federal administrations, both Democratic and Republican, scientists who worked in government agencies were allowed to present their honest conclusions. As of this writing, scientists at government agencies can no longer count on having this freedom. Scientists on the faculties of public universities, or who use federal funds for scientific research, may also be at risk of having their conclusions predetermined. When Lysenkoism dominated Soviet science, Russian genetic and agricultural research gained an international reputation for unreliability. It is not impossible that scientific research in the United States today may be similarly discredited by scientists in other countries if current trends continue.

Scientific research must adhere to scrupulous standards of honesty. Scientists are among the most honest people in the world. Why? The scientific method itself constrains people, who might otherwise be no more or less honest than others, to observe high standards of honesty. Success in business depends on what a businessperson can get people to buy; for preachers and politicians, success depends on whatever they can get people to believe. But for scientists, success depends on the reliability, therefore the honesty, of the research. Dishonest science is science that ultimately fails. In the short term, however, some scientists have pursued dishonest practices. Examples include:

- Outright fabrication of data
- Altering or omitting a few data, which may alter the conclusion
- Unnecessary duplication of publication

A scientist may be tempted to omit a few inconvenient data, especially if the scientist convinces himself or herself that those particular data might be erroneous and need to be omitted. In such a case, the scientist may legitimately omit data, if he or she admits it and presents the reasons for it in the resulting reports and publications. Duplication of publication can make the research appear to be more extensive than it is.

Temptations to be dishonest can have motivations such as the following:

- *Follow the money.* Most cases of scientific dishonesty have involved expensive research in the biomedical sciences. The most notorious recent example was the claim by medical researcher Woo-suk Hwang to have produced human embryonic stem cell lines that contained nuclei transferred from other human cells, a breakthrough that would have greatly advanced biotechnology and medicine. He was, very briefly, the national hero of South Korea and very popular among top scientists in the United States. When collaborators discovered that Hwang had not been honest about the sources of the human egg cells used to create the stem cell lines, they began to investigate his other claims. They discovered in 2005 that his stem cell lines were fraudulent; he had used computer image manipulation to produce the photographs that were published by *Science,* one of the leading journals in the world.
- *Prejudices.* Some scientists have fabricated data that confirm their prejudices. The most famous example is Sir Cyril Burt, whose data demonstrated a genetic and racial component to intelligence, but which was later discovered to be fabricated. Another example is PILTDOWN MAN, a fossil discovery that appeared to confirm the northern European origin of human intelligence.
- *Tenure and promotion.* Successful research, and numerous publications, increase the chances that a scientist will have academic success and research grants, which enhance the chances of academic success even more.

If scientists ran the world, it would be more honest than it is. At least the scientific method provides a system, however imperfect, for verifying any one scientist's assertions. In contrast, no one can verify a preacher's statement that "God told me so."

Science is beautiful. Scientific investigations, besides being generally more reliable than other ways of knowing, are also beautiful. Beauty is something people generally associate with art, but when a scientific hypothesis, confirmed by experiment, provides a simple explanation for what had previously seemed a complex set of disconnected facts, the result is something that professional scientists (as well as science educators and amateur scientists) experience as beautiful. Perhaps most beautiful of all is what William Whewell, a broadly trained scholar in the first half of the 19th century (and also the man who defined the modern use of the word *science*), called "consilience": When a hypothesis explains, at the same time, the facts of several previously unconnected fields of study, then it is very likely to be true. (This is not exactly the same as Wilson's use of the word, as described earlier.) Whewell applied the concept of consilience to Newtonian physics, which brought together the facts of physics, mathematics, and astronomy into a unified explanation. A creationist, Whewell refused to apply it to evolution, after he read the book written by his former student Charles Darwin. Most scientists today consider evolutionary science to be the supreme consilience: it explains the facts of geology, paleontology, ecology, genetics, and embryology all at once. How could a false theory explain so many things so well? Consilience is consistent with the ancient definition of beauty given by Greek philosopher Democritus: "unity in diversity."

Relatively few scholars have excelled in both science and the humanities, but those who have, such as entomologist-novelist Volodya (Vladimir) Nabokov, testify to the beauty of science. Nabokov said that there is "no science without fancy, no art without facts."

Science is a community. It would be unreasonable to expect all scientists to operate in the same way. Some classes and textbooks present a detailed and numbered list of the scientific method, but most scientists could not recite such a list. Ask any two scientists, and one will get two different lists.

Not everything that scientists do in the enterprise of science is strictly scientific research. In particular, there is a place in the scientific community for crazy geniuses. Many scientists would place astronomer Sir Fred Hoyle and chemist Sir Francis Crick in this category (not crazy in the clinical sense). The imagination of scientists, especially of the crazy geniuses, draws connections and generates ideas that the more pedestrian of humans could not have guessed. Scientists must be free to do this, but being wrong is a real occupational hazard of the scientific enterprise, especially from these geniuses. Crick, at the Cavendish Laboratory in England, was 35 years old and still had not finished with his Ph.D., when he elucidated the structure of DNA, one of the major breakthroughs in science. He did not work alone: he used data from other people, such as chemist Rosalind Franklin; and he worked with geneticist James Dewey Watson. But Crick has also been wrong: His hypothesis of the comma-less genetic code was wrong, and few scientists accept his speculations about directed panspermia (see ORIGIN OF LIFE). Fred Hoyle teamed up with Crick to proclaim directed panspermia. Ordinary scientists accept such insights with gratitude, then test them individually, keeping the ones that work. Some scientists are specialists in their narrow fields; others are generalists, working mostly as educators and writers; some scientists study the ethical and social implications of scientific research and theory. These are some of the things science can do because it is a community of scientists.

Because science limits itself to physical causation and hypothesis testing, it can reach definite conclusions. This is why biologist Peter Medawar has described science as "the art of the soluble," and Stephen Jay Gould (see GOULD, STEPHEN JAY) called science "an enterprise dedicated to posing answerable questions."

Some philosophers have claimed that science is merely a Western social convention. Most scientists, however, strongly believe that the scientific method is the uniquely correct way to truth about physical processes. Science cannot be dismissed as a mere belief system that is no better than other ways of thinking. Other ways of thinking can, of course, be valid for nonphysical concepts.

Finally, one cannot help but wonder if science has already made all of the major discoveries of the universe. In the past century and a half, scientists have discovered the main outlines of the history of the universe and life, and the chemical basis of life. It seems unlikely that any new breakthroughs are coming comparable to those of Darwin or Einstein. Some have pointed out that all the easy research has been done; only the hard questions remain, involving the very small, the very distant, and the very old. Scientific research may start yielding diminishing returns, very soon. Will society still be willing to pay for it?

Further Reading

Bowler, Peter J., and Iwan Rhys Morus. *Making Modern Science: A Historical Survey.* Chicago: University of Chicago Press, 2005.

Center for Science, Mathematics, and Engineering Education. *Every Child a Scientist: Achieving Scientific Literacy for All.* Washington, D.C.: National Academies Press, 1998. Available online. URL: http://books.nap.edu/catalog/6005.html. Accessed May 5, 2005.

Environmental Protection Agency. "Global warming—climate—uncertainties." Available online. URL: http://yosemite.epa.gov/oar/globalwarming.nsf/content/climateuncertainties.html. Accessed October 7, 2005.

Feynman, Richard P. *The Pleasure of Finding Things Out: The Best Short Works of Richard P. Feynman.* New York: Basic Books, 2005.

Gould, Stephen Jay. *The Hedgehog, the Fox, and the Magister's Pox: Mending the Gap Between Science and the Humanities.* New York: Harmony Books, 2003.

———. *Rocks of Ages: Science and Religion in the Fullness of Life.* New York: Ballantine Books, 2002.

Horgan, John. *The End of Science: Facing the Limits of Knowledge in the Twilight of the Scientific Age.* New York: Little, Brown, 1997.

Jenkins, Stephen H. *How Science Works: Evaluating Evidence in Biology and Medicine.* New York: Oxford University Press, 2004.

Kuhn, Thomas. *The Structure of Scientific Revolutions,* 3rd ed. Chicago: University of Chicago Press, 1996.

Lightman, Alan. *A Sense of the Mysterious: Science and the Human Spirit.* New York: Pantheon, 2005.

Losee, John. *Theories on the Scrap Heap: Scientists and Philosophers on the Falsification, Rejection, and Replacement of Theories.* Pittsburgh, Pa.: University of Pittsburgh Press, 2005.

Menuge, Angus. *Agents under Fire: Materialism and the Rationality of Science.* Lanham, Md.: Rowman and Littlefield, 2004.

Moody, Chris. *The Republican War on Science.* New York: Basic Books, 2005.

Moore, John A. *Science as a Way of Knowing: The Foundations of Modern Biology.* Cambridge, Mass.: Harvard University Press, 1993.

Ruse, Michael. *Taking Darwin Seriously: A Naturalistic Approach to Philosophy.* Oxford, U.K.: Blackwell, 1986.

Union of Concerned Scientists. *Scientific Integrity in Policymaking: An Investigation into the Bush Administration's Misuse of Science.* March 2004. Available online. URL: http://www.ucsusa.org/scientific_integrity/. Accessed October 7, 2005.

Waxman, Henry A. "About politics & science: The state of science under the Bush Administration." U.S. House of Representatives. Available online. URL: http://democrats.reform.house.gov/features/politics_and_science/index.htm. Accessed October 7, 2005.

Wigley, Tom M., et al. "Temperature Trends in the Lower Atmosphere: Understanding and Reconciling Differences." Available online. URL: www.climatescience.gov/Library/sap/sap1-1/finalreport/default. htm. Accessed July 11, 2006.

Wilson, Edward O. *Consilience: The Unity of Knowledge.* New York: Random House, 1999.

Working Group on Teaching Evolution, National Academy of Sciences. *Teaching About Evolution and the Nature of Science.* Washington, D.C.: National Academies Press, 1998.

Scopes Trial In 1925 a high school teacher in Dayton, Tennessee, John T. Scopes, was put on trial for teaching evolution, which was prohibited by Tennessee law. Since Scopes clearly violated state law, the judge found him guilty, but the punishment was only a small fine.

The Tennessee law upon which the Scopes Trial was based was not the first antievolution legislation in the United States. That honor goes to Oklahoma, where an antievolution amendment was added to a state law regarding free textbooks in 1923. This law was repealed shortly after the 1925 Scopes Trial.

The Scopes Trial has become the symbol of religious opposition to the teaching of evolution, for several reasons. First, it was the only example of prosecution of an instructor for teaching evolution. Second, because it was widely viewed as the flashpoint of the struggle between God and secularism, it drew national attention, particularly when two of the most famous lawyers in the country came to Dayton: William Jennings Bryan, to prosecute Scopes, and Clarence Darrow to defend him (see top photo on page 362). Bryan was a famous politician: a former secretary of state under Woodrow Wilson,

The two principal figures in the 1925 Scopes Trial in Dayton, Tennessee, were Clarence Darrow (left) and William Jennings Bryan. ***(Courtesy of the Granger Collection)***

three-time presidential candidate, and renowned speaker and defender of Christian orthodoxy. Clarence Darrow was just as famous as a lawyer who was willing to work for unpopular defendants. He had just finished defending Nathan Leopold and Richard Loeb, two young men who had murdered another young man, in a trial that gave Darrow national notoriety. The attention that the trial received was enhanced by the reports written by the famous antireligious writer H. L. Mencken.

Several subsequent court cases regarding the teaching of evolution have gained national attention, but none have involved the prosecution of a teacher. In 1968 science teacher Susan Epperson sued the state of Arkansas on account of its law that prohibited her from teaching evolution (see photo below). The case eventually went to the Supreme Court, which ruled in her favor. Arkansas again came to national attention in 1981 in the case of *McLean v. Arkansas Board of Education*, in which a science teacher sued the state because of a law that mandated equal time for the teaching of evolution and CREATIONISM. Arkansas federal judge William R. Overton ruled in favor of the science teacher in early 1982. Most recent activity regarding the mandating of alternatives to evolutionary science in the classroom has involved decisions by state boards of education, as in Kansas in 1999 and 2005, and the imposition of stickers on textbooks warning students to not believe the evolution contained therein. None of these events has had the drama of the original Scopes Trial, which was viewed as suppression of modern science by old-time religion and was inevitably compared with the ecclesiastical trial of Italian astronomer Galileo Galilei for his belief that the Earth revolved around the Sun. The legacy of the Scopes Trial for evolutionary science and education in the United States has far exceeded the actual consequences for the people directly involved or the community of Dayton. John T. Scopes spent a successful career as a geologist, and life in Dayton quickly returned to normal.

Most people know about the Scopes Trial from *Inherit the Wind*, a stage and screenplay written by Jerome Lawrence and Robert E. Lee. This encyclopedia entry makes extensive reference to *Inherit the Wind*, not just because it remains a famous movie but because it allows an analysis of the Scopes Trial. The title of the movie refers to a biblical passage in the Proverbs of Solomon: "He who troubles his own house shall inherit the wind," and strongly implies that fundamentalist antagonism toward modern science, as expressed by the citizens of Hillsboro (the fictitious name for Dayton), is futile and destructive. This play and movies subsequently made from it have been immensely popular and remained in continuous production.

The play was finished in 1950, in the midst of the frenzy associated with the anticommunist interrogations by U.S. Senator Joseph McCarthy, but it did not debut until 1955, in the Dallas Theatre. It ran as an immensely successful Broadway play, followed by a major motion picture by MGM in 1960, directed by Stanley Kramer. Its immediate popularity resulted largely from the fear of possible McCarthy-era state control of scientific and social thought. In the movie trailer, Kramer makes it clear that *Inherit the Wind* is important for far more than its entertainment value.

The 1960 movie won many awards worldwide (such as the Berlin Festival, where German teenagers honored it). Its fame resulted not only from the excellent screenplay and striking cinematography but also from the all-star cast: Fredric March played Matthew Harrison Brady, the fictitious William Jennings Bryan; Spencer Tracy played Henry Drummond, the fictitious Clarence Darrow; Gene Kelly played E. K. Hornbeck, the fictitious H. L. Mencken; Dick York (later

Susan Epperson (right) won a Supreme Court case in 1968 against an Arkansas law that prohibited the teaching of evolution. Writer Jerry R. Tompkins arranged for Susan Epperson to meet John T. Scopes (left) in 1969. ***(Courtesy of Jerry R. Tompkins)***

Comparison of *Inherit the Wind* with the Actual Scopes Trial

In the actual Scopes Trial:

1. The American Civil Liberties Union (ACLU) sponsored the defense as a test case for intellectual freedom. The ACLU placed an advertisement in Tennessee newspapers offering to pay legal expenses for any teacher who was willing to challenge the Butler Law, which prohibited the teaching of evolution. Although the ACLU arranged for the defense, Clarence Darrow offered his services for free (the only time he ever did so). One of the local trial organizers, George Rappleyea, was supportive of the defense.
2. Bryan befriended Darrow as much as circumstances allowed.
3. Darrow had recently defended two young men who are generally considered to have been guilty of a horrendous crime.
4. No one had observed Scopes teaching evolution. In fact, Scopes himself was unable to recall, under oath, whether he had actually taught evolution in a manner that contradicted Tennessee law. The indictment against Scopes was largely a staged event to bring attention, and business, into the community. It worked: To this day, Dayton, Tennessee, is a tourist stop primarily because of this trial that occurred 80 years ago.
5. In the trial, the local citizenry took little interest in what Scopes was or was not teaching, particularly since he was not a science teacher. He was a coach, who substituted for the regular science teacher (W. F. Ferguson) during a two-week illness. Scopes was never formally arrested and spent no time in jail. Occasionally he showed up late for the trial, which started without him. Clearly, the Scopes trial was not primarily about Scopes.
6. The events in the other box 6 are all (except Cates and Mrs. Brady) fictitious characters. All the events are fictitious, although the story of the little boy was based on a real, though unconnected, occurrence.
7. Both Bryan and Darrow were welcomed warmly. Both were famous men, which was perfect for the real purpose of the trial: putting Dayton on the map. Bryan had no animosity toward Scopes and even offered to pay his fine. It was H. L. Mencken who was not appreciated by the locals.
8. Bryan enjoyed and supported science. He had read Darwin's *Origin of Species.* He even approved the teaching of evolution, with the exception of human evolution. He did not attempt to defend a recent creation of the Earth.
9. Twelve scientists and theologians testified for the defense.
10. Bryan admitted that some passages of the Bible were not intended for literal interpretation.
11. The trial was the end of Scopes's teaching career, because he chose to go on to graduate school (he became a geologist).
12. Bryan died five days after the trial.
13. Perhaps most importantly, Darrow asks for a guilty verdict. The main reason for this was so that the case could go to a higher court.

In *Inherit the Wind*:

1. The ACLU is not mentioned. Instead, the newspaper that sent Hornbeck (Mencken) to the trial paid for the defense. All of the local organizers are militant fundamentalists defending Hillsboro against atheism imported from the big city.
2. Brady (Bryan) and Drummond (Darrow) had been friends many years previously, but ended up in fierce battle.
3. Drummond (Darrow) had done nothing more than defend some sexually explicit literature, thus making his entire career seem to consist of defending intellectual freedom.
4. The preacher and local dignitaries stalked into Scopes's classroom and observed the teaching of human evolution. The trial was, for them, a crusade, even though they recognized that "the whole world is laughing at us." The town leaders did recognize the economic benefits that national notice would bring, but only after Matthew Harrison Brady (William Jennings Bryan) entered the scene.
5. In the play and movie, the local citizenry took intense interest: They marched, shouted, burned Cates (Scopes) in effigy, and threw a rock through the window of the prison in which he was kept (although the jailer let him out long enough to play cards and meet with his fiancée). They sing "Gimme That Ol' Time Religion" numerous times, making it the only annoying aspect of an otherwise monumental movie. The movie clearly depicted Cates (Scopes) as a martyr.
6. Cates (Scopes) had a fiancée, none other than the daughter of the community's minister, Rev. Jeremiah Brown. She wanted Cates to recant but remained by his side even as Rev. Brown damned Cates (most memorably in a public sermon) and, indirectly, her. The minister is represented as unquestionably abusive to his family and community, as when he proclaimed that the "little Stebbins boy" would go to hell because he missed Sunday school before he drowned. Mrs. Brady (Bryan) slaps the young woman.
7. Brady (Bryan) is welcomed with a parade and cheers, while Drummond (Darrow) is virtually unnoticed upon his arrival; Drummond is later confronted by placards accusing him of bringing Satan into the God-fearing community.
8. Brady (Bryan) considers science to be ungodly and would not consider reading Darwin's book. He claims the Earth was created in 4004 B.C., which allowed Drummond (Darrow) to make him look foolish.
9. No such testimony was allowed in the fictional version.
10. Brady (Bryan) proclaims a literal belief in every passage of the Bible.
11. Scopes' teaching career was finished because he was fired.
12. Brady died while trying to deliver what he considered his most important speech, and to which nobody was listening.
13. Drummond (Darrow) defends the teacher's innocent plea.

the husband in *Bewitched*) played Bertram T. Cates, the fictitious John Scopes; and Henry Morgan (later Sergeant Joe Friday's *Dragnet* sidekick) was the judge. Interestingly, both March (in 1931) and Tracy (in 1941) had starred in movie versions of *Dr. Jekyll and Mr. Hyde,* a story that also explores the uneasy connection between human and animal natures. In 1960 *Inherit the Wind* became the world's first "in-flight movie," shown by TWA to first-class passengers on some flights.

As is always the case in historical fiction, the events of the Scopes Trial were altered by the authors of *Inherit the Wind* in such a way as to focus attention on what they considered the important themes of the event. In particular, the play and movie depict the trial as a dramatic battle between hate-filled, benighted fundamentalists and the calm saints of scientific light and truth. Examples are provided in the accompanying box. Probably every researcher and teacher of evolutionary science has experienced fanatical opposition from fundamentalist creationists (as has the author of this encyclopedia and many of his acquaintances). Such vehement opposition, as depicted in the play and movie, is in fact less common than many anti-creationists suppose. The author of this encyclopedia teaches evolutionary science in Bryan County, Oklahoma, squarely in the Bible Belt and named after none other than William Jennings Bryan. The overwhelming majority of people who oppose evolutionary science have been courteous to the author, even when he puts on his Charles Darwin costume to teach. *Inherit the Wind,* by depicting creationists as benighted, makes a point worth considering, but those who see the movie should recognize that not all creationists should be painted with the same brush.

In particular, William Jennings Bryan was not the antiscientific dogmatist portrayed in *Inherit the Wind.* Though he would be considered a fundamentalist by most observers today, his main political and personal motivation was not the squashing of evolutionary science but the defense of individual freedoms against the threat of scientific EUGENICS and social Darwinism, and the perceived effect of secularism on demeaning the worth of the individual. Creationism itself began as an organized movement in the 20th century, not in the 19th, ignited largely by fears of the same secularism that concerned Bryan.

NBC produced TV versions of *Inherit the Wind* in 1965 and again in 1988. In 1999 MGM released its second movie version, starring George C. Scott as Bryan and Jack Lemmon as Darrow, both in nearly the final roles of their careers. The later productions did not portray the citizens of Hillsboro as being quite as vehement in their fundamentalism, although whether this resulted from the fear of conservative backlash or from a more accurate fairness is difficult to say. The 1999 version, like all modern movies based on actual incidents, carries a disclaimer that "any resemblance to real persons, living or dead, is accidental."

The scene that perhaps remains most in people's minds is the final one. Drummond (Darrow) leaves the empty courtroom, holding first the Bible, then the *Origin of Species,* and deciding to take both of them with him.

Further Reading

Davis, Edward B. "Science and religious fundamentalism in the 1920s." *American Scientist* 93 (2005): 253–260.

Larson, Edward J. *Summer for the Gods: The Scopes Trial and America's Continuing Debate over Science and Religion.* Cambridge, Mass.: Harvard University Press, 1998.

Lawrence, Jerome, and Robert E. Lee. *Inherit the Wind.* New York: Bantam Books, 1955.

Moore, Randy A. "Creationism in the United States, VII. The Lingering Impact of *Inherit the Wind.*" *American Biology Teacher* 61 (1999): 246–250.

Overton, William R. "McLean v. Arkansas Board of Education." Available online. URL: http://www.talkorigins.org/faqs/mclean-v-arkansas.html. Accessed May 10, 2005.

United States Supreme Court. "Epperson v. Arkansas." Available online. URL: http://www.talkorigins.org/faqs/epperson-v-arkansas.html. Accessed May 10, 2005.

seedless plants, evolution of Seedless plants are *bryophytes* (mosses and their relatives) and *pteridophytes* (ferns, horsetails, and club mosses, and their relatives). (Algae are sometimes considered to be seedless plants, but they are actually photosynthetic protists; see EUKARYOTES, EVOLUTION OF). Seedless plants must live in moist environments on land, for two reasons:

- *Vascular tissue* is the plumbing system of higher plants. Vascular tissue consists of *xylem,* which brings water up from the soil to the leaves, and *phloem,* which carries nutrients (dissolved in water) throughout the plant. Bryophytes do not have vascular tissue; because they must soak up water rather than transporting it in vascular tissue, they must live in moist environments. Pteridophytes have vascular tissue.
- Both bryophytes and pteridophytes reproduce sexually by releasing sperm or similar cells that swim or are carried to eggs. This form of sexual reproduction can only occur in a wet environment, and this is why pteridophytes also must live in moist environments. When plants that produced seeds evolved, sexual reproduction could be carried out in the absence of a wet environment (see GYMNOSPERMS, EVOLUTION OF; ANGIOSPERMS, EVOLUTION OF).

Because life began in water, the earliest plants were aquatic. The first plants that lived on land were seedless plants similar to those that are in moist terrestrial environments today. An analogous situation occurred with animals, in which the earliest animals on the Earth were aquatic, and the first animals that lived on land were restricted to moist environments. As with plants, this was due both to a reduced ability to survive dry conditions, and to the inability to sexually reproduce in dry conditions. In both plants and animals, evolution has produced progressively more species that live in dry environments.

The green algae (Chlorophyta) are considered to be the ancestors of all land plants, even though they are not the most complex algae. Many are unicellular, some form sheets, and others form more complex structures, but none of them rival the gigantic kelp in size or complexity. However, they have important similarities to land plants on the cellular level:

- Green algae include many freshwater species. Almost all brown and red algae, for example, are marine, as are many of the green algae. Land plants evolved from freshwater, not marine, ancestors.
- The cells of green algae are chemically and structurally similar to those of land plants. They have the same kinds of chlorophyll, store the same kinds of carbohydrates, and have the same kinds of cell walls.

The green algal origin of land plants, long known from the fossil record, has been confirmed by DNA studies (see DNA [EVIDENCE FOR EVOLUTION]) in which green algae of the family Charophyceae have been found to be more similar to land plants than are any other algae.

Freshwater algae often live in shallow water, and shallow water frequently dries up. Algae that had adaptations that allowed them to continue to function when water became mud would be favored by evolution. The first land plants, as seen in the fossil record, were small plants in wetlands, consistent with this explanation.

Nearly all organisms have a sexual life cycle in which a haploid phase alternates with a diploid phase. Cells with unpaired chromosomes are haploid, cells with paired chromosomes are diploid. A type of cell division known as MEIOSIS separates chromosome pairs and produces haploid cells; in turn, some haploid cells come back together, in a process called fertilization, to restore the original diploid number (see SEX, EVOLUTION OF). In animals, the haploid phase consists only of eggs and sperm, but in plants, the haploid phase is multicellular. As green algae evolved into land plants, they diverged:

- Some of them evolved into bryophytes (see figure), in which the haploid multicellular phase is relatively large and independent, while the diploid multicellular phase is relatively small and depends on the haploid cells to supply their needs. They reproduce when sperm swim through a film of water to eggs. None of these plants grow taller than a few centimeters. These plants were already established on the Earth by at least 350 million years ago.
- Some of them evolved into plants in which the diploid multicellular phase was relatively large and independent, while the haploid multicellular phase was relatively small and sometimes dependent on the diploid cells to supply their needs. These plants were the earliest known land plants. They lived in marshlike habitats during the SILURIAN PERIOD 435 million years ago, although a few fossil spores have been found from the previous ORDOVICIAN PERIOD. Some of their fossils have been so well preserved that the vascular pipes can actually be seen in rock cross sections. They were the ancestors of pteridophytes and seed plants.

The earliest land plants, in the swamps of the Silurian period, were only about three feet (a meter) tall or less, and were hardly more than branching stems with reproductive structures at the tips. They had no roots or leaves. They included:

Mosses do not produce seeds. They produce spores, which are single-celled reproductive structures, in the capsules shown in this photo. Each spore grows into a new moss plant. *(Photograph by Stanley A. Rice)*

- the rhyniophytes such as *Cooksonia*. The rhyniophytes apparently became extinct without descendants. Even though they looked very much like modern whiskferns, they are unrelated to them; the whiskferns are more closely related to the ferns and apparently evolved from a more complex into a simpler form in more recent times.
- the zosterophylls. Some zosterophylls evolved into the ancestors of the club mosses.
- the trimerophytes. Some trimerophytes evolved into the ancestors of horsetails, whiskferns, and true ferns. These plants were common during the Silurian and Devonian periods. Their evolutionary descendants included large forest trees of the CARBONIFEROUS PERIOD (see below) as well as smaller plants, which are today's ferns and fernlike plants.

By the Carboniferous period, many of these plants had grown very tall. Some of them, such as *Lepidodendron,* were relatives of modern club mosses, while others such as *Calamitales* were relatives of modern horsetails. Then as now some ferns reached the size of small trees. Their leaves were long, narrow, and very simple. Their water pipes allowed them to grow tall, but their upward growth was limited by their inability to continue producing new wood. They dominated the Carboniferous swamps. They could only live in moist environments, such as swamps, because their sexual reproduction required sperm to swim through water, just as seedless plants do today. Therefore, the forests of the Carboniferous period were found only in swamps; the hillsides and dry plains were most likely barren. Fortunately for them, swamp conditions were extremely widespread on the Earth during the Carboniferous. When the plants died, they did not completely decompose. The piles of organic matter that they produced constitute today's coal deposits. Although coal itself contains no fossils, mineral inclusions *(coal balls)* often do (see FOSSILS AND FOSSILIZATION). These coal balls provide most of the information that scientists have about these plants.

Beginning in the Carboniferous period, seed plants started their evolutionary diversification and eventually dominated most of the Earth. In very few places on land do the seedless plants dominate (though the ocean is still the realm of seaweeds, including forests of giant kelp). Seedless plants, however, still have a fair measure of species diversity: While there are only six species of whiskferns and 15 species of horsetails, there are over a thousand species of club mosses, and over 11,000 species of ferns. Though they primarily hide in the shade of the seed plants, they are still important to the BIODIVERSITY of the living world.

Further Reading

Graham, Linda E., Martha E. Cook, and J. S. Busse. "The origin of plants: Body plan changes contributing to a major evolutionary radiation." *Proceedings of the National Academy of Sciences USA* 97 (2000): 4,535–4,540.

Kenrick, Paul, and Peter Crane. "The origin and early evolution of plants on land." *Nature* 389 (1997): 33–39.

Niklas, Karl J. *The Evolutionary Biology of Plants*. Chicago: University of Chicago Press, 1997.

Shear, William A. "The early development of terrestrial ecosystems." *Nature* 351 (1993): 283–289.

selfish genetic elements Evolution occurs in populations because individual organisms reproduce more or less than other individuals; individuals are the units of NATURAL SELECTION. Natural selection therefore favors individuals that have the greatest reproductive success, not the individuals that best benefit the population (see GROUP SELECTION); that is, natural selection favors efficiently selfish individuals. The term *selfish* as used by biologists does not refer to deliberate or conscious selfishness, but only the processes and adaptations that favor individuals at the expense of populations or species. The term *conflict of interest* denotes a situation in which some individuals benefit from processes or adaptations that are harmful to other individuals.

Sometimes, helping other individuals can enhance an individual's evolutionary success; therefore even ALTRUISM is fundamentally selfish, in evolutionary terms. In contrast, the organs, tissues, and cells of a multicellular organism cannot reproduce or survive on their own. Their success is entirely dependent upon the success of the individual of which they are a part. The body as a whole controls the replication of its component cells. Cells that escape bodily control of replication can become cancer. Because of this, it would not seem possible that cells or any components of them could act selfishly without natural selection destroying the individual that contains them. However, in some cases, such apparently selfish behavior on the part of genetic elements has been documented.

Genetic components within cells can sometimes spread at the expense of other components, or damage other components, within individuals. Sometimes the selfish activities of genetic components may harm individuals; sometimes their selfish activities may benefit the individual. This entry considers several examples of selfish genetic elements.

Selfish noncoding DNA. Genes, or segments of NONCODING DNA, can be replicated and spread by the activities of enzymes such as reverse transcriptase. Genes can be copied, and the copy or copies can be inserted in the same or other chromosomes. This process is called *gene duplication.* Some segments of DNA, called *transposons* or *jumping genes,* can be copied and inserted into new locations. Sometimes short segments of DNA, meaningless in terms of genetic information, can be replicated over and over. These can all be considered selfish genetic elements. During the history of eukaryotes, duplicated genes and noncoding DNA have spread so much that they now comprise more than 90 percent of the DNA in the cells of many organisms, including humans. The accumulation of selfish DNA can have negative, neutral, or positive effects on cells:

- *Negative effects.* The insertion of a copied chunk of DNA inside an existing gene can disrupt that gene; the cell may die without the activity of that gene. An inserted chunk of DNA can also disrupt the PROMOTER and other sequences that control the way the cell uses that gene, or the DNA that encodes the proteins that bind to promoter. In such cases, the gene, even if itself intact, may be expressed too much, or not expressed. Cells in which these negative effects occur may die; or (if cancer results) natural selection may eliminate them because the individual that contains them dies.
- *Neutral effects.* Having lots of extra noncoding DNA may have no effect on the cell, if the genetic DNA continues to function normally. This may be much more likely to happen in plant cells, which can often tolerate polyploidy, than in animal cells (see HYBRIDIZATION). An organism whose cells produce 10 times as much DNA as is necessary may seem to be wasteful with its resources and therefore inferior. Despite this, almost all eukaryotic organisms have chromosomes chock full of noncoding DNA.
- *Beneficial effects.* Some cell biologists such as Thomas Cavalier-Smith have suggested that the extra DNA makes the nucleus bigger, which has the effect of increasing the rate at which the genes can be used by the cell. In this case, the extra DNA would be favored by natural selection.

Homing endonucleases. Some fungi and algae produce endonuclease enzymes that cut DNA with a certain sequence that is about 20 nucleotides in length. This sequence is found only in chromosomes that do not carry the gene for the enzyme. Therefore, in heterozygous individuals, the enzyme encoded by one chromosome of a homologous pair cuts the other chromosome of the pair; the chromosome with the endonuclease gene is then used as a template for repair, with the result that both chromosomes in the pair will then carry the endonuclease gene. In this way, the endonuclease gene can spread in these populations.

Competition of maternal and paternal alleles in animal embryos. An embryo receives two alleles for each gene, one from the mother and one from the father (see MENDELIAN GENETICS). This creates a situation in which a conflict of interest is possible. Both of the parents benefit from the successful development, birth, and nurture of the offspring. The interests of the mother are best served if the offspring are successful, without impairing her ability to successfully reproduce in the future. The interests of the father are best served if the offspring are successful even if the mother's ability to successfully reproduce in the future is impaired. This would not make any sense in a monogamous relationship, but most animals (including humans) are not strictly monogamous. The father's fitness interests are therefore best served by having more offspring either by the same female or by some other. The father's fitness benefits if the embryo that carries his genes commandeers resources from the mother even at the expense of her long-term health! In contrast, the mother's fitness benefits if she is in control of how much nurture to provide to the embryo, that is, if her alleles can counteract the selfish alleles of the father. This creates a conflict of interest between mother and father, expressed in the development of the embryo they create.

The only way that maternal and paternal alleles can differentially affect fetal development is if there is some way of distinguishing between them. There is such a way. In *genetic imprinting,* either the maternal or the paternal allele is chemically inactivated, at least temporarily, in the embryo. Therefore, while the embryo contains alleles from both parents for any given gene, the allele from only one of the parents may be functional. This is an example of *epigenetics* (see DNA [RAW MATERIAL OF EVOLUTION]). The imprinting may be accomplished by methylation of the cytosine nucleotides of the inactivated allele. The effect of genomic imprinting on conflict of interest between selfish maternal and paternal genetic elements has been explored by evolutionary biologist David Haig.

The conflict of interest between maternal and paternal alleles has been investigated in laboratory mice, in which alleles can be substituted in the fertilized egg cell, an experiment that is technically (and for humans ethically) impossible with most other animal species. One gene, Igf2 (insulin-like growth factor 2), produces a protein that causes the mother to devote more nutrients to the embryo but also causes higher blood pressure. In the placenta, the maternal allele for Igf2 is inactivated. The paternal gene for the receptor of Igf2, however, is inactivated. The Igf2 gene acts in the interest of the father, the receptor gene acts in the interest of the mother. Experiments with mice indicate that the mouse embryo develops normally if it has both the paternal Igf2 gene and the maternal receptor, or neither—but not if both of the genes are represented by the alleles from one parent. Igf2 and its receptor are just two of the 40 genes in mice that are imprinted. Humans have most of these same genes. Some medical conditions associated with pregnancy, such as high blood pressure, may therefore be best interpreted from an evolutionary, conflict of interest perspective (see EVOLUTIONARY MEDICINE).

Cytoplasmic male sterility factors in plants. Cytoplasmic genes are found in the DNA of mitochondria and chloroplasts (see SYMBIOGENESIS). The mitochondria and chloroplasts cannot survive without the cell, therefore it would seem that their genes would not harm the cell in which they reside. However, mitochondria are passed on only in female reproductive cells. Most plants are either hermaphroditic or monoecious—they produce both pollen and ovules. In dioecious species of plants, however, there are separate male and female plants (see REPRODUCTIVE SYSTEMS). Nuclear genes can be passed on either through either pollen or through ovules with equal effectiveness. However, the mitochondria and the genes they contain are passed on to the next generation more effectively if female reproduction occurs much more than male reproduction. A mitochondrion that produces a chemical that inhibits the production of male cells, therefore, may be passed on into the next generation more than a mitochondrion that does not produce this chemical. Such a chemical is called a cytoplasmic male sterility factor. Numerous such factors have been found in plants. If a cytoplasmic male sterility factor spreads through a population of plants, it will cause some of the plants to be female (by failing to produce pollen). In flowering plants, male sterility factors may also be found in chloroplasts, since the chloroplasts are passed on only through the female cells. This does not happen in conifers, since their chloroplasts are passed on through the pollen.

As cytoplasmic male sterility spreads in a population, and females become more common, the reproductive success of the male function may increase by frequency-dependent selection; if so, there will be an advantage to male sexual reproduction. But what is to stop the spread of cytoplasmic male sterility before it causes the whole population to become female, and then extinct? Mutations occur in the nucleus, which cause genes to produce *silencing factors* that counteract the effects of the cytoplasmic sterility factors. In some populations, sexual reproduction represents a balance between cytoplasmic sterility factors that inhibit male function and nuclear silencing factors that restore it. Populations of plants may have several different cytoplasmic male sterility factors and different nuclear silencer genes.

Flowering plants had ancestors that produced hermaphroditic flowers. Female plants are frequently plants whose male function has been inhibited by sterility factors. In some cases, hermaphroditic plants come from lineages in which cytoplasmic male sterility has never occurred; in other cases, hermaphroditic plants are individuals in which the male function has been lost and then restored by silencing factors. One cannot tell by looking at a hermaphroditic plant whether or not it is hermaphroditic because of a balance between sterility and silencing factors. This entire phenomenon of sterility and silencing was discovered only after DNA technology allowed the actions of the genes to be determined.

Sex chromosomes. In many species, the gender of an individual is determined by sex chromosomes. In humans, females have two X chromosomes, while males have an X and a Y chromosome, although in other species the pattern can be quite different (see SEX, EVOLUTION OF). The chromosome associated with males (in humans, the Y) is usually much smaller than the chromosome associated with females (in humans, the X). Consider what would happen if the X chromosome encodes a gene that produces a chemical that destroys the Y chromosome. An X chromosome that destroyed Y chromosomes would cause the reproductive cells to more commonly carry the X than the Y chromosome—resulting in a disproportionately large number of female offspring. From the viewpoint of the X chromosome, this could be advantageous. Since X chromosomes are inside of female cells two-thirds of the time and male cells only one-third of the time, an X chromosome that killed Y chromosomes might be passed on more efficiently into the next generation than an X chromosome that did not do so. The result of the spread of the dangerous X chromosome would be a population that consisted mostly of females, with few males.

What process might be able to stop the spread of the dangerous X? Once again, it is possible that silencing factors in the nucleus might be involved. Another process, however, has apparently been the reduction in size of the Y chromosome. Many of the genes in the Y chromosome have migrated to other chromosomes, with the result that the Y chromosome (in humans) now has very few genes. The Y chromosome, being smaller, has fewer sites upon which chemicals encoded by genes in the X chromosome can bind—that is, the Y chromosome has evolved to become a smaller target.

In wood lemmings, some X chromosomes (denoted X*) are dominant feminizers: X*Y are female, not male. When an X*Y female mates with a normal XY male, one-fourth of the offspring will be YY and will never develop. As a result, two-thirds of the surviving offspring will carry the X* chromosome, not the expected one-half. This allows the X* chromosome to spread through the rodent population. Once again, the frequency-dependent reproductive advantage of rare males counteracts this trend.

Meiotic drive. Meiotic drive can occur in two ways. First, it may occur when a chromosome that originated from either the mother or the father ends up in more than half of the gametes produced by the offspring. Second, it may occur when gametes with a chromosome either from the mother or the father are more successful at fertilization.

The offspring of the F_1 generation receive alleles from both the mother and the father. The pattern expected from Mendelian genetics is that half of the gametes produced by MEIOSIS in these F_1 individuals will carry the paternal allele, and half will carry the maternal allele. This produces the familiar 3:1 ratio in the F_2 generation. In contrast, meiotic drive can cause the paternal allele or the maternal allele to be present in more than half of the gametes.

One way this can happen is through what have been called "selfish centromeres." During meiosis, protein strands pull the chromosomes apart so that each resulting cell receives a full set of chromosomes. *Centromeres* are segments of DNA in a chromosome to which these protein strands attach. A larger centromere might be able to link with more strands, allowing its chromosome to be preferentially pulled to one of the resulting cells. There is no advantage associated with this in the production of sperm or pollen, since spermatocytes in animals and microsporocytes in higher plants typically produce four sperm or pollen grains. That is, both the maternal and paternal chromosomes are assured of ending up in a male sex cell. In the production of eggs or ovules, however, a conflict of interest can arise that creates a situation that may favor selfish centromeres. During meiosis of oocytes in animals and megasporocytes in higher plants, only one egg or ovule is produced; the other two or three cells *(polar cells)* disintegrate. The chromosome with a bigger centromere (a selfish centromere) may be more likely to end up in the egg or ovule rather than being lost in a polar cell. Research by evolutionary biologist Lila Fishman has shown that, in hybridization between two species of monkeyflower (*Mimulus guttatus* and *M. nasutus*), the *M. guttatus* chromosomes get passed on in the ovules of hybrids much more effectively than do the *M. nasutus* chromosomes, perhaps because the *M. guttatus* chromosomes have better centromeres. About 10 percent of species have "B chromosomes," which end up in reproductive cells more often than other chromosomes.

Meiotic drive may also occur if the sperm or pollen with a chromosome from one of the original parents are more successful at fertilizing eggs or ovules than are the sperm or pollen that contain the chromosome from the other parent. This amounts to competition between sperm as they swim or pollen tubes as they grow through female cone tissue or the style of a flower. This is not likely to occur with eggs or ovules, since eggs and ovules are largely stationary. Paternal or maternal alleles may also promote the death of zygotes that contain the other allele.

Gene-centered view of selection. Evolutionary biologist Richard Dawkins (see DAWKINS, RICHARD) explains natural selection in terms of selfish genes. Genes are using cells as their vehicles for reproduction, and natural selection favors the most efficiently selfish genes.

This viewpoint has been much criticized. Critics have pointed out that in very few cases can genes express themselves individually in the phenotype of the organism. Genes act in complex developmental pathways. When natural selection acts upon individuals (causing some to reproduce more, some less, some not at all) it has at best an extremely indirect effect on favoring one gene over another.

At least two responses to these criticisms are possible. First, one of the ways in which selfish genes can get themselves most effectively passed on into the next generation is by cooperating with other genes. On the level of the individual, this happens frequently when the COEVOLUTION of species produces symbiosis, and when it progresses as far as symbiogenesis. If individuals and species can cooperate with one another even though they are selfish, then genes should be able to do the same thing. Second, as indicated above, conflicts of interest do frequently occur among the genes of an organism. Natural selection eliminates the individuals whose genes are in too great a conflict.

Genes and other replicative elements of the DNA often operate in a manner that human observers call selfish. Natural selection favors the ability of other genes to suppress the selfishness of these elements. Evolutionary biologist Egbert Leigh has called this a "parliament of genes," in which selfish interests of different constituencies balance one another, allowing the work of the cell usually to get done.

Further Reading

Burt, Austin, and Robert Trivers. *Genes in Conflict: The Biology of Selfish Genetic Elements.* Cambridge, Mass.: Harvard University Press, 2006.

Cavalier-Smith, Thomas. "Economy, speed and size matter: Evolutionary forces driving nuclear genome miniaturization and expansion." *Annals of Botany* 95 (2005): 147–175.

Charlesworth, Deborah, and Valérie Laporte. "The male-sterility polymorphism of *Silene vulgaris:* Analysis of genetic data from two populations and comparison with *Thymus vulgaris.*" *Genetics* 150 (1998): 1,267–1,282.

Dawkins, Richard. *The Selfish Gene.* New York: Oxford University Press, 1976.

Doolittle, W. Ford, and C. Sapienza. "Selfish genes, the phenotype paradigm and genome evolution." *Nature* 284 (1980): 601–603.

Fishman, Lila, and John H. Willis. "A novel meiotic drive locus almost completely distorts segregation in *Mimulus* (monkeyflower) hybrids." *Genetics* 169 (2005): 347–353.

Frank, S. A. "Sex allocation theory for birds and mammals." *Annual Review of Ecology and Systematics* 21 (1990): 13–55.

Haig, David. *Genomic Imprinting and Kinship.* New Brunswick, N.J.: Rutgers University Press, 2002.

Herbst, E. W., et al. "Cytological identification of two X-chromosome types in the wood lemming *(Myopus schistocolor).*" *Chromosoma* 69 (1978): 185–191.

Jones, Steve. *Y: The Descent of Men.* New York: Houghton Mifflin, 2003.

Leigh, Egbert G. J. "How does selection reconcile individual advantage with the good of the group?" *Proceedings of the National Academy of Sciences USA* 74 (1977): 4,542–4,546.

Pardo-Manuel de Villena, F., and C. Sapienza. "Nonrandom segregation during meiosis: The unfairness of females." *Mammalian Genome* 12 (2001): 331–339.

Sykes, Bryan. *Adam's Curse: A Future without Men.* New York: Norton, 2004.

sex, evolution of Sex is the major biological process by which genes are recombined among members of the same species. Almost all species have some form of genetic recombination.

Sexual Recombination in Different Life-forms

Even bacteria have genetic recombination. Within many bacterial species, one cell grows a tube toward another cell, through which a small circle of DNA, called a *plasmid,* travels from one cell to the other. Alternatively, under certain conditions bacteria can absorb plasmids that have been released into their environments. Bacteria of different species, even different genera, can exchange plasmids. This is the main way in which resistance to antibiotics can spread from one species of bacteria to another; resistance to the antibiotic vancomycin spread from relatively harmless intestinal bacteria to harmful staph bacteria in this way (see RESISTANCE, EVOLUTION OF).

Many biologists define species in terms of the ability to exchange genetic information (see ISOLATING MECHANISMS; SPECIATION). If this concept is applied literally, then there may be, as one biologist has claimed (see MARGULIS, LYNN), only one gigantic worldwide species of bacteria. The biological species concept is also difficult to apply to many plants, in which cross-pollination can occur among species in a genus, or even between genera. Whether complete or partial, genetic isolation is necessary for species to diverge.

Among eukaryotes (see EUKARYOTES, EVOLUTION OF) sexual reproduction occurs by the alternation of MEIOSIS and fertilization. In most eukaryotic cells, chromosomes occur in pairs; these cells are called *diploid.* Under certain conditions, some of these diploid cells undergo a special kind of cell division known as meiosis, in which the chromosome pairs are separated, producing cells that are *haploid* instead of diploid. Haploid cells, therefore, have only half the number of chromosomes that diploid cells have. In humans, for example, diploid cells (which are nearly all of the cells of the body) have 46 chromosomes, consisting of 23 pairs; the haploid cells (ova and sperm cells) have only 23 chromosomes, with no pairs. The haploid cells of eukaryotes either fuse together or else they grow into structures that produce other haploid cells that fuse together (see below). The haploid cells that fuse (eggs and sperm) are called *gametes.* When gametes unite, *fertilization* has occurred (see MENDELIAN GENETICS). In eukaryotic species, meiosis alternates with fertilization, resulting in an alternation between haploid and diploid. Each such cycle is called a generation.

In fungi, many protists, and some plants, the gametes are the same size (the species are *isogamous*), therefore neither gamete may be called male or female. However, in plants and animals, meiosis produces both large and small reproductive cells. The large reproductive cells are considered female, and the small ones are considered male. The most likely advantage for the evolution of male vs. female reproductive cells is specialization. Isogamous reproductive cells are not particularly good at moving or at nourishing the embryo that develops from them. In contrast, small male reproductive cells can move efficiently. Because they are small they can be numerous, therefore some of these male reproductive cells can reach their target. Large female reproductive cells can efficiently nourish the embryo.

In plants, meiosis produces *spores.* In most plants, large female spores are called *megaspores* (from the Greek for large) while small male spores are called *microspores.* The megaspores grow into multicellular haploid female structures; the microspores grow into multicellular haploid male structures. The multicellular haploid structures produce gametes: Females produce egg cells or nuclei, males produce sperm cells or nuclei. In mosses and ferns, the multicellular haploid structures grow on moist soil; the male structures release sperm that swim through a film of water to the egg that remains in a female structure. In seed plants (such as pine trees and flowering plants), the multicellular haploid structures do not grow on the ground. Instead, the female structures remain protected and fed inside of the immature seed, where they produce egg nuclei. The male haploid structures are *pollen grains,* which are carried through the air (by wind or by animals) to the female structures. The pollen grain grows a tube, through which its sperm nucleus travels toward the egg nucleus (see ANGIOSPERMS, EVOLUTION OF; GYMNOSPERMS, EVOLUTION OF).

In animals, the sexual life cycle is simpler. Meiosis directly produces gametes. In animals with *external fertilization,* the male animal releases sperm into the environment, where they swim toward the eggs, which the female has also laid in the environment. The environment must be moist for this to succeed. In animals with *internal fertilization,* the male releases sperm into the body of the female. The sperm must still swim to the egg, but they do so in the protected internal environment of the female's body.

Sex determination (the determination of whether an individual is male or female or some combination of the two) can be environmental or genetic. With environmental sex determination, for example in reptiles, environmental conditions such as temperature determine the sex of the individual. There is no consistent pattern among reptile species as to which incubation temperature causes the differentiation of which gender.

With genetic sex determination, the genes of the individual determine the sex into which the embryo develops. In many animals, gender is determined not just by genes but by sex chromosomes. In humans, females have two X chromosomes while males have an X and a Y; females are the *homogametic* sex in humans. In birds, butterflies, snakes, fishes, and some plants, males have two Z chromosomes while females have a Z and a W, making females the *heterogametic* sex.

There is no universally accepted explanation as to why there should be separate sex chromosomes. Sexual differentiation of chromosomes, as with X and Y, effectively prevents crossing over (exchange of chunks during meiosis). This allows the two chromosomes to specialize on genes specific to each sex. The fitness interests of the two sexes are not the same. For example, a gene that controls calcium metabolism might be used for antlers by males and for milk by females.

In humans, not all male genes are on the Y chromosome. Most genes that determine male sexual characteristics are on other chromosomes. The Y chromosome does carry the gene that switches on the other male genes. Conversely, there is a female-determining gene on the X chromosome. Sex determination can, however, be complex:

- Male-determining genes on the Y chromosome switch on the production of juvenile and adult forms of testosterone. Therefore the presence of a Y chromosome usually results in maleness. In order for testosterone to produce male characteristics, however, there must be a testosterone receptor protein. Some individuals have a defective testosterone receptor, therefore the testosterone has no effect on their physical characteristics. These XY individuals develop into females. Normal XX females have and use a little bit of testosterone, but the XY individuals with defective receptors do not, and they develop into what have been called super-females.
- Some XY individuals cannot produce the juvenile form of testosterone, yet can produce and respond to the adult form. These individuals grow up as little girls, then at puberty they turn into boys (see POPULATION GENETICS).
- XO individuals (with one X chromosome and no Y; Turner's syndrome) are female but sterile, and while often of normal intelligence they have certain impairments of social understanding.
- In most XY individuals, the male-determining factor overpowers female characteristics. But if the female determining factor is doubled on the X chromosome, an XY individual (genetically male) develops as a female.

As evolutionary biologist Ashley Montagu said, Y is not equal to X. The X chromosome carries many traits, essential to males as well as females, while the Y chromosome is very small and consists mostly of NONCODING DNA. Why is the Y small? The explanation that is usually presented is a conflict of interest between male and female genes (see SELFISH GENETIC ELEMENTS). Suppose a mutation occurred on the X chromosome that damaged the Y chromosome. This would not be good for males, but since the X chromosome spends two-thirds of its time in females, it may be selected anyway. This would lead to males becoming rare, an occurrence that seems to have happened in some species of insects. In response, the Y could lose genes, evolving to be a smaller target for destructive genes produced by the X chromosome.

Evolutionary Reasons for Sex

Scientists have long puzzled over the evolution of sex. Why do organisms go through all the trouble of producing eggs and sperm, when they could just produce copies of themselves?

Some plants produce new plants from specialized structures such as runners, rhizomes, bulbs, or corms. Such *asexual propagation* is a common way for humans to make cuttings from plants. Individuals from asexual propagation are much more likely to survive than seeds. A runner from a plant, for example, is much larger than a seed, and thus more likely to survive; moreover, the runner can remain attached to the parent plant and be fed by it until it is large enough that its survival is virtually assured. *Oxalis pes-caprae* is a little yellow wildflower that grows in the springtime in the mild environments of the Mediterranean, Australia, and California. It is pentaploid, which means that its chromosomes come in groups of five rather than in diploid pairs. Meiosis cannot occur in these plants, because very early in meiosis the chromosomes are divided into two equal groups, and an odd number of chromosomes cannot divide into two equal groups. Therefore a pentaploid plant is sexually sterile. It cannot produce seeds; it propagates by means of bulbs. Some plants produce seeds asexually (without meiosis or fertilization). Asexual propagation by means of diploid eggs or ovules, which are essentially clones of the mother, is called *parthenogenesis* ("partheno" refers to the Greek goddess Athena who supposedly grew out of Zeus's head). One example is the common dandelion, *Taraxacum oficinale.* It is triploid, which means that its chromosomes come in groups of three rather than in diploid pairs. Both of these plant species produce flowers as if they were reproducing sexually, as their immediate ancestors did and as their close relatives still do. Most plants, however, reproduce sexually either along with or instead of asexual propagation.

Parthenogenetic animal species are even more uncommon than asexually propagated plants. Most parthenogenetic animal populations appear to be evolutionary dead ends, on a few of the outermost twigs of the TREE OF LIFE. However, a few animal species, such as some mites and the bdelloid rotifers, appear to have persisted for a long time without sexual reproduction. The bdelloid rotifers have been asexual for a long enough evolutionary time that they have evolved into 360 asexual species. Genetic analysis confirms that this lineage has probably been asexual throughout its evolutionary history. Some animals, such as many species of aphids, propagate themselves asexually during the summer and undergo sexual reproduction in the autumn. Within the fungus kingdom, most species have sexual reproduction, although in many fungi sexual reproduction has never been observed.

Parthenogenesis is not possible in some lineages. Mammals cannot be parthenogenetic. Because some maternal alleles are methylated, only the father's allele is functional, thus a parthenogen will be missing these genes and cannot survive. Conifers cannot be parthenogenetic because the chloroplasts are passed on through pollen.

One important difference between asexual propagation and sexual reproduction is that sexually produced offspring have a different combination of genes than the parents. In asexual propagation, the "parents" and the propagules are genetically identical to one another, and all the propagules are identical to one another. In contrast, each sexually produced offspring is genetically unique. This occurs because diploid organisms have pairs of chromosomes, one member of each pair coming from the mother, the other coming from the father. When meiosis separates the pairs, the chromosomes that came from the father and those that came from the mother are usually distributed randomly among the spores or gametes. Therefore the genes on the different chromosomes experience *independent assortment.* In humans, with 23 pairs of chromosomes, meiosis can distribute the chromosomes into different sperm cells or different ova in 2 to the 23rd power, or more than eight million, different ways. When an egg and a sperm come back together in fertilization, any of these eight million kinds of sperm could unite with any of the eight million kinds of eggs. The genes carried in the cells of one individual can mix with those carried by another individual. Although each individual can carry, at most, two different versions (or *alleles*) of a gene, a population can have many versions of the gene; and sexual reproduction mixes them all together. Although new mutations *(somatic mutations)* can produce new alleles in the cells of asexually propagated organisms, they cannot produce new combinations of alleles.

But what are the advantages of such sexual genetic mixing? There must be a strong advantage to sexual reproduction, since nearly every species has it. The advantages must be major, to compensate for the disadvantages of sexual reproduction:

- Sexual reproduction seems hopelessly vulnerable. Sperm or pollen must find eggs or ovules, and most of them fail.
- The average individual in the population can produce only half as many offspring sexually as it would be able to asexually.
- Good combinations of genes are broken up; this is called *recombinational load.*
- Most plants are capable of a great deal of developmental *plasticity,* in which they adjust their gene expression and their growth to the prevailing environment. Even genetically identical plants, therefore, can have much larger leaves if grown in moist, shady conditions than in dry, sunny conditions (see ADAPTATION). Animals have less plasticity than plants. Nevertheless, plasticity in both animals and plants would seem to assure that only major genetic variations would make an immediate difference to survival of the offspring in the next generation. Most of the genetic variation among offspring of one parent is minor, rather than major, genetic variation. Therefore, in the short term, the asexual species would seem to have an advantage over the sexual species.

Where, then, is the tremendous advantage that sexual reproduction confers upon the species that possess it, as nearly all do? The following categories of advantage have been proposed:

Variable offspring in unpredictable physical and biological environments. When the physical environment is unpredictable, as nearly all environments are, it is impossible to know which combination of genes will be best in the next generation. By having a great diversity of combinations, a population is more likely to persist. The problem with this explanation is that natural selection does not favor the best groups but the best individuals (see NATURAL SELECTION; GROUP SELECTION). But the same reasoning could be applied to individuals: The individuals that produce the greatest genetic diversity of offspring are the most likely to have at least some of those offspring survive in the unpredictable future. If an organism is perfectly adapted to its current environment, its clonal propagules would not be perfectly adapted to the changed environment of the future. As evolutionary biologist George C. Williams points out, this would be especially true under conditions of intense competition—conditions that nearly every lineage of organisms encounters from time to time. In fact, in many natural populations, nearly all of the offspring die. The very few winners are more likely to be the lucky recipients of superior genes from sexual recombination than clonal copies of their parents.

The physical environment (for example, climate) is not the only or even the most important factor in the success of genetically varied offspring. Evolutionary biologists such as Robert M. May and William D. Hamilton have proposed that the diverse offspring produced by sexual reproduction have an advantage in responding to the biological environment, in particular to parasites. The genetic diversity of sexual offspring is minor, from the viewpoint of survival in different climatic conditions; much of this minor genetic diversity consists of proteins and other chemicals that confer resistance to specific diseases. As Hamilton points out, parasites almost always evolve faster than their hosts. Bacteria can have one generation every 20 minutes, while it may take 20 years for a human generation. Evolution proceeds rapidly in most parasites, a fact that has made medical doctors finally take notice of the process of evolution (see EVOLUTIONARY MEDICINE). An asexually reproducing species is therefore at great risk of being killed off by parasites that can quickly evolve the perfect adaptations to infect it. According to this view, dandelions and pentaploid oxalis are either just lucky (which is why such examples are rare), or else the parasite populations are kept under control by climatic conditions. The few asexual species of animals may persist because they are continually migrating and parasites do not easily locate them. A species of snails in New Zealand has both sexual and parthenogenetic forms. The sexual forms are more common in habitats where infection by parasitic worms is common. This proposal overlaps with the RED QUEEN HYPOTHESIS of evolutionary biologist Leigh Van Valen.

Genetic diversity in resisting infection must be of crucial importance in natural populations, for it is of crucial importance in agricultural populations. Plant breeders continually search through wild relatives of crop plants for new genes to introduce either by crossbreeding or by genetic engineering. Some of these genes are for faster growth or improved yield, but often the genes that the plant breeders are looking for are genes that confer resistance against viruses, bacteria, and fungi. When plant breeders have ignored this genetic diversity, fungus blights have broken out and killed vast acreages of crops. This occurred in the early 1970s when Southern Corn Leaf Blight killed many corn plants that all shared the same ineffective resistance genes. A similar argument applies to the breeding of livestock, though livestock breeders usually cross different lines of livestock rather than seeking genes from wild populations. Plant and animal breeders have discovered, sometimes through multimillion-dollar mistakes, that genetic diversity in crop populations is essential to keep diseases under control; it seems certain, therefore, that wild populations need genetic diversity, therefore sex, for exactly the same reason.

Elimination and avoidance of bad mutations. August Weismann, a cell biologist of the late 19th century, proposed that sexual reproduction not only allows good combinations of genes to succeed but allows bad MUTATIONS to be partially eliminated. This idea was expanded by geneticists R. A. Fisher (see FISHER, R. A.) and Hermann Muller, and evolutionary biologist William D. Hamilton. Most mutations are harmful. Many of them are *recessive,* which means that they can be hidden by the functional *dominant* alleles. In a diploid individual, a pair of genes may consist of one good allele and one bad one; if the bad genes are recessive, the good allele will hide the bad one. This process could go on and on, causing a buildup of bad alleles. However, in sexual recombination, the pairing of good and bad alleles is broken. Some of the offspring (the *heterozygotes*) will receive one good and one bad allele; some (the *homozygous dominant* individuals) will receive two good ones; some (the *homozygous recessive* individuals) will receive two bad ones. The offspring with two bad alleles will probably die. The death of the homozygous recessive offspring partially cleans out the bad genes from the population. The process is never complete, because bad genes can always hide in the heterozygotes, but, according to this theory, it is better than nothing, which is what asexual propagation would do (see POPULATION GENETICS).

Sexual reproduction not only allows bad mutations to be eliminated but also allows them to be avoided before being

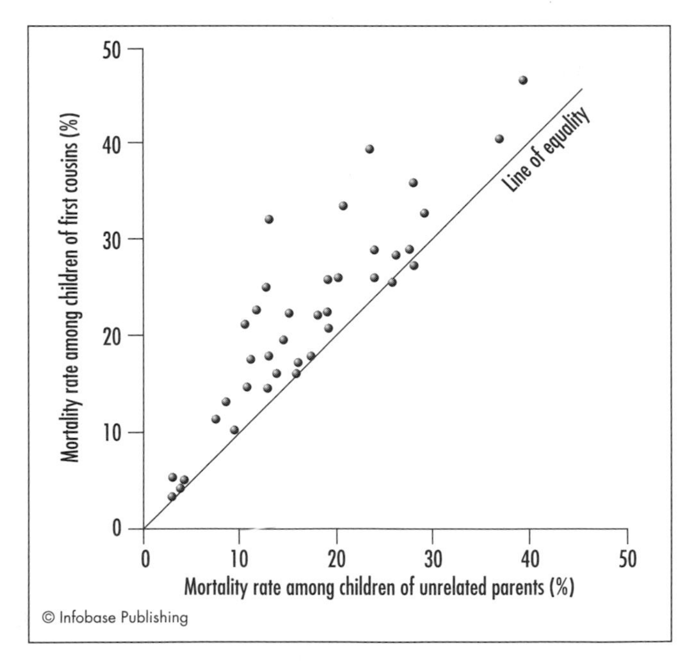

Each point on this graph represents a human population. Childhood mortality rates vary greatly among populations. The horizontal axis is the mortality rate for children whose parents are unrelated. The vertical axis is the mortality rate for children whose parents are first cousins, in the same population. The line represents equal mortality rates for children of unrelated and first-cousin marriages. The points show a predominantly greater mortality rate for first-cousin marriages within these populations. *(Adapted from Bittles and Neel)*

eliminated—that is, if sexual recombination occurs between individuals that are not genetically similar to one another. *Consanguineous* matings (between close relatives, who may share many of the same bad mutations) may produce a relatively large number of offspring that are homozygous for those mutations, a result frequently called *inbreeding depression.* In humans it is well known that incest leads to a high incidence of deleterious mutations (see figure on page 372). This would seem to be the case for all species, but it apparently is not. Some lineages of plants exhibit inbreeding depression, and some do not. Inbreeding depression in plants is most noticeable when individuals experience stressful conditions or competition. Depression in plants may show up more clearly in the second generation of inbreeding; the effects of inbreeding may have been hidden during the first year by maternal resources provided in the seeds. Sibling and other consanguineous matings occur routinely in some invertebrates. For example, there is a wasp *(Scleroderma immigrans)* in which the females not only mate with their own sons but their own grandsons. Their populations appear to have very few deleterious mutations. In the case of the wasps, this may be because the males are haploid, under which condition all deleterious genes are expressed and eliminated (see REPRODUCTIVE SYSTEMS).

It is possible that sexual reproduction has not been selected at all. Paleontologist Niles Eldredge explains that sexual species tend to produce new species more quickly than do asexual species, a process called species sorting. A lineage that produces many new lineages will persist over time, just by chance, more than a lineage that produces few or no new lineages. Random events of EXTINCTION can readily eliminate asexual lineages, which speciate little, but not sexual lineages. To Eldredge, then, the evolution of sex may represent species sorting rather than natural selection.

Several plausible reasons for the evolutionary advantage of sexual reproduction have been proposed. The major difficulty, according to some evolutionary biologists, is not to explain why sexual reproduction exists but rather to explain why it is obligate in so many species.

The evolution of sex provides the perfect illustration that natural selection favors individuals, rather than the whole species. Since male animals (and the male function of plants) can produce many more offspring than females, it would seem that a population would not require very many males. Indeed, in agricultural settings, farmers and ranchers keep only a few choice males as sources of pollen (in trees such as date palms) or sperm. The sex ratio of agricultural populations is therefore heavily weighted toward the females. However, in wild populations, the sex ratio is usually close to 1:1. Suppose that one begins with a wild population that consists, like an agricultural population, of many females and a few males. The few males would leave many offspring. Any mutation that caused a female to produce more male offspring would therefore be favored by natural selection. This process of frequency-dependent selection would continue until males constituted half of the population. Departures from a 1:1 sex ratio can occur as a result of cytoplasmic sterility factors (see SELFISH GENETIC ELEMENTS).

There is little evidence regarding the origin of the process of meiosis, which produces sex cells, or of fertilization, which brings sex cells together. Lynn Margulis and Dorion Sagan point out that sexual fusion is not very different from other kinds of cellular fusion that merge the genetic components of the two cells (see SYMBIOGENESIS).

Natural selection favors the sexual production of diverse offspring, mainly because it cleanses bad mutations from the population and produces a great diversity of genes that resist parasites; these benefits accrue to individuals, not just to populations.

Further Reading

Bell, Graham. *The Masterpiece of Nature: The Evolution and Genetics of Sexuality.* Berkeley: University of California Press, 1982.

Bittles, A. H., and J. V. Neel. "The costs of human inbreeding and their implications for variations at the DNA level." *Nature Genetics* 8 (1994): 117–121.

Cosmides, Leda M., and John Tooby. "Cytoplasmic inheritance and intragenomic conflict." *Journal of Theoretical Biology* 89 (1981): 83–129.

Hamilton, William D., R. Axelrod, and R. Tanese. "Sexual reproduction as an adaptation to resist parasites." *Proceedings of the National Academy of Sciences USA* 87 (1990): 3,566–3,573.

———. *Narrow Roads of Gene Land, volume 2: Evolution of Sex.* New York: Oxford University Press, 2002.

Hurst, Laurence D. "Selfish genetic elements and their role in evolution: the evolution of sex and some of what that entails." *Philosophical Transactions of the Royal Society of London B* 349 (1995): 321–332.

———. "Why are there only two sexes?" *Proceedings of the Royal Society of London B* 263 (1996): 415–422.

Lively, Curtis M. "Evidence from a New Zealand snail for the maintenance of sex by parasitism." *Nature* 328 (1987): 519–521.

Margulis, Lynn, and Dorion Sagan. *Mystery Dance: On the Evolution of Human Sexuality.* New York: Simon and Schuster, 1991.

Otto, Sarah P., and Scott L. Nuismer. "Species interactions and the evolution of sex." *Science* 304 (2004): 1,018–1,020.

Paland, Susanne, and Michael Lynch. "Transitions to asexuality result in excess amino acid substitutions." *Science* 311 (2006): 990–992.

Ridley, Matt. *The Red Queen: Sex and the Evolution of Human Nature.* New York: HarperCollins, 1993.

Takebayashi, Naoki, and Peter L. Morrell. "Is self-fertilization an evolutionary dead end? Revisiting an old hypothesis with genetic theories and a macrorevolutionary approach." *American Journal of Botany* 88 (2001): 1,143–1,150.

West, Stuart A., Curtis M. Lively, and Andrew F. Read. "A pluralist approach to the evolution of sex and recombination." *Journal of Evolutionary Biology* 12 (1999): 1,003–1,012.

Williams, George C. *Sex and Evolution.* Princeton, N.J.: Princeton University Press, 1975.

sexual selection Sexual selection is the selection of characters that enhance reproductive success rather than (even at the expense of) survival. The individuals in which the characteristic is well developed will reproduce more often, and leave more offspring, than individuals in which the characteristic is absent or less well developed.

Sexual selection is similar to NATURAL SELECTION, in that a characteristic evolves because it causes the individuals that possess the trait, or that possess it in a more highly developed state, to leave more offspring than individuals that do not possess it. The difference is that sexual selection favors characteristics that directly enhance reproductive success rather than characteristics that enhance it indirectly through survival or obtaining resources. Charles Darwin (see DARWIN, CHARLES) proposed natural selection as the major mechanism of evolution in his world-changing 1859 book *On the Origin of Species By Means of Natural Selection.* He also proposed sexual selection in his 1871 book *The Descent of Man and Selection in Relation to Sex* (see *ORIGIN OF SPECIES* [book]; *DESCENT OF MAN* [book]).

Sexual Selection in Animals

In most animal species, males can produce more offspring than can females (see REPRODUCTIVE SYSTEMS). This is partly because sperm cells are small and cheap, while egg cells are large and expensive. In animals with external fertilization, such as most fishes, this may be almost the entire difference in reproductive cost between males and females. In species with internal fertilization (such as most terrestrial vertebrates), at least part of the embryonic development occurs within the body of the female. Bird and reptile eggs are relatively large by the time the female lays them. In terrestrial species with live birth *(vivipary),* it is the female in which embryonic development occurs. In most animal species in which newborns receive parental care, it is the female that provides a disproportionately large amount of that care.

Sexual selection can operate at the level of the reproductive cell, or of the reproductive individual:

Sexual selection upon sperm. In many animal species, males produce millions of sperm while females produce far fewer eggs. This can lead to an intense selection process occurring upon the sperm themselves. In humans, very few of the many millions of sperm reach the egg. This is why a sperm count of 50 million, although that sounds like a large number, may indicate infertility.

- The acidic environment of the vagina kills many sperm.
- The sperm must also be strong swimmers, to climb up through the cervix into the uterus.
- Since the sperm are genetically different from the cells of the woman's body, the woman's immune system begins to launch an attack on the sperm.

Selection of strong sperm may also result in the selection of strong offspring, because many of the metabolic genes that promote sperm success also promote the vigor of the offspring. Sexual selection has not only favored the production of strong sperm but also the production of chemicals in the semen that partially suppress the woman's immune response. Selection for successful sperm is a very important part of reproductive success. Each sperm is the product of a long series of cell divisions, each one an opportunity for MUTATIONS. There are therefore many mutant sperm that need to be "weeded out." This explanation seems inadequate to explain all of the characteristics of sperm cells in all animal species. No convincing explanation has yet been offered for why the sperm of *Drosophila bifurca,* a species of fruit fly that is about one-twentieth of an inch (1 mm) long, can have tails almost two inches (6 cm) long.

Sexual selection at the individual level. Since one male can produce so many sperm, and since in many animal species the act of mating is the male's only contribution to the success of the offspring, a female animal cannot typically increase her reproductive success by mating more often. Therefore it is usually males that compete with other males for access to females; and females that choose which males they allow to fertilize their eggs.

Females may choose among males on the basis of genetically based characteristics that have no connection with the quality of the male. According to the "sexy sons" hypothesis, females in a species may choose a male with any characteristic, perhaps an outlandish color pattern. The sons of these females will also have this characteristic, which will enhance their chances of being chosen by females in the next generation. Evolutionary theorist R. A. Fisher (see FISHER, R. A.) first pointed out that this could result in a positive feedback loop of runaway sexual selection.

Females may choose among males on the basis of genetically based characteristics that indicate the quality of the male:

- The best genes may be those that allow males to obtain resources most effectively.
- The best genes may be those that allow resistance to parasites and diseases. Because parasites can evolve so much more rapidly than their hosts, most animal species are in a continual race to keep ahead of their parasites (see RED QUEEN HYPOTHESIS).

There are many *fitness indicators* that females can use to choose the best males. A fitness indicator is any trait that reliably displays the strength of the state of health of the male. A healthy male is likely to be superior in resource acquisition and in resistance to parasites and therefore likely to have higher potential fitness. A fitness indicator must, therefore, be expensive—and the more expensive it is, the better a fitness indicator it is. This is the "handicap principle" proposed by evolutionary biologist Amotz Zahavi. Evolutionary biologists William Hamilton and Marlene Zuk proposed that the most valuable fitness indicators are those that show the male to be free from parasites. Fitness indicators include:

- A male that defends a large territory in prime quality habitat not only has more resources to offer to the female but is probably healthier and therefore able to drive away other males. It is not only expensive for the male to obtain and defend the territory but to advertise it to females. A male mockingbird may sing all day for several weeks, which uses a lot of energy and makes him noticeable to predators.
- Another fitness indicator is bodily ornamentation. Consider a bird species that uses colored feathers as fitness indicators. Any male, even an inferior one, can (potentially) produce a few small colored feathers, but only a healthy

one, free of parasites, is likely to produce many long colorful feathers. High levels of testosterone, which enhances aggression in birds and mammals, also decreases the function of the immune system. The males that manage to have both the longest, most colorful feathers as well as the most aggressive territorial defense may be able to leave the most offspring. Feathers will continue to evolve to be longer and more colorful until the feathers become too expensive or burdensome. In many bird species, male ornamentation consists of only a few feathers or a crest; but in some, such as peacocks and birds of paradise, the ornamentation is so excessive that it constitutes a significant drain of resources and puts the male in greatly enhanced danger of attack by predators. Researchers have experimentally augmented the feathers of some males and found that the males with artificially long feathers have more offspring.

- In many bird species, males congregate in *leks* to display their plumage and their mating dances, and females stroll around and choose from among them.
- Symmetry is a fitness indicator in many species, including humans, who show a preference for faces and bodies that are symmetrical. Poor health often results in asymmetry between the right and left sides of a vertebrate.

Expensive fitness indicators among mammals include antlers (which in the now extinct Irish elk could reach over six feet, or two meters, in expanse). Natural selection can put a stop to sexual selection, but not before sometimes bacchanalian excesses of ornamentation are produced. The more expensive the better, up to a point, because if the ornamentation were cheap, less healthy males could cheat by producing an ornament that merely made it look healthy. Aside from being expensive, and being truthful representations of health, fitness indicators may be arbitrary, as reflected in the tremendous diversity of feather ornamentation and courtship dances among the almost 9,000 bird species.

Once fitness indicators come into existence, as a result of sexual selection, they can serve as reproductive ISOLATING MECHANISMS that help to separate one population into two. This can lead to SPECIATION, when males with novel traits are chosen by females with a preference for those same novel traits.

In many animal species, sexual selection has produced particularly marked characteristics that facilitate competition between males:

- *Size dimorphism.* In many animal species, the male is larger than the female even though it is the females, not the males, that have most of the costs of reproduction. Large size gives males a competitive advantage in fighting with one another.
- *Weapons.* The same antlers that serve as fitness indicators for female choice also serve as weapons for fighting other males.
- *Impulsiveness.* In most species, it is the males that are more impulsive, and sometimes more violent. This is why, as Darwin noted in 1871, it is usually males (even among the invertebrate animals) that are, as he said in his inimitable Victorian prose, more "pugnacious" than females. In their hurry to mate, males can sometimes be absurdly undiscriminating: Male turkeys, for example, will attempt to mate with a replica of a female turkey head.
- *Interference with the reproduction of other males.* There are numerous instances in which males interfere with the opportunity for other males to inseminate females. In some stick insects such as *Necroscia sparaxes,* the male will copulate with the female for weeks at a time, thus keeping rival males away. Whether in flies or in primates, males may have very large reproductive organs, which deliver a large amount of sperm. This helps to outcompete the sperm that a previous male may have left. Male voles produce more sperm if there are other males in the vicinity than they do if they are alone. The males of some species have penes with bristles, which are believed to scrub away sperm left by other males. In other cases, males produce secretions that plug up the female's reproductive tract after he has mated with her, thus preventing the sperm of other males from entering. However, the males have also evolved the ability to remove the plugs inserted by other males.
- *Stealth.* In many cases, males have evolved to become stealthy in their approach to females. In some salamanders such as *Plethodon jordani,* the female follows the male until the male drops a spermatophore, a structure that contains sperm; the female then sits on the spermatophore and receives the sperm. Sometimes a third male will intrude between the couple, waddle along as if he is a female, then when the first male drops his spermatophore, the second runs off with the female. In some fishes, some males imitate females, infiltrate themselves into the harems controlled by dominant males, and mate with some of the females. Perhaps the ultimate example is when certain male bats *(Myotis lucifugus)* creep around amid hibernating females and inseminate them.
- *Chemical components of semen.* Some male houseflies even produce semen that contains proteins that reduce the female's interest in copulation—after he has finished with her, of course. This reduces the chance that she will accept the advances of another male. In some cases, the compounds in the semen are even slightly poisonous. In one experiment, in which male flies were allowed to evolve in competition but female flies were not, the semen from the male flies actually killed the females.

Male-male competition may in some cases have promoted speciation. In dragonflies and damselflies, males clasp females behind their heads to mate with them, thus keeping other males away. Their penes scrub out the sperm of previous males. This process can be so violent that the females get severely injured. But the post-reproductive survival of the female does not matter to the male. According to evolutionary biologist Ola Fincke, the female evolutionary response to this conflict of interest may have been new shapes and colors, which make it more difficult for the males to recognize the females.

In animals with either monogamous or harem mating systems, in which the male can be reasonably confident of his exclusive access to the female, males have small reproductive organs. Gorilla males, for example, have penes even smaller than those of some ducks.

Because, in these instances, the male reproductive investment may be completed upon successful copulation, the male may be expendable. In many spiders and in praying mantises *(Mantis religiosa),* females often eat the males—in the case of the mantis, even while copulation is occurring.

There are cases in which the males have the larger reproductive investment, and females choose among them. Females may provide eggs to males, who take care of them; often, males receive eggs from more than one female:

- In the rhea *(Rhea americana),* a South American bird similar to the ostrich, the male incubates the eggs and takes care of the hatchlings.
- In some wading birds, such as the red-necked phalarope *(Phalaropus lobatus),* it is the males that raise the young. The red-necked phalarope is also one of the few bird species in which the females are more brightly colored than the males.
- In some stickleback fishes such as *Gasterosteus aculeatus,* females prefer to provide eggs to males who already have eggs in the nests they have built. This has led to the evolution of stealing behavior, in which males will steal eggs from the nests built by other males.
- Males of the fish species *Cyrtocara eucinostomus,* though only 10 centimeters in length, build enormous sand castles a meter wide at the base, one mouthful of sand at a time. The female chooses the male that constructs the best castle.
- In some frogs, and fishes such as the sea horse, the female lays the eggs and then the male scoops them up into his mouth to protect them while they hatch, a process called *mouth-brooding.* This activity benefits the male in that he can be certain the offspring are his. However, during this prolonged period of time, the male cannot fertilize any other eggs. It therefore represents a considerable expense for the male. Since it typically takes more than one male to mouth-brood the eggs produced by one female, it is the females that compete with one another for access to mouth-brooding males.

Therefore, while sperm are individually cheap, male reproductive expenditure can be expensive.

In some instances, the females may be so widely dispersed that a male is lucky to ever find one, and a female is unlikely to be found by more than one male. In these cases, there is no advantage for a male to be large. In some species of invertebrates (such as some barnacles), as well as some fishes, the males are small and even merge into the flesh of the females so that they resemble an organ or a lump on the female.

Sexual selection has operated strongly during human evolution. Men have evolved numerous fitness indicators to influence female choice. Some of these fitness indicators are visible characteristics such as muscularity; sexual selection may have caused men to become more muscular than they need to be as a result. Most male fitness indicators have less to do with appearance than with function. The human penis is larger than that of some other primates. There may be two reasons for this. One reason is, as noted above, large penes are associated with promiscuous mating systems, as in chimpanzees, while small penes are associated with harem mating systems, as in gorillas; humans are intermediate. Another reason is that large penis size may have been selected because of tactile pleasure that it provides to the females, who would then choose to keep company with the man who could provide this pleasure. This explanation would make no sense if pregnancy often occurred after a single copulation (as in cats), because by the time the female determined that a certain male gave less pleasure than others it would be too late to reject him. However, because numerous copulations are typically necessary for pregnancy to occur in humans, pleasure is a useful piece of information on which a woman could partly base her decision. Therefore female choice that maintains a relationship, as well as the choice that initiates it, has been important in human evolution.

As noted above, males sometimes select among competing females. This is especially the case in animal species that have expensive offspring, such as humans. Human reproduction is so expensive that there is a definite survival benefit to the offspring if both parents help to raise them (see LIFE HISTORY, EVOLUTION OF). The genetic fitness of the men, as well as the women, would be enhanced if the man stays with the woman and helps with the provision and protection of the offspring. If the woman has fitness indicators that will attract the man and entice him to stay, her reproductive success may be enhanced, as well as his. The man, in turn, may want his genes to be carried by a woman who is most likely to raise the children to adulthood. Women have, therefore, also evolved fitness indicators that show that they are healthy and well-nourished. A primary indicator of health is the face. A face that is symmetrical and unblemished is an indicator of past and current health. Another indicator is fat accumulation. Fat accumulation serves not only as a fitness indicator but as a direct contributor to the ability to raise children, through nursing and later the ability to survive food shortages during which feeding the child or children receives top priority. Fat accumulation throughout the body would be less obvious than fat accumulation in visible places such as breasts and buttocks. A fitness indicator, in order to be effective, must be expensive and show variation among individuals. Faces, breasts, and buttocks not only indicate health and nutrition but also age. If skin did not wrinkle and breasts did not sag, with advancing age, they would not have served as reliable indicators of female fitness.

Evolutionary scientists have been slow to catch on to Darwin's insight that sexual selection has been paramount in human evolution. They have instead focused on the more practical survival aspects of life as paramount in human evolution, for example language as a means of communicating the locations of resources or sorting out social hierarchies; tools as means of obtaining resources, or as weapons against competing tribes; and the relative hairlessness of the human body as an adaptation that allows warm-blooded mammals to cool off in the hot African savannas. In general, all of the many adaptations that are collectively called intelligence have been attributed to natural selection for the serious businesses of resource acquisition and war (see INTELLIGENCE, EVOLUTION OF). In many cases, these explanations were and are believable. Tools certainly have enhanced the ability of humans to exploit natural resources, and complex social interactions would be impossible without language (even though chimpanzee societies are quite complex despite the lack of true language). In other cases, the explanations sound like just-so stories. Why should humans have evolved hairlessness, when other hot-blooded mammals of the African savanna have kept their hair?

Even the believable explanations seem insufficient. Consider, for example, language, which some scientists consider the major adaptation that distinguishes human from animal intelligence (see LANGUAGE, EVOLUTION OF). Theories proposed early in the 20th century attempted to explain the origin of language in terms of practical natural selection. What should have been obvious even at that time is that languages are far more complex than they need to be for such simple practical requirements. Just how complex does language have to be in order to say "Lion!" or "Time for lunch!"?

Perhaps the best explanation for the origin of language is that it was an indicator of the quality of intelligence and of the brain. People who could not just talk but speak with complex grammar and poetical allusions obviously had good brains. Many other characteristics of human intelligence, such as the production of art and music, serve as clear indicators of the quality of the brain and could therefore function as sexual fitness indicators. Evolutionary biologist Geoffrey Miller explains that the human brain itself is a fitness indicator, since it is such an expensive organ (consuming tremendous amounts of energy and oxygen), and at least half of the genes are involved in brain development. Therefore practically anything that the brain does can be a fitness indicator. As such, it can be the subject of sexual selection. By choosing intelligent mates, both men and women could be reasonably certain that their mate had good genes.

Another example of a characteristic often considered to have evolved for its practical benefits is hunting. The practical explanation of hunting is that men bring home the meat while women gather nuts, berries, and grains. While plant materials may provide most of the calories, the meat brought home by the men provides essential nutrients that would be missing from a vegetarian diet. Even if it is true that a purely vegetarian diet would be inadequate, this does not explain why men hunt big game in nearly every tribe that has been studied and continue to do so today in civilized nations. Men could provide adequate meat by hunting small game. In fact, they could provide more meat in this manner, and with considerably less danger to themselves. Simply put, hunting is not a practical way for a man to provide for his family. It appears that hunting, in all societies, has been and is a sport. Its purpose is to be a fitness indicator—only very strong males with good brains can show off their qualities by undergoing the physical exertion and having the intelligence to track down, or trick, dangerous game animals, and then kill them. By choosing good hunters, the women could be reasonably sure that their mates were strong and intelligent.

Many human characteristics, not the least of which is the large human brain, are part of a pattern called NEOTENY in which juvenile characteristics are retained into adulthood. This explains many unique human features as part of a suite of neotenous traits but does not explain why neoteny occurred in the first place. What, if anything, caused neoteny to operate during human evolution? It is possible that the retention of juvenile characteristics was selected because they made older people look younger, thus deceptively enhancing fitness indicators of youth in both sexes?

Sexual Selection in Plants

Mate choice, resulting in sexual selection, can occur in organisms other than animals. Evolutionary biologists Mary Willson and Nancy Burley pointed out that male plants, or the male function of bisexual plants, compete with one another in terms of sperm or pollen production, and that female plants, or the female function of bisexual plants, may be able to discriminate among sperm or pollen. Male competition in plants occurs for the same reason that it does in animals: A plant can produce more sperm, or more pollen, than a female plant can produce eggs, or ovules; and the female plant, or female function within the plant, not only produces eggs or ovules but also nourishes the resulting embryo, often within a seed, and provides for its dispersal, often within a fruit. Instances of sexual selection in plants may include the following:

- Male competition can cause plants to undergo male reproduction earlier than female reproduction, since the first male plant to produce sperm or pollen can fertilize numerous female plants. Botanist Stanley Rice and his students have shown that in cottonwoods *(Populus deltoides)*, which have separate male and female trees, the male trees open their flower buds earlier, on the average, than female trees—although of course their flowering times must overlap or else pollination would fail. Male trees that open earlier may pollinate many female flowers before other male trees get a chance to do so. This could be interpreted as sexual selection among male trees.
- Plants do not copulate. Pollen must be carried from the male parts of one plant to the female parts of another. Plants rely on the wind, or on many species of pollinators, to accomplish this. The plants that more effectively attract pollinators are more effective at sexual reproduction. All of the characteristics of plants related to pollination (see COEVOLUTION) can be considered the products of sexual selection. In animal-pollinated dioecious plant species, in which some plants are male and others are female, male flowers are usually larger and more colorful than female flowers. This can be interpreted as male-male competition for sperm delivery and therefore sexual selection.
- Once a flower has been pollinated, the pollen must grow a tube down into the flower, making contact with an ovule, which may be a great distance away. In maize, for example, the tube from a pollen grain must grow up to eight inches (20 cm), a thousand times the length of the pollen grain, to reach the ovule. The sperm nucleus then descends

through this tube to complete the act of fertilization. The structure through which the pollen tube must grow, the style, can be said to exist for the sole purpose of "weeding out" inferior pollen grains. Botanist Allison Snow has found that, in the bush *Hibiscus moscheutos,* pollen grains from some plants grow tubes faster than pollen grains from other plants. Since flowers often receive more pollen than they need, there is often a race between the different pollen tubes to be the first ones to arrive at the immature seeds—which may constitute sexual selection among the pollen tubes.

- In terms of differences among pollen grains in the rate of growth, the style functions just as the food source through which the tube grows, but a pistil is more than just a food source for the pollen tube. The pistil itself may chemically encourage the growth of some pollen tubes and discourage others. This is difficult to demonstrate. It has been shown that there is greater variation in pollen tube growth rate in the pistils of healthy plants than of plants that are experiencing stresses such as drought. This implies but does not prove that a healthy plant chemically distinguishes among pollen tubes. The haploid cells surrounding the egg cell, in the immature seed, have genes that control the release of sperm from the pollen tubes. This implies the possibility of sexual selection in which the female cells are in control of which sperm are allowed to fertilize the immature seeds.
- Botanist Diane Marshall found that radish plants may provide more food to the developing fruits that contain seeds that are more genetically variable (the result of a greater diversity of pollen donors) than to fruits that contain genetically more uniform seeds. Botanists K. B. Searcy and M. R. McNair found that monkeyflower plants *(Mimulus guttatus)* that grow in soil that contains mildly toxic levels of copper selectively abort seeds that had been fertilized by pollen from donors that were not resistant to copper. These may be examples of sexual selection in which the maternal plant selects among pollen sources.
- The endosperm is the food supply for the embryo inside the seed. It is triploid, having formed from two sets of chromosomes from the plant that produces it but only one set from the pollen. Several researchers, such as botanist Mark Westoby, have interpreted this as sexual selection in which the maternal plant gains more control than the paternal plant in controlling the growth of the embryo.

Therefore while natural selection explains the diversity of adaptations that allow organisms to survive and reproduce in different environments, and in response to other species, it is sexual selection that primarily explains much of the seemingly arbitrary characteristics of animals, and perhaps many characteristics of plants as well. Natural selection has made the world of organisms efficient; sexual selection has made it wildly beautiful.

Further Reading

Albert, Arianne Y. K., and Sarah P. Otto. "Sexual selection can resolve sex-linked sexual antagonism." *Science* 310 (2005): 119–121.

Andersson, Malte. "Evolution of classical polyandry: Three steps to female emancipation." *Ethology* 111 (2005): 1–23.

Bernasconi, G., et al. "Evolutionary ecology of the prezygotic stage." *Science* 303 (2004): 971–974.

Cuervo, J. J., and R. M. De Ayala. "Experimental tail shortening in barn swallows *(Hirundo rustica)* affects haematocrit." *Functional Ecology* 19 (2005): 828–835.

Darwin, Charles. *The Descent of Man, and Selection in Relation to Sex.* London: John Murray, 1871. Reprinted with introduction by James Moore and Adrian Desmond, New York: Penguin Classics, 2004.

DelBarco-Trillo, Javier, and Michael H. Ferkin. "Male mammals respond to a risk of sperm competition conveyed by odours of conspecific males." *Nature* 431 (2004): 446–449.

Diamond, Jared. *Why Is Sex Fun?* New York: Basic Books, 1997.

Haig, David, and Mark Westoby. "Parent-specific gene expression and the triploid endosperm." *American Naturalist* 134 (1989): 147–155.

———. "Seed size, pollination costs and angiosperm success." *Evolutionary Ecology* 5 (1991): 231–247.

Hamilton, William D., and Marlene Zuk. "Heritable true fitness and bright birds: A role for parasites?" *Science* 218 (1982): 384–387.

Heinsohn, Robert, Sarah Legge, and John A. Endler. "Extreme reversed sexual dichromatism in a bird without sex role reversal." *Science* 309 (2005): 617–619.

Judson, Olivia. *Dr. Tatiana's Sex Advice to All Creation.* New York: Henry Holt, 2002.

Marshall, Diane L., and M. W. Folsom. "Mate choice in plants: An anatomical to population perspective." *Annual Review of Ecology and Systematics* 22 (1991): 37–63.

———, and Pamela K. Diggle. "Mechanisms of differential pollen donor performance in the wild radish, Raphanus sativus (Brassicaceae)." *American Journal of Botany* 88 (2001): 242–257.

Miller, Geoffrey. *The Mating Mind: How Sexual Choice Shaped the Evolution of Human Nature.* New York: Doubleday, 2000.

Miller, Paige M., et al. "Sexual conflict via maternal-effect genes in ZW species." *Science* 312 (2006): 73.

Nielsen, M. G., and W. B. Watt. "Interference competition and sexual selection promote polymorphism in *Colias* (Lepidoptera, Pieridae)." *Functional Ecology* 14 (2000): 718–730.

Pomiankowski, A., and Y. Iwasa. "Runaway ornament diversity caused by Fisherian sexual selection." *Proceedings of the National Academy of Sciences USA* 95 (1998): 5,106–5,111.

Searcy, K. B., and M. R. McNair. "Developmental selection in response to environmental conditions of the maternal parent in *Mimulus guttatus.*" *Evolution* 47 (1993): 13–24.

Snow, Allison A. "Postpollination selection and male fitness in plants." *American Naturalist* 144 (1994): 569–583.

———, and Timothy Spira. "Pollen-tube competition and male fitness in Hibiscus moscheutos." *Evolution* 60 (1996): 1,866–1,870.

Trivers, Robert. "Parental investment and sexual selection." Pages 136–179 in Campbell, B., ed., *Sexual Selection and the Descent of Man.* Chicago: Aldine, 1972.

Van Oort, H., and R. D. Dawson. "Carotenoid ornamentation of adult male common redpolls predicts probability of dying in a salmonellosis outbreak." *Functional Ecology* 19 (2005): 822–827.

West, Peyton M. "The lion's mane." *American Scientist* 93 (2005): 226–235.

Willson, Mary F., and Nancy Burley. *Mate Choice in Plants: Tactics, Mechanisms, and Consequences.* Princeton, N.J.: Princeton University Press, 1983.

Zahavi, Amotz. *The Handicap Principle: A Missing Piece of Darwin's Puzzle.* New York: Oxford University Press, 1997.

Silurian period The Silurian period (440 million to 410 million years ago) was the third period of the PALEOZOIC ERA (see GEOLOGICAL TIME SCALE). It followed the ORDOVICIAN PERIOD and preceded the DEVONIAN PERIOD.

Climate. The climate of the continents was generally warm and moist, except the landmasses directly over the South Pole, during the Silurian period. Some southern glaciers from the Ordovician period melted and raised the sea level.

Continents. As in the preceding and following periods, most of the landmass was near the South Pole (Gondwana). Other landmasses that today constitute Eurasia, Siberia, and China were near the equator.

Marine life. During the Silurian period, aquatic animals continued expanding from shallow marine waters into deeper waters and into freshwater. Fishes, the first vertebrates, continued to evolve, both jawless fishes and the first fishes with jaws (see FISHES, EVOLUTION OF).

Life on land. Fishes first adapted to freshwater habitats on the continents during the Silurian period. The most noticeable advance of life during the Silurian period was onto dry land. It did not look like much: A few invertebrates lived in the mud, and very small plants (a few inches tall) grew from mud up into the air rather than living completely underwater. But it was a major event in the history of life. The arthropods (see INVERTEBRATES, EVOLUTION OF) were mostly carnivores and detritivores. The plants contained lignins and toxins; the animals could not generally digest the plant material until it had been softened and detoxified by fungi. While today the food chain is based largely upon the consumption of fresh green plant tissue, the Silurian food chain depended indirectly on plants. These early plants (such as *Cooksonia*) had xylem cells, which are the plumbing that allows plants to transport water upward from the ground to the ends of their stems. They resembled modern plants such as club mosses and horsetails (see SEEDLESS PLANTS, EVOLUTION OF).

Extinctions. The Silurian period began after the Ordovician extinction, which was one of the five MASS EXTINCTIONS in the history of the Earth.

Further Reading

Museum of Paleontology, University of California, Berkeley. "The Silurian." Available online. URL: http://www.ucmp.berkeley.edu/silurian/silurian.html. Accessed May 13, 2005.

Simpson, George Gaylord (1902–1984) American *Paleontologist* George Gaylord Simpson was one of the architects of the MODERN SYNTHESIS which brought together Darwinian NATURAL SELECTION and MENDELIAN GENETICS. The modern synthesis was a multidisciplinary effort: Based on the mathematical framework of Fisher, Haldane, and Wright (see FISHER, R. A.; HALDANE, J. B. S.; WRIGHT, SEWALL), it took form with the research of Dobzhansky (see DOBZHANSKY, THEODOSIUS) with fruit flies, Mayr (see MAYR, ERNST) with birds, Stebbins (see STEBBINS, G. LEDYARD) with plants, and Simpson with fossil vertebrates. It was Simpson who supplied the paleontological information, especially regarding the evolution of horses, that was necessary for the modern synthesis.

Born June 16, 1902, Simpson had an insatiable interest in all things; when he received a copy of the *Encyclopaedia Britannica,* he read it straight through. Growing up in Denver, he hiked in the Rocky Mountains with friends and family. When he enrolled at the University of Colorado at Boulder, he wanted to be a creative writer. As a sophomore he took a geology course, which determined his career. He never did give up his interest in writing, however; he wrote a novel, *The Dechronization of Sam Magruder,* that explored some scientific concepts related to the time dimension. Some of his science books, such as *Attending Marvels* and *The Meaning of Evolution,* became classic best sellers partly because of their literary merit.

He transferred to Yale to continue his studies in paleontology and graduated in 1923. He remained at Yale, where he completed a Ph.D. about the fossil mammals of the American West. He spent some time at the British Museum of Natural History until in 1927 he joined the American Museum of Natural History in New York as a curator of fossil vertebrates. In 1942 he joined the U.S. Army and served as an intelligence officer until illness brought him back to the United States. In 1958 Simpson moved to the Museum of Comparative Zoology at Harvard, where he completed most of the rest of his career. His two major expeditions into the field were to Patagonia and to the Brazilian Amazon.

Simpson strongly believed that the dispersal of mammals could readily explain their distribution patterns in the fossil record and on the Earth today, and that it was not necessary to invoke the CONTINENTAL DRIFT theory of Alfred Wegener. As continental drift became more widely accepted, Simpson held out in his opposition, until the amount of evidence that supported PLATE TECTONICS as the mechanism of continental drift finally convinced him. The willingness and ability to change even the most fundamental theoretical framework, as Simpson did, is one of the characteristics that makes the SCIENTIFIC METHOD differ from most aspects of religion. Simpson died October 6, 1984.

Further Reading

Laporte, Léo. "George Gaylord Simpson: Paleontologist and Evolutionist, 1902–1984." Available online. URL: http://people.ucsc.edu/~laporte/simpson/Index.html. Accessed May 12, 2005.

Simpson, George Gaylord. *Attending Marvels: A Patagonian Journal.* New York: Macmillan, 1934.

———. *Concession to the Improbable: An Unconventional Autobiography.* New Haven, Conn.: Yale University Press, 1978.

———. *The Major Features of Evolution.* New York: Columbia University Press, 1953.

———. *The Meaning of Evolution.* New Haven, Conn.: Yale University Press, 1949.

———. "One hundred years without Darwin are enough." *Teachers College Record* 60 (1961): 617–626.

———. *Tempo and Mode in Evolution.* New York: Columbia University Press, 1944.
———. *This View of Life: The World of an Evolutionist.* New York: Harcourt, Brace and World, 1964.

Smith, William (1769–1839) British *Geologist, Engineer* William Smith was the geologist who figured out that fossils could be used to correlate sedimentary rock layers (see FOSSILS AND FOSSILIZATION). If a sedimentary layer in one location contained an assemblage of fossils, and a sedimentary layer in another location contained the same or similar assemblage of fossils, the layers in both locations (even if separated by hundreds of miles) were almost certainly formed at the same time. In this way, the relative ages of fossil deposits can be determined, although their absolute ages cannot. Smith's insight allowed geologists in the 19th century to reconstruct the geological record of the Earth's entire history, although they had no idea of how many millions of years the geological record spanned (see AGE OF EARTH; RADIOMETRIC DATING).

Born March 23, 1769, William Smith received little formal education but spent much time collecting fossils as he grew up. He studied mapping and became a surveyor. Excelling in this skill, Smith spent six years supervising the digging of the Somerset Canal in southwestern England. Smith noticed that the fossils within a vertical section of sedimentary rocks were always in the same order from the bottom to the top of the section, and the sedimentary rock sections on one side of England matched those on the other side. Not every species was useful for correlating the strata of different locations; species that had very widespread distributions or that persisted for a long time were not suitable for precise correlation.

Smith was not the first to construct geological maps but was the first to do so by fossils and not by the mineral composition of the rocks. When Smith finally secured money from private investors in 1812, he began a fossil-based geological map of all of England and Wales, a task he finished in 1815. Even though this map was initially disregarded by scientists, it is difficult to overestimate its importance. It was an essential base of information that allowed UNIFORMITARIANISM to replace CATASTROPHISM and that allowed the insights of later geologists (see LYELL, CHARLES) and, eventually, evolutionary scientists (see DARWIN, CHARLES). Smith died August 28, 1839.

Further Reading

Winchester, Simon, and Soun Vannithone. *The Map That Changed the World: William Smith and the Birth of Modern Geology.* New York: Perennial, 2002.

Snowball Earth Periods of ice formation, global in extent, occurred during PRECAMBRIAN TIME. The Sturtian Glaciation ended about 700 million years ago, the Marinoan Glaciation about 635 million years ago, and the Gaskiers Glaciation about 580 million years ago. There was an earlier period of ice formation, about 2.4 billion years ago, about which much less is known, because there are far fewer deposits that have survived from that distant time. The ice formed on both land and sea, not only in polar and temperate latitudes but even in the tropics, which would have made the Earth look very much like a snowball. Because there may have been little open ocean, there would have been very little evaporation, hence very little rain or snow, once the snowball conditions formed.

Calculations by Russian scientist Mikhail Budyko in the 1960s suggested that the Earth could never have had such a period of cold temperatures, because if this had occurred, the Earth would have remained frozen forever. This would occur because of the albedo effect: The sunlight would have reflected from the white snow and ice, directly back into outer space, without being absorbed; the only sunlight that creates warmth is the sunlight that is absorbed. (This is why a stick lying on top of the snow on a sunny day will melt its way into the snow; the stick absorbs sunlight and becomes warm, while the snow does not.) Budyko's calculations did not include the effects of volcanic eruptions. Eruptions would have created numerous unfrozen spots in the ocean and would have released carbon dioxide into the atmosphere. Carbon dioxide contributes to the GREENHOUSE EFFECT by absorbing infrared photons that are emitted by the Earth. Very little infrared light would have been emitted by a Snowball Earth, but as carbon dioxide levels built up over millions of years, enough heat might have accumulated to begin melting the ice. Two major processes by which carbon dioxide is removed from the air are the photosynthesis of organisms and the weathering of rocks. With so few organisms, and with the rocks and oceans covered by ice, there would have been almost nothing to remove carbon dioxide from the air, thus allowing it to accumulate. These processes would have allowed the Earth to emerge from its snowball condition.

Snowball Earth is an example of a truly radical theory that has become widely accepted, in various forms, by the scientific community. Its principal proponents are geologists Paul Hoffman and Daniel Schrag of Harvard University, basing their studies on earlier work by geologists Joseph Kirschvink and Brian Harland.

- One line of evidence suggesting that there was a Snowball Earth is the massive accumulation of tillite *dropstones* in late Precambrian deposits. Dropstones are a diverse mixture of rocks that have been transported far from their places of origin and dropped onto the bottom of the ocean. It is generally agreed that only icebergs can carry dropstones for such distances. The large deposits of dropstones indicate that there must have been a lot of icebergs in the late Precambrian. These deposits are found from all over the world: Spitsbergen, China, Australia, Russia, Norway, Namibia, Newfoundland, and the Rocky Mountains. Therefore, the glaciation was global.
- Another line of evidence is the presence of late Precambrian ironstones. When oxygen first began to be produced by photosynthesis of microscopic organisms, it reacted with iron in the ocean water and precipitated to the bottom of the sea, forming iron deposits. This stopped happening when the iron was removed from the ocean water,

and it is not happening extensively today. However, there was a period of ironstone formation in the late Precambrian. Researchers claim that it occurred because the ice of Snowball Earth isolated the oceans from the air, allowing a volcanic buildup of iron in the ocean which could not react with atmospheric oxygen. As the ice melted, iron deposits again began to form.

- Iridium is rare on Earth but common in cosmic dust. It would have accumulated on the Snowball Earth ice surface, then deposited quickly in a concentrated layer when the ice melted. Geologists have found a high concentration of iridium in deposits from the end of the Marinoan Glaciation.

During the late Precambrian, there were not only a lot of icebergs but a lot of tropical icebergs. The evidence for this is that the magnetic fields of the deposits usually have a nearly horizontal orientation. The magnetic field of a rock is produced by the orientation of iron crystals (see PALEOMAGNETISM). After the rock has hardened, the iron crystals cannot move, and the magnetic field is frozen into place. (Frequently, conditions such as heat or the flow of mineralized water can cause the original magnetic field to be altered and even overprinted by another orientation; geologists are careful to check for this possibility.) The Earth's magnetic field is oriented not only north-to-south, but the field lines are nearly vertical at the poles and horizontal at the equator.

If the tropical seas were largely covered with icebergs, the rest of the world must have been covered with ice, making the whole planet a snowball (with the exception of openings in the ice over volcanoes). Life would not have become extinct under these conditions—organisms could survive near the volcanic openings—but there must have been a drastic reduction in the number of species and in population sizes. There is broad agreement among evolutionary scientists that these glaciations occurred; the disagreement regards how completely the Earth was covered with ice—was it really a snowball, or were some large areas free of ice cover?

Little agreement has been reached about what might have caused the Snowball Earth. One suggestion is that photosynthesis by green bacteria and protists removed so much carbon dioxide from the air that the greenhouse effect was almost completely canceled. This allowed ice to accumulate, which reflected more sunlight away from the Earth, which caused more cooling, which allowed more ice to accumulate, and so on in a self-reinforcing (positive feedback) loop. Other scientists have suggested that all of the continents were near the equator at that time, which reduced the ability of ocean currents to redistribute heat into the polar regions, in which an inexorable buildup of ice began.

Once the Snowball began to melt, the change was astonishingly rapid. The accumulation of volcanic and meteoric dust on the ice may have reduced albedo enough to allow melting to begin. High concentrations of atmospheric carbon dioxide caused a huge greenhouse effect; the world underwent transition from its coldest to its warmest temperatures. Once rain began to fall, it contained dissolved carbon dioxide, in the form of carbonic acid. Once the land was exposed, then minerals in the rocks could react with the acid rain, producing carbonates that accumulated in the oceans. The accumulation of carbonate was so massive that it produced a thick whitish layer of rock immediately above (after) the layer of dropstones. The dropstones indicated the Snowball; the carbonate layer indicated the rapid melting of the Snowball. Carbon isotope ratios (see ISOTOPES) suggest that this carbonate was of inorganic origin—directly from atmospheric carbon dioxide, not from organisms. There were not yet any land plants to absorb carbon dioxide through photosynthesis, but photosynthetic protists (such as seaweeds) began to grow abundantly in the oceans (which were fertilized by the high concentrations of iron). These cells absorbed carbon dioxide, and the newly exposed ocean water itself could also absorb carbon dioxide, thus bringing the greenhouse effect slowly under control.

Some researchers believe that the last of the Snowball events stimulated the evolution of complex life. The EDIACARAN ORGANISMS and the rapid evolution of early Cambrian animals (see CAMBRIAN EXPLOSION) both occurred soon after this Snowball event. It has been suggested that because the Precambrian world was dominated by single-celled organisms, an extinction event such as Snowball (as well as the buildup of oxygen that immediately followed it) was actually necessary to allow the evolution of new adaptations, such as the multicellular body. As paleontologist Andrew Knoll points out, in the world of retreating ice after the last Snowball, even poorly functioning new multicellular animal forms had all the real estate to themselves.

One consequence of this discovery is both unsettling and reassuring. The Earth has not always been a nice place to live. The climate is apparently capable of going wild; and if it did so before, it may do so again. The very presence of life itself on this planet may spare the Earth a repeat of such drama. Today, unlike the Precambrian, the continents are covered with plants, which help keep both global cooling and global warming under control. Some scientists have formalized this concept as the GAIA HYPOTHESIS. Conservation efforts aimed at maintaining plant cover and diversity may not just be a good idea and the right thing to do, but may be essential to the future of the planet.

Further Reading

Bodiselitsch, Bernd, et al. "Estimating duration and intensity of Neoproterozoic snowball glaciations from Ir anomalies." *Science* 308 (2005): 239–242. Summarized by Richard A. Kerr, "Cosmic dust supports a Snowball Earth." *Science* 308 (2005): 181.

Hoffman, Paul F. "Snowball Earth." Available online. URL: http://www.snowballearth.org. Accessed July 11, 2006.

Jenkins, Gregory S., Mark A. S. McMenamin, Christopher P. McKay, and Linda Sohl, eds. *The Extreme Proterozoic: Geology, Geochemistry, and Climate.* Washington, D.C.: American Geophysical Union, 2004.

Knoll, Andrew H. *Life on a Young Planet: The First Three Billion Years of Evolution on Earth.* Princeton, N.J.: Princeton University Press, 2003.

Walker, Gabrielle. *Snowball Earth: The Story of a Great Global Catastrophe That Spawned Life as We Know It.* New York: Crown Publishers, 2003.

social behavior *See* ALTRUISM; BEHAVIOR, EVOLUTION OF.

Social Darwinism *See* EUGENICS.

sociobiology Sociobiology is the application of evolutionary biology to human and other animal behavior. Charles Darwin wrote of the relevance of evolution to understanding human and other animal behavior (see DARWIN, CHARLES), but the formalization of sociobiology as a distinct science, and the invention of the name, date from the 1975 book by Edward O. Wilson (see WILSON, EDWARD O.). Wilson explained the evolutionary basis of the behavior of a wide range of different animals, then in the last chapter applied these principles to humans. He expanded human sociobiology into a best-selling 1978 book *On Human Nature.*

Wilson identified a number of human behaviors that appear to be universal. If these behaviors were determined entirely by the environment in which individuals developed and on their personal decisions, they would be unlikely to be universal. Wilson did not claim that environment and personal decision had no effect on human behavior, but that they did not influence them completely (see essay, "How Much Do Genes Control Human Behavior?"). Wilson then provided explanations of how each of those behaviors would have proven beneficial to the fitness of humans, and therefore would have been favored by NATURAL SELECTION (see BEHAVIOR, EVOLUTION OF). The environment in which these behaviors would have proven beneficial is not the modern environment but the *environment of evolutionary adaptedness,* during the hundred thousand years that *HOMO SAPIENS* existed prior to agriculture and civilization, and quite likely the environments of earlier *Homo* and earlier HOMININ species. These behaviors include:

- *Incest avoidance.* Human societies differ markedly in reproductive systems, from (near) monogamy to the formation of harems to (near) promiscuity. Societies differ markedly in which behaviors they consider moral and immoral. All human societies consider close consanguineous matings (incest) to be immoral. The evolutionary benefit of the "incest taboo" is that incest produces many offspring that display detrimental genetic traits (see SEX, EVOLUTION OF).
- *Recognition of discrete colors.* Colors do not really exist. Photons have a continuous range of wavelengths; wavelengths between about 0.000015 inch (400 nm) and about 0.000028 inch (700 nm) are visible to humans. The recognition of discrete colors results in part from the three kinds of cones in the human retina (responding to wavelengths humans recognize as red, green, and blue) and the interpretation of nerve impulses by the human brain. Humans of all cultures have brains that interpret very similar sets of colors. Evolutionary advantages of color recognition include the ability to recognize food sources and to use color as a medium of cultural communication.
- *Face pattern recognition.* Humans of all cultures have brains that recognize individual faces and do so on the basis of mostly the same characteristics. The evolutionary advantages include the ability to keep track of which individual humans are friends and which are rivals within a society.
- *Facial expressions.* Humans of all cultures produce and respond to the same facial expressions in the same way, including anger, terror, surprise, and happiness. Smiling is not only universal but even individuals blind from birth, and who could not possibly be imitating someone else, smile. The advantages for social communication and cohesion can hardly be overestimated.
- *Specific sugar preferences.* All human culture groups like sugar and distinguish in a similar manner among different kinds of sugar. Existing variation in sugar preference often disappears when highly sugared modern food becomes available to cultures to which it was previously unavailable. Humans that liked sugar sought out and consumed higher quality food than those that did not. As it turns out, this universal trait is maladaptive under conditions of modern civilization. Modern conditions began too recently to influence human biological evolution.
- *Fear of strangers (xenophobia).* Every culture group has feared and loathed people from other cultures. The advantage that this provided was that people were willing to fight against other cultures and acquire their land and resources. The appreciation that the resulting conflicts are counterproductive, and the growing sense that hatred of people from other cultures is immoral, is a recent and cultural development in the human species. The fear of strangers within one's own culture is also a recent phenomenon. For most of the time the human species has existed, people lived in small groups within which there were no strangers.
- *Phobias.* Phobias are fears that have no rational basis and are seldom chosen or understood by the people who experience them. Examples include the fear of spiders, snakes, and heights. Wilson points out that these phobias correspond to ancient, not modern, dangers. Phobias of guns, for example, are not well documented. Spiders, snakes, and falling out of trees (and later off of cliffs) were real and continual dangers for the ancestors of modern humans. The evolutionary advantage of phobias was that the person could take immediate action without having to figure out, or learn, the appropriate response—a behavior pattern particularly valuable for young children.

Other researchers have added other apparently universal behavior patterns. For example, men prefer women who are young and pretty. Youth is an indicator that a woman has a long period of potential reproduction ahead of her, and beauty may be a reflection of health and therefore the ability to resist parasites. These researchers also noticed that women often prefer men who are older and as a result have more resources. The difference in mate preferences between men and women has been interpreted as an example of SEXUAL SELECTION in which men maximize their fitness by inseminating more females but women maximize their fitness by access to more resources. Psychologists Leda Cosmides and John Tooby say that humans are better able to figure out social than logical situations, and have the ability to detect nuances of character that may indicate dishonesty—and that this, too, is a genetically based universal trait. As more supposedly universal characteristics were added to the list, it became clear that most aspects of human thoughts, feelings, and behavior may have been influenced by evolution. Therefore sociobiology has come to incorporate *evolutionary psychology*.

Sociologists had long assumed that all human behavior was learned and culturally conditioned. One of the most extensive sets of observations that appeared to confirm this came from anthropologist Margaret Mead, who studied Samoan societies, in which moral standards that Western observers assumed to be universal were, in fact, not. Mead's work has been criticized, particularly by sociologist Derek Freeman, because it was based on a relatively small number of informants and had problems of statistical analysis. Freeman even claims that some of Mead's informants were playing tricks on her.

One nearly unavoidable problem with the detection of most universal behaviors is that no truly primitive societies remain. Even the Samoans that Mead studied in the early 20th century, and especially the American Samoans studied later by Freeman, had been in contact with Western cultures for centuries before being studied. Did women of these cultures prefer men with resources because of innate behavior patterns selected by evolution, or because their culture had been influenced by Western cash societies?

Probably the best way around this problem has been to study the native cultures of Amazonia, such as the Yanomamö studied by anthropologist Napoleon Chagnon and others. Many of these cultures had no known direct contact with civilization until the 20th century. However, indirect contact, in which they may have heard about civilizations from neighboring tribes, may have influenced their cultures. Recent evidence further suggests that these allegedly primitive tribes descended from ancestors who lived in cities until their populations were ravaged by European diseases, which arrived before the Europeans did.

Many critics of sociobiology have pointed out shortcomings, including:

- Natural selection (and probably sexual selection as well) undoubtedly favored the evolution of the gigantic human brain (see INTELLIGENCE, EVOLUTION OF). This huge brain is now capable of a tremendous range of behaviors. The potential for these behaviors evolved, but not the behaviors themselves. These critics admit that some behaviors such as smiling and the preference for certain sugars may be influenced by genes, but this is not much of a basis for deriving a full-scale evolutionary psychology. One evolutionary biologist (see GOULD, STEPHEN JAY) wrote that sociobiologists "... reify the human repertory of behaviors into 'things' (aggression, xenophobia, homosexuality), posit selective advantages for each item (usually by telling a speculative story), and complete a circle of invalid inference by postulating genes, nurtured by natural selection, 'for' each trait."
- Even genetically identical twins are not identical in their social development. Geneticist Richard Lewontin (see LEWONTIN, RICHARD) cites the example of quintuplets whose parents capitalized upon them by enforcing the same style of dress and hair, and took them on tour. At adolescence, each child was allowed to choose her own path in life. The quintuplets promptly followed different paths. Three married and had families, two did not; two went to college, three did not; three were attracted to a religious vocation, but only one made it a career, and she died in a convent at age 20—of epilepsy, which the others did not have. Lewontin says that "... each of their unhappy adulthoods was unhappy in its own way." This line of reasoning does not show that there is no genetic influence upon behavior but reminds all observers that social forces are important and can in some cases overwhelm genetic proclivities.

Future work in sociobiology will undoubtedly be based upon brain studies made possible by such techniques as positron-emission tomography (PET scans), which reveal, in three dimensions, the parts of the brain that are active during various mental processes and emotions. So far, these studies have revealed both broad and seemingly universal patterns as well as individual differences.

Further Reading

Barash, David P. *Madame Bovary's Ovaries: A Darwinian Look At Literature*. New York: Random House, 2006.

Beckstrom, John H. *Sociobiology and the Law: The Biology of Altruism in the Courtroom of the Future*. Urbana: University of Illinois Press, 1985.

Berreby, David. *Us and Them: Understanding Your Tribal Mind*. New York: Little, Brown, 2005.

Buller, David J. *Adapting Minds: Evolutionary Psychology and the Persistent Quest for Human Nature*. Cambridge, Mass.: MIT Press, 2005.

Cosmides, Leda, and John Tooby. *What Is Evolutionary Psychology? Explaining the New Science of the Mind*. New Haven, Conn.: Yale University Press, 2005.

Darwin, Charles. *The Expression of Emotions in Man and Animals*. London: John Murray, 1872. Reprinted with introduction by Paul Ekman. New York: Oxford University Press, 2002.

De Waal, Frans. *Our Inner Ape: A Leading Primatologist Explains Why We Are Who We Are*. New York: Penguin, 2005.

Francis, Richard C. *Why Men Won't Ask for Directions: The Seductions of Sociobiology*. Princeton, N.J.: Princeton University Press, 2006.

Freeman, Derek. *Margaret Mead and the Heretic: The Making and Unmaking of an Anthropological Myth*. New York: Penguin, 1997.

Gould, Stephen Jay. "The ghost of Protagoras." Chap. 4 in *An Urchin in the Storm: Essays about Books and Ideas*. New York: Norton, 1987.

Lewontin, Richard, Steven Rose, and Leon J. Kamin. *Not in Our Genes*. New York: Pantheon, 1985.

Mead, Margaret. *Coming of Age in Samoa: A Psychological Study of Primitive Youth for Western Civilization*. New York: William Morrow, 1928. Reprint, New York: Penguin, 2001.

Olsson, Andreas, et al. "The role of social groups in the persistence of learned fear." *Science* 309 (2005): 785–787.

Wilson, Edward O. *Sociobiology: The New Synthesis*. Cambridge, Mass.: Harvard University Press, 1975.

———. *On Human Nature.* Cambridge, Mass.: Harvard University Press, 1978.

solar system *See* UNIVERSE, ORIGIN OF.

speciation Speciation occurs when two or more species diverge from a single ancestral population. NATURAL SELECTION and SEXUAL SELECTION can produce evolutionary change within lineages of organisms over time. However, as many writers have pointed out, Darwin's book (see DARWIN, CHARLES; *ORIGIN OF SPECIES* [book]) did not actually explain how two or more species could evolve from one species, but only how one species could change into another. Something more is needed, besides natural selection, to explain speciation. A certain amount of *reproductive isolation* is necessary. If one population of organisms is separated into two populations, the two groups cannot interbreed. This could result from *geographical isolation,* in which the two populations are isolated from one another by impassible barriers such as deserts or oceans. Or it could result from characteristics of the organisms themselves: ISOLATING MECHANISMS are negative processes that prevent interbreeding between groups, and *specific mate recognition systems* are positive processes that encourage breeding within groups.

Once even partial reproductive isolation occurs, the genes of the two populations no longer mix completely together. The two populations may eventually become two species because new genetic combinations, and new mutations, that occur in one population will be different from those in the other. The two populations could then diverge along different evolutionary lines:

- Natural selection may favor different characteristics in each group strongly enough to compensate for interchange of genes *(gene flow)* between the populations. After the populations have begun to diverge, natural selection may favor isolating mechanisms and specific mate recognition systems that complete the process of speciation.
- Different and random genetic changes *(genetic drift)* may occur in each group (see FOUNDER EFFECT). Even if natural selection favors the same characteristics in both populations, they will probably diverge because they will not have the same genes upon which natural selection can act.

Evolutionary biologists distinguish species on the basis of either a biological or a phylogenetic species concept. The *biological species concept* (see MAYR, ERNST) recognizes the potential to interbreed under natural conditions as the definition of species membership. Even though this concept has some difficulties (HYBRIDIZATION between recognized species occurs frequently, and investigators cannot know the interbreeding potential of geographically isolated populations or fossil species), it is the concept that most closely reflects the evolutionary process itself. The *phylogenetic species concept* recognizes that species are distinct if experts can distinguish them.

Natural or sexual selection may occur rapidly after reproductive isolation, as the incipient species adjusts by directional selection to the new environmental conditions or the new set of species with which it is in contact. This may be followed by a long period of stabilizing selection. This has occurred frequently enough, according to many paleontologists, to produce a pattern of PUNCTUATED EQUILIBRIA in the fossil record.

Speciation can occur in different ways, as described in the following sections.

Allopatric speciation. Allopatric speciation occurs when populations are geographically or ecologically isolated. This has occurred frequently on islands, where dispersal of plants and animals from the mainland is rarely successful, and once it occurs, the island population is unlikely to disperse back to the mainland. This is the process that has produced the vast number of unique *endemic* species found only on certain islands (see BIOGEOGRAPHY). The presence of many endemic species on islands was one of Darwin's chief evidences that evolution has, in fact, occurred. Most species of nettles are small, stinging herbs; but in Hawaii, a stingless nettle bush has evolved, in isolation from mainland nettles. Hawaii has or had many species of birds, snails, insects, plants, and many other species found nowhere else in the world. Not surprisingly, Hawaii has the longest list of threatened and endangered species of any state in the United States (see BIODIVERSITY). Given long enough time, the geographic separation of populations almost inevitably produces new species through allopatric speciation.

Peripatric speciation. Peripatric speciation is similar to allopatric speciation but involves the geographical isolation of small populations from the main population. Small populations can be important in speciation, since the genetic makeup of small populations can change more rapidly than that of large populations (see POPULATION GENETICS). A classic experimental demonstration of rapid evolution in small populations is the study of evolutionary change, over the course of a year and a half, in fruit flies (see DOBZHANSKY, THEODOSIUS). Dobzhansky established 20 large, and 20 small, populations of fruit flies, each with a certain chromosome at 50 percent frequency. In most populations, this chromosome became less frequent. In all of the large populations, the chromosome became less frequent to about the same extent; they all evolved similarly. In the small populations, some hardly evolved at all, and some evolved a great deal; the small populations had a much greater range of evolutionary outcomes than the large populations.

Parapatric speciation. A large population may evolve into two or more species by specialization on different habitats within the geographical range of the ancestral species.

Sympatric speciation. Reproductive isolation can also occur between two populations that live in the same location. Examples include:

- If two species of plants open their flowers in different seasons (for example, the autumn-flowering vs. the spring-flowering species of alder trees of the genus *Alnus*), they cannot interbreed even if they live in the same habitat.
- If two species of plants depend upon different pollinators (such as the hummingbird-pollinated vs. the bee-pollinated

monkeyflowers of the genus *Mimulus*), they will not interbreed even if they are right next to one another. Genera of plants with bilaterally symmetrical flowers tend to have more species than genera that are otherwise similar but have radially symmetrical flowers, perhaps because bilaterally symmetrical flowers often have more specialized relationships with pollinators.

- If two species of animals specialize on different food plants (especially if they live on and mate on these food plants as well), they will not interbreed very often even if their respective food plants are right next to one another. It has been reported that some European corn borer larvae now consume hops. This has resulted in adult insects that emit different mating pheromones. Adult borers whose larvae ate hops preferentially mate with other such adults, rather than adults whose larvae ate corn.
- If two species of animals have different mating calls or rituals, they will not interbreed very often even if they are in contact. This has been observed with DARWIN'S FINCHES. Indigobirds are nest parasites, which means that they lay their eggs in the nests of other bird species. Cladistic analysis indicates that indigobirds evolved into several species more recently than did the host species. This apparently happened because the young indigobirds learned songs and habits from their hosts, and when they grew up, they tended to mate with other indigobirds that had similar songs and habits.
- If two species of plants can cross-pollinate, but the pollen grains do not successfully fertilize the egg nuclei; or, if two animal species can mate with one another but the sperm from one population has inferior performance in the female reproductive tracts of the other; then the proximity of the two populations does not matter.
- In some cases, SYMBIOGENESIS may cause reproductive isolation, if one population that consists of individuals in which a mutualistic microbe is living cannot crossbreed with a population whose individuals lack the microbe.

Isolating mechanisms can separate two populations and allow them to evolve into two species. The two populations might as well be separated by an ocean. Given long enough time, the reproductive separation of populations almost inevitably produces new species through sympatric speciation.

Speciation by hybridization. Hybrids between species may result in offspring with incompatible chromosomes. If the chromosomes within the cells of the offspring are unable to match up with one another during meiosis, the offspring are sterile (see MENDELIAN GENETICS). Among animals, hybrids are generally the end of the evolutionary line. A strong mule, for example, cannot produce offspring. Nevertheless, an occasional hybridization can be an important source of new genes for a species. The gene for organophosphate insecticide resistance (see RESISTANCE, EVOLUTION OF) originated in the mosquito species *Culex pipiens,* but it spread worldwide, even across mosquito species boundaries, probably because of rare but important hybridizations.

However, among plants, sterile individuals produced by hybridization may have options that are not available to animals. Many plants reproduce without sex, a process called *apomixis.* The embryo produced inside the seed is an exact genetic copy of the parent, and the plant that grows from it is a clone of the parent (see SEX, EVOLUTION OF). An example is the wildflower *Oxalis pes-caprae* in the shamrock family. Many individuals in this species have chromosomes in groups of five rather than in groups of two; they produce beautiful flowers, but because MEIOSIS cannot handle unpaired chromosomes, the flowers cannot produce either pollen or egg nuclei. They reproduce only by making underground bulbs by which they survive the dry season. Each lineage of these plants is reproductively isolated from all the rest. Mutations may occur, and if the mutations are passed on through the bulbs, each lineage can evolve into what looks like a different species. In their native South Africa, there are a few *Oxalis pes-caprae* with evenly paired chromosomes, which can reproduce sexually, but they are reproductively isolated from the sterile *Oxalis* individuals. To call each sterile lineage of *Oxalis* a separate species would be maddeningly complex, and yet each sterile lineage does function as if it were a separate species. Dandelions *(Taraxacum oficinale)* have chromosomes in groups of three and are also unable to reproduce sexually. They produce seeds, but the seeds contain copies of the parental genes. Pollination is necessary to stimulate seed production but does not mix the genes of pollen and egg nuclei. Each clump of dandelions, then, functions as if it were a separate species. A single population of dandelions can contain dozens of distinct genetic lineages, which may have arisen by the multiple evolution of apomixis, or by somatic mutation, or both.

Apomixis is not the only way that evolution can allow sterile plants to reproduce. When the reproductive cells of animals experience spontaneous chromosome doubling, the cell usually dies. Most animals tolerate only the most minor changes in the number of chromosomes; even then, as in Down syndrome (trisomy of chromosome 21) in humans, a single extra chromosome results in sterility and retardation. Plants, in contrast, frequently undergo spontaneous doubling in the number of chromosomes. A plant egg cell with doubled chromosomes may survive. A sexually sterile plant, if it undergoes chromosome doubling, may produce offspring that are not sterile because the doubled chromosomes now form pairs, which allows meiosis to occur. Each chromosome can now match up with a copy of itself. The population of plants with doubled chromosomes, however, may not be able to interbreed with either of the ancestral populations. The hybrid is then instantly isolated from both parents, and has characteristics different from both; it becomes an "instant species." New species have arisen by hybridization in the goatsbeard *Tragopogon* and in the sunflower *Helianthus.* This form of speciation is common in the plant kingdom. The study of chromosomes suggests that almost half of all plant species, and 95 percent of fern species, originated by chromosome doubling following hybridization.

Speciation has been investigated as it occurs, as described in the following sections.

Experimental speciation. Speciation is difficult to observe in the wild or to demonstrate experimentally. Perhaps the best example is from fruit flies of the genus *Drosophila.* Researchers

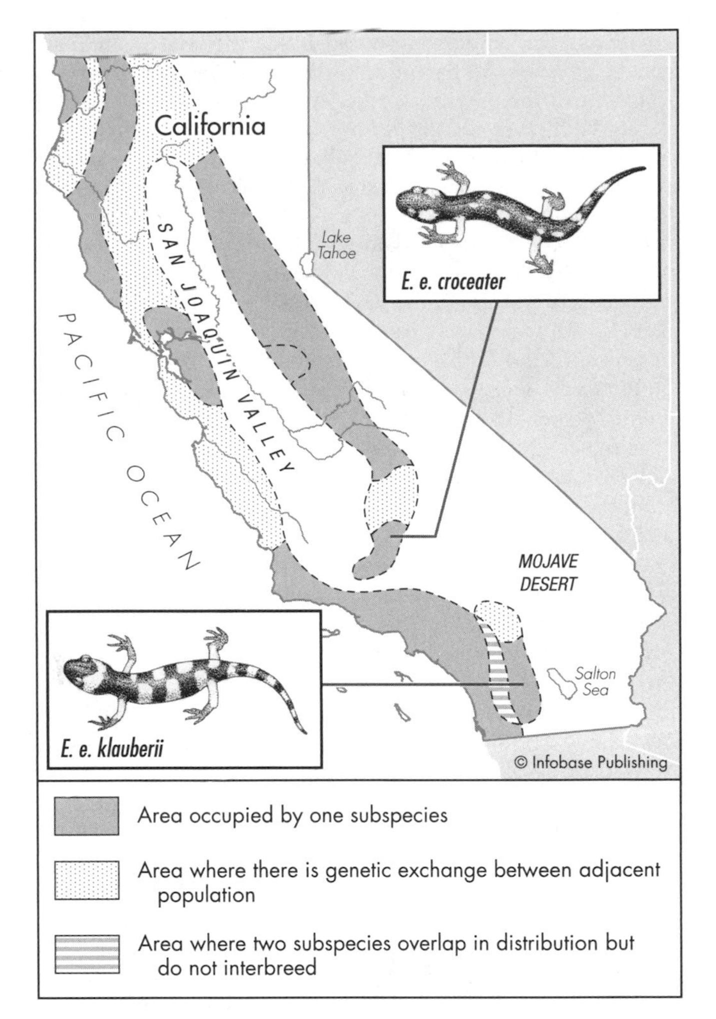

***Ensatina eschscholtzii* is an amphibian that forms a ring species. Each subspecies overlaps with other subspecies, but the subspecies at the southern ends of the distribution are unable to interbreed with one another. (*Klauberi and Croceater*)**

can raise many generations of fruit flies under laboratory conditions. Beginning with a single population of flies, they separated them into two populations, one kept in warm conditions, the other kept in cool conditions. After many generations, the two populations of flies had diverged enough that they did not interbreed when they were placed back together. This may have occurred partly from the gradual accumulation of genetic differences but may also have involved symbiogenesis, since one of the populations lost microbes that had lived in the digestive system of its ancestors. In another fruit fly experiment, researchers kept some flies in the dark and others in the light. Not surprisingly, the flies in the dark lost the genetic ability to use visual cues for mating; they evolved mating cues based upon touch and vibration. The visual flies could not interbreed with the tactile flies.

Reconstructing speciation in the wild. In natural populations where speciation may be impossible to observe in action, researchers have frequently found convincing evidence regarding how the speciation occurred. Researchers Douglas Schemske and H. D. Bradshaw have found the traits that were almost certainly responsible for the evolution of two species of *Mimulus* from a single ancestor, and Loren Rieseberg has identified the two parental species of sunflowers whose chromosomes have come together to form the hybrid species *Helianthus paradoxa*. Peter and Rosemary Grant have carefully measured divergence in beak size in two groups of Darwin's finches which, if it continues for a long time, may result in the formation of new species.

Speciation occurring in the wild. In some cases, two incipient species (not quite yet differentiated enough to be considered separate species) can exist in adjacent habitats. If there is a series of such species, each one grading into the next, it is possible that the species at the two ends of the series cannot interbreed. This would constitute a *ring species*. Examples include gulls of the genus *Larus* across the northern continents, salamanders of the genus *Ensatina* in the mountains of California (see figure), and the greenish warbler *Phylloscopus trochiloides* around the Himalayas.

Evolutionary scientists have found numerous examples of speciation that is occurring in wild populations (see table). Evolution has produced far more species than any human observer would consider necessary. For example, there are well over 350,000 species of beetles. If evolution produced species only in response to environmental conditions, there would be far fewer species of organisms in the world: a few species of animals and plants best suited to moist, cool conditions, a few other species best suited to hot, dry conditions, etc. Species frequently evolve in response to one another (see COEVOLUTION) and as a result of the population fragmentation that results from geographic and reproductive isolation.

A Few Examples of Speciation

Organism	Divergence that is occurring
Flowering plants	Coevolution with pollinators
Hawaiian plants	Coevolution with birds for seed dispersal
Fireweeds	Polyploidy
Soapberry bugs	Different host plants
Goldenrod gall insects	Different host plants
Heliconius butterflies	Different warning colorations
Field crickets	Male courtship song
Stickleback fish	Bottom-dwelling vs. open-water-dwelling populations
Sockeye salmon	Populations that return from the sea v. remained in lakes
Blue tits	Populations that breed in evergreen v. deciduous oaks
Bowerbirds	Male courtship patterns

Further Reading

Abrahamson, Warren G., et al. "Gall flies, inquilines, and goldenrods: A model for host-race formation and sympatric speciation." *American Zoologist* 41 (2001): 928–938.

Arnqvist, G., et al. "Sexual conflict promotes speciation in insects." *Proceedings of the National Academy of Sciences USA* 97 (2000): 10,460–10,464.

Blondel, Jacques, et al. "Selection-based biodiversity at a small spatial scale in a low-dispersing insular bird." *Science* 285 (1999): 1,399–1,402.

Carroll, Scott P., Hugh Dingle, and Stephen P. Klassen. "Genetic differentiation of fitness-associated traits among rapidly evolving populations of the soapberry bug." *Evolution* 51 (1997): 1,182–1,188.

Dreckmann, U., and M. Doebeli. "On the origin of species by sympatric speciation." *Nature* 400 (1999): 354–357.

Freeman, Scott, and Jon C. Herron. "Mechanisms of speciation." Chap. 15 in *Evolutionary Analysis,* 3rd ed. Upper Saddle River N.J.: Pearson Prentice Hall, 2004.

Futuyma, Douglas. *Evolution.* Sunderland, Mass.: Sinauer Associates, 2005.

Grant, Verne. *Plant Speciation,* 2nd ed. New York: Columbia University Press, 1981.

Gray, D. A., and W. H. Cade. "Sexual selection and speciation in field crickets." *Proceedings of the National Academy of Sciences USA* 97 (2000): 14,449–14,454.

Howard D. J., and Stuart H. Berlocher, eds. *Endless Forms: Species and Speciation.* New York: Oxford University Press, 1998.

Irwin, Darren E., et al. "Speciation by distance in a ring species." *Science* 307 (2005): 414–416.

Jiggins, Chris D., et al. "Reproductive isolation caused by colour pattern mimicry." *Nature* 411 (2001): 302–305.

Martin, Robert A. *Missing Links: Evolutionary Concepts and Transitions through Time.* Sudbury, Mass.: Jones and Bartlett, 2004.

Mayr, Ernst. *What Evolution Is.* New York: Basic Books, 2001.

Ortíz-Barrientos, Daniel, and Mohamed A. F. Noor. "Evidence for a one-allele assortative mating locus." *Science* 310 (2005): 1,467.

Palumbi, Stephen R. *The Evolution Explosion: How Humans Cause Rapid Evolutionary Change.* New York: Norton, 2001.

Price, Jonathan P., and Warren L. Wagner. "Speciation in Hawaiian angiosperm lineages: Cause, consequence, and mode." *Evolution* 58 (2004): 2,185–2,200.

Rundle, H. D., et al. "Natural selection and parallel speciation in sympatric sticklebacks." *Science* 287 (2000): 306–308.

Sabara, Holly. "Evolution of reproductive isolation in polyploids." Available online. URL: http://www.uoguelph.ca/botany/research/evollab/RI.htm. Accessed May 13, 2005.

Sargent, Risa D. "Floral symmetry affects speciation rates in angiosperms." *Proceedings of the Royal Society of London B* 271 (2004): 603–608.

Schilthuizen, Menno. *Frogs, Flies, and Dandelions: The Making of Species.* New York: Oxford University Press, 2001.

Uy, J. A. C., and G. Borgia. "Sexual selection drives rapid divergence in bowerbird display traits." *Evolution* 54 (2000): 273–278.

Via, Sara. "Sympatric speciation in animals: The ugly duckling grows up." *Trends in Ecology and Evolution* 16 (2001): 381–390.

Wake, D. B. "Incipient species formation in salamanders of the *Ensatina* complex." *Proceedings of the National Academy of Sciences USA* 94: 7,761–7,767.

Wood, Chris C., and C. J. Foote. "Genetic differentiation of the sympatric anadromous and non-anadromous morphs of sockeye salmon *(Oncorhynchus nerka).*" *Evolution* 50 (1996): 1,265–1,279.

Spencer, Herbert (1820–1903) British *Evolutionary philosopher* Herbert Spencer popularized evolutionary ideas in England and the United States. His vision of evolution was vague and, because it incorporated a Lamarckian mechanism (see LAMARCKISM), wrong, but it was, for a long time, more popular than Darwin's evolutionary theory (see DARWIN, CHARLES; *ORIGIN OF SPECIES* [book]). Spencer was lecturing about evolutionary ideas before Darwin published the *Origin of Species,* and Darwin derived part of his theory—if only a single phrase—from Spencer.

Born April 27, 1820, Spencer grew up in an intellectual environment (son of a schoolmaster) and spent his life writing a great outpouring of philosophical works. Evolution was at the core of all of his works. He proclaimed that the evolutionary history of all things produced greater complexity and greater interdependence. Even before Darwin published the *Origin of Species,* Spencer wrote about the "survival of the fittest," the phrase that Darwin later adopted to describe natural selection. Spencer had come close to the idea of natural selection, but he included it in a Lamarckian framework—that the efforts of organisms, including humans, could produce new characteristics that could be incorporated into their biological framework and passed down to future generations. Spencer was not the careful scientist that Darwin was; his scientific examples were few and vague. Since progress resulted from effort, then rich people were rich because they had superior abilities, and the poor were poor because they were lazy, or incompetent, or both, according to Spencer's view. Spencer wrote in 1851:

> Blind to the fact that under the natural order of things, society is constantly excreting its unhealthy, imbecile, slow, vacillating, faithless members, these unthinking, though well-meaning men advocate an interference which not only stops the purifying process but even increases the vitiation—absolutely encourages the multiplication of the reckless and incompetent by offering them an unfailing provision, and discourages the multiplication of the competent and provident by heightening the prospective difficulty of maintaining a family.

Spencer's evolution proceeded by effort, while Darwin's seemed to be founded upon luck. When Spencer visited America in 1882, the year Darwin died, he was immensely popular in this nation of newfound wealth resulting from the efforts of independent people exploiting a huge wilderness of resources they had recently expropriated from the Native Americans. When religious leaders such as Henry Ward Beecher embraced evolution within their theology, it was Spencer's version, not Darwin's. Spencer's views encouraged the growth of social Darwinism and EUGENICS that were used as justification of the oppression of minorities within countries, and the aggression of one country against another—activities that Darwin himself, a hater of slavery, would almost certainly have condemned.

It can be argued that Spencer opened the minds of many people to evolutionary thinking, which was a valuable work even though scientists no longer accept his ideas about how evolution works. But his work also helped contribute to a fundamentalist backlash the following century (see CREATIONISM).

Politician Robert Reich has pointed out that many Christian fundamentalists, while rejecting evolution, have embraced a "survival of the fittest" mentality for commerce and government that would have pleased Herbert Spencer. Evolutionary biologist Ernst Mayr said that Spencer's positive contributions to the development of science were negligible. Just about the only component of Spencerian evolution that scientists still use, and that only figuratively, is his phrase *survival of the fittest.* This phrase, however catchy, is more trouble than it is worth, and science educators spend a lot of time explaining what it does not mean (see NATURAL SELECTION). Unfortunately, no truly accurate phrase has been invented that has the popular appeal of Spencer's aphorism. Spencer died December 8, 1903.

Further Reading

Reich, Robert B. "The two Darwinisms." *The American Prospect,* 20 December, 2005.

statistics *See* SCIENTIFIC METHOD.

Stebbins, G. Ledyard (1906–2000) American *Botanist, Evolutionary biologist* George Ledyard Stebbins, Jr., was one of the major botanical contributors to the MODERN SYNTHESIS, which unified Darwinian NATURAL SELECTION with MENDELIAN GENETICS. Like Stebbins, other biologists provided information crucial to the modern synthesis: They worked with experimental populations of flies (see DOBZHANSKY, THEODOSIUS), wild populations of birds (see MAYR, ERNST), and fossils (see SIMPSON, GEORGE GAYLORD). In particular, Stebbins showed that several processes that contribute to evolution, such as HYBRIDIZATION and polyploidy, occur much more commonly in plants than in animals.

Born January 6, 1906, Stebbins entered Harvard University, intending to study law, but he became interested in botany, staying for a Ph.D. in 1931. When Stebbins joined the faculty of the University of California at Berkeley, he began studying the genetics of wild plants. He moved to the University of California at Davis, where he remained until retirement. He became acquainted with and was influenced by Dobzhansky, who was at Davis during his retirement. Stebbins won numerous awards, including the National Medal of Science.

Stebbins was an active conservationist. He went beyond just defending the habitats of species but called for the protection of natural areas in which many species had evolved in a small area, of which there are many in California. Stebbins died January 19, 2000.

Further Reading

Smocovitis, V. B. "G. Ledyard Stebbins, Jr. and the evolutionary synthesis (1924–1950)." *American Journal of Botany* 84 (1997): 1,625–1,637.

Stebbins, G. Ledyard. *Variation and Evolution in Plants.* New York: Columbia University Press, 1950.

survival of the fittest *See* NATURAL SELECTION; SPENCER, HERBERT.

symbiogenesis Symbiogenesis is the generation of evolutionary novelty through symbiosis. Symbiosis is the physiological association of members of two or more distinct kinds of organism and results from COEVOLUTION. When populations of two species form a close symbiotic partnership, NATURAL SELECTION may favor genetic changes that enhance the partnership of the two species. If these genetic changes are associated with reproductive isolation (see ISOLATING MECHANISMS), then the partners may be considered new species distinct from their immediate ancestors. Further, the partners may merge their activities so closely, as when photosynthetic bacteria became chloroplasts or when algae merge with fungi to become lichens, that they are often named as a single new species. As a result, symbiogenesis is the genesis of new species by merger of some populations of previously independent species. The principal modern scientist who has promoted awareness of symbiogenesis is Lynn Margulis of the University of Massachusetts at Amherst (see MARGULIS, LYNN).

The principal image that has dominated Darwinian evolutionary theory is the TREE OF LIFE, where ancestral species branch out to form new species. One of the earliest images of the tree of life is the single illustration that is found in Charles Darwin's landmark 1859 book (see *ORIGIN OF SPECIES* [book]). In this image, the branches grow farther and farther apart but never merge. When symbiogenesis occurs, however, the tree of life experiences grafts, sometimes between branches that are very far apart on the tree.

The earliest statements of the idea of symbiogenesis were made by two Russian botanists, Konstantin S. Merezhkovsky of Kazan University and Andrey S. Famitsyn of the Academy of Sciences in St. Petersburg, prior to the Russian Revolution. After Merezhkovsky died in 1918 and Famitsyn in 1921, another Russian botanist, Boris M. Kozo-Polyansky of Voronezh State University, kept the idea alive. It was never, however, a major theory in the Soviet Union, especially in the Stalinist era when the theories of LYSENKOISM dominated genetics and evolution and remained practically unknown in the United States until recently, when their ideas were promoted by Margulis and science writer Dorion Sagan. Independently of the Russian scientists, an American anatomist, Ivan E. Wallin (at the University of Colorado Medical Center in Denver) formulated an idea of symbiogenesis which he called symbionticism in a 1927 book. His ideas were ignored and he finished his career without pursuing symbiosis any further.

Lynn Margulis investigated the inheritance patterns of some genetic traits that appeared to be passed on not through the nucleus but through the cytoplasm of the cell. In sexual reproduction, most of the cytoplasm of the fertilized egg cell comes from the unfertilized egg, rather than from the sperm; the cytoplasmic factors showed a pattern of inheritance that was passed on solely through the maternal line. This indicated that there was DNA in the cytoplasm, not just in the nucleus (see DNA [RAW MATERIAL OF EVOLUTION]), and she had an idea where that DNA was located.

All independently living cells (and most viruses as well) have genetic information encoded in DNA. In contrast, organelles (specialized structures within the cell) generally do not have their own DNA; their production and operation are dictated by the DNA in the cell's nucleus. Margulis wrote a paper in 1967 in which she proposed that chloroplasts and mitochondria had and used their own DNA (see PHOTOSYNTHESIS, EVOLUTION OF; RESPIRATION, EVOLUTION OF). The DNA was there because these organelles are the evolutionary descendants of bacteria that had merged with ancestral eukaryotic cells (see EUKARYOTES, EVOLUTION OF). The bacteria remained, without killing their host cells; and the host cell lineages evolved processes that accommodated and made use of the bacteria. The bacteria became symbionts. If chloroplasts and mitochondria were merely cell structures, one would not expect them to have their own DNA. The chloroplasts evolved from photosynthetic bacteria (often called blue-green algae; see BACTERIA, EVOLUTION OF) while the mitochondria evolved from aerobic bacteria that consumed organic materials.

These symbiotic events happened more than a billion years ago. Frequently in cells, segments of DNA can move from one location to another (see HORIZONTAL GENE TRANSFER). Many of the genes of the mitochondria and chloroplasts were transposed from the symbiotic bacteria to the nuclei of the host cells. The simplified symbionts were now incapable of surviving on their own, and they became the chloroplasts and mitochondria that are today found only in eukaryotic cells.

Though not accepted by many scientists at first, Margulis's proposal of symbiogenesis quickly advanced in the scientific community. As better chemical and microscopic techniques became available, convincing evidence of the bacterial origin of chloroplasts and mitochondria emerged. Chloroplasts and mitochondria:

- have their own DNA, which resembles bacterial DNA more than it resembles the DNA of eukaryotic chromosomes.
- have ribosomes, which are structures that use genetic information in DNA to produce proteins. As a result, mitochondria and chloroplasts produce some of their own proteins. In the case of an important photosynthetic enzyme called rubisco, part of the enzyme (the large subunit, rbcL) is produced by the chloroplasts while the small subunit is produced by the ribosomes out in the chloroplast, using information now found in the nucleus.
- reproduce themselves, rather than being constructed by the cell.

In addition, some marine invertebrates consume algae but keep the algal chloroplasts alive in some of their cells. This strongly suggests that chloroplasts and mitochondria may have arisen in a similar way, as symbiotic bacteria that lost most of their genetic independence—but not quite all of it, thus betraying their symbiogenetic origins.

As techniques became available to actually determine the sequence of nucleotides within DNA and other nucleic acids, it became possible to make comparisons between the DNA of different species to determine their evolutionary relationships. Following the initial lead of Carl R. Woese of the University of Illinois (see WOESE, CARL R.), biologists constructed a tree of life showing the branching patterns of evolution (see DNA [EVIDENCE FOR EVOLUTION]). The tree was based solely upon living species, from which DNA and RNA could be obtained. Mitochondrial genes turned out to be very close to aerobic bacteria on the tree of life, and not at all similar to the genes in the nuclei of the cells in which mitochondria live. Chloroplast genes turned out to more closely resemble cyanobacteria on the tree of life than the genes in the nuclei of the plant cells in which they now live. This is precisely what would be expected as confirmation of Margulis's symbiogenetic theory of their origins.

While nearly all biologists accept Margulis's symbiogenetic explanation of the origin of chloroplasts and mitochondria, some of her subsequent proposals have not met with widespread acceptance. Perhaps the most famous example is her proposal that the motile structures of eukaryotic cells were originally symbiotic bacteria as well.

Margulis uses the term *undulipodia* ("undulating feet" in Greek) to denote structures in eukaryotic cells that are usually called *cilia* and eukaryotic *flagella.* Cilia are hairlike structures on the outside of many eukaryotic cells, for example free-living protists, and the cells that line animal respiratory passages, and that have the power of movement. Cilia assist in the movement of protists such as *Paramecium,* and some small aquatic animals. Flagella are whiplike structures with which many eukaryotic cells can swim, for example protists like *Euglena,* and sperm cells. Many bacteria also have flagella with which they propel themselves, but the smaller rotary structure of bacterial flagella is totally unlike the more complex structure of eukaryotic flagella, which is why Margulis does not use the same term for both of them.

Margulis has also pointed out similarity in composition between undulipodia and the mitotic spindle apparatus in eukaryotic cells. Before mitosis begins, chromosomes replicate; in mitosis, small protein strands in the spindle move the chromosomes into two groups, which then form the nuclei of two cells. Margulis points out structural similarities between undulipodia and the spindle apparatus; and that eukaryotic cells do not have undulipodia and a spindle apparatus at the same time. This implies that, even though they now look and act very differently, undulipodia and the spindle apparatus evolved from the same ancestral structure.

Margulis claims that the protein strands within undulipodia and within the spindle apparatus are the remnants of spirochete bacteria. How might this have occurred? Today, there are examples of spirochete bacteria that burrow partway into protist cells. The protruding parts of the bacteria flap in a coordinated rhythm, propelling the protist cell. These protists, when viewed under the microscope, look just like ciliated or flagellated cells; one must look closely, and watch the process of their formation, to realize that they are not cilia at all, but bacteria! In other words, spirochetes are, today, doing the very thing that Margulis claimed they did back when undulipodia came into existence.

One piece of evidence that is missing here, and which was so important in gaining the acceptance of the symbiogenetic origins of chloroplasts and mitochondria, is DNA. It has not been established beyond doubt that undulipodia (in particular, the structures from which they grow) and the spindle apparatus (in particular, the structures from which they grow) contain any DNA. Some studies have shown that there may be DNA in these structures. If the DNA is present, the conclusion that undulipodia and spindles have a symbiogenetic origin will be hard to resist. The absence of DNA will not, however, disprove the theory, because it is possible that all of the DNA that the ancestral spirochetes once possessed may have moved to the nuclei of the host cells. In the absence of DNA evidence, Margulis and others are seeking to determine whether the proteins found in spirochetes and the proteins found in undulipodia and spindles have structural similarities that are too great to be explained by chance.

Another structure that may have had a symbiogenetic origin is the hydrogenosome, a structure that produces hydrogen and is found in some protist cells. In some cases the genes that control hydrogenosome activity are all together in the host cell nucleus and resemble bacterial rather than eukaryotic genes, as if they were transposed to the nucleus from the ancestral hydrogenosome, which may have been a mitochondrion or a bacterium.

Margulis reconstructs the history of eukaryotic cells in four stages, which she calls Serial Endosymbiotic Theory:

1. The nucleus had a bacterial origin. The most likely ancestor for the nucleus is an archaebacterium (see ARCHAEBACTERIA) similar to modern *Thermoplasma.* Some archaebacteria have proteins that resemble the histones associated with the DNA in eukaryotic chromosomes. The ancestral cell was now a eukaryote, with a nucleus.
2. Spirochetes associated with this cell and eventually became undulipodia and the spindle.
3. Aerobic bacteria moved into some of these cells and eventually became mitochondria. Some of the protists that had mitochondria became the ancestors of animals, fungi, and plants.
4. Photosynthetic cyanobacteria were consumed, but not digested, by some cells that already had mitochondria, and eventually became chloroplasts. Some of the protists that had chloroplasts became the ancestors of plants.

One of the strongest proofs of symbiogenesis is the fact that the structures now called chloroplasts have originated more than once. Most chloroplasts, like those of green algae and of all terrestrial plants, resemble bacteria and are surrounded by a simple membrane. But the chloroplasts of some protists, such as dinoflagellates, have multiple membranes. This is also true of the apicoplast of the *Plasmodium falciparum* protist that causes malaria, which is a degenerate, multiple-membraned chloroplast. The *Plasmodium* does not use its apicoplasts for photosynthesis. The multiple membranes of these chloroplasts and apicoplasts suggest that they evolved when photosynthetic eukaryotic cells invaded other eukaryotic cells, stayed there, and simplified. In other words, regular chloroplasts are degenerate cells inside a cell; dinoflagellate chloroplasts and plasmodial apicoplasts represent degenerate cells inside of degenerate cells inside of a cell! There is even microscopic evidence that these chloroplasts have nucleomorphs, which appear to be little, vestigial nuclei (see VESTIGIAL CHARACTERISTICS).

A dramatic experimental demonstration of symbiogenesis (albeit unintended) occurred when biologist Kwang Jeon of the University of Tennessee at Knoxville noticed that his cultures of the protist *Amoeba proteus* looked sick. Under the microscope he could see that each amoeba was infected with thousands of rod-shaped bacteria. The amoebae usually digested bacteria, but these bacteria could resist the digestive enzymes of the amoebae. Most of the amoebae died from the bacterial infection. However, a few amoebae could resist the bacteria and survive. Jeon cultured the survivors over and over. He was artificially selecting for the amoebae that could best resist the bacterial infection, as well as for the bacteria that had the mildest effects on the amoebae. After about a year and a half, the amoebae and bacteria had developed a mutually beneficial relationship. After several years, the coevolution of amoebae and bacteria had progressed so far that neither of the species could survive without molecules supplied by the other. They had become an essential part of each other's environment. Might the amoebae now be considered a new species of protist? Might the bacteria now be considered a new species as well? Perhaps they could even be considered one new species of organism?

Another example of a symbiogenetic process in action involves the bacterium *Buchnera* that infects some insects such as aphids. Like an organelle, the *Buchnera* bacterium can be transmitted from one generation to the next through eggs. Because *Buchnera* manufactures certain amino acids that aphids cannot get from the sap that they eat, it can be considered mutualistic. Did mitochondria begin their evolution in a similar way?

Evolutionary biologists Noriko Okamoto and Isao Inouye have found evidence of a secondary symbiosis that appears to be in progress. There is a green algal symbiont that lives in certain protist host cells. The symbiont still has a nucleus, mitochondria, and a large green plastid, but the flagella, cytoskeleton, and internal membranes are gone. When the symbiont divides, it produces two cells, one with the plastid and one without. The cell without the green plastid then develops a feeding apparatus and engulfs the cell with the green plastid. This may represent a direct observation of a "missing link" in the evolution of chloroplasts by symbiosis.

Recent experimental evidence has reconstructed some other possible symbiogenetic pathways. Evolutionary biologist Joel Sachs raised two kinds of bacteriophage (viruses that live inside of bacteria) in laboratory bacteria. He used antibiotics to eliminate any bacteria that did not contain both of the bacteriophages. Therefore, he was imposing selection for those bacteriophages that could live together in the same host. He got a surprise. Not only did the bacteriophages live together in some of the bacteria, but they even fused together into a single bacteriophage.

The symbiogenetic event that produced chloroplasts and other similar structures was not a single, freak event; symbiogenesis cannot, therefore, be safely ignored in evolutionary theory as a weird aberration. Symbiogenesis has occurred, and continues to occur, a lot. In fact, according to Margulis, it occurs so often that Earth should be considered a "symbiotic planet." The worldwide tendency of organisms to associate with one another may, according to Margulis, help to explain why the Earth operates as a single system (see GAIA HYPOTHESIS).

Symbiogenesis can occur more than once even in a single cell. One of the best examples is a protist that lives in the intestines of termites in Australia. This protist appears to have several different types of undulipodia. Actually, they are four different kinds of spirochetes that form a symbiotic association with the protist. And the symbiosis does not end there. The protist itself is a symbiont inside of termite intestines, and it helps the termites to digest wood. Termites could not survive if they did not have protists living in their guts that aid with the digestion of wood.

Whenever coevolution produces symbiosis, new opportunities for evolution result. Small mutualists living inside of larger cells may lose some of their genes to the host cell. When this occurs, symbiogenesis has led to a fusion rather than a separation of branches on the tree of life.

Further Reading

Akhmanova, A., et al. "A hydrogenosome with a genome." *Nature* 396 (1998): 527–528.

Boxma, B., et al. "An anaerobic mitochondrion that produces hydrogen." *Nature* 434 (2005): 74–79.

Chapman, Michael J., Michael F. Dolan, and Lynn Margulis. "Centrioles and kinetosomes: Form, function, and evolution." *Quarterly Review of Biology* 75 (2000): 409–429.

Douglas, Susan, et al. "The highly reduced genome of an enslaved algal nucleus." *Nature* 410 (2001): 1,091–1,096.

Dyall, Sabrina, Mark T. Brown, and Patricia J. Johnson. "Ancient invasions: From endosymbionts to organelles." *Science* 304 (2004): 253–257.

Falkowski, Paul G., et al. "The evolution of modern eukaryotic phytoplankton." *Science* 305 (2004): 354–360.

Jeon, K. W. "Integration of bacterial endosymbionts in amoebae." *International Review of Cytology* 14 (1983): 29–47.

Margulis, Lynn. *Symbiotic Planet: A New Look at Evolution.* New York: Basic Books, 1998.

———. "Serial endosymbiotic theory (SET) and composite individuality: Transition from bacterial to eukaryotic genomes." *Microbiology Today* 31 (2004): 172–174.

———, and Dorion Sagan. "The beast with five genomes." *Natural History,* June 2001, 38–41.

———, and ———. *Acquiring Genomes: A Theory of the Origins of Species.* New York: Basic Books, 2002.

———, and R. Fester, eds. *Symbiosis as a Source of Evolutionary Innovation.* Cambridge Mass.: MIT Press, 1991.

Okamoto, Noriko, and Isao Inouye. "A secondary symbiosis in progress?" *Science* 310 (2005): 287.

taxonomy *See* CLADISTICS; LINNAEAN SYSTEM.

technology Technology is the use of tools, which are objects that are not produced by the body of the animal that uses them. In a more restricted sense, tools are objects that have been modified from their natural state before use.

There are numerous examples of animal species using technology in the broad sense:

- Some birds use sticks to pry insects from holes in trees, and chimpanzees use sticks to remove termites from nests.
- Chimpanzees use rocks to crack nuts.
- Gorillas have been observed to use sticks to check the depth of water through which they wade.

Technology in the narrow sense has only developed within the human lineage.

The first evidence of the primate use of modified tools is the production of stone tools by *HOMO HABILIS* and/or related species. This was the beginning of the Paleolithic ("old stone") Age. The toolmakers, of the Oldowan technology phase (named for Olduvai Gorge in Tanzania), struck stones, removing flakes and creating sharp surfaces on one side of the remaining core (unifacial tools) (see table on page 394). These HOMININS may have used both the flakes and the cores as tools. Oldowan tools were also made by early *HOMO ERGASTER* and by Asian *HOMO ERECTUS*. Later *H. ergaster,* and *HOMO HEIDELBERGENSIS,* made tools by producing finer flakes, and cores that were modified on two sides (bifacial tools). The cores of the Acheulean technology phase (named after St. Achuel, in France) had much more cutting surface per weight of stone than did the older Oldowan tools (see table). When *H. heidelbergensis* evolved into NEANDERTALS, a more advanced technology phase, the Mousterian (named after Le Mouster in France) emerged. Mousterian tools had more cutting surface, and showed more diversity within a set of tools, than Acheulean tools. Mousterian tools were produced by the Levallois technique (named after yet another archaeological site in France) in which the toolmaker determined the general shape of the flake as soon as he or she removed it from the core, then modified the edge. Mousterian tools had very limited geographical variability.

Some advanced tools, such as an almost 90,000-year-old bone harpoon, were made in Africa (see *HOMO SAPIENS*). Most anthropologists consider that these inventions occurred only in *H. sapiens.* Until about 30,000 years ago, progress in tool technology occurred slowly. For almost 50,000 years, Neandertals and *H. sapiens* coexisted in what is now Israel and Palestine, and both used Mousterian tools.

Advances in toolmaking occurred in *H. sapiens* populations in all parts of the world. What is often called an explosion of technological invention occurred when *H. sapiens* reached Europe. Anthropologists speculate that *H. sapiens* developed these tools, along with more art and more complex social structure, in response to the most recent of the ICE AGES as well as to the Neandertals who were already present in Europe. *H. sapiens* in Europe went through a succession of technological phases that was rapid in comparison to the entire previous history of hominins (see table). Finer tools, a much more diverse toolkit, and geographical differences in toolkits resulted. The last Neandertals used a toolkit that was similar to the Aurignacian tools of *H. sapiens.* These Châtelperronian tools may have been copied from *H. sapiens,* or stolen from them. The possibility that *H. sapiens* derived the ideas for Aurignacian tools from Neandertals cannot be discounted.

Scientific knowledge of toolmaking is restricted to stone tools, since bone, wood, and other media decompose. Wooden spears, made 450,000 years ago by *H. heidelbergensis,* have been recovered from deposits where moisture and lack of oxygen restrict decomposition. *H. sapiens* used stone tools during the Mesolithic ("middle stone") Age, as they hunted, gathered, and formed more complex social groups. In the Middle East, they set stone or bone into wooden handles for harvesting wild grains. *H. sapiens* continued using

Technological and Cultural Periods

Culture	Beginning (Thousand years ago, approximate)	Species	Length of cutting edge (cm) per kg of stone	Characteristics
Lower Paleolithic:				
Oldowan[1]	2,500	*H. habilis*[2] *H. erectus* Early *H. ergaster*	11	Simple, one worked surface Uniform throughout species range
Acheulean	1,500	Late *H. ergaster* *H. heidelbergensis*[3] Early Neandertal	45	Simple, two worked surfaces Uniform throughout species range
Middle Paleolithic:				
Mousterian	75	Neandertal Early *H. sapiens*	220	Prepared cores Uniform throughout species range
Upper Paleolithic:				
Aurignacian	33	Early *H. sapiens*	650	Advanced stone tools Diverse toolkit Geographical diversity
Châtelperronian	33	Late Neandertal	650	Similar to Aurignacian Limited geographic range
Gravettian	27	*H. sapiens*	1,300	More advanced stone tools More diverse toolkit Geographic diversity
Solutrean	21	*H. sapiens*	2,000	
Magdalenian	17	*H. sapiens*	2,500	
Mesolithic:	10	*H. sapiens*		Tribal village life
Neolithic:	8	*H. sapiens*		Agricultural village life
Bronze Age	5	*H. sapiens*		Civilization
Iron Age	2	*H. sapiens*		

[1]Paleolithic, Mesolithic, and Neolithic periods in Europe correspond closely to the Early, Middle, and Late Stone Ages of Africa but are not defined for Africa or Asia, therefore the terms do not strictly apply to *H. habilis, H. erectus,* and *H. ergaster.*
[2]Includes *H. rudolfensis*
[3]Includes *H. antecessor*

stone tools in the Neolithic ("new stone") Age, with which they began to clear forests and grow crops (see AGRICULTURE, ORIGIN OF). The earliest civilizations frequently made implements and weapons from bronze, while later civilizations used iron. Human technology accelerated when the Industrial Revolution began. The Bronze and Iron Ages arose at different times in different parts of the world. Few truly Stone Age cultures remain in the 21st century, as most tribal societies have selectively adopted modern technologies.

Further Reading

Shreeve, James. "Mystery on Mount Carmel." Chap. 7 in *The Neandertal Enigma: Solving the Mystery of Modern Human Origins.* New York: William Morrow, 1995.

Teilhard de Chardin, Pierre (1881–1955) French *Anthropologist, philosopher* Pierre Teilhard de Chardin was a priest and paleontologist who was passionate about incorporating an evolutionary viewpoint into Christian theology. This got him into trouble both with the Catholic Church and with scientists. For the most part, both his version of evolution and his version of theology have been abandoned. Especially in the middle of the 20th century, many lay people derived their ideas of evolution from reading Teilhard de Chardin's books.

Born May 1, 1881, Teilhard de Chardin grew up as a collector and amateur naturalist. He entered the Catholic clergy as a Jesuit priest. He completed studies in England, taught in Egypt, then returned to England, where his reading of philosopher Henri Bergson inspired him to integrate his understanding of evolutionary science into his theological studies. Bergson championed the idea that evolution was propelled by a life force (*élan vital*) rather than by NATURAL SELECTION. While working at the paleontology laboratory at the Natural History Museum in Paris, Teilhard de Chardin became interested in human prehistory, and he participated in the study of newly discovered CRO-MAGNON caves. His studies were interrupted by World War I. Teilhard de Chardin was a stretcher-bearer at the front lines. War experiences caused him to think even more about his philosophy. Afterward, Teilhard de Chardin continued his paleontological studies and earned a science doctorate from the Catholic Institute of Paris in 1922. In 1923 he visited China to participate in paleontological excavations, then returned to France. When his writings met with church hierarchy disapproval, he went back to China, where he remained (except for brief visits to Europe and the United States) almost 20 years, to focus on paleontology. He contributed to the development of geological maps and began to write book manuscripts, such as *The Phenomenon of Man* and *The Divine Milieu,* for which he is most remembered. In 1929 he participated in the discovery of Peking man (see *HOMO ERECTUS*) in the caves of Choukoutien (now Zhoukoudian) near Peking. Teilhard de Chardin's work contributed to an understanding of the technology of Peking man and the relationship of this species to Java man.

Teilhard de Chardin did not publish his books, largely because the Catholic Church refused to condone his evolutionary beliefs about human origins. They were published after his death and became very popular. He presented evolution as a goal-directed process, aiming toward an "omega point" which he identified with Jesus Christ—a concept rejected by both scientists and theologians. He also predicted the formation of a "noosphere" of worldwide human mental interconnectedness. Some observers claim that the Internet is the fulfillment of Teilhard de Chardin's noosphere proposal. Since many of Teilhard de Chardin's proposals are unclear (biologist Peter Medawar called them a "bouquet of aphorisms"), this claim cannot be tested.

Teilhard de Chardin's early interest in paleontology put him at the site at which PILTDOWN MAN was discovered. Piltdown man was later revealed as a hoax, and Teilhard de Chardin's involvement in designing the hoax, if any, has not been determined. Teilhard de Chardin died April 10, 1955.

Further Reading

Gould, Stephen Jay. "Teilhard and Piltdown." Section 4 in *Hen's Teeth and Horse's Toes.* New York: Norton, 1983.

Medawar, Peter B. "Teilhard de Chardin and *The Phenomenon of Man.*" In *The Art of the Soluble.* London: Methuen, 1967.

Teilhard de Chardin, Pierre. *The Phenomenon of Man.* New York: Perennial, 1976.

terminal Cretaceous event *See* CRETACEOUS EXTINCTION.

Tertiary period The Tertiary period (65 to two million years ago) was the first period of the CENOZOIC ERA (see GEOLOGICAL TIME SCALE). It followed one of the MASS EXTINCTIONS which occurred at the end of the CRETACEOUS PERIOD (see CRETACEOUS EXTINCTION) and preceded the QUATERNARY PERIOD, which is the current period of Earth history. The Cenozoic era is also known as the Age of Mammals, because after the extinction of the DINOSAURS the mammals had opportunity to evolve into tremendous diversity, including many large forms.

Many evolutionary scientists and geologists now divide the Cenozoic era into the Paleogene (Paleocene, Eocene, Oligocene) and the Neogene (Miocene, Pliocene, Pleistocene, Holocene) rather than the traditional Tertiary and Quaternary periods.

Climate. Warm moist conditions that allowed extensive forests were widespread in the early Tertiary period (Paleogene). Forests grew near the North Pole. Scientists estimate that conditions were warm enough that the trees did not need to be deciduous to avoid snowfall in winter. They were deciduous, scientists have concluded, because of the half year of darkness rather than because of cold temperatures. During the second half of the Tertiary period (Neogene), climatic conditions were cooler and drier than they had been for most of previous Earth history. The world's first extensive deserts and grasslands developed during the Tertiary period. Periodic ICE AGES began with the Quaternary period.

Continents. The northern and southern continents were largely separate from one another at the beginning of the Tertiary period. As the continents continued to move, connections began to form between the northern and southern continents. The subcontinent of India collided with the Eurasian continent, forming the Himalayas (see CONTINENTAL DRIFT; PLATE TECTONICS). Northern Europe and North America were just beginning to separate at the beginning of the Tertiary. Tree species that lived in both areas now evolved into separate species (see BIOGEOGRAPHY).

Marine life. All modern groups of marine organisms existed during the Tertiary period, including the first aquatic mammals (see WHALES, EVOLUTION OF).

Life on land. The Cretaceous extinction left a world in which many organisms had died, and much space and many resources were available for growth. Not only had the dinosaurs become extinct but also numerous lineages within the birds, reptiles, and mammals (see BIRDS, EVOLUTION OF; REPTILES, EVOLUTION OF; MAMMALS, EVOLUTION OF). The conifers that had dominated the early Mesozoic forests came to dominate only the forests of cold or nutrient-poor mountainous regions in the Cenozoic (see GYMNOSPERMS, EVOLUTION OF).

- *Plants.* The flowering plants evolved in the Cretaceous period but proliferated during the Tertiary period, into the forest trees that dominate the temperate and tropical regions, and many shrubs and herbaceous species (see ANGIOSPERMS, EVOLUTION OF). The explosive speciation of flowering plants resulted from COEVOLUTION with insect groups such as bees, butterflies, and flies. During the middle of the Tertiary period, dry conditions allowed the spread of grasses and other plants with adaptations to aridity, and the spread of grasslands and deserts.
- *Animals.* When mammals began to proliferate after the Cretaceous extinction, they evolved into many specialized forms, including many that still exist (such as bats and whales) as well as many that are now extinct. Some modern orders of mammals had much larger body sizes than they do today. There were rodents the size of rhinoceroses, rhinoceroses as tall as a house, and raccoons that evolved to the size and ferocity of bears. All lineages of birds except one probably became extinct near the end of the Cretaceous period, and all modern birds evolved from this one lineage. Some birds were larger than any that exist today: Some carnivorous birds were more than 10 feet (almost 2 m) tall. Grazing animals evolved from browsing ancestors, taking advantage of the grass food base (see HORSES, EVOLUTION OF).

Extinctions. There were no mass extinctions during the Tertiary period, although many animals became extinct as the climate became cooler and drier during the Paleogene-Neogene transition, and as the period of ice ages began when the Tertiary period ended.

thermodynamics Thermodynamics is the science of the movement, transformation, and availability of energy and its effects upon matter within a defined system. The laws of thermodynamics underlie all of the physical and chemical events in the universe and therefore underlie the evolutionary process.

There are two major laws of thermodynamics:

The First Law

A simplified version of the *First Law of Thermodynamics* states that energy in a system cannot be created or destroyed. Also called the *conservation of energy,* this principle indicates that all energy must come from and go somewhere when events occur. Energy can move from one place to another or change from one form to another, but the total amount of energy remains constant.

The exception to the simplified version of the First Law presented above is that matter and energy can be transformed into one another. All matter began as energy, shortly after the big bang (see UNIVERSE, ORIGIN OF). After the big bang, most of the energy in the universe has come from nuclear fusion, in which matter has been transformed back into energy. The transformation of matter into light energy, as four hydrogen atoms fuse into a helium atom inside the Sun, continually produces energy inside of stars. Three other sources of energy in the solar system are:

- The heat that remains from the origin of the solar system. Earth and Venus still have molten cores, but the core of Mars has apparently lost its heat.
- The decay of radioactive atoms that were formed in the supernova that preceded the solar system (see ISOTOPES; RADIOMETRIC DATING). Most of the heat of the molten cores of Earth and Venus come from radioactive decay of some of the atoms within them.
- The heat that results from the effects of gravity. Much of the energy that creates volcanic eruptions on the moons of Jupiter results from the gravitational pull of the planet.

Energy comes in several forms. Strictly speaking, thermodynamics deals only with the first two of these forms:

Kinetic energy results from the movement of atoms and molecules. The temperature of matter results from the kinetic energy of its atoms. At absolute zero, there is no kinetic energy. As kinetic energy increases, atoms in its solid state vibrate more and more. When kinetic energy (temperature) increases to the melting point, the atoms or molecules begin to slide past one another, forming a liquid. When kinetic energy increases to the boiling point, the atoms or molecules of a gas move in straight lines until they collide with other matter, or until gravity restrains them.

Potential energy is stored energy that is not currently causing anything to happen. The classic example is a rock at the top of a hill, which is not currently moving but could at any moment roll down the hill. A coiled spring, waiting to expand, and a stretched rubber band, waiting to contract, contain potential energy. Perhaps the most common example is the potential energy that is stored within the bonds of molecules. Some molecules contain a lot of potential energy; some of these molecules are flammable or explosive, under the right circumstances. Potential energy can also be stored by an imbalance of particles, atoms, or molecules. A battery has potential energy; the electrons are not moving, but they can, as soon as the circuit is closed. When one side of a biological membrane has more atoms or molecules of a certain kind than the other side, there is a gradient of potential energy. If the atoms or molecules are allowed to move across the membrane, they will. This is what happens in the impulse of a nerve cell.

Electrical energy results from the movement of electrons, usually through atoms or molecules called conductors. The movement of electrons creates an electrical current, and also creates a magnetic field.

Radiant energy results from the movement of photons, which are particles but have no mass and also function as waves. High energy (short wavelength) photons include X-rays; low energy (long wavelength) photons include radio waves. The spectrum of visible light is in the medium range of photon energy and wavelength. Blue or violet light has higher energy and shorter wavelengths than red light. Photons that are just beyond the violet end of the visible spectrum (ultraviolet) are even higher in energy (which is why they can kill bacteria and cause sunburns and MUTATIONS). Photons that are just beyond the red end of the visible spectrum (infrared) are even lower in energy, and humans usually perceive them as heat rather than light.

The Second Law

A simplified version of the *Second Law of Thermodynamics* is that, whenever events occur, the amount of entropy increases. Entropy can best be understood as disorder, or, as one of the founders of thermodynamics, chemist J. Willard Gibbs, described it, "mixedupness." The natural tendency is for orderliness to decay into disorder. This occurs because, for any system, there are far more possible disordered states than there are ordered states. The process of diffusion allows the Second Law to produce many of its effects. *Diffusion* occurs from the individual movements of atoms or molecules toward a less ordered, or more uniform, arrangement.

In most events with which humans are familiar, both the First and Second Laws operate. In nearly every event, energy changes from one form to another and entropy increases. When no energy input occurs from outside a system, events tend to occur in one direction only: They continue until equilibrium is reached in which energy is uniform throughout the system (First Law), and maximum disorder has been reached (Second Law). Consider the following examples:

Diffusion of heat. Heat diffuses from regions of higher temperature to regions of lower temperature. Warm molecules move faster (have more kinetic energy) than cold molecules and can transfer their energy to the cold molecules by colliding into them. As a result, heat energy diffuses from regions of high temperature to regions of low temperature. Diffusion of heat is also called *conduction. Convection* occurs when a mass parallel movement of molecules, such as those in the air, carry heat from a warm region to a cool region. If conduction and convection go to completion, an equilibrium of lukewarm molecules will result. This is what happens when a cup of hot coffee, or a recently dead mammal, cools off to environmental temperature. (The room actually becomes slightly warmer from the heat lost by the coffee cup.) The temperature of an object can increase when heat is conducted to it from another source that is warmer (see figure). Energy must be expended, for instance by a refrigerator, to make a relatively cool place even cooler; the coils in the back of the refrigerator disperse the heat, from the space inside the refrigerator and from the machinery, into the air. The First Law indicates that the total amount of energy is unchanged, and the Second Law indicates that the energy has reached a maximum state of disorder: The energy is no longer concentrated in any one location.

Movement of air. Air moves from regions of high pressure to regions of low pressure. Gas molecules in air that has high pressure (high potential energy) flow toward regions of air that have lower pressure, producing wind. Because wind involves the parallel movement of many gas molecules, it is not an example of diffusion. Air movement continues until an equilibrium is reached in which all regions have equal pressure. Since warm air has a lower pressure than cool air, temperature differences can cause air to move. Energy must be expended, by a fan or a pump, to force air to move in the absence of pressure differences. Other fluids, such as water, also move from regions of high to regions of low pressure. The First Law indicates that the total amount of energy has remained unchanged, even though it changed from potential to kinetic forms; and the Second Law indicates that pressure has reached maximum uniformity, when equilibrium is reached.

Diffusion of molecules. Molecules diffuse from locations in which they are more concentrated toward locations in which they are less concentrated. For example, a concentrated mass of sugar molecules is dropped into water. This is an orderly arrangement of molecules, with all of the sugar molecules in one place, and the water molecules in another. Both kinds of molecules move randomly, as a result of kinetic energy. They become less orderly as the sugar and water molecules mix together, until an equilibrium arrangement is reached in which both kinds of molecules have the same concentration everywhere. The molecules are very unlikely to

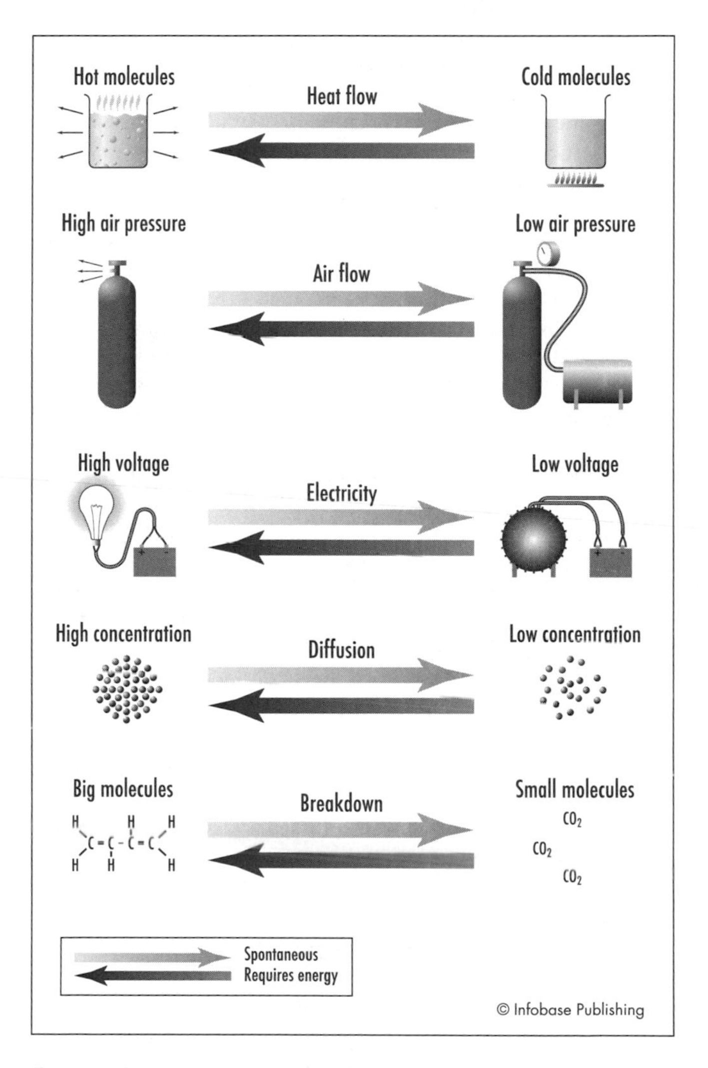

Energy and matter change from high energy and highly structured states to lower energy and disordered states. Heat flows from hot to cold places, resulting in a uniform temperature; air flows from high to low air pressure, resulting in a uniform pressure; electricity flows from high to low voltage, resulting in a uniform voltage; molecules diffuse from high to low concentration, equalizing their concentration; and big molecules with much stored energy break down into small molecules with less stored energy.

return to their original orderly arrangement simply by random movements.

Electricity. Electricity flows from regions of high voltage to regions of lower voltage. Electric current occurs when electrons flow from regions in which they have high potential energy (high voltage) to regions in which electrons have low voltage. This process continues until the voltage reaches equilibrium, as when a battery runs down. Energy must be expended, as in an electric generator, to raise the voltage of some electrons and thus create an electric current. The First Law indicates that the total amount of energy has remained constant, although it changed form from potential to electrical, and the Second Law indicates that the electrons have reached maximum uniformity, when equilibrium is reached.

Chemical reactions. Chemical reactions (in which atoms change molecular arrangements) also follow this pattern. A great deal of potential energy is stored within the chemical bonds of molecules, particularly in large biological or other organic molecules. If the reactant or reactants at the beginning of the reaction have more potential energy than the product or products have at the end of the reaction, then the energy has been released into another form. It could be released as kinetic energy (the reaction can release heat). Some reactions release light, while others release electricity. The First Law indicates that during these energy-releasing reactions, potential energy has changed into other forms of energy, but the total amount of energy remains constant, and the Second Law indicates that disorder has increased as a large, orderly molecule has broken down into smaller, less orderly molecules. Large molecules tend to fall apart into smaller components because large molecules contain more potential energy and are more orderly than small molecules. The total amount of energy remains the same, though it was transformed from potential to kinetic forms.

In other chemical reactions, the reactant or reactants have less potential energy than the product or products. These reactions will not occur unless energy is put into the reaction. In some of these reactions, kinetic energy of molecules becomes potential energy within bonds: The reactions cause their environment to become colder. In other reactions, light is absorbed by the electrons. In organisms, the energy that is released from one reaction can be used as an energy source by another reaction. The First Law indicates that during these energy-absorbing reactions, various forms of energy become potential energy, but the total amount of energy remains constant, and the Second Law indicates that entropy has actually decreased, as (usually) smaller, disorderly molecules have come together to form a larger, orderly molecule. Biological processes can assemble small molecules into larger ones. This process requires an input of energy and represents a decrease in entropy. For reasons described below, events that build small molecules into large ones do not represent violations of the Second Law.

For any of the events listed above to go in the reverse direction, inputs are necessary. It has already been noted that inputs of energy are necessary. However, a simple addition of energy is usually not enough to reverse the tendency toward disorder. If one puts kinetic energy into a sample of small molecules by heating them, one will not necessarily cause the small molecules to assemble into large molecules. It would probably just make them hotter and even more disorderly. A process that uses some kind of information is necessary to direct the use of the energy. Organisms have enzymes by which energy is directed to assemble small molecules into large ones and to make molecules to move from regions of lower concentration to regions of higher concentration. This represents an input of information. The information source need not be complex. Many inorganic catalysts are quite simple but can allow simple molecules to react into more complex forms.

What does all this have to do with evolution? Although evolution does not require an increase in complexity (see PROGRESS, CONCEPT OF), the history of the Earth has been characterized by the emergence of more and more complex organisms. This means that, at least in some cases, entropy has decreased during evolution. Some critics (see CREATIONISM) have claimed that evolution therefore violates the Second Law. This is not the case. A decrease in entropy requires an input of information. In biological systems, most of this information is in the form of enzymes, which control chemical reactions, and which are produced using information contained in nucleic acids (see DNA [RAW MATERIAL OF EVOLUTION]). If mutations accumulated without any restraints, information would degrade and entropy would increase. However, NATURAL SELECTION removes deleterious mutations, causing an accumulation of neutral or even beneficial mutations, which represent new information and a potential decrease in entropy. Sometimes, as in gene duplication (followed by mutation), horizontal gene transfer, or SYMBIOGENESIS, the information in a cell can suddenly increase.

Thermodynamics is especially relevant to an understanding of the origin of the first biological molecules and living systems (see ORIGIN OF LIFE). Life originated when small molecules assembled into large organic molecules. There was no preexisting life that could provide information (for example, in the form of enzymes) to direct this process. Yet experimental evidence indicates that entropy can decrease even without an input of complex information. Miller's classic experiment (see MILLER, STANLEY) brought small molecules and an energy source together to produce larger molecules. Subsequent experimental simulations of the origin of life have used clay surfaces that nonrandomly orient the smaller molecules and promote their synthesis into large molecules. The processes that would break large molecules down into smaller ones would overwhelm the processes that build small molecules into large ones, according to the Second Law, if it were not for the clay surfaces, which represent a source of information. The clay surfaces allow a spatial separation of the large molecules, so that they are no longer in circulation and exposed to the processes that would break them down.

The total amount of energy in the universe remains constant, but the total amount of disorder is always increasing. Eventually all the energy in the universe will be uniformly distributed and very weak, producing an equilibrium of maximum disorder from which nothing from within the universe itself can lift it.

Thomson, William *See* KELVIN, LORD.

tree of life The classification of all organisms into a single system has been depicted as a tree in which the branch patterns represent evolutionary relationships among species. Traditionally, taxonomists (scientists who study the classification of organisms) specialize on small groups of organisms, therefore the large-scale classification of all life-forms has often been unclear and controversial. For example, in earlier decades, it was unclear whether oomycetes and slime molds are or are not fungi (see EUKARYOTES, EVOLUTION OF). Constructing an overall tree of life, therefore, was a speculative enterprise at best.

Two developments have allowed the construction of a realistic tree of life. First was the invention of CLADISTICS. In this approach, species are bracketed together on the basis of their similarities. This approach allows the construction of a literally tree-shaped classification diagram for organisms (called a cladogram). A major limitation of this approach for a tree of life for all organisms is finding traits that all organisms possess. Botanists could construct cladograms on the basis of leaf and flower characteristics, mammalogists could use skeletal characteristics, but on what basis could both plants and animals be merged into a single cladogram? The second advance was the study of nucleotide sequences (see DNA [EVIDENCE FOR EVOLUTION]). The evolutionary closeness of two species was directly related to the similarity of their DNA. DNA is something that all species have in common. The problem is that comparisons must be made between the same genes or other DNA segments in the different species. Are there any genes that all species possess? There are a few, associated with the basic steps of metabolism. All organisms, for example, possess a gene for phosphoglucose isomerase. This enzyme, and the DNA that encodes it, have frequently been used in cladistic analyses. Some researchers suggest the cytochrome oxidase I gene, and still others a ribosomal RNA gene, as a basis for classifying all organisms. Others are skeptical that any single gene can be used as a definitive classification of all organisms.

The evolutionary history of organisms can be studied through three independent sources of information: fossils, cladograms based upon anatomical data, and cladograms based on molecules such as DNA. None of these three sources of data is derived from any of the others. All three sources of information have converged upon more or less the same outline of evolutionary history. The groupings of organisms, and their relative times of divergence, are about the same whether the cladograms are based upon anatomical features or upon DNA, and they correspond closely to the order with which they appear in the fossil record. This result could not have happened if evolution had not occurred.

Prodigious computing power is necessary for constructing anything beyond the simplest cladograms. Not surprisingly, a cladogram of all of the world's almost two million known species (or even the subset of them from which DNA is available) remains a dream. Cladograms based upon representative species from major groups of organisms have been produced.

Patterns of evolutionary divergence. Among the evolutionary patterns that have resulted from constructing the tree of life are the following:

- Traditional distinctions between major groups were confirmed, for example, the distinction between prokaryotes and eukaryotes and between plants and animals. Within these larger groups, the traditional classifications were often confirmed: The arthropod clade included traditionally recognized groups such as arachnids, insects, and decapods, and the vertebrate clade contained traditionally recognized groups such as birds and mammals. The analysis confirmed the evolutionary closeness of arthropods to roundworms, of segmented worms to mollusks, and of chordates to echinoderms (see INVERTEBRATES, EVOLUTION OF). Botanists had long maintained that plants evolved from green algae. Tree of life analyses confirmed this and further indicated that the organisms similar to modern charalean green algae were the ancestors of plants.
- Much evidence indicates that chloroplasts in plant cells evolved from cyanobacteria that took up residence in ancestral plant cells, and that mitochondria evolved from proteobacteria that took up residence in ancestral eukaryotic cells (see SYMBIOGENESIS). In this tree of life, cyanobacteria and chloroplasts clustered together, as did mitochondria and proteobacteria, thus confirming their symbiogenetic origin. Chloroplasts and mitochondria bear little genetic resemblance to their hosts.
- Some distinct evolutionary lineages of organisms had previously been lumped together but turn out to be very different. Water molds are not fungi (they cluster together with brown algae). Slime molds turn out to be a composite group, some of them closely related to fungi, some not. And while monocot plants form a single clade, the dicot plants represent distinct lineages (see ANGIOSPERMS, EVOLUTION OF).
- Some surprises resulted from observing which large groups were more similar to which others. ARCHAEBACTERIA turn out to resemble eukaryotes more closely than they do the eubacteria; this suggests that the nucleus of the eukaryotic cell may have had an archaebacterial origin (see EUKARYOTES, ORIGIN OF). Fungi turn out to be more similar to animals than they are to plants. Both fungi and animals use chitin and collagen as structural materials. These results did not contradict previous evolutionary concepts so much as filling in a void of knowledge.

Assigning dates to evolutionary divergences. It is one thing to determine the series of events in evolutionary history and quite another to determine when these events occurred. Paleontologists expect eventual agreement between the ages of evolutionary lineages as determined by a MOLECULAR CLOCK and those determined from the fossil record. As evolutionary biologists Michael J. Benton and Francisco J. Ayala point out, one should expect that the fossil record will always provide an underestimate, and molecules an overestimate, of the age of the evolutionary lineage:

- *Fossils provide underestimates of the age of a lineage.* When an evolutionary divergence first occurs, the two or more resulting lineages are indistinguishable on the basis of fossils. If one saw the two lineages, the lineages would at this point look alike. By the time their fossils become noticeably different, evolutionary divergence has already been going on for some time. Fossils may not appear in the record until a particular innovation, for example the evolution of hard skeletons, occurs.

- *Molecules provide overestimates of the age of a lineage.* Benton and Ayala point out that the branch point indicated by molecular cladograms may represent the origin of a molecular polymorphism, rather than an actual evolutionary branch point. The two lineages of molecules may have diverged even while they were still mixed together in the same population, constituting a polymorphism within a population rather than a divergence into different populations. It is well known that some genes mutate faster than others, meaning that each of these molecular clocks ticks at a different rate. Molecular estimates of the time at which bilateral animals diverged from the sponges and cnidarians range from 1.2 billion to only about 600 million years ago, depending on which and how many segments of DNA or other molecules were used.

Therefore the actual evolutionary branch point will probably be somewhere between the anatomical and the molecular estimates. The anatomical and the molecular branch points may represent two different evolutionary events, rather than two estimates of the same event.

In some cases, the correspondence between fossils and evolutionary lineages may not be clear. For example, molecular evidence suggests that modern birds diversified during the CRETACEOUS PERIOD. Birds such as *ARCHAEOPTERYX* were already in existence by the earlier JURASSIC PERIOD. The explanation for this is that modern birds are probably not the descendants of *Archaeopteryx,* which proved to be an evolutionary dead end. Birds diversified even before *Archaeopteryx,* but most of the branches of bird evolution became extinct.

Because of horizontal gene transfer among bacteria, among archaebacteria, and even between the two lineages, the prokaryotic tree of life might more closely resemble a web, from which the eukaryotic tree of life emerges. This might make it difficult or impossible to find "LUCA," or the *last universal common ancestor* of all extant life-forms.

As evolutionary scientists continue to construct the tree of life, they will learn answers to many questions that have puzzled biologists for centuries, particularly with regard to which organisms are more closely related to which others.

Further Reading

Benton, Michael J., and Francisco J. Ayala. "Dating the tree of life." *Science* 300 (2003): 1,698–1,700.

Ciccarelli, Francesca D., et al. "Toward automatic reconstruction of a highly resolved tree of life." *Science* 311 (2006): 1,283–1,287.

Tree of Life Web Project. "Explore the tree of life." Available online. URL: http://www.tolweb.org/tree. Accessed May 13, 2005.

Zimmer, Carl. "Did DNA come from viruses?" *Science* 312 (2006): 870–872.

Triassic period The Triassic period (250 million to 210 million years ago) was the first period of the MESOZOIC ERA (see GEOLOGICAL TIME SCALE). It followed the PERMIAN PERIOD, which ended with the PERMIAN EXTINCTION. The Mesozoic era is also known as the Age of Dinosaurs, because DINOSAURS were the largest land animals during that time.

Climate. Because the continent of Pangaea was so large, the middle of the continent had arid conditions.

Continents. A single worldwide continent, Pangaea, had formed during the Permian period (see CONTINENTAL DRIFT). Pangaea was centered at the equator. The Panthalassian ("world ocean") Sea surrounded it, and the Tethys Sea penetrated into it. The widening of the Tethys Sea separated Pangaea into the northern Laurasia and the southern Gondwana by the end of the Triassic period.

Marine life. Marine invertebrates and vertebrates began to evolve into many new forms after the Permian extinction had eliminated some groups (such as TRILOBITES) and reduced the diversity of others (such as brachiopods; see INVERTEBRATES, EVOLUTION OF). All modern groups of marine organisms existed during the Triassic, except aquatic mammals. Large aquatic reptiles began to evolve in the oceans during the Triassic.

Life on land. Because of the Permian extinction, the Triassic period began with few but very common species. Scientists speculate that the dry conditions in the middle of Pangaea provided an advantage to seed plants over plants that reproduced by spores, and to reptiles over amphibians and mammals.

- *Plant life.* The seed fern *Glossopteris* dominated vast forest areas at the beginning of the Triassic. Seed plants later dominated over seedless plants, forming the first extensive forests of conifers, which had first evolved in the late Paleozoic era (see GYMNOSPERMS, EVOLUTION OF; SEEDLESS PLANTS, EVOLUTION OF).
- *Animal life.* One of the most common vertebrates of the early Triassic was the reptile *Lystrosaurus.* All vertebrates were small at the beginning of the Triassic period. Some moderately large forms evolved by the end of the period, including the first dinosaurs, although the largest dinosaurs did not evolve until the Jurassic and Cretaceous periods.

Further Reading

Kazlev, M. Alan. "The Triassic period." Available online. URL: http://www.palaeos.com/Mesozoic/Triassic/Triassic.htm. Accessed May 16, 2005.

trilobites Trilobites were among the most abundant and diverse arthropods in the oceans of the PALEOZOIC ERA. Although they have been described as the cockroaches of the ancient seas, they represent an evolutionary lineage distinct from that of insects; they were more closely related to modern arachnids such as the horseshoe crab and spiders (see INVERTEBRATES, EVOLUTION OF). Besides being diverse and abundant, they have been marvelously preserved as fossils, since they were the first arthropods to have external skeletons (even though just their upper surfaces) reinforced with calcite. They

Trilobites (such as this *Phacops rana* from the Devonian) were extremely common arthropods that lived in the Paleozoic oceans. Notice the large compound eyes and the body segments. ***(Courtesy of Wim van Egmond/ Visuals Unlimited)***

were abundant for 270 million years, becoming extinct only at the PERMIAN EXTINCTION. For all four reasons they have been studied extensively by paleontologists as providing a window into how evolution works over long periods of time.

They are called trilobites because of the three lobes, or lengthwise sections, that constitute their bodies behind the head segment (see figure). The numerous segments of the middle part of the body were flexible and allowed trilobites to curl up lengthwise, at least a little bit. They had numerous legs, and each leg had a branch of delicate gills which, protected underneath the shell, allowed them to breathe oxygen from the water.

Trilobites had tremendous diversity. More than 5,000 genera have been recognized. They ranged in size from being as big as a dinner plate to as small as a pea. They had many different ways of obtaining food. Some were predators, eating annelid worms, while others were eaten by such larger predators as *Anomalocaris* during the CAMBRIAN PERIOD (see BURGESS SHALE) and mollusks and the earliest fishes during the ORDOVICIAN PERIOD. Some were scavengers, and some had enormous head brims with which they may have filtered particles of food from the ocean water. They had many different ways of moving about with their two rows of numerous legs: Some scurried about on the seafloor, while others swam at different depths. Some were covered with spines, others were smooth. Most trilobites had eyes made of calcite crystals unlike those of any other animal before or since. Some had clusters of many small eyes (sometimes in fantastic towerlike structures), and some had smaller clusters of large eyes. Some species secondarily lost their eyes during evolution in darkness.

Further Reading

Fortey, Richard A. *Trilobite: Eyewitness to Evolution.* New York: Knopf, 2000.

———. "The lifestyles of the trilobites." *American Scientist* 92 (2004): 446–453.

Parker, Andrew. *In the Blink of an Eye.* New York: Perseus, 2003.

U

unconformity An unconformity is a geological feature in which two adjacent deposits do not show an unbroken sequence of layers in the correct time order in which they were originally formed. Examples include *disconformities* in which there are missing layers between the upper and lower layers of sedimentary deposits; *nonconformities* in which sediments were deposited over rocks of volcanic origin or rocks that were metamorphosed by heat and pressure; and *angular unconformities* in which the lower layers of deposits were pushed into a non-horizontal angle, and upper layers were then deposited upon them. In many cases, unconformities occur because of the movement of the Earth's crust (see PLATE TECTONICS). In most of these cases, something similar to this sequence of events occurred:

1. Deposition was interrupted. For example, the continental shelf on which layers of sediment had been accumulating was raised above the ocean by the movement of the Earth's crust, or the ocean level dropped.
2. The lower layers of sediments were transformed into rock.
3. The upper layers of sediments eroded away.
4. The layers were then submerged again, by continued crustal movements or because ocean levels rose again. Crustal movement could bend or push the layers to a different angle.
5. New layers of sediments were deposited on top of the transformed lower layers.

In one example, from the Black Hills of South Dakota, vertical layers of rock were once horizontal sediments but were turned into rock and turned on their sides. Then, on top of them, new layers of sediment were deposited, which then also turned into rock. Both deposits are Precambrian; there is a 1.2-billion-year gap represented by the line between the vertical and horizontal layers (see photo).

More than a billion years separates the deposition of the horizontal layers of sedimentary rock on the top and the vertical layers of sedimentary rock on the bottom of this unconformity in the Black Hills of South Dakota. *(Photograph by Stanley A. Rice)*

Unconformities can be produced by a number of processes. For example, inverted layers can result when movements of the crust cause a buckling, which pushes upper layers up and over, leaving them upside down. Alternatively, when two formations collide, one of them (even if it is older) can be thrust over the other (even if it is younger). The result of either of these processes would be older rocks on top of younger rocks.

Unconformities are among the many things that make the interpretation of the geological record challenging. Creationists (see CREATIONISM) often cite unconformities as examples of the failure of the scientific interpretation of billions of years of Earth history and of the evolutionary science based upon it. However, creationists are unable to explain how unconformities could have occurred if all of the deposits were produced during a single, worldwide Noachian Flood. How could horizontal layers be deposited upon vertical layers, if they had all been mud at the same time during the Flood? In fact, it was the angular unconformity at Siccar Point that led James Hutton (see HUTTON, JAMES) to reject the prevailing model of CATASTROPHISM and paved the way toward the acceptance of UNIFORMITARIANISM.

uniformitarianism Uniformitarianism is an assumption and a procedure that underlies much of modern geological science. It began mainly with the geological studies of Sir Charles Lyell (see LYELL, CHARLES) and his predecessors such as James Hutton (see HUTTON, JAMES), although Lyell called it simply *uniformity*.

Lyellian uniformitarianism can be summarized as "The present is the key to the past." Uniformitarianism, in part, attempts to explain all past geological events in terms of processes that are now in operation, rather than by special events, whether natural or supernatural. Lyell proposed it as an alternative to CATASTROPHISM, in which some scientists (see CUVIER, GEORGES) explained the geological past as a series of worldwide catastrophic events that separated disjunct ages of Earth history during which very little (geologically speaking) happened. Catastrophism did not demand solely supernaturally caused catastrophes, although catastrophists considered the biblical Flood of Noah to be the most recent of the worldwide catastrophes.

Lyell assigned four meanings to uniformity, some of which are considered unquestionable today and others of which are now rejected by most scientists.

Uniformity of law. The laws of physics and chemistry that underlie geological processes have remained unchanged through time. Scientists do not question this assumption, for to do otherwise would make the scientific investigation of the past impossible.

Uniformity of process. Past geological events should be interpreted as the result of processes that we can see in operation today. All geologists accept this. Lyell favored the interpretation that scientists should explain the past only in terms of processes that are happening all the time, at any given moment—processes such as volcanic activity, earthquakes, erosion, sedimentation, compression, uplift, and (something not known in Lyell's time) PLATE TECTONICS. Geologists today would include processes that happen only rarely but which have been observed within recent history, such as the impact of extraterrestrial objects upon the Earth. When Lyellian uniformitarianism dominated geological science in the early to middle 20th century, such impacts were ignored. What may have been a comet exploded near the ground in Siberia in 1908, and large meteorites have apparently struck the Earth in the recent past, such as the impact that created the Barringer Crater in Arizona only 50,000 years ago (see ASTEROIDS AND COMETS). The work of astronomer Gene Shoemaker (now famous as codiscoverer of the Shoemaker-Levy comet that slammed into Jupiter in 1995) in the 1960s on what is now recognized as the Ries Crater in Germany opened the door for geologists to consider meteorite and other impacts upon the Earth. The most famous extraterrestrial impact was the asteroid that caused the Chicxulub Crater in Yucatán 65 million years ago and contributed to the extinctions of many species, including all dinosaurs (see CRETACEOUS EXTINCTION). Lyell would not have been pleased at this. Modern geologists continue to reject supernatural causation for past geological events.

Uniformity of rate. This meaning, now called *gradualism,* demands that geological processes operated in the past at the same rate that they do today—in other words, that there were no periods in the past in which volcanic eruptions occurred on a greater scale than they do today. Geologists now recognize that this is not strictly true; massive volcanic eruptions, far exceeding anything to occur within historical times, produced the Deccan Traps in what is now India, starting just before the end of the CRETACEOUS PERIOD.

Uniformity of state. This is the meaning that is not accepted in modern geology—that directional change (progress) has not occurred during Earth history. Uniformity of state maintains that there have been changes, but no permanent ones. Volcanoes erupt, sea levels go up and down, glaciers advance and retreat, but the Earth, according to this view, always goes back to the way it was before. In geological sciences, this is now known not to be strictly true. The early Earth, during the Archaean eon, was very different, its atmosphere without oxygen, and with massive tides caused by a close moon (see PRECAMBRIAN TIME). Lyell extended his philosophy to include life—he believed that species may disappear but they will return; EXTINCTION, in his view, is not forever. This assumption was what led Lyell (at first) to reject any form of evolution, even from his friend Charles Darwin. It was the catastrophists such as Cuvier who insisted that the history of life had a direction, and that changes occurred over time, not the uniformitarians. Today, scientists reject uniformity of state, not only because of the triumph of evolutionary science but also because they know Earth history has been characterized by very different assemblages of organisms through time: lots of marine but barren landscapes in the Cambrian, forests dominated by seedless vascular plants in the Carboniferous, the age of dinosaurs, the age of mammals, the ice ages, to name a few.

It is the last meaning, the extreme form of uniformitarianism, that invoked some of the sarcasm for which writer Mark Twain was famous. Twain pointed out that the Mississippi River is getting shorter each year because new river courses cut off oxbows. From this, he said, scientists may conclude that a million years ago the Mississippi River stuck out over the Caribbean like a fishing rod, and a million years from now, the Mississippi River will shorten, causing New Orleans and St. Louis to be in the same place and have a joint board of aldermen.

Uniformitarianism played a valuable role in allowing geological science to develop the tools of investigating the past using knowledge from the present. Only recently, however, has the strict grip of uniformitarianism loosened enough to allow exceptions to be understood.

Further Reading

Benton, Michael. "The Death of Catastrophism." Chap. 3 in *When Life Nearly Died*. London: Thames and Hudson, 2003.

universe, origin of Although the universe includes everything, this discussion is limited to the origin of galaxies, stars, and planets. Although some scholars use the term EVOLUTION for the processes and changes in the history of the universe, the processes of origin and change in the universe have no counterparts to genetics or natural selection.

The modern understanding of the origin of the universe would have been practically unthinkable to people even a century ago. For one thing, they had no concept of the size of the universe. Astronomers knew that the universe was, for them, incalculably vast. It might have been possible to use triangulation to calculate the distance of a star, using the diameter of Earth's orbit around the Sun as the base of the triangle, but the stars are so far away that the angle opposite the base was effectively zero. Before large modern telescopes, it was not clear whether the nebulae were clouds of gas or clusters of stars (there are many of each). It was not until 1923 that astronomer Edwin Hubble was able to focus on the Andromeda galaxy well enough to discern individual stars. If those stars appeared so small, then the galaxy must be incredibly distant. There was a suspicion that nebulae and stars went through the equivalent of life cycles, but there was no direct observation that showed that anything changed in the universe, except for occasional supernova explosions.

With improved techniques, it became possible to estimate, though not directly measure, the distance of galaxies. Astronomers Henrietta Swan Leavitt and Harlow Shapley, at the Harvard College Observatory, determined that Cepheid variable stars had a very reliable correlation between the periodicity of their variation and their absolute luminosity. The intensity of light decreases with the square of its distance from the observer. When Edwin Hubble looked at the individual stars in the Andromeda galaxy, he found some that changed their luminosity in the same way as a Cepheid variable star. From the correlation calculated by Leavitt and Shapley, Hubble calculated the absolute luminosity; and by comparing the absolute with the observed luminosity, he calculated the distance of the galaxy. The resulting distance could be meaningfully expressed only in light-years, the distance that light can travel in a year, as it travels at 186,000 miles (almost 300,000 km) per second. The Andromeda galaxy was 900,000 light-years away.

Astounding as this discovery was, Hubble's main breakthrough was his discovery that the universe is expanding. Ever since German chemist Joseph Frauenhofer discovered that chemical elements block certain wavelengths of light, astronomers had been able to analyze the chemical composition of stars and planets by studying these lines of darkness in the spectrum of light from a star or planet. Hubble discovered that the Frauenhofer lines displayed a marked *redshift*—the absorbance lines were further toward the red (long wavelength) end of the spectrum than they should be—and that different stars had different degrees of redshift. When an object is coming toward the observer, waves emitted from the object are shortened; when an object is moving away from the observer, waves emitted from the object are lengthened. This *Doppler effect* explains why the whistle of a train coming toward an observer has a higher pitch than when the train has passed the observer: Sound consists of waves of air molecules. The redshift of stars and galaxies suggested that they are moving away from the Earth (or, more properly, that the space between them and the Earth is expanding). This suggested an expanding universe.

Hubble and other astronomers noticed something else. The more distant galaxies had a greater redshift than the closer galaxies. There was a constant relationship between speed and distance, a constant now called the *Hubble constant*. This meant:

- For galaxies too distant for Cepheid variable stars to be observed, the Hubble constant allowed a calculation of distance, once the redshift is measured. The redshift of some galaxies was huge: In some cases, the hydrogen band that should be at 0.000005 inch (122 nm, in the ultraviolet part of the spectrum) is shifted all the way over to 0.00003 inch (720 nm, in the infrared part of the spectrum). But the relationship between redshift and distance had to be reliable, unless the laws of nature were not reliable. Some of the galaxies observed by the orbiting telescope named after Edwin Hubble are so distant that they actually appear red in the photographs.
- All the galaxies began at a single point at a single time. This was the origin of the modern *big bang* theory. An explosive origin of the universe had been proposed by the Belgian priest and physicist Georges Lemaître earlier in the century, but the redshift was the first evidence for it. The universe had to have begun from a gigantic explosion that produced literally everything. Finally, the inverse of the Hubble constant even provided an estimate of when this happened, sometime less than 15 billion years ago. One recent estimate by the National Aeronautics and Space Administration (NASA) indicates that the universe began 13.7 ± 1 billion years ago.

The very idea that the universe even had a beginning was not automatically accepted by all astronomers. Sir Fred Hoyle, a famously free-thinking British astronomer, defended an alternative view, which he called the *steady state* model.

Yes, the universe is expanding, but it has been doing so forever. New matter forms spontaneously between existing galaxies and eventually forms new galaxies. Ironically, the name "big bang" began as a humorous term that Hoyle used for the theory he did not accept. Almost all astronomers have now accepted the big bang theory.

Astronomers could then extrapolate backward, to the time when all the matter and energy of the universe was in one place, at the beginning. Just how much could matter be compressed? That is, how big was the primordium from which the universe exploded? The masses and forces involved would exceed anything that could be studied on Earth, therefore mathematical models had to be used. Astronomers calculated back to a time when the universe quite literally had infinite density and zero volume. This condition is called, with cosmic understatement, a singularity.

From this point of infinite density, the universe exploded. Why did it wait until a certain point in time, about 14 billion years ago, to explode? Space and time are linked (the space-time continuum was one of the consequences of Einstein's general theory of relativity). Since time itself began at that point, it is a meaningless question; there is no "before." It would be like asking what God was doing before God began to create the universe.

When the universe was less than 10^{-43} seconds old, the almost infinitely dense universe experienced some quantum fluctuations, which determined the whole future course of universal history. When the universe was one millisecond old, it was denser than the nucleus of an atom, consisting neither of energy nor matter as we know them today. As the universe expanded and cooled, particles, then hydrogen atoms, formed. Almost equal amounts of matter and antimatter were produced and canceled each other out; the slight excess of matter (one part in 100 billion) that survived is what the universe is made of. Nuclear fusion occurred, but only for a few seconds, so the sphere consisted mostly of energy and hydrogen atoms. This superheated sphere continued to expand and cool. It took a half-million years to cool down to about 4,700°F (3,000°K). At this point, the energy level was low enough that a human observer would not have been able to see it. Astronomers refer to this period as the Dark Age, even though the darkness is defined relative to the human visible spectrum.

As the universe expanded for another half-million years, there were slight heterogeneities in the arrangements of the atoms. Some were closer to each other and began to attract one another. The expanding sphere of gas and energy became a sphere of gas balls. Gravity made these balls ever denser. Finally, when the universe was one million years old, thermonuclear explosions began to occur, turning many of the balls of gas into stars. Soon whole galaxies were lit up by their stars. Scientists now know that there are billions of galaxies. The Hubble telescope, in orbit above the Earth's atmosphere, used a 10-day exposure to photograph thousands of galaxies in a small space (the equivalent of the intersection area of two gun scope crosshairs at arm's length) in which no galaxies at all had previously been detected.

The big bang was over, but some of its energy lingered. As the universe continued to expand, the temperature of this energy decreased. Today, there is still a little bit of it. Calculations by astronomer George Gamow suggested that the energy should be in the microwave range, with a wavelength of about three inches (7 cm), and give outer space a temperature of three degrees above absolute zero. This background radiation was measured by astronomers Arno Penzias and Robert Wilson in 1964 using an antenna that had been developed for communication with satellites. They detected a signal at 2.9 inches (7.35 cm) that was so faint that it could barely be distinguished from the random noise of the atmosphere and of their instruments. They made measurements of just how much of the noise was produced by their instruments and made correction for it. They still detected the microwave radiation, and it was coming from the entire sky. If the radiation had been produced by the atmosphere, there would have been more of it where radiation would have to penetrate more of the atmosphere, that is, nearer to the horizon. If it had been produced by stars, there would have been more of it in the plane of the Milky Way than away from it. There remained one possible factor that could have caused spurious results. Pigeons had been nesting in the radio telescope, coating it with what Penzias called "a white dielectric material." But even after cleaning out the telescope, the signal remained. The wavelength of 2.9 inches (7.35 cm) corresponded to a temperature of about six degrees F (2.5 to 4.5°C) above absolute zero. Penzias and Wilson had measured the energy left over from the big bang. Sir Martin Rees, the British Astronomer Royal, calls it "the 'afterglow' of a pregalactic era when the entire universe was hot and dense and opaque."

Stars go through what can be called "life cycles." Their main source of energy is the fusion of four hydrogen atoms to form one helium atom. A small amount of the matter is transformed into a huge amount of energy. This is described by the equation $E = mc^2$, also worked out by Einstein: The coefficient that relates matter to energy is the speed of light multiplied by itself, which is a huge number. The inside of a star is hot and dense, and fusion reactions occur. Inside the larger stars, additional fusion reactions produced atoms larger than helium. These stars gradually expand and become dimmer. At some point, most stars explode. The largest stars explode into supernovae. One supernova can release as much energy as an entire galaxy. The middle of a supernova has greater energy than anyplace in the universe had since the beginning, and fusion reactions produced even the largest atoms. Some supernovae are so powerful that they produce black holes, which are so dense and have such strong gravitational pull that even light cannot escape from them. There may be black holes at the centers of many galaxies.

One large star "lived" for 10 billion years and exploded in a supernova. It left behind a nebula that contained a lot of unused hydrogen but also a lot of heavier elements. Gravity began to pull these atoms together again in a flat, circular, swirling mass. The hydrogen in the dense center of the circle ignited once again in fusion reactions about five billion years ago. This was the birth of the Sun. The atoms away from the center formed clusters (planetesimals), some of gases, others of heavier elements. The gaseous clusters never produced sufficient pressure to ignite; they became the gas giant planets Jupiter, Saturn, Uranus, and Neptune. The clusters of heavier elements became the remaining planets, including Earth. This explains why the solar system is only four and a half billion years old, in a 14-billion-year-old universe: The Sun is a second-generation star. New stars continue to form in nebulae: Stars in the Orion Nebula are less than 100,000 years old. More than 150 planets have been detected revolving around other stars. In most cases, the presence of the planet is detected by the movement of the star caused by the gravity of a large planet, or by changes in light intensity caused by a large planet moving between the star and human observers. In 2005 the first photograph of a planet around a star other than the Sun was published.

The "life span" of the sun will be about 10 billion years. About half of this time has passed. Martin Rees uses the analogy of walking across the United States from coast to coast: Each step would represent two thousand years in the Sun's lifetime. All of human history would fit into three or four steps in the middle of Kansas.

Eventually the Sun will cool to a red color and expand, becoming a red giant. Life on Earth will have ended before this. About the time, five billion years from now, that the Sun explodes, the Milky Way galaxy will collide with the Andromeda galaxy. Although galaxies are mostly empty space, gravitation will draw stars together in many colorful explosions—none of which humans will see.

But what will happen to the universe? Will it expand forever, reaching a uniform deadness of absolute zero? If the average density of the universe exceeds three atoms per cubic meter, the gravitational force will be sufficient to draw the atoms back together, creating yet another big bang. The visible matter of the universe is 50 times less than what is required for this to happen. However, the vast majority of matter in the universe may be "dark matter," dispersed between the galaxies and stars, and not reflecting any light. Furthermore, particles called neutrinos have almost no mass (they weigh one billionth as much as a hydrogen atom), but there may be so many of them that they constitute a considerable part of the mass of the universe. The density of the universe may be great enough, after all, to cause it to coalesce and renew.

Further Reading

Davies, Paul. *The Last Three Minutes: Conjectures about the Ultimate Fate of the Universe.* New York: Basic Books, 1997.

Ferris, Timothy. *The Whole Shebang: A State of the Universe Report.* New York: Simon and Schuster, 1998.

Guth, Alan H., and David I. Kaiser. "Inflationary cosmology: Exploring the universe from the smallest to the largest scales." *Science* 307 (2005): 884–890.

Lineweaver, Charles H., and Tamara M. Davis. "Misconceptions about the Big Bang." *Scientific American,* March 2005, 36–45.

Rees, Martin. *Our Cosmic Habitat.* Princeton, N.J.: Princeton University Press, 2003.

Schilling, Govert. "Picture-perfect planet on course for the history books." *Science* 308 (2005): 771.

Smolin, Lee. *The Life of the Cosmos.* New York: Oxford University Press, 1997.

Tyson, Neil deGrasse, and Donald Goldsmith. *Origins: Fourteen Billion Years of Cosmic Evolution.* New York: Norton, 2004.

Weinberg, Steven. *The First Three Minutes.* New York: Basic Books, 1977.

vestigial characteristics Vestigial characteristics are traits of organisms that no longer have a useful primary function but are leftovers or vestiges of characteristics that had a useful primary function in evolutionary ancestors. They usually result from a process such as the following:

1. A characteristic has an important function in an organism. For example, stamens in flowers produce pollen.
2. As circumstances change, the characteristic is no longer important for the organism. NATURAL SELECTION no longer eliminates individuals that lack the full development of the characteristic. For example, flowers (at least, those with male function) need stamens but may not need all of the stamens that they have.
3. Mutations may cause incomplete development of the characteristic. For example, many flowers produce staminodes, which are sterile stamens—shafts that produce no pollen. This structure is now considered vestigial.
4. If the vestigial structure represents no significant cost to the organism, then natural selection will not eliminate it. Also, in many cases, natural selection may be in the process of eliminating a vestigial characteristic but has not completed this process.

It is always risky to call a characteristic useless because it is always possible to discover a use for the characteristic later. Critics of evolutionary science (see CREATIONISM) like to cite the example of the German anatomist Robert Wiedersheim who, in 1895, listed about a hundred organs which he considered vestigial in humans. Many of these organs turned out to have important functions (for example, tonsils) even though a person could in fact survive without them. The fact that a person can survive with only one leg (or none) does not prove that legs are useless. However, *vestigial* and *useless* are not the same thing; and it is not possible to so easily dispose of vestigial characteristics as evidence for evolution as the following examples indicate.

Vestigial characteristics can be found at all levels of organism structure and function:

At the biochemical level. The DNA of eukaryotic organisms contains a large, sometimes prodigious, amount of NONCODING DNA. In most eukaryotes, well over half of the DNA is noncoding—that is, it does not result in the production of a protein. While some of this noncoding DNA has important functions, much of this DNA is vestigial. Pseudogenes, for example, are very similar in structure to true genes, but they do not have a PROMOTER, therefore they are not used. They are old genes that the cell no longer uses. The cell has deleted the gene the same way a computer deletes a file—not by eliminating it entirely, but by deleting the information about how to find and use it (see DNA [RAW MATERIAL OF EVOLUTION]). Much human DNA consists of instructions for making reverse transcriptase—an enzyme not used by humans but which is the remnant of the past activity of retroviruses.

At the cellular level. Some cellular organelles, such as chloroplasts and mitochondria, are the simplified descendants of symbiotic bacteria (see SYMBIOGENESIS). Chloroplasts and mitochondria still have some of their own DNA and genes, and their own ribosomes, which allow them to produce proteins from their own genes. As they no longer have all of the genes that they need for survival, they cannot live independently of the host cell. Many cell structures such as the endoplasmic reticulum did not arise by symbiosis; they function just fine without their own DNA or ribosomes, because the nucleus of the cell has all the genes, and the cytoplasm of the cell has all the ribosomes, necessary for the synthesis and activity of the endoplasmic reticulum. Chloroplasts and mitochondria are a transitional form—they have lost enough genes that they cannot live on their own, but they have not lost all of them. While their genes are, in fact, useful, these genes do not need to be in the chloroplasts and mitochondria themselves. Perhaps even more striking is the fact that the chloroplasts of some protists (such as dinoflagellates) are the evolutionary descendants not of bacteria but of eukaryotic algae. The chloroplasts of dinoflagellates not only have extra membranes but even have nucleomorphs, which are vestigial nuclei left over from the eukaryotic algal ancestors!

Examples at the organ level are almost innumerable.

At the organ level, in plants. Staminodes, mentioned previously, are one example—or are many examples, since they have evolved independently in several different plant families. Also, grasses have flowers, but as these flowers are wind-pollinated, they have no use for petals. Grass flowers do, however, have tiny vestigial petals called lodicules.

At the organ level, in animals. Various animal species have structures that are usually considered vestigial. Examples include:

- The appendix in humans is a remnant of the cecum, a pouch of the intestine that has an important function in animals that have a high food intake of plant materials.
- Flightless birds have wing remnants. Ostriches have wings, but they are small and not used for flight; kiwis have vestigial wings so small that they have no important function.
- Some snakes, and some whales, have vestigial hip and leg bones, even though they do not have legs (see WHALES, EVOLUTION OF). The ancestors of horses had several toes, while modern horses have just one (the hoof); occasionally, modern horses develop extra toes, because they still have the vestigial, unused genes for these extra toes.
- Some toothless animals have teeth during their embryonic stages (for example, baleen whale embryos develop teeth, which are then reabsorbed before birth) or have the genes for teeth but do not use them (for example, chickens).
- Male mammals have nipples. Male nipples are not themselves so much vestigial as they are a side effect of the development of nipples in female mammals (see DEVELOPMENTAL EVOLUTION). The fact that nipple production is not prevented during male development can be considered a vestigial characteristic.
- Blind cave fishes may have no eyes but still have eye sockets. Tissue transplants, and crosses between subspecies of blind fish, result in small but still functional eyes. This demonstrates that blindness evolved recently in these fish populations.

Another example of a vestigial characteristic is patterns of seasonal activity that make sense only when one considers the evolutionary past. Deciduous trees open their buds in the springtime. Some tree species (for example, elms and maples) open their buds early, when there is still some danger of frost; other tree species (for example, persimmons) do not open their buds until nearly all danger of frost is past. These trees deal with the danger of frost in two different ways: The trees that open their buds early must tolerate the frost, while the trees that open their buds late are avoiding the frost. *Tolerance* requires the trees to produce protective chemicals within their buds, but in return, they are able to utilize the sunshine of early spring. *Avoidance* requires the trees to lose the opportunity to utilize early spring sunshine, but in return, they do not need to produce protective chemicals. In general, trees that have evolved in cold climates (such as northern North America) tolerate frost, while trees that have evolved in warm climates (such as southern North America) avoid frost. Trees that are native to the middle latitudes of North America (for example, in Oklahoma) have a mixture of tree species that tolerate and avoid frost; both methods should work equally well. It just so happens that the tree species native of Oklahoma that tolerate frost are members of plant families that evolved in the north, and trees species native to Oklahoma that avoid frost are members of plant families that evolved in the south, during the early CENOZOIC ERA in the northern continents. Some trees tolerate, some avoid, frost, not because of current climatic conditions, but because they have inherited the evolutionary adaptations of their ancestors. This evolutionary pattern can be considered vestigial, left over from the evolutionary past.

The primary assumption behind creationist attacks on the concept of vestigial characteristics is that an organism designed by a higher intelligence cannot have useless characteristics (see INTELLIGENT DESIGN); biologist Michael Denton went so far as to say that there can be no exceptions. Therefore they have attempted to prove that characteristics which evolutionary scientists have ever identified as vestigial are, in fact, useful. The problems with their approach include:

- They assume that, if they demonstrate a function for the vestigial characteristic, they have shown it to be useful. Some creationists have argued that the human appendix is part of the lymphatic system. While the appendix does have a great deal of lymphatic tissue, this does not make it an important part of the lymphatic system. The lymphatic tissue of the appendix is important because the appendix traps bacteria and is prone to infection; the lymphatic tissue is therefore important because the appendix is vestigial, not in spite of it. Creationists also argue that nipples on male mammals are, in fact, useful, because under some circumstances males have been known to lactate (for example, from a surge of estrogens resulting from food intake following near starvation). Yes, this is a function, but can it really be considered useful? Some creationists have also made the claim that nipples in male mammals are useful because they contribute to erotic stimulation. While this is true, erotic stimulation is not the major function of a nipple, nor is such a complex structure necessary for erotic stimulation. Staminodes in flowers can be colorful, like petals, and help to attract pollinators, but this cannot be a primary or necessary function. Of course, with enough special pleading, a scenario of usefulness can be constructed for almost anything.
- They assume that, if they make a credible argument for one component of a category, they have proved the usefulness of the entire category. Perhaps the major example of this is their argument that the usefulness of some of the noncoding DNA demonstrates that none of the noncoding DNA is vestigial.
- In some cases, the potential usefulness of the vestigial characteristic is for evolution itself. So-called junk DNA may be very useful as a potential future source of genetic vari-

ability. In fact, some transposons are activated at times of environmental stress, and they can sometimes activate pseudogenes when they come to rest in a new location. Thus, during times of environmental challenge, a population of organisms can generate greater genetic diversity, which may allow more rapid evolution. But how can a creationist say that all noncoding DNA is useful, because it might be needed for future evolution, if they do not even accept evolution?

Vestigial characteristics, even when they have some utility, therefore represent evidence either of the remnants of previous evolutionary history or of evolution still in the process of occurring.

Further Reading

Espinasa, Luis, and Monika Espinasa. "Why do cave fish lose their eyes? A Darwinian mystery unfolds in the dark." *Natural History,* June 2005, 44–49.

Gould, Stephen Jay. "Hen's teeth and horse's toes." Chap. 14 in *Hen's Teeth and Horse's Toes.* New York: Norton, 1983.

Rice, Stanley A. "South with the spring: A study of evolution and tree buds." *National Center for Science Education Reports* 23 (2003): 27–32.

Wallace, Alfred Russel (1823–1913) British *Evolutionary biologist* Alfred Russel Wallace is famous principally as the codiscoverer, along with Charles Darwin (see DARWIN, CHARLES) of NATURAL SELECTION as the principal mechanism of evolution. However, he was famous also for his travel writings, his contributions to the study of what is now called BIODIVERSITY, and his social activism. He was perhaps the most famous British naturalist of the late 19th century.

Born January 8, 1823, Wallace came from an intellectually active and sometimes financially challenged family. He worked with his brother as a surveyor and then taught surveying skills at a small British college. He found increasingly more time for the study of local natural history (insects, plants, geology), especially after he began a friendship with another young naturalist (see BATES, HENRY WALTER).

Wallace and Bates decided to travel to South America to collect plant and animal specimens. Since neither was rich, they planned to sell the specimens to pay for the expedition. They went separate ways in South America. Both amassed large collections. Bates remained many years, studying the butterflies that were the first and still most famous example of MIMICRY, a type still called Batesian mimicry. Wallace spent four years in South America and traveled further in some of the river systems of South America than any European had ever gone. His studies in the Amazon basin formed the basis of his developing thoughts about BIOGEOGRAPHY and about the process of evolution. He lived among, and closely studied, the ways and languages of the native peoples. In so doing, he developed an appreciation that tribal peoples have an intellectual capacity equal to that of civilized peoples, a belief that few European intellectuals held at that time. As he returned to England with his collections, the ship caught fire, and he lost everything except some of his drawings and notes. Upon returning home, he published and spoke about some of his observations to scientific societies and the general public.

Wallace's accomplishments as a naturalist were not yet famous but were sufficient for him to obtain support from the Royal Geographic Society for a journey to Indonesia and New Guinea (at that time called the Malay Archipelago). He spent nearly eight years there, from 1854 to 1862. He traveled 14,000 miles and collected 125,660 biological specimens (including more than a thousand new to science). The book that he wrote upon his return, *The Malay Archipelago,* ranks as one of the most popular pieces of travel writing from the 19th century, partly because he published some of the earliest information about birds of paradise, orangutans, and the native peoples of New Guinea.

Wallace's work in Indonesia would have been sufficient to guarantee him lifelong fame as a naturalist; he certainly had a lifetime of work awaiting him, upon his return, just to analyze and write about his specimens and travels. But it was an event that occurred at the midpoint of his journey, in 1858, that changed the course of scientific history. He had contracted malaria, and was lying ill with it, when his fevered mind came up with an explanation for how the process of evolution could work, a process that we today call natural selection. When he was well enough, he wrote his ideas in an article entitled "On the Tendency of Varieties to Depart Indefinitely from the Original Type." It was the word "indefinitely" (meaning without limit) that made Wallace's proposal distinct from orthodox scientific views. Wallace was ready to send the article back to England for publication. Wallace knew, but did not much care, that his views might be controversial. He decided, however, to let another, more experienced, scientist review his article before publication. He sent his article to Charles Darwin.

Wallace could not have known what an effect his article would have. Like Wallace, Darwin had traveled extensively in tropical regions of the world and had spent much time (while ill) thinking about evolution. Like Wallace, Darwin had come up with natural selection as a way of explaining how evolution could occur. However, Darwin had done all of this almost exactly 20 years earlier. Darwin had not published any of his hypotheses on evolution, however, because he knew they would be controversial, and he wanted to amass all available evidence into a huge book that would answer every possible objection and settle the question. Therefore Darwin delayed 20 years, while Wallace was ready to publish right away.

Darwin was astounded that, after this long delay, another scientist had come up with the same ideas, even using the same term to describe the process. Darwin said, "I never saw a more striking coincidence. If Wallace had my manuscript sketch written out in 1842, he could not have made a better short abstract." In consultation with the famous geologist Sir Charles Lyell (see LYELL, CHARLES), Darwin decided that his 1842 summary of natural selection should be presented and published jointly with Wallace's article. Darwin was ill, so Lyell presented the joint Darwin-Wallace paper to the Linnaean Society in 1858. Strangely, the scientists in attendance took little notice of the paper, nor did the president of the Linnaean Society who summarized the events of 1858 as unremarkable. Wallace, still in Indonesia, knew nothing of this.

While in Indonesia, Wallace noticed that the animals on the island of Lombok differed strikingly from the animals of Bali, and those of Borneo differed from those of Sulawesi. The mammals of Lombok and Borneo included many marsupials (see MAMMALS, EVOLUTION OF) and few large predators, while the animals of Bali and Sulawesi were mostly placental mammals and included large predators such as tigers. He also noticed a difference in the birds. In some cases, the islands with different animals were almost within sight of one another. The line he drew on the map to separate the two faunas is now known as Wallace's Line. Wallace interpreted this, correctly, as resulting from the separate evolution of the two groups of animals. He could not have known that a deep ocean trench separated the two groups, which were on different continental plates (see CONTINENTAL DRIFT) and which had therefore had little contact throughout geological history. Wallace's line is considered one of the earliest major contributions to the science of biogeography.

After returning to England in 1862, Wallace spent three years organizing his specimens, presenting papers to scientific meetings, and writing. He earned money from his famously interesting public presentations; one speaking tour took him as far as California. His striking appearance (he was tall, had a long beard, and snow-white hair for the last 40 years of his life) contributed to his public success. He married a woman almost 20 years younger than he, but she shared his passion for gardening and natural history. Later, they journeyed overseas on botanizing expeditions. Although his collections and writings proved profitable, his investments did not. He had to take on small editing and speaking assignments for pay and move his family progressively further from London to find cheaper accommodations. Eventually, Darwin intervened to convince the British government to give Wallace an annual stipend in recognition of his services to science.

Wallace continued his studies of evolution and advanced into other areas of science. His paper, "The Origin of Human Races and the Antiquity of Man Deduced from the Theory of Natural Selection," presented in 1864, preceded Darwin's book *The Descent of Man* (see DESCENT OF MAN [book]) by seven years. Wallace published a paper in 1863 that gave an evolutionary explanation for the hexagonal construction of the cells in beehives, a topic Darwin incorporated into later editions of the *Origin of Species*. Wallace also published a theory of glaciation and evolutionary explanations of mimicry, protective coloration, and color vision in animals, bird migration, and the necessity of aging. He was one of the first to present evidence that the PLEISTOCENE EXTINCTION may have been caused by overhunting by humans. His writings were among the earliest that contrasted different kinds of life history in animals (see LIFE HISTORY, EVOLUTION OF). His ideas about the evolution of reproductive isolation (see ISOLATING MECHANISMS) have even been called the "Wallace Effect." He wrote in 1907 that the ice caps of Mars were carbon dioxide (dry ice), not water.

Wallace also expanded his studies into many different areas throughout the remainder of his life. He began to give serious time and attention to social issues, in which he defended the causes of the working man and the oppressed. His views may be mainly attributed to his experiences with Amazonian, Indonesian, and Papuan peoples, and to his exposure to socialism when he worked as a surveyor. Among the social issues he championed are:

- *Land nationalization.* Wallace called for the protection of rural lands and historical monuments, the formation of greenbelts and parks, for the preservation of nature but also for the recreation of the lower and middle classes.
- *Labor reform.* Rather than calling for strikes, Wallace in 1899 called for workingmen to lay aside a portion of their salaries into a fund that would allow them to buy stock in the company and thereby partly or wholly control the company. He also called for double overtime pay rates. Later in life he fully converted to socialism.
- *Opposition to vaccination.* This issue puzzles many modern scientists: Why would one of the most prominent scientists in Europe oppose vaccination? Wallace had both scientific and social reasons for it. Scientifically, he could demonstrate with epidemiological data that the incidence of contagious diseases began to decline before the introduction of vaccination. Today scientists know that he was correct: Advances in public hygiene have proved at least as important as vaccination in improving public health. Socially, Wallace saw vaccination as the government's way of trying to improve the lot of the poor without having to spend the money necessary for public sanitation. It was much cheaper and easier to jab poor people in the arm than to build sewers and provide safe drinking water.
- *Women's suffrage.* He was one of the earliest proponents of voting rights for women in male intellectual society.
- *Consumer safety.* Wallace proposed in 1885 that all manufactured items have labels that identify their component materials, and that the government regulate standards for these materials.
- *Eugenics.* Wallace criticized the attempt, then popular among scientists, including Darwin's cousin (see GALTON, FRANCIS), to attribute social problems to the supposed

genetic inferiority of the lower classes and nonwhite races (see EUGENICS). Wallace saw this as an abuse of science and an attempt to justify oppression.

In the late 1860s, Wallace became an adherent of spiritualism, with which he remained for the rest of his life. As a scientist he sought experimental confirmation but was satisfied with demonstrations that more skeptical scientists would not accept. He wrote more than 100 publications on spiritualism. His lecture on life after death, delivered mainly during his California speaking tour, was one of his most popular. His outspoken association with spiritualism, however, did not prevent him from receiving a continuous stream of scientific awards and honors.

Related to his belief in spiritualism was his acceptance of what is now called the ANTHROPIC PRINCIPLE, in which he believed that the universe was designed for man—a perfect universe in which humans were intended to evolve. The rebuttal written by Samuel Clemens (Mark Twain), "Was the World Made for Man?" has become much more famous than Wallace's original writings on the subject.

The publication of Wallace's first article about natural selection jointly with Charles Darwin in 1858 created the impression that Darwin and Wallace believed the same things about evolution. Wallace's publications about evolution were to prove that this was not the case at least with human evolution. Wallace distinguished between the material process of natural selection that produced the human body, and a spiritual one that produced the human mind, a distinction Darwin never made. Today, nearly all evolutionary scientists explain the origin of human behavior patterns in terms of evolutionary processes acting upon the brain (see ALTRUISM; SOCIOBIOLOGY). Many scholars assume that Wallace's rejection of natural selection as an explanation for the origin of the human mind resulted from his irrational attraction to spiritualism. However, evolutionary biologist Stephen Jay Gould and others have pointed out a different possibility. Both Darwin and Wallace were aware that tribal peoples had intelligence equal to that of civilized peoples—Darwin knew this from his experience with the Fuegians, and Wallace from the Amazonians and Indonesians. But Wallace pointed out, further, that the tribal peoples did not need or use the high levels of intelligence of which they proved themselves capable when given European education. Wallace reasoned that if natural selection were the sole explanation of the origin of the human mind, then scientists would see tribal peoples with low intelligence. Why did these people have so much more intelligence than they used? (This argument is now rejected by anthropologists, who recognize that tribal peoples have very complex languages, customs, and oral traditions.) Wallace, in other words, was taking natural selection more seriously, not less seriously, than Darwin. Of course, Wallace's acceptance of spiritualism made his rejection of the evolution of mind much easier.

Wallace remained intellectually active well into his old age. Between 1898 and 1910, mostly in his ninth decade, he published more than 4,000 pages of material. Wallace died November 7, 1913.

Further Reading

Brackman, Arnold C. *A Delicate Arrangement: The Strange Case of Charles Darwin and Alfred Russel Wallace.* New York: Times Books, 1980.

Camerini, Jane, ed. *The Alfred Russel Wallace Reader: A Selection of Writings from the Field.* Baltimore, Md.: Johns Hopkins University Press, 2001.

Fichman, Martin. *An Elusive Victorian: The Evolution of Alfred Russel Wallace.* Chicago: University of Chicago Press, 2004.

Raby, Peter. *Alfred Russel Wallace: A Life.* Princeton, N.J.: Princeton University Press, 2002.

Smith, Charles H. "The Alfred Russel Wallace Page." Available online. URL: http://www.wku.edu/~smithch/index1.htm. Accessed May 13, 2005.

Wallace's line *See* BIOGEOGRAPHY; WALLACE, ALFRED RUSSEL.

whales, evolution of Whales and other *cetaceans* (such as dolphins) evolved from land-dwelling mammalian ancestors (see MAMMALS, EVOLUTION OF). Cetaceans are spectacularly well adapted to life in the oceans, so well that they were considered fishes until the last couple of centuries. Among the adaptations that were necessary in the evolution of whales are:

- *Front flippers and a tail fluke.* Pinnipeds such as seals and walruses crawl onto land for mating and childbirth, but cetaceans spend their entire lives in water. An animal that spends all of its time in the water does not need to walk, and hind legs are only a hindrance. Flippers provide a large surface area with which the front limbs can propel the cetacean through water, and the horizontal fluke on the tail also allows propulsion. Cetaceans move their tails up and down in typical mammalian fashion, unlike the side-to-side motion of the tails of fishes.
- *Nostrils on the top of the head.* Cetaceans, being mammals, need to breathe air. It is much more convenient to emerge from the ocean for breathing if the nostrils are on the top of the head.
- *Change in eating.* While many cetaceans (such as killer whales and dolphins) have teeth and pursue prey, some of the largest whales live off of plankton, which they strain from the water with *baleen* or whalebone, which they have instead of teeth.
- *Ability to hold its breath.* When diving deeply for long periods of time, whales must take as much oxygen with them as possible. Whenever they emerge, they can exchange up to 90 percent of the air in their lungs with the atmosphere, compared to the typical human breath that exchanges only 50 percent. Their muscles contain a great deal of myoglobin, a protein similar to hemoglobin that releases oxygen into muscle tissue.
- *Changes in ear structure.* The structure of the ear that is best for hearing in air is quite different from that which is best for hearing underwater.
- *Changes in birth.* Most mammals are born headfirst, to allow the newborn to begin breathing as soon as possible.

This would, however, cause a baby whale to drown. Baby whales are born tailfirst.

All of these evolutionary transitions had to occur quickly, as whales that resemble modern ones were in existence by the Eocene epoch, the second epoch of the TERTIARY PERIOD, following the CRETACEOUS EXTINCTION. The extinction of the large aquatic reptiles sufficiently reduced competition for this style of life, which allowed whales to evolve rapidly. Another possible reason for the rapid evolution of whales is that, once they stopped coming onto land during any stage of their life cycles, they were freed from the structural constraints imposed by gravity. Nearly a complete set of intermediate links has been found:

- The earliest whale ancestor that has been discovered is *Pakicetus*. Only the skull has been preserved, so scientists know little about the rest of the skeleton. How valuable it would be to know whether this species primarily swam, or primarily walked! The ear structure was not suitable for underwater hearing, but the eyes were near the top of the head. The bones were found in association with land snails, therefore this species must have foraged in the water but primarily stayed near land.
- Perhaps the best of the MISSING LINKS (no longer missing) between whales and their terrestrial ancestors is *Ambulocetus natans*, of which most of the skeleton has been found. *Ambulocetus* means "walking whale," and that is a good description of it. It had front and hind legs, and a tail without a fluke. It was found associated with shallow marine organisms, so it probably spent more of its time out in the water than did *Pakicetus*.
- *Rodhocetus* had less of an ability to walk on land than did *Ambulocetus* but still retained some skeletal characteristics of land mammals. In particular, the vertebral column appears to have a mixture of terrestrial and aquatic traits.
- *Basilosaurus* and *Dorudon* clearly had skeletal features that were intermediate between *Ambulocetus* and modern whales. With very small pelvis and hind limbs, they were unable to walk. "Saurus" means lizard; this species was misnamed by paleontologist R. Harlan in 1834, before being correctly identified as a primitive whale by Sir Richard Owen in 1842 (see OWEN, RICHARD).

This series of organisms represents intermediates between terrestrial whale ancestors and modern whales. The hind femur (upper limb bone) of *Ambulocetus* was similar to that of related terrestrial mammals; the hind femur of *Basilosaurus* was very small; the hind femur of *Rodhocetus* was intermediate between the two. In terms of habitat, *Pakicetus* lived in shallow waters near the mouths of rivers; *Ambulocetus* inhabited shallow marine waters; *Rodhocetus* and *Basilosaurus* were fully marine.

Not only is there an essentially complete series of species illustrating the stages of whale evolution, but modern whales frequently produce VESTIGIAL CHARACTERISTICS that are left over from earlier evolutionary stages. This apparently happens because the genes for the structures still exist in modern whales but are normally not expressed, unless very early in embryonic development. Some modern whales have vestigial hind limb structures; others produce hind limb buds during fetal development. Toothless baleen whales have tooth buds as embryos, but the buds are resorbed during development.

Further Reading

Gingerich, Philip D., et al. "Hind limbs of Eocene *Basilosaurus:* Evidence of feet in whales." *Science* 249 (1990): 154–157.

———, et al. "New whale from the Eocene of Pakistan and the origin of cetacean swimming." *Nature* 368 (1994): 844–847.

Gould, Stephen Jay. "Hooking Leviathan by its past." Chap. 28 in *Dinosaur in a Haystack: Reflections in Natural History.* New York: Harmony, 1995.

Rose, Kenneth D. "The ancestry of whales." *Science* 293 (2001): 2,216–2,217.

Sutera, Raymond. "The origin of whales and the power of independent evidence." *National Center for Science Education Reports* 20 (2001): 33–41.

Thewissen, J. G. M., and M. Aria. "Fossil evidence for the origin of aquatic locomotion in archaeocete whales." *Science* 263 (1994): 210–212.

———, and E. M. Williams. "The early radiations of Cetacea (Mammalia): Evolutionary pattern and developmental correlations." *Annual Review of Ecology and Systematics* 33 (2002): 73–90.

Zimmer, Carl. *At the Water's Edge.* New York: Touchstone, 1998.

Wilson, Edward O. (1929–) American *Evolutionary biologist* Edward Osborne Wilson (see photo on page 417) has been at the leading edge of breakthroughs in evolutionary science, entomology, biogeography, and the study of the diversity of species on this planet. He even helped to invent some of the terms that are now central to these subjects: SOCIOBIOLOGY, island BIOGEOGRAPHY, BIODIVERSITY, and BIOPHILIA, which is an innate, genetically based love of nature that, Wilson maintains, is universally present in our species. He has also been an immensely popular writer on these subjects: *On Human Nature* and *The Ants* (with entomologist Bert Hölldobler) have won Pulitzer Prizes, and *The Diversity of Life* and *The Future of Life* have been immensely popular books about biodiversity.

Born June 10, 1929, in Birmingham, Alabama, Wilson grew up primarily in Alabama and Florida. Very early he became fascinated with the natural world, and as a Boy Scout he spent a lot of time exploring the marshes and swamps, collecting insects. He became an amateur expert on things that most people did not see even if they looked straight at them. It was Wilson who, at about 12 years of age, was the first to report the presence of the red imported fire ant, which is now a major problem, in the United States (see INVASIVE SPECIES). His youthful passion for understanding the diversity of the natural world was the force that impelled his entire subsequent career.

He earned bachelor's and master's degrees at the University of Alabama and worked as an entomologist for the Alabama Department of Conservation, primarily studying the fire ants whose invasion he had discovered. It was at the university that he learned about evolutionary science, and from that point onward he questioned the religious foundations he had embraced as a youth. He completed his Ph.D. at Harvard University in 1955. He was invited to join the Harvard faculty in 1956, and he is still there, as Pellegrino University Professor Emeritus. He works as Honorary Curator in Entomology at the Museum of Comparative Zoology.

Receipt of a Harvard fellowship allowed him to travel soon after he began his faculty appointment in 1956. He studied ants in New Guinea and other Pacific Islands. He was the first outsider to climb Mt. Boana, the crest of the central Sarawaget Mountains of New Guinea. He also traveled in the American tropics. He collected lots of ants, always with the goal of understanding the evolutionary history of the entire ant family Formicidae.

Wilson's observations also made him think about how species diversity developed on islands. Curious as always, when he returned to Harvard he sat in on mathematics classes, recognizing that he needed to know more math if he was to develop general theories of population biology. He shared his understanding with the world in *A Primer of Population Biology* (with geneticist William H. Bossert), which is still one of the fundamental texts on the subject. With ecologist Robert H. MacArthur of Princeton University, Wilson developed the theory of island biogeography, which tied species immigration and extinction to the area of islands and their distance from the mainland.

Edward O. Wilson is one of the leading evolutionary scientists of modern times. ***(Courtesy of Science Photo Library)***

But is it possible to study island biogeography experimentally? In the late 1960s Wilson, with ecologist Daniel Simberloff, removed all of the animals from mangrove islands of various sizes and distances from the mainland in the Florida Keys—a difficult task, especially when a hurricane is coming—then documented the colonization of these islands by animals. His work transformed the study of island species from stories to a science and made island biogeography theory useful to conservation efforts.

Wilson's studies of ants created breakthroughs of understanding in how insects communicate. He and collaborators worked out the principles of how insects communicate by chemicals. Subsequently, many researchers have found examples of chemical communication among vertebrates and even among plants.

Wherever he traveled, at home or abroad, he could not help but notice that human activity was destroying natural habitats and, in the process, replacing high-diversity ecological communities with low-diversity artificial habitats. He could see, before most scientists had thought about it, that species were being driven into extinction faster than anyone could even recognize them, let alone study them. He knew that there was yet a lot to be learned from wild species of ants. For example, ants have high population densities yet they hardly ever experience epidemic diseases. This is because their metapleural glands and some symbiotic bacteria produce chemicals (which some people call antibiotics) that kill bacteria. Humanity needs new antibiotics (see RESISTANCE, EVOLUTION OF), yet humans are driving many ant species into extinction. Wilson's 1984 *Biophilia* argued that humans have a psychological need for contact with the natural world. In *The Diversity of Life* and *The Future of Life,* Wilson explained the causes and consequences of the sixth of the MASS EXTINCTIONS, which humans are bringing upon the world, and what humans might be able to do to slow or stop it. Not content with writing, Wilson has been active with the American Museum of Natural History, Conservation International, The Nature Conservancy, and the World Wildlife Fund. The destruction of biodiversity, he writes, "is the folly our descendants are least likely to forgive us."

Wilson was largely responsible for synthesizing the modern understanding of insect behavior in his 1971 book *The Insect Societies.* He applied these concepts across the entire animal kingdom in his 1975 book *Sociobiology: The New Synthesis.* There was only a little bit, the last chapter, about humans, but it sparked a vigorous controversy about the role of evolutionary biology in human behavior and created a field of study now sometimes called evolutionary psychology. Wilson did not hesitate to make his message clear to the general public in his 1978 book *On Human Nature.* Strong disagreement came from other scientists, including some Harvard colleagues (see GOULD, STEPHEN JAY; LEWONTIN, RICHARD). Outside the academic world many activists believed that sociobiology provided an intellectual justification for oppression. At a 1978 meeting of the American Association for the Advancement of Science in Washington D.C., Wilson was one of the platform scientists at a discussion of human evolution. Protesters chanted, "Racist Wilson you can't hide, we charge you with genocide!" A protester jumped up on stage with a pitcher of water and poured it all over Wilson, saying, "Wilson, you are all wet!" Evolutionary anthropologist Napoleon Chagnon helped push the protester off the stage, and Stephen Jay Gould took the microphone to denounce the protesters' tactics. Meanwhile Wilson just wiped off the water and continued with the meeting. No one who has met Wilson can believe that he in any way condones the misuse of sociobiology as a rationalization for political oppression.

If anyone could handle the challenge of trying to bring together all fields of knowledge into a single unified structure, it would be Edward Wilson. His 1998 *Consilience* called for a return to original Enlightenment ideals for bridging the sciences and the humanities. College education is a smorgasbord of largely disconnected subjects that leaves undergraduates more annoyed than enlightened; the typical divisions of a newspaper (news, business, sports, leisure) are even more disconnected. Rather than to have science as just one area of study, equal perhaps to music, or as an occasional page in a newspaper, Wilson wants science, particularly evolutionary science, to be understood as the foundation of all human history and activity. Naturally this approach has drawn attacks from scholars in the humanities who do not want their field to be seen as a subset of science. Wilson writes, "My truths, three in number, are the following: first, humanity is ultimately the product of biological evolution; second, the diversity of life is the cradle and greatest natural heritage of the human species; and third, philosophy and religion make little sense without taking into account these first two conceptions." Although Wilson admits his approach to consilience may be wrong, it is widely recognized as one of the most concise and organized attempts ever made.

Wilson has received many of the highest awards that are available. There is no Nobel Prize for evolutionary biology, but Wilson has received the Crafoord Prize, the award given by the Royal Swedish Academy of Sciences for areas not covered by the Nobel Prize, as well as the National Medal of Science in the United States. In addition to his two Pulitzer prizes, Wilson has received prestigious prizes in Japan, France, Italy, and Saudi Arabia and from world conservation organizations. As much as he cherishes these prizes, he is still very grateful for the teaching awards he has received from the students of Harvard University.

Even though most people his age have decided to take it easy, Wilson still has major projects. His primary goal is to facilitate the formation of an "Encyclopedia of Life," which will use modern technology to speed up the process of documenting the Earth's rapidly disappearing biodiversity. His motto might be reflected in this quote, "Love the organisms for themselves first, then strain for general explanations, and, with good fortune, discoveries will follow. If they don't, the love and the pleasure will have been enough." Even though he is one of the most respected scientists of modern times, he will not hesitate to crouch down on the ground to look at an ant. He is still, at heart, a Boy Scout working on an insect merit badge.

Further Reading

Cowley, Geoffrey. "Wilson's world." *Newsweek,* 22 June 1998, 58–62.

MacArthur, Robert H., and Edward O. Wilson. *Theory of Island Biogeography.* Princeton, N.J.: Princeton University Press, 1967.

Wilson, Edward O. *Biophilia.* Cambridge, Mass.: Harvard University Press, 1984.

———. *Consilience: The Unity of Knowledge.* New York: Vintage Press, 1998.

———. *The Diversity of Life.* Cambridge, Mass.: Harvard University Press, 1992.

———. "The encyclopedia of life." *Trends in Ecology and Evolution* 18 (February 2003): 77–80.

———. *The Future of Life.* New York: Vintage Press, 2002.

———. *Nature Revealed: Selected Writings, 1949–2006.* Baltimore, Md.: Johns Hopkins University Press, 2006.

———. *Naturalist.* Washington, D.C.: Island Press, 1994.

———. *On Human Nature.* Cambridge, Mass.: Harvard University Press, 1978.

———. *Sociobiology: The New Synthesis.* Cambridge, Mass.: Harvard University Press, 1975.

———, and William H. Bossert. *A Primer of Population Biology.* Sunderland, Mass.: Sinauer Associates, 1971.

———, and Bert Hölldobler. *The Ants.* Cambridge, Mass.: Harvard University Press, 1990.

Woese, Carl R. (1928–) American *Microbiologist, Evolutionary scientist* Carl R. Woese (see photo on page 419) is one of the few scientists whose research has transformed the way scientists think about life and evolution. His work laid the foundation for the molecular study of evolution and for the TREE OF LIFE. Born July 15, 1928, Woese studied physics as an undergraduate at Amherst College, and biophysics for his doctorate at Yale. He joined the microbiology faculty at the University of Illinois at Urbana-Champaign in 1964.

Woese's first major contribution was to demonstrate the importance of using DNA to test hypotheses of evolutionary relatedness, rather than using visible anatomical structures (see DNA [EVIDENCE FOR EVOLUTION]). This is especially important for comparisons among prokaryotes such as BACTERIA, which have relatively few visible differences, and for comparing prokaryotes with more complex organisms such as humans. Evolutionary change leaves a record in the DNA, even when almost all visible differences between two species (for example, between a bacterium and a human) have vanished in evolutionary time. Woese was not the only scientist to think of using DNA to reconstruct evolutionary history, but very few other scientists attempted it, because in the 1970s the techniques were very laborious and slow. Techniques that had to be done by hand and required weeks of work are now automated and can be done overnight. It

Carl R. Woese pioneered the use of nucleotide sequence comparisons to reconstruct evolutionary history of all organisms. ***(Courtesy of Bill Wiegand/University of Illinois News Bureau)***

was necessary to amplify the DNA of a particular gene, then to separate it into fragments on a thin gel, a process called electrophoresis. Radioactively labeled segments exposed a sheet of photographic film, producing a barcode pattern that allows the researcher to determine the sequence of nucleotides in the DNA (see BIOINFORMATICS). Woese's insight was to choose a gene that all organisms possess: the gene for a type of ribosomal RNA. Not only is this gene universal, but it changes very slowly over evolutionary time. Woese worked 8–12 hours at a stretch, looking at the photographic sheets in a darkroom, nearly every day for 10 years. The first gene sequence took him nearly a year. Similar work can now be done in a few days. By 1976 Woese had sequenced the ribosomal RNA gene from 60 kinds of bacteria. Woese was also one of the first scientists to analyze relationships among species by producing a phylogenetic tree, a process used today in the study of CLADISTICS.

Then in 1976 a colleague suggested that Woese should analyze a methanogenic bacterium. Methanogens ("methane producers") live in anaerobic swamp conditions. When Woese completed the nucleotide sequence of this methanogen, he found that it resembled none of the other prokaryotes. Its ribosomal RNA was as different from those of the other prokaryotes as from more complex organisms such as plants and humans. Woese repeated his analysis to make sure of his results. As he analyzed other prokaryotes that live in conditions of extreme heat, acidity, and salinity, he found that many of them clustered into a group separate from the bacteria. He called this cluster ARCHAEBACTERIA (now called Archaea). Based on ribosomal RNA sequences, Woese constructed the first version of the now famous tree of life, which classified all life-forms into three main branches: the archaebacteria, the bacteria, and eukaryotes (see EUKARYOTES, EVOLUTION OF). This tree revealed that almost all of the genetic diversity was found among microbes; the plants, animals, and fungi formed three tiny twigs of the tree of life. Two archaebacteria that look similar to human observers might be as different from one another, on the genetic level, as a mouse and a spider. While this was at first surprising, in retrospect it made sense. Almost 80 percent of the evolutionary history of life occurred during PRECAMBRIAN TIME, when most life-forms were microbial.

For the decade after Woese's 1977 discovery of Archaea, most microbiologists paid little attention to his work. One scientist even cautioned the colleague who had given Woese the methanogen to distance himself from Woese's work. Major new concepts in science often meet with considerable resistance before being accepted (see SCIENTIFIC METHOD). Woese shared this experience with other evolutionary scientists who proposed bold new concepts: PUNCTUATED EQUILIBRIA (see GOULD, STEPHEN JAY; ELDREDGE, NILES), SOCIOBIOLOGY (see WILSON, EDWARD O.), and SYMBIOGENESIS (see MARGULIS, LYNN). His work gradually moved from acceptance to become the new standard view. By 2000 Woese had won the top award in microbiology, and the National Medal of Science, an award also won by Wilson and by Margulis. In 2003 Woese won the Crafoord Prize from the Royal Swedish Academy of Sciences, the award given to scientists in areas of research that are not covered by the Nobel Prize.

Woese has also contributed creative insights into the ORIGIN OF LIFE and the origin of the genetic code (see DNA [RAW MATERIAL OF EVOLUTION]). Woese continues to do research at the University of Illinois, focusing on the effects that HORIZONTAL GENE TRANSFER may have had on the evolution of the earliest life-forms.

Further Reading

Woese, Carl R. "Interpreting the universal phylogenetic tree." *Proceedings of the National Academy of Sciences USA* 97 (2000): 8,392–8,396.

———, and G. E. Fox. "Phylogenetic structure of the prokaryotic domain: The primary kingdoms." *Proceedings of the National Academy of Sciences USA* 74 (1977): 5,088–5,090.

Wright, Sewall (1889–1988) American *Evolutionary geneticist* Sewall Wright was an American geneticist who contributed greatly to the understanding of population genetics and evolution, principally by explaining the process of genetic drift (see FOUNDER EFFECT). He also has had a major and continuing impact on many fields of scientific research, even more in the social than in the natural sciences, by developing the statistical method known as path analysis, a breakthrough method that allowed correlations to be studied in complex systems. He also made significant contributions to philosophy.

Born December 21, 1889, Wright grew up in Illinois in an intellectual family. Very early in life, he showed great mathematical ability and a passion for the natural world. His father's printing business published his pamphlet "The Wonders of Nature" when Wright was seven years old. He studied mathematics at Lombard College, where his father taught. In graduate school at the University of Illinois and at Harvard, Wright studied biology and genetics. He worked for the USDA, studying the genetics of animal breeding. This is where he first began to study the effects of inbreeding and the changes that can occur in small populations.

He is best remembered for developing the theoretical framework for understanding genetic drift, in which random events in small populations can alter the direction of breeding and of evolution. A related area of study was inbreeding, in which recessive mutations can show up more frequently in small than in large populations, and this is therefore a major problem in endangered species (see EXTINCTION; MENDELIAN GENETICS; POPULATION GENETICS). He is considered one of the three scientists who laid the theoretical foundation for the MODERN SYNTHESIS (see also FISHER, R. A.; HALDANE, J. B. S.). His views contrasted sharply with those of the other two, who emphasized the role of NATURAL SELECTION in maintaining genetic variation in natural populations. Wright's emphasis upon evolutionary events in small populations, however, allied him more closely with the Japanese evolutionary geneticist Motō Kimura, who claimed that most genetic variation in populations was neutral in its effects on the organisms.

Wright continued to make major contributions to science well after his mandatory retirement from the University of Chicago at age 65. His major work, *Evolution and the Genetics of Populations,* was four volumes in length, the final volume of which was released when he was 98 years old. Wright won many scientific awards. His first paper was published in 1912, his last in 1988, a span of 76 years. He died March 3, 1988.

Further Reading

Provine, William B. *Sewall Wright and Evolutionary Biology.* Chicago: University of Chicago Press, 1986.

APPENDIX

DARWIN'S "ONE LONG ARGUMENT": A SUMMARY OF *ORIGIN OF SPECIES*

Evolutionary sociologist Ashley Montagu said, "Next to the Bible, no work has been quite as influential, in virtually every aspect of human thought, as *The Origin of Species.*" When it was published in 1859, it sold out on the first day (see *ORIGIN OF SPECIES* [book]). In it, Darwin presented convincing evidence that evolution had occurred, and an explanation of how evolution occurs. Countless scientists and science educators have marveled at the powerful simplicity of Darwin's arguments for the fact and the process of evolution. Many have had the same experience that T. H. Huxley (see HUXLEY, THOMAS HENRY) had upon learning of this theory: How stupid to not have thought of it before!

Many people have learned about *Origin of Species* not from reading it but from reading what others have written about it. Many of evolutionary biologist Stephen Jay Gould's writings have elucidated the development of specific portions of Darwin's arguments and thoughts (see GOULD, STEPHEN JAY). Many histories of scientific thought (most notably evolutionary biologist Ernst Mayr's *The Growth of Biological Thought*) summarize it (see MAYR, ERNST). Ernst Mayr specifically explained Darwin's reasoning in his book *One Long Argument.* More recently, geneticist Steve Jones "updated" *Origin of Species* in his book *Darwin's Ghost*—he used the same chapter divisions and borrowed a little of Darwin's beautiful prose, but it consists almost wholly of new material.

Origin of Species is not only the foundation of modern evolutionary biology. It is also the foundation of ecology: Chapter 3 is one of the earliest examples of ecological science. Darwinian evolution is, in fact, now considered the foundation of all branches of biology; as geneticist Theodosius Dobzhansky said, "Nothing in biology makes sense except in the light of evolution" (see DOBZHANSKY, THEODOSIUS).

Because of its length and complexity, most scientists and science educators have not read *Origin of Species* in its entirety. This is exactly parallel to the phenomenon that the Bible is largely unread by Christian congregations, many of which would be surprised at its contents. Even though Darwin wrote this book in the passion of rushing to press, because of Alfred Russel Wallace's independent discovery of natural selection (see WALLACE, ALFRED RUSSEL), and made it the "briefest abstract" of his ideas, it is still more than 400 pages in length. Its prose is delightful but thick, and its complex sentences are more easily accessible to Victorian than to modern American readers. Darwin assumed that his readers had immense facility with biological concepts; for example, he just expected them to know what cirripedes are (they are barnacles).

The Bible is accompanied by an entire publishing industry of commentaries, devotionals, and various translations that attempt to make the original text accessible to everyone from young people to truck drivers. *Origin of Species,* however, exists only in referenced chunks and in complete form. The appendix of this encyclopedia consists of a summary of this book, in which the original force of the "one long argument" can be apparent to the modern reader.

One may be astonished at how much Darwin knew. Many creationist arguments still in tiresome use today were already answered in Darwin's book. Many pieces of evidence that many students of biology assume to be the product of 20th-century science are already documented in *Origin of Species.* This summary uses the sixth edition, the one in most common circulation; in it, Darwin answered many of the objections raised by his critics to the contents of the previous editions. The word *evolution* is freely used in this summary, even though Darwin did not use the word *evolved* until the very end of the book. At the time, the word EVOLUTION still carried some connotations of the unfolding of a predetermined history, an impression Darwin labored to avoid. This summary avoids the use of the words *gene* and *genetic,* concepts of which Darwin was unaware (see MENDELIAN GENETICS).

The *Origin of Species* is not the Bible of modern biology. There is none. Modern biologists are perfectly free to disagree with those (admittedly few) instances in which Darwin was in error.

The following summary is not meant as a replacement for the original, any more than a tract replaces the Bible. The author hopes that this summary will inspire many to read *Origin of Species* in its entirety.

Charles Darwin may be the most fascinating person you never met. When you read *Origin of Species,* you will also get to know Charles Darwin as a person—passionate for truth, humble in response to his critics, continually in awe of the natural world.

The author has used the following format to summarize *Origin of Species:*

- The author uses the same chapters and chapter titles as *Origin of Species.*
- The author has condensed Darwin's text but maintained much of the style and terminology. This summary largely restricts itself to the concepts and knowledge that Darwin had. Occasionally it has been necessary to insert updated material, in brackets. The author has retained Darwin's use of the first person in this summary.
- The exact quotations from the sixth edition of *Origin of Species* are contained within quotation marks or, for longer passages, in indented blocks.
- The author has inserted cross-references to entries in this encyclopedia.

Chapter 1. Variation under Domestication

Perhaps the best way to see what evolution can do is to study domesticated plant and animal species (see ARTIFICIAL SELECTION).

First, artificial selection demonstrates that wild populations, from which crops and livestock were domesticated, contain a tremendous amount of heritable variation. Breeders can still improve old varieties of, and produce new varieties from, even the oldest cultivated plant species and domesticated animal species. Breeders sometimes try to maintain the purity of the stock by breeding only the best animals in their herds. Even while they are trying to prevent change from occurring, they end up causing it to occur: These superior stocks become even better over the decades as a result. There is no end in sight for heritable variation in these species.

Second, plant and animal breeding shows what selection is capable of doing. Consider the pigeon. There are many breeds that are astonishingly different from one another in anatomy and behavior. Yet they all came from the wild rock dove. Scientists know this because if one crosses these breeds with one another, the offspring after two generations once again look like rock doves. The characteristics of the wild rock doves still persist in their highly modified descendants. These breeds are so different from one another that if one encountered them in the wild one would consider them different species or even different genera. They are known to be varieties of one species only because breeders produced them within historical times. Nature provides the raw material of variation; human selection adds them up in a direction that humans find useful. Some of this selection is intentional; but some of it is unintentional. When breeders select for one trait, other traits will be selected along with them.

This is artificial selection in the sense that humans select the characteristics in the plants and animals that are most favorable to human purposes, even to the detriment of the plants and animals: Many domesticated plants and animals can no longer survive in a natural state. Artificial selection demonstrates that wild populations contain a tremendous amount of variation; and it is an example of what selection can do.

Chapter 2. Variation under Nature

Natural populations contain a great deal of heritable variation (see POPULATION GENETICS). It is difficult even for experts to decide which groups of organisms constitute species, and which constitute mere varieties; species delineations are often arbitrary. Because of this, there is no reason to believe that varieties evolved but species did not.

Large genera are experiencing rapid evolution, and varieties are on their way to evolving into separate species. This explains (1) why the species within large genera are more similar to one another than are species in small genera, and (2) why the species within large genera have more varieties than do species within the smaller genera.

Chapter 3. Struggle for Existence

Natural populations experience competition. There are hardly ever enough resources to allow all of the offspring of any species to survive and reproduce. "We behold the face of nature bright with gladness," for example, the singing birds, not realizing that these birds survive only because of the destruction of the seeds and insects that they eat, and that at frequent intervals these birds themselves suffer starvation and death.

Malthus explained how human populations grow exponentially (see MALTHUS, THOMAS; POPULATION). This is also true of every plant and animal species. Some species are increasing in number, but not all can do so, "for the world would not be able to hold them." Populations will, therefore, always outrun resources, unless and until some disaster reduces their numbers. Even a single pair of elephants, with perhaps the slowest reproductive rate known, could produce millions of offspring in a few hundred years, if all survived. "Lighten any check to population growth, mitigate the destruction of offspring ever so little," and the population size of any species will quickly increase.

This demonstrates that it is the struggle for existence rather than the rate of reproduction that determines the size of a natural population. The population size of a species does not depend upon how many offspring its individuals can produce; one species of fly may produce a thousand eggs at a time, another just one, but both are equally abundant. Parasitic flies can make some livestock species incapable of surviving in certain regions. Rare plants are abundant in the few places in which they live. Even in a desert or tundra, organisms compete for the scarce resources that are available. When a plant produces a thousand seeds, these seeds go out into a world already thickly clothed with plants, and the young seedlings must compete with them. Most of these young seedlings die.

I have demonstrated this with my own experiments. In one case, I marked 357 young seedlings; 295 of them were killed within the year, mostly because slugs and insects ate them. In another case, a plot of turf in which livestock were allowed to graze had 20 plant species, while in an ungrazed plot nine of these species died out because the more vigorous species outgrew them. I have also seen how the exclusion of grazing allows conifers to grow in what is normally heathlands. Everywhere in nature, the struggle for existence determines which species live or die.

Not all natural interactions are struggles. The very survival of most species depends upon their interactions with other species. I have found that, when pollinators (mostly bumblebees) are excluded from clover flowers, the flowers set no seed. Clover, therefore, depends upon bees. Mice destroy beehives; cats destroy mice; therefore the abundance of clover may depend upon the abundance of cats!

Consider an entangled riverbank of vegetation, and how complex are the interactions among the species. But this riverbank was at one time barren. All these interactions have formed as the vegetation grew onto the barren riverbank. "Throw up a handful of feathers, and each one will fall to the ground according to" the laws of physics; but it would be easier to calculate the trajectories of each of these feathers than to calculate the course of events as plant species grow back in an area that was once barren!

This description makes Nature seem like a violent place. But we can reflect that the struggle for existence is not always violent. When one animal kills another, "no fear is felt, death is generally prompt ... and the vigorous, the healthy, and the happy individuals survive and multiply."

Chapter 4. Natural Selection

Populations contain much heritable variation; many more offspring are born than can possibly survive; can it therefore be doubted that individuals having any advantage, however slight, over others would have the best chance of surviving and of reproducing their kind? Similarly, any heritable variation in the least degree injurious would be eliminated. This is NATURAL SELECTION. It is an unconscious process; the organisms need have no more volition or awareness of it than the planets have of the gravitation that keeps them in their orbits. Nor is it necessary for a new place to be opened up in the landscape in order for a new species to evolve into it; for a new species can evolve and replace a species that is already present.

Nowhere in the world are all, or any, of the species so perfectly adapted that improvement is not possible. If artificial selection by humans can produce such great diversity of crops and livestock in such a short time, how much greater diversity of forms can be produced by natural selection over vast ages! Natural selection is "daily and hourly scrutinizing, throughout the world, the slightest variations; rejecting those that are bad, preserving and adding up all that are good; silently and insensibly working, whenever and wherever opportunity offers," to improve organisms in their physical environments and in their interactions with other organisms. Even differences we would consider slight—for example, a little bit more fuzz on a peach—might make a big difference in resisting infection by fungus.

Natural selection has been very effective, producing new species and bringing about the extinction of others. The only things natural selection cannot do are: (1) to produce results for which heritable variation is not available; (2) to select one species solely for the benefit of another species.

SEXUAL SELECTION is a type of natural selection, in which individuals are selected not for how well they survive in their physical environments or in their interactions with other species but by a "struggle between the individuals of one sex, generally the males, for the possession of the other sex." From complex mammals to simple insects, males fight, whirling around as at a war dance. This competition is especially intense among polygamous species, in which one male has several to many female mates. It is not always a violent fight; in many birds, for example, the males compete with one another to show off to the females, performing strange antics and with gorgeous plumage.

Two interacting species can influence one another's evolution. For example, within a population of plants, those with flowers that produced more nectar might be chosen by pollinators more than those with flowers that produced less nectar; and within a population of pollinators, those that could obtain nectar from the flowers more rapidly would survive and reproduce better than those that obtained nectar more slowly. Natural selection can, in this case, produce two species most perfectly adapted to one another. [This is now called COEVOLUTION.]

One puzzling question in the study of evolution is, how could the sexes have evolved? It would seem that, if each individual simply produced a copy of itself, rather than having to crossbreed with another individual, each individual could produce more offspring. More offspring would be produced if each individual had one, rather than two, parents. Two responses can be made to this (see REPRODUCTIVE SYSTEMS; SEX, EVOLUTION OF).

First, it appears that crossbreeding usually produces:

- more vigorous offspring. Inbred lines of plants and animals are usually inferior in quality.
- more variable offspring. A greater variety of offspring from one pair of parents means that there is a greater chance for at least some of the offspring to succeed in the heterogeneous conditions of life into which they enter.

For these reasons, practically every known species of plant and animal requires cross-fertilization: Even hermaphroditic animals require mates, and flowers often require cross-pollination even though each flower may have both male and female parts. In some plants, the flowers have both male and female parts but they are not active at the same time. Even in flowers in which self-fertilization is possible, as demonstrated by my own experiments with cabbages, the pollen that comes from a different plant produces more seeds than pollen from the same plant. If self-fertilization of flowers were beneficial, flowers would never take the risk of opening!

Second, individuals may benefit from specializing upon the male function, or the female function, thus performing that function more efficiently than could an individual that attempts to do both. A plant species with flowers that have both male (stamens) and female (pistils) parts can evolve into one in which some of the plants specialize on pollen production and others on fruit and seed production. They can eventually evolve into a species with separate male individuals (the flowers of which have only stamens) and female individuals (the flowers of which have only pistils).

The following two objections that have been raised against the effectiveness of natural selection:

First, such a large number of offspring die that any good, new variation that might appear among them would be very likely to disappear. However, natural selection would still favor the beneficial variations among the survivors.

Second, any new variation would get swamped by intermixing with the mediocre majority with which the new variant crossbreeds. However, a new variation, when it arises, interbreeds primarily with close neighbors; the new trait is therefore not likely to get lost in the big pool of the whole population. It would become established as a local variety. It is not uncommon to find two varieties of animal which can interbreed, but which nevertheless remain distinct.

Small, isolated islands usually produce their own, unique species. These species are seldom able to compete against the species that have evolved on large continents. For example, continental placental mammals are displacing the marsupial mammals of Australia. In fact small, isolated islands often have species that appear to have not changed for long periods of time (see LIVING FOSSILS). For reasons described above, it is important for a new variety to be partially isolated in order for natural selection to begin favoring it (see ISOLATING MECHANISMS). This isolation need not be on an island; a large continent has isolated habitats (forests, lakes, etc.) that are as effective as islands in producing new species. Perhaps most effective of all is a large continental area that experiences fluctuations of conditions (such as changes in sea level, or changes in climate): First, species in isolated habitats evolve uniquely in each habitat, then when conditions change these species spread, come in contact with each other, and evolve into even more species in response to one another, then they become isolated yet again and evolve into yet more species. The possible lack of isolation on a continent, compared to an island, is more than compensated by its larger population sizes, with its accompanying greater heritable variability.

Varieties within a species can evolve into separate species as a result of the following:

- Competition is stronger between varieties of one species than between species, because they are competing for a more similar set of resources. This is indirectly demonstrated by the fact that the species that successfully invade a new continent are usually members of genera that are not already present on that continent.
- Natural selection will favor the individuals within two varieties that are most different from one another, specializing on different resources.

This results in what I call *divergence of character* (see CHARACTER DISPLACEMENT).

While divergence has occurred throughout the history of life, CONVERGENCE has also occurred: When two species, very different in structure, begin to adapt to a common environment, natural selection favors the same characteristics in each, causing them to become more similar to one another.

Natural selection has also led to advancement in complexity of organisms over time. Natural selection favors whatever traits confer advantages, whether the traits are more complex, or less. There are more ways for a complex trait to prevail over a simple trait than for the reverse to happen.

The result is an evolutionary diversification of species that resembles a vast bush, with many branches from a common ancestral trunk. Many of its branches have been pruned away by extinction, the remaining ones have diverged, adapting in different ways to different environments. Some branches diverge more, some less; some produce more branchlets, some fewer; some go nowhere, into extinction; we behold only the outermost twigs and try to figure out the history of branching that has produced their pattern.

Someday people will recognize that species have arisen through the slow work of natural selection, rather than sudden creation, just as they now realize that geological formations are the result of gradual processes such as erosion and uplift, rather than due to huge, brief floods. [Darwin was overly optimistic about both of these; see CREATIONISM.]

Chapter 5. Laws of Variation

Many factors determine the development of an organism's characteristics. The direct effect of the environment can make individual organisms acclimatize to new conditions, but heritable variation that is not directly related to environmental conditions also arises. Natural selection acts only on the heritable variation. There are constraints upon this heritable variation. For example, some traits (such as albinism with deafness in cats) are correlated for unknown reasons. [This is now called linkage.] Other traits are correlated because of the "economy" of the organism's growth: Greater growth of one part necessitates lesser growth of another (see ALLOMETRY). Therefore the degeneration of some organs may result from natural selection. For example, cave organisms do not need their eyes, but what harm would it be if they had eyes anyway? The answer is, eyes are easily damaged and infected, and if they are unnecessary, the organism benefits from not having them, and natural selection gets rid of them.

There is more variability in those parts of organisms that are (1) multiple, (2) unspecialized, (3) rudimentary, (4) rapidly evolving. Unspecialized organs need flexibility, since they cannot specialize on just one use. Rudimentary organs, not being needed by the organism, are not controlled as much by natural selection. Rapidly evolving organs have more variability because they are still in the process of evolving. Those traits that differ more between species also differ more within them: This is a pattern that makes sense if the traits are evolving, but makes no sense in terms of special creation.

Traits can reappear after even hundreds of generations, available for natural selection, and providing evidence for evolutionary ancestry. Ancestral characteristics are frequently seen in pigeons, and among species of the horse genus. To say that God put those characteristics into each species, to make it look as if they had a common ancestor when in fact they did not, "makes the works of God a mockery and a deception."

Chapter 6. Difficulties of the Theory

By this point, many difficulties of the theory must have occurred to the reader. They used to bother me, too, until I carefully thought them through, and I am now prepared to answer them.

Transitional forms. If all species have arisen by gradual changes from common ancestors, why do we not see innumerable intergradations among organisms today? In fact, we see distinct species instead—and this is also what we see in the fossil record. But consider these facts. First, the crust of the earth is a vast fossil museum—but a very imperfect one (see below). Second, when a new variety begins to form, it will be in direct competition with the parental stock, or with other varieties forming alongside it from the same parent stock. Competition is strongest between and among the most closely related forms. This will lead to the diminution, even extinction, of most of the closely related forms and result in what I have previously called divergence of character. The rarity of intermediate forms does not disprove natural selection; it is the result of natural selection. This is in fact what we see; when varieties within species form even today, the intermediate forms are less common than those that differ the most from one another.

Complex adaptations. Many people have pointed out [and continue to do so; see INTELLIGENT DESIGN] that in complex adaptations, the incipient forms would have experienced no benefit; of what use is a partway-evolved adaptation? Of what use is a part of a wing, or a part of an eye? But, as a matter of fact, throughout the animal kingdom, we see some very good examples of intermediate forms that benefit from their seemingly imperfect adaptations. Some single-celled animals have simple eyespots that allow them only to distinguish light from darkness, and to orient themselves toward the light; they cannot, in the strict sense of the word, see, but they benefit immensely from their ability to detect light. Flying fish do not, in the strict sense of the word, fly, but they can launch themselves above the waves long enough to escape from predators. Flying squirrels cannot really fly, but their ability to glide can and often does save their lives, when pursued by a predator or when they fall out of a tree. While we cannot demonstrate, from living animals or from fossils, a complete series of transitional forms within a given lineage, we can find examples of intermediate forms of the general class of adaptations—which shows us of what natural selection is capable.

Occasionally, an animal's habits can evolve without a corresponding change in anatomy. But the habit is just as much a part of its adaptation as is its anatomy (see BEHAVIOR, EVOLUTION OF; GENE-CULTURE COEVOLUTION). While it may seem incredible to us that something as complex as the vertebrate eye, or the flight of birds, could evolve by means of natural selection, we have the evidence to demonstrate that such a thing is possible. [One cannot use what one evolutionary biologist (see DAWKINS, RICHARD) has called the "argument from personal incredulity," which implies that if one cannot imagine it to have happened, it cannot have happened.] If there were any example of an adaptation that could not have been formed by small steps in natural selection, "my theory would utterly break down; but I can find no such case."

There is another pattern that makes sense only from the viewpoint of evolution. Complex eyes are found not just in vertebrates but in invertebrates as well—the eyes, for example, of cephalopods such as the squid, and of arthropods such as insects. But in each case, the eye has a fundamentally different structure. In fact, there are at least three different organs, all of which we call by the same name, "eye." Insects have compound eyes. Vertebrates have eyes in which the nerves come out in front of the light sensors and converge in the optic nerve. Cephalopods have eyes in which the nerves come out in back of the light sensors. It appears that these "eyes" have evolved along different pathways, and converged upon similar but not identical structures: structures that still retain vestiges of their different evolutionary pasts. Other examples of convergence include organs for the production of electricity in various aquatic animals, and the wings of birds vs. bats vs. insects.

> In all cases of beings, far removed from each other in the scale of organization, which are furnished with similar and peculiar organs, it will be found that although the general appearance and function of the organs may be the same, yet fundamental differences between them can always be detected ... natural selection would have had different materials or variations to work on, in order to arrive at the same functional result; and the structures thus acquired would almost necessarily have differed. On the hypothesis of separate acts of creation the whole case remains unintelligible.

A new adaptation does not have to arise from scratch; it can arise from the modified use of a previously existing organ. Lungs, for example, develop in the same location as the swim bladders that fish use for flotation and have many structural similarities to them; lungs could therefore have evolved from swim bladders. [It is now considered more likely that swim bladders evolved from lungs; see FISHES, EVOLUTION OF.] In some cases, the corresponding structures, which had been present in the adult or juvenile forms, might have been lost if the adult or juvenile phase of the life cycle had been lost (as has happened in some animals), making it look as if the structure had appeared out of nowhere (see NEOTENY).

> I have been astonished how rarely an organ can be named, toward which no transitional grade is known to lead ... Nature is prodigal in variety, but [stingy] in innovation.

Why, on the theory of Creation, should there be so much variety and so little real novelty?

Organs of seemingly little importance. Sometimes complex organs exist which seem to be of little importance to the organisms—how could natural selection have produced them? [Why would a Creator have put them there, for that matter?] But this is more a problem of our knowledge than of evolutionary process. For why should any such organ, even if specially created, continue to exist, if it were really useless? Surely it must have some function, or else have had some value to the ancestors of the species that now possesses it (see VESTIGIAL CHARACTERISTICS).

Some authors have claimed that complex organs exist merely for the sake of beauty in the eyes of human beholders. Natural selection tells us that this is never the case. Why did beautiful seashells exist in the Eocene epoch, then become extinct—simply so that fossil shell hunters could find them and enjoy them? Why were ancient diatoms so beautiful—just so that students of fossils could admire them under the microscope? Why are flowers beautiful? Wind-pollinated flowers of oak and nut trees are not beautiful, because the wind is not attracted by their beauty. Other flowers are beautiful, not for our sake but because they attract insect and bird pollinators. Many male birds are beautiful, not to impress bird-watchers but to impress female birds. Do not press this argument too far—if God put beauty into the world just for us, then why did he put into the world the ugliness which parasites inflict upon their hosts?

Nature will never produce absolute perfection, but natural selection produces adaptations that are only as good as they need to be to outdo the competition. There are, in fact, examples of imperfection in the natural world; this is an understandable result of natural selection, not a criticism of God.

Chapter 7. Miscellaneous Objections

Many of the criticisms that my work has occasioned are not worth answering, because they have been made by people who have not bothered to study my theory, or natural history. For example, one critic said that evolution must produce new species, each of which has a longer life span than its predecessors, which is ridiculous, because a shorter life span might very well have greater adaptive benefits to survival, as some short-lived weeds are more successful than trees in disturbed habitats. But to those that are worthy of an answer, I continue to respond in this chapter.

The natural world simply must contain some kind of evolution, since if each species were originally created to be perfectly adapted to its original conditions, they would have to change in order to adjust to new conditions.

I here respond to three important objections:

1. When we compare modern livestock with those depicted in Egyptian hieroglyphics, we see that no evolutionary changes have occurred during that span of time. This is hardly surprising, as a couple of thousand years is nothing compared to an evolutionary time span.
2. Varieties and species do not differ from one another in single characters but in many ways. This is not a problem for natural selection, which acts upon organisms, with all of their characteristics.
3. How could natural selection have produced characteristics that are of no use to the organisms that possess them? Some naturalists have claimed that such characteristics are common. However, we must be very careful when we say that a characteristic is of no use, simply because we do not know what it is used for. Consider the dimorphic and trimorphic flowers, in which different plants within a species have flowers with different lengths of stamens and pistils. At first this arrangement appeared to have no function, but we now know that it promotes cross-pollination. In some cases, differences among species have resulted from the accidents of ancestry. For example, feathers and hair might be equally effective in holding in warmth, but birds always have feathers and mammals have hair—not because it is necessarily better for birds to have feathers than hair, and mammals to have hair rather than feathers, but simply because they have inherited these characteristics.

One must admit the art and force of the objections published by St. George Jackson Mivart in his 1871 book, *The Genesis of Species.* But he is wrong. Some of his objections arise from his mistakes. For example, he says that mammary glands could not have evolved, since a rudimentary gland producing only a drop of fluid would not have helped a baby animal in a transitional species. But, in fact, while such an arrangement might not work in an early mammal with external mammae, such an arrangement would have been beneficial in marsupials, in which the mammae are protected inside a pouch—a factor that Mivart seems to have overlooked. However, other objections raised by Mivart deserve answers.

1. One of his objections involves the giraffe. He asks if having a long neck is such an advantage to the giraffe, why haven't all the other African grazing mammals evolved long necks? But once giraffes began to evolve long necks, allowing them to reach higher into the trees for leaves to eat, it would not be an advantage for any other animals to do so, unless they could outreach the newly evolving giraffes. [It now appears that the neck of the giraffe evolved by sexual selection; see ADAPTATION).] Other authors have raised similar objections, such as, why haven't ostriches evolved the power of flight? I think the answer is obvious—ostriches are doing so well with running, and the energy and modifications of structure required for flight are so great, that it simply would not be worth it and would not be selected by the evolutionary process. Flight is not always better for birds. This is why there are no islands on which seals have evolved into terrestrial forms or bats into ground-dwelling forms—the advantages of staying as they are have outweighed the advantages of changing. Why haven't apes [other than humans] evolved high intelligence? Intelligence is over-

rated: It has been an advantage in our evolution but is not necessarily worth the cost in all situations (see INTELLIGENCE, EVOLUTION OF). In their particular circumstances, intelligence would not be an advantage to apes. Cows just walk around and moo and eat grass; an intelligent cow would not be able to do this much better than a stupid cow. Furthermore, natural selection can act only upon the combinations of characteristics that happen to have been provided to it, randomly, by nature.

2. Mivart also said that if heritable variations occurred in all directions, would they not neutralize each other? They would, in the absence of natural selection; but natural selection preserves the good variations and disposes of the bad ones, causing a directionality of change even from an initial wellspring of random variation.
3. Mivart also considered several examples of complex adaptations which, he maintains, could not have evolved by small steps.
   - Mivart mentions the baleen whale, with its mouth of huge plates that filter plankton out of the water for food. Of what use would the earliest stages of evolution of baleen plates have been? When we consider the various kinds of whales, we cannot find direct evidence of all of the intermediate stages of baleen development. But when we consider birds, many of which also subsist by filtering plankton out of the water, we can in fact find all intermediate stages of development of filtering structures in their beaks, all the way from the rudimentary devices used by the common duck to the complex sieves of the shoveler duck and the flamingo. The common duck occasionally filters plankton out of the water as food, while the shoveler duck and flamingo subsist almost exclusively from filtering plankton. In all cases, the birds do in fact find these adaptations to be of use to them: The common duck spends a lot of time sieving plankton out of the water, even though its anatomy appears to us inferior for this purpose. [Fossils of intermediate forms in the evolution of whales have now been found; see WHALES, EVOLUTION OF.]
   - A similar response is possible concerning the evolution of the prehensile tail in monkeys; many mammals have tails that are just a little bit prehensile, and this helps them, when very young, to cling to their mothers, even though they cannot use these tails to help them swing through the trees.
   - A similar response is also possible regarding the evolution of certain muscular forceps in echinoderms such as starfish, to the pollination adaptations of orchids, and to the evolution of climbing in vines. The stems of some plants, which are not climbers, nevertheless rotate during growth, which helps them grow toward sunlight but also provides a launching pad for the evolution of the climbing habit of vines. [This process, known as circumnutation, is one of the many discoveries Darwin made in fields other than evolutionary science—in this case, plant physiology.]

   In most cases, we can find the entire range of intermediate structures, each of which is of plain service to the species that possesses it.
4. Mivart also says, how could something as strange as the flounder have evolved in one big step? Flounders lie on one side on the bottom of the sea, with both of their eyes on the top side. I quite agree that this adaptation could not have arisen in a single step. But the pleuronectid family of fishes, of which the flounder is the most extreme member, exhibits a whole range of intermediate adaptations to lying on their sides. We do not have to imagine these intermediate forms; they exist.

Almost all naturalists admit evolution in some form; even Mivart proposes an evolutionary theory. His theory, however, requires the sudden appearance of fully developed forms, rather than natural selection from among slightly developed forms. But what do we see occurring in the natural world? We observe the slight variations that my theory requires; we never see the sudden appearance of fully developed forms, as Mivart's theory requires. We see, in embryos, the evidence of gradual modifications that have occurred in the past. To insist that large steps have taken place during evolution, and have left no trace of their occurrence in the embryo, is to "enter the realms of miracle, and to leave those of Science."

Chapter 8. Instinct

Many instinctive animal behaviors seem too wondrously complex to have arisen gradually by natural selection (see BEHAVIOR, EVOLUTION OF). But I will now explain how instincts could have evolved. Instincts are behaviors that animals perform, as it were, mindlessly, sometimes to an extent that appears to us ridiculous [as when a crow attacks its own reflection in a window, over and over, for months] and which we would think the animal would eventually figure out. But instinct and reason are two very different things.

Instincts are adaptations that are just as important in natural selection as are any features of anatomy, and they have a heritable basis as surely as do the anatomical features. Species have adapted to one another most marvelously. For example, aphids drink the sap of plants and excrete the fluid which they have only partly digested. But some of them will not excrete until an ant touches them. This is because the ants and aphids form a mutualism, in which the ants protect the aphids, and the aphids feed the ants with the excreted, sugary fluid. The aphids do not learn this behavior; they are born with it. The behavior we see in dogs, for example, pointers that stand motionless in the direction of their quarry to point the hunter toward it, have been inherited from wild behaviors: in this case, the motionlessness of the dog just prior to pouncing on its prey.

Could adaptive behaviors, however complex, have arisen by gradual natural selection? Consider these examples:

1. Some cuckoos are nest parasites, laying their eggs in the nests of other bird species. The cuckoo hatchlings, even

when young and blind, eject the other hatchlings out of the nest, and the parent birds, of a different species, raise the young cuckoo. Cowbirds also exhibit such behavior; and among cowbird species, we see all the intermediate stages ranging from occasionally laying the eggs in the nests of other bird species, to depending upon this strategy entirely. In one example, the cowbird seems to employ an intermediate strategy that does not work very well.

2. Some ant species enslave other ant species by raiding their nests and stealing pupae, then raising the ants of the other species as slaves. We see here, also, the entire range of adaptations. In some ant species, the masters do much of the work, forcing the slaves to help them; for example, when it comes time to move to a new nest, the masters carry the slaves. In some of these, the slaves accompany the masters on food-gathering expeditions; in others, they stay at home. At the other end of the spectrum, there are ant species in which the masters are entirely dependent upon the work of the slaves and cannot even feed themselves without the help of the slaves. At moving time, the slaves carry the masters.
3. Honeybees make the wax chambers of their nests entirely by instinct. These chambers appear complex: The hexagonal form of the chambers is the mathematically perfect compromise of strength and economy of space. "He must be a dull man who can examine the exquisite structures of a comb, so beautifully adapted to its end, without enthusiastic admiration." Although the comb-building instinct appears complex, it results from a few basic instincts, and the intermediate stages of its evolution can still be found represented in other species of bees. "Let us look to the great principle of gradation, to see whether Nature does not reveal to us her method of work."
   - We find, among bee species, the whole range of honeycomb complexity, from the simple round chambers of bumblebees to the complex hexagonal chambers of honeybees, with other species intermediate.
   - The complexity of beehives results from the repeated application of a few basic behaviors, which I demonstrated by experiments in which I provided different starting conditions for the bees to build their chambers and observed what they did. It is quite simple, actually. "The work of construction seems to be a sort of balance struck between many bees, all instinctively standing at the same relative distance from each other, all trying to sweep equal spheres, and then building up, or leaving ungnawed, the planes of intersection between these spheres." The bees are not intelligent architects; they follow simple instincts and in so doing build up complex structures. [This is an example of what is now called EMERGENCE.]

The intermediate stages of honeycomb evolution would have proved advantageous; for wax is a very expensive material for the bees to make, and any adaptation that helped them economize its use would be selected. "Thus, as I believe, the most wonderful of all known instincts, that of the hive-bee, can be explained by natural selection having taken advantage of numerous, successive, slight modifications of simpler instincts."

One objection to my theory at first appears fatal. In some social insects—including honeybees—the worker individuals are sterile. How could natural selection favor characteristics in the worker bees that they themselves cannot transmit to the next generation? Clearly the answer is that the characteristics are transmitted by the queen bee and drones. Natural selection works on families, not just individuals. A livestock breeder who finds that a certain animal has desirable traits, but discovers this only after slaughtering the animal, can still breed for those traits by using the dead animal's closest relatives as the breeding stock. A division of labor within a society of ants, even if this division includes some of the members being sterile, is beneficial to the ants as surely as a specialization of labor is of benefit within human society. [Today this is recognized as *inclusive fitness;* see ALTRUISM.]

Chapter 9. Hybridism

Scientists commonly believe that crosses between varieties within a species (as of different breeds of pigeon) produce fertile offspring, whereas the crosses between different species (as when horse and donkey cross to make a mule) produce sterile offspring (see HYBRIDIZATION). If this is an unbreakable rule of nature, then natural selection could never transform varieties into species! But this pattern, while commonly observed, is not a rule, for these reasons:

1. Crosses between varieties, and between species, produce all different gradations of fertility and sterility. It is not true to say that crosses between varieties are perfectly fertile while those between species are completely sterile. In some plants, interspecific crosses produce offspring even more vigorous and fertile than crosses within a species! The details of which varieties or species can be interbred with which others, and which cannot, seem unrelated to how different or similar the varieties or species appear to one another. In some cases, crossing two species yields different results, if one rather than the other species is used as the father (compare, for example, mules and hinnies). Sometimes two species can be easily crossbred but they produce few fertile offspring, while sometimes two species can be crossbred only with difficulty, but the resulting offspring are vigorous and fertile. There is no simple rule that varieties can cross and species cannot. You simply cannot tell which crosses will yield good offspring and which will not, until you try the crosses yourself.
2. Some of the studies of crossbreeding are flawed because they are performed with livestock or with crop plants that have experienced many generations of inbreeding, which itself results in reduced reproductive vigor.
3. It is not surprising that, according to some naturalists, varieties can cross while species cannot, because they define species that way! If two forms cannot cross, they are defined as different species. This is circular reasoning. [For a discussion of the complexities surrounding what is now called the "biological species concept," see SPECIATION.]

Chapter 10. On the Imperfection of the Geological Record

I have already explained why it is that we do not today see numerous intermediate forms between species: Natural selection favors the forms that are most distinct from the parents, and from one another, in a group of newly evolving varieties. Now I will explain why these innumerable intermediate forms seem to be absent from the fossil record as well. "… this, perhaps, is the most obvious and serious objection which can be urged against [my] theory."

The geological record is highly imperfect; indeed, it could hardly be otherwise. We know that different breeds of pigeon were bred from a single ancestral species; yet there is very little if any record of this in the fossils.

Go look for yourself, along a shoreline, and see the erosion that is there taking place, and think about how slowly and irregularly erosion is occurring there, to be deposited in the shallow oceans nearby. Yet this erosion has occurred, in the past, enough to produce a total of almost 14 miles of depth in geological deposits, and these 14 miles represent only the period of time in which deposition was occurring, during times of the subsidence of the crust; at least half of the time, the crust was rising, causing erosion instead of subsidence. After all, if you have deposition in one location, you must have erosion somewhere else. You will not find all 14 miles in any one location; this figure is calculated by adding up the deposits in one location that correspond to those of another location [a procedure developed by earlier geologists; see SMITH, WILLIAM].

During my journeys I saw, in one location, a mass of conglomerate rock more than 10,000 feet thick. This rock was formed of rounded pebbles imbedded in metamorphic rock that had been transformed by heat; this conglomerate rock, therefore, was formed from earlier rocks. How old the Earth must be! Just how long is a million years? Some professors have demonstrated this to their students by making a mark, one-tenth of an inch in diameter, on a strip of paper 83 feet 4 inches long; the mark represents a century, the strip of paper a million years! Yet the Earth has existed many millions of years.

Here are just some of the reasons that the geological strata are imperfect as a record of past life on Earth:

1. Only a few places on the Earth have been thoroughly explored by geologists.
2. The only organisms that are preserved as fossils are those that die:
    - during periods of deposition. Those that die during periods of erosion are unlikely to get covered with sediments and preserved.
    - at a time and place where deposition is occurring rapidly enough to bury them before they decompose. In most parts of the ocean, sediments are not accumulating; this is why most of the ocean is blue.
    - underwater, or where they can be transported to a riverbed, delta, or continental shelf.
3. The very times when species are more likely to become extinct than to evolve new forms—the times when the crust is subsiding, and terrestrial regions are being flooded and plunged into dark depths—are the very times when fossils are most likely to form; conversely, the times when species are most likely to evolve—the times when the crust is rising, and new terrestrial surfaces (as well as intertidal and subtidal surfaces) are being formed—are the times when fossils are least likely to form.

Therefore if a species, or whole group, evolved under conditions when fossils were not forming, and were then fossilized, they would appear in the fossil record as if they were specially created. While positive evidence is trustworthy—we can trust a fossil when we find it—negative evidence (the absence of fossils) tells us nothing worthwhile. [This is one of the assertions most vigorously debated regarding PUNCTUATED EQUILIBRIA.]

While this may explain why the geological record, on a worldwide scale, does not preserve evidence of innumerable transitional forms, it does not explain why innumerable transitional forms are not preserved within any one formation. In particular, some species seem to appear suddenly in the fossil record, without predecessors leading up to it, even when there has been no interruption of deposition. The appearance of a species in the fossil record is much more likely to be the result of immigration from a different region, where it had already evolved, than of its sudden origin in that location. In some cases a species seems to disappear, then reappear, then disappear, then reappear; this is not because it re-evolved several times, but because it vanished, then migrated back in from another area, several times. Even within any one formation, deposition may not have been uninterrupted; each formation probably represents a whole series of switches between erosion and deposition.

A more serious problem is what appears to be the sudden appearance of a great variety of multicellular lifeforms at the beginning of the Cambrian period (see CAMBRIAN EXPLOSION). My theory, if true, should explain the origin of all these forms, gradually, from a common ancestor; therefore the world must have swarmed with creatures even before this time. There is even a good question as to whether the Earth has existed long enough for the diversity of forms seen in the Cambrian deposits to have evolved, if William Thomson's calculations are correct. [It turned out that they were not; see KELVIN, LORD.] I can give no satisfactory explanation for the absence of fossils in Precambrian rocks. The presence of organic matter in these deposits indicates that life existed throughout that time, a fact also demonstrated by the fossil of Eozoon. [It turned out that Eozoon was not a fossil; however, numerous unicellular fossils, and a few multicellular ones, have been found in Precambrian deposits since Darwin's time. Darwin was right that life, albeit primarily single-celled, existed throughout that time.] The case must at present remain inexplicable; and it may be truly urged as a valid argument against the views here entertained. [The discovery of the slightly more ancient EDIACARAN ORGANISMS has not helped to, even now, solve the problem of the relatively rapid origin of multicellular life.]

Chapter 11. On the Geological Succession of Organic Beings

Species evolve very slowly; and even the ancient deposits contain a few species that are very similar to those that exist today. Once a species becomes extinct, however, it never comes back. Species become rare before they become extinct; to be surprised at the extinction of a species makes no more sense than to be surprised at its rarity. The history of life is recorded in the geological strata; each new set of species represents not a new creation but "only an occasional scene, taken almost at hazard, in an ever slowly changing drama."

The pattern of fossils in the sedimentary rocks is not random, as one might expect if they were deposited by a great flood. In fact, the species contained within each layer closely resemble, and are intermediate between, those of the layer immediately above and immediately below. These similarities, furthermore, are found in strata of a particular age throughout the world. It has long been known that the more ancient a fossil deposit is, the more greatly its species differ from those found on the Earth today. I wrote in 1839 and 1845 about the close relationship, on each continent, between living and extinct forms, even when the continents differ greatly among themselves. For example, most marsupials today are found in Australia, which is also where most of the fossil marsupials are found. There is nothing special about the climate of Australia that makes it particularly friendly to marsupials; in fact, placental mammals have largely displaced them. I have now explained why this pattern has occurred.

Groups of organisms that today appear entirely distinct can often be linked by the fossils of species now extinct; that is, we can find many examples of "missing links" that are no longer missing. For example, fossilized animals have been found intermediate between whales and terrestrial mammals. The *Archaeopteryx* and *Compsognathus* link birds to their dinosaurian reptile ancestors. We cannot, however, expect the geological record to provide evidence of all the links. "Thus, on the theory of descent with modification, the main facts with respect to the mutual affinities of the extinct forms of life to each other and to living forms, are explained in a satisfactory manner. And they are wholly inexplicable in any other way."

Has progress occurred over evolutionary time? This question cannot be answered, for we have no way of defining which life-forms are higher than which other forms. The only true experimental test would be to allow the ancient species to compete with the modern ones, and see whether the modern ones win; but this can never happen. We cannot even guess the outcome. For example, who would ever have predicted, just from studying their anatomy, that European placental mammals would have so completely displaced the native Australian marsupials? There has undoubtedly been much progress over evolutionary time, but natural selection favors only the amount of progress that is beneficial to a species; after that point, no more progress is necessary. Therefore the persistence of species that closely resemble very ancient forms is not at all surprising.

Chapters 12–13. Geographical Distribution

The geographical distribution of organisms can be reasonably explained only in terms of descent from a common ancestor, followed by evolutionary modification (see BIOGEOGRAPHY).

Everyone has noticed how different are the animal and plant species of the different continents, especially the Old vs. New Worlds, even when they share very similar climates. For example, the animals of South America are much more closely related to one another than to the animals of either Australia or Africa, despite having similar Southern Hemisphere latitudes and range of climates. For example, South America, Africa, and Australia all have large flightless birds, but they are in three different groups: the rheas, the ostriches, and the emus, respectively. There are entirely different sets of species among the blind animals of caves in the different continents. Just as on land, there is a similar biogeographical pattern found in the oceans. This pattern cannot be explained in terms of the independent creation of species suited to each type of habitat [nor by dispersal of species after a gigantic flood, a theory to which many modern creationists cling but which had been nearly abandoned by the creationists of Darwin's time]. "On the principle of inheritance with modification we can understand how it is that ... whole genera, and even families, are confined to the same areas, as is so commonly and notoriously the case." [Darwin believed, like all scientists prior to the 20th century, that the continents had not moved. CONTINENTAL DRIFT makes an evolutionary explanation of biogeography even more convincing.]

Dispersal, followed by descent with modification, also explains many facts regarding island biogeography, such as the following:

1. Nearly everywhere in the world, animal species on islands most closely resemble the animal species on the continents nearest to them. "Why should this be so? Why do the species which are supposed to have been created in the Galápagos Archipelago, and nowhere else, bear so plainly the stamp of affinity to those created in America?" The climate of the Galápagos, off the coast of South America, is very similar to that of the Cape Verde Islands, off the coast of Africa; yet their species are entirely different.
2. Islands usually have fewer species than corresponding areas of continents. This has occurred because they have not existed as long as the continents, and not as many species have evolved there.
3. Islands almost always have a high proportion of endemic species—species found nowhere else in the world. This has occurred because the endemic species have evolved in isolation. The islands that do not have very many endemic species are those in which isolation has been incomplete: in which organisms frequently disperse from the mainland (as in the birds of Bermuda) and intermix with their native relatives. The Galápagos Islands have their own species of animals, for example mockingbirds; this is because each species evolved on its own island, and then, if one of the island species dispersed from one island to another, it lost

out in competition with the species that was already present on the island.

4. On many islands, mammals are nearly absent. On the Galápagos Islands, large turtles have taken the place in the economy of nature that grazing mammals would have filled; in New Zealand, it is gigantic wingless birds that did this. The exception proves the rule: On many islands, the only native mammals are bats! Amphibians are also frequently absent from islands. In both cases, it is not because the animals cannot survive on the islands—continental mammals and amphibians thrive when humans introduce them to islands—but because they cannot disperse well over great distances. The theory of creation gives us no reason why islands should be so deficient in amphibians and in large mammals. Some naturalists say that oceanic islands have not existed long enough for mammals to have been created. Although some volcanic islands are young, many are old, and yet they do not have mammals either. [Darwin was responding to the idea of progressive creationism, rather than the six-day creationism espoused by many modern creationists, who could not possibly have raised this point.]
5. On many islands, plants whose mainland relatives are herbaceous grow into trees (as in the tree-sunflowers of St. Helena). This has occurred because herbaceous plants are frequently much better at dispersing over wide distances than are trees; once they have dispersed to islands, however, some of these herbaceous plants find it advantageous to evolve into trees.

These patterns make no sense if the species of each island were independently created.

Many facts of biogeography make sense when we consider the effects of recent Ice Age glaciations, as revealed by the research of Louis Agassiz (see AGASSIZ, LOUIS; ICE AGES):

1. Mountaintops within a continent, for example within North America, often have the same alpine tundra species upon them, even though these species cannot grow in the intervening lowlands. They are also very similar to the northern arctic tundra species of the same continent. How could the same species travel among the arctic tundra and the alpine tundras of these different mountaintops? They got there because they retreated both northward and up the mountains, where the alpine plants are now stranded, as the weather became warmer and the glaciers retreated. So far, this has little to do with evolution. But here is the fact that can be explained only by evolution: Each continent has a different set of alpine tundra species, each most closely resembling the arctic tundra species of its own continent. Mountains are [ecological] islands of cool climate surrounded by warm lowland, just as geological islands are surrounded by water.
2. The forest tree species are different on the different continents, although related: For example, there are oak forests in Europe, Asia, and North America, but each of these places has its own species of oaks (see ADAPTIVE RADIATION). How did oaks originally get to these three continents? The arctic regions of the world form a nearly continuous landmass, separated only by narrow channels (e.g., on either side of Greenland, and the Bering Strait). In earlier periods, when the weather was warmer, oaks and other forest trees grew far north of where they are now found and could have dispersed freely among the three continents. After the glaciations began, however, the oaks of the three continents have been separated and have evolved into different species on each continent. [Scientists would now add that the continents have also drifted apart.]
3. Some species similarities among distant regions in the Southern Hemisphere may be explained by dispersal through Antarctica, which today is covered with ice but in earlier ages was warmer and had forest species, the fossils of which we have found.

In order to understand geographical distribution we must understand dispersal—the ways that organisms get from one place (as I believe, their point of origin) to new locations. Among the factors that influence dispersal are the following:

1. Islands that are now separated by oceanic waters may not always have been so in the past. When ocean levels were lower, some regions that are now separate islands (the islands separated by what are now shallow seas) were mountains in a plain, allowing the organisms to freely travel among them. However, other islands (the ones now separated by deep oceans) have been isolated for long periods. We would expect, by this theory, that the organisms found on islands separated by shallow seas should more closely resemble one another than the organisms found on islands separated by deep oceans. This is in fact the case. A very good example [and still the most famous] is provided by Mr. Wallace from the Malay Archipelago: The islands near Asia, separated from Asia and from one another by shallow seas, have Asian mammals, all placental; the islands near Australia, separated from Australia and from one another by shallow seas (but from the Asian islands by a deep trench) have Australian mammals, many of them marsupials. Also, species that are found in widely separated places may have once had a more continuous range, and the individuals in the intervening areas have become extinct.
2. Organisms may have astonishing powers of dispersal. My own experiments have shown that many species of seeds (provided their fruits are mature) can survive in, and float upon, saltwater for weeks—which is long enough to allow them to disperse to distant islands on ocean currents. Other seeds, which would otherwise die in ocean water, occasionally disperse to islands within clods of dirt lodged in driftwood. Others can germinate after being carried in, and expelled from, the crops or the intestines of birds—which have frequently been known to fly to distant islands. Waterfowl could easily carry seeds in the mud on their feet, thus dispersing freshwater plant seeds between

freshwater habitats separated by vast expanses of saltwater. Most botanists are unaware of how many seeds a little bit of pond mud can contain: From three tablespoonfuls of pond mud, I grew 534 plants! My own experiments have also shown that seeds can even be carried inside of predatory birds that have eaten prey that had eaten fruits and seeds! Incredibly enough, even locusts can carry seeds, when they are blown great distances in the wind. Some plant parts have even been carried on icebergs.

The freshwater species are remarkably similar in widely separated regions. How did this happen? Since lakes do not last forever, and rivers shift course, freshwater species have to disperse from one location to another, in order to persist. Because freshwater species are unusually good at dispersing from one freshwater location to another, it is not surprising that they have dispersed widely in the world. Although freshwater fishes usually cannot survive in saltwater, nor saltwater fishes in fresh, there are many groups of fishes that contain both fresh and saltwater species; evolutionary adaptation to differences in salinity must not, therefore, be very difficult. Indeed, some individual fish (such as salmon) can acclimatize to changes in saltwater vs. freshwater. Freshwater fishes, then, could have dispersed even through saltwater. Ducks fly over great distances and could have dispersed freshwater snails from one place to another. As every child knows, salt kills terrestrial snails; but freshwater snails can even survive floating in saltwater, so long as they have sealed themselves into their shells with a membrane.

Natural selection therefore explains not just the origin of species but of groups of species throughout the world.

Chapter 14. Mutual Affinities of Organic Beings

"From the most remote period in the history of the world organic beings have been found to resemble each other in descending degrees, so that they can be classed in groups under groups. This classification is not arbitrary like the grouping of stars in constellations." Rather than a confusion of types, like a sky full of stars, organisms occur in recognizable groups, such as the many species of lizards, of oak trees, of carnivores. The species in each of these groups are adapted to live in a great variety of environmental conditions and parts of the world. This fact is so familiar that we hardly stop to ask ourselves why it should be so. This taxonomic pattern, I will explain, results from the operation of natural selection over the entire history of the Earth.

As explained in earlier chapters, most species eventually become extinct. The few survivors are the ones that produce the species that come later. The groups that contain the largest number of species, and the species that contain the most individuals, are the ones that are least likely to become extinct. As a result, to him who has will more be given: The largest groups become even greater in importance. For example, there are millions of species of animals, but they fall into just a couple of dozen groups [phyla], and the many species of flowering plants fall into just two groups, the monocots and the dicots. This is because all of the animals and flowering plants are the evolutionary descendants of just the few ancient populations that did not become extinct. A further example of this is the discovery of Australia, which has resulted in the discovery of an enormous number of new and unique species, and in the discovery of no new insect classes and very few new plant families.

Biologists classify organisms not just to keep track of all the information but also in an attempt to understand their relationships to one another, based on similarities and differences. Biologists could use an artificial system of classification—for instance, to classify all blue organisms together, as in blue whales, bluebirds, and bluebonnet flowers. But biologists find such arbitrary classifications exceedingly unsatisfying and strive instead to produce a natural system of classification which closely resembles lineages of descent. These natural systems are used by virtually all taxonomists, not just those who would agree with me about evolution. I have shown that these lineages of descent are not metaphorical, that they do not reflect some unknowable "plan of the Creator," but have really resulted from a community of descent. Biologists recognize, rather than invent, the patterns of relationship that define the genera of organisms: As Linnaeus said, the characteristics do not define the genus, but the genus defines the characteristics [see LINNAEAN SYSTEM].

It might seem obvious that the most important characteristics for the survival of an organism would also be the most important characteristics for its classification, but this is not the case. Biologists have long recognized that fishes and dolphins are very different on the inside, however similar their external adaptations to swimming may appear. Among plants, the leaves are extremely important as organs of survival, but their shapes are virtually useless in classification—no two leaves even on the same tree have exactly the same shape.

Sometimes we classify organisms on the basis of characteristics that appear not to be important to the function and survival of these organisms. This actually makes sense, from an evolutionary viewpoint, because natural selection will have acted upon, and wrought great diversity of form in, those characteristics that are important; the unimportant characteristics, such as the rudimentary and now useless remnants of petals in the flowers of grasses, are those that will have survived without modification from ancient times and represent evidence of ancestry. Often, these rudimentary characteristics are found in the embryonic stage of the organism's development. The flowering plants are divided into monocots and dicots on the basis of the number of leaves possessed by the embryo, inside the seed. If we base our classification upon a single characteristic, we may err; but when we base it on sets of characteristics, shared by all the species in a group, we can have confidence in the resulting pattern.

Evolutionary ancestry not only explains the taxonomic arrangement of groups but also the amount of difference among the various groups. Families, or genera, of organisms that differ more greatly from one another are simply those that have diverged, through evolution, for a longer period of time since their common ancestor.

The evolutionary classification of species is a process almost exactly parallel to the evolutionary classification of languages. Linguists classify all Romance languages—Spanish, French, Italian, Portuguese, and Romanian—together based on their structural, grammatical characteristics and similarity of words, just as they also classify Germanic languages—English, German, Dutch, Norwegian, Swedish, and Danish. No linguist doubts that all the Romance languages have evolved from Latin, which fact is historically attested; nor that all the Germanic languages evolved from the tongue of the Teutonic tribes. [In the 20th century, Luigi Luca Cavalli-Sforza demonstrated that not only the process of classification but the actual patterns of human evolution and of human language were very similar to one another (see LANGUAGE, EVOLUTION OF).]

Evolution also explains the difference between homology and analogy. The hands of humans and the flippers of whales have homological resemblance. Despite a vast difference in external appearance and the use to which they are put, hands and flippers evolved from the limb of the same ancestral animals, as is indicated by the interconnections of the bones within them. The placental dogs of the northern continents and the marsupial thylacines of Australia, however, have an analogical resemblance: Quite different mammalian ancestors evolved separately and independently into doglike organisms. This fact, obscured by external similarities, is revealed by internal differences, for example of the number and arrangement of teeth, in which the thylacines resemble not dogs but other marsupials. No one would dream of classifying the dray horse and racehorse into separate groups despite their differences in structure and function, nor would anyone dream of classifying racehorses and greyhounds together, even though both of them are sleek and fast.

Here are some examples of homology: "What can be more curious than that the hand of a man, formed for grasping, that of a mole for digging, the leg of a horse, the paddle of the porpoise, and the wing of the bat, would all be constructed on the same pattern, and should include similar bones, in the same relative positions?" Among marsupials,

> ... the hind-feet of the kangaroo, which are so well fitted for bounding over the open plains,—those of the climbing, leaf eating koala, equally well fitted for grasping the branches of trees,—those of the ground-dwelling, insect or root-eating, bandicoots,—and those of some other Australian marsupials,—should all be constructed on the same extraordinary type, namely with the bones of the second and third digits extremely slender and enveloped within the same skin, so that they appear like a single toe furnished with two claws. Notwithstanding this similarity of pattern, it is obvious that the hind feet of these several animals are used for as widely different purposes as it is possible to conceive ... what can be more different than the immensely long spiral proboscis of the sphinx-moth, the curious folded one of a bee or bug, and the great jaws of a beetle?—yet all these organs, serving for such widely different purposes, are formed by ... modifications of an upper lip, mandibles, and two pairs of maxillae.

Despite their great diversity, all the different flowers develop from the same four parts: sepals, petals, stamens, and pistils. "On the ordinary view of the independent creation of each being, we can only say that so it is;—that it has pleased the Creator to construct all animals and plants in each great class on a uniform plan; but this is not a scientific explanation." It is not an explanation at all; it merely says, it is so because it is so, so there. But the reason that all vertebrates have vertebrae is that the ancestral species from which they evolved had them; arthropods all have jointed appendages and external skeletons because their ancestors did. Despite tremendous evolutionary diversification, vertebrates and arthropods still "retain ... plain traces of their original or fundamental resemblances" to their ancestors.

In fact, many of the aforementioned embryonic characteristics reveal homology, both within and among species. Here is an example of homology of origin of different organs within a species. It appears to us that the sepals, petals, stamens, and pistils of flowers actually constitute a spiral of leaves that have been modified for reproduction. This appearance is confirmed when we study the embryonic development of flowers, inside of their buds, in which rudimentary leaflike structures that under some conditions do in fact develop into leaves, will under other conditions develop into flowers instead of into leaves. Here is an example of homology between species: All vertebrates, during embryonic development, have gill slits with looplike blood vessels in them, which serve no purpose for birds and mammals, but which are remnants of the evolutionary past. Another example is that some insects retain, in their embryonic stage, a resemblance to their annelid (earthworm-like) ancestors—as in caterpillars. [Annelids are no longer classified closely with insects; see INVERTEBRATES, EVOLUTION OF.]

Speaking of embryos, they themselves offer considerable support for evolution. However different the adult forms of vertebrates, from fishes to chickens to humans, it is sometimes difficult to distinguish their embryos, especially at very early stages. This has occurred because natural selection has not caused the adaptation of these organisms to different conditions during their embryonic stage. It is the adults of these organisms that have very different methods of survival, from swimming to flying to running; the embryos float in a protected fluid, whether in an egg or in the uterus. The study of embryos "rises greatly in interest, when we look at the embryo as a picture, more or less obscured, of the progenitor" of the species.

The evolution of [what is now called] MIMICRY represents another example of evolutionary homology vs. analogy. Mr. Bates (see BATES, HENRY WALTER) has observed many examples of butterflies in which one species, without particularly marked means of self-defense, resembles another species, that has well-developed defenses. The mockers [mimics] are rare, and the mocked [models] are abundant. [An example familiar to many readers will be the resemblance of the rare, nonpoisonous viceroy butterfly to the abundant and poisonous monarch.] The mockers closely resemble the mocked in external coloration and form, but the details of their internal and reproductive anatomy reveal that they are not related to the mocked, but to one another! Sometimes, several mockers, quite different in external appearance, are all in the same butterfly species! Clearly, the resemblance of the rare to the abundant butterfly is an example of analogy, in which the rare butterfly has evolved coloration and structure that will allow it to, as it were, hide behind the protection of the abundant species—sometimes so effectively that even scientists could not at first tell the difference between the mockers and the mocked; but the resemblances of the rare individuals or species to one another are true homology of close evolutionary relationship. "Now if a member of one of these persecuted and rare groups were to assume a dress so like that of a well-protected species that it continually deceived the practiced eye of an entomologist, it would often deceive predaceous birds and insects, and thus often escape destruction. Mr. Bates may almost be said to have actually witnessed the process" of evolution in action.

Finally, rudimentary organs speak clearly of evolutionary ancestry. Hardly any animal can be named that does not possess rudimentary organs, such as the functionless mammae of male mammals, or the teeth that are present in but not used by fetal whales and absent in the adults, which use plates of baleen rather than teeth to obtain their food. Aquatic salamanders have gills, but terrestrial ones do not—except the unborn salamanders, which have very well-developed gills, which they do not need. Most flowers have both male (stamens) and female (pistils) parts, but some flowers have only one or the other. It is common to find rudimentary, useless pistils inside of male flowers. Some snapdragons have a rudimentary, useless fifth stamen. Natural selection would get rid of costly, unnecessary organs, but would not necessarily get rid of them completely. When these organs have simplified to the point of being rudimentary and at the same time no longer costly, there has been no advantage in getting rid of them completely. Rudimentary organs provide the same evidence for the evolution of organisms as do silent letters for the evolution of languages: The letters, retained in spelling, are vestiges of ancestral pronunciations.

Why should rudimentary organs exist at all, if the Creator made everything perfect? Some writers have remarked that the Creator put them there "for the sake of symmetry" or "to complete the scheme of Nature." But this is, as described earlier, not an explanation; nor is it even consistent with itself: For why would the Creator have given rudimentary [vestigial] hips to boa constrictors but not to other snakes?

> ... the several classes of facts which have been considered in this chapter, seem to me to proclaim so plainly, that the innumerable species, genera and families, with which this world is peopled, are all descended ... from common parents ... that I should without hesitation adopt this view, even if it were unsupported by other facts ...

Chapter 15. Recapitulation and Conclusion

As this book represents one long argument, I will recapitulate that argument, then draw some conclusions that emerge from an evolutionary understanding of the world.

"That many and serious objections may be advanced against ... natural selection, I do not deny." But even the most complex adaptations can be explained from a few simple starting premises, each supported by evidence. We must be very cautious in claiming that any adaptation that we observe could not have been produced by many small steps, each selected by nature. All we have to do is look at the tremendous diversity and advancements that have been produced during the evolution of our crops and livestock. "There is no reason why the principles which have acted so efficiently under domestication should not have acted under nature." Natural selection seems inevitable. Even the slightest individual differences can make the difference in natural selection: "A grain in the balance may determine which individuals shall live and which shall die." Because of this, "if there has been any variability under nature, it would be an unaccountable fact if natural selection had not come into play." Further, "I can see no limit to" the power of natural selection "in slowly and beautifully adapting each form to the most complex relations of life," so long as the heritable variability is available in the populations. That it has done so is indicated by the fact that we cannot distinguish species, which many suppose to be independent creations, from mere varieties, which everyone admits have a natural origin. The fact that most groups become extinct and a few of them diversify is inevitable, for if all species survived and diversified "the world could not hold them," and this also explains why the larger groupings [phyla, orders, etc.] are so few in number. It also explains why nature has so much variety but so little innovation, as each lineage pursues its own course of adaptation: "The same general end is gained through an almost infinite diversity of means."

This theory even explains why there is so much beauty in nature—the beauty of birds and flowers is the result of sexual selection or selection by pollinators. But it also explains why not everything is beautiful, for the faces of "hideous bats with a distorted resemblance to the human face" can evolve as readily as something that we consider beautiful. We ought not "to marvel if all the contrivances of nature be not, as far as we can judge, absolutely perfect, as in the case even of the human eye; or if some of them be abhorrent to our ideas of fitness," or appear to us wasteful, for natural selection produces adaptations, not perfection.

This theory has an elegant simplicity. For example, the occasional appearance of stripes on the legs of juvenile members of the horse genus is simply explained as a vestige of their common ancestors and is inexplicable on the basis of independent creation of the species in that genus. Evolution explains all the examples of characteristics that "bear the plain stamp of inutility." With elegant simplicity evolution explains the patterns observed in the fossil record, both the appearance and the permanent disappearance of species, and the similarity of the species in any one layer to those in the layers immediately above and below that layer. With elegant simplicity this theory explains what every traveler has seen, that the animals and plants are very different on the different continents, even when conditions are nearly the same, and that islands have relatively few species but those few species are frequently unique.

My theory has been much maligned. The "power of steady misrepresentation" is great, but truth eventually prevails. It is doubtful that a false theory could explain so much, so well, as does this theory. Someday, criticisms of evolution by means of natural selection will seem as quaint as Leibniz accusing Newton of invoking mystical forces to explain physics. My theory has been accused of undermining religion; but Leibniz criticized Newton's theory of gravitation as subversive of religion. Despite attacks from several clergymen, I have received the praise of several others, including one who believed that special creation of each species was unworthy of the Creator, as if this Creator had to specially make new species to fill in the gaps created by the operation of the Creator's very own laws! How disparaging to the Creator to think that many species "have been created with plain but deceptive, marks of descent from a single parent." For the Creator to use natural means in the birth and death of species is no stranger than for this Creator to use natural means in the ordinary birth and death of each individual.

It is not surprising that people would find the origin of species by gradual transformations hard to believe. Humans always have a hard time believing transformations "of which we do not see the steps." Despite the resistance of older scholars, I look to the younger scientists who will grow up not having been indoctrinated in the old ways of thinking.

Do I believe that evolution explains the origin of all species from a common ancestor? "I believe that animals are descended from at most only four or five progenitors, and plants from an equal or lesser number." But the number of original species, from which all others evolved, may have been smaller yet. [Though scientists knew little about biochemistry at the time, Darwin knew that] the same poison can in some cases affect every kind of cell. Sexual reproduction, on a cellular level, is very similar in all species. For these reasons, belief in the evolution of all species from a single origin is not out of the question.

When the concept of evolution is finally grasped throughout the scientific world, we can "foresee that there will be a considerable revolution in natural history." At the very least, what a sigh of relief will come from naturalists no longer having to worry about whether related forms are separate species or mere varieties; "we shall at last be free from the vain search for the undiscovered and undiscoverable essence of the term species ... We will no longer look at an organic being as a savage looks at a ship, as something wholly beyond his comprehension;" but see it as the product of natural selection in the same way "as any great mechanical invention is the summing up of the labor, the experience, the reason, and even the blunders of numerous workmen; when we thus view each organic being, how far more interesting—I speak from experience—does the study of natural history become!" An evolutionary understanding can revolutionize geology, since the degree of evolutionary change can be roughly used as a measure of time over which the rock layers were formed. Even the study of psychology can be based upon evolution (see SOCIOBIOLOGY). "Much light will be thrown on the origin of man and his history."

This theory even gives us some comfort. As we consider the unbroken lineages that have led to the present-day diversity of species, we know that "the ordinary succession by generation has never once been broken, and no cataclysm has desolated the whole world. Hence we may look with some confidence to a secure future of great length." Evolution leads all organisms toward perfection, even if none reach it.

> It is interesting to contemplate a tangled bank, clothed with many plants of many kinds, with birds singing on the bushes, with various insects flitting about, and with worms crawling through the damp earth, and to reflect that these elaborately constructed forms, so different from each other, and dependent upon each other in so complex a manner, have all been produced by laws acting around us ... Thus, from the war of nature, from famine and death, the most exalted object of which we are capable of conceiving, namely, the production of the higher animals, directly follows. There is grandeur in this view of life, with its several powers, having been originally breathed by the Creator into a few forms or into one; and that, whilst this planet has gone cycling on according to the fixed law of gravity, from so simple a beginning endless forms most beautiful and most wonderful have been, and are being evolved.

Further Reading

Darwin, Charles. *On the Origin of Species by Means of Natural Selection,* 6th ed. New York: New American Library, 1958.

Jones, Steve. *Darwin's Ghost: The Origin of Species Updated.* New York: Random House, 2000.

Mayr, Ernst. *The Growth of Biological Thought: Diversity, Evolution, and Inheritance.* Cambridge, Mass.: Harvard University Press, 1982.

———. *One Long Argument: Charles Darwin and the Genesis of Modern Evolutionary Thought.* Cambridge, Mass.: Harvard University Press, 1991.

Further Resources

FURTHER READING: GENERAL AUDIENCES

Bryson, Bill. *A Short History of Nearly Everything*. New York: Broadway Books, 2003.

Darwin, Charles. *On the Origin of Species by Means of Natural Selection,* 6th ed. Reprinted with introduction by Julian S. Huxley. New York: Times Mirror, 1958.

———. *The Descent of Man, and Selection in Relation to Sex*. 2nd ed. London: John Murray. Reprinted with introduction by James Moore and Adrian Desmond. New York: Penguin Books, 2004.

Dawkins, Richard. *The Ancestor's Tale: A Pilgrimage to the Dawn of Evolution*. New York: Houghton Mifflin, 2004.

Desmond, Adrian, and James Moore. *Darwin: The Life of a Tormented Evolutionist*. New York: Warner Books, 1991.

Dupré, John. *Darwin's Legacy: What Evolution Means Today*. New York: Oxford University Press, 2003.

Freeman, Scott, and Jon C. Herron. *Evolutionary Analysis,* 3rd ed. Upper Saddle River, N.J.: Pearson Prentice Hall, 2004.

Gould, Stephen Jay, ed. *The Book of Life: An Illustrated History of the Evolution of Life on Earth*. New York: Norton, 1993/2001.

Jones, Steve. *Darwin's Ghost: The Origin of Species Updated*. New York: Random House, 2000.

Knoll, Andrew H. *Life on a Young Planet: The First Three Billion Years of Evolution on Earth*. Princeton, N.J.: Princeton University Press, 2003.

Margulis, Lynn, and Dorion Sagan. *Acquiring Genomes: A Theory of the Origins of Species*. New York: Basic Books, 2002.

Mayr, Ernst. *The Growth of Biological Thought: Diversity, Evolution, and Inheritance*. Cambridge, Mass.: Harvard University Press, 1982.

———. *One Long Argument: Charles Darwin and the Genesis of Modern Evolutionary Thought*. Cambridge, Mass.: Harvard University Press, 1991.

———. *What Evolution Is*. Cambridge, Mass.: Harvard University Press, 2001.

Pagel, Mark, ed. *Encyclopedia of Evolution*. New York: Oxford University Press, 2002.

Quammen, David. "Was Darwin Wrong? No." *National Geographic,* November 2004, 2–35.

Southwood, Richard. *The Story of Life*. New York: Oxford University Press, 2003.

FURTHER READING: YOUNGER AUDIENCES

Hosler, Jay. *The Sandwalk Adventures: An Adventure in Evolution Told in Five Chapters*. Columbus, Ohio: Active Synapse, 2003.

Sís, Peter. *The Tree of Life: A Book Depicting the Life of Charles Darwin, Naturalist, Geologist, & Thinker*. New York: Farrar Straus Giroux, 2003.

GENERAL WEB SITES

Museum of Paleontology. University of California, Berkeley. "Understanding Evolution." Available online. URL: http://evolution.berkeley.edu. Accessed October 18, 2005.

National Center for Science Education. "National Center for Science Education." Available online. URL: http://www.ncseweb.org. Accessed October 18, 2005.

Tree of Life Web Project. "Explore the Tree of Life." Available online. URL: http://www.tolweb.org/tree/. Accessed October 18, 2005.

White, Toby. "Palaeos: The Trace of Life on Earth." Available online. URL: http://www.palaeos.com. Accessed October 18, 2005.

ORGANIZATIONS

National Center for Science Education. Website: http://www.ncseweb.org.

Society for the Study of Evolution. Website: http://www.evolutionsociety.org.

Index

Note: **Boldface** page numbers indicate main entries, *italic* page numbers indicate illustrations, *m* refers to maps, *t* refers to tables.

B

D

E

H

J

K

L

M

N

O

P

Q

R

S

T

U